A FRENCH-ENGLISH DICTIONARY

FOR

CHEMISTS

BY

AUSTIN M. PATTERSON, Ph.D.

Vice-President and Professor of Chemistry, Antioch College

TOTAL ISSUE TWENTY THOUSAND

NEW YORK

JOHN WILEY & SONS, Inc.

LONDON: CHAPMAN & HALL, LIMITED

Printed in U. S. A.

TECHNICAL COMPOSITION CO.
CAMBRIDGE, MASS., U. S. A.

3–39

PREFACE

The present work is the response to a demand (not at all anticipated) for a French companion to the author's "German-English Dictionary for Chemists." It follows the plan of the earlier book, which seems to have met with general favor.

Prof. F. W. Tanner of the University of Illinois has read the galley proof and added a number of biological terms, and Prof. Frank Vogel of the Massachusetts Institute of Technology has read the page proofs and the manuscript of the Introduction and has also contributed the general statement prefacing the "Conjugation of French Verbs." The author is likewise under obligation to Prof. C. W. Foulk, Mr. E. J. Crane and Dr. Maurice L. Dolt, and to numerous other friends who have given opinions or answered queries.

Reference works chiefly consulted: *General*, Clifton-Grimaux-McLaughlin, French-English Dictionary; the Larousse dictionaries, especially the latest, "Larousse pour tous." *Technical*, Répertoire technologique des noms d'industries et de professions (by French Government officials); Willcox, French-English Military Technical Dictionary; De Méric, Dictionnaire des termes de médecine. The material has been rather scattered, consequently a large proportion has been obtained by the reading of books, catalogs and journals.

A really good dictionary is a growth. No one can realize as well as the author how incomplete the vocabulary of this one still is. The very helpful correspondence regarding errors and omissions in the German-English dictionary leads to the hope that readers will go to the same trouble in this case.

AUSTIN M. PATTERSON

XENIA, OHIO, January, 1921

ABBREVIATIONS

a.	adjective.	*masc.*	masculine.
abbrev.	abbreviation.	*Math.*	Mathematics.
adv.	adverb.	*Mech.*	Mechanics.
Agric.	Agriculture.	*Med.*	Medicine.
Anat.	Anatomy.	*Metal.*	Metallurgy.
Arith.	Arithmetic.	*Micros.*	Microscopy.
Astron.	Astronomy.	*Mil.*	Military.
Bact.	Bacteriology.	*Min.*	Mineralogy.
Biol.	Biology, Biological.	*n.*	noun.
Bot.	Botany.	*Obs.*	obsolete.
Calico.	Calico Printing.	*Org.*	Organic.
cap.	capital.	*p.a.*	participial adjective (and participle).
Ceram.	Ceramics.		
cf.	compare.	*p.def.*	past definite.
Chem.	Chemistry.	*Petrog.*	Petrography.
Com.	Commerce.	*Pharm.*	Pharmacy.
cond.	conditional.	*Photog.*	Photography.
conj.	conjunction.	*Physiol.*	Physiology.
Cryst.	Crystallography.	*pl.*	plural.
Dist.	Distilling.	*p.p.*	past participle.
Elec.	Electricity.	*p.pr.*	present participle.
esp.	especially.	*pr.*	present.
Expl.	Explosives.	*prep.*	preposition.
f.	feminine noun.	*pron.*	pronoun.
fem.	feminine.	*Pyro.*	Pyrotechnics.
fig.	figuratively.	*sing.*	singular.
fut.	future.	*sp.*	species.
Geol.	Geology.	*specif.*	specifically.
Geom.	Geometry.	*subj.*	subjunctive.
imp.	imperfect.	*syn.*	synonym.
imper.	imperative.	*Tech.*	Technical.
indic.	indicative.	*Thermochem.*	Thermochemistry.
interj.	interjection.	*v.i.*	intransitive verb.
m.	masculine noun.	*v.r.*	reflexive verb.
Mach.	Machinery.	*v.t.*	transitive verb.
		Zoöl.	Zoölogy.

INTRODUCTION

Thanks to the many words which French and English have in common and to the wonderful clarity of French expression, the beginner usually finds it easier to learn to read French than German, especially if he has studied Latin to some account. Separable prefixes, involved sentence order and a certain well known " longwordformingtendency " are not there to trouble him. Chemical names do not present great difficulty. Nevertheless there are pitfalls, such as the numerous idiomatic phrases, the irregular verb forms and the words which, tho posing as old friends, do not mean just what they ought to! These matters the author has tried to keep in mind in the preparation of the dictionary. The following notes may be of assistance in the use of the vocabulary.

Phrases. A large number of phrases has been included. As a rule, a phrase is entered under the first important word which occurs in it, prepositions, conjunctions and articles being passed by (*en détail* under *détail, tout d'un coup* under *tout*, etc.).

Verbs. Conjugations of typical regular verbs have been placed at the end of the Introduction in the hope that this may save references to a grammar. Irregular verb forms which are likely to appear in scientific literature are given where they are most needed, i.e., in the vocabulary under their own spellings, and not safely concealed under the parent verb.

General Words. A general vocabulary superior to that of many pocket dictionaries is included with the technical words. The special chemical meaning of a general word is ordinarily placed first.

Words Common to French and English. Words of the same or nearly the same spelling in the two languages are defined even when the meaning is exactly the same as in English. Manifestly, these words are only a selection from among those which might have been included. Especially is this true of the names of chemical compounds.

Derived Words. As a space-saving device the meanings of many words are merely indicated, reference being made to another word which is fully defined. For example, several senses of the verb *brasser* (to stir) are given. A listing of all the corresponding meanings of the noun *brassage* is avoided by using for the latter word the form " stirring, etc. (see brasser)." Participles and participial adjectives are often treated in the same way.

Feminine Forms. These are not given except in special cases. The usual sign of the feminine is the letter *e*; thus, *gris* (gray) becomes *grise* in the feminine singular, while *neutre* (neutral) which ends in *e* in the masculine form remains unchanged in the feminine. The following irregularities should be noted:

-ve (fem.) corresponding to	**-f** (masc.);	as,	*actif, active.*
-se	"	" **-x**;	as, *nitreux, nitreuse.*
-se	"	" **-r**;	as, *trompeur, trompeuse.*
-che	"	" **-c**;	as, *blanc, blanche.*
-que	"	" **-c**;	as, *turc, turque.*
-gue	"	" **-g**;	as, *long, longue.*
-trice	"	" **-teur**;	as, *indicateur, indicatrice.*

Doubling of the final consonant; as, *gros, grosse.*

Plural Forms. The regular formation of the plural is by addition of *s* to the singular. Departures from this rule are:

-aux (pl.) corresponding to	**-al, -ail** or **-au** (sing.);	as,	*égal, égaux.*
-eaux	"	" **-eau**;	as, *nouveau, nouveaux.*
-eux	"	" **-eu**;	as, *feu, feux.*
-oux	"	" **-ou**;	as, *caillou, cailloux.*

Nouns ending in *-s*, *-x* and *-z* and masculine adjectives ending in *-s* and *-x* in the singular are unchanged in the plural.

Nouns Used as Adjectives. French differs from English in its adjectival use of nouns. In some instances the use is the same; as, *odeur éther*, ether odor; *solution lilas*, lilac solution. For the most part, however, French nouns used as adjectives might be construed as in apposition; as, *chimiste ingénieur*, chemical engineer; *tube séparateur*, separating tube. On the other hand such English expressions as " sodium chloride " (*chlorure de sodium*) and "gun steel " (*acier à canon*) require the use of a preposition when rendered into French.

Inorganic Nomenclature. Most inorganic names are readily translated from French into English as soon as one learns that *-ure* corresponds to the English *-ide*; *-ique* to *-ic*; and *-eux* (fem. *-euse*) to *-ous*.

Names of hydracids differ in construction from the English names; as, *chlorhydrique*, hydrochloric.

Names of salts containing the preposition *de* should not invariably be translated by using "of." For example, while *acétate de cuivre* is correctly translated "acetate of copper " the form " copper acetate " is more frequent.

Apparently, the older fashion of naming salts from the bases instead of the metals (as *carbonate de soude*, carbonate of soda) is more persistent in French than in English, hence in many cases it would be well to use the modern form (in this case, sodium carbonate) in translating them.

Organic Nomenclature. The use of adjectives in French organic names, instead of a combining form as in English and German, may be noted; thus the compound $C_6H_4Cl_2$ (dichlorobenzene) is often called *benzène* (or *benzine*) *dichloré*, although *dichlorobenzène* and even the Germanism *dichlorbenzol* also occur.

Names of esters are found in which the name of the alcohol radical is prefixed to the name of the acid; as, *éther amylacétique*. This should be rendered "amyl acetate" rather than "amylacetic ester" or "amylacetic ether" if the compound $CH_3CO_2C_5H_{11}$ is meant, and other cases should be treated similarly.

Words ending in *-ique* are used in a general descriptive sense more frequently than the corresponding English words in *-ic*; *berbérinique*, for instance, would be properly translated "berberinic" if designating an acid, but it often means simply "of berberine" or "berberine"; as, *sels berbériniques*, berberine salts.

German names ending in *-insäure* may usually be translated into French and English in two ways, that is, as *acide -inique* (-inic acid) or as *acide -ique* (-ic acid); and it sometimes happens that the one alternative is chosen in French, the other in English. For example the compound named Levulinsäure is usually called levulinic acid in English while in French the form *acide lévulique* seems to be the more common. Such differences are not numerous.

In order to have names which are in accord with the usage of the American Chemical Society as embodied in the "Directions to Abstractors" the following rules (which also harmonize well with the best usage in Great Britain) should be observed. They are not necessarily to be applied to names of pharmaceuticals and dyes, and possibly some exceptions should be allowed for commercial use only, as glycer*in* instead of glycer*ol*. Differences between trade and scientific usage are always to be regretted.

(1) Translate *oxy-* by *hydroxy-* when it designates hydroxyl, as is commonly the case in organic names. When *oxy-* designates the ketonic group (CO) it is preferably translated *keto-*.

(2) Translate names of compounds the chief function of which is alcoholic or phenolic so that the name ends in *-ol*; as, glycerol, resorcinol, mannitol, pinacol (not pinacone).

(3) When the French ending *-ol* does not indicate hydroxyl it should be translated *-ole* (as, anisole, indole), or in the case of a few hydrocarbons *-ene* (as *benzol*, benzene; *toluol*, toluene; *styrol*, styrene).

(4) The ending *-ine* or *-in* should be translated *-ine* in the case of basic substances and *-in* elsewhere; as, aniline, glycine, palmitin, albumin. Exceptions are *benzine* (meaning benzene) and the names of alcohols and phenols (see (2)).

(5) The form *amido-* should be so translated only when it denotes combination with an acid group. Usually it is to be translated *amino-*; as, *acide amidopropionique*, aminopropionic acid; *amidophénol*, aminophenol. The same holds for imido-, anilido-, etc.

(6) In such combining forms as bromo-, cyano-, chloro-, nitro-, etc., when they denote substituting radicals, the connective *o* is to be used invariably; as

chlorobenzène or *chlorbenzène*, chlorobenzene; *acide chloracétique*, chloroacetic acid. This usage is by no means universal, but those who cannot reconcile themselves to such spellings as "bromoacetic" should at least avoid the German forms bromphenol, acetphenetidine, etc., by using the connective *o* before consonants. (French literature does not contain so many of these objectionable forms as does the German.)

(7) The French ending *-ane* should be translated *-ane* if it is the name of a hydrocarbon (or parent heterocyclic compound) which is *fully saturated;* otherwise, *-an.* Examples: methane, menthane, tolan, furan, pentosan.

(8) Names of acids ending in *-carbonique* are translated *-carboxylic,* not -carbonic.

CONJUGATION OF FRENCH VERBS

1. General Statement

(By Professor Frank Vogel)

In general the tenses of *all* regular verbs and of most irregular verbs can be formed or recognized from the five principal parts, viz.: (1) present infinitive, (2) present participle, (3) past participle, (4) present indicative, (5) past definite indicative.

(1) From the present infinitive, by adding directly *-ai, -as, -a, -ons, -ez, -ont,* the future indicative is formed.   Likewise by adding *-ais, -ais, -ait, -ions, -iez, -aient* the conditional present is obtained, but infinitives in *-re* always drop the *-e.*

(2) From the present participle, by changing *-ant* into *-ais, -ais, -ait, -ions, -iez, -aient,* we obtain the imperfect indicative; and by changing *-ant* into *-e, -es, -e, -ions, -iez, -ent* we obtain the present subjunctive.

(3) The past participle with the auxiliary *avoir* for all transitive verbs (but with *être* for intransitives denoting motion, change of condition or position, and for reflexives) forms the compound tenses in the active voice.   However, with the past participle and the required tense form of *être* the tenses of the passive voice are formed.

(4) The present indicative gives the imperative by dropping the pronoun of the second person singular and of the first and second person plural, also final *s* of the second person singular in the first (*-er*) conjugation is dropped except when followed by the pronouns *-y* or *-en.*

(5) The past definite gives the imperfect subjunctive by changing the first person singular final *i* or *s* to *-sse, -sses, -ˆt, -ssions, -ssiez, -ssent.*   Note that the third person singular always has a circumflex (ˆ) accent over the vowel (*a, i, u*) before the *t.*

2. Active Voice (and Intransitive Verbs)

Laver, brunir and *rompre* are used as types of the three regular conjugations in *-er, -ir* and *-re*. See the note about certain intransitive verbs in paragraph (3) of the General Statement.

laver, *to wash*

Simple Tenses	*Compound Tenses*

Present Infinitive
  laver, to wash

Perfect Infinitive
  avoir lavé, to have washed

Present Participle
  lavant, washing

Perfect Participle
  ayant lavé, having washed

Past Participle
  lavé, washed

Present Indicative
  je lave, I wash, am washing
  tu laves, you wash
  il lave, he washes
  nous lavons, we wash
  vous lavez, you wash
  ils lavent, they wash

Past Indefinite
  j'ai lavé, I have washed, I washed
  tu as lavé
  il a lavé
  nous avons lavé
  vous avez lavé
  ils ont lavé

Imperfect Indicative
  je lavais, I was washing, used to wash, washed
  tu lavais
  il lavait
  nous lavions
  vous laviez
  ils lavaient

Pluperfect Indicative
  j'avais lavé, I had washed, had been washing
  tu avais lavé
  il avait lavé
  nous avions lavé
  vous aviez lavé
  ils avaient lavé

Past Definite
  je lavai, I washed
  tu lavas
  il lava
  nous lavâmes
  vous lavâtes
  ils lavèrent

Past Anterior
  j'eus lavé, I had washed
  tu eus lavé
  il eut lavé
  nous eûmes lavé
  vous eûtes lave
  ils eurent lavé

Future

je laverai, I shall wash
tu laveras
il lavera
nous laverons
vous laverez
ils laveront

Future Anterior

j'aurai lavé, I shall have washed
tu auras lavé
il aura lavé
nous aurons lavé
vous aurez lavé
ils auront lavé

Present Conditional

je laverais, I should wash
tu laverais
il laverait
nous laverions
vous laveriez
ils laveraient

Conditional Anterior

j'aurais lavé, I should have washed
tu aurais lavé
il aurait lavé
nous aurions lavé
vous auriez lavé
ils auraient lavé

Imperative

lave, laves,* wash (sing.)
qu'il lave, let him wash
lavons, let us wash
lavez, wash (pl.)
qu'ils lavent, let them wash

Present Subjunctive

que je lave, that I may wash

que tu laves
qu'il lave
que nous lavions
que vous laviez
qu'ils lavent

Perfect Subjunctive

que j'aie lavé, that I may have washed
que tu aies lavé
qu'il ait lavé
que nous ayons lavé
que vous ayez lavé
qu'ils aient lavé

Imperfect Subjunctive

que je lavasse, that I might wash

que tu lavasses
qu'il lavât
que nous lavassions
que vous lavassiez
qu'ils lavassent

Pluperfect Subjunctive

que j'eusse lavé, that I might have washed
que tu eusses lavé
qu'il eût lavé
que nous eussions lavé
que vous eussiez lavé
qu'ils eussent lavé

---

* Only when -*y* or -*en* follows.

brunir, *to brown*   **rompre,** *to break*

(Only the simple tenses of these verbs are given, the compound tense forms being exactly like those of *laver* with the substitution of *bruni* and *rompu* for *lavé*.)

Present Infinitive

brunir, to brown   rompre, to break

Present Participle

brunissant, browning   rompant, breaking

Past Participle

bruni, browned   rompu, broken

Present Indicative

je brunis, I brown, am browning	je romps, I break, am breaking
tu brunis	tu romps
il brunit	il rompt
nous brunissons	nous rompons
vous brunissez	vous rompez
ils brunissent	ils rompent

Imperfect Indicative

je brunissais, I was browning, used to brown, browned	je rompais, I was breaking, used to break, broke
tu brunissais	tu rompais
il brunissait	il rompait
nous brunissions	nous rompions
vous brunissiez	vous rompiez
ils brunissaient	ils rompaient

Past Definite

je brunis, I browned	je rompis, I broke
tu brunis	tu rompis
il brunit	il rompit
nous brunîmes	nous rompîmes
vous brunîtes	vous rompîtes
ils brunirent	ils rompirent

Future

je brunirai, I shall brown	je romprai, I shall break
tu bruniras	tu rompras
il brunira	il rompra
nous brunirons	nous romprons
vous brunirez	vous romprez
ils bruniront	ils rompront

Present Conditional

je brunirais, I should brown	je romprais, I should break
tu brunirais	tu romprais
il brunirait	il romprait
nous brunirions	nous romprions
vous bruniriez	vous rompriez
ils bruniraient	ils rompraient

Imperative

brunis, brown (sing.)	romps, break (sing.)
qu'il brunisse, let him brown	qu'il rompe, let him break
brunissons, let us brown	rompons, let us break
brunissez, brown (pl.)	rompez, break (pl.)
qu'ils brunissent, let them brown	qu'ils rompent, let them break

Present Subjunctive

que je brunisse, that I may brown	que je rompe, that I may break
que tu brunisses	que tu rompes
qu'il brunisse	qu'il rompe
que nous brunissions	que nous rompions
que vous brunissiez	que vous rompiez
qu'ils brunissent	qu'ils rompent

Imperfect Subjunctive

que je brunisse, that I might brown	que je rompisse, that I might break
que tu brunisses	que tu rompisses
qu'il brunît	qu'il rompît
que nous brunissions	que nous rompissions
que vous brunissiez	que vous rompissiez
qu'ils brunissent	qu'ils rompissent

3. Passive Voice (Conjugation of être)

The passive is formed from the past participle (which agrees with the subject in gender and number) by using with it the appropriate tense form of *être*.

Examples: *il est lavé*, he (or it) is washed; *elle a été lavée*, she (or it) has been washed; *ils seraient brunis*, they would be browned; *ayant été rompu*, having been broken. The conjugation of *être* is as follows:

Infinitives

être, to be avoir été, to have been

Participles

étant, being été, been ayant été, having been

Present Indicative

je suis, I am, I am being
tu es, you are
il est, he is
nous sommes, we are
vous êtes, you are
ils sont, they are

Past Indefinite

j'ai été, I have been, I was
tu as été, you have been
il a été, he has been
nous avons été, we have been
vous avez été, you have been
ils ont été, they have been

Imperfect Indicative

j'étais, I was, was being, used to be
tu étais, you were
il était, he was
nous étions, we were
vous étiez, you were
ils étaient, they were

Pluperfect Indicative

j'avais été, I had been
tu avais été
il avait été
nous avions été
vous aviez été
ils avaient été

Past Definite

je fus, I was
tu fus, you were
il fut, he was
nous fûmes, we were
vous fûtes, you were
ils furent, they were

Past Anterior

j'eus été, I had been
tu eus été
il eut été
nous eûmes été
vous eûtes été
ils eurent été

Future

je serai, I shall be
tu seras
il sera
nous serons
vous serez
ils seront

Future Anterior

j'aurai été, I shall have been
tu auras été
il aura été
nous aurons été
vous aurez été
ils auront été

Present Conditional	*Conditional Anterior*
je serais, I should be	j'aurais été, I should have been
tu serais	tu aurais été
il serait	il aurait été
nous serions	nous aurions été
vous seriez	vous auriez été
ils seraient	ils auraient été

Imperative

sois, be (sing.)
qu'il soit, let him be
soyons, let us be
soyez, be (pl.)
qu'ils soient, let them be

Present Subjunctive	*Perfect Subjunctive*
que je sois, that I may be	que j'aie été, that I may have been
que tu sois	que tu aies été
qu'il soit	qu'il ait été
que nous soyons	que nous ayons été
que vous soyez	que vous ayez été
qu'ils soient	qu'ils aient été

Imperfect Subjunctive	*Pluperfect Subjunctive*
que je fusse, that I might be	que j'eusse été, that I might have been
que tu fusses	que tu eusses été
qu'il fût	qu'il eût été
que nous fussions	que nous eussions été
que vous fussiez	que vous eussiez été
qu'ils fussent	qu'ils eussent été

4. Reflexive Verbs

In the simple tenses, reflexive verbs are conjugated like the corresponding active forms, but with the use of the pronouns *me, te, se, nous, vous, se* (as the person and number may be). In the compound tenses, in addition to the use of the above pronouns the auxiliary *être* (instead of *avoir* as in the active forms) is employed; thus, *laver* (active), *se laver* (reflexive); *avoir lavé* (active), *s'être lavé* (reflexive).

Reflexive verbs are translated:

(1) Reflexively; as, *se laver*, to wash oneself.

(2) Reciprocally; as, *se laver*, to wash each other or one another. (The above two meanings are not usually given in the vocabulary.)

(3) Passively; as, *se publier*, to be published. (This meaning is often given, especially when it is a customary one.)

(4) Intransitively; as, *l'or se fond*, the gold melts, gold melts. (Such meanings have been included as far as possible.)

The following inflections must serve as examples for the reflexive verbs:

Present Indicative	*Perfect Indicative*
je me lave, I wash myself, I wash	je me suis lavé, I have washed myself, I have washed
tu te laves	tu t'es lavé
il se lave	il s'est lavé
nous nous lavons	nous nous sommes lavés
vous vous lavez	vous vous êtes lavés
ils se lavent	ils se sont lavés

FRENCH-ENGLISH DICTIONARY
FOR CHEMISTS

A

a, *pr. 3 sing.* of avoir. — *abbrev.* (asymétrique), asymmetric(al).

à, *prep.* to; at; on, in; of; by; from; according to. (When *à* is used between substantives the second modifies the first, as *canne à sucre*, sugar cane. Phrases commencing with *à* are placed under the first important word in the phrase.)

abaissement, *m.* lowering, depression; dropping (of a perpendicular); reduction (of equations); fall, subsidence, decline, diminution; abasement; debasement.
— *du point de congélation*, freezing-point lowering.
— *moléculaire*, molecular lowering (or depression).

abaisser, *v.t.* lower; drop, let fall (a perpendicular); reduce (an equation); abate, reduce, diminish, depress; abase; degrade. — *v.r.* become lower, fall, subside; demean oneself.

abandon, *m.* abandonment, relinquishment, abandon.

abandonner, *v.t.* leave; let go of; abandon, give up, give over.

abaque, *m.* graphic table; abacus.

abasourdir, *v.t.* deafen; astound.

abasourdissant, *p.a.* deafening; astounding.

abat, *m.* killing, slaughter; slaughtered animal; (*pl.*) certain inferior parts of slaughtered animals; downpour. — *pr. 3 sing.* of abattre.

abatage, *m.* killing, slaughter; destruction, demolition; throwing down; lowering.
— *à la poudre*, blasting.
— *hydraulique*, hydraulic mining or excavating.

abâtardir, *v.t.* render degenerate, deteriorate.

abatis, *m.* offal (of animals); material cut or pulled down; slaughter; cutting or pulling down.

abat-jour, *m.* shade (for a lamp or the eyes); window (of certain kinds); skylight; abat-jour; awning.

abattage. = abatage.

abattement, *m.* prostration; dejection; faintness.

abattis. = abatis.

abattoir, *m.* abattoir, slaughterhouse.

abattre, *v.t.* knock down, throw down, cut down, pull down; kill, slaughter; depress; prostrate; cut off, cut away; soak (skins). — *v.r.* fall, fall down; abate; stoop; become discouraged.

abcès, *m.* abscess.

abcisse, *f.* abscissa.

abdiquer, *v.t.* give up, renounce; abdicate.

abducteur, *a.* leading away; (*Physiol.*) abducent.
tube —, delivery tube.

abécédaire, *a.* alphabetical; elementary.

abeille, *f.* bee.

abélite, *f.* nitroglycerin.

abêtir, *v.t.* stupefy, render stupid.

abhorrer, *v.t.* abhor, detest.

abiétine, *f.* abietin.

abiétinée, *f.* (*Bot.*) a tree of the fir genus (*Abies*); broadly, any of various conifers.

abiétique, *a.* abietic.

abîme, *m.* abyss, depth, deep; (*Candles*) tallow vat.

abîmer, *v.t.* injure, spoil; ruin, destroy, engulf, overwhelm. — *v.r.* spoil; sink, be engulfed.

abiogénèse, *f.* abiogenesis.

ablation, *f.* ablation, removal.

ablativo, *adv.* in confusion, promiscuously.

ablette, *f.* bleak (a fish with silvery scales).

abluant, *a. & n.* abluent, detergent.

abluer, *v.t.* wash, wash off.

aboi, *m.* bark, barking; bay, (last) gasp.

abolir, *v.t.* abolish; cancel, annul, repeal.

abolissement, *m.*, **abolition,** *f.* abolishment, abolition, etc. (see abolir).

abondamment, *adv.* abundantly; copiously, amply.

abondance, *f.* abundance; plentifulness, plenty.
d'—, extemporaneously.

abondant, *p.a.* abundant; abounding, plentiful; fluent.

abonder, *v.i.* abound, be abundant.

abonné, *m.*, **abonnée,** *f.* subscriber.

abonnement, *m.* subscription; contract.

abonner, *v.r.* subscribe; contract.

abonnir, *v.t.* improve. — *v.i.* & *r.* become good or better, improve.

abord, *m.* approach.

d'—, first, at first.

d'— *que*, as soon as.

dès l'—, from the first.

aborder, *v.t.* approach; accost; take up, commence on (a topic).

abortif, *a.* & *m.* abortive.

abouchement, *m.* joining (of tubes or the like), junction; (*Biol.*) anastomosis; interview, conference.

aboucher, *v.t.* join end to end (as tubes); bring together. — *v.r.* join; (*Biol.*) inosculate; confer.

about, *m.* end, butt end.

abouter, *v.t.* join end to end.

aboutir, *v.i.* abut, end, terminate; converge; result; have result, succeed; (of an abscess) come to a head; (*Bot.*) bud.

aboutissant, *p.a.* abutting, etc. (see aboutir). — *m.* circumstance.

aboutissement, *m.* result, outcome; suppuration.

abrasif, *m.* abrasive.

abre, *m.* (*Bot.*) Abrus.

abrégé, *m.* abridgment, summary, résumé, abstract. — *p.a.* abridged, brief.

abréger, *v.t.* abridge, abbreviate, shorten. — *v.i.* be short, be brief.

abreuvage, abreuvement, *m.* soaking, etc. (see abreuver).

abreuver, *v.t.* soak, saturate; water; irrigate; (*Painting*) prime, size; (*Soap*) lye.

abréviation, *f.* abbreviation.

abri, *m.* shelter, cover, refuge; specif., dugout.

à l'— *de*, protected from, sheltered from, out of contact with, away from.

abricot, *m.* apricot.

abricotier, *m.* apricot tree.

abriter, *v.t.* shelter, cover, protect.

abroger, *v.t.* abrogate; repeal.

abscisse, *f.* abscissa.

absenter, *v.r.* absent oneself, be absent.

absinthe, *f.* wormwood (*Artemisia*, esp. *A. absinthium*); absinth.

— *marine*, sea wormwood (*A. maritima*).

— *pontique*, Roman wormwood (*Artemisia pontica*).

grande —, common wormwood.

absinthé, *a.* absinthiated.

absolu, *a.* absolute; peremptory.

absolument, *adv.* absolutely; peremptorily, positively.

absolvant, *p.pr.* of absoudre.

absolvent, *pr. 3 pl.* of absoudre.

absorbant, *a.* absorbent, absorptive, absorbing, absorption. — *m.* absorbent; (*Explosives*) dope.

absorber, *v.t.* absorb. — *v.r.* be absorbed.

absorbeur, *m.* absorber. — *a.* absorbing.

absorption, *f.* absorption.

absorptivité, *f.* absorptivity, absorptive power.

absoudre, *v.t.* absolve, acquit, excuse.

absous, *p.p.* of absoudre.

absout, *pr. 3 sing.* of absoudre.

abstenir, *v.r.* abstain, refrain.

absterger, *v.t.* cleanse, clean.

abstersif, *a.* abstergent, detergent.

abstient, *pr. 3 sing.* of abstenir.

abstractivement, *adv.* abstractively; abstractly.

abstraire, *v.t.* abstract, take away.

abstrait, *a.* abstract; abstracted.

abstraitement, *adv.* abstractly; abstractedly.

abstrayant, *p.pr.* of abstraire.

abstrich, *m.* = crasse noire, under *crasse*.

abstrus, *a.* abstruse.

absurde, *a.* absurd.

absurdement, *adv.* absurdly.

absurdité, *f.* absurdity, absurdness.

abus, *m.* abuse, misuse; error, mistake.

abuser, *v.i.* (with *de*) make ill use of, abuse. — *v.t.* deceive. — *v.r.* be deceived, be mistaken.

abusif, *a.* abusive, excessive.

abyssin, *a.* Abyssinian.

abzug, *m.* (*Lead*) scum which forms at the beginning of cupellation. Cf. crasse noire, under *crasse*.

acabit, *m.* quality.

académie, *f.* academy (in its French senses).

académique, *a.* academic.

acajou, *m.* mahogany; cashew (*Anacardium occidentale*).

— *à noix*, — *à pommes*, cashew.

acanthe, *f.* acanthus.

accablant, *a.* overwhelming, oppressive, annoying.

accabler, *v.t.* load, weigh down, crush, overwhelm.

accaparement, *m.* monopolization, monopoly.

accaparer, *v.t.* monopolize; accumulate, amass.

accéder, *v.i.* accede, agree.

accélérateur, *m.* accelerator. — *a.* accelerating, accelerative.

accéléré, *p.a.* accelerated, quick.

accélérer, *v.t.* accelerate, hasten.

accentué, *p.a.* accentuated; accented.

accentuer, *v.t.* accentuate; accent.

acceptation, *f.* acceptance.

accepter, *v.t.* accept.

accepteur, *m.* acceptor (as of oxygen); acceptor.

acception, *f.* regard, preference; sense, meaning (of words).

accès, *m.* access; attack (of fever,.etc.).

accessoire, *a.* & *m.* accessory.

accessoirement, *adv.* accessorily, in an accessory or subordinate way.

accident, *m.* accident.

par —, by accident, accidentally.

accidentel, *a.* accidental, adventitious.

accidentellement, *adv.* accidentally.

accise, *f.* excise.

accolade, *f.* (*Printing*) brace.

accolement, *m.* uniting, joining, union.

accoler, *v.t.* unite, join; place side by side; (*Printing*) brace; embrace. — *v.r.* become stuck together (as a rubber tube).

accommodant, *a.* accommodating.

accommodement, *m.* accommodation, arrangement, adjustment.

accommoder, *v.t.* accommodate, adapt; arrange; suit. — *v.r.* adapt oneself; agree; put up (with).

accompagnement, *m.* accompaniment.

accompagner, *v.t.* accompany; suit, harmonize with.

accompli, *p.a.* accomplished; perfect.

accomplir, *v.t.* accomplish, effect; complete; fulfil.

accomplissement, *m.* accomplishment, etc. (see accomplir).

accord, *m.* accord; agreement; chord.

d'—, agreed; in accord.

accorder, *v.t.* accord, grant, concede; reconcile; harmonize. — *v.r.* be in accord or harmony (with); agree (to).

accoter, *v.t.* prop up; lean. — *v.r.* lean.

accouchement, *m.* (*Med.*) delivery, accouchement.

accouplé, *p.a.* coupled, joined.

accouplement, *m.* coupling, pairing; joining, uniting; (*Mach.*) coupling, connection.

accoupler, *v.t.* couple, pair; join, unite, connect.

accourcir, *v.t.* shorten.

accoutumé, *p.a.* accustomed; usual.

accoutumer, *v.t.* accustom.

accrocher, *v.t.* hang, hang up; hook; catch; snag; delay, arrest. — *v.r.* catch, cling.

accroissement, *m.* increase; accretion, growth.

accroître, *v.t.* increase, enlarge. — *v.r. & i.* increase, grow.

accru, *p.p.* of accroître.

accueil, *m.* reception; welcome.

accueillir, *v.t.* receive; welcome; (*Com.*) honor.

accul, *m.* corner, blind alley, impasse.

acculer, *v.t.* corner.

accumulateur, *m.* accumulator; (*Elec.*) accumulator, storage battery, secondary battery.

accumuler, *v.t.* accumulate.

accusé, *p.a.* indicated, etc. (see accuser); pronounced, clearly indicated. — *m.* accused; acknowledgment.

— *de réception,* (*Com.*) acknowledgment.

accuser, *v.t.* indicate, show; accuse, charge; declare; acknowledge (receipt).

acénaphtène, *m.* acenaphthene.

acérage, *m.* steeling, etc. (see acérer).

acérain, *a.* relating to or of the nature of steel, steel, steely.

acerbe, *a.* (of tastes) sharp, harsh, acrid.

acerbité, *f.* sharpness, harshness, acerbity.

acerdèse, *f.* (*Min.*) manganite.

acéré, *p.a.* steeled, edged or tipped with steel; sharp, keen.

acérer, *v.t.* steel, edge, point or overlay with steel; temper; render keen.

acéreux, *a.* (*Bot.*) acerose.

acescence, *f.* acescence.

acétal, *m.* acetal.

acétamide, *f.* acetamide.

acétanilide, *f.* acetanilide.

acétate, *m.* acetate.

— *d'alumine,* aluminium acetate.

— *d'ammoniaque,* ammonium acetate.

— *d'ammoniaque liquide,* (*Pharm.*) solution of ammonium acetate.

— *de chaux,* calcium acetate.

— *de cuivre brut,* crude copper acetate, verdigris.

— *de plomb,* lead acetate.

— *de potasse,* potassium acetate.

— *de soude,* sodium acetate.

— *ferrique liquide,* (*Pharm.*) solution of ferric acetate.

— *neutre de plomb,* (*Pharm.*) lead acetate.

— (*sous-*) *de plomb liquide,* (*Pharm.*) solution of lead subacetate.

acéteux, *a.* acetous.

acétification, *f.* acetification, acetifying.

acétifier, *v.t.* acetify.

acétimètre, *m.* acetimeter.

acétimétrie, *f.* acetimetry.

acétine, *f.* acetin.

acétique, *a.* acetic.

acide — *concentré,* concentrated acetic acid, specif., glacial acetic acid.

acide — *cristallisable,* glacial acetic acid.

acide — *monochloré,* monochloroacetic acid.

acide — *trichloré,* trichloroacetic acid.

acétol, *m.* ketol; acetol; (*Pharm.*) acetolatum.

acétomel, *m.* (*Pharm.*) oxymel.

acétomellé, *a.* treated or mixed with oxymel.

acétomètre, *m.* acetimeter, acetometer.

acétométrie, *f.* acetimetry, acetometry.

acétone, *f.* ketone; acetone.

acétonémie, acétonhémie, *f.* acetonemia.

acétonique, *a.* ketonic; acetonic; of or containing acetone.

acétonitrile, *m.* acetonitrile.

acétonurie, *f.* acetonuria.

acétophénone, *f.* acetophenone.

acétoselle, *f.* wood sorrel, oxalis, esp. *Oxalis acetosella.*

acétosité, *f.* acetosity, acetous quality.

acétoxime, *m.* ketoxime; acetoxime.

acet-phénétidine, *f.* (*Pharm.*) phenacetin, acetphenetidin (*p*-acetophenetide).

acétylacétate, *m.* acetoacetate, acetylacetate.

acétylacétique, *a.* acetoacetic, acetylacetic.

acétylacétone, *f.* acetylacetone (2,4-pentanedione).

acétylant, *p.a.* acetylating.

acétyle, *m.* acetyl.

acétylène, *m.* acetylene.

acétylénique, *a.* of or pertaining to acetylene.

acétyler, *v.t.* acetylate.

acétylure, *m.* acetylide.

achaine, *m.* (*Bot.*) achene.

acharné, *p.a.* intense, furious, desperate.

acharnement, *m.* ardor, fury, tenacity.

acharner, *v.t.* excite, incite, infuriate. — *v.r.* be bent or set; be infuriated.

achat, *m.* purchase, purchasing.

ache, *f.* any plant of the genus *Apium* (of which celery is one species); parsley.

— *d'eau,* water parsley.

— *de montagne,* skirret.

— *des marais,* smallage, wild celery.

acheminé, *p.a.* sent, forwarded; (of plate glass) rough-polished.

acheminement, *m.* preliminary, step, progress.

acheminer, *v.t.* send, forward; prepare for, promote. — *v.r.* advance, progress; start, turn (toward).

acheter, *v.t.* buy, purchase.

acheteur, *m.* buyer, purchaser.

achevage, *m.* finishing.

achevé, *p.a.* finished, perfect; worn-out.

achèvement, *m.* completion, finishing; finish.

achever, *v.t.* finish; complete; conclude. — *v.r.* end, be completed, conclude.

acheveur, *m.* finisher.

achevoir, *m.* finishing tool; finishing room.

achillée, *f.* any plant of the genus *Achillea* (including milfoil or yarrow).

— *sternutatoire,* sneezewort.

acholie, *f.* acholia, lack of bile.

achoppement, *m.* obstacle; stumbling.

achopper, *v.i. & r.* stumble.

achromatique, *a.* achromatic.

achromatiser, *v.t.* achromatize.

achrome, *a.* achromic, colorless.

achromie, *f.* achromatism.

achromique, *a.* achromic, colorless.

aciculaire, *a.* acicular, needle-shaped.

acide, *a. & m.* acid.

— *aminé,* amino acid.

— *au soufre,* (*Elec.*) sulfuric acid of 26° B.

— *Broenner,* Brönner acid (6-amino-2-naphthalenesulfonic acid).

— *cyané,* cyano acid.

— *de Freund,* Freund's acid (a disulfo derivative of α-naphthylamine).

acide *des chambres,* chamber acid.

— *du citron,* citric acid.

— *du tartre,* (dextro)tartaric acid.

— *G,* G acid.

— *gras,* fatty acid.

— *H,* H acid.

— *marin,* hydrochloric acid.

— *minéral,* mineral acid, inorganic acid.

— *oxygéné,* oxygen acid, oxacid.

— *phénique,* phenol, carbolic acid.

— *sulfonitrique,* a mixture of sulfuric and nitric acids.

— *thymique,* thymic acid (often used, esp. in *Pharm.*, as a synonym of thymol but also applied to a nucleic acid hydrolysis product).

(For the meaning of other phrases of *acide* see the respective adjectives.)

acide-acétone, *m.* ketonic acid, acid ketone.

acide-alcool, *m.* hydroxy acid, alcohol-acid.

acide-cétone, *m.* ketonic acid, acid ketone.

acideur, *m.* (*Bleaching*) acid-liquor man.

acidifiant, *p.a.* acidifying. — *m.* acidifier.

acidificateur, *m.* acidifier; specif., an apparatus for separating the fatty acids from fats.

acidifier, *v.t.* acidify. — *v.r.* become acid; turn sour.

acidifleur, *m.* acidifier.

acidimétrie, *f.* acidimetry.

acidimétrique, *a.* acidimetric, -metrical.

acidité, *f.* acidity.

acidose, *f.* acidosis.

acidule, *a.* acidulous, subacid.

acidulé, *p.a.* acidulated; acidulous.

aciduler, *v.t.* acidulate.

acier, *m.* steel.

— *à canon,* gun steel.

— *adouci,* annealed steel.

— *affiné,* fined steel.

— *à outils,* tool steel.

— *à ressort,* spring steel.

— *au bore,* boron steel.

— *au chrome,* chrome steel.

— *au creuset,* crucible steel.

— *au réverbère,* reverberatory furnace steel.

— *boursouflé,* blister steel.

— *bronze,* steel bronze.

— *brut,* crude steel, raw steel.

— *cémenté,* cement steel.

— *chromé,* chrome steel.

— *corroyé,* shear steel, weld steel.

— *coulé,* cast steel, ingot steel.

— *damassé,* damask steel.

— *de Damas,* Damascus steel, damask steel.

— *de fonte,* furnace steel.

— *de forge,* forge steel, fined steel.

— *de fusion,* cast steel, ingot steel.

— *de l'ingot,* ingot steel.

— *des Indes,* wootz, Indian steel.

acier *deux fois corroyé*, double-shear steel.
— *diamant*, tool steel.
— *double-chromé*, double-chrome steel.
— *doux*, mild steel, soft steel.
— *dur*, — *durci*, hard steel.
— *ferreux*, half steel.
— *fondu*, cast steel, ingot steel.
— *fondu au creuset*, crucible steel.
— *forgé*, forged steel.
— *indien*, wootz, Indian steel.
— *laminé*, rolled steel.
— *malléable*, malleable steel, forge steel.
— *manganésé*, manganese steel.
— *marchand*, rolled steel.
— *mi-dur*, semi-hard steel.
— *moulé*, cast steel.
— *naturel*, natural steel.
— *nickelé*, nickel steel; nickel-plated steel.
— *par reaction*, reaction steel.
— *poli*, polished steel.
— *poule*, blister steel.
— *puddlé*, puddled steel.
— *raffiné*, refined steel.
— *raffiné à deux marques*, double-shear steel.
— *raffiné à une marque*, single-shear steel.
— *recuit*, reheated steel, tempered steel.
— *sauvage*, wild steel.
— *soudable*, welding steel.
— *soudé*, weld steel.
— *spécial*, special steel.
— *supérieur*, hard steel.
— *tendre*, soft steel.
— *trempé*, hardened steel.
— *une fois corroyé*, single-shear steel, shear steel.
— *vif*, hard steel.
aciérage, *m.* steeling.
aciération, *f.* steeling; conversion into steel; steel making.
aciérer, *v.t.* steel (plate, overlay, edge or point with steel); convert into steel. — *v.r.* become steel.
aciéreux, *a.* relating to or resembling steel, steely.
aciérie, *f.* steel mill, steel plant, steel works.
acier-nickel, *m.* nickel steel.
aclaste, *a.* aclastic.
acompte, *m.* payment, installment.
aconit, *m.* aconite.
— *napel*, monkshood (*Aconitum napellus*).
aconitine, *f.* aconitine.
aconitique, *a.* aconitic.
acore, *m.* (*Bot.*) Acorus.
— *odorant*, — *vrai*, calamus, sweet flag (*A. calamus* or its root).
Açores, *n.* Azores.
acotylédone, acotylédoné, *a.* acotyledonous.
à-coup, *m.* jerk, start, sudden movement or halt.
acoustique, *a.* acoustic. — *f.* acoustics.

acquérir, *v.t.* acquire; attain, obtain; purchase; ascertain (a fact). — *v.i.* improve.
acquiert, *pr. 3 sing.* of acquérir.
acquis, *p.a.* acquired, etc. (see acquérir).
— *m.* acquirement, attainment; experience, knowledge. — *p. def. 1 & 2 sing.* of acquérir.
acquit, *m.* payment; receipt (of a bill). — *p. def. 3 sing.* of acquérir.
pour —, (on a bill) paid.
acquit-à-caution, *m.* permit to move bonded goods.
acquittement, *m.* payment; acquittal.
acquitter, *v.t.* pay, discharge, perform (an obligation); receipt; acquit.
âcre, *a.* acrid.
âcreté, *f.* acridness, acridity; acrimony.
acridine, *f.* acridine.
acridinique, *a.* acridine, of acridine.
acridique, *a.* acridic, acridinic.
acrogéne, *m.* acrogen.
acroléine, *f.,* **acrol,** *m.* acrolein.
acroléique, *a.* acrylic.
acrylique, *a.* acrylic.
acte, *m.* act; transaction (of a society); deed; document, certificate.
prendre — *de*, receive a certificate of; make a note of.
actée, *f.* (*Bot.*) baneberry (*Actæa*).
— *à grappes*, black cohosh (*Cimicifuga racemosa*).
acteur, *m.* actor.
actif, *a.* active. — *m.* assets.
— *social*, assets of a company or of a firm.
à l'— *de*, to the credit of, in favor of.
actinium, *m.* actinium.
action, *f.* action; share (of stock); gesticulation; animation (in speaking, etc.).
— *de priorité*, preference share (not synonymous with preferred share).
— *ordinaire*, ordinary share (not necessarily synonymous with common share).
— *privilégiée*, = action de priorité.
— *protectrice*, protective action.
hors d'—, out of action; out of gear.
actionnable, *a.* movable; (*Law*) actionable.
actionnaire, *m.* shareholder, stockholder.
actionné, *p.a.* set in motion, operated; busy, occupied; sued.
actionner, *v.t.* set in motion, start, run, operate; excite; sue.
activant, *p.a.* activating, etc. (see activer).
activement, *adv.* actively.
activer, *v.t.* activate; accelerate (motion); intensify, make hotter (a fire or flame); hasten, urge on, push.
activité, *f.* activity; (*Mach.*) operation.
actualité, *f.* actuality.
actuel, *a.* actual; present; at that time.

actuellement, *adv.* actually; now, at present; then, at that time.

acuité, *f.* acuteness.

acutangle, *a.* acute-angled.

acutesse, *f.* sharpness, pointedness.

acyclique, *a.* acyclic.

acylé, *p.a.* acylated.

acyler, *v.t.* acylate.

adamantin, *a.* adamantine.

adamine, *f.* (*Min.*) adamite.

adapter, *v.t.* fit, adjust; adapt, suit. — *v.r.* fit; suit; apply.

additif, *a.* additive.

additionnel, *a.* additional.

additionner, *v.t.* mix, treat (as, *additionné d'acide,* mixed or treated with acid, to which acid has been added); add, add up.

adduction, *f.* conveying, piping (as of water).

adénine, *f.* adenine.

adepte, *m. & f.* adept.

adhérence, *f.* adherence, adhesion.

adhérent, *a.* adherent, adhering.

adhérer, *v.i.* adhere, cling, stick.

adhésif, *a.* adhesive.

adhésion, *f.* adhesion.

adhésivement, *adv.* adhesively.

adiabatique, *a.* adiabatic.

adiabatiquement, *adv.* adiabatically.

adiabatisme, *m.* adiabatic state or phenomenon.

adiante, *m.* (*Bot.*) adiantum, maidenhair fern.

adiaphorèse, *f.* adiaphoresis.

adipeux, *a.* adipose.

adipique, *a.* adipic.

adipocire, *f.* adipocere.

adipogénie, *f.* (*Physiol.*) production of fat.

adipose, *f.* (*Med.*) adiposis.

adiposité, *f.* adiposeness, adiposity.

adjectif, *a.* adjective.

adjoignant, *p.p.* of adjoindre.

adjoignit, *p.def. 3 sing.* of adjoindre.

adjoindre, *v.t.* adjoin, attach.

adjoint, *p.a.* adjoined, attached; assistant. — *m.* adjunct, associate; assistant, deputy. — *pr. 3 sing.* of adjoindre.

adjonction, *f.* adjunction, adjoining, addition.

adjuger, *v.t.* adjudge; award; sell (at auction).

adjuvant, *a. & m.* adjuvant.

admettre, *v.t.* admit; accept (as true); receive, allow, concede; assume, suppose.

administrateur, *m.* administrator, director (of a company), manager, trustee, etc.

administrer, *v.t.* administer, manage, direct, govern.

admirablement, *adv.* admirably, excellently.

admirent, *p.def. 3 pl.* of admettre. — *pr. 3 pl.* of admirer.

admirer, *v.t.* admire; wonder at.

admis, *p.a.* admitted, etc. (see admettre). — *p.def. 1 & 2 sing.* of admettre.

admissibilité, *f.* admissibility.

admit, *p.def. 3 sing.* of admettre.

admixtion, *f.* admixture, admixtion.

adonner, *v.r.* devote oneself, apply oneself (to), occupy oneself (with).

adopter, *v.t.* adopt.

adorer, *v.t.* adore.

ados, *m.* slope, talus; sloping bed (in gardens); (*Agric.*) ridge, or tillage in ridges.

adossé, *p.a.* having the back against something, supported at the back (said, e.g., of a table fastened to a wall).

adosser, *v.t.* set the back of (against), support at the back, lean, rest.

adoucir, *v.t.* soften; sweeten; smooth; polish; round off; (*Metal.*) anneal; make milder, mollify, soothe, relieve, moderate. — *v.r.* become soft, sweet or mild, be relieved or moderated.

adoucissage, *m.* smoothing, polishing; polish (of metals, etc.); polishing powder; softening (of colors).

adoucissant, *a.* softening; emollient, demulcent. — *m.* polishing substance; (*Med.*) emollient.

adoucissement, *m.* softening, sweetening, etc. (see adoucir); first polish (of metals).

adoucisseur, *m.* smoother, polisher.

adoux, *m.* woad.

adragant(h), *a. & m.* tragacanth.

adrénaline, *f.* adrenaline.

adrénalinique, *a.* adrenaline, of adrenaline.

adresse, *f.* address; dexterity, skill, cleverness.

adresser, *v.t.* address; send, forward. — *v.r.* address oneself; be addressed; apply; tend.

adroit, *a.* adroit, expert, clever.

adroitement, *adv.* adroitly, expertly, cleverly.

adsorber, *v.t.* adsorb.

adsorption, *f.* adsorption.

adulte, *a. & m. or f.* adult.

adultérant, *p.a.* adulterating, adulterant.

adultération, *f.* adulteration.

adultérer, *v.t.* adulterate.

aduste, *a.* burnt, adust.

advenir, *v.i.* happen, befall.

adventice, *a.* adventitious.

adventif, *a.* adventive; adventitious.

advenu, *p.p.* of advenir.

advers, *a.* adverse, opposing, opposite.

adversaire, *m.* adversary, opponent, competitor.

adversité, *f.* adversity.

advient, *pr. 3 sing.* of advenir.

aérage, *m.* ventilation; airing.

aérateur, *m.* aërator; fan, ventilator.

aération, *f.* aëration; ventilation.

aéré, *p.a.* aërated; ventilated; aired, airy.

aérer, *v.t.* aërate; ventilate; air.

aérien, *a.* aërial; of the air, air; airy.

aérifère, *a.* air-conducting, air-.

aérifier, *v.t.* aërify.

aériforme, *a.* aëriform, gaseous.

aériser, *v.t.* aërify.

aérobie, *a.* aërobic. — *m.* aërobe.

aérodynamique, *f.* aërodynamics.

aérogène, *a.* See gaz aérogène, under *gaz.*

aérolithe, *m.* aërolite, aërolith.

aérolithique, *a.* aërolitic.

aéromètre, *m.* aërometer.

aérométrie, *f.* aërometry.

aérostat, *m.* aërostat (used in French only of lighter-than-air craft), balloon.

aérostation, *f.* art of constructing and navigating balloons, aërostation, aëronautics.

aérostatique, *a.* aërostatic. — *f.* aërostatics.

æsculine, *f.* esculin, æsculin.

aétite, *f.* eaglestone, ætites.

affadir, *v.t.* render insipid or flat, flatten, deaden; disgust. — *v.r.* become insipid or flat; pall.

affadissant, *p.a.* insipid, flat, vapid; tiresome.

affadissement, *m.* insipidity, flatness, loss of flavor.

affaibli, *p.a.* weakened, etc. (see affaiblir); dilute.

affaiblir, *v.t.* weaken; dilute; reduce, diminish; debase (coinage). — *v.i.* become weak, lose strength.

affaiblissement, *m.* weakening, etc. (see affaiblir); weakness; debility.

affaiblisseur, *a.* weakening; (*Photog.*) reducing. — *m.* (*Photog.*) reducer.

affaire, *f.* affair; thing, matter, question, etc.; (also *pl.*) business; occupation; undertaking; fortune; difficulty, danger; quarrel; battle, engagement; (legal) case; need, occasion.

avoir — à, have to do with, be concerned with; have business with.

affaissement, *m.* weighing down, sinking, etc. (see affaisser).

affaisser, *v.t.* (cause to bend or sink under a weight) weigh down, sink, settle; deject, depress. — *v.r.* sink, sag, settle, subside; collapse, give way; succumb.

affamé, *p.a.* famished, starved; eager.

affecter, *v.t.* affect; destine, allot.

affectionné, *a.* affectionate, devoted; beloved.

affectionner, *v.t.* love, like; interest.

afférent, *a.* (*Physiol.*) afferent; (*Law*) belonging, proper.

affermer, *v.t.* lease, rent.

affermir, *v.t.* make firm, firm; harden; strengthen, consolidate; fasten, fix; confirm. — *v.r.* become firm, hard, fast or strong; be consolidated.

affermissement, *m.* making firm, etc. (see affermir).

affichage, *m.* posting, etc. (see afficher).

affiche, *f.* bill, poster, placard.

afficher, *v.t.* post, post up; announce, publish; parade; expose.

affilage, *m.* sharpening.

affilé, *p.a.* sharpened, sharp.

affiler, *v.t.* sharpen.

affilié, *p.a.* affiliated. — *m.* member, associate.

affilier, *v.t.* affiliate. — *v.r.* join (an organization).

affin, *a.* similar; related, at-law.

affinage, *m.* refining, etc. (see affiner). — *anglais,* (*Metal.*) puddling.

affinement, *m.* refining, etc. (see affiner).

affiner, *v.t.* refine; fine; ripen (cheese); heckle (flax); point (needles, etc.). — *v.r.* become pure, become refined; (of cheese) ripen; (of weather) clear.

affinerie, *f.* refinery.

affineur, *m.* refiner.

affinitaire, *a.* of or pertaining to affinity.

affinité, *f.* affinity.

affirmativement, *adv.* affirmatively.

affirmer, *v.t.* affirm.

affleurage, *m.* (*Paper*) thinning the pulp.

affleuré, *p.a.* level, flush; properly mixed.

affleurement, *m.* leveling; (*Geol. & Mining*) outcropping, outcrop.

affleurer, *v.t.* make level or flush; be even or level with; come to the level of; bring to the proper level; mix in proper proportion (as flours). — *v.i.* be level or flush; (of strata) crop out.

affleurie, *f.* finest flour.

affligé, *p.a.* afflicted.

affligeant, *p.a.* afflicting, distressing.

affliger, *v.t.* afflict, distress.

affluence, *f.* influx, inflow, affluence.

affluer, *v.i.* flow in, flow, run; throng; abound.

affolé, *p.a.* (of the magnetic needle) disturbed, deranged; infatuated.

affouage, *m.* (*Iron*) fuel supply.

affouillement, *m.* undermining, caving in, erosion; exfoliation (by rust); cavity; pitting.

affouiller, *v.t.* undermine, erode.

affranchir, *v.t.* free, liberate, deliver, exempt; pay, prepay; fire (casks); trim, trim off.

affranchissement, *m.* freeing, etc. (see affranchir).

affréter, *v.t.* charter (a vessel).

affreusement, *adv.* frightfully, horribly.

affreux, *a.* frightful, fearful; hideous; detestable.

affront, *m.* affront; disgrace.

affronter, *v.t.* place on the same level, end to end or face to face, join flush; front, face; cheat.

affusion, *f.* pouring on, affusion.

affût, *m.* watch, lookout; hiding place; carriage (of a gun).
— *à l'— de*, on the lookout for.

affûtage, *m.* sharpening.

affûter, *v.t* sharpen.

affûteur, *m.* sharpener.

afin de. in order (to).

afin que. in order that, so that.

africain, *a.* African.

Afrique, *f.* Africa.
— *du Sud*, South Africa.

agame, *a.* (*Bot.*) agamous.

agaric, *m.* agaric.
— *amadouvier*, tinder agaric (*Polyporus igniarius*).
— *blanc*, — *purgatif*, white agaric, purging agaric (*Polyporus officinalis* or a preparation of it).

agate, *f.* agate.

agaté, *a.* agaty, containing agate.

agatifier, agatiser, *v.t.* agatize.

agave, agavé, *m.* agave.

âge, *m.* age.

âgé, *a.* aged, old.

agence, *f.* agency.

agencement, *m.* arrangement; (*pl.*) fittings, fixtures.

agencer, *v.t.* arrange; dispose, adjust, order; fit up.

agenda, *m.* memorandum book, notebook.

agenouiller, *v.r.* kneel.

agent, *m.* agent; medium.
— *condensateur*, condensing agent.
— *de liaison*, means of union, (*Org. Chem.*) condensing agent.
— *réducteur*, reducing agent.
— *révélateur*, (*Photog.*) developer.
— *transmetteur*, carrying agent, carrier.

agglomérant, *m.* agglomerant, agglomerating agent; specif., (*Ceram.*) a substance which imparts hardness to the paste.

agglomérat, *m.* agglomerate.

agglomération, *f.* agglomeration.
grand —, (of people) crowded center, city.

aggloméré, *p.a.* agglomerated. — *m.* agglomerate, mass; specif.: fuel made into the form of briquets or lumps from coal dust or the like; unrefined sugar molded into briquets; mass composed of a mixture of manganese dioxide and coke for Leclanché cells.
— *combustible*, fuel briquet.
— *magnésien*, magnesite brick, magnesite block.

agglomérer, *v.t. & r.* agglomerate, form or collect into a mass.

agglutinage, *m.* agglutination; fritting, sintering.

agglutinant, *a. & m.* agglutinant.

agglutinatif, *a.* agglutinant, agglutinative.
— *m.* agglutinant.

agglutiner, *v.t.* agglutinate. — *v.r.* be agglutinated.

agglutinine, *f.* agglutinin.

aggraver, *v.t.* aggravate.

agir, *v.i.* act; be in action; take action.
— *en*, act as, behave as.
il s'agit de, it is a question of, it concerns.

agissant, *p.a.* acting; active.

agitant, *p.a.* agitating.

agitateur, *m.* stirrer, specif. stirring rod; shaker; agitator.
— *à bouton*, stirring rod with a disk-shaped end.
— *mécanique*, mechanical stirrer or shaker; (*Brewing*) mashing apparatus.
— *ordinaire*, ordinary glass stirring rod.
— *va-et-vient*, shaker with a reciprocating motion.

agitation, *f.* agitation (stirring, shaking, etc.).

agiter, *v.t.* agitate (stir, shake, disturb, etc.).
— *v.r.* be agitated, be in motion.

aglomérant, aglutinage, etc. See agglomérant, agglutinage, etc.

agneau, *m.* lamb.

agneline, *a.* In *laine agneline*, lamb's wool.

agrafe, *f.* fastening (of various kinds), fastener, clamp, cramp, clasp, clip, hook, catch, snap.

agrafer, *v.t.* fasten (with an *agrafe*), clamp, cramp, clasp, clip, hook, lock.

agrandir, *v.t.* enlarge; extend; exaggerate; aggrandize, exalt. — *v.r.* enlarge, become larger; increase, grow, extend.

agrandissement, *m.* enlargement, etc. (see agrandir).

agréable, *a.* agreeable, pleasant, pleasing.

agréablement, *adv.* agreeably, pleasantly.

agréer, *v.t.* accept, receive favorably; approve; permit. — *v.i.* be pleasing, be acceptable.

agrégat, *m.* aggregate.

agrégation, *f.* aggregation; aggregate; admission (into a society or company); (university) fellowship, admission to a fellowship.

agrégé, *m.* a person who has been declared fitted to teach in a secondary school or in a university, fellow, temporary professor.

agréger, *v.t.* aggregate; admit (into a society).

agrément, *m.* consent, approval; agreeableness, pleasantness, charm; pleasure; ornament.

agrès, *m.pl.* apparatus, equipment, implements, fittings.

agressivement, *adv.* aggressively.

agricole, *a.* agricultural. — *m.* agriculturist.

agriculteur, *m.* agriculturist. — *a.* agricultural.

agripaume, *f.* motherwort (*Leonurus cardiaca*).

agronome, *m.* agronomist. — *a.* agronomic.

agronomique, *a.* agronomic.

agrostide, *f.* bent grass (*Agrostis*).

aguerrir, *v.t.* inure, harden, season.

ai, *pr. 1 sing.* of avoir (to have).

aidant, *p.a.* aiding. — *m.* aid, helper.

aide, *f.* aid, help, assistance. — *m. & f.* assistant, helper, aid.

— *de laboratoire,* laboratory assistant.

à l'— de, with the aid of.

aider, *v.t.* assist, help, aid. — *v.r.* (with *de*) make use (of); help oneself or one another; exert oneself.

— *en s'aidant de,* with the aid of, with the employment of.

aide-souffleur, *m.* (*Glass*) blower's assistant.

aïeul, *m.* grandfather; (*pl.*) ancestors.

aigle, *m.* eagle.

aigre, *a.* sour, tart, sharp; (of metals) brittle, cold-short; piercing, keen, harsh. — *m.* sourness, tartness, etc.

sentir l'—, smell sour.

tourner à l'—, turn sour.

aigre-doux, *a.* between sweet and sour, subacid, sourish.

aigrelet, *a.* sourish, subacid, slightly sour.

aigremoine, *f.* (*Bot.*) agrimony.

aigremore, *m.* powdered soft wood charcoal, used in pyrotechnics.

aigret, *a.* slightly sour, sourish, subacid.

aigrette, *f.* aigrette, egret; (*Zoöl.*) egret.

aigretté, *a.* having an aigrette.

aigreur, *f.* sourness, tartness, acidity; (of metals) brittleness, cold-shortness; harshness; (*Med., pl.*) acidity of the stomach.

avoir de l'—, be sour.

aigri, *p.a.* soured, etc. (see aigrir).

aigrir, *v.t.* sour, turn sour; irritate; aggravate. — *v.r.* sour, turn sour; become irritated or aggravated.

aigrissement, *m.* souring, etc. (see aigrir).

aigu, *a.* pointed; acute; (of needles, etc.) sharp; (of sounds) shrill.

aiguayer, *v.t.* rinse, wash, water.

aigue-marine, *f.* aquamarine.

aiguille, *f.* needle; pointer, index, hand; spire, pinnacle, peak; (*Railways*) switch, points.

— *à coudre,* sewing needle.

— *aimantée,* magnetic needle.

— *à tricoter,* knitting needle.

— *de boussole,* compass needle.

— *de fusil,* firing pin.

— *de pin,* pine needle.

— *en fil de platine,* platinum needle.

aiguillé, *a.* needle-shaped, acicular.

aiguillette, *f.* long, narrow slice or strip; (of firearms) firing pin; (of costumes) aiguillette.

aiguillon, *m.* thorn, spine, prickle; sting (of bees, etc.); goad; sharp peak.

aiguillonner, *v.t.* goad; sting.

aiguisage, aiguisement, *m.* acidification; sharpening.

aiguiser, *v.t.* acidify; sharpen.

ail, *m.* any species of *Allium*; usually, garlic (*Allium sativum*).

— *d'Ascalon,* shallot.

aile, *f.* wing (in various senses); flange; blade (of a propeller); sail (of a windmill); tooth (of a pinion); ale.

— *de moulin,* (*Glass*) a warped form assumed by cylinder glass when the annealing furnace is too cold.

ailé, *a.* winged.

aileron, *m.* blade, paddle (as on a mechanical stirrer or a water wheel); pinion (end of a bird's wing); fin; (of an airplane) aileron.

ailette, *f.* small wing, blade, fin; lug, flange, stud, tenon.

aille, *pr. 1 & 3 sing. subj.* of aller (to go).

aillent, *pr. 3 pl. subj.* of aller.

ailleurs, *adv.* elsewhere, somewhere else, anywhere else.

d'—, moreover, besides, further, in addition; otherwise.

par —, by another way, by other means.

aimable, *a.* lovely, amiable, kind.

aimant, *m.* magnet; loadstone.

— *en fer à cheval,* horseshoe magnet.

aimantaire, *a.* magnetic.

aimantation, *f.* magnetization.

aimanté, *p.a.* magnetized, magnetic.

aimanter, *v.t.* magnetize. — *v.r.* be magnetized, become magnetic.

aimantin, *a.* magnetic; adamantine.

aimer, *v.t.* love, like, be fond of.

— *mieux,* like better, prefer.

s'— à, like, be fond of.

aine, *m.* groin.

-aine. A suffix used in forming nouns from numeral adjectives; as, une cinquant*aine,* fifty, half a hundred (not so definite as cinquante).

aîné, *a.* eldest, elder, senior.

ainsi, *adv.* so, thus. — *conj.* so, thus; therefore.

— *que,* as, as well as.

par —, therefore, consequently.

pour — dire, so to say, as it were.

air, *m.* air; (in furnaces) blast.

— *ambiant,* atmosphere.

— *carburé,* carburetted air.

— *comprimé,* compressed air.

— *fixe,* (*Old Chem.*) fixed air (carbon dioxide).

— *inflammable,* (*Old Chem.*) inflammable air (hydrogen gas).

— *vicié,* vitiated air, specif. choke damp.

— *vital,* (*Old Chem.*) vital air (oxygen gas).

airain, *m.* bronze; brass (in the old indefinite sense of copper alloy).

airain *de Corinthe*, Corinthian bronze, Corinthian brass (alloy of copper, gold and silver).

aire, *f.* area; (flat) surface; floor, floor space; platform; hearth, sole (of a furnace); bed; face (as of a hammer); aerie.
— *de grillage*, roasting hearth, roasting furnace.
— *de vent*, point of the compass.

airelle, *f.* whortleberry, blueberry (*Vaccinium*).

ais, *m.* board; (of a cask) stave; shelf.
— *de carton*, pasteboard.

aisance, *f.* ease; (*pl.*) water closet.

aise, *f.* ease; (*pl.*) comforts; joy, enjoyment.
— *a.* glad, content.

aisé, *a.* easy; well off.

aisément, *adv.* easily.

aissante, *f.*, **aisseau,** *m.* shingle.

aisselle, *f.* (*Anat.*, etc.) axilla.

ajointer, *v.t.* join end to end.

ajouré, *p.a.* containing openings, perforated.

ajourer, *v.t.* make openings in, perforate.

ajournement, *m.* adjournment, etc. (see ajourner).

ajourner, *v.t.* adjourn; postpone; assign, summon. — *v.r.* adjourn, be adjourned; be postponed.

ajouté, *m.* addition.

ajouter, *v.t. & n.* add. — *v.r.* be added.
— *foi à*, put faith in, believe.

ajoutoir, *m.* = ajutage.

ajustage, *m.* adjusting, fitting, finishing.

ajustement, *m.* adjustment, etc. (see ajuster).

ajuster, *v.t.* adjust; fit; fit up, arrange; tune; aim, aim at; attire, adorn; (*Mach.*) fit.
— *v.r.* be adjusted; be fitted; fit; agree; be reconciled; dress.

ajusteur, *m.* adjuster; fitter; sizer (of coins).

ajustoir, *m.* coin balance.

ajutage, *m.* outlet tube, efflux tube, discharge tube or pipe, ajutage; (of a burette) tip.

ajutoir, *m.* = ajutage.

alabandine, *f.* (*Min.*) alabandite.

alambic, *m.* still; (*Old Chem.*) alembic.
— *à siphon*, siphon still.
— *d'essai*, testing still, still for analytical work.

alambiquer, *v.t.* distill; make too subtle; puzzle.

alandier, *m.* alandier (fireplace placed at the base of certain special furnaces or kilns and fired from the outside).

alanine, *f.* alanine.

alarmant, *p.a.* alarming.

alarme, *f.* alarm.

alarmer, *v.t.* alarm. — *v.r.* be alarmed; be uneasy.

albâtre, *m.* alabaster; a kind of white flint glass.
— *calcaire*, calcareous alabaster.

albâtre *gypseux*, gypseous alabaster.
— *oriental*, Oriental alabaster (calcareous alabaster).

alberge, *f.* clingstone peach (**or** apricot).

albinisme, *m.* albinism.

albinos, *m. & a.* albino.

albugineux, *a.* albugineous.

albumen, *m.* albumen.
— *corné*, horny albumen.

albumine, *f.* albumin.

albuminé, albumineux, *a.* albuminous; (of paper) albuminized, albumen.

albuminimètre, *m.* albuminimeter.

albuminoïde, *a. & m.* albuminoid.

albuminurie, *f.* albuminuria.

albuminurique, *a.* albuminuric.

albumoïde, *m.* albumoid.

albumose, *n.* albumose.

alcalamide, *f.* (*Org. Chem.*) alkalamide (an *N*-acyl derivative of an amine, whether aliphatic or not).

alcalescence, *f.* alkalescence, alkalescency.

alcalescent, *a.* alkalescent.

alcali, *m.* alkali.
— *carbonaté*, alkali carbonate.
— *volatil*, volatile alkali (old name for ammonia).
— *volatil concret*, ammonium carbonate (old name).

alcalifiant, *a.* alkalifying.

alcaligène, *a.* alkaligenous. — *m.* alkaligen (an old name for nitrogen).

alcalimètre, *m.* alkalimeter.

alcalimétrie, *f.* alkalimetry.

alcalimétrique, *a.* alkalimetric(al).

alcalin, *a.* alkaline. — *m.* alkaline substance, alkali; alkali metal.

alcalinisation, *f.* alkalization.

alcaliniser, *v.t.* render alkaline, alkalize.

alcalinité, *f.* alkalinity.

alcalino-barytique, *a.* (*Ceram.*) alkalino-barytic, containing alkali metal and barium.

alcalino-plombifère, *a.* (*Ceram.*) alkalino-plumbiferous, containing alkali metal and lead.

alcalino-salé, *a.* (*Soap*) alkalino-saline, containing alkali and salt.

alcalino-siliceux, *a.* alkalinosiliceous, containing alkali and silica.

alcalino-terreux, *a.* alkaline-earth. — *m.* alkaline-earth metal.

alcalisation, *f.* alkalization.

alcaliser, *v.t.* alkalize, render alkaline.

alcaloïde, *a.* alkaloid, alkaloidal. — *m.* alkaloid.
— *animal*, animal alkaloid.
— *végétal*, vegetable alkaloid, plant alkaloid.

alcaloïdique, *a.* alkaloid, alkaloidal.

alcarazas, alcarraza, *m.* alcarraza (water cooler of porous earthenware).

alchimie, *f.* alchemy.
alchimique, *a.* alchemical, alchemic, alchemistic(al).
alchimiste, *m.* alchemist.
alcool, *m.* alcohol.
— *absolu,* absolute alcohol.
— *allylique,* allyl alcohol.
— *amylique,* amyl alcohol.
— *amylique de fermentation,* amyl alcohol of fermentation (a mixture of 2- and 3-methylbutanol).
— *anisique,* anisyl alcohol.
— *benzylique,* benzyl alcohol.
— *butylique,* butyl alcohol.
— *campholique,* borneol.
— *camphré,* spirit of camphor.
— *caprique,* decyl alcohol.
— *caproïque,* hexyl alcohol.
— *caprylique,* octyl alcohol.
— *carburé,* carburetted alcohol.
— *cérylique* (or *cérotique*), ceryl alcohol.
— *cinnamique,* cinnamic alcohol.
— *coniférylique,* coniferyl alcohol.
— *de betteraves,* beet alcohol, beetroot spirit.
— *de cœur,* in the rectification of alcohol, the third portion to distill over, following the *moyens goûts* and preceding the *bon goût.*
— *de fécule,* starch spirit, fecula brandy.
— *de menthe,* spirit of peppermint.
— *dénaturé,* denatured alcohol.
— *dilué,* diluted alcohol, dilute alcohol.
— *d'industrie,* industrial alcohol.
— *éthalique,* cetyl alcohol, ethal.
— *éthylénique,* glycol, ethylene glycol (1, 2-ethanediol).
— *éthylique,* ethyl alcohol.
— *fin,* in the rectification of alcohol, the purest portion of the distillate, Cologne spirit. It is the first part of the *bon goût.*
— *furfurolique,* furfuryl alcohol (2-furan-carbinol).
— *hexylique,* hexyl alcohol.
— *industriel,* industrial alcohol, alcohol for use in the arts.
— *mauvais goût,* amyl alcohol of fermentation (a mixture of 2- and 3-methyl-1-butanol).
— *mélissique,* melissyl (myricyl) alcohol.
— *méthylique,* methyl alcohol.
— *myricique,* myricyl alcohol.
— *ordinaire,* common or ordinary alcohol (ethyl alcohol; specif., a distillate of about 80 to 95 per cent strength).
— *phényléthylique,* phenethyl alcohol.
— *pipéronylique,* piperonyl alcohol.
— *primaire,* primary alcohol.
— *propargylique,* propargyl alcohol.
— *propylique,* propyl alcohol.
— *secondaire,* secondary alcohol.
— *subérique* (or *subérylique*), suberyl alcohol (cycloheptanol).

alcool *sulfuré,* mercaptan, thiol.
— *supérieur,* higher alcohol.
— *terpénique,* terpene alcohol.
— *tertiaire,* tertiary alcohol.
— *thionylique,* see thionylique.
— *vanillique,* vanillic alcohol, vanillyl alcohol.
— *vinique,* vinic alcohol, ethyl alcohol.
alcoolase, *f.* alcoholase (zymase).
alcoolat, *m.* (*Pharm.*) spirit (strictly, one made from alcohol and a plant by distillation; one made simply by dissolving the volatile principle in alcohol is properly called *teinture d'essence*).
— *ammoniacal aromatique,* aromatic spirit of ammonia.
— *ammoniacal fétide,* fetid spirit of ammonia.
— *antiscorbutique,* compound spirit of horse-radish.
alcoolate, *m.* (*Org. Chem.*) alcoholate.
alcoolature, *f.* (*Pharm.*) tincture of a fresh herb or herbs, alcoholature.
alcoolé, *m.* (alcoholic) tincture.
(For the meaning of phrases see the corresponding phrases under *teinture.*)
alcoolification, *f.* conversion into alcohol, alcoholification.
alcoolique, *a.* alcoholic.
alcooliquement, *adv.* alcoholically.
alcoolisable, *a.* alcoholizable.
alcoolisation, *f.* alcoholization.
alcooliser, *v.t.* alcoholize.
alcoolisé, *p.a.* alcoholized; containing alcohol.
alcoolisme, *m.* alcoholism.
alcoolomètre, *m.* alcoholometer.
alcoolométrie, *f.* alcoholometry.
alcoolyse, *f.* alcoholysis.
alcoolyser, *v.t.* alcoholyze.
alcoomètre, *m.* alcoholometer.
alcoométrie, *f.* alcoholometry.
alcooscopie, *f.* alcoholoscopy, alcoholometry.
alcoosol, *m.* alcosol.
alcoylation, *f.* alkylation.
alcoyle, *m.* alkyl.
alcoylé, *p.a.* alkylated; alkyl (as, *radical alcoylé,* alkyl radical).
alcoyler, *v.t.* alkylate.
aldéhydate, *m.* compound with an aldehyde.
— *d'ammoniaque,* aldehyde ammonia.
aldéhyde, *m.* aldehyde.
— *acétique,* acetic aldehyde, acetaldehyde.
— *allylique,* acrolein.
— *anisique,* anisic aldehyde, anisaldehyde.
— *benzoïque,* benzoic aldehyde, benzaldehyde.
— *butylique,* butyl aldehyde (either normal butyraldehyde or isobutyraldehyde).
— *cinnamique,* cinnamic aldehyde, cinnam-aldehyde.
— *crotonique,* crotonic aldehyde, crotonalde-hyde.

aldéhyde *cuminique*, cumic aldehyde, cumaldehyde, cuminaldehyde.
— *éthalique*, palmitaldehyde.
— *éthylique*, ethyl aldehyde (acetaldehyde).
— *formique*, formaldehyde, formic aldehyde.
— *glycolique*, glycolaldehyde, glycolic aldehyde.
— *œnanthylique*, enanthaldehyde, enanthole, œnanthol.
— *protocatéchique*, protocatechualdehyde, protocatechuic aldehyde.
— *salicylique*, salicylaldehyde, salicylic aldehyde.
— *toluique*, tolualdehyde, toluic aldehyde.
— *trichloré*, trichloroacetaldehyde, chloral.
— *valérique*, valeraldehyde, valeric aldehyde.
— *vinique*, acetaldehyde.
aldéhydique, *a.* aldehydic, aldehyde (as, *groupement aldéhydique*, aldehyde group).
aldohexose, *f.* aldohexose.
aldol, *m.* aldol.
aldolisation, *f.* aldolization.
aldoliser, *v.t.* aldolize.
aldonique, *a.* aldonic.
aldose, *n.* aldose.
aldoxime, *m.* aldoxime.
ale, *f.* ale.
aléa, *m.* chance, hazard.
aléatoire, *a.* hazardous, uncertain.
alène, *f.* awl.
alentour, *adv.* around, about. — *m.pl.* environs.
alerte, *a.* alert, quick. — *f.* alert, alarm.
— *au gaz*, (*Mil.*) gas alarm.
alésage, *m.* bore, internal diameter; boring smooth.
aléser, *v.t.* bore, bore smooth, drill.
aléseuse, *f.* drill, drilling machine.
alésoir, *m.* boring tool or machine, borer, broach, reamer, bit.
alésure, *f.* borings.
alétris, *m.* (*Bot.*) Aletris.
— *farineux*, colic root (*A. farinosa*).
aleuromètre, *m.* aleurometer.
aleurone, *f.* aleurone.
alexine, *f.* alexin.
alfa, *m.* esparto, alfa (*Stipa tenacissima*).
alfénic, *m.* alphenic, sugar candy.
alfénide, *m.* alfenide.
algarobille, *n.* algarobilla.
algaroth, *m.* powder of Algaroth, algaroth.
algèbre, *f.* algebra.
algébrique, *a.* algebraic(al).
algébriquement, *adv.* algebraically.
Alger, *n.* Algiers.
Algérie, *f.* Algeria.
algérien, *a.* Algerian, Algerine.
algue, *f.* (*Bot.*) alga (*pl.* algæ).
alidade, *f.* alidade.

aliéné, *p.a.* alienated; insane.
aliéner, *v.t.* alienate; sell; transfer. — *v.r.* be alienated; become insane.
alignement, *m.* alignment; line, straight line.
aligner, *v.t.* align; line up, arrange, provide.
aliment, *m.* food; nourishment; sustenance; (*Mach.*) material supplied, supply, feed; (*Bact.*) medium.
— *d'épargne*, protein-sparing food.
— *respiratoire*, respiratory food.
alimentaire, *a.* alimentary, for or pertaining to food; (*Mech.*) feed, feeding.
alimentateur, *m.* (*Mech.*) feeder, feed. — *a.* alimentary, nourishing.
alimentation, *f.* alimentation, nutrition; supplying, supply; (*Mech.*) feeding, feed.
alimenter, *v.t.* feed; supply; nourish; maintain.
alimenteux, *a.* nourishing, nutritive, alimental.
alinéa, *m.* paragraph; break.
aliphatique, *a.* aliphatic.
aliquote, *a.* aliquot. — *f.* aliquot part.
alizarine, *f.* alizarin.
alkékenge, *m.* alkekengi (*Physalis alkekengi*).
alkermès, *m.* (*Old Pharm.*) alkermes.
alkyler, *v.t.* alkylate.
alkylurée, *f.* alkylurea.
alla, *p.def. 3 sing.* of aller (to go).
allaitement, *m.* (infant) feeding, suckling.
allantoïde, *f.* (*Anat.*) allantois.
allantoïne, *f.* allantoin.
allé, *p.p.* of aller (to go).
allégement, *m.* lightening, easing, relief.
alléger, *v.t.* lighten, ease, relieve, alleviate.
allégir, *v.t.* reduce in size or thickness.
allègre, *a.* brisk, lively; vivacious; gay.
allègrement, *adv.* briskly; gaily.
allégresse, *f.* briskness, liveliness; vivacity.
alléluia, *m.* wood sorrel (*Oxalis acetosella*).
Allemagne, *f.* Germany.
allemand, *a. & m.* German.
allène, *m.* allene.
allénique, *a.* allene (as, *hydrocarbure allénique*. allene hydrocarbon).
aller, *v.i.* go; be in motion; extend; progress; continue, last; tend; amount; fit; be becoming, harmonize; (with infinitive) be going, be about (to). — *v.r.* (with *en*) go away, leave, disappear; escape; leak; evaporate; die; be nearly; (with infinitive) be going, be about (to). — *m.* going, journey.
— *à la lessive*, (of fabrics) stand lye, wash.
— *au-devant de*, meet, go to meet.
— *au feu*, stand fire, endure fire.
— *chercher*, see under *chercher*.
— *de soi*, go without saying, be self-evident.
— *ensemble*, go together, harmonize, make a set.

aller *trouver*, seek out.
allons, come! let's go!
en —, go away, disappear, etc. (see aller, *v.r.*); turn out, happen.
faire —, make go, start, drive, run.
s'en —, see aller, *v.r.*
y —, go at it, proceed; be a question (of).
alliacé, *a.* alliaceous.
alliage, *f.* alloy; (*Arith.*) alligation.
— *de Wood*, Wood's alloy, Wood's metal.
— *de Rose*, Rose's alloy, Rose's metal.
alliance, *f.* alliance.
allié, *p.a.* alloyed; allied. — *m.* (and **alliée**, *f.*) ally; relation by marriage.
allier, *v.t.* alloy; unite; blend; ally; marry.
— *v.r.* alloy, form an alloy; unite; become allied; harmonize, blend; intermarry.
allocation, *f.* allowance, grant.
allocinnamique, *a.* allocinnamic, *allo*-cinnamic.
allonge, *f.* adapter; percolator, extraction tube; the jacket of a certain type of upright condenser; extension piece, eking piece; (*Mach.*) coupling rod.
— *à déplacement*, percolator.
allongement, *m.* lengthening, elongation, prolongation, extension, stretching, stretch.
allonger, *v.t.* lengthen, elongate, prolong; stretch out, extend; draw out, (of metals) wiredraw. — *v.r.* lengthen, grow longer; extend; be extended.
allons, *pr. indic. & imper. 1 pl.* of aller (to go).
allophane, *f.* (*Min.*) allophane.
allotropie, *f.* allotropy.
allotropique, *a.* allotropic.
allotropisme, *m.* allotropism.
allouer, *v.t.* allow, grant.
alloxane, *m.* alloxan.
alloxanique, *a.* alloxanic.
alloxantine, *f.* alloxantin.
alluchon, *m.* (*Mach.*) cog, tooth.
allumage, *m.* lighting; ignition (as in explosion engines).
allume-feu, *m.* kindling, piece of kindling.
allume-gaz, *m.* gas lighter.
allumer, *v.t.* light, kindle, set fire to; fire (a furnace, or the like); excite. — *v.r.* light, take fire, light up; be excited.
allumette, *f.* match.
— *à friction*, friction match.
— *amorphe*, safety match.
— *chimique*, chemical match.
— *congrève*, Congreve match.
— *de sûreté*, safety match; fusee.
— *soufrée*, sulfur match, brimstone match.
— *phosphorique*, phosphorus match.
— *suédoise*, Swedish match; safety match.
allumette-bougie, *f.* wax match, vesta.
allumeur, *m.* igniter; primer; lighter.
allumière, *f.* match factory; match box.

allumoir, *m.* lighter. — *a.* lighting.
allure, *f.* course, direction; working (of a furnace or kiln); gait, pace; conduct.
— *crue*, (*Metal.*) poor working, irregular working.
alluvial, alluvien, *a.* alluvial.
alluvion, *m.* alluvium, alluvion.
alluvionnaire, *a.* alluvial.
alluvionnement, *m.* alluviation.
allyle, *m.* allyl.
allylé, *a.* allyl.
allylène, *m.* allylene.
allylénure, *m.* a metallic derivative of allylene, allylenide (as, *allylénure de mercure*).
allylique, *a.* allyl.
almanach, *m.* almanac.
almandine, *f.* (*Min.*) almandite, almandine.
aloès, aloës, *m.* aloes, aloe.
— *caballin*, caballine aloes, horse aloes.
— *dépuré*, purified aloes.
— *hépatique des Barbades*, Barbados aloes.
— *socotrin* (or *sucotrin*), Sokotrine aloes.'
aloétine, *f.* aloëtin.
aloétique, *a. & m.* aloetic.
aloi, *m.* standard fineness, standard; alloy; quality.
aloïne, *f.* aloin.
aloïque, *a.* aloetic.
alonge, etc. See allonge, etc.
alors, *adv.* then, at that time, in that case.
— *que*, when; even if, altho.
jusqu' —, till then.
alouette, *f.* lark.
alourdir, *v.t.* render heavy; increase (weight); load (fabrics, paper, etc.); render dull.
alourdissement, *m.* rendering heavy, etc. (see alourdir); heaviness, dullness.
aloyage, *m.* bringing to standard fineness; alloying; alloy.
aloyer, *v.t.* bring (gold or silver) to standard fineness; alloy.
alpaca, alpaga, *m.* alpaca.
Alpes, *f.pl.* Alps.
alpestre, *a.* Alpine; (*Bot.*) alpestrine, subalpine.
alphabétique, *a.* alphabetic(al).
alphabétiquement, *adv.* alphabetically.
alphabétiser, *v.t.* alphabetize, alphabet.
alphénic, alphœnix, *m.* alphenic, sugar candy.
alphénol, *m.* a compound which is both an alcohol and a phenol, alphenol.
alpin, *a.* Alpine, alpine.
alquifoux, *m.* (*Ceram.*) alquifou.
alsacien, *a.* Alsatian.
altérabilité, *f.* alterability.
altérable, *a.* alterable; unstable; (of foods) perishable.
altérant, *p.a.* alterant; causing thirst; (*Med.*) alterative. — *m.* alterant; (*Med.*) alterative.

altératif, *a.* alterative.

altération, *f.* alteration, adulteration, etc. (see altérer).

altéré, *p.a.* altered, adulterated, etc. (see altérer).

altérer, *v.t.* alter; specif., change for the worse, impair, spoil, adulterate (food), debase (coin), disturb; make thirsty. — *v.r.* alter, be altered, undergo change, deteriorate, spoil. — *v.i.* excite thirst.

alternance, *f.* alternation.

alternant, *p.a.* alternating; alternate.

alternateur, *m.* alternator.

alternatif, *a.* alternate; alternative; (*Elec.*) alternating; (*Mach.*) reciprocating.

alternation, *f.* alternation; (of crops) rotation.

alternative, *f.* alternation; alternative.

alternativement, *adv.* alternately, in succession.

alterne, *a.* (*Geom. & Bot.*) alternate.

alterner, *v.i.* alternate; rotate crops.

althée, *m.* (*Bot.*) althea.

althéine, *f.* altheine (asparagine).

alude, *f.* (*Leather*) aluta.

aludel, *m.* aludel.

aluine, *f.* wormwood (old name).

alumelle, *f.* blade (of a knife or the like).

alumen, *m.* (*Pharm.*) alumen, alum.

aluminage, *m.* alumination; specif., (*Dyeing*) treatment with an aluminium mordant.

aluminaire, *a.* aluminiferous.

aluminate, *m.* aluminate.

alumine, *f.* alumina.

aluminé, *a.* aluminous.

aluminer, *v.t.* aluminate, aluminize.

alumineux, *a.* aluminous.

aluminium, *m.* aluminium, aluminum.

aluminium-éthyle, *m.* aluminium ethyl.

aluminiumferrosilicium, *m.* an alloy of aluminium, iron and silicon.

aluminium-méthyle, *m.* aluminium methyl.

alumino-ferrique, alumino-ferrugineux, *a.* of or pertaining to aluminium and iron, alumino-ferric.

alumino-métallurgie, *f.* aluminometallurgy.

alumino-thermie, *f.* aluminothermy.

alumino-thermique, *a.* aluminothermic.

alun, *m.* alum.

— *ammoniacal,* ammonium alum, ammonia alum.

— *calciné,* — *brûlé,* burnt alum.

— *d'ammoniaque,* ammonia alum, ammonium alum.

— *de chrome,* chrome alum.

— *de fer,* iron alum.

— *de fer ammoniacal,* ammonium iron alum.

— *de plume,* feather alum (halotrichite, alunogen or epsomite).

— *de potasse,* potash alum, potassium alum.

— *de roche,* rock alum (Roman alum).

alun *de Rome,* Roman alum.

— *de soude,* soda alum.

— *desséché,* burnt alum.

— *ordinaire,* common alum (potassium alum).

— *romain,* Roman alum.

alunage, *m.* aluming.

alunation, *f.* preparation of alum.

aluné, *p.a.* alumed, containing alum.

aluner, *v.t.* alum.

alunerie, *f.* alum works.

alunière, *f.* alum pit; alum works.

alunifère, *a.* aluminiferous.

alunite, *f.* (*Min.*) alunite.

alunogène, *m.* (*Min.*) alunogen.

alute, *f.* (*Leather*) aluta.

alvéolaire, *a.* alveolar.

alvéole, *m. & f.* cell, small cavity; socket; seat (mechanical sense); alveolus.

alysson, alysse, *m.* (*Bot.*) Alyssum.

Am, *symbol* for ammonium.

amabilité, *f.* amiability, kindness.

amadou, *m.* amadou, punk, (German) tinder.

amadouerie, *f.* amadou factory.

amadouvier, *m.* tinder fungus; specif., tinder agaric (*Polyporus igniarius*).

amaigri, *p.a* thin, lean.

amaigrir, *v.t.* make thin, thin, reduce; impoverish (soil); (*Ceram.*) thin, lean (clay). — *v.r.* become thin or lean; shrink.

amaigrissant, *p.a.* making thin, etc. (see amaigrir). — *m.* (*Ceram.*) thinning agent, leaning material.

amaigrissement, *m.* growing thin or lean; loss of flesh, emaciation.

amalgamation, *f.* amalgamation.

amalgame, *m.* amalgam.

amalgamer, *v.t. & i.* amalgamate.

amalgameur, *m.* amalgamator.

amande, *f.* almond; kernel; nucleus.

— *amère,* bitter almond.

— *douce,* sweet almond.

en —, almond-shaped.

amandé, *a.* containing or treated with almonds. — *m.* almond milk, milk of almonds.

amandine, *f.* a cosmetic having an almond base; (*Org. Chem.*) amandin.

amanite, *f.* (*Bot.*) amanita.

amant, *m.,* **amante,** *f.* lover.

amarante, *f.* amaranth.

amarescent, *a.* bitterish, slightly bitter.

amarine, *f.* amarine.

amas, *m.* mass; heap; accumulation; crowd; (*Mining*) pipe, pipe vein.

amasser, *v.t.* amass, collect, gather, accumulate. — *v.r.* collect, gather.

amateur, *m.* lover; amateur. — *a.* fond (of), interested (in).

amati, *p.a.* mat, unpolished, deadened, dead.

amatir, *v.t.* mat, render mat, deaden, blanch.

ambiant, a. atmospheric; ambient, surrounding.

ambigu, a. ambiguous.

ambiguïté, f. ambiguity, ambiguousness.

ambigument, adv. ambiguously.

ambitionner, v.t. aspire to.

ambre, m. amber.

— *blanc,* spermaceti.

— *gris,* ambergris.

— *jaune,* yellow amber, ordinary amber.

ambré, a. amber-colored, amber; ambered (perfumed with ambergris).

ambrette, f. amber seed, musk seed, ambrette (seed of the abelmosk).

ambroïde, m. ambroid.

ambroïne, f. ambroïn (an insulating material).

ambrosie, f. (Bot.) ragweed (Ambrosia).

âme, f. core; (of firearms) bore; (of bellows) valve; soul.

— *de chargement,* loading tube.

amélioration, f. improvement, betterment, amelioration; (Metal.) a treatment of ore to improve its condition, as by exposure to the weather or to water.

améliorer, v.t. improve, ameliorate; purify (metals); correct. — v.r. improve.

amenage, m. (Com.) carriage.

aménagement, m. management, etc. (see aménager).

en —, in preparation, under construction.

aménager, v.t. manage; prepare, fit up; saw (timber).

amende, f. fine.

amendement, m. improvement, amendment; (Agric.) amendment (a substance which improves the soil by modifying its physical properties rather than by supplying plant food; according to one authority *amendement* has been used to mean the amount of fertilizer necessary to replace the material removed by a crop).

amender, v.t. improve, amend, reclaim; improve (soil) by the use of amendments (see amendement). — v.i. improve, mend.

amenée, f. bringing, etc. (see amener); inlet, inflow, entrance.

amener, v.t. bring, lead, conduct, bring in, introduce, admit; lower (a flag).

amenuiser, v.t. make thin, reduce in size.

amer, a. bitter. — m. bitter substance; (pl.) bitters; gall.

— *de bœuf,* ox gall.

— *de Welter,* Welter, picric acid.

américain, a. & m. American.

Amérique, f. America.

— *du Nord, du Centre* (or *centrale*), *du Sud,* North, Central, South America.

amertume, f. bitterness.

améthyste, m. amethyst.

ameublir, v.t. (of soil) mellow, loosen.

ameublissement, m. mellowing, loosening (of soil).

ami, m. friend. — a. friendly.

amiantacé, a. amianthoid, amianthine.

amiante, m. (fibrous) asbestos, amianthus.

— *platiné,* platinized asbestos.

amianté, p.a. covered or treated with asbestos; as, toile métallique *amiantée* (asbestos wire gauze).

amiantin, a. asbestine, asbestos.

amiantine, f. asbestos cloth.

amibe, f. (Zoöl.) ameba, amœba.

amiboïde, a. (Biol.) ameboid, amœboid.

amidase, f. amidase.

amide, f. amide.

— *acétique,* acetamide, acetic amide.

— *acide,* acid amide.

amidé, a. amido; (when the substance is of the nature of an amine and not an amide) amino; (less correctly) ammino (as, *composés amidés du chrome,* ammino compounds of chromium).

amide-alcali, m. alcalamide.

amidification, f. conversion into an amide, amidification.

amidine, f. amidine (an aminoimino base); amidin (from starch).

amido-. amino-, amido- (should be translated amido- only when the compound is of amide nature, otherwise amino-).

amidoazoïque, a. aminoazo.

amidogène, m. amidogen (the group NH_2).

amidol, m. (Photog. & Chem.) amidol.

amidon, m. starch (especially in its industrial uses; cf. fécule).

— *nitré,* nitrated starch, nitro starch.

— *soluble,* soluble starch.

amidonné, p.a. starched; starched.

amidonner, v.t. treat with starch, starch.

amidonnerie, f. starch factory, starch works.

amidonnier, m. starch maker, starch manufacturer; starch dealer; a kind of grain similar to spelt.

amidonnière, f. a kind of kneading trough used in starch manufacture.

amidophénol, m. aminophenol.

amidure, m. (metallic) amide; any compound (as Hg_2NH_2Cl) in which ammoniacal hydrogen is replaced by a metal.

aminci, p.a. made thin or thinner, thin; slender. — m. any thin part of a device or mechanism.

amincir, v.t. make thin or thinner, reduce in thickness. — v.r. become thin.

amincissement, m. making thin, etc. (see amincir).

amine, f. amine.

— *acide,* amino acid, acid amine.

aminé, a. aminated, amino.

aminogène, m. aminogen, amidogen.

aminonaphtol, *m.* aminonaphthol.

aminophénol, *m.* aminophenol.

aminurée, *f.* substituted urea (formed from an amine).

amitié, *f.* friendship; favor; attraction; (of colors) harmony.

amméline, *f.* ammeline.

ammonal, *m.* (*Expl.*) ammonal.

ammoniac, *a.* ammoniacal; ammoniac.

ammoniacal, *a.* ammoniacal.

ammoniacalité, *f.* ammoniacal quality.

ammoniacaux, *pl. masc.* of ammoniacal.

ammoniacé, *a.* ammoniated, ammoniacal.

ammoniaco-. = ammonico-.

ammoniaco-cobaltique, *a.* of or pertaining to ammonium and cobalt.

ammoniaco-magnésien, *a.* of or pertaining to ammonium and magnesium (as, *phosphate ammoniaco-magnésien*, ammonium magnesium phosphate).

ammoniadynamite, *f.* (*Expl.*) ammonia dynamite.

ammoniagélatine, *f.* (*Expl.*) ammonia gelatin.

ammoniaque, *f.* ammonia.

— *composée,* (*Org. Chem.*) ammonia derivative.

— *liquide,* liquid ammonia.

— *liquide officinale,* (*Pharm.*) ammonia water, aqua ammoniæ.

ammoniate, *m.* ammoniate.

ammonico-. A combining form used in compound adjectives to denote *ammonium*.

ammonicomanganeux, *a.* of or pertaining to ammonium and bivalent manganese (as, *phosphate ammonicomanganeux*, ammonium manganous phosphate).

ammonicosodique, *a.* of or pertaining to ammonium and sodium.

ammonié, *a.* ammoniated.

ammonio-. ammonio-, of ammonia or ammonium.

ammonio-iridique, *a.* of or pertaining to ammonia and iridium.

ammonio-métallique, *a.* of or pertaining to ammonium and a metal.

ammonio-platinique, *a.* of or pertaining to ammonia and platinum.

ammonium, *m.* ammonium.

ammoniure, *m.* ammoniate.

amniotique, *a.* amniotic.

amoindrir, *v.t. & r.* diminish, decrease.

amollir, *v.t. & r.* soften; weaken.

amome, *m.* (*Bot.*) Amomum.

amonceler, *v.t. & r.* pile up, accumulate, amass.

amorçage, *m.* priming, etc. (see amorcer).

amorce, *f.* priming, primer; fuse; beginning, opening, entrance; bait; allurement; (*Drawing*) sketching, indication. — *a.* priming, starting, (of a burner) pilot.

— *à étincelles,* high-tension fuse.

amorce *à friction,* friction primer.

— *à percussion,* percussion primer.

— *cirée,* wax primer.

— *d'âme,* central primer.

— *de quantité,* quantity fuse, low-tension fuse.

— *de tension,* high-tension fuse.

— *périphérique,* rim-fire primer.

— *voltaïque,* low-tension fuse.

amorcement, *m.* priming, etc. (see amorcer).

amorcer, *v.t.* prime (a gun, a siphon, etc.); start (as a pump, a dynamo, crystallization); (*Tech.*) make ready in some way; bait.

amorceur, *m.* primer; starter; baiter. — *a.* priming, starting.

amorçoir, *m.* primer box; an instrument for beginning holes, as a center bit.

amordancer, *v.t.* mordant.

amorphe, *a.* amorphous.

amorphie, *f.* amorphism.

amorti, *p.a.* deadened, etc. (see amortir).

amortir, *v.t.* deaden; damp (vibrations); reduce in intensity, tone down, diminish, weaken; pay, pay for, liquidate, amortize.

amortissement, *m.* deadening, etc. (see amortir); amortization; finishing ornament, coping, finial.

amortisseur, *m.* deadener, damper, etc. (see amortir); (*Elec.*) dimmer; shock absorber.

amouille, *f.* colostrum of cows, beestings.

amouler, *v.t.* grind (on a grindstone).

amour, *m.* love.

amour-propre, *m.* self-love, self-esteem.

amovible, *a.* removable; revocable.

ampérage, *m.* (*Elec.*) amperage.

ampère, *m.* (*Elec.*) ampere.

ampère-heure, *m.* (*Elec.*) ampere hour.

ampèremètre, *m.* (*Elec.*) ammeter, ampere-meter.

amphibie, *a.* amphibious. — *m.* amphibian.

amphibole, *f.* (*Min.*) amphibole.

amphibolite, *f.* (*Petrog.*) amphibolite.

amphigène, *a.* (*Bot. & Old Chem.*) amphigenous. — *m.* amphigen; (*Min.*) leucite.

amphithéâtre, *m.* auditorium, lecture hall; amphitheater.

amphore, *f.* amphora.

amphotère, *a.* amphoteric.

amplement, *adv.* amply, fully.

ampleur, *f.* ampleness, breadth, fullness.

ampliation, *f.* copy, duplicate.

pour —, true copy (with an official certification).

amplifiant, *p.a.* amplifying; (*Optics*) magnifying.

amplificateur, *a.* amplifying; enlarging; exaggerating. — *m.* amplifier; (*Photog.*) enlarging camera.

amplificatif, *a.* amplifying; (*Optics*) magnifying.

amplification, *f.* amplification; magnification.

amplificatrice, *fem.* of amplificateur.

amplifier, *v.t.* amplify; magnify; exaggerate.

amplitude, *f.* amplitude.

ampoule, *f.* bulb; a bulbous vessel (see the phrases below); (*Pharm.*) ampoule; blister; ampulla.

— *à brome,* bromine funnel, dropping funnel.

— *à décantation,* — *à décanter,* separating funnel.

— *à robinet,* dropping funnel; separating funnel.

— *compte-gouttes,* medicine dropper.

— *de Crookes,* Crookes tube.

ampoulé, *a.* blistered; (of style) turgid.

ampouler, *v.t.* blister.

ampoulette, *f.* Diminutive of ampoule.

amusant, *p.a.* amusing, entertaining.

amusement, *m.* amusement; pretext; trifling.

amuser, *v.t.* amuse; deceive; waste.

amygdale, *f.* tonsil.

amygdalin, *a.* amygdaline (of or pertaining to almonds); tonsillar.

amygdaline, *f.* amygdalin (the glucoside).

amygdalique, *a.* amygdalic.

amygdaloïde, *f.* amygdaloid, amygdaloidal.

amylacé, *a.* amylaceous, starchy.

amylacétique, *a.* In the phrase *éther amyl-acétique,* amyl acetate.

amylamine, *f.* amylamine.

amylase, *f.* amylase.

amylazoteux, *a.* In the phrase *éther amylazoteux,* amyl nitrite.

amylbromhydrique, *a.* In the phrase *éther amylbromhydrique,* amyl bromide.

amylchlorhydrique, *a.* In the phrase *éther amylchlorhydrique,* amyl chloride.

amyle, *m.* amyl.

amylène, *m.* amylene.

amylformique, *a.* In the phrase *éther amylformique,* amyl formate.

amyliodhydrique, *a.* In the phrase *éther amyliodhydrique,* amyl iodide.

amylique, *a.* amyl; (of fermentation) amylic.

amylnitreux, *a.* = amylazoteux.

amyloïde, *f.* amyloid. — *a.* amyloid, amyloidal.

amylolytique, *a.* amylolytic.

amylopectine, *f.* amylopectin.

amylose, *f.* amylose.

an, *m.* year.

anacarde, *m.* fruit of the cashew, including the hypocarp (cashew apple) and nut (cashew nut); anacardium.

anacardier, *m.* (*Bot.*) anacardium.

anaérobie, *a.* (*Bact.*) anaërobic. — *m.* anaërobe.

analgésie, analgie, *f.* analgesia, analgia.

analgésine, analgine, *f.* analgesine (antipyrine).

analgésique, analgique, *a. & m.* analgesic, analgic.

analogie, *f.* analogy.

analogique, *a.* analogic, analogical.

analogiquement, *adv.* analogically.

analogue, *a.* analogous. — *m.* analog.

analysable, *a.* analyzable.

analyse, *f.* analysis.

— *chimique,* chemical analysis.

— *élémentaire,* elementary analysis, ultimate analysis.

— *immédiate,* proximate analysis.

— *spectrale,* spectrum analysis.

analyser, *v.t.* analyze.

analyseur, *m.* analyzer; analyst.

analyste, *m.* analyst. — *a.* analytical.

analytique, *a.* analytic, analytical. — *f.* analytics.

analytiquement, *adv.* analytically.

ananas, *m.* pineapple, ananas.

anaphylactique, *a.* anaphylactic.

anaphylatoxine, *f.* anaphylatoxin.

anaphylaxie, *f.* anaphylaxis.

anastigmat(e), -matique, *a.* anastigmatic.

anastigmat, *m.* anastigmat.

anastomose, *f.* anastomosis.

anastomoser, *v.r.* anastomose.

anate, *f.* annatto(?).

anatomie, *f.* anatomy.

anatomique, *a.* anatomical, anatomic.

ancêtres, *m.pl.* ancestors, forefathers.

anchuse, *f.* (*Bot.*) any plant of the genus *Anchusa,* bugloss, oxtongue.

ancien, *a.* ancient; old; former, late, ex-; old-fashioned. — *m.* ancient; old person; senior.

anciennement, *adv.* anciently, of old; formerly.

ancienneté, *f.* age; seniority; antiquity.

ancolie, *f.* (*Bot.*) columbine.

ancrage, *m.* anchorage.

ancre, *f.* anchor.

ancrer, *v.t.* anchor; brace; fix, establish.

andalous, andalousien, *a.* Andalusian.

andalousite, *f.* (*Min.*) andalusite.

andésite, *f.* (*Petrog.*) andesite.

andorite, *f.* (*Min.*) andorite.

andropogon, *m.* (*Bot.*) Andropogon, andropogon.

-ane. (*Org. Chem.*) in the names of *saturated* hydrocarbons and heterocyclic parent compounds, -ane; in other cases, -an; as, *éthane,* ethane; *furfurane,* furan, furfuran; *uréthane,* urethan.

âne, *m.* ass, donkey.

anéantir, *v.t.* annihilate, destroy, abolish, annul; prostrate.

anéantissement, *m.* annihilation, etc. (see anéantir).

anélectrique, *a.* anelectric.

anémie, *f.* anemia, anæmia.

anémique, *a.* anemic, anæmic.

anémone, *f.* (*Bot.*) anemone.

anéroïde, *a.* & *m.* aneroid.

ânesse, *f.* she-ass.

lait d'—, ass's milk.

anesthésiant, *a.* anesthetic.

anesthésie, *f.* anesthesia.

anesthésier, *v.t.* anesthetize.

anesthésique, *a.* & *m.* anesthetic.

aneth, anet, *m.* (*Bot.*) Anethum.

— *odorant,* dill (*Anethum graveolens*).

anéthol, *m.* anethole.

ange, *m.* angel; angel fish.

angélique, *a.* angelic. — *f.* angelica (*Angelica archangelica*).

angine, *f.* (*Med.*) angina.

anglais, *a.* English. — *m.* English (language).

Anglais, *m.* Englishman.

angle, *m.* angle.

— *aigu,* acute angle.

— *critique,* critical angle.

— *d'équerre,* right angle.

— *dièdre,* dihedral angle.

— *droit,* right angle.

— *optique,* optic angle.

— *plan,* plane angle.

— *rapporteur,* protractor.

— *relevé,* observed angle.

— *trièdre,* trihedral angle.

anglésite, *f.* (*Min.*) anglesite.

Angleterre, *f.* England.

angoissant, *p.a.* distressing.

angoisse, *f.* anguish; spasm.

anguille, *f.* eel.

angulaire, *a.* angular; corner.

angulairement, *adv.* angularly.

angulé, *a.* angular, angulated.

anguleux, *a.* angular.

angusture, *f.* angustura bark, angustura.

anhydrate, *m.* anhydride.

anhydre, *a.* anhydrous.

anhydride, *m.* anhydride.

— *acétique,* acetic anhydride.

— *azoteux,* nitrous anhydride.

anyhdrifier, *v.r.* become anhydrous.

anhydrisation, *f.* dehydration.

anhydrite, *f.* (*Min.*) anhydrite.

anhygrométrique, *a.* not hygroscopic.

anilidure, *m.* anilide.

aniline, *f.* aniline.

animal, *a.* & *m.* animal.

animalisation, *f.* animalization.

animalisé, *p.a.* animalized; animal.

animaliser, *v.t.* animalize.

animalité, *f.* animality, animal nature or life.

animateur, *a.* animating.

animatrice, *fem.* of animateur.

animaux, *pl. masc.* of animal.

animé, *f.* animé, gum animé. — *p.p.* of animer.

animer, *v.t.* animate; set or keep in motion, impel, drive; heighten (colors); incite.

anion, *m.* anion.

anis, *m.* anise.

— *étoilé,* star anise.

— *vert,* anise, aniseed.

anisateur, *m.* an apparatus used in making a distilled liquor flavored with anise.

aniser, *v.t.* anisate, flavor with anise.

anisette, *f.* anisette.

anisine, *f.* anisine.

anisique, *a.* anisic; (of the alcohol) anisyl.

anisol, *m.* anisole.

anisotrope, *a.* anisotropic.

annales, *f.pl.* annals.

Ann. chim. phys., *abbrev.* Annales de chimie et de physique.

anneau, *m.* ring; link, hoop, band, collar, (circular) rim, (ring-shaped) handle; loop; coil; ringlet; (*Elec.*) ring armature.

à —, with ring, ringed.

— *benzénique,* benzene ring.

— *de sûreté,* safety ring.

— *flotteur,* floating ring, ring float.

année, *f.* year.

annelet, *m.* small ring, ringlet, annulet.

annexe, *f.* annex; appendage; attachment; fitting. — *a.* annexed, attached.

annexer, *v.t.* annex.

annihiler, *v.t.* annihilate, destroy, annul.

anniversaire, *a.* & *m.* anniversary.

annonce, *f.* announcement; notice; advertisement; sign.

annoncer, *v.t.* announce; tell, declare, publish; advertise; show, indicate.

annoter, *v.t.* note down; annotate.

annuaire, *m.* yearbook, annual; directory

annuel, *a.* annual, yearly.

annuellement, *adv.* annually, yearly.

annulaire, *a.* annular; ring (finger).

annuler, *v.t.* annul, cancel; nullify, paralyze.

anode, *f.* anode.

anodin, *a.* anodyne, soothing; tame, spiritless. — *m.* anodyne.

anodique, *a.* (*Elec.*) anodic, anode.

anomal, *a.* anomalous.

anomalie, *f.* anomaly.

anone, *f.* (*Bot.*) Annona.

anonym, *a.* anonymous; joint-stock (company).

anormal, *a.* abnormal.

anormalement, *adv.* abnormally.

anorthique, *a.* (*Min.*) anorthic (triclinic).

anotto, *m.* annatto.

anoxhémie, *f.* (*Med.*) anoxemia.

anse, *f.* handle (attached at both ends); loop; small bay, inlet; (*Mach.*) slot, keyway.

— *à vis,* screw eye.

— *de panier,* basket handle; a curve composed of arcs of three circles.

anser, *v.t.* furnish with a handle.
ansette, *f.* small handle.
antagoniste, *a.* antagonistic.
antécédemment, *adv.* previously, antecedently.
antenne, *f.* antenna; catch, hook; (of a torpedo) striker.
antérieur, *a.* anterior, front, fore; anterior, prior, previous.
antérieurement, *adv.* previously, anteriorly.
antériorité, *f.* priority, anteriority.
anthelminthique, *a. & m.* anthelmintic.
anthocyane, *m.* anthocyanin, anthocyan.
anthocyanine, *f.* anthocyanin.
anthracène, *m.* anthracene.
anthracène-carbonique, *a.* anthracenecarboxylic.
anthracénique, *a.* anthracene, of anthracene.
anthracéno-disulfoné, *a.* anthracenedisulfonic.
anthracéno-sulfoné, *a.* anthracenesulfonic.
anthracine, *f.* anthracene.
anthracifère, *a.* anthraciferous.
anthracite, *m.* anthracite.
anthraciteux, *a.* anthracitic, anthracitous.
anthracitifère, *a.* anthracitiferous, anthraciferous.
anthracnose, *f.* (*Bot.*) anthracnose.
anthracomètre, *m.* anthracometer.
anthragallol, *m.* anthragallol.
anthranilique, *a.* anthranilic.
anthranol, *m.* anthranol.
anthraquinone, *f.* anthraquinone.
anthrarufine, *f.* anthrarufin.
anthrol, *m.* anthrol.
antiacide, *a. & m.* antacid.
antialcalin, *a. & m.* antalkaline.
antiasphyxiant, *a.* antiasphyxiating.
anticalcaire, *m.* boiler compound, scale remover.
anticathode, *f.* anticathode.
antichlore, *m.* antichlor.
anticipation, *f.* anticipation; (*Com.*) advance.
anticiper, *v.t.* anticipate.
anticombustible, *a. & m.* incombustible.
anti-congélant, *a.* antifreezing. — *m.* antifreezing agent.
anticorps, *m.* antibody.
antidéperditeur, antidéperditif, *a.* (*Med.*) diminishing waste due to lack of assimilation.
antidérapant, *a.* nonskidding, nonskid. — *m.* a nonskid tire or device.
antifébrine, *f.* antifebrin (acetanilide).
antiferment, *m.* antiferment.
antifermentescible, *a.* antifermentative; nonfermenting. — *m.* antifermentative; nonfermenting substance.
anti-floculant, *a.* antiflocculating.

antifondant, *m.* (*Ceram.*) a substance which checks or prevents fusion, anti-flux.
antifriction, *a.* antifriction. — *f.* antifriction alloy, antifriction metal.
antigène, *m.* antigen.
antihypnotique, *a. & m.* antihypnotic.
anti-incrustant, *a.* scale-preventing.
antimammoniaque, *f.* stibine.
antimoine, *m.* antimony.
— *cru,* (*Pharm.*) antimony sulfide.
— *diaphorétique,* (*Pharm.*) diaphoretic antimony (an acid potassium antimonate forming a white, insoluble powder).
— *sulfuré,* antimony sulfide.
antimonial, *a. & m.* antimonial.
antimoniate, *m.* antimonate, antimoniate.
— *grenu de potasse,* dihydrogen potassium pyroantimonate (a granular powder, $H_2K_2Sb_2O_7 . 6H_2O$).
antimonide, *a. & m.* antimonial.
antimonié, *a.* containing antimony, antimonial; antimoniuretted (hydrogen).
antimonieux, *a.* antimonious, antimonous.
antimonine, *m.* antimonin (a lactate of calcium and antimony used in tanning).
antimonique, *a.* antimonic.
antimonite, *n.* antimonite.
antimoniure, *m.* antimonide.
antinéphrétique, *a. & m.* antinephritic.
antinevralgique, *a. & m.* antineuralgic.
antiparasitaire, *a. & m.* antiparasitic.
antiphlogistique, *a. & m.* antiphlogistic.
antiplastique, *a.* antiplastic.
antiputride, *a.* antiputrefactive, antiseptic. — *m.* an antiputrefactive substance.
antipyrétique, *a. & m.* antipyretic.
antipyrine, *f.* antipyrine.
antique, *a.* antique; antiquated.
antiquité, *f.* antiquity.
antirhumatisant, *a. & m.* antirheumatic.
anti-rouille, *a.* anti-rust.
antisepsie, *f.* antisepsis.
antiseptique, *a. & m.* antiseptic.
antisérum, *m.* antiserum.
antithermique, *a.* (*Med.*) antithermic, antipyretic.
antithèse, *f.* antithesis.
antitoxine, *f.* antitoxin.
antitoxique, *a.* antitoxic.
antivénéneux, *a.* counteracting poison, antitoxic, alexipharmic. — *m.* antipoison, antidote.
antivénérien, *a.* antivenereal, antisyphilitic.
antivermineux, *a. & m.* anthelmintic, vermifuge.
antizymique, *a. & m.* antizymotic, antizymic.
antracène, *m.* anthracene.
antre, *m.* cave, cavern, den; (*Anat.*) antrum.
anurie, *f.* (*Med.*) anuria.
Anvers, *n.* Antwerp.

anxiété, *f.* anxiety.

anxieux, *a.* anxious.

aorte, *f.* (*Anat.*) aorta.

août, *m.* August; harvest.

aoûter, *v.t., r. & i.* ripen.

apaiser, *v.t.* appease, pacify, calm, allay.

Apalaches, *f.pl.* Appalachians.

apercevoir, *v.t. & r.* perceive, notice, see.

aperçoit, *pr. 3 sing.* of apercevoir.

aperçu, *p.a.* perceived, etc. (see apercevoir).
— *m.* glance, (rapid) view, rough estimate,
outline, sketch.

apériodique, *a.* aperiodic.

apéritif, *a.* stimulating the appetite; (*Med.*)
aperient, aperitive. — *m.* appetizer, apéritif; (*Med.*) aperient, aperitive.

aphanite, *f.* (*Petrog.*) aphanite.

aphlogistique, *a.* aphlogistic, flameless.

aphrodisiaque, *a. & m.* (*Med.*) aphrodisiac.

apiol, *m.* apiole, apiol.

aplanir, *v.t.* make smooth or level, smooth.
— *v.r.* become smooth or level, be leveled.

aplanissement, *m.* making smooth, etc. (see
aplanir); smoothness, evenness.

aplastique, *a.* aplastic.

aplat, *m.* flat tint.

aplati, *p.a.* flattened, flat; oblate (spheroid);
debased.

aplatir, *v.t.* flatten; debase. — *v.r.* be flattened,
become flat, flatten; fall, sink.

aplatissage, *m.* flattening.

aplatissement, *m.* flattening; flatness.

aplatisserie, aplatissoire, *f.* flatting mill, rolling mill.

aplomb, *m.* perpendicular, perpendicularity,
plumb; equilibrium, balance; self-posses-
sion.
d'—, upright, plumb; perpendicularly, in
plumb; in equilibrium.

apoatropine, *f.* apatropine.

apomorphine, *f.* apomorphine.

aposorbique, *a.* aposorbic.

apothicaire, *m.* apothecary.

apothicairerie, *f.* apothecary's shop; dispensary.

apôtre, *m.* apostle.

apozème, *m.* (*Pharm.*) infusion.

apparaître, *v.i.* appear; be clear, be evident.

appareil, *m.* apparatus; machine, device,
equipment; (ceremonious) preparation;
pomp; (surgical) dressing; style of construction; course (of stone, etc.); (of
bricks) bond.
— à *chargement,* charging apparatus.
— à *chlore,* chlorine apparatus.
— à *colonne,* (*Distilling*) column apparatus,
column.
— à *flammes,* (*Mil.*) flame projectors.
— à *pluie,* shower apparatus, sprinkling
apparatus.

appareil à *prélèvement de gaz,* (*Mil.*) gas
respirators or masks.
— à (or *de*) *ruissellement,* trickling apparatus.
— *broyeur,* crushing (or grinding) apparatus,
crusher, grinder.
— *casseur,* breaking (or crushing) apparatus,
breaker, crusher, stamp mill.
— *concasseur,* = appareil casseur.
— *coupeur,* cutting apparatus, cutting machine, cutter.
— *d'alimentation,* (*Mach.*) feed gear.
— *d'allumage,* lighting apparatus; igniting
gear.
— *débourbeur,* cleansing apparatus, washing
apparatus, washer.
— *de chauffage,* heating apparatus, heater.
— *de cinglage,* (*Metal.*) shingling machine,
shingler.
— *d'éclairage,* lighting apparatus; lighting
plant.
— *de coulée,* casting (or founding) apparatus
or equipment.
— *de découpage,* cutting apparatus, cutting
machine, cutter.
— *de grenage,* granulating apparatus or machine, granulator.
— *de lavage,* washing apparatus, washing
machine, washer.
— *de lessivage,* leaching apparatus, lixiviation
apparatus.
— *de succion,* suction apparatus.
— *de sûreté,* safety device.
— *d'extinction,* fire extinguisher.
— *dialyseur,* dialyzing apparatus.
— *distillatoire,* distilling apparatus.
— *étalon,* calibrating apparatus, calibrator.
— *frigorifique,* refrigerating apparatus, refrigerator.
— *générateur,* generating apparatus, generator.
— *graisseur,* lubricating apparatus or device,
lubricator.
— *malaxeur,* an apparatus for mixing or
working pasty material, mixer, beater, malaxator, etc.
— *moteur,* (*Mach.*) driver.
— *photographique,* photographic apparatus;
specif., camera.
— *producteur,* (for gases) generating apparatus, generator.
— *propulseur,* propelling apparatus, propeller.
— *protecteur,* protecting (or protective) device or apparatus.
— *récolteur,* collecting apparatus.
— *Schilt,* (*Mil.*) a kind of flame projector.
— *solidificateur,* solidifying apparatus, solidifier.
— *Strombos,* (*Mil.*) Strombos horn, Strombos
gas alarm.

appareil *trieur*, sorting apparatus, sorter, classifier.

appareillage, *m.* equipment; matching; preparation; bonding, bond (of bricks, etc.).

appareiller, *v.t.* match; prepare; dress (cloth).

apparemment, *adv.* apparently.

apparence, *f.* appearance; probability; trace.

apparent, *a.* apparent; conspicuous; plausible.

apparition, *f.* appearance; apparition.

appartement, *m.* apartment, suite of rooms.

appartenant, *p.a.* belonging, appertaining.

appartenir, *v.i.* belong; concern, be the province (of), be the duty (of).

appartient, *pr. 3 sing.* of appartenir.

apparu, *p.p.* of apparaître.

apparut, *p.def. 3 sing.* of apparaître.

appauvri, *p.a.* impoverished, exhausted.

appauvrir, *v.t.* impoverish, exhaust. — *v.r.* become poor or poorer; deteriorate.

appauvrissement, *m.* impoverishment, exhaustion.

appel, *m.* call; appeal; tension, pull; (of air) draft; ventilation; roll call; challenge. — *par vibrateur*, — *vibré*, *(Elec.)* buzzer.

appelé, *p.a.* called, etc. (see appeler).

appeler, *v.t.* call; call for, require; draw, drive (air, etc.); announce. — *v.i.* appeal; call; (of a chimney) draw. — *v.r.* be called.

appendice, *m.* appendix; appendage, projection, stud, etc.

appesantir, *v.t.* make heavy, weigh down; dull. — *v.r.* become heavy; be oppressed; dwell, insist (upon).

appétissant, *a.* appetizing, tempting.

appetit, *m.* appetite.

applaudissement, *m.* applause; praise.

applicable, *a.* applicable.

applicateur, *m.* *(Wine)* pasteurizer.

application, *f.* application; posting (of bills).

applique, *f.* something applied or affixed, mounting, setting, reinforce, inlay, etc.; *(Lighting)* bracket fixture, bracket.

appliqué, *p.a.* applied; attentive.

appliquer, *v.t.* apply; post (bills). — *v.r.* be applied; apply; apply oneself. — *v.i.* require application.

appoint, *m.* balance, remainder, difference, amount over; *(Explosives)* priming charge.

appointements, *m. pl.* salary, pay.

appointer, *v.t.* point, sharpen to a point; tack, stitch; salary.

apport, *m.* contribution (as to the capital of a company); deposit (of documents).

apporter, *v.t.* bring; contribute; adduce, cite; apply, use.

apposer, *v.t.* affix, append, add.

apposition, *f.* affixing, etc. (see apposer); apposition.

appréciable, *a.* appreciable.

appréciateur, *m.* estimator, valuer, appreciator; specif., a kind of hydrometer.

appréciation, *f.* estimation, measurement, valuation; appreciation.

apprécier, *v.t.* estimate, value; appreciate.

apprenant, *p.pr.* of apprendre.

apprendre, *v.t.* learn; teach; inform.

apprennent, *pr. 3 pl. indic. & subj.* of apprendre.

apprenti, *m. & f.* apprentice.

apprentissage, *m.* apprenticeship; practice.

apprêt, *m.* dressing, finishing, finish; sizing; priming; preparation; affectation.

apprêtage, *m.* dressing, finishing.

apprêter, *v.t.* dress, finish; prepare.

apprêteur, *m.* dresser, finisher.

apprêteuse, *f.* dresser, finisher; *(Paper)* calenders, surfacing rolls.

appris, *p.p* of apprendre.

approbation, *f.* approval, approbation.

approchant, *a.* like, somewhat like, near, approximate. — *adv.* nearly, about. — *p.pr.* of approcher.

approche, *f.* approach.

approché, *p.a.* approximate; brought near, etc. (see approcher)

approcher, *v.t.* bring or place near or nearer, bring up; approach. — *v.r. & i.* approach, be near.

approfondi, *p.a.* deepened, etc. (see approfondir); profound.

approfondir, *v.t.* deepen; go deeply into, study profoundly, fathom. — *v.r.* deepen; study deeply.

approfondissement, *m.* deepening; profound study.

approprié, *p.a.* appropriate, suitable; appropriated, etc. (see approprier).

approprier, *v.t.* appropriate; adapt, suit; put in order, fix up.

approuver, *v.t.* approve, approve of, sanction.

approvisionnement, *m.* supplying, supply, stock.

approvisionner, *v.t.* supply.

approximatif, *a.* approximate.

approximation, *f.* approximation.

approximativement, *adv.* approximately.

approximer, *v.t.* approximate.

appui, *m.* support (in various senses); prop, stay, rest; fulcrum; balustrade; rail; sill; emphasis, stress; pressure, impress.

appuie, *pr. 1 & 3 sing.* of appuyer.

appui-main, *m.* support for the hand, hand rest.

appuyer, *v.t.* support; prop, lean, rest; base. — *v.r.* be supported, lean, rely (on). — *v.i.* be supported, rest; bear, press (on); dwell, insist (on).

âpre, *a.* acrid, tart; harsh, rough; fierce, violent; greedy.

âprement, *adv.* acridly, etc. (see âpre).

après, *prep.* after; behind, beyond; about, at; against, upon. — *adv.* after, afterward. *d'*—, according to, after; following, next.

après-demain, *adv.* day after tomorrow.

après-guerre, *n.* (period) after the war.

après-midi, *f. & m.* afternoon.

âpreté, *f.* acridity, etc. (see âpre).

apside, *f.* apsis; (*Arch.*) apse.

apte, *a.* capable; qualified, fit.

aptitude, *f.* capability, capacity, aptitude.

apyre, *a.* infusible; incombustible.

aquarelle, *f.* water-color painting.

aqueduc, aquéduc, *m.* aqueduct.

aqueux, *a.* aqueous; watery.

aquifère, *a.* water-bearing, aquiferous.

aquosité, *f.* aqueousness; wateriness, aquosity.

arabe, *a.* Arabic; Arabian.

Arabie, *f.* Arabia.

arabine, *f.* arabin (arabic acid).

arabinose, *n.* arabinose.

arabique, *a.* arabic.

arabite, *f.* arabitol, arabite.

arabonique, *a.* arabonic.

arac, *m.* arrack.

arachide, *f.* peanut (*Arachis hypogæa* and related species); (*Bot.*) Arachis.

arachidique, *a.* arachidic.

arachine, *f.* arachin.

arack, *m.* arrack.

aragonite, *f.* (*Min.*) aragonite.

araignée, *f.* spider; spider's web.

aralie, *f.* (*Bot.*) Aralia.

— *épineuse,* Hercules'-club (*A. spinosa*).

— *nue,* wild sarsaparilla (*Aralia nudicaulis*).

araser, *v.t.* level; be flush with.

aratoire, *a.* farming, agricultural.

arbacine, *f.* arbacine.

arbitraire, *a.* arbitrary. — *m.* arbitrariness.

arbitrairement, *adv.* arbitrarily.

arborer, *v.t.* raise, hoist.

arborescent, *a.* arborescent; branched (chain)

arborisation, *f.* arborization, dendrite.

arborisé, *a.* arborized.

arbouse, *f.* arbutus fruit, arbutus berry.

arbousier, *m.* (*Bot.*) arbutus.

— *commun,* strawberry tree (*A. unedo*).

arbre, *m.* tree; (*Mach.*) shaft, spindle, axis, axle, arbor; mast.

— *à cames,* cam shaft.

— *à cire,* wax tree, esp. the wax myrtle (*Myrica*).

— *à manivelle,* crank shaft.

— *à noyau,* (*Founding*) core spindle.

— *à perruque,* fustet (*Cotinus cotinus*).

— *à suif,* tallow tree; wax myrtle.

— *à vermillon,* kermes, kermes oak (*Quercus coccifera*).

— *coudé,* cranked shaft, cranked axle.

arbre *creux,* hollow shaft; hollow tree.

— *debout,* upright shaft, vertical shaft or spindle.

— *de couche,* horizontal shaft; main shaft.

— *de Diane,* arbor Dianæ, arborescent silver.

— *de haute futaie,* forest tree, timber tree.

— *de Saturne,* arbor Saturni, arborescent lead.

— *de tour,* lathe spindle, mandrel.

— *du communicateur,* gearing shaft.

— *du tiroir,* eccentric shaft.

— *fruitier,* fruit tree.

— *intermédiaire,* countershaft.

— *moteur,* driving shaft, main shaft.

— *philosophique,* arbor Dianæ.

arbrisseau, *m.* shrub, bush (of considerable size).

arbue, *f.* clay used as a flux.

arbuste, *m.* shrub, bush (of small size).

arbutine, *f.* arbutin.

arc, *m.* arc; arch; bow.

— *voltaïque,* voltaic arc, electric arc.

arcade, *f.* curved part, arch, bow (as of a saddle); arcade.

arcane, *m.* arcanum.

arcanne, *f.* ruddle, red ocher.

arcanson, *m.* rosin.

arc-boutant, *m.* buttress, stay, prop; flying buttress.

arc-bouter, *v.t.* buttress, prop, brace.

arceau, *m.* arch, small arch.

arc-en-ciel, *m.* rainbow.

archal, *m.* brass (in the phrase *fil d'archal,* brass wire).

archangélique, *f.* (*Bot.*) angelica.

arche, *f.* fire chamber, furnace, arch; arch (of other kinds); ark.

— *à pots,* (*Glass*) a furnace in which the pots are given a preliminary heating.

— *à tirer,* (*Glass*) annealing furnace for cylinders.

archée, *f.* (*Alchemy*) archeus.

archet, *m.* curved part, bow, arch; (violin) bow.

archi-. arch-.

archives, *f.pl.* archives; record office.

arctique, *a.* arctic.

ardemment, *adv.* ardently.

ardent, *a.* burning, blazing, glowing; ardent; (of colors) bright, vivid; (of hair) red, sandy; spirited. — *m.* ignis fatuus.

ardeur, *f.* ardor; (of colors) vividness; (*Med.*) sensation of heat; mettle.

— *d'estomac,* heartburn.

— *d'urine,* ardor urinæ, scalding of the urine.

ardillon, *m.* tongue (of a buckle).

ardoise, *f.* slate; slate color.

ardoisé, *p.a.* slate-colored; slated.

ardoiser, *v.t.* render slate-colored; slate.

ardoiseux, ardoisier, *a.* of or like slate, slaty.

ardoisière, *f.* slate quarry.

ardoisin, *a.* resembling slate.
ardu, *a.* arduous, steep, difficult.
arduité, *f.* arduousness.
are, *m.* are (100 square meters or 119.6 sq. yds.)
aréca, aréca, *m.* (*Bot.*) areca.
aréfaction, *f.* drying, desiccation.
arénacée, *a.* arenaceous, sandy.
arénation, *f.* sanding; (*Med.*) arenation.
arène, *f.* sand; arena; (*pl.*) amphitheater.
aréner, *v.i. & r.* sink, sag.
aréneux, *a.* sandy, arenose.
aréolaire, *a.* areolar.
aréole, *f.* areola.
aréomètre, *m.* hydrometer, areometer.
— *chercheur,* exploring hydrometer, exploring spindle (to indicate which of a set of spindles is to be used).
aréométrie, *f.* hydrometry, areometry.
aréométrique, *a.* hydrometric, areometric.
arête, *f.* edge; ridge; awn, beard (of grain); (fish) bone; (*Arch.*) arris.
— *de poisson,* (*Tech.*) herring bone.
— *mousse,* rounded edge.
— *vive,* sharp edge.
à vives arêtes, sharp-edged.
argémone, *f.* (*Bot.*) prickly poppy (*Argemone*).
argent, *m.* silver; money.
— *corné,* (*Min.*) horn silver (cerargyrite).
— *fulminant,* fulminating silver; specif., silver fulminate.
— *raffiné,* refined silver.
argentage, *m.* silvering, silver plating.
argental, *a.* containing silver, argental (as, *mercure argental,* argental mercury, a native silver amalgam).
argentan, *m.* argentan, German silver (an alloy of copper, nickel and zinc).
— *au manganèse,* an alloy of copper, manganese and zinc.
argenté, *p.a.* silvered; silvery, silver-white.
argenter, *v.t.* silver.
argenterie, *f.* silver ware, silver plate.
argenteur, *m.* silverer.
argentifère, *a.* argentiferous, containing silver.
argentin, *a.* silvery, silver; Argentine.
argentine, *f.* argentine (in various senses); specif., powdered tin, used to give a silvery coating on fabrics.
argentique, *a.* of silver, silver, argentic.
argenton, *m.* an alloy of copper, nickel and tin
argenture, *f.* silvering, silver plating.
— *électro-chimique,* — *galvanique,* electrosilvering, silver electroplating.
argilacé, *a.* argillaceous, clayey.
argile, *f.* clay.
— *à façonner,* potter's clay.
— *à fougère,* slate clay.
— *à porcelaine,* kaolin, porcelain clay.
— *pauvre,* lean clay.
— *plastique,* plastic clay, potter's clay.

argile *réfractaire,* refractory clay, fire clay.
— *schisteuse,* slate clay.
— *smectique,* fuller's earth.
argileux, *a.* clayey, clay, argillaceous.
argilière, *f.* clay pit.
argilifère, *a.* argilliferous.
argilite, *f.* argilite, clay slate.
argillacé, *a.* argillaceous, clayey.
argilo-alcalin, *a.* containing clay and alkali, argillo-alkaline.
argilo-calcaire, *a.* containing clay and lime, argillo-calcareous.
argilo-glaiseux, *a.* clayey, of stiff clay.
argilo-siliceux, *a.* containing clay and silica, argillo-siliceous.
argiloïde, *a.* like clay, argilloid.
argilolithe, *m.* clay stone.
arginine, *f.* arginine.
argon, *m.* argon.
argue, *f.* a machine for drawing ingots of gold, silver, etc.
arguer, *v.t.* rough-draw (gold, silver, etc.); argue, conclude.
arguérite, *f.* arguerite (a native amalgam, AgsHg).
argumenter, *v.i.* argue. — *v.t.* argue with.
argyrose, *f.* (*Min.*) argentite (Ag₂S).
argyrythrose, *f.* (*Min.*) pyrargyrite (Ag₃SbS₃).
aricine, *f.* aricine.
aride, *a.* arid.
aridité, *f.* aridity.
aristol, *m.* (*Pharm.*) aristol.
aristoloche, *f.* (*Bot.*) Aristolochia.
arithmétique, *f.* arithmetic. — *a.* arithmetical.
arithmétiquement, *adv.* arithmetically.
armagnac, *m.* a well known French brandy.
armature, *f.* a strengthening part or system (usually metallic), frame, truss, braces, clamps, sheathing, facing; fitting, mounting; (*Magnetism*) armature.
arme, *f.* arm, weapon; arm (of the service).
— *à feu,* firearm.
— *blanche,* a cutting or piercing weapon.
— *de guerre,* military weapon.
— *de jet,* ballistic weapon.
— *petite,* side arm.
— *rayée,* rifle.
armé, *p.a.* armed, etc. (see armer).
armée, *f.* army; forces. — *a. fem.* of armé.
— *de mer,* sea forces; navy, fleet.
— *de terre,* land forces.
armement, *m.* armament; arming; equipment.
armer, *v.t.* arm; equip, fit out; reinforce (cement or glass); insert a fuse in (a shell); cock (a firearm); wind (a dynamo); man.
armoire, *f.* closet, cupboard, press, wardrobe.
— *à dessiccation,* drying closet.

armoise, *f.* (*Bot.*) Artemisia, wormwood; = armoisin.
— *absinthe,* — *des champs,* wormwood (*A. absinthium*).
— *commune,* mugwort (*A. vulgaris*).
armoisin, *m.* a kind of silk made at Lyons.
armure, *f.* frame, braces, etc. (see armature); armature (of a magnet); pole piece (of a dynamo); armor.
arnicine, *f.* arnicine.
arnique, arnica, *f.* arnica.
aromate, *m.* aromatic.
aromaticité, *f.* aromatic quality, aromaticness.
aromatique, *a.* aromatic.
aromatiquement, *adv.* aromatically.
aromatisation, *f.* aromatization.
aromatiser, *v.t.* aromatize.
arome, *m.* aroma.
arpentage, *m.* land measurement.
arquer, *v.t., i. & r.* curve, bend, arch.
arquifoux, *m.* (*Ceram.*) alquifou.
arrachage, arrachement, *m.* pulling up, etc. (see arracher).
arrache-pied. In the phrase *d'arrache-pied,* uninterruptedly, consecutively.
arracher, *v.t.* pull up, dig up, pull out, tear out; tear away, pull off, strip off; take away (violently); eradicate; extort.
arrak, *m.* arrack.
arrangement, *m.* arrangement, etc. (see arranger); economy; order, method.
arranger, *v.t.* arrange; adjust; suit, please; trim, trim up, dress. — *v.r.* be arranged; place oneself; make arrangements; manage.
arrêt, *m.* stopping, stop, stoppage; (*Tech.*) stop, check, catch, arrest, stay; judgment, sentence; decree, order; arrest; (legal) attachment; steadfastness.
arrêtage, *m.* (*Mach.*) stop, check, arrest, catch.
arrête-gaz, *m.* gas check.
arrêter, *v.t.* stop; check; fasten, fix; arrest; detain; seize, attach (legally); decide; engage (lodging, etc.). — *v.r. & i.* stop, cease; stay; dwell (on).
arrêtoir, *m.* (*Mach.*) stop, catch, check.
arrière, *adv.* back, behind, in the rear, astern.
— *m.* rear, hind part, stern.
en —, back, backward, behind, behindhand.
en — *de,* behind.
arriéré, *p.a.* delayed, deferred; backward; behindhand.
arrière-goût, *m.* aftertaste.
arriérer, *v.t.* delay, defer, postpone. — *v.r.* be, get or stay, behind.
arrimer, *v.t.* stow, pack.
arrivage, *m.* arrival.
arrivée, *f.* admission tube or pipe, feed pipe; inlet; arrival.
— *d'eau,* water feed pipe.
d'—, at first; at once; on arrival.

arriver, *v.i.* arrive; approach; come; succeed; happen, occur.
arrondi, *p.a.* rounded; round.
arrondir, *v.t.* round, make round; round off. — *v.r.* become or grow round.
arrondissage, *m.* rounding, making round.
arrondissement, *m.* rounding; roundness; district, arrondissement.
arrosage, *m.* wetting, etc. (see arroser).
arrosement, *m.* wetting, etc. (see arroser); (*Ceram.*) application of a glaze by pouring it on the object.
arroser, *v.t.* wet, moisten; sprinkle; water.
arrosoir, *m.* sprinkler, watering pot.
A.R.S., *abbrev.* of appareil respiratoire spécial (the name of a French gas mask).
arsendiméthyle, *m.* arsenic dimethyl (cacodyl).
arséniate, *m.* arsenate, arseniate.
— *de fer,* iron arsenate.
— *de soude,* sodium arsenate.
— *de soude desséché,* (*Pharm.*) exsiccated sodium arsenate.
— *ferreux,* ferrous arsenate.
arséniaté, *a.* arsenate of.
arsenic, *m.* arsenic.
— *blanc,* white arsenic (arsenious oxide).
arsenical, *a.* arsenical. — *m.* arsenical compound or substance, arsenical.
arsenicaux, *pl. masc.* of arsenical.
arsénié, *a.* arseniuretted; arsenicated; arsenical.
arsénieux, *a.* arsenious.
arsénifère, *a.* arseniferous.
arsénio-sulfure, *m.* (*Min.*) sulfarsenide.
arsénio-vanadique, *a.* arsenovanadic, arseniovanadic.
arsénique, *a.* arsenic.
arsénite, *m.* arsenite.
arséniure, *m.* arsenide.
— *d'hydrogène,* hydrogen arsenide, arsine.
arsénolithe, *m.* (*Min.*) arsenolite.
arsinate, *m.* arsinate; arsonate (incorrect use).
arsine, *f.* arsine.
art, *m.* art; (*pl.*) fine arts.
— *d' Hermès,* Hermetic art, alchemy.
artémisie, artémise, *f.* artemisia, wormwood.
artère, *f.* artery; (*Elec.*) feeder.
artérialiser, *v.t.* arterialize.
artériel, artérieux, *a.* arterial.
artériosclérose, *f.* (*Med.*) arteriosclerosis.
artésien, *a.* Artesian.
artichaut, *m.* artichoke.
— *des Indes,* sweet potato.
— *des toits,* houseleek.
article, *m.* article; joint; link; matter, subject.
articulaire, *a.* articular.
articulation, *f.* articulation, joint, jointing
— *sphérique,* ball-and-socket joint.

articulé, *p.a.* articulated; jointed; articulate. — *m.* (*pl.*, *Zoöl.*) Articulata.

articuler, *v.t.* articulate; assert, allege.

artifice, *m.* artifice; firework, fireworks, pyrotechnic composition.
— *à signaux,* pyrotechnic signaling device.
— *d'éclairage,* any illuminating firework, flare, rocket, etc.
— *de guerre,* military fireworks; strategem.
— *de rupture,* any explosive device used for demolition.
— *de signaux,* signal fireworks.
— *éclairant,* = artifice d'éclairage.
— *incendiare,* incendiary firework, incendiary.

artificiel, *a.* artificial.

artificiellement, *adv.* artificially.

artificier, *m.* fireworks maker, pyrotechnist; (*Mil.*) artificer (maker of shells, fuses, etc.).

artificieux, *a.* artful.

artillerie, *f.* artillery.

artisonné, *a.* moth-eaten, worm-eaten.

artiste, *m. & f.* artist; adept.

artistement, *adv.* artistically; skillfully.

Arts et Métiers, *abbrev.* of École nationale d'arts et métiers (national school of arts and trades), any one of a class of engineering schools.

as, *m.* ace. — *pr. 2 sing.* of avoir (to have).

asaret, *m.* (*Bot.*) Asarum; specif., asarabacca (*A. europæum*).
— *du Canada,* wild ginger (*A. canadense*).

asarone, *f.* asarone (1, 2, 4-trimethoxy-5-propenylbenzene).

asbeste, *m.* asbestos.

asbolane, *m.* (*Min.*) asbolite; (wrongly) graphite.

asboline, *f.* (*Pharm.*) asbolin.

ascendant, *a.* ascending, rising; return, reflux (condenser). — *m.* ascendant, ascendency; ancestor; propensity.

ascenseur, *m.* elevator, lift, hoist.

ascension, *f.* rising, rise, ascent; ascension; (of a piston) upstroke.

ascensionnel, *a.* ascensional, upward.

ascidie, *f.* (*Zoöl.*) ascidian.

ascite, *f.* (*Med.*) ascites.

asclépiade, *f.*, **asclépias,** *m.* (*Bot.*) any of several asclepiadaceous plants, specif. swallowwort (*Cynanchum vincetoxicum*).

ascomycète, *a.* (*Bot.*) ascomycetous. — *m.* ascomycete.

asepsie, *f.* asepsis.

aseptique, *a. & m.* aseptic.

aseptiquement, *adv.* aseptically.

aseptiser, *v.t.* asepticize, aseptify.

aseptol, *m.* (*Pharm.*) aseptol.

asiatique, *a.* Asiatic.

Asie, *f.* Asia.
— *Mineure.* Asia Minor.

asparagine, *f.* asparagine.

asparaginique, *a.* asparaginic (aspartic).

aspartique, *a.* aspartic.

aspect, *m.* sight; aspect.

asperge, *f.* asparagus; asparagus shoot.

aspergement, *m.* = aspersion.

asperger, *v.t.* sprinkle, sparge.

aspergille, *f.* (*Bot.*) Aspergillus.

aspérité, *f.* roughness, unevenness; harshness; asperity.

aspersion, *f.* sprinkling; aspersion.

aspersoir, *m.* sprinkler, sparger.

asphaline, *f.* (*Explosives*) asphaline.

asphalte, *m.* asphalt, asphaltum.

asphodèle, *m.* (*Bot.*) asphodel.

asphyxiant, *p.a.* asphyxiating, suffocating.

asphyxie, *f.* asphyxia.

asphyxier, *v.t.* asphyxiate, suffocate.

asphyxique, *a.* (*Med.*) asphyxial.

aspic, *m.* spike lavender, aspic (*Lavandula spica*); asp.

aspirail, *m.* vent hole, air hole.

aspirant, *p.a.* suction, sucking, aspirating; aspiring. — *m.* aspirant, candidate.

aspirateur, *m.* aspirator; exhauster; ventilator. — *a.* aspirating, exhausting; inhaling.

aspiration, *f.* aspiration, etc. (see aspirer); (*Mach.*) intake, admission, suction.

aspirer, *v.t.* aspirate, draw by suction, suck up, suck in, exhaust; inhale. — *v.i.* aspire; breathe.

asque, *m.* (*Bot.*) ascus.

assa dulcis, assa-dulcis, *f.* asa dulcis.

assa fœtida, assa-fœtida, *f.* asafetida.

assaillir, *v.t.* assail, assault, attack.

assainir, *v.t.* render healthy, purify.

assainissement, *m.* making healthy, purification, sanitation.

assaisonnant, *p.a.* seasoning, flavoring.

assaisonnement, *m.* seasoning, flavoring; condiment.

assaisonner, *v.t.* season, flavor.

assamar, assamare, *m.* assamar.

assaut, *m.* assault, attack.

assèchement, *m.* drainage.

assécher, *v.t.* drain.

asse-fétide, *n.* asafetida.

assemblage, *m.* assemblage, assembling; joint, coupling; (*Elec.*) connection, joining up; (*Masonry*) bond.
— *à vis,* screw coupling.
— *douille,* socket joint.
— *en quantité,* (*Elec.*) connection in parallel.
— *en tension* (or *en série*), (*Elec.*) connection in series.
— *mixte,* (*Elec.*) connection in series and parallel.

assemblée, *f.* assembly; meeting.

assembler, *v.t.* assemble, collect, put together; (*Elec.*) connect, join up. — *v.r.* assemble, meet.

assentir, *v.i.* assent.

asseoir, *v.t.* seat; set, put, place; found, base. — *v.r.* sit down, be seated.

asservir, *v.t.* (*Mach.*) put or have under control.

asseyant, *p.pr.* of asseoir.

assez, *adv.* enough, sufficiently; rather, passably. — *m.* enough, sufficiency. — *bien,* fairly well.

assied, *pr. 3 sing.* of asseoir.

assiéger, *v.t.* besiege; beset.

assiéra, *fut. 3 sing.* of asseoir.

assiette, *f.* plate, dish; position; horizontal position; state (of mind); assessment.

assignation, *f.* assignment; check; assignation.

assigner, *v.t.* assign; appoint; charge (on or to).

assimilabilité, *f.* assimilability.

assimilateur, *a.* assimilating, assimilative.

assimilatif, *a.* assimilative.

assimilation, *f.* assimilation.

assimiler, *v.t.* assimilate.

assis, *p.a.* seated, etc. (see asseoir).

assise, *f.* stratum, layer, bed, course; base.

assistance, *f.* assistance; audience; attendance.

assister, *v.t.* assist; attend. — *v.n.* (with *à*) be present at, attend, witness.

association, *f.* association.

associé, *p.a.* associated, etc. (see associer). — *m.* associate; partner; stockholder.

associer, *v.t.* associate; make (a person) an associate, partner or member. — *v.r.* become an associate; share, join; associate; harmonize.

assoit, *pr. 3 sing.* of asseoir.

assolement, *m.* rotation of crops.

assombrir, *v.t.* darken, obscure, make gloomy. — *v.r.* become dark or gloomy.

assommer, *v.t.* knock down, kill (with a blow); beat; stun, overwhelm; bore.

assomptif, *a.* assumptive, assumed.

assomption, *f.* assumption.

assorti, *p.a.* assorted; matched; stocked.

assortiment, *m.* assortment.

assortir, *v.t.* assort; match; stock, supply. — *v.r. & i.* harmonize, match.

assoupir, *v.t.* render drowsy; allay; quiet. — *v.r.* become drowsy; doze, nap; be quieted.

assoupissant, *p.a.* rendering drowsy, etc. (see assoupir); soporific.

assouplir, *v.t.* supple, make supple or pliable. — *v.r.* become supple or pliable.

assouplissage, *m.* suppling, etc. (see assouplir).

assourdir, *v.t.* deafen; muffle; tone down (colors).

assouvir, *v.t.* satiate, sate, satisfy.

assoyant, *p.pr.* of asseoir.

assuje(t)tir, *v.t.* subject; make fast, make tight, fasten; confine, bind; wedge in. — *v.r.* submit.

assumer, *v.t.* assume.

assurance, *f.* assurance; insurance; security.

assuré, *p.a.* assured; insured; secured; sure, certain; firm, steady.

assurément, *adv.* assuredly, certainly; securely; confidently.

assurer, *v.t.* assure; insure; secure; assert. — *v.r.* assure, secure, etc. oneself; satisfy oneself; rely, trust; take possession.

astérisque, *m.* asterisk.

astéroïde, *m.* asteroid; meteorite.

asthme, *m.* asthma.

asti, *m.* Asti (wine); a kind of silk; = astic.

astic, *m.* polisher, polishing stick; a polishing paste of whiting, soap and alcohol.

astiquer, *v.t.* polish, smooth.

astragale, *m.* astragal; (*Bot.*) Astragalus.

astre, *m.* (any) heavenly body.

astreignant, *p.pr.* of astreindre.

astreindre, *v.t.* restrict, limit; compel, force.

astreint, *p.a.* restricted, etc. (see astreindre) — *pr. 3 sing.* of astreindre.

astrictif, *a.* astrictive, astringent.

astringence, *f.* astringency.

astringent, *a. & m.* astringent.

astronome, *m.* astronomer.

astronomie, *f.* astronomy.

astronomique, *a.* astronomical.

astronomiquement, *adv.* astronomically.

asymétrie, *f.* asymmetry.

asymétrique, *a.* asymmetric(al).

asymétriquement, *adv.* asymmetrically.

asymptote, *f.* asymptote.

asymptotique, *a.* asymptotic(al).

atakamite, *f.* (*Min.*) atacamite.

atelier, *m.* workroom; workshop; factory; working party, gang; studio.
— *d'artificiers,* (*Mil.*) (pyrotechnic) laboratory.
— *de fonte,* foundry; smeltery; melting house.
— *de granulation,* (*Powder*) graining room.
— *de précision,* shop where accurate fitting, etc., is done.
— *de teinture,* dyehouse.
— *mécanique,* machine shop.

athanor, *m.* (*Old Chem.*) athanor.

athermane, *a.* athermanous.

athérome, *m.* (*Med.*) atheroma.

atmidomètre, *m.* atmometer, atmidometer.

atmolyse, *f.* atmolysis.

atmomètre, *m.* atmometer.

atmosphère, *f.* atmosphere.

atmosphérique, *a.* atmospheric(al).

atome, *m.* atom.

atomicité, *f.* atomicity.
atomisme, *m.* atomism.
atomiste *m.* atomist.
atomistique, *a.* atomistic(al). — *f.* atomistics.
a-t-on. See -t-.
atoxique, *a.* atoxic, nonpoisonous.
atoxyl, *m.* (*Pharm.*) atoxyl.
atrabilaire, *a.* (*Med.*) atrabiliary.
âtre, *m.* hearth.
-âtre. -ish; as, *rougeâtre,* reddish, *verdâtre,* greenish, etc.
atrolactique, atrolactinique, *a.* atrolactic.
atromarginé, *a.* (*Paper*) black-bordered.
atropamine, *f.* atropamine.
atrophier, *v.t.* atrophy.
atropine, *f.* atropine.
atropique, *a.* atropic.
attachant, *p.a.* attaching, etc. (see attach).
attache, *f.* attachment; fastening (of any kind), tie, cord, band, brace, rivet, clamp, clip, etc.; assent, consent.
attaché, *p.a.* attached, etc. (see attacher). — *m.* attaché; adherent.
attachement, *m.* attachment.
attacher, *v.t.* attach; fasten (in any way), bind, tie, pin, stick, etc.; interest, please. — *v.r.* attach oneself, be attached, cling, stick.
attaquable, *a.* attackable, liable to attack.
peu —, refractory.
attaque, *f.* attack.
— *aux gaz,* — *gazeuse,* — *par gaz,* (*Mil.*) gas attack.
— *par vagues de gaz,* (*Mil.*) gas-wave attack, cloud-gas attack.
attaquer, *v.t.* attack; commence on; address.
attarder, *v.t.* delay.
atteignant, *p.pr.* of atteindre.
atteindre, *v.t.* attain; reach; strike, hit; injure; overtake; equal, come up to.
atteint, *p.p. & pr. 3 sing.* of atteindre.
atteinte, *f.* reach; attack, blow, stroke, hit; pain, harm, shock.
hors d'—, out of reach, beyond reach; irreproachable.
attenant, *a. & prep.* adjoining. — *adv.* near by.
attendant, *p.pr.* of attendre.
en —, meanwhile, meantime.
en — *que,* till, until.
attendre, *v.t.* wait for, await; look for, expect; keep (wine, etc.). — *v.r.* expect; rely. — *v.i.* wait.
attendrir, *v.t.* soften, make soft, make tender. — *v.r.* become tender or soft, be softened.
attendrissement, *m.* softening, making tender; tenderness, softness.
attendu, *p.a.* waited for, etc. (see attendre). — *prep.* on account of, considering.

attendu *que,* considering that, since.
attente, *f.* waiting; expectation.
attenter, *v.t.* attempt.
attentif, *a.* attentive.
attention, *f.* attention.
attentivement, *adv.* attentively.
atténuant, *p.a.* attenuating, attenuant; extenuating. — *m.* (*Med.*) attenuant.
atténuation, *f.* attenuation; extenuation.
atténuer, *v.t.* attenuate, diminish; extenuate. — *v.r.* become attenuated or diminished.
attiédi, *p.a.* made tepid, etc. (see attiédir); tepid, lukewarm.
attiédir, *v.t.* make tepid, make lukewarm (whether by warming or cooling); cool. — *v.r.* become tepid or lukewarm.
attiédissement, *m.* making tepid, etc. (see attiédir); lukewarmness, tepidness; coolness.
attique, *a.* Attic; attic. — *m.* attic.
attirable, *a.* attractable.
attirail, *m.* apparatus, equipment, requisites, appliances, implements, goods, plant.
attirant, *p.a.* attractive, attracting.
attirer, *v.t.* attract; lead to. — *v.r.* attract each other or one another.
attiser, *v.t.* stir, stir up, poke (a fire).
attiseur, *m.* stoker.
attisoir, attisonnier, attisonnoir, *m.* poker, fire rake.
attouchement, *m.* contact; touch, touching.
attoucher, *v.t.* touch.
attracteur, *a.* attracting.
attractif, *a.* attractive, attracting.
attraction, *f.* attraction.
attractivement, *adv.* with attraction, attractively.
attraire, *v.t.* attract, entice.
attrait, *m.* attraction; bait. — *p.p.* of attraire.
attrape, *f.* crucible tongs; catch, trap.
attraper, *v.t.* catch; get; strike; cheat, dupe.
attrayant, *p.pr.* of attraire.
attrempage, *m.* heating gradually, etc. (see attremper).
attremper, *v.t.* heat gradually, anneal (glass), temper (steel), subject (glass pots, furnaces, etc.) to a preliminary heating.
attribuable, *a.* attributable.
attribuer, *v.t.* attribute; assign. — *v.r.* claim; be attributed.
attrition, *f.* attrition.
au. Contraction of *à le.*
aube, *f.* dawn; float, paddle, bucket (of a wheel).
aubépine, *f.*, **aubépin,** *m.* hawthorn (*Cratœgus*).
aubier, *m.* sap wood, sap, (*Bot.*) alburnum.
aucun, *a.* any; (with *ne,* expressed or implied) no, not any. — *pron.* any, any one, (*pl.*) some; (with *ne,* expressed or implied) no one, not one, none.

aucunement, *adv.* in any way, at all; (with *ne,* expressed or implied) in no way, not at all; (less commonly) to a certain extent.

audace, *f.* audacity.

audacieux, *a.* audacious, daring.

au deçà, au dedans, etc. See deçà, dedans, etc.

au-dessous, *adv.* below, beneath.
— *de,* under, beneath, below.

au-dessus, *adv.* above, upwards.
— *de,* above, over; (fig.) beyond.

au-devant, *adv.* in front, before.
— *de,* before; to meet.

audience, *f.* audience; sitting; court.

audition, *f.* hearing, audition; attendance (on lecture courses); auditing.

auditoire, *m.* auditorium; audience.

auge, *f.* trough; bucket (of a wheel); hod.
— *à goudron,* tar bucket.
— *de roulage,* bed (of an incorporating mill for gunpowder).
— *de trempage,* (*Expl.*) dipping trough.

augée, *f.* troughful; hodful.

auget, *m.* small trough; bucket (of a wheel or a conveyor); groove; cartridge carrier.

augmentation, *f.* increase, augmentation.

augmenter, *v.t.* increase, augment, (of prices) raise or advance. — *v.i.* increase; (of prices) rise. — *v.r.* increase, be increased.

augure, *m.* omen; augur.

aujourd'hui, *adv.* today; nowadays, at present.

aulne, *m.* & *f.* = aune.

aulx, *pl.* of ail.

aune, *m.* (*Bot.*) alder (*Alnus*); alder wood.
— *f.* ell (old measure, 1.188 meters).

aunée, *f.* (*Bot.*) elecampane (*Inula helenium*).

auparavant, *adv.* previously, before.

auprès, *adv.* near, near by.
— *de,* near, near to; about, with (persons); in comparison with; to.

auquel. Contraction of à lequel.

aura, *fut 3 sing.* of avoir (to have).

auramine, *f.* auramine.

aurantia, *n.* (*Dyes*) aurantia.

aurantiacées, *f.pl.* (*Bot.*) Aurantiaceæ.

auréole, *f.* areola (small area); specif. (as in blowpipe analysis), ring; (*Astron.*) corona; aureole, areola.

auréosine, *f.* (*Dyes*) aureosin.

aureux, *a.* aurous.

aurez, *fut. 2 pl.* of avoir (to have).

auriculaire, *a.* auricular. — *m.* little finger.

aurifère, *a.* auriferous.

aurifique, *a.* aurific.

aurique, *a.* auric.

auront, *fut. 3 pl.* of avoir (to have).

aurore, *f.* dawn; aurora. — *a.* & *m.* the color of the sky at dawn (used variously to mean pink, reddish yellow, or golden yellow).

aurore *boréale,* aurora borealis.

aurure, *m.* auride.

auryle, *m.* auryl.

aussi, *adv.* also, too; so. — *conj.* therefore, so, and so.
— *bien,* as well; so well; indeed, in fact, for.
— *bien que,* as well as.
— . . . *que,* as . . . as.

aussitôt, *adv.* immediately, at once; (usually with *que*) as soon as, whenever.

austère, *a.* (of taste) harsh, rough; austere, severe.

austérité, *f.* harshness, etc. (see austère).

Australasie, *f.* Australasia.

Australie, *f.* Australia.

australien, *a.* Australian.

autant, *adv.* as much, as many, as far.
— . . . —, as much . . . so much.
— *de,* so much, so many.
— *que,* as much, as many as; as far as, in so far as.
— . . . *que,* as . . . as.
d'—, by as much, to the same extent.
d'— *moins,* the less so, all the less.
d'— *plus,* the more so, all the more.
d'— *plus* (or *moins*) *que,* all the more (or less) as, so much the more (or less) because.
d'— *que,* in as much as, since, as; as far as.

autel, *m.* altar; (*Metal.*) altar: either fire bridge (*grand autel*) or flue bridge (*petit autel*); (*Astron.*) Ara, the Altar.

auteur, *m.* author; authority.

authentique, *a.* authentic. — *f.* original (text).

auto-ballon, *m.* dirigible.

auto-camion, *m.* auto truck, motor truck.

autoclave, *m.* & *a.* autoclave.

autocopiste, *a.* (self-)copying.

auto-excitateur, *a.* (*Elec.*) self-exciting.

auto-excitation, *f.* (*Elec.*) self-excitation.

autogène, *a.* autogenic, autogenous.

autolyse, *f.* autolysis.

autolytique, *a.* autolytic.

automate, *m.* automatic; automaton.

automatie, automaticité, *f.* automatic quality or power.

automatique, *a.* automatic.

automatiquement, *adv.* automatically.

automne, *m.* & *f.* autumn, fall.

automobilisme, *m.* automobile manufacture; use of automobiles, automobilism.

automoteur, *a.* self-acting, self-moving.

automotrice, *fem.* of automoteur.

autonome, *a.* autonomous.

autopsie, *f.* autopsy.
— *cadavérique,* post-mortem, autopsy.

auto-régulateur, *a.* self-regulating.

autorisation, *f.* authorization.

autoriser, *v.t.* authorize; commission. — *v.r* be authorized; obtain authority.

autorité, *f.* authority.
autour, *adv.* around, round.
— **de,** around, round, about.
autre, *a. & pron.* other.
autrefois, *adv.* formerly, in former times.
autrement, *adv.* otherwise.
Autriche, *f.* Austria.
Autriche-Hongrie, *f.* Austria-Hungary.
autrichien, *a.* Austrian.
autruche, *f.* ostrich.
autrui, *m.* other, others, other people.
auvergne, *f.* tan liquor.
auvergner, *v.t.* soak in tan liquor.
auvernat, *m.* auvernat (a red wine).
aux. Contraction of *à les,* to the, at the, etc.
auxdits. Contraction of *à les dits,* to the said things or persons, to the above, etc.
auxiliaire, *a. & m.* auxiliary.
auxiliairement, *adv.* in an auxiliary way.
auxomètre, *m.* (*Optics*) auxometer.
auxquelles. Contraction of *à lesquelles.*
auxquels. Contraction of *à lesquels.*
avait, *imp. 3 sing.* of avoir (to have).
aval, *m.* downstream direction; guaranty.
— *adv.* downstream.
en —, downstream, down.
en — *de,* below.
avaler, *v.t.* swallow; lower (into a cellar), cellar; guarantee. — *v.r.* be swallowed; hang down; go downstream.
avalies, *f.pl.* wool from slaughtered sheep.
avance, *f.* advance; projection; (*Mach.*) lead; (*Soap*) a lye of density 25° B.
d' —, *par* —, in advance.
avancé, *p.a.* advanced, etc. (see avancer); (of meat, etc.) spoiled; forward, early; late.
avancement, *m.* advancement; projection.
avancer, *v.t.* advance; hasten; gain. — *v.i.* advance; progress; project; be fast, gain time. — *v.r.* advance; progress; project.
avant, *prep.* before. — *adv.* far; before. — *m.* front, fore part, head.
— *de,* — *que,* — *que de,* before.
— *J.-C.,* before Christ, B.C.
en —, on, onward, forward, ahead.
en — *de,* in advance of, before, beyond.
avantage, *m.* advantage; odds; donation.
avantageusement, *adv.* advantageously.
avantageux, *a.* advantageous; becoming; presuming.
avant-bras, *m.* forearm.
avant-chauffeur, *m.* preheater.
avant-creuset, *m.* (*Metal.*) forehearth.
avant-dernier, *a. & n.* last but one, next to last.
avant-garde, *f.* advance guard.
avant-goût, *m.* foretaste.
avant-guerre, *m.* (period) before the war.
d' —, before-the-war, ante-bellum.
avant-hier, *adv. & m.* day before yesterday.
avant-pièce, *n.* = pièce maîtresse, under *pièce.*

avant-propos, *m.* preface; preamble; introduction.
avant-vin, *m.* = vin doux, under *vin.*
avare, *a.* avaricious; sparing; greedy. — *m. & f.* miser.
avarie, *f.* damage; (in maritime commerce) average.
avarier, *v.t.* damage, spoil. — *v.r.* spoil
avec, *prep.* with. — *adv.* with it.
d' —, from.
avelanède, *f.,* acorn cup of the valonia oak, (*pl.*) valonia.
aveline, *f.* filbert.
avenant, *a.,* pleasing; comely.
à l' —, in proportion, in keeping.
avènement, *m.* coming, arrival, accession.
avenir, *m.* future. — *v.i.* happen, occur.
à l' —, in the future, for the future.
aventure, *f.* adventure; venture; chance; fortune.
à l' —, haphazard, at random.
d' —, *par* —, by chance, perchance.
aventurer, *v.t.* venture, risk. — *v.r.* venture, run risk.
avenu, *p.p.* of avenir.
avérer, *v.t.* prove, establish.
avertir, *v.t.* inform, notify; warn.
avertissement, *m.* preface (to books); notice; admonition, warning, hint.
avertisseur, *m.* alarm, annunciator, call bell, indicator; one that notifies or warns. — *a.* warning, indicating, notifying.
avet, *m.,* **avette,** *f.* silver fir (*Abies picea*).
aveu, *m.* consent; acknowledgment, avowal.
aveuglant, *p.a.* blinding.
aveugle, *a.* blind. — *m. & f.* blind person.
aveugler, *v.t.* blind; stop up, stop.
avez, *pr. 2 pl.* of avoir (to have).
avide, *a.* avid, desirous, greedy, grasping.
— *de,* (*Chem.*) having strong affinity for.
avidité, *f.* avidity.
avient, *pr. 3 sing.* of avenir.
aviez, *imp. 2 pl.* of avoir (to have).
avilir, *v t.* depreciate; debase, degrade. — *v.r.* (of goods) depreciate; (of prices) fall.
avilissement, *m.* debasement, etc. (see avilir).
aviner, *v.t.* season, soak or stain with wine.
avion, *m.* airplane, aëroplane.
avions, *imp. 1 pl.* of avoir (to have).
aviron, *m.* oar.
avis, *m.* notice; preface; opinion; information, intelligence; warning; counsel, advice; decision; vote.
à mon —, in my opinion.
— *autorisé,* official notice.
aviser, *v.t.* advise; inform; perceive, see. — *v.i. & r.* think.
— *de,* think of, see to.
s' — *de,* think of; presume to.
s' — *à,* see to.

avivage, *m.* brightening, etc. (see aviver).

aviver, *v.t.* brighten (in various senses); brighten, heighten, freshen (colors); polish, burnish; quicken (with mercury, as in making mirrors or in electroplating); cut to sharper angles; brisk up, make hotter (a fire); enliven. — *v.r.* brighten, become brighter; revive; (of pain) become sharper.

avocat, *m.* advocate; barrister, lawyer.

avocatier, *m.* avocado, alligator pear (tree).

avoine, *f.* oats; oat plant; (*pl.*) oat crop.

avoir, *v.t.* have. — *m.* property, possessions; credit.

— *affaire à,* have business with, have to do with, be dealing with.

— *dû faire,* must have done (and so for other infinitives than *faire*).

— *lieu,* take place, occur.

— *raison,* be right.

— *tort,* be wrong.

— *vent de,* get wind of.

il y a, there is, there are; (of time) it is, ago.

il y a eu, there has been, there have been.

il y avait, there was, there were.

avoisinant, *p.a.* adjacent (to), neighboring.

avoisiner, *v.t.* be adjacent to, border on. — *v.r.* be adjacent, be near.

avons, *pr. 1 pl.* of avoir (to have).

avortement, *m.* abortion.

avoué, *p.a.* avowed; approved. — *m.* attorney.

avouer, *v.t.* acknowledge, avow; approve.

avrelon, *m.* mountain ash (*Sorbus aucuparia*).

avril, *m.* April.

avrillet, *m,* spring wheat.

axe, *m.* axis; axle, arbor, spindle, shaft.

axial, *a.* axial.

axillaire, *a.* axillary.

axiomatique, *a.* axiomatic.

axiome, *m.* axiom.

axonge, *f.* semisolid fat, grease, especially lard; (*Pharm.*) lard, adeps.

— *balsamique,* — *benzoinée,* (*Pharm.*) benzoinated lard.

axuel, *a.* axial.

ay, *m.* a kind of sparkling wine from Ay.

ayant, *p. pr.* having (from *avoir*).

Az. Symbol for *azote* (nitrogen).

azédarac, azédarach, *m.* (*Bot.*) China tree, azedarach, margosa.

azélaïque, *a.* azelaic.

azimidé, *a.* azimido, azimino.

azine, *f.* azine.

azinique, *a.* azine.

azoamidé, azoaminé, *a.* aminoazo.

azobenzène, *m.* azobenzene.

azobenzoïque, *a.* azobenzoic.

azodiphénylène, *m.* azodiphenylene (phenazine).

azoïque, *a.* azo. — *m.* azo compound.

azophénolique, *a.* azophenolic, hydroxyazo.

azotate, *f.* nitrate.

— *d'ammoniaque,* ammonium nitrate.

— *d'argent,* silver nitrate.

— *de baryte,* barium nitrate.

— *de chaux,* calcium nitrate.

— *de cuivre,* copper nitrate.

— *de fer,* ferric nitrate.

— *de fer liquide,* (*Pharm.*) solution of ferric nitrate.

— *de lithine,* lithium nitrate.

— *de magnésie,* magnesium nitrate.

— *de plomb;* lead nitrate.

— *de potasse,* potassium nitrate, nitrate of potash.

— *de soude,* sodium nitrate, nitrate of soda.

— *d'urane,* uranium nitrate.

— *mercurique liquide,* (*Pharm.*) solution of mercuric nitrate.

azote, *m.* nitrogen.

— *restant,* residual nitrogen.

azoté, *a.* nitrogenous; nitrogenized, azotized.

azoter, *v.t.* nitrogenize, azotize.

azoteux, *a.* nitrous.

azothydrique, *a.* hydrazoic, triazoic.

azotine, *f.* azotine (a nitrogenous fertilizer; also, an explosive of the gunpowder type).

azotique, *a.* nitric.

azotite, *m.* nitrite.

— *d'amyl,* amyl nitrite.

azotomètre, *m.* azotometer, nitrometer.

azotoxyhémoglobine, *f.* nitric oxide hemoglobin.

azoture, *m.* nitride.

— *de plomb,* lead nitride.

azoturie, *f.* (*Med.*) azoturia.

azotyle, *m.* nitryl, nitro group.

azoxique, *a.* azoxy.

azoxybenzène, *m.* azoxybenzene.

azur, *m.* azure, (sky) blue; smalt.

— *de cuivre,* blue verditer.

azurage, *m.* bluing, etc. (see azurer).

azuré, *a.* azure.

azurer, *v.t.* blue, color blue, azure.

azurin, *a.* pale blue.

azyme, *a.* unleavened. — *m.* azym, unleavened bread.

azymique, *a.* azymic.

B

babeurre, *m.* buttermilk.
babiole, *f.* toy, bauble, trifle.
babouin, *m.* baboon.
bac, *m.* back, vat, tank; (of a storage battery) jar; ferryboat.
— *à carbonater,* — *à carbonatation,* (*Sugar*) carbonating vat, saturator.
— *à cristalliser,* crystallizing tank, crystallizer.
— *d'attente,* (*Sugar*) crystallizer.
— *décanteur,* decanting tank.
— *jaugeur,* (*Sugar*) a container for measuring the proper amount of massecuite into the centrifugal.
— *mesureur,* (*Sugar*) a container for measuring beet juice.
— *refroidisseur,* cooling tank.
bâche, *f.* cistern, tank; cover (of canvas or the like), tarpaulin.
bachelier, *m.* bachelor.
bacillaire, *a.* bacillar, bacillary.
bacille, *m.* bacillus.
bacille-virgule, *m.* comma bacillus, cholera bacillus.
bactéricide, *a.* bactericidal.
bactérie, *f.* bacterium; (*pl.*) bacteria.
bactéries aérobies, aërobic bacteria.
bactérien, *a.* bacterial, (less often) bacterian.
bactériologie, *f.* bacteriology.
bactériolyse, *f.* bacteriolysis.
bactériolysine, *f.* bacteriolysin.
bactrioles, *f.pl.* gold beatings.
badamier, *m.,* **badamie,** *f.* (*Bot.*) Terminalia.
Bade, *n.* Baden (duchy); Baden-Baden (watering place).
badiane, *f.,* **badianier,** *m.* star anise (*Illicium*).
badigeon, *m.* (*Painting*) whitewash, or wash of any color; badigeon; putty.
badigeonnage, *m.* painting, etc. (see badigeonner).
badigeonner, *v.t.* paint, coat (with iodine, with tar, etc.); whitewash, wash (in any color).
badiner, *v.i.* have play, hang loose; jest, play.
badois, *a.* of Baden.
bagace, *f.* = bagasse.
bagage, *m.* baggage, luggage.
bagasse, *f.* spent substance (as indigo waste, the marc of grapes or of olives, etc.); specif., (*Sugar*) bagasse, cane trash; (*Bot.*) fruit of the *bagassier*.

bagassier, *m.* (*Bot.*) an ulmaceous tree of Guiana.
bagassière, *f.* (*Sugar*) a trough which conducts sugar cane to the rolls.
bague, *f.* ring; hoop, collar, sleeve, band.
— *d'appui,* supporting ring or collar.
— *d'assemblage,* collar, sleeve.
— *de garniture,* packing ring, piston ring.
baguenaude, *f.* bladder senna pod.
baguenaudier, *m.* (*Bot.*) bladder senna (*Colutea arborescens*).
baguette, *f.* rod; stick, wand, switch; ramrod; (of a rocket) stick; drumstick.
— *de nettoyage,* cleaning rod.
— *de verre,* — *en verre,* glass rod.
baie, *f.* berry; bay.
— *de genévrier,* — *de genièvre,* juniper berry.
baignage, *m.* bathing, etc. (see baigner).
baigner, *v.t.* bathe, lave; wet. — *v.i.* float, swim; dip; soak, be immersed.
baignoire, *f.* bath tub; (in a theater) box.
bail, *m.* lease.
bâiller, *v.i.* yawn, gape, be open, show crevices.
bain, *m.* bath (in various senses); bathing.
— *à sensibiliser,* (*Photog.*) sensitizing bath.
— *d'air,* air bath.
— *d'argent,* (*Photog.*) silver bath.
— *de Barèges artificiel,* a bath imitating the sulfur water of Barèges, made by adding a prepared mixture of salts to ordinary water.
— *de cendres,* ash bath.
— *de développement,* (*Photog.*) developing bath.
— *de fixage,* (*Photog.*) fixing bath.
— *de fonte,* bath of melted cast iron.
— *de mortier,* bed of mortar.
— *de plomb,* lead bath.
— *de sable,* sand bath.
— *de sel,* salt bath.
— *de teinture,* dye bath, liquor.
— *de trempe,* tempering bath.
— *de vapeur,* steam bath; (*Med.*) vapor bath.
— *de virage,* (*Photog.*) toning bath.
— *d'huile,* oil bath.
— *fixateur,* (*Photog.*) fixing bath.
— *réducteur,* reducing bath.
— *révélateur,* (*Photog.*) developing bath.
en —, melted, fused.
mettre en —, melt, fuse.

31

bain-marie, *m.* water bath.
baïonette, *f.* bayonet.
baiser, *v.t. & r.* kiss. — *m.* kiss, kissing.
baisse, *f.* fall, decline.
  — *du change,* fall of exchange.
baisser, *v.t.* lower; bow (the head). — *v.i.* fall, sink, decline; deteriorate. — *v.r.* stoop; fall; be dejected. — *m.* setting (of the sun).
baissière, *f.* bottom, foot, part just above the lees (as of wine in a cask).
baissoir, *m.* (*Salt*) brine tank.
balai, *m.* broom; (*à laver*) mop; (*Elec.*) brush.
  — *de communication,* commutator brush.
  — *électrique,* dynamo brush.
balaie, *pr. 3 sing.* of balayer.
balais, balai, *a.* In *rubis balai(s),* balas ruby.
balance, *f.* balance.
  — *à court fléau,* short-beam balance.
  — *à ressort,* spring balance.
  en —, in suspense, in the balance.
balancé, *p.a.* balanced; swayed, rocked; undecided; weighed, considered.
balancement, *m.* balancing, etc. (see balancer); balance; (*Astron.*) libration.
balancer, *v.t.* balance; swing, rock; weigh, examine. — *v.r.* swing, sway; balance.
balancerie, *f.* balance factory, scales factory; balance making.
balancier, *m.* beam, lever; pendulum; press (for stamping, punching or the like); fly, flyer (of a wheel); balance, balance wheel (of a watch); rod (of a pump); balance maker.
  — *à vis,* fly press, screw press.
  — *découpoir,* cutting press.
balançoire, *f.* swing; seesaw; (*Cement*) an apparatus for sifting the ground cement.
balaustier, *m.* pomegranate tree.
balayage, balayement, *m.* sweeping.
balayer, *v.t.* sweep.
balayette, *f.* small broom or brush.
balayure, balayures, *f. sing. & pl.* sweepings.
balbutier, *v.i.* stammer, lisp. — *v.t.* stammer.
bale, bâle, *f.* glume, husk, chaff (of seeds).
Bâle, *f.* Basel, Bâle (Swiss canton and town).
Baléares, *n.pl.* Balearic isles.
baleine, *f.* whale; whalebone, (less commonly) baleen; rib, stay (as of an umbrella).
baleinier, *a.* whaling. — *m.* whaler; whalebone dealer.
balèvre, *f.* projection (on a casting); lower lip.
baline, *f.* a coarse woolen packing fabric, baline.
balistique, *f.* ballistics. — *a.* ballistic.
balistite, *f.* (*Expl.*) ballistite.
baliveau, *m.* young tree, sapling, staddle.
balkanique, *a.* Balkan.
ballage, *m.* (*Metal.*) balling.
ballant, *a.* swinging, loose.

ballast, *m.* ballast.
ballastage, *m.* ballasting.
balle, *f.* ball; bale; pack; bullet; (*Bot.*) = bale.
  — *à éclairer,* (*Pyro.*) light ball.
  — *à feu,* (*Pyro.*) fire ball.
  — *à fumée,* (*Pyro.*) smoke ball.
  — *ardente,* (*Pyro.*) light ball.
  — *conique,* conical bullet.
  — *d'artifice,* pyrotechnic ball, fire ball.
  — *de remplissage,* shrapnel bullet.
  — *luisante,* (*Pyro.*) light ball.
  — *multiple,* fragmentation bullet.
baller, *v.t.* (*Metal.*) ball.
ballon, *m.* balloon; specif., a spherical or round-bottomed flask, balloon flask (but also applied rather loosely to certain other flasks); carboy; inflated ball (as a football; top (of an automobile); hood.
  — *à col court,* short-necked balloon.
  — *à distillation,* — *à distiller,* balloon flask for distillation.
  — *à filtrer,* round-bottomed filtering flask.
  — *à fond plat,* flat-bottomed flask (usually a Florence flask).
  — *à gaz,* gas balloon, specif., a flexible gas holder with delivery tube and stopcock, for laboratories.
  — *à long col,* long-necked balloon, bolthead.
  — *d'artifices,* fire balloon.
  — *de Pasteur,* Pasteur flask.
  — *d'extraction,* balloon flask for extraction (short wide neck, flat bottom).
  — *diluteur,* dilution flask (used in the analysis of air).
  — *réserve,* a special bottle (resembling a wash bottle) for preserving a sterile liquid and delivering it in the form of a jet.
ballonner, *v.t.* distend, inflate. — *v.i.* bulge.
ballot, *m.* (small) bale, package.
ballote, *f.* (*Bot.*) Ballota.
ballotte, *f.* small ball; a wine vessel; = ballote.
ballotter, *v.t.* toss, toss about, shake; pack in bales; ballot for. — *v.i.* shake, wobble, tumble.
bâlois, *a.* of Basel or Bâle.
balsamier, *m.* balsam tree.
balsamifère, *a.* balsamiferous.
balsamique, *a.* balsamic.
bambou, *m.* bamboo.
banaliser, *v.t.* render banal or commonplace.
banane, *f.* banana.
bananier, *m.* banana (plant).
banc, *m.* bed, stratum; bench; (*Tech.*) frame, bed; bank (of sand, etc.).
  — *à étrangler,* (*Pyro.*) choking frame, choking press.
  — *d'artifices,* (*Pyro.*) laboratory table.
bandage, *m.* tire; hoop; bandage; bandaging.
  — *caoutchouc,* rubber tire.
  — *creux,* penumatic tire.

bandage *de roue,* tire.
— *plein,* solid tire.
— *pneumatique,* pneumatic tire.
bande, *f.* band; strip; slip; strap; belt; bandage; tire; side, shore.
bandé, *p.a.* bandaged, etc. (see bander).
— *m.* compression (of a spring).
bandelette, *f.* narrow band, small strip; flat bar iron; small bandage.
— *agglutinative,* adhesive strip.
bander, *v.t.* bandage, bind up; compress, set, wind up (a spring); tighten (a cord); bend (something elastic); cock (a gun). — *v.i.* be tight.
bandine, *f.* buckwheat flour; buckwheat.
bandoir, *m.* spring, mainspring.
bandoulière, *f.* shoulder belt.
banlieue, *f.* outskirts, environs, suburbs.
banne, *f.* hamper; awning; tarpaulin; (*Mining*) corf, corb.
bannière, *f.* banner.
bannir, *v.t.* banish.
banque, *f.* bank; banking.
banqueroute, *f.* bankruptcy.
banquier, *m.* banker.
baquet, *m.* tub; bucket; trough; tray; (*Glass*) a water bucket for cooling the blowpipe.
barattage, *m.* churning.
baratte, *f.* churn.
baratte-malaxeur, *m.* a combined churn and kneader.
baratter, *v.t.* churn.
Barbade, *f.* Barbados, Barbadoes.
barbare, *a.* barbarian; barbarous. — *m.* barbarian.
barbaresque, *a.* of Barbary.
barbe, *f.* beard; rough edge, burr (of castings, etc.); (*Paper*) deckle; barb; wattle, gill; fin.
— *de plume,* feather of a quill.
barbeau, *m.* cornflower, bluebottle (*Centaurea cyanus*); barbel (a fish). — *a.* cornflower (blue), light blue.
barbelé, *a* barbed.
barbillon, *m.* barb; barbel.
barbotage, *m.* bubbling, etc. (see barboter); mash (as of bran and water).
barboter, *v.i.* (of gases) bubble; paddle, dabble; mumble. — *v.t.* mix, stir; dip; splash.
barboteur, *m.* bubbler (vessel in which a gas delivery tube dips into a liquid, *e.g.* a gas-washing bottle or a potash bulb); mixer, stirrer, agitator; duck. — *a.* bubbling; stirring, mixing.
— *à palettes,* paddle mixer.
barbotin, *m.* sprocket wheel, chain pulley.
barbotine, *f.* (*Ceram.*) slip, slop; (*Pharm.*) santonica.
barbotteur, *m.* = barboteur.
barbouiller, *v.t.* daub, smear, soil; scribble; bungle; mumble.

barbu, *a.* bearded; (*Bot.*) barbate.
bardeau, *m.* shingle; slate (for roofing).
bardelle, *f.* (*Glass*) bullion bar (an arm or bar on which the glass is rested).
baril, *m.* barrel; barrel-shaped vessel.
barille, *f.* barilla (either alkali or plant).
— *douce,* the first quality of Spanish crude soda.
— *mélangée,* second-grade Spanish crude soda.
barillet, *m.* small barrel, keg; (*Gas*) a horizontal washing apparatus, hydraulic main; (*Mach.*) barrel, drum, cylinder; (of a fuse) case.
bariolage, *m.* medley; variegation.
bariolé, *p.a.* variegated, party-colored, motley.
barique, *f.* See barrique.
baritel, *m.* hoisting machine, whim.
barium, *m.* barium.
barle, *f.* (*Mining*) fault.
barlong, *a.* of unequal sides, lopsided; oblong.
baromètre, *m.* barometer.
— *à cadran,* wheel barometer.
— *à cuvette,* cup barometer.
— *à mercure,* mercury barometer.
— *à siphon,* siphon barometer.
— *enregistreur,* recording barometer, barograph.
— *portatif,* portable barometer.
barométrique, *a.* barometric(al).
barométriquement, *adv.* barometrically.
barométrographe, *m.* barometrograph, barograph.
baronne, *f.* baroness.
baroque, *a.* (of pearls) irregular, baroque; odd, grotesque, baroque.
barque, *f.* (small) boat, skiff.
barquieux, *m.pl.* (*Soap*) tanks having perforated false bottoms and used for the preparation of lye.
barrage, *m.* dam; barrier; (*Mil.*) barrage.
barras, *m.* barras, galipot.
barre, *f.* bar (in various senses); line, stroke.
— *à T,* T-bar, T-iron.
— *d'aller,* (*Elec.*) positive bus bar.
— *d'appui,* hand rail.
— *de fourneau,* — *de grille,* fire bar, grate bar.
— *de retour,* (*Elec.*) negative bus bar.
— *d'excentrique,* eccentric rod.
— *directrice,* guide bar.
— *du foyer,* fire bar, grate bar.
— *témoin,* (*Metal.*) trial bar, tap bar.
barreau, *m.* (small) bar; bar (legal sense).
— *aimanté,* bar magnet.
— *de grille,* grate bar.
— *d'essai,* test bar.
barrer, *v.t.* bar; dam; strike out, cross out.
barrette, *f.* (small) bar; rib, stay, strip.
— *d'essai,* test bar, test strip.
barrière, *f.* barrier.

barrique, *f.* barrel, cask (esp. for liquids).

baryte, *f.* baryta.

— *caustique,* caustic baryta, barium oxide.

— *hydratée,* hydrated baryta, barium hydroxide.

barytine, *f.* (*Min.*) barite, barytes.

barytique, *a.* of or pertaining to barium or baryta, barytic; as, *sulfate barytique,* barium sulfate.

baryum, *m.* barium.

bas, *a.* low; lower; shallow. — *m.* bottom, lower part, foot; stocking. — *adv.* low; down, off.

à —, down; low, off with; down with.

à — *de,* down from, off.

— *foyer,* (*Metal.*) any broad, low furnace (as contrasted with shaft furnace), hearth furnace, shallow hearth.

basse pression, low pressure.

en —, below; downstairs; downward.

en — *de,* at the bottom of.

par —, below; downward.

par en —, downward; at the bottom; from below.

basalte, *m.* basalt.

basaltique, *a.* basaltic.

basane, *f.* tanned sheepskin, sheep, basil.

basané, *p.a.* tanned, sunburnt.

bascule, *f.* a piece or apparatus having a rocking or seesaw motion, rocker, lever, plier, swipe, bascule, balance (esp. of the steelyard kind); specif., a conveyor in the form of a trough or pipe which is rocked forward and backward; reciprocating or seesaw motion.

à —, rocking.

— *romaine,* Roman balance.

basculer, *v.t.* move back and forth about a pivot, rock, swing, sway, tilt, dip, drop.

basculeur, *m.* rocker; (*Mach.*) rocking lever.

base, *f.* base; basis.

— *antimoniée,* (*Org. Chem.*) any of various substituted stibines.

— *arséniée,* (*Org. Chem.*) any of various substituted arsines.

— *cadavérique,* cadaveric base, ptomaine.

— *nucléinique,* nuclein base.

— *névrinique,* neurine base.

— *oxygénée,* (*Org. Chem.*) a base containing oxygen, as a hydroxy amine.

— *phosphorée,* (*Org. Chem.*) any of various substituted phosphines.

— *purique,* purine base.

— *pyridique,* pyridine base.

— *quinoléique,* quinoline base.

— *xanthique,* xanthine base.

baser, *v.t.* base, ground. — *v.r.* be based.

se — *sur,* be based on, take as a basis.

bas-fond, *m.* low ground, bottom; shoal water, shoal; (rarely) deep water.

bas-foyer, *m.* See bas foyer, under *bas.*

basicité, *f.* basicity.

basification, *f.* basification, basifying.

basilic, *m.* (*Bot.*) basil, esp. sweet basil (*Ocimum basilicum*); basilisk.

basique, *a.* basic.

basophile, *a.* (*Biol.*) basophilic, basophilous.

basse, *a. fem.* of bas. — *f.* shoal; (*Music*) bass.

basse-cour, *f.* farmyard; back yard.

basse-étoffe, *f.* base metal, esp. an alloy of lead and tin.

basse-mer, *f.* low water, low tide.

basserie, *f.* (*Leather*) plumping liquor.

bassesse, *f.* lowness; baseness, meanness.

bassin, *m.* basin (in various senses); pan, specif. scale pan; dock; (*Anat.*) pelvis.

— *de repos,* settling basin, settling tank.

— *filtrant,* filtering basin, filter bed.

bassine, *f.* a variously shaped open vessel (usually circular in form but either shallow or deep and either metal or pottery), pan, jar.

— *conique,* a pail-shaped vessel, widening from bottom to top and having the handles on the sides.

— *cylindrique,* a cylindrical vessel or jar, usually with handles on the sides.

bassiner, *v.t.* dampen, moisten, sprinkle; bathe.

bassinet, *m.* (small) basin; (of a fuse) cup.

bassorine, *f.* bassorin.

bassotin, *m.* (*Dyeing*) indigo vat.

bastrinque, *m.* a kind of black-ash furnace.

bas-ventre, *m.* lower part of the belly.

bat, *pr. 3 sing.* of battre.

bataille, *f.* battle; battle array.

batailler, *v.i.* battle, struggle.

bataillon, *m.* battalion.

bâtard, *a.* bastard, mongrel, hybrid.

batate, *f.* sweet potato, batatas.

batavique, *a.* In the phrase *larme batavique,* Rupert's drop.

bateau, *m.* boat.

— *à vapeur,* steamboat, steamer.

bateau-citerne, *m.* tank boat, tank steamer.

batée, *f.* pan (for washing gold).

bâti, *p.a.* built; (*Sewing*) basted. — *m.* frame, framework; support, bed, stand; basting.

bâtiment, *m.* building, structure; vessel, ship.

— *de graduation,* (*Salt*) graduation house, graduation tower.

bâtir, *v.t.* build; (*Sewing*) baste. — *v.i.* build. — *v.r.* be built; build for oneself.

bâton, *m.* stick; rod, cane, staff, wand, baton.

à bâtons rompus, by snatches.

bâtonnage, *m.* casting into stick form.

bâtonner, *v.t.* beat with a stick; strike out.

bâtonnet, *m.* small stick, small rod, rodlet.

batracien, *m.* (*Zoöl.*) batrachian.

battage, *m.* beating; stamping; churning; thrashing; driving (of piles).

battant, *a.* beating; swinging (door); fighting; (*Mach.*) going, working. — *m.* leaf (of a door); clapper (of a bell); batten (of a loom).

batte, *f.* beater, beetle, rammer; (of a churn) beater, plunger.

battement, *m.* beating; beat; clapping, flapping, stamping.

batterie, *f.* battery; fighting, fray.
— *électrique,* electric battery.

batteur, *m.* beater.
— *d'or,* gold beater.

batteuse, *f.* beater; churn; thrashing machine.

battitures, *f.pl.* hammer scale, iron scale (chiefly Fe_3O_4).

battoir, *m.* beater, rammer.

battre, *v.t.* beat; pound, strike, hammer, batter, ram, thrash, drive (piles), coin (money). — *v.i.* beat; throb, palpitate; clap, stamp; (*Mach.*) be loose.
— *son plein,* be in full blast.

battu, *p.a.* beaten, etc. (see battre).

batture, *f.* gold lacquer.

baudruche, *f.* goldbeater's skin.

bauge, *f.* a mortar of clay and straw.

baume, *m.* balsam; balm.
— *calédonien,* (*Pharm.*) a solution of kauri gum in an equal weight of 90 per cent alcohol.
— *de Canada,* Canada balsam.
— *de Carthagène,* balsam of Tolu.
— *de cheval,* (*Bot.*) horse balm (*Collinsonia*).
— *de copahu,* copaiba, copaiba balsam.
— *de Giléad,* balm of Gilead.
— *de la Mecque,* balm of Gilead, Mecca balsam.
— *de Pérou,* — *de Pérou noir,* — *des Indes,* — *de Sonsonate,* balsam of Peru.
— *de Tolu* (or *tolu*), balsam of Tolu (or tolu).
— *du commandeur,* (*Pharm.*) an old remedy essentially the same as compound tincture of benzoin.
— *ophthalmique rouge,* (*Pharm.*) ointment of red mercuric oxide.
— *tranquille,* (*Pharm.*) an olive oil solution of several narcotic and aromatic plant substances, used as an anodyne.
— *vert,* (*Bot.*) spearmint.

baumier, *m.* balsam tree.

baux, *pl.* of bail.

bauxite, *f.* (*Min.*) bauxite.

bavard, *a.* talkative; gossiping.

bavarois, *a.* Bavarian.

bave, *f.* slaver, slobber, slime, foam.

baver, *v.i.* slaver, slobber, dribble, run, (of ink) blot.

bavette, *f.* bib; plate.

Bavière, *f.* Bavaria.

bavoché, *p.a.* blurred, printed indistinctly.

bavure, *f.* rough edge, seam, burr, beard, mold mark.

bayer, *v.i.* gape.

bdellium, *m.* bdellium.

beau, *a.* beautiful, handsome, fine, elegant, fair, pleasant; proper, fitting; (in compound words) step-, -in-law. — *m.* best; beauty; beautiful; beau, fop.

beaucoup, *adv.* much, greatly. — *m.* much, a great deal, a good deal.
à — *près,* by far, by a great deal.
— *de,* much, many, a great many.
de —, by far, by much.

beauté, *f.* beauty.

beaux-arts, *m.pl.* fine arts.

bec, *m.* burner; lip (of a vessel); beak, bill; neck (of a still or retort); any of various projecting parts, nose, spout, stud, catch, arm, jaw (of a vise), mouthpiece, tip.
à — (of vessels) with lip, lipped.
— *à gaz,* gas burner.
— *à incandescence,* incandescent gas burner.
— *à tirage,* burner for creating a draft.
— *Bunsen,* Bunsen burner.
— *cacheteur,* a burner for sealing letters, etc.
— *Carcel,* Carcel lamp.
— *de cigogne,* stork's bill; herb Robert (*Geranium robertianum*).
— *de sûreté,* safety burner.
— *d'étain,* (*Min.*) a twinned cassiterite crystal.
— *de tirage,* = bec à tirage.
— *papillon,* fishtail burner(?).
— *renversé,* inverted burner.

bec-de-grue, *m.* (*Bot.*) crane's-bill (*Geranium*).
— *tacheté,* spotted crane's-bill (*G. maculatum*).

bêche, *f.* spade.

becherglas, *m.* beaker.

bégayer, *v.i. & t.* stammer, stutter, lisp.

beige, *a.* (of wool) raw, natural; of the color of, or made from, raw wool.

bel, *a.* Used instead of *beau* before vowels and in certain phrases. See beau.

belge, *a.* Belgian.

belge-silésien, *a.* (*Zinc*) designating or pertaining to the combined Belgian and Silesian process.

Belgique, *f.* Belgium.

bélier, *m.* ram.

belladone, *f.* belladonna.

belladonine, *f.* belladonnine.

belle, *a. fem.* of beau. — *f.* belle, fair one.

belliqueux, *a.* warlike, martial, bellicose.

bellite, *f.* (*Expl.*) bellite.

ben, *m.* ben, ben nut, ben tree (*Moringa*).

bénédiction, *f.* blessing, benediction, consecration.

bénéfice, *m.* benefit; (*Com.*) profit, gain; (*Gem Cutting*) diamond waste.
au — *de,* to the benefit of.

bénéficiel, *a.* beneficial.

bénéficier, *v.i.* make a profit. — *v.t.* work with profit.

bénévole, *a.* kind, benevolent; (of students) special, not candidate for a degree.

bengale, *m.* (*Pyro.*) Bengal light; (*cap.*) Bengal.

benigne, *a. fem.* of benin.

bénin, *a.* benign, benignant, kind, mild.

bénir, *v.t.* bless.

bénit, béni, *p.a.* blessed, consecrated, holy.

benjoin, *m.* benzoin (either resin or plant).

benne, *f.* a receptacle in which material is transported in factories, mines, etc.: basket, cage, cradle, car, skip, bucket.

benoîte, *f.* (*Bot.*) avens, bennet (*Geum*).

— *aquatique,* — *des ruisseaux,* water avens (*G. rivale*).

bénoléique, *a.* behenolic.

benzaldoxime, *m.* benzaldoxime.

benzène, *m.* benzene.

— *diazimide,* triazobenzene.

— *sulfoxyde,* phenyl sulfoxide.

benzènedicarbonique, *a.* benzenedicarboxylic.

benzène-monosulfoné, *a.* benzene(mono)-sulfonic.

benzène-sulfoné, *a.* benzenesulfonic.

benzènetricarbonique, *a.* benzenetricarboxylic.

benzènetrisulfoné, *a.* benzenetrisulfonic.

benzénique, *a.* of or pertaining to benzene, benzene, aromatic.

benzéno-disulfonique, *a.* benzenedisulfonic.

benzéno-sulfonate, *m.* benzenesulfonate.

benzéno-sulfone, *m.* phenyl sulfone.

benzéno-sulfonique, *a.* benzenesulfonic.

benzényle, *m.* benzenyl.

benzhydrol, *m.* benzohydrol, benzhydrol.

benzidine, *f.* benzidine.

benzidinique, *a.* benzidine; as, *transposition benzidinique,* benzidine rearrangement.

benzile, *m.* benzil.

benzine, *f.* benzene (or, *Com.* benzole); benzine; turpentine.

— *colophanée,* turpentine (the oleoresin).

— *de la houille,* benzene (or benzole).

— *légère,* light benzole.

— *lourde,* heavy benzole.

— *rectifiée,* oil of turpentine.

— *régie,* heavy benzole.

benzoate, *m.* benzoate.

benzobleu, *m.* benzo blue (a dye).

benzoflavine, *f.* (*Dyes*) benzoflavine.

benzofurfurane, *m.* benzofuran, benzofurfuran.

benzoïne, *f.* benzoïn.

benzoïné, *a.* benzoinated.

benzoïque, *a.* benzoic.

benzol, *m.* benzole, benzol (the commercial product); benzene (the compound).

benzonaphtol, *m.* benzonaphthol (β-naphthyl benzoate, used as an antiseptic).

benzonitrile, *m.* benzonitrile.

benzophénone, *f.* benzophenone.

benzopurpurine, *f.* (*Dyes*) benzopurpurin.

benzothiophène, *m.* benzothiophene (thionaphthene).

benzoylacétique, *a.* benzoylacetic (β-ketohydrocinnamic).

benzoyle, *m.* benzoyl.

benzoylformique, *a.* benzoylformic (keto-α-toluic).

benzylamine, *f.* benzylamine.

benzyle, *m.* benzyl.

benzylique, *a.* benzyl, benzylic.

béquille, *f.* crutch; handle (of a cock); prop, support.

berbérine, *f.* berberine.

berbérinique, *a.* berberine, of berberine.

berce, *f.* (*Bot.*) cow parsnip (*Heracleum*).

berceau, *m.* cradle (in various senses); barrel vault; (*Founding*) a furnace for drying molds.

bercement, *m.* rocking, rocking motion.

bercer, *v.t.* rock; deceive; (*Engraving*) cradle.

bergamote, *f.* bergamot, bergamot orange; bergamot orange tree.

bergamotier, *m.* bergamot orange tree (*Citrus aurantium bergamia*).

berger, *m.* shepherd.

béribéri, *m.* beriberi.

béril, *m.* beryl.

Bermudes, *f.pl.* Bermudas.

béryl, *m.* beryl.

béryllium, *m.* beryllium.

besaigre, *a.* turning sour, sour.

besicles, *f.pl.* spectacles.

besogne, *f.* work; task, labor, business, job.

besogner, *v.i.* work.

besoin, *m.* want, need, necessity.

au — at need, if necessary.

avoir — *de,* have need of, need, want.

il est —, it is necessary.

bestiaux, *m.pl.* live stock, cattle.

bétail, *m.* live stock, cattle.

bétaïne, *f.* betaïne.

bête, *f.* beast, brute, animal; simpleton; aversion. — *a.* silly, stupid.

bêtise, *f.* stupidity, folly, nonsense.

bétoine, *f.* betony (*Betonica*).

béton, *m.* concrete; (rarely) colostrum.

— *aggloméré,* ordinary cement concrete.

— *armé,* reinforced concrete, armored concrete, ferroconcrete.

— *au quart,* — *au septième,* etc., concrete with one volume of cement to four, seven. etc. of other material.

béton *Coignet,* a kind of concrete made by mixing sand with cement or hydraulic lime, and adding water.

— *gras,* concrete rich in cement.

— *maigre,* concrete poor in cement.

bétonnage, *m.* concrete work, concrete; laying of concrete.

bétonner, *v.t.* construct, or cover, with concrete.

bétonneur, *m.* concrete worker.

bétonnière, *f.* concrete mixer.

bette, *f.* any plant of the genus *Beta,* including beets and chard.

betterave, *f.* beet, beetroot.

— *à sucre,* sugar beet.

— *fourragère,* forage beet.

— *potagère,* beet for cooking.

— *sucrière,* sugar beet.

beurre, *m.* butter (in various senses).

— *de cacao,* cacao butter (less properly called cocoa butter).

— *de coco,* coconut butter, coconut oil.

— *de muscade,* nutmeg butter.

— *de vache,* cow butter, ordinary butter.

— *frais,* fresh butter, unsalted butter.

beurrerie, *f.* place where butter is made, creamery.

bevilacque, *n.* (*Bot.*) Indian pennywort.

bézoard, *m.* bezoar.

bézoardique, *a.* bezoardic.

acide —, old name for ellagic acid.

biacétyle, *m.* biacetyl, 2, 3-butanedione.

biacide, *a.* biacid, diacid.

biais, *m.* slope, inclination, slant; bias, bent; way of looking or proceeding, shift. — *a.* slanting, oblique, skew, askew, awry.

de —, *en* —, obliquely, aslant, askew, on the bias.

biaiser, *v.i.* slant, slope, slope off.

biatomique, *a.* diatomic.

bibarytique, *a.* containing two atoms of barium, dibarium.

bibasique, *a.* bibasic, dibasic.

bibelot, *m.* small ornament, knickknack.

bibliographe, *m.* bibliographer.

bibliographie, *f.* bibliography.

bibliographique, *a.* bibliographic(al).

bibliothécaire, *m.* librarian.

bibliothèque, *f.* library; bookcase.

bibromer, *v.t.* (*Org. Chem.*) introduce two atoms of bromine into.

bibromhydrique, *a.* (*Org. Chem.*) of the nature of a dibromohydrin (thus, ethylene chloride is called the *éther bibromhydrique* of glycol).

bibromure, *m.* dibromide.

bicalcique, *a.* dicalcium, dicalcic.

bicarbonate, *m.* bicarbonate.

— *de potasse,* potassium bicarbonate.

— *de soude,* sodium bicarbonate.

bicarboné, *a.* bicarburetted.

bicarbure, *m.* dicarbide, bicarbide, (formerly) bicarburet.

bicarré, *a.* (*Math.*) raised to the fourth power.

bichet, *m.* annatto.

bichloré, *a.* dichloro, dichloro-.

bichlorhydrique, *a.* of the nature of a dichlorohydrin. See bibromhydrique.

bichlorure, *m.* dichloride, bichloride.

bichromate, *m.* dichromate, bichromate.

— *de potasse,* potassium dichromate.

— *de soude,* sodium dichromate.

biconcave, *a.* biconcave, concavo-concave.

biconvexe, *a.* biconvex, convexo-convex.

bidon, *m.* can, tin-plate vessel (often holding about 5 liters) for oil, gasoline or the like; large jug or pitcher of wood or metal; canteen.

— *à huile,* oil can.

— *incendiare,* (*Mil.*) incendiary can.

bielle, *f.* (*Mach.*) rod, arm, link, connecting rod, side rod.

— *d'accouplement,* coupling rod.

— *motrice,* driving rod.

bien, *adv.* well; very; very much; much; quite, entirely. — *m.* good; good thing, blessing; property, estate, possession.

à —, successfully.

— *de,* much, many.

— *entendu,* well understood; be it well understood; certainly.

— *entendu que,* it being understood that, provided that.

— *peu,* very little.

— *plus,* more than that, what is more.

— *que,* altho, tho.

— *venu,* welcome.

en —, honorably; favorably.

si — *que,* so that; however, however much.

bien-être, *m.* welfare, comfort, comforts.

bienfaisant, *a.* beneficent, beneficial, helpful.

bienfait, *m.* benefit.

biennal, *a.* biennial.

bientôt, *adv.* soon, shortly, presently.

bière, *f.* beer; bier.

— *de Brunswick,* mum.

— *de conserve,* — *de garde,* lager beer.

— *de gingembre,* ginger beer.

— *de malt,* malt liquor.

biffer, *v.t.* erase, cancel, strike out.

bifurcateur, *m.* dividing piece, forked tube.

— *en forme de T,* T tube.

— *en forme d' Y,* Y tube.

bifurqué, *p.a.* bifurcated, bifurcate, forked; (of chains of atoms) branched.

bigarade, *f.* bitter orange.

bigarré, *a.* mottled, party-colored.

biiodhydrique, *a.* of the nature of a diiodohydrin. See bibromhydrique.

biiodure, *m.* diiodide, biiodide, biniodide.

— *de mercure,* mercuric iodide.

bijon, *m.* (*Pharm.*) pine resin.

bijou, *m.* jewel.

bijouterie, *f.* jewelry; jeweler's trade or shop.

bilan, *m.* balance sheet.

bile, *f.* bile, gall.

— *de bœuf,* oxgall, ox bile.

biliaire, *a.* biliary, bile.

bilieux, *a.* bilious.

billard, *m.* billiards; billiard table or room.

bille, *f.* small ball (of metal or other material); billiard ball; marble; log; billet; sucker, shoot; cutting (for propagation).

billet, *m.* note, bill, ticket, ballot, certificate.

— *de banque,* bank note.

— *simple,* promissory note; (*Railways*) single ticket.

billon, *m.* coin bronze; billon (debased alloy formerly used for money); (*Agric.*) ridge.

billot, *m.* block.

biloupe, *f.* a magnifying glass with two lenses.

bimagnésique, *a.* containing two atoms of magnesium, dimagnesium.

bimétallique, *a.* bimetallic; containing two atoms of a metal, dimetallic.

binaire, *a.* binary.

binitré, *a.* dinitrated, dinitro. — *m.* dinitro compound.

biochimie, *f.* biochemistry.

biochimiste, *m.* biochemist.

biographe, *m.* biographer; biograph.

biographie, *f.* biography.

biologie, *f.* biology.

biologique, *a.* biologic(al).

biologiste, biologue, *m.* biologist.

bioxyde, *m.* dioxide, (formerly) bioxide.

— *d'argent,* silver dioxide.

— *d'azote,* nitrogen dioxide.

— *d'étain,* tin dioxide, stannic oxide.

bipolaire, *a.* bipolar.

bipotassé, *a.* containing two atoms of potassium, dipotassium.

biprimaire, *a.* biprimary (said, *e.g.,* of a compound containing two primary alcohol groups).

biprisme, *m.* biprism.

biréfringent, *a.* birefringent.

bis, *a.* grayish brown, brown, swarthy. — *adv.* a second time, twice (used with numbers; e.g. *10 bis* in a series is equivalent to 10A or 10½). — *interj.* encore!

bisaccharide, *m.* bisaccharide (disaccharide).

bisage, *m.* second dyeing.

bisaille, *f.* coarse flour; mixture of peas and vetches.

bisannuel, *a.* biennial.

bisazoïque, *a.* disazo. — *m.* disazo compound or dye.

biscuit, *m.* biscuit; (*Ceram.*) biscuit, bisque. — *a.* twice baked, twice burned.

biscuit *de porcelaine,* biscuit porcelain, biscuit, bisque.

biscuité, *a.* (of bread) baked longer than usual.

bisé, *p.a.* redyed, dyed a second time.

biseau, *m.* bevel; bezel.

biseautage, *m.* beveling.

biseauter, *v.t.* bevel.

bisecondaire, *a.* bisecondary. Cf. biprimaire.

biser, *v.t.* redye, dye a second time. — *v.i.* (of grain) darken, deteriorate.

bismuth, *m.* bismuth.

bismuthifère, *a.* bismuthiferous.

bismuthine, *f.* (*Min.*) bismuthinite, bismuthine.

bismuthique, *a.* bismuthic.

bismuth-triéthyle, *m.* bismuth triethyl, triethylbismuthine.

bissecter, *v.t.* bisect.

bissecteur, *a.* bisecting.

bissection, *f.* bisection.

bissectrice, *a. fem.* of bissecteur. — *f.* bisectrix.

bissexuel, *a.* bisexual.

bistorte, *f.* (*Bot.*) bistort.

bistortier, bistotier, *m.* a wooden pestle.

bistre, *m.* bister, bistre.

bistrer, *v.t.* color with bister; brown. ,

bi-sublimé, *p.a.* twice sublimed, resublimed.

bisubstitué, *a.* disubstituted, bisubstituted.

bisulfite, *n.* bisulfite, acid sulfite.

— *de soude,* sodium bisulfite.

bisulfitique, *a.* of or pertaining to a bisulfite, bisulfitic; bisulfite.

bisulfure, *m.* disulfide, bisulfide.

bitartrate, *m.* bitartrate, acid tartrate.

— *de potasse,* potassium bitartrate.

bitertiaire, *a.* bitertiary. Cf. biprimaire.

bitter, *m.* bitters (the liquor, esp. made from gin).

bitumage, *m.* covering with bitumen.

bitume, *m.* bitumen.

— *de Judée,* Jew's pitch, bitumen of Judea.

bitumer, *v.t.* bituminize.

bitumeux, *a.* bituminous.

bituminer, *v.t.* bituminize.

bitumineux, *a.* bituminous.

bituminifère, *a.* bituminiferous.

bituminisation, *f.* bituminization.

bituminiser, *v.t.* bituminize.

bizarre, *a.* bizarre, fantastic.

blafard, *a.* dull, dim, pale.

blaireau, *m.* brush (of badger's hair or other soft hair); badger.

blâme, *m.* blame.

blanc, *a.* white; blank; clean. — *m.* white; blank space, blank; white heat; white man, white boy; white sauce.

— *ancien,* — *antique,* (*Glass*) white glass with a slight bluish green tint, cathedral glass.

— *d'antimoine,* antimony white.

blanc *d'argent*, a fine variety of white lead.
— *de baleine*, spermaceti.
— *de baryte*, baryta white.
— *de bismuth*, bismuth white.
— *de Bougival*, a form of whiting.
— *de céruse*, white lead, lead white.
— *de Champagne*, whiting, whitening.
— *de champignon*, mushroom spawn.
— *de chaux*, whitewash, coat of whitewash.
— *de Chine*, Chinese white.
— *de fard*, pearl white (basic bismuth nitrates).
— *de Hambourg*, Hamburg white.
— *de Hollande*, Dutch white.
— *de Krems*, Kremnitz white, krems white.
— *de lait*, milky white.
— *de lune*, (*Metal.*) white heat, welding heat.
— *de Meudon*, a form of whiting.
— *de Mulhouse*, lead sulfate.
— *de neige*, snow white (zinc white).
— *de perle*, pearl white (bismuth oxychloride).
— *de plomb*, white lead, lead white.
— *d'Espagne*, Spanish white.
— *de Troyes*, a form of whiting.
— *de Venise*, Venetian white.
— *de zinc*, zinc white.
— *d'œuf*, egg white, white of egg.
— *fixe*, blanc fixe (barium sulfate).
— *mat*, dead white, dull white.
— *minéral*, whiting, whitening.
— *permanent*, permanent white.
— *sale*, dingy white, grayish or yellowish white.
— *soudant*, welding heat, white heat.
en —, in white; unpainted; in blank.
blanchâtre, *a.* whitish.
blanche, *a. fem.* of blanc.
blanchet, *m.* cloth filter; a kind of white wool cloth.
blancheur, *f.* whiteness.
blanchi, *p.a.* whitened, etc. (see blanchir).
blanchiment, *m.* whitening, etc. (see blanchir).
— *chimique*, chemical bleaching.
— *naturel*, — *sur pré*, grass bleaching.
blanchir, *v.t.* whiten, bleach, blanch; wash; clean, polish (metals); refine (cast iron); tin; dress, trim, rough-grind, smooth; scald, parboil; wash for; exonerate. — *v.i.* turn white, whiten, bleach, blanch, pale, fade; (of hair) turn gray; graze. — *v.r.* turn white, whiten, bleach.
— *à la chaux*, whitewash.
blanchissage, *m.* washing, laundering; whitewashing; refining (of sugar).
blanchissant, *p.a.* whitening, etc. (see blanchir); foaming, foamy, frothy.
blanchisserie, *f.* bleachery; bleaching trade; laundry, place where clothes are washed.
blanchisseur, *m.* bleacher; launderer. — *a.* bleaching; laundering.

blanchisseuse, *fem.* of blanchisseur.
blanchoyer, *v.i.* have a white reflection.
blanc-soudant, *m.* welding heat, white heat.
blankett, *m.* (*Mining*) blanket.
blankite, *f.* blankite.
blanquette, *f.* blanquette (a kind of kelp or crude soda); a white sparkling wine of the Midi; (*Dist.*) low wines.
blé, *m.* wheat; (with qualifying words) grain, corn (including wheat, rye, barley, maize, etc.); wheat field, grain field.
— *barbu*, bearded wheat.
— *cornu*, ergot.
— *de printemps*, spring wheat.
— *de Turquie*, maize, (Indian) corn.
— *d'hiver*, winter wheat.
— *méteil*, mixed wheat and rye.
— *noir*, buckwheat.
blende, *f.* (*Min.*) blend, blende.
blendeux, *a.* blendous, of or containing blend.
blesser, *v.t.* wound, hurt, injure; offend.
blessure, *f.* wound, hurt, injury; grief.
blet, *a.* bletted, over-ripe, half-decayed.
blettir, *v.i.* blet, become over-ripe.
bleu, *a.* blue. — *m.* blue; bruise.
— *azoïque*, azo blue.
— *azur*, — *céleste*, azure, sky blue.
— *cœruleum*, ceruleum (greenish blue pigment).
— *Coupier*, Coupier's blue (induline).
— *d'acier*, steel blue.
— *d'alizarine*, alizarin blue.
— *d'aniline*, aniline blue.
— *d'azur*, smalt.
— *de Berlin*, Berlin blue.
— *de Brème*, Bremen blue.
— *de Chine*, Chinese blue.
— *de ciel*, sky blue.
— *de cobalt*, cobalt blue.
— *de diphénylamine*, diphenylamine blue.
— *de gallamine*, gallamine blue.
— *d'émail*, smalt.
— *de Meldola*, Meldola's blue.
— *de méthylène*, methylene blue.
— *de montagne*, mineral blue (blue copper carbonate).
— *d'empois*, smalt.
— *de naphtol*, naphthol blue.
— *de Nicholson*, Nicholson's blue.
— *de Nil*, Nile blue.
— *de nuit*, night blue.
— *de Paris*, Paris blue.
— *de phénylène*, phenylene blue.
— *de Prusse*, Prussian blue.
— *de résorcine*, resorcin blue.
— *de roi*, king's blue.
— *de safre*, zaffer blue, smalt.
— *de Saxe*, Saxon blue, Saxony blue.
— *de Sèvres*, Sèvres blue.
— *de toluylène*, toluylene blue.

bleu *d'indigo*, indigo blue, indigotin.
— *d'outremer*, ultramarine blue.
— *égyptien*, Egyptian blue, copper glass.
— *hydron*, hydron blue.
— *lumière*, light blue, bleu lumière (the dye); bleu lumière (the pigment).
— *minéral*, mineral blue.
— *Nil*, Nile blue.
— *outremer*, ultramarine blue, ultramarine.
— *patenté*, patent blue.
— *sombre*, dark blue; (*Dyes*) night blue.
— *turquin*, a dark, mat blue.
— *Victoria*, Victoria blue.

bleuâtre, *a.* bluish.

bleu-bleu, *m.* a kind of blue ultramarine.

bleuir, *v.t.* blue, make blue. — *v.i.* turn blue, become blue, blue.

bleuissage, *m.* bluing.

bleuissant, *p.pr.* of bleuir.

bleuissement, *m.* turning blue, bluing.

bleu-lapis, *m.* (*Glass*) a blue color obtained by the use of cobalt oxide.

bleuté, *p.a.* tinged with blue, bluish.

bleuter, *v.t.* tinge with blue, blue slightly.

blindage, *m.* armor, armor plating; casing, protection; (*Mil.*) blindage.

blinder, *v.t.* armor; case; (*Mil.*) blind.

bloc, *m.* block; lump, mass, lot; log.
— *de plâtre*, (*Bact.*) gypsum block.
en —, in a lump, in the lump.

blocage, *m.* stopping, etc. (see bloquer).

blocus, *m.* blockade.

blond, *a.* light-colored (light yellow or light brown, distinguished from *blanc*, white); blond, fair. — *m.* light color; blond.

blondeur, *f.* light color.

blondir, *v.i.* turn light-colored, turn light yellow, (of grain) whiten.

blondoyer, *v.i.* have a light yellow reflection.

bloom, *m.* (*Metal.*) bloom.

bloquer, *v.t.* stop, arrest, block, lock, jam; clamp, fix, set; blockade; enclose (in masonry); dub out, fill out (with plaster).

blouse, *f.* blouse, frock, smock.
— *de laboratoire*, laboratory coat (esp. a long, full overgarment).

bluette, *f.* spark, sparkle; flake or scale of hot iron.

bluetter, *v.i.* spark, sparkle.

blutage, *m.* bolting.

bluter, *v.t.* bolt (flour, cement, etc.).

bluterie, *f.* bolting mill, bolting house.

blutoir, *m.* bolter, bolting machine, bolting cloth.

Bo. Symbol for boron.

bobèche, *f.* socket (of a lamp or candlestick); sconce (of a candlestick); steel edge (on a knife or chisel).

bobinage, *m.* spooling, winding; wiring, coil.

bobine, *f.* bobbin, spool, drum, reel; (*Elec.*) bobbin, coil.

bobine *à réaction*, reactance coil, choking coil.
— *de dérivation*, shunt coil.
— *de résistance*, resistance coil.
— *de Ruhmkorff*, Ruhmkorff coil.
— *d'induction*, — *inductrice*, induction coil.
— *induite*, secondary coil.
— *primaire*, primary coil.
— *secondaire*, secondary coil.

bobineur, *m.* winder, spooler.

bocage, *m.* scrap (of metals); grove, coppice.

bocal, *m.* wide-mouthed bottle; any of various wide-mouthed vessels, vase, globe, jar; mouthpiece.

bocard, *m.* stamp, stamp mill, stamping mill, crusher.

bocardage, *m.* stamping, crushing.

bocarder, *v.t.* stamp, crush (ores, etc.).

bocaux, *pl.* of bocal.

bœuf, *m.* ox; (*pl.*) cattle; beef.
— *salé*, salt beef, corned beef.

boghead, *m.* Boghead coal, boghead.

boguette, *f.* buckwheat.

bohé, *m.* bohea (a kind of black tea).

Bohême, *n.* Bohemia; Bohemian glass.

Bohémien, *a.* Bohemian.

boire, *v.t.* drink; absorb, drink in, suck up, soak up; swallow (an insult). — *v.i.* drink; (of paper) blot. — *m.* drink, drinking; creek, inlet, channel.
faire —, soak (skins, etc.).

bois, *m.* wood; timber.
— *amer*, bitterwood, quassia.
— *à ouvrer*, timber, lumber.
— *blanc*, soft wood (nonresinous).
— *cochon*, a balsamaceous tree known as *Tetragastris panamensis* or *Hedwigia balsamifera*.
— *colorant*, dyewood.
— *d'absinthe*, quassia.
— *d'acajou*, mahogany.
— *d'aigle*, eaglewood, agalloch.
— *d'amarante*, mahogany.
— *d'aubier*, sapwood.
— *débité*, wood sawed into planks, boards, etc., timber, lumber.
— *de Brésil*, brazilwood.
— *de brin*, whole timber (not split).
— *de campêche*, logwood, campeachy wood; (*Pharm.*) hematoxylon.
— *de charpente*, timber, lumber.
— *de chauffage*, firewood.
— *de construction*, timber, lumber.
— *de fer*, ironwood.
— *de fer blanc*, (*Bot.*) Sideroxylon cinereum or its wood.
— *de fil*, wood cut with the grain.
— *de frêne*, ash wood, ash.
— *de gaïac*, guaiacum wood, lignum vitæ.
— *de haute futaie*, forest timber; timber forest.
— *de paille*, straw board.

bois de Panama, soapbark, quillaja bark, Panama bark.

— de quassie, quassia, quassia wood.

— de réglisse, licorice root.

— de rose, rosewood (the cabinet wood); rhodium wood, rosewood (containing a fragrant oil).

— de sang, logwood.

— de santal, sandalwood.

— de sappan, sapan wood (a red dyewood).

— de service, timber, lumber.

— des îles, brazilwood (?).

— desséché, dry wood, seasoned wood.

— de teinture, dyewood.

— de travail, timber, lumber.

— d'Inde, logwood.

— d'œuvre, timber, lumber.

— doux, soft wood; cascarilla; licorice root.

— droit, straight-grained wood.

— du Brésil, brazilwood.

— du centre, heartwood.

— dur, hard wood, hardwood.

— durci, artificial wood, wood pulp.

— fin, fine wood, close-grained wood.

— gentil, (Pharm.) mezereum.

— gras, soft wood.

— jaune, yellowwood, specif., the wood of Morus tinctoria.

— parfait, heartwood.

— rouge, redwood.

— soufré, sulfurated wood.

— tendre, soft wood.

— vif, green timber, green wood.

boisage, m. woodwork, framing, casing, planking.

boisé, a. wooded; wainscoted, paneled.

boiserie, f. wainscoting, paneling.

boiseux, a. woody, ligneous.

boisseau, m. the hole or hollow part into which the key or plug of a stopcock or tap fits, plug hole, shell; a section of earthenware pipe (specif., without a qualifying word, a tile of square cross section as distinguished from boisseau rond which is cylindrical); bushel.

boisson, f. drink, beverage, liquor; specif., piquette; drinking, drunkenness.

— fermentée, fermented liquor, fermented beverage.

boit, pr. 3 sing. of boire.

boite, f. (of wine, etc.) state of fitness, maturity, ripeness; piquette.

boîte, f. box; case, chest, canister, etc.

— à feu, fire box.

— à fumée, smoke box.

— à noyau, (Founding) core box.

— à réactifs, reagent box.

— à soupape, valve chest.

— à vent, (of a furnace) wind box; (of an **organ**) wind chest.

boîte de distribution, (Elec.) connecting box, switchboard.

— de Pétri, Petri dish.

— de résistance, (Elec.) resistance box.

— d'origine, original box or case.

— gainerie, case, box (for keeping instruments or the like).

— perdue, a box or case not intended to be returned when emptied.

boitement, m. limping; (Mach.) irregular action.

boiteux, a. lame, limping.

boivent, pr. 3 pl. indic. & subj. of boire.

bol, m. bole; bolus, large pill; bowl.

bolaire, a. bolar, bolary.

bolet, m. (Bot.) boletus.

Bolivie, f. Bolivia.

bolivien, a. Bolivian.

bolonais, a. Bolognese.

bombage, m. rendering convex, etc. (see bomber).

bombagiste, m. maker of convex glass.

bombardement, m. bombardment.

bombarder, v.t. bombard; shell.

bombe, f. bomb.

— éclairante, light ball, illuminating bomb.

— fumigène, smoke bomb.

— incendiaire, incendiary bomb.

bombé, p.a. rendered convex, bulged, convex.

bombement, m. bulging, convexity.

bomber, v.t. render convex, cause to bulge or swell out; (Glass) bend to convex form. — v.i. bulge, swell out, become convex.

bombeur, m. one who renders convex.

— de verre, maker of convex glass.

bombiste, m. bomb maker.

bombonne, f. = bonbonne.

bon, a. good; simple, foolish. — adv. good, well. — m. good; good part, best; comfort; order, check, draft; security, bond, certificate.

à — compte, cheaply.

à — marché, cheap; cheaply.

à la bonne heure, very well, good, so be it.

— goût, (in the rectification of alcohol) the fraction which follows the alcool de cœur and is followed by the moyen goût (it includes the alcool fin and some alcohol of second quality); a superior quality of certain other chemicals, e.g. acetic acid.

— marché, cheap.

bonne première, (Soap) a salt lye of 28° B.

bonne seconde, (Soap) a salt lye of 26° B.

— teint, fast dye, fast color; fast.

de bonne heure, soon, early.

bonbon, m. bonbon, sweetmeat.

bonbonne, f. carboy; large bottle such as those used in distilling nitric acid; oil can.

bond, m. bound.

bonde, *f.* opening, vent, outlet, inlet, (of a cask) bunghole; stopper, bung; sluice.
— *trop-plein,* overflow.
— *de remplissage,* opening for putting in material, charging hole.

bonder, *v.t.* fill very full, cram.

bondir, *v.i.* bound.

bondon, *m.* bung, stopper; (less often) bunghole; bondon cheese, bondon.

bondonner, *v.t.* bung.

bonheur, *m.* good fortune, luck; welfare; success; happiness.

bonhomme, *m.* good man; fellow; puppet; mullen; bolt, catch.

boni, *m.* surplus.

bonification, *f.* improvement; allowance, rebate.

bonifier, *v.t.* improve; make up, make good.
— *v.r.* improve.

bonne, *a. fem.* of bon. — *f.* (servant) maid.

bonnement, *adv.* simply, sincerely; exactly.

bonnet, *m.* cap; end of a window-glass cylinder to which the blowpipe is attached.

bonneterie, *f.* knitted goods, knitting, hosiery.

bonnette, *f.* shade (on certain instruments).

bonté, *f.* goodness, kindness.

boracique, *a.* boracic (boric).

boracite, *f.* (*Min.*) boracite.

borate, *m.* borate.
— *de chaux,* borate of lime (calcium borate).
— *de soude,* borate of soda (sodium borate), specif. borax.

boraté, *a.* borated.

borax, *m.* borax.

bord, *m.* edge, border, margin; rim; brim; bank (of a river); shore; binding, border; side (of a ship or a road).

bordeaux, *m.* Bordeaux wine, bordeaux; (*Dyes*) bordeaux.

bordelaise, *f.* a cask for Bordeaux wine, holding 225–230 liters; a special bottle for Bordeaux wine.

border, *v.t.* border, edge, line; sheathe.

bordereau, *m.* memorandum, statement.

bordure, *f.* border, edging, edge, binding.

borduré, *a.* bordered.

bore, *m.* boron.

boré, *a.* containing boron, of boron.

boréal, *a.* north, northern, boreal.

bore-méthyle, *m.* boron methyl, boron trimethyl.

boréthyle, *m.* borethyl, boron triethyl.

borgne, *a.* one-eyed; (of things) deficient in some way.

boricisme, *m.* boricism, boric acid poisoning.

borique, *a.* boric.

boriqué, *a.* treated with or containing boric acid.

borne, *f.* boundary, limit; post (to mark a point or limit); landmark; (*Elec.*) terminal.
— *d'attache,* (*Elec.*) binding post.

bornéol, *m.* borneol.

borner, *v.t.* limit, confine, restrict; bound.

bornylamine, *f.* bornylamine.

bornyle, *m.* bornyl.

bornylène, *m.* bornylene.

boro-tungstique, *a.* borotungstic.

borure, *m.* boride.

bossage, *m.* bossing, boss, swelling.

bosse, *f.* bump, boss, protuberance, prominence, hump; (*Art*) relief.

bosseler, *v.t.* emboss; dent.

bossuer, *v.t.* dent.

botanique, *f.* botany. — *a.* botanical.

botaniquement, *adv.* botanically.

botaniste, *m.* botanist.

botryoïde, *a.* botryoidal.

botte, *f.* bundle, bunch, (of wire) coil, (of yarn) hank; boot; butt (large cask); (*Zinc*) adapter.

bottillon, *m.* bunch, small bundle.

bottine, *f.* high shoe, half-boot.

bouc, *m.* goat, he-goat; goatskin bottle, skin.

boucage, *m.* anise.

boucanage, *m.* smoking (of meat or fish).

boucanager, *v.t.* smoke, cure by smoking.

boucaut, *m.* hogshead.

bouchage, *m.* stoppering, etc. (see boucher).
— *à l'émeri,* making of ground stoppers, stopper grinding.

bouche, *f.* mouth; opening, orifice.
— *à feu,* gun, cannon.

bouchement, *m.* filling, plugging, stopping.

boucher, *v.t.* stopper, stop, stop up, plug, plug up, cork, lute, close. — *m.* butcher.
— *à l'émeri,* provide with a ground stopper.

boucherie, *f.* butchery; butcher trade.

bouche-trou, *m.* stopgap.

bouchon, *m.* stopper; plug, stopple, cork, tampon, wad, wadding; cap, cover, lid.
— *à deux trous,* two-holed stopper.
— *à l'émeri,* ground stopper.
— *à vis,* screw stopper.
— *d'amorce,* fuse plug, primer plug.
— *de sûreté,* safety plug.
— *de vidange,* drip cock.
— *en caoutchouc,* rubber stopper; rubber plug.
— *en liège,* cork stopper, cork.
— *en verre,* glass stopper.

bouchon-écrou, *m.* screw plug.

bouchonnier, *m.* maker of, or dealer in, corks or stoppers.

boucle, *f.* loop, ring, eye; bend (as of a U-tube); elbow, turn; buckle; curl (of hair).

boucler, *v.t.* buckle; ring; curl (hair).

bouclier, *m.* buckler, shield.

bouder, *v.i.* pout; sulk; be backward.

boudin, *m.* (stuffed) pudding, sausage; something resembling a sausage, as a helical spring, a fuse, a roller, a glass bulb, a roll of opium, or a cylinder of soap as it issues from the milling machine.

boudinage, *m.* (*Soap*) pressing (of toilet soap) into a homogeneous cylinder.

boudine, *f.* (*Glassblowing*) bull's-eye.

boudineuse, *f.* a machine for pressing toilet soap into cylinders.

boue, *f.* mud; slime; mire, dirt; pus.

bouée, *f.* buoy.

boueux, *a.* muddy; slimy; foul, dirty.

bouffée, *f.* puff; whiff, breath; outburst.

bouffer, *v.i.* puff, puff out, bulge; (of bread) rise.

bouffi, *p.a.* puffed, puffy, swollen, bloated.

bouffir, *v.t. & i.* puff up, puff out, bloat.

bouge, *m.* bulge, swell, convexity; closet.

bouger, *v.i.* budge, stir.

bougie, *f.* candle, taper (esp., a candle made from wax or fatty acids, in distinction from *chandelle*); cylindrical or candle-shaped filter (esp. one for sterilizing liquids), filter candle; (*Med.*) bougie; (*Mach.*) spark plug.

— *Berkefeld,* Berkefeld filter.

— *Chamberland,* Chamberland filter.

— *décimale,* decimal candle, bougie décimale.

— *de cire,* wax candle.

— *de stéarine,* stearin candle (made of stearic acid).

— *d'oribus,* resin torch.

— *électrique,* electric-light carbon.

— *normale,* standard candle.

— *stéarique,* = bougie de stéarine.

bougie-heure, *f.* candle-hour.

bougran, *m.* buckram.

bouillage, *m.* boiling.

bouillaison, *f.* fermentation (of beer, cider, etc.).

bouillait, *imp. 3 sing.* of bouillir.

bouillant, *p.a.* boiling; scalding hot.

bouillent, *pr. 3 pl. indic. & subj.* of bouillir.

bouillerie, *f.* distillery.

bouilleur, *m.* distiller; still; boiler; boiler tube; heater (of a boiler).

— *d'eau-de-vie,* brandy distiller or (in general) distiller.

— *de cru,* one who distils only from certain fruits of his own raising, such as grapes, apples, pears, plums.

— *de profession,* a distiller who uses material raised by others, or materials not allowed the *bouilleur de cru*, such as peaches, apricots, potatoes, sugar beets, cereals.

— *d'os,* bone boiler.

bouilli, *p.a.* boiled. — *m.* boiled meat.

bouillie, *f.* pap, gruel, thickened milk; hence, paste, mash, pulp; (*Paper*) pulp.

bouillie *bordelaise,* Bordeaux mixture.

— *bourguignonne,* Burgundy mixture.

bouillir, *v.i. & t.* boil.

bouilloire, *f.* vessel for boiling water, boiler, teakettle.

bouillon, *m.* bubble (in a hot liquid or in the solid formed by cooling it), bleb; bubbling; broth, stock; ripple, wave; (*Agric.*) liquid manure; (*Glass*) an air bubble formed in gathering the molten glass.

— *blanc,* (*Bot.*) mullein.

bouillonnement, *m.* bubbling, effervescence.

bouillonner, *v.i.* bubble, bubble up, effervesce; gush out, gush forth.

bouillonneux, *a.* containing bubbles, blebby.

bouillotte, *f.* boiler, kettle.

bouillotter, *v.i.* boil gently, simmer.

boulage, *m.* (*Sugar*) the expressing of sugar beets; also, the expressed beets.

boulange, *f.* unbolted flour.

boulanger, *m.* baker. — *v.t.* bake (bread).

boulangerie, *f.* baking; bakery, bakeshop.

boule, *f.* ball; bulb.

— *à gaz,* gas connection (having a ball-shaped part to which one or more stopcocks are attached).

— *à robinets,* a ball-shaped connection (for gas or the like) to which stopcocks are attached, stopcock fixture.

— *à teinture,* dye ball.

— *blanche,* white ball, specif. a ball of whiting.

— *de bleu,* blue ball, ball of bluing.

— *de noir,* ball of black pigment or dye, black ball.

— *de verre,* glass bulb.

boules d'acier, boules de Mars, boules de Nancy, (*Pharm.*) balls of iron and potassium tartrate.

bouleau, *m.* birch, birch tree, birch wood.

boulée, *f.* cracklings of tallow, greaves.

bouler, *v.i.* swell, puff out, (of bread) rise; roll like a ball.

boulet, *m.* ball; specif., cannon ball.

boulette, *f.* small ball, pellet.

bouleur, *m.* maker of glass bulbs or globes.

bouleversement, *m.* overthrow; commotion; panic.

bouleverser, *v.t.* overturn, upset, unsettle, throw into confusion, demolish.

bouloir, *m.* stirrer, beater.

boulon, *m.* bolt, pin.

— *à bout percé,* eyebolt.

— *à clavette,* keyed bolt.

— *à écrou,* screw bolt, bolt and nut.

— *à goupille,* pinned bolt.

— *à œillet,* eyebolt.

— *à oreilles,* wing bolt, fly bolt.

— *à vis,* screw bolt.

— *taraudé,* threaded bolt, screw bolt.

boulonner, *v.t.* bolt.

bouquet, *m.* bunch, cluster, clump, tuft, wisp; bouquet.

bourdaine, *f.* (*Bot.*) alder buckthorn (*Rhamnus frangula*).

bourde, *f.* poorest quality of Spanish soda.

bourdonner, *v.i.* hum, buzz, murmur.

bourg, *m.* market town.

bourgade, *f.* small town.

bourgène, *f.* = bourdaine.

bourgeois, *a.* private; plain (and good); civilian; middle-class, bourgeois.

bourgeon, *m.* bud; shoot; (*pl.*) long-staple wools.

bourgeonnement, *m.* budding.

bourgeonner, *v.i.* bud.

Bourgogne, *f.* Burgundy.

bourgogne, *m.* Burgundy wine, burgundy.

bourguignon, *a.* Burgundian, Burgundy.

bourlet, *m.* = bourrelet.

bourrache, *f.* borage (*Borago officinalis*); any plant of the genus *Borago.*

bourrage, *m.* (*Mining*) tamping.

bourre, *f.* (a fibrous mass or material) wad, wadding, stuffing, padding, pad, plug; short hair; down; tag wool; floss; silk waste; trash, refuse; (*Mining*) tamping.

bourré, *p.a.* packed, etc. (see bourrer).

bourrelé, *p.a.* flanged; tormented.

bourrelet, *m.* padding, pad; swelling; flange.

bourrer, *v.t.* pack, wad, stuff, pad, tamp, ram.

bourroir, *m.* tamper.

bourru, *a.* (of wine) unfermented; rough; crabbed.

bourse, *f.* purse; pouch, bag; stock exchange.

— *à pasteur,* (*Bot.*) shepherd's purse (*Bursa bursa-pastoris*).

boursouflé, *p.a.* swollen, swelled, inflated.

boursouflement, *m.* swelling, expansion, inflation; blister, bleb, bubble (in metals, etc.).

boursoufler, *v.t., r. & i.* swell, expand, puff up, inflate.

boursouflure, *f.* swelling, swell.

bousage, *m.* (*Calico*) dunging.

bouse, *f.* ox dung, cow dung.

boussole, *f.* compass; galvanometer.

— *des sinus,* sine galvanometer.

— *des tangentes,* tangent galvanometer.

— *électrique,* galvanometer.

bout, *m.* end; extremity; tip; bit. — *pr. 3 sing.* of bouillir.

boute, *f.* hogshead; barrel; tub.

boutefeu, *m.* (*Expl.*) igniter.

bouteille, *f.* bottle; (for compressed gases) cylinder; any of various elongated vessels; bubble.

— *de Leyde,* Leyden jar.

— *de savon,* soap bubble.

bouteiller, *v.i.* (*Glass*) become full of bubbles.

bouter, *v.i.* (of liquids) become thick or ropy.

boutique, *f.* shop, store; stock, outfit.

bouton, *m.* button; knob, stud, lug, handle; plug; bud; pimple, pustule.

— *de bielle,* crank pin.

— *de commutateur,* (*Elec.*) switch plug.

— *de sonnerie,* push button.

boutonner, *v.t.* button. — *v.i.* bud; button.

bouture, *f.* slip, cutting; sucker, shoot.

bovin, *a.* bovine, pertaining to cattle.

boyau, *m.* intestine, gut; hose, (flexible) tubing; catgut; narrow passage.

boyauderie, *f.* gut works.

bracelet, *m.* circular band, ring; bracelet.

bractée, *f.* (*Bot.*) bract.

brai, *m.* pitch; tar; rosin.

— *gras,* soft pitch, moist pitch (e.g., the residue left when coal tar is distilled to not over 200°).

— *liquide,* tar.

— *sec,* hard pitch, dry pitch (e.g., the residue left when coal tar distillation is pushed as far as is practicable).

braise, *f.* wood coals (either glowing or not), embers, bed of coals; overburned, friable charcoal, breeze; braised meat.

— *chimique,* tinderbox paste.

bran, *m.* bran; sawdust.

brancard, *m.* handbarrow; litter, stretcher.

branche, *f.* branch; arm, division, subdivision, etc.; (of a tripod) leg.

branchement, *m.* branching; branch tube or pipe.

brancher, *v.t.* branch; connect (tubes, wires).

branchette, *f.* small branch, branchlet.

branderie, *f.* brandy distillery.

brandevin, *m.* brandy (made from wine)

brandevinier, *m.* brandy distiller or seller.

branle, *m.* swing, swinging; motion, impulse.

branler, *v.t.* swing; shake (the head). — *v.i.* swing, shake, sway, move.

branloire, *m.* handle, lever (of a bellows), rock-staff; seesaw.

braquer, *v.t.* point, direct, turn, fix (in a certain direction).

bras, *m.* arm; handle; (*Mach.*) lever, rod, bolt, etc.; brace; hand.

à —, by hand, hand-.

brasage, *m.* brazing, hard-soldering.

brase, *f.* wood coals (see braise).

brasement, *m.* = brasage.

braser, *v.t.* braze, solder with hard solder.

brasier, *m.* fire of wood coals; brazier.

brasiller, *v.i.* shine, sparkle.

brasque, *f.* (*Metal.*) brasque, lining (as a paste of powdered charcoal and clay).

brasquer, *v.t.* brasque, line with brasque.

brassage, *m.* mixing, etc. (see brasser).

brasser, *v.t.* mix, stir, stir up, rabble, puddle; mash (malt, etc.); brew (beer, etc.).

brasserie, f. brewery; beer shop, café.

brasseur, m. brewer; mixer, rabbler, puddler.
— *mécanique,* mechanical stirrer (or rabbler or puddler).

brassin, m. mash tub, mash vat; quantity brewed at one time, brewing; quantity of soap boiled at one time, boiling; the process of making soft soap.

brasure, f. brazing; brazed joint or seam.

braunite, f. (*Min.*) braunite.

brave, a. brave, gallant, fine, worthy, kind.

braye, f. puddle (moist clay used for lining).

brayer, v.t. pitch, tar.

brebis, f. ewe, (female) sheep.

brecciolaire, a. brecciated.

brèche, f. gap, breach, hole, notch (in an edge); (*Geol.*) breccia.

bref, a. brief, short. — *adv.* in short, in brief. en —, in brief.

Brême, n. Bremen.

Brésil, m. Brazil.

brésil, m. brazilwood, brazil.

brésilien, a. Brazilian.

brésiline, f. brasilin.

brésiller, v.t. dye red (with brazilwood); shatter, shiver. — v.i. fall to powder.

brésillet, m. braziletto, brasiletto.

brésilleur, m. brazilwood dyer.

Bretagne, f. Brittany.

bretelle, f. strap, sling; brace, suspender.

breton, a. Breton, of Brittany.

bretté, a. toothed, jagged.

breuvage, m. beverage, drink; draft; drench.

brève, a. fem. of bref.

brevet, m. patent; certificate; diploma; commission; (*Dyeing*) a decoction of bran and woad.
— *de corps,* a patent obtained on a chemical compound or other substance.
— *de perfectionnement,* patent for improvement (of an invention).
— *d'importation,* patent for importation.
— *d'invention,* patent for an invention.

brevetable, a. patentable.

brevetage, m. (*Alum*) addition of potassium sulfate.

breveté, p.a. patented, etc. (see breveter).
— m. patentee.
— *S.G.D.G.,* patented without guarantee by the government (the regular label on patented articles).

breveter, v.t. patent; grant a patent (or certificate) to.

brévium, m. brevium, thorium X_2.

bride, f. clamp, cramp; strap, stay, band; flange; bridle; rein, reins.

brider, v.t. bridle, check, restrain.

brie, m. Brie cheese.

brier, v.t. knead.

brièvement, adv. briefly; soon, shortly.

brièveté, f. brevity.

brigade, f. gang, party, body, band; **brigade.**

brigadier, m. foreman; corporal.

brillament, adv. brilliantly.

brillant, a. brilliant. — m. brilliancy; polish; bright work.

brillanter, v.t. cut as a brilliant; honor.

brillement, m. brilliancy, luster.

briller, v.i. shine; be bright, be brilliant.

brin, m. bit, small piece; strand; staple, fiber (of wool, etc.); blade, shoot, sprig, twig.

brinasse, f. hemp combings, inferior tow.

brindille, f. slender branch, branchlet.

brique, f. brick; (of soap) bar.
— *blanche,* fire brick; white brick.
— *creuse,* hollow brick.
— *crue,* unburnt brick.
— *de liège,* cork brick.
— *de parement,* face brick, facing brick.
— *de scories,* slag brick, slag stone.
— *émaillée,* enameled brick, glazed brick.
— *hollandaise,* Dutch brick, Dutch clinker.
— *pilée,* brick dust.
— *réfractaire,* fire brick.
— *vernissée,* glazed brick.
— *vitrifiée,* vitrified brick.

briquet, m. steel (for striking fire); tinder box.
— *chimique,* chemical tinderbox, automatic lighter.

briquetage, m. brickwork.

briqueté, p.a. bricked; made of bricks; brick-colored, (*Med.*) lateritious; imitating brick.

briqueter, v.t. brick.

briqueterie, f. brickmaking; brick factory; match factory.

briqueteur, m. bricklayer.

briquetier, m. brick maker (or seller); match maker (or seller); bricklayer.

briquetier-fumiste, m. brick-chimney builder.

briquettage, m. briquetting.

briquette, f. briquette, briquet.
— *de poussier,* coal-dust briquette.
— *de tourbe,* turf briquette, turf cake.

bris, m. breaking, shattering, smashing.

brisant, p.a. breaking, etc. (see briser); (of explosives) shattering, disruptive, sudden, brisant.

brisé, p.a. broken, etc. (see briser); jointed, folding.

brise-circuit, brise-courant, m. circuit breaker.

brise-jet, m. an attachment to a tap for lessening the force of the stream.

brisement, m. breaking, etc. (see briser).

briser, v.t. break; shatter, smash; break off, interrupt; weary, wear out. — v.i. break. — v.r. break; (of light) be refracted; fold.

bristol, m. Bristol board, bristol.

brisure, *f.* break; crack, flaw; broken part; folding joint.

brittanique, *a.* British.

brocatelle, *f.* brocatel, brocatelle.

broche, *f.* spindle; broach (in various senses), spike, peg, pin; spigot.

broché, *p.a.* (of books) stitched only, in paper binding; (of fabrics) figured.

brochette, *f.* pin, peg; skewer.

brochure, *f.* stitching (of books); brochure, pamphlet; inwoven pattern.

broder, *v.t.* embroider.

broderie, *f.* embroidery; insignia.

broie, *pr. 3 sing.* of broyer.

broiement, broiment, *m.* = broyage.

bromacétique, *a.* bromoacetic, bromacetic.

bromal, *m.* bromal.

bromalizarine, *f.* bromoalizarin, bromalizarin.

bromate, *m.* bromate.

brome, *m.* bromine; (*Bot.*) brome grass.

bromé, *p.a.* brominated, containing bromine, bromo.

bromer, *v.t.* brominate.

bromhydrate, *m.* hydrobromide.

bromhydrine, *f.* bromohydrin, bromhydrin.

bromhydrique, *a.* hydrobromic.

bromique, *a.* bromic.

bromo-aurate, *m.* bromaurate.

bromo-aurique, *a.* bromauric.

bromoforme, *m.* bromoform.

bromo-ioduré, *a.* treated with or containing bromide and iodide.

bromonitré, *a.* bromonitro.

bromuration, *f.* bromination.

bromure, *m.* bromide.

— *de calcium,* — *de chaux,* calcium bromide.

— *de fer,* iron bromide, specif. (esp. in *Pharm.*) ferrous bromide.

— *de magnésie,* magnesium bromide.

— *de méthyle,* methyl bromide, bromomethane.

— *d'éthyle,* ethyl bromide, bromoethane.

— *ferreux,* ferrous bromide.

bromuré, *a.* brominated; containing bromine.

bronche, *f.* (*Anat.*) bronchus, bronchial tube.

broncher, *v.i.* stumble, trip, falter.

bronchique, *a.* bronchial.

bronchite, *f.* bronchitis.

bronzage, *m.* bronzing; browning (of steel).

bronze, *m.* bronze.

— *à canon,* gun metal.

— *aciéré,* steely bronze.

— *à cloches,* bell metal.

— *au manganèse,* manganese bronze.

— *d'aluminium,* aluminium bronze.

— *de nickel,* nickel bronze.

— *des monnaies,* coin bronze.

— *dur,* — *durci,* hard bronze, hardened bronze.

— *en poudre,* bronze powder.

bronze *moulu,* bronze powder.

— *naturel,* common bronze.

— *phosphoreux,* phosphor bronze.

— *siliceux,* silicon bronze.

bronzé, *p.a.* bronzed; browned; bronze, bronzy.

bronze-acier, *m.* steel bronze.

bronzer, *v.t.* bronze; brown (iron or steel).

broquette, *f.* tack.

brossage, *m.* brushing.

brosse, *f.* brush.

— *douce,* soft brush.

— *en fil de métal,* wire brush.

— *passe-partout,* small dusting brush.

— *rude,* stiff brush.

brosser, *v.t.* brush.

brosserie, *f.* brushmaking; brush factory.

brossier, *m.* brush maker.

brossure, *f.* (*Leather*) color brushed on.

brou, *m.* hull, husk (of a nut, specif. walnut).

brouet, *m.* gruel, porridge, broth.

brouette, *f.* wheelbarrow.

brouettée, *f.* barrowful, wheelbarrow load.

brouetter, *v.t.* wheel in a barrow.

brouetteur, brouettier, *m.* barrowman.

broui, *m.* blowpipe (of an enameler). — *a.* blighted.

brouillard, *m.* fog, mist.

brouillement, *m.* mixing, etc. (see brouiller).

brouiller, *v.t.* mix, mix up, mingle, jumble, confuse, beat (eggs), embroil. — *v.i.* blunder, bungle, make trouble.

brouillerie, *f.* discord, quarrel, trouble.

broussaille, *f.* brushwood, underbrush.

brouter, *v.i.* browse; (of tools) cut unevenly.

broutille, *f.* small twigs or shoots; trifle.

brownien, *a.* Brownian.

broyage, *m.* grinding, etc. (see broyer).

broyé, *p.a.* ground, etc. (see broyer).

broyement, *m.* grinding, etc. (see broyer).

broyer, *v.t.* grind, pulverize, powder, crush, pound, bray, bruise, break.

broyeur, *m.* grinder, etc. (see broyer).

— *à boulets,* ball grinder, ball mill.

— *pulvérisateur,* crusher and pulverizer.

broyeuse, *f.* grinder, etc. (see broyer).

broyeuse-mélangeuse, *f.* grinder and mixer.

broyeuse-sécheuse, *f.* grinder and drier.

brucelles, *f.pl.* (spring) forceps, (spring) pincers.

brucine, *f.* brucine.

bruire, *v.i.* make noise, rustle, rattle, hum, etc.

bruissant, *p.a.* noisy, roaring, humming, rattling.

bruissement, *m.* noise: rumbling, humming, rattling, etc.

bruit, *m.* noise; sound; rumor; reputation.

— *p.p. & pr. 3 sing.* of bruire.

brûlable, *a.* burnable.

brûlage, *m.* burning (specif. of weeds, **brush**, etc.); hole in a furnace fire.

brûlant, *p.a.* burning, etc. (see brûler).

brûlé, *p.a.* burned, burnt, etc. (see brûler).
— *m.* burn; burning.

brûlement, *m.* burning.

brûler, *v.t.* burn; roast; distill (wine, etc.).
— *v.i.* burn. — *v.r.* burn oneself; be burnt.
— *en dedans,* (of a burner) burn at the base.

brûlerie, *f.* distilling; distillery, esp. brandy distillery.

brûleur, *m.* burner; heater; distiller (esp. of brandy); roaster (as of coffee).
— *à gaz,* gas burner.
— *ambulant,* traveling distiller.
— *à pétrole,* oil burner, petroleum burner.
— *de Bunsen,* Bunsen burner.
— *d'os,* bone calciner.
— *électrique,* electric heater.

brûlot, *m.* burnt brandy; half-burned charcoal.

brûlure, *f.* burn; scald; burning; (*Agric.*) blight.

brume, *f.* thick fog or mist.

brumeux, *a.* foggy, misty.

brun, *a. & m.* brown.
— *acide,* acid brown.
— *Bismarck,* Bismarck brown.
— *châtain,* chestnut brown, chestnut.
— *clair,* light brown.
— *de Bruxelles,* Brussels brown (a pigment made by charring bone).
— *de momie,* mummy brown.
— *de montagne,* umber.
— *de Prusse,* Prussian brown.
— *foncé,* dark brown.
— *marron,* chestnut brown, chestnut, maroon.
— *rouge,* brown ocher.
— *Van Dyck,* Vandyke brown.

brunâtre, *a.* brownish.

brune, *a. fem.* of brun. — *f.* dusk.

bruni, *m.* burnish, polish. — *p.a.* browned, etc. (see brunir).

brunir, *v.t.* brown; burnish, polish. — *v.i.* brown, turn brown.

brunis, *m.* burnish, polish.

brunissage, *m.* burnishing; browning.

brunissement, *m.* browning; burnishing.

brunisseur, *m.* burnisher (person).

brunissoir, *m.* burnisher (tool).

brunissure, *f.* browning; burnishing, polishing.

brun-mat, *a.* dull brown, mat brown.

brun-rouille, *a.* rusty brown, rust-colored.

brusque, *a.* sudden, abrupt; brusque, blunt.

brusquement, *adv.* suddenly, abruptly; brusquely.

brut, *a.* crude; raw; gross; rough, unpolished, uncut, unfinished, etc.; (of land) uncultivated; inorganic; brute, brutish; un-educated.
— *de coulée,* — *de fonte,* rough-cast; as cast.

Bruxelles, *n.* Brussels.

bruyant, *p.a.* noisy, rattling, humming, etc.

bruyère, *f.* heath, heather; heath, moor.

bryone, *f.* (*Bot.*) bryony.

bryonine, *f.* bryonin.

bu, *p.p.* of boire.

buandier, *m.* bleacher; launderer.

bube, *f.* pimple.

bubon, *m.* (*Med.*) bubo.

bubonique, *a.* bubonic.

bucail, *m.,* **bucaille,** *f.* buckwheat.

bucco, *n.* (*Pharm.*) buchu.

bûche, *f.* (of wood) billet, stick, log, block; (of coal) lump; (*Wiredrawing*) block.

bûcher, *v.t.* hew, cut. — *m.* woodhouse, woodshed; woodpile.

bûchette, *f.* small stick (of wood).

bûchilles, *f.pl.* borings.

buée, *f.* damp vapor, moisture; (formerly) lye.

buer, *v.i.* give off vapor, steam, reek.

buffle, *m.* buffalo; buff (leather); buff stick.

buffleterie, *f.* the chamois process applied to ox hide, also the leather so produced, and articles made of it.

buglosse, *f.* (*Bot.*) bugloss (*Anchusa*), esp. *A. italica* (in English *A. officinalis* is the one most often referred to).

buis, *m.* (*Bot.*) box, boxwood, box tree.

buisson, *m.* bush; thicket.

bulbe, *m. & f.* bulb.
— *de colchique,* — *de safran bâtard,* (*Pharm.*) colchicum corm.

bulbeux, *a.* bulbous, bulbose.

bulle, *f.* bubble; blister, bleb, vesicle; droplet, drop; bulla; bull (edict).

bulletin, *m.* bulletin, report, notice; ticket; receipt.
— *d'essai,* assay report, report of analysis.

Bull. assoc. chim. sucr. dist., *abbrev.* Bulletin de l'Association des Chimistes de Sucrerie et de Distillerie.

Bull. sci. pharmacol., *abbrev.* Bulletin des Sciences pharmacologiques.

Bull. soc. chim., *abbrev.* Bulletin de la Société chimique de France.

Bull. soc. encour. ind. nat., *abbrev.* Bulletin de la Société d'Encouragement pour l'Industrie nationale.

bulleux, *a.* vesicular.

bureau, *m.* office; table, desk, counter; bureau; officers of an organization.

burent, *p.def. 3 pl.* of boire.

burette, *f.* burette, buret; cruet; (*Mach.*) oil can, oiler.
— *à huile,* oil can, oiler.
— *anglaise,* Bink's burette (having the bottom closed and the tip branching off just below the top).
— *à pince,* burette with pinchcock (Mohr burette).
— *à robinet,* burette with stopcock (Geissler burette).

burette *de Mohr*, Mohr burette (pinchcock burette).

burgos, *m.* (*Ceram.*) a kind of gold luster.

burinage, *m.* engraving, graving, cutting (of metal); specif., (*Metal.*) cutting out of cracks from ingots.

buriner, *v.t.* engrave, etc. (see burinage).

buse, *f.* nozzle (as of a blowpipe or bellows); blast pipe (of a furnace); twyer; pipe, ventilating pipe; channel; buzzard.
— *de soufflet*, blast pipe.

bussent, *imp. 3 pl. subj.* of boire.

busserole, *f.* (*Bot.*) red bearberry (*Arctostaphylos uva-ursi*).

but, *m.* object, purpose, aim, end, goal; butt.
— *p.def. 3 sing.* of boire.

butalanine, *f.* butalanine.

buté, *p.a.* fixed, determined; butted.

butée, *f.* fixed piece, lug, stop, shoulder, support; abutment; striking, tapping.

buter, *v.i.* aim (at); hit the mark. — *v.t.* butt; oppose; buttress.

butte, *f.* small hill, hillock, knoll.

buttoir, butteur, *m.* (*Mach.*) catch, stop, projection, stud, tappet, driver, dog.

butyle, *m.* butyl.

butylène, *m.* butylene.

butylénique, *a.* butylene, of butylene.

butylique, *a.* butyl.

butyracé, *a.* butyraceous.

butyrate, *m.* butyrate.

butyreux, *a.* of butter, buttery, butyraceous.

butyrine, *f.* butyrin.

butyrique, *a.* butyric.

butyroamylique, *a.* In the phrase *éther butyroamylique*, amyl butyrate.

butyromètre, *m.* butyrometer.

buvable, *a.* drinkable, fit to drink.

buvant, *p.pr.* of boire.

buvard, *a.* (of paper) blotting, bibulous, absorbent. — *m.* blotter.

C

c., *abbrev.* (centimètre) centimeter.

c', ç', *abbrev.* of ce.

C.A., *abbrev.* (coefficient d'abaissement) coefficient of (freezing-point) lowering.

ça, *pron.* Abbreviated form of *cela.*

çà, *adv.* here.

caballin, *a.* caballine.

cabestan, *m.* capstan, winch.

cabinet, *m.* cabinet; closet; study; office; business.

câble, *m.* cable, rope.

— *de transmission,* power cable.

— *double,* (*Elec.*) two-wire cable.

— *métallique,* wire cable, wire rope.

— *sous-marin,* submarine cable.

câbler, *v.t.* lay up, twist into a cable; cable.

cabochon, *m.* (*Jewelry*) cabochon.

cabrillon, *m.* goat's-milk cheese.

cabron, *m.* kid skin; burnishing tool.

cacao, *m.* cacao; cocoa.

— *soluble,* soluble cocoa.

cacaotier, -tière. = cacaoyer, -yère.

cacaoyer, *m.* cacao tree (*Theobroma cacao*).

cacaoyère, *f.* cacao plantation.

cachalot, *m.* sperm whale, cachalot.

caché, *p.a.* hidden, unseen, secret.

cachemire, *m.* cashmere.

Cachemire, *n.* Kashmir, Cashmere.

cacher, *v.t.* hide, conceal, secrete.

cachet, *m* seal; stamp; ticket; (*Pharm.*) cachet.

— *de pain,* (*Pharm.*) cachet (wafer of unleavened bread).

cacheter, *v.t.* seal.

cacheteur, *m.* sealer. — *a.* sealing.

cachexie, *f.* (*Med.*) cachexia.

cachibou, *m.* cachibou (resin).

cacholong, *m.* (*Min.*) cacholong.

cachon, *m.* (*Glass*) a box in which blowpipes are kept.

cachou, *m.* catechu.

— *de Laval,* (*Dyes*) cachou de Laval.

cachoutannique, *a.* catechutannic.

cachouter, *v.t.* dye or treat with catechu.

cacodylate, *m.* cacodylate.

cacodyle, *m.* cacodyl.

cacodylique, *a.* cacodylic.

cacothéline, *f.* cacotheline.

cactacée, cactée, *f.* cactus (any plant of the family Cactaceae).

cactier, cactus, *m.* cactus (plant of the genus *Cactus*).

c.-à-d., *abbrev.* (c'est-à-dire) that is to say, that is, i.e.

cadavérine, *f.* cadaverine.

cadavérique, *a.* cadaveric.

cadavre, *m.* cadaver, corpse.

cade, *m.* (*Bot.*) cade (*Juniperus oxycedrus*).

cadeau, *m.* present, gift.

cadence, *f.* cadence.

en —, in cadence; at the same time.

cadet, *a.* younger, youngest, junior. — *m.* younger brother, younger son, junior; cadet.

cadette, *f.* hewn paving stone; younger daughter (or sister).

cadetter, *v.t.* pave with hewn stones.

Cadix, *n.* Cadiz.

cadmie, *f.* cadmia.

cadmique, *a.* of or containing cadmium, cadmic.

cadmium, *m.* cadmium.

cadran, *m.* dial, face; (*Mach.*) index plate.

— *indicateur,* indicating dial.

cadre, *m.* frame, border; framework; outline, plan, arrangement; compass, limits; (*Mil.*) skeleton organization.

cadrer, *v.i.* agree, tally, harmonize, suit.

caduc, *a.* decaying, falling, declining; null, void; (*Bot.*) caducous.

cæsium, *m.* cesium, cæsium.

café, *m.* coffee.

— *au lait,* coffee and milk, or the light brown color of this.

— *de grains,* cereal coffee.

— *de Moka,* Mocha coffee.

— *du Soudan,* kola nuts.

— *en grains,* whole coffee.

— *en poudre,* ground coffee.

— *torréfié,* roasted coffee.

— *vert,* green coffee, unroasted coffee.

caféier, *m.* coffee tree.

caféière, *f.* coffee plantation.

caféine, *f.* caffeine.

caféique, *a.* caffeic.

caféone, *f.* caffeone.

cafétannique, *a.* caffetannic.

caffût, *m.* scrap iron, old iron.

cafier, *m.* coffee tree.

cafique, *a.* caffeic.

49

cage, *f.* case, casing; frame, framework; cage.

cahier, *m.* blank book, record book, notebook, account book, etc.; specifications.

— *d'analyse,* analytical notebook.

— *de charges,* specifications.

cahot, *m.* jolt, jolting; obstacle.

cahoter, *v.t. & i.* jolt.

caïeu, *m.* (*Bot.*) offset bulb; (of garlic) clove.

caille, *f.* quail.

caillé, *p.a.* coagulated, curdled, clotted. — *m.* curd, curds, coagulated part of milk.

caillebottage, *m.* curdling, clotting.

caillebotte, *f.* curdled milk; curd, curds.

caillebotté, *p.a.* curdled, clotted; curdy.

caillebotter, *v.t. & r.* curdle, clot.

caille-lait, *m.* (*Bot.*) cheese rennet, bedstraw (*Galium,* esp. yellow bedstraw, *G. verum,* called also *caille-lait jaune*).

caillement, *m.* coagulation, curdling, clotting.

cailler, *v.t. & i.* coagulate, curdle, clot.

caillette, *f.* rennet.

caillot, *m.* coagulum, curd, clot.

caillou, *m.* pebble; flint, flint stone.

— *roulé,* rolled pebble.

cailloutage, *m.* (*Ceram.*) flintware; pebble work.

caillouteux, *a.* pebbly; flinty.

cailloutis, *m.* road metal, macadam.

cailloux, *pl.* of caillou.

caïnça, *f.* (*Pharm.*) cahinca root, cahinca.

caire, *m.* coir (coconut fiber).

caisse, *f.* case, box, chest; (for liquids) tank; safe; bank, cashier's office; cash; fund; shell (of a drum or a pulley); body (of a car, etc.); drum.

— *à claire-voie,* crate.

— *à eau,* water tank.

— *à feu,* fire box.

— *à outils,* tool chest, tool box.

— *de cémentation,* cementation box.

— *de résistance,* (*Elec.*) resistance box.

— *en fonte,* cast-iron box or tank.

cajeput, cajeputier, *m.* (*Bot.*) cajuput, cajeput.

cal, *m.* callosity, callus.

calage, *m.* leveling, etc. (see caler).

calambac, calambart, calambour, *m.* Agalloch wood.

calament, *m.* (*Bot.*) calamint.

calaminaire, *a.* In the phrase *pierre calaminaire,* lapis calaminaris, calamine.

calamine, *f.* (*Min.*) calamine.

calamité, *f.* calamity.

calamiteux, *a.* calamitous.

calandrage, *m.* calendering.

calandre, *f.* calender; (*Sugar*) the upper division of one of the cylindrical units of a multiple-effect evaporator.

calandrer, *v.t.* calender.

calandreur, *m.* calenderer.

calant, *p.a.* leveling, etc. (see caler).

calcaire, *a.* calcareous, or containing lime. — *m.* limestone.

— *carbonifère,* Carboniferous limestone.

— *coquillier,* shell limestone.

— *hydraulique,* hydraulic limestone.

calcarelle, *f.* a kind of kiln for extracting sulfur.

calcarifère, *a.* calciferous.

calcarone, *m.* calcarone (large sulfur kiln).

calcédoine, *f.* chalcedony.

calcédonieux, *a.* chalcedonic.

calcifère, *a.* calciferous.

calcilithe, *f.* compact limestone.

calcinable, *a.* calcinable.

calcinage, *m.* calcination, calcining.

calcination, *f.* calcination.

calcine, *f.* calcine (calcined product); (*Enameling*) a powder containing lead and tin oxides (lead stannate) in varying proportions; (*Soda*) the strongly heated second hearth of a furnace for converting sodium chloride into sodium sulfate.

calciner, *v.t.* calcine, burn (as lime).

calcinure, *n.* (*Glass*) minute surface crack.

calcique, *a.* of or containing calcium or lime; as, *chlorure calcique,* calcium chloride.

calcite, *f.* (*Min.*) calcite.

calcium, *m.* calcium.

calcul, *m.* calculus; calculation; arithmetic.

— *differentiel,* differential calculus.

calculant, *p.a.* calculating.

calculateur, *m.* calculator. — *a.* calculating.

calculatoire, *a.* calculatory, calculating.

calculer, *v.t. & i.* calculate. — *v.r.* be calculated.

calculeux, *a.* (*Med.*) calculous, calculary.

cale, *f.* (a piece to level, steady or support an object), chock, block, wedge, key, quoin, prop; hold (of a ship); cove.

calé, *p.a.* leveled, etc. (see caler).

calebasse, *f.* calabash.

caleçon, *m.* (pair of) drawers.

calédonien, *a.* Caledonian, Scotch; of or pertaining to New Caledonia (as, *minerais calédoniens*).

calendrier, *m.* calendar.

calepin, *m.* memorandum book, notebook.

caler, *v.t.* level, set, adjust, support, chock, prop, wedge, key; lower (a sail); (of a boat) draw. — *v.i.* jam, stick.

calfater, *v.t.* calk, caulk.

calfeutrer, *v.t.* stop up the chinks of, make tight, close, stuff.

calibrage, *m.* calibration, gaging.

calibre, *m.* caliber; caliper; gage; template; mold; model, pattern (of a molding); mandrel.

calibrer, *v.t.* calibrate, gage, measure, size.

calice, *m.* cup; chalice; (*Bot.*) calyx.

caliche, *f.* (*Min.*) caliche.

calicot, *m.* calico.

Californie, *f.* California.

califourchon, *m.* hobby; hobbyhorse.
— à —, astride, astraddle.

calin, *m.* calin (alloy for tea canisters).

calleux, *a.* callous, indurated, horny.

callosité, *f.* callosity.

calmant, *p.a.* calming, soothing; (*Med.*) calmative. — *m.* (*Med.*) calmative.

calme, *a. & m.* calm.

calmer, *v.t.* calm, soothe, quiet.

calomel, *m.* calomel.

calorie, *f.* calorie, calory.

calorifère, *m.* heater, heating apparatus, heating furnace, heating stove. — *a.* conveying heat; (of pipes) hot-air, hot-water, steam (as the case may be).
— à air, hot-air heater.
— à eau, hot-water heater.
— à vapeur, steam heater.

calorifique, *a.* calorific.

calorifuge, *a.* heat-insulating. — *m.* heat insulation, heat insulator.

calorimètre, *m.* calorimeter.

calorimétrie, *f.* calorimetry.

calorimétrique, *a.* calorimetric(al).

calorique, *m.* heat; (formerly) caloric.

calorisateur, *m.* (*Sugar*) calorisator.

calot, *m.* block, wedge. Cf. cale.

calotte, *f.* calotte, skull cap; hence, anything of similar shape: cap, calotte, capping, knob, head, cup, segment of a sphere (esp. one less than a hemisphere), cupola or its roof or lining, end of a window-glass cylinder opposite the blowpipe, case of a watch, (*Anat.*) brain cap.

calquage, *m.* tracing, copying.

calque, *m.* tracing, copy. — *a.* tracing, copying.

calquer, *v.t.* trace, copy (by tracing).
— sur, copy after.

calquoir, *m.* tracing point.

calus, *m.* callosity; callus.

camaïeu, *m.* = camée.

camarade, *m.* comrade, fellow.

cambouis, *m.* oil or grease blackened by use in machinery, coom, coomb; grease for polishing, lubricating, or the like.

cambre, *m.* camber, convexity.

cambré, *p.a.* cambered, curved, bowed.

cambrement, *m.* cambering, curving.

cambrer, *v.t.* camber, bend, curve, bow, warp.

cambrien, *a.* (*Geol.*) Cambrian.

cambrure, *f.* camber, cambering, curvature.

came, *f.* cam, wiper, lifter, tappet; cog.

camée, *m.* cameo; (*Painting*) camaïeu.

caméléon, *m.* chameleon.
— minéral, chameleon mineral.

caméline, *f.* wild flax, gold-of-pleasure (*Camelina sativa*).

camelle, *f.* (*Salt*) a heap of sea salt.

camion, *m.* truck, dray, camion.

camionnage, *m.* truckage, drayage, cartage.

camionner, *v.t.* transport by truck or dray.

camomille, *f.* camomile, chamomile.
— commune, = camomille ordinaire.
— d'Allemagne, German camomile (*Matricaria chamomilla*).
— de Perse, Persian insect powder.
— ordinaire, wild camomile (*Matricaria* species, esp. *M. chamomilla*).
— puante, stinking camomile (*Anthemis cotula*).
— pyréthre, pellitory (*Anacyclus pyrethrum*).
— romaine, Roman camomile (*Anthemis nobilis*).

camoufler, *v.t.* camouflage, disguise.

camourlot, *m.* cement for filling the joints between flagstones.

campagne, *f.* campaign (e.g., in metallurgy); season; country; field; voyage; country-seat.
en —, in operation; in the field.

campanien, *a.* Campanian.

campêche, *m.* logwood, campeachy wood.

camphène, *m.* camphene.

camphine, *f.* camphine.

camphocarbonique, *a.* camphocarboxylic.

camphoglycuronique, *a.* camphoglucuronic, camphoglycuronic.

camphol, *m.* camphol (borneol).

campholique, *a.* campholic.
alcool —, borneol.

camphorate, *m.* camphorate.

camphored, *n.* a sort of wide-mouthed bottle.

camphorifère, *a.* producing camphor.

camphorique, *a.* camphoric.

camphoroïde, *a.* resembling camphor.

camphorone, *f.* camphorone.

camphoronique, *a.* camphoronic.

camphre, *m.* camphor.
— anisique, fenchone.
— de matricaire, Matricaria camphor (l-camphor).
— de menthe, mint camphor (menthol).
— du Japon, Japan camphor, ordinary camphor.
— monobromé, bromocamphor, (*Pharm.*) monobromated camphor.

camphré, *a.* camphorated; camphoraceous.

camphre-poudre, *m.* camphorated powder.

camphrer, *v.t.* camphorate.

camphrier, *m.* camphor tree. (*Cinnamomum camphora*).

can, *m.* edge.

canadien, *a.* Canadian.

canaigre, *f.* canaigre (*Rumex hymenosepalus*).

canal, *m.* canal; pipe, tube, conduit; channel; flue; drain; sluice; groove; fluting; bed (of a river).

canal d'aérage, ventilating flue or channel.
— d'amenée, inlet pipe or tube, inflow channel.
— d'amorce, detonator hole.
— de coulée, (Founding) runner, gate, sprue.
— de fuite, efflux pipe or tube, waste pipe, waste channel.

canalisation, f. piping, etc. (see canaliser); system of pipes or tube; pipe line; connection, connections (as for steam or gas); sewerage; (Elec.) wiring, esp. conduit work.
— de lumière, electric light wiring.

canaliser, v.t. pipe (gas, etc.); unite, connect; centralize; canalize; (Elec.) wire.

canamelle, f. sugar cane.

canard, m. duck (specif., drake); hoax; (Mining) air passage; (Pyro.) water rocket.

canaux, pl. of canal.

cancer, m. (Med.) cancer.

cancereux, a. cancerous.

cancre, m. crab.

candélabre, m. chandelier, branched fixture; candelabrum.

candi, p.a. candied; candy. — m. candy, esp. rock candy; candied fruit.

candidat, m. candidate.

candir, v.i. & r. candy, become candied.

candiserie, f. candy (esp. rock candy) manufacture.

cane, f. (female) duck; = canne.

canette, f. duckling; (Weaving) cop.

canevas, m. canvas; outline, skeleton.

canif, m. penknife.

canin, a. canine.

caniveau, m. gutter, channel, conduit, drain, sewer.

cannaie, f. cane field; canebrake.

canne, f. cane; (Glass) blowpipe.
— à sucre, sugar cane.

cannelé, p.a. channeled, grooved, fluted.

canneler, v.t. channel, groove, flute.

cannelier, m. cinnamon tree (Cinnamomum).

cannelle, f. cinnamon; faucet, tap; groove, channel.
— de Ceylan, Ceylon cinnamon.
— de Magellan, (Pharm.) Winter's bark.
— de Saïgon, Saigon cinnamon.

cannellé, a. cinnamon-colored.

cannelure, f. groove, channel, fluting.

cannette, f. = canette; = cannelle.

canon, m. cannon, gun; (collectively) guns, artillery; cylindrical part or object, as the barrel of a rifle or a syringe, the nozzle or pipe of a bellows, a roll or stick of sulfur, a reel for yarn, a cylindrical jar; (Zinc) a crucible of oval section; canon.
— de campagne, field gun.
— de fusil, gun barrel, rifle barrel.
— rayé, rifled gun, rifle; rifled barrel.

canot, m. small boat, launch; canoe.

cantharide, f. cantharis, Spanish fly. — a. iridescent like the cantharis.

cantharider, v.t. cantharidate.

cantharidine, f. cantharidin.

cantharidique, a. cantharidic.

cantine, f. chest, box, case; canteen.

canton, m. canton (in France, subdivision of an arrondissement).

canule, f. small tube, cannula, (of a syringe) nozzle.

caoutchène, m. caoutchene.

caoutchouc, m. rubber, caoutchouc; (rubber) tire.
— à eau (or à raccord), rubber tubing made from cut sheet.
— creux, single-tube tire.
— cuir, imitation leather made from rubberized cloth.
— des huiles, a rubber substitute made from certain oils.
— de synthèse, synthetic rubber.
— dilaté, fine-cut sheet rubber.
— durci, hard rubber.
— en feuilles, sheet rubber.
— factice, artificial rubber.
— mélangé, mixed rubber.
— pneumatique, pneumatic tire.
— régénéré, reclaimed rubber.
— sulfuré, vulcanisé, vulcanized rubber.

caoutchoutage, m. treating with rubber, rubberizing.

caoutchouter, v.t. treat with rubber, rubberize.

caoutchoutier, m. rubber plant; rubber worker.

caoutchoutifère, a. producing rubber.

cap, m. cape; head.
— de Bonne-Espérance, Cape of Good Hope.

capable, a. capable.

capacité, f. capacity; expert.

capcique, n. capsicum.

capillaire, a. capillary. — m. capillary, capillary tube; maidenhair fern, capillaire.

capillarité, f. capillarity.

capitaine, m. captain.

capital, a. capital. — m. capital; chief thing.
— d'établissement, fixed capital.
— social, capital of a firm or company, social capital.

capital-action(s), m. capital stock.

capitale, f. capital (either city or letter).

capitalement, adv. capitally; essentially; absolutely.

capitaliser, v.t. capitalize.

capitaliste, m. capitalist.

capital-obligations, m. liabilities.

capitel, m. (Soap) a lye of ashes and lime.

capiteux, a. (of liquors) heady, intoxicating.

capituler, v.i. capitulate.

capnofuge, a. smoke-preventing.

caporal, *m.* corporal.

capot, *m.* cover, hood.

câpre, *f.* (*Bot.*) caper.

capricieux, *a.* capricious.

caprin, *a.* caprine, of or pertaining to a goat.

caprinate, *m.* caprate.

caprinique, *a.* capric.

caprique, *a.* capric; decyl (in the phrase *al-cool caprique*).

caproique, *a.* caproic; hexyl (in the phrase *alcool caproique*).

capryle, *m.* capryl; (less properly) octyl.

caprylique, *a.* caprylic; octyl (in the phrase *alcool caprylique*).

capsulage, *m.* capping (as of bottles).

capsulaire, *a.* capsular.

capsule, *f.* an evaporating dish or dish of similar shape, even when large; small dish or cup, capsule; capsule (in other senses); cap; (*Explosives*) cap, primer, detonator; (*Mach.*) ring gage.

— *à trous,* perforated dish, strainer, colander.

— *avec manche,* dish with handle, (in some cases) casserole.

— *conique,* a dish having the form of an inverted frustum of a cone.

— *d'évaporation,* — *évaporatoire,* evaporating dish.

— *fulminante,* fulminating cap, percussion cap.

— *médicamenteuse,* medicinal capsule.

— *surrénale,* suprarenal capsule.

captage, *m.* conducting of water (as by piping) to the place where it is to be used.

captation, *f.* catching, collecting, collection.

capter, *v.t.* catch, collect, capture; make available; conduct (water) to the place of use; captivate.

captif, *a.* captive. — *m.* captive, prisoner.

captiver, *v.t.* captivate; subdue; confine.

capuchon, *m.* cap (as for a bottle or a test tube); (*Bot.,* etc.) hood.

capuchonné, *a.* capped, hooded.

capucine, *f.* nasturtium; band (of a firearm).

capuline, *f.* a black fabric made of wool, goat's hair and horsehair, used in the purification of fatty acids by expression.

caput-mortuum, *m.* (*Old Chem.*) caput mortuum.

caquage, *m.* barreling.

caque, *f.* a kind of barrel.

caquer, *v.t.* barrel.

car, *conj.* for, because, as, inasmuch as.

caraba, *m.* cashew nut oil, cardol.

carabé, *m.* a variety of amber.

caracoli, *m.* caracoli, caracoly (an alloy).

caractère, *m.* character; characteristic.

— *d'imprimerie,* type.

caractérisant, *p.a.* characterizing, -teristic.

caractériser, *v.t.* characterize; distinguish.

caractéristique, *a. & f.* characteristic.

carafe, *f.* wash bottle, washing bottle; carafe, decanter.

caramel, *m.* caramel.

caramélique, *a.* pertaining to, or of the nature of, caramel.

caraméliser, *v.t. & r.* caramelize, turn into caramel; treat or mix with caramel.

carat, *m.* carat; small diamond.

carature, *f.* gold alloy; alloying (of gold).

— *blanche,* alloying of gold with silver, or the alloy so produced.

— *rouge,* alloying of gold with copper, or the alloy so produced.

carbamate, *m.* carbamate.

carbamide, *f.* carbamide, urea.

carbamique, *a.* carbamic.

carbanile, *n.* carbanil, phenyl isocyanate.

carbazide, *f.* carbazide.

carbazol, *m.* carbazole.

carbazotique, *a.* carbazotic (picric).

carbimide, *f.* carbimide, carbonimide.

carbinol, *m.* carbinol.

carbinolique, *a.* carbinol, pertaining to, or of the nature of, a carbinol.

carboazotine, *f.* carboazotine (a kind of safety blasting powder).

carbodynamite, *f.* carbodynamite.

carbolique, *a.* carbolic.

carbonado, *m.* carbonado.

carbonatage, *m.* carbonation.

carbonatation, *f.* carbonation, carbonatation.

carbonate, *m.* carbonate.

— *d'ammoniaque,* carbonate of ammonia, ammonium carbonate.

— *de chaux,* carbonate of lime, calcium carbonate.

— *de fer,* iron carbonate.

— *de lithine,* lithium carbonate.

— *de magnésie,* magnesium carbonate.

— *de plomb,* lead carbonate.

— *de potasse,* carbonate of potash, potassium carbonate.

— *de soude,* carbonate of soda, sodium carbonate.

— *de soude potassé,* potassium sodium carbonate.

— *de strontiane,* strontium carbonate.

— *lithique,* lithium carbonate.

— *niccolique,* nickel carbonate.

carbonaté, *p.a.* carbonated.

carbonater, *v.t.* carbonate. — *v.r.* become carbonated.

carbonateur, *m.* carbonator; (*Sugar*) carbonation (or carbonatation) man.

carbone, *m.* carbon.

— *combiné,* combined carbon.

— *de recuit,* annealing carbon.

— *de trempe,* temper carbon.

— *fixe,* fixed carbon.

carbone *graphitique,* graphitic carbon.

carboné, *a.* of, pertaining to or containing carbon, carbonaceous; carburetted.

carboneux, *a.* In the phrase *acide carboneux,* an old name for oxalic acid.

carbonide, *m.* carbonide (rarely used), carbide.
— *a.* resembling carbon.

carbonifère, *a.* carboniferous; (*Geol.*) Carboniferous.

carbonimètre, *m.* carbonimeter (a buret for determining carbon dioxide).

carbonimide, *f.* carbonimide, carbimide.

carbonique, *a.* carbonic; (*Org. Chem.,* esp. in combination) carboxylic.

carbonisation, *f.* carbonization.

carboniser, *v.t.* carbonize.

carboniseuse, *f.* (*Textiles*) an apparatus for conducting the carbonization process.

carbonite, *f.* (*Expl.*) carbonite.

carbonoïde, *a.* resembling carbon.

carbonométrie, *f.* carbonometry.

carbonyl-amidé, *a.* containing both carbonyl and amidogen or, specif., an acid amide group, $CONH_2$.

carbonyle, *m.* carbonyl.

carbonylé, *a.* containing carbonyl, carbonyl.

carborundum, *m.* carborundum.

carbosulfure, *m.* carbon disulfide.

carbosulfureux, *a.* containing carbon disulfide.

carboxyhémoglobine, *f,* carbonyl hemoglobin, carbon monoxide hemoglobin.

carboxyle, *m.* carboxyl.

carboxylpyrazinique, *a.* pyrazinecarboxylic.

carboxyphénolique, *a.* phenolcarboxylic (designating aromatic hydroxy acids).

carburant, *p.a.* carburet(t)ing. — *m.* carburetant; fuel (for explosion engines).

carburateur, *m.* carburet(t)or. — *a.* carburet(t)ing.

carburation, *f.* carburization, carburation.

carbure, *m.* carbide; specif., hydrocarbon.
— *acétylénique,* acetylene hydrocarbon.
— *allénique,* allene hydrocarbon.
— *à noyau fermé,* ring hydrocarbon.
— *aromatique,* aromatic hydrocarbon.
— *d'hydrogène,* hydrogen carbide, hydrocarbon.
— *diéthylénique,* diethylenic hydrocarbon.
— *ditérébénique,* diterpene.
— *éthylénique,* ethylene hydrocarbon.
— *métallique,* metallic carbide.
— *monotérébénique,* terpene proper ($C_{10}H_{16}$).
— *polytérébénique* (or *multitérébénique*), polyterpene.
— *saturé,* saturated hydrocarbon.
— *sesquitérébénique,* sesquiterpene.
— *térébénique,* terpene.

carburé, *p.a.* carburized; carburetted; containing carbon.

carburer, *v.t.* carburize; carburet.

carburet, *m.* carburet (old term for carbide).

carbylamine, *f.* carbylamine, carbamine, isonitrile.

carbyle, *m.* carbyl.

carcaise, *f.* annealing kiln for plate glass.

carcas, *m.* refined cast iron.

carcasse, *f.* skeleton; frame, framework; case (of a torpedo); carcass; body; (*Mil.*) carcass, incendiary shell.

carcel, *m.* (*Photometry*) carcel.

carcinomateux, *a.* (*Med.*) carcinomatous.

carcinome, *m.* (*Med.*) carcinoma.

cardage, *m.* carding; teaseling.

cardamome, cardamone, *m.* cardamom, cardamon.

cardan, *f.* Cardan joint, Cardan jointing.
à la —, Cardanic, in the manner or of the nature of a Cardan joint.

carde, *f.* card, carding machine; teasel; chard.

carder, *v.t.* card (wool, etc.); teasel (cloth).

cardiaire, *f.* (*Bot.*) motherwort (*Leonurus cardiaca*). — *a.* of the heart, cardiac.

cardialgie, *f.* (*Med.*) cardialgia.

cardiaque, *a. & m.* (*Med.*) cardiac. — *f.* = cardiaire.

cardinal, *a. & m.* cardinal.

cardol, *m.* cardol (cashew-nut oil).

carence, *f.* deficiency, lack (as in diet); (*Law*) absence of assets.

cargaison, *f.* cargo; shipping.

caricine, *f.* caricin (papain).

carie, *f.* (*Med.,* etc.) caries.
— *sèche,* (of wood) dry rot.

carillon, *m.* rod (of iron); chimes; noise.

carinthien, *a.* Carinthian.

carkhèse, *f.* = carcaise.

carline, *f.* (*Bot.*) carline thistle (*Carlina*).

carmin, *m. & a.* carmine.
— *aluné,* (*Micros.*) alum carmine.
— *boracique,* (*Micros.*) borax carmine.
— *d'indigo,* (*Dyes*) indigo carmine.

carminatif, *a. & m.* (*Med.*) carminative.

carminé, *p.a.* carminated; carmine-colored.

carminer, *v.t.* treat or mix with carmine; color carmine.

carminique, *a.* carminic.

carné, *a.* flesh-colored; of meat, meat, flesh.

carneau, carnau, *m.* the flue or pipe connecting a furnace with the chimney, vent, smoke pipe, (in a boiler) fire tube.

carnet, *m.* notebook, memorandum book; account book.
— *journalier,* journal.

carnine, *f.* carnine.

carnivore, *a.* carnivorous. — *m.pl.* Carnivora.

carotide, *a. & f.* (*Anat.*) carotid.

carotine, *f.* carotin.

carotte, *f.* carrot; trick, ruse.

carottine, *f.* carotin.

caroube, *f.* carob bean, carob.

caroubier, *m.* carob tree, carob.

carouge, *f.* = caroube.

carpobalsame, *m.* carpobalsamum. *Obs.*

carquaise, *f.* = carcaise.

carre, *f.* thickness; crown (of a hat); back and shoulders; corner.

carré, *p.a.* squared; square; perpendicular; square-built; (*Paper*) of the size called *carré* (see below). — *m.* square; something square or rectangular, as an iron bar of square cross section, a garden bed, a staircase landing, etc.; (*Paper*) a size about 45 by 56 cm. (18 by 22 in.), medium, demy.

carreau, *m.* tile; pane (of glass); paving brick; floor; ground; square; check, checker; a coarse file of square cross section; garden bed; tailor's goose; square cushion; (*Masonry*) stretcher; (*Cards*) diamond; thunderbolt.

— *de poêle,* stove tile.

— *encaustique,* encaustic tile.

— *en ciment,* cement slab, cement block.

carrelage, *m.* paving with tile, tiling.

carreler, *v.t.* pave with tile, tile.

carrelet, *m.* cubical grain (of explosives); packing needle; (*Zoöl.*) plaice.

carrelier, *m.* tile maker.

carrément, *adv.* squarely, square.

carrer, *v.t.* square.

carrière, *f.* quarry; career; race course; arena; course; scope, play.

carrosserie, *f.* coach (or carriage) building.

carrossier, *m.* coach builder, carriage builder.

carroyage, *m.* checkering, squares.

carte, *f.* cardboard; card; map, chart; bill, account; bill of fare.

— *couchée,* glazed cardboard.

— *de visite,* visiting card.

— *postale,* postal card, post card.

cartel, *m.* syndicate, trust; (*Mil.*) cartel.

cartelle, *f.* veneer.

carter, *m.* (*Mach.*) a protecting piece or cover, casing, case.

carthame, *m.* safflower (*Carthamus tinctorius*); any plant of the genus *Carthamus.*

carthamine, *f.* carthamin.

carthamique, *a.* carthamic.

cartilage, *m.* cartilage.

cartilagineux, *a.* cartilaginous.

carton, *m.* pasteboard, (paper) board; a piece of pasteboard; pasteboard case or box, carton; portfolio; cartoon; (*Printing*) a double leaf, or four pages.

— *bitumé* (or *bituminé*), tar board, tar paper.

— *Bristol,* Bristol board.

— *couché,* glazed board.

— *couvert,* covered pasteboard.

— *cuir,* leatherboard.

— *d'amiante,* asbestos board.

carton *de collage,* board made by gluing sheets of paper together, pasteboard in the proper sense.

— *de deuxième moulage,* board made from old paper.

— *de montagne,* asbestos board.

— *de moulage,* board made from paper pulp by molding or pressing into sheets, (in part) millboard.

— *de paille,* straw board.

— *de papier-mâché,* — *de pâte secondaire,* board made from old paper.

— *de premier moulage,* board made, like paper, from original materials.

— *de tourbe,* peat board.

— *en pâte de bois,* wood board.

— *glacé,* glazed board.

— *minéral,* asbestos board.

— *ondulé,* corrugated board.

— *porcelaine,* enameled board.

en —, (of books) in boards.

carton-cuir, *m.* leatherboard.

cartonnage, *m.* binding in boards; making of paper-board objects; paper-board boxes (collectively).

cartonné, *p.a.* bound in boards.

cartonner, *v.t.* bind in boards.

cartonnerie, *f.* pasteboard (or paper board) manufacture; paper-board factory.

cartonneux, *a.* resembling pasteboard, stiff.

carton-paille, *m.* strawboard.

carton-pâte, *m.* papier-mâché, paper pulp.

carton-pierre, *m.* a hard variety of paper board containing clay, cement, etc.

cartouche, *f.* thimble, shell (for extractions); cartridge. — *m.* case; frame, cartouche (for a name, title or the like).

— *à blanc,* — *à poudre,* — *blanche,* — *sans balle,* blank cartridge.

— *à fumée,* smoke cartridge.

cartoucherie, *f.* cartridge factory.

carvacrol, *m.* carvacrol.

carvène, *m.* carvene.

carvi, *m.* (*Bot.*) caraway (*Carum carui*).

carvol, *m.* carvol (old name for carvone).

caryophyllées, *f. pl.* (*Bot.*) Caryophyllaceae (Silenaceae).

caryophylline, *f.* caryophyllin.

cas, *m.* case; position, state; matter, affair; value.

en — *que,* in case, if.

en tout —, in any case.

cascade, *f.* cascade.

en —, in cascade, in a (descending) series.

cascarille, *f.* cascarilla.

case, *f.* compartment; hut.

caséase, *f.* casease.

caséate, *m.* caseate.

caséation, *f.* caseation.

caséeux, *a.* caseous, cheesy.

caséification, *f.* caseation.
caséifier, *v.t.* convert into cheese.
caséiforme, *a.* like cheese, cheesy.
caséine, *f.* casein.
caséinerie, *f.* cheesemaking; cheese factory.
caséinogène, *a.* casein-producing. — *m.* casein-ogen.
caséique, *a.* caseic.
caser, *v.t.* place; put away; put in order.
caserette, *f.*, **caserel, caseret,** *m.* cheese mold.
caserne, *f.* barracks.
casette, *f.* little case; (*Ceram.*) sagger, seggar.
caséum, *m.* caseum (paracasein).
casier, *m.* filing case, set of pigeonholes; compartmented case or rack.
casilleux, *a.* brittle.
caspien, *a.* Caspian.
  la Caspienne, the Caspian Sea.
casque, *m.* helmet; (*Bot.*) hood, helm.
casquette, *f.* cap.
cassable, *a.* breakable.
cassage, *m.* breaking, crushing.
cassant, *p.a.* breaking, etc. (see casser); brittle, short; peremptory.
  — *à chaud,* hot-short, red-short.
  — *à froid,* cold-short.
cassave, *f.* cassava.
casse, *f.* breaking, breakage; ladle; pan, basin, boiler; case (esp. for type); (*Bot. & Pharm.*) cassia; (*Metal.*) crucible; (*Wines*) casse.
  — *aromatique,* cassia bark, Chinese cinnamon (*Cinnamomum cassia,* etc.).
  — *du Brésil,* = casse vraie.
  — *en bâtons,* cassia pods.
  — *en bois,* (*Pharm.*) cassia bark.
  — *mondée,* cassia pulp.
  — *vraie,* purging cassia, cassia pods (*Cassia fistula*).
cassé, *p.a.* broken, etc. (see casser); worn out, weak; (of wine) affected with casse. — *m.* (*Paper, pl.*) casse paper, cassie paper; (*Candy*) crack, snap (stage reached at about 127° C.).
casse-coke, *m.* an instrument for breaking up large pieces of coke.
casse-diable, *m.* (*Bot.*) St.-John's-wort (*Hypericum,* esp. *H. perforatum*).
cassement, *m.* breaking, etc. (see casser).
casse-pierre, casse-pierres, *m.* stone breaker, stone crusher; stone hammer; (*Bot.*) saxifrage.
casse-poitrine, *m.* fiery liquor.
casser, *v.t.* break; crush; crack (nuts); annul.
  — *v.i. & r.* break.
  — *le vide,* break the vacuum, interrupt the vacuum.
casserole, *f.* casserole.
casseur, *m.* breaker. — *a.* breaking, crushing.

cassiée, *f.* (*Bot.*) any member of a tribe of cæsalpiniaceous plants including the genus *Cassia.*
cassier, *m.* cassia tree, cassia plant.
cassis, *m.* cassis, black currant (*Ribes nigrum*) or a liqueur made from it.
cassitérite, *f.* (*Min.*) cassiterite.
casson, *m.* fragment (as of glass or pottery); broken or imperfect sugar loaf; broken cocoa nibs.
cassonade, *f.* raw or muscovado sugar, cassonade.
cassure, *f.* fracture.
  — *à éclats,* splintering fracture.
  — *à fibres,* fibrous fracture.
  — *à grain* (or *à grains*), granular fracture.
  — *à nerf,* fibrous fracture.
  — *conchoïde,* conchoidal fracture.
  — *de bois,* woody fracture.
  — *fibreuse,* fibrous fracture.
  — *granulaire,* granular fracture.
  — *lamelleuse,* lamellar fracture.
  — *métallique,* metallic fracture.
  — *résineuse,* resinous fracture.
  — *terreuse,* earthy fracture.
  — *vitreuse,* vitreous fracture.
castillan, *a.* Castilian.
castine, *f.* (*Metal.*) limestone, used as a flux.
castoréum, *m.* castoreum.
castorine, *f.* castorin.
casuel, *a.* casual, incidental.
cataire, *f.* catnip (*Nepeta cataria*). — *a.* purring.
catalan, *a.* Catalan, (less often) Catalonian.
catalase, *f.* catalase.
catalepsie, *f.* (*Med.*) catalepsy.
Catalogne, *f.* Catalonia.
catalogue, *m.* catalog(ue), list.
cataloguement, *m.* catalogging, cataloguing.
cataloguer, *v.t.* catalog(ue).
catalyse, *f.* catalysis.
catalyser, *v.t.* catalyze, catalyse.
catalyseur, *m.* catalyzer, catalyst.
catalyte, *f.* catalyst, catalyzer.
catalytique, *a.* catalytic.
cataphorèse, *f.* cataphoresis.
cataphorétique, *a.* cataphoretic.
cataplasme, *m.* cataplasm, poultice.
catapuce, *f.* caper spurge (*Euphorbia lathyris*).
catéchine, *f.* catechol, catechin.
catéchique, *a.* catechuic, pertaining to catechu.
catéchu, caté-chu, *m.* catechu.
catégorie, *f.* category.
catégorique, *a.* categorical.
cathartine, *f.* cathartin (cathartic acid).
cathartique, *a. & m.* (*Med.*) cathartic.
cathédrale, *a. & f.* cathedral.
cathérétique, *a. & m.* (*Med.*) catheretic.
cathète, *f.* (*Geom.*) cathetus, perpendicular.
cathétomètre, *m.* cathetometer.

cathode, *f.* cathode.
cathodique, *a.* cathodic, cathode.
catholicon, *m.* (*Med.*) catholicon, panacea.
cati, *p.a.* glossed, lustered. — *m.* gloss, luster.
catir, *v.t.* gloss, luster (cloth by pressing).
catissage, *m.* glossing, lustering (of cloth).
catoptrique, *f.* catoptrics. — *a.* catoptric(al).
Caucase, *m.* Caucasus.
caucasien, caucasique, *a.* Caucasian.
cauchemar, *m.* incubus, nightmare.
caucher, *m.* parchment book for gold leaf.
causant, causatif, *a.* causative, causing.
causativement, *adv.* causatively.
cause, *f.* cause; (*Law*) consideration.
  à — *de,* because of, on account of.
  à — *que,* because, for the reason that.
  — *d'erreur,* cause of error, source of error.
causé, *p.a.* caused; (*Law*) for value received.
causer, *v.t.* cause. — *v.i.* talk, chat.
causticité, *f.* causticity.
caustificateur, *m.* causticizer.
caustification, *f.* causticization.
caustifier, *v.t.* causticize, render caustic.
caustique, *a. & m.* caustic. — *f.* caustic curve,
  caustic.
  — *au chlorure de zinc,* (*Pharm.*) Canquoin's
  paste.
  — *de Vienne,* (*Pharm.*) Vienna caustic,
  Vienna paste.
caustiquement, *adv.* caustically.
cautère, *m.* cautery, cauterization; cautery,
  cauterizing agent; ulcer.
cautérisant, *p.a.* cauterizing. — *m.* cauterizing agent, cauterant.
cautérisation, *f.* cauterization.
cautériser, *v.t.* cauterize.
caution, *f.* security, surety, guarantee, bail.
cautionné, *p.a.* guaranteed, warranted; bailed.
cautionnement, *m.* security, deposit, bail.
cautionner, *v.t.* guarantee; give security for.
cavalerie, *f.* cavalry.
cavalier, *m.* rider (as of a balance); staple;
  horseman; cavalier; (*Paper*) a size corresponding to printing demy (about 17½ by
  22½ in.).
  — *curseur,* sliding rider, adjustable rider.
cave, *f.* cellar. — *a.* hollow, concave.
caveau, *m.* small cellar, vault.
  — *à glace,* ice box.
caver, *v.t.* hollow, hollow out, excavate; dig,
  mine, undermine; stake, bet.
caverne, *f.* cave, cavern; cavity.
caverneux, *a.* cavernous, hollow.
cavité, *f.* cavity, hollow.
cazette, *f.* (*Ceram.*) sagger, seggar.
ce, *a.* this; that. — *pron.* this, that, he, she, it.
  (For the pl. see *ces.*)
  à — *que,* from what, according to what; by
  the fact that; that, in order that.

céanothe, céanote, *m.* (*Bot.*) Ceanothus;
  specif., (*Pharm.*) New Jersey tea (*C. ameri-
  canus*).
ceci, *pron.* this.
cécité, *f.* blindness.
  — *des couleurs,* color blindness.
cédat, *m.* natural steel.
céder, *v.t.* give up, part with; yield, surrender;
  cede, transfer; sell. — *v.i.* yield, give in,
  give way; be inferior.
cédrat, *m.* citron (*Citrus medica,* specif. the
  variety *genuina,* or its fruit).
cédraterie, *f.* citron plantation.
cédratier, *m.* citron tree.
cèdre, *m.* cedar (wood or tree); specif., any tree
  of the genus *Cedrus;* citron.
  — *blanc,* white cedar.
  — *de l'Himalaïa,* Himalaya cedar, deodar.
  — *des Bermudes,* Bermuda cedar (*Juniperus
  bermudiana*).
  — *de Sibérie,* Siberian cedar, stone pine
  (*Pinus cembra*).
  — *de Virginie,* red cedar (*Juniperus virginiana*).
  — *du Liban,* cedar of Lebanon (*Cedrus libani*).
  — *rouge,* red cedar.
cédrel, *m.,* **cédrèle,** *f.* any tree of the genus
  Cedrela.
cédrie, *f.* cedar resin.
cédron, *m.* (*Bot.*) cedron (*Simaba cedron*).
ceignant, *p.pr.* of ceindre.
ceignit, *p.def. 3 sing.* of ceindre.
ceindre, *v.t.* surround; put on (as a sword).
ceint, *p.a.* surrounded, inclosed, encircled.
  — *pr. 3 sing.* of ceindre.
ceinture, *f.* belt, band, girdle.
cela, *pron.* that; that person, he, she, him, her,
  they, them.
  à — *près,* with that exception; for all that.
céladon, *m. & a.* pale green.
celer, *v.t.* conceal, hide.
céleri, *m.* celery.
célérité, *f.* celerity.
céleste, *a.* celestial, heavenly.
  bleu —, sky blue, azure.
célestine, *f.* (*Min.*) celestite, celestine.
celle, *pron.* she, her, one, the one, that.
celle-ci, *pron.* this, this one, she, her; the latter,
  the last.
celle-là, *pron.* that, that one, she, her; the
  former, the first.
celles, *pron.* they, them, those. (*fem.* of ceux.)
celles-ci, *pron.* they, them, these; the latter,
  the last.
celles-là, *pron.* they, them, those; the former,
  the first.
cellier, *m.* a cool place for making wine or for
  keeping wine, vegetables, or the like (distinguished from *cave* in not being underground or vaulted).

cellose, *f.* cellose.
cellulaire, *a.* cellular.
cellule, *f.* cell; cellule, small cell.
— *à compter,* — *à fond divisé,* counting chamber.
— *composée,* composite cell.
— *de Thoma,* Thoma counting chamber.
cellulé, *a.* cellular, cellulated.
celluleux, *a.* cellular.
cellulite, *f.* cellulite; cellulitis.
celluloïd, celluloïde, *m.* celluloid.
cellulose, *f.* cellulose.
— *sulfitique,* sulfite cellulose.
cellulosique, *a.* cellulosic, cellulose.
celui, *pron.* he, him, one, the one, that.
celui-ci, *pron.* this, this one, he, him; the latter, the last.
celui-là, *pron.* that, that one, he, him; the former, the first.
cembre, cembro, *m.* stone pine (*Pinus cembra*).
cément, *m.* cement, specif. the powder used in cementation. (Cf. ciment.)
cémentation, *f.* (*Metal.*) cementation.
cémentatoire, *a.* cementatory; (*Mining*) cement (see eau cémentatoire, under *eau*).
cémenter, *v.t.* (*Metal.*) subject to cementation, cement, caseharden, face-harden. — *v.i.* undergo cementation.
cémenteux, *a.* of the nature of cement.
cendrage, *m.* (*Founding*) ashing over of molds.
cendraille, *f.* ashy débris from lime kilns.
cendre, *f.* ash, ashes.
— *bleue,* saunders blue (either ultramarine ash or a basic copper carbonate).
— *bleue naturelle* (also *pl.*), a blue pigment composed of powdered azurite.
— *de houille,* coal ashes.
— *de plomb,* dust shot, very fine shot.
— *d'os,* bone ash.
— *d'outremer,* ultramarine ash.
— *fixe,* permanent ash (one that remains when glowed).
cendres de bois, wood ashes.
cendres d'orfèvre goldsmith's ash, precious-metal residues.
cendres gravelées, a grayish potassium carbonate (pearlash) made from wine lees.
cendres perlées, pearlash.
— *verte,* green verditer.
mettre en cendres, reduce to ashes, burn to ashes, incinerate.
cendré, *a.* ash-colored, ashen.
cendrée, *f.* lead ashes, litharge; dross, refuse; dust shot, very fine shot.
— *de Tournay,* a hydraulic cement made from coal dust and lime.
cendrer, *v.t.* mix or treat with ashes; give an ashen color to.

cendreux, *a.* ashy, covered with or full of ashes; (*Metal.*) weak or flawy from ash spots.
cendrier, *m.* receptacle for ashes, ash pit, ash box, ash pan, ash bucket.
cendrière, *f.* peat; a woman that sells ashes.
cendrure, *f.* a rough pitted condition of metal, esp. steel, due to ashes; ash spot, cinder spot.
censé, *a.* supposed, regarded, considered.
censeur, *m.* censor; proctor. — *a.* censorious.
censure, *f.* censorship; censure.
cent, *a. & m.* hundred.
pour —, per cent.
cent., *abbrev.* centime (⅒ of a cent); (centième) hundredth.
centaine, *f.* hundred (vaguer than *cent*).
centaurée, *f.* (*Bot.*) centaury.
— *américaine,* American centaury (*Sabbatia angularis*).
cent. cub., *abbrev.* (centimètre cube, centimètres cubes), cubic centimeter(s).
centenaire, *a.* centennial, centenary, centenarian. — *m.* centennial, centenary; centenarian.
centennal, *a.* centennial.
centésimal, *a.* centesimal; percent; centigrade.
centiare, *m.* centiare, centiar (1 sq. meter).
centième, *a. & m.* hundredth.
alcool à 25 centièmes, 25 percent alcohol.
en centièmes, in hundredths, in per cent.
centigrade, *a.* centigrade.
centigramme, *m.* centigram.
centilitre, *m.* centiliter (10 c.c.).
centime, *m.* centime (about ⅕ cent or r⅒ d.).
centimètre, *m.* centimeter.
— *carré,* square centimeter.
— *cube,* cubic centimeter.
centrage, *m.* centering.
central, *a.* central.
centrale, *f.* central establishment or plant.
centralement, *adv.* centrally.
centralisateur, *a.* centralizing.
centralisation, *f.* centralization.
centraliser, *v.t.* centralize.
centration, *f.* centering.
centre, *m.* center.
centrer, *v.t.* center.
centreur, *m.* centering apparatus, centerer.
centrifugateur, *m.* = centrifugeur.
centrifugation, *f.* centrifugalization, centrifugation.
centrifuge, *a.* centrifugal.
centrifuger, *v.t.* centrifugalize, centrifugate, centrifuge.
centrifugeur, *m.,* **centrifugeuse,** *f.* centrifugal machine, centrifugal, centrifuge.
centripète, *a.* centripetal.
centrosome, *m.* (*Biol.*) centrosome.

centuple, *a. & m.* centuple, hundredfold.

centupler, *v.t.* centuple, increase a hundred-fold.

cep, *m.* stock.

cépage, *m.* variety (of grapevine).

cependant, *conj.* nevertheless, however, still. — *adv.* meanwhile, in the mean time.

céphalalgie, *f.* cephalalgia, headache.

céphalique, *a.* cephalic.

cephœline, *f.* cephaëline.

céracée, *f.* a kind of Swiss cheese.

cérame, *m.* an earthenware vessel. — *a.* pertaining to, or suitable for making, earthenware vessels.

céramique, *a.* ceramic. — *f.* ceramics. — *n.* ceramic ware, earthenware.

céramiste, *n.* ceramist. — *a.* ceramic.

céramographie, *f.* ceramography.

céramographique, *a.* ceramographic.

cérasine, *f.* cerasin.

cérat, *m.* (*Pharm.*) cerate; (loosely) ointment. — *cosmétique,* cold cream, ointment of rose water. — *de Goulard,* Goulard's cerate, cerate of lead subacetate. — *de résine anglais,* rosin cerate. — *de résine composé,* compound rosin cerate, Deshler's salve. — *de savon,* soap cerate. — *épulotique,* ointment of zinc oxide, zinc ointment. — *saturné,* = cérat de Goulard. — *simple,* simple cerate, cerate.

cératoïde, *a.* ceratoid, horny (or hornshaped).

cerce, *f.* circle, ring, specif.: the frame of a circular sieve; (*Ceram.*) an earthenware ring used as a support for articles in the kiln.

cerceau, *m.* hoop.

cerclage, *m.* encircling; hooping.

cercle, *m.* circle; hoop, ring, (annular) band, tire. *en cercles,* in circles; (of liquors) in the wood, in casks.

cercler, *v.t.* surround with hoops or bands; encircle; ring; limit.

cercueil, *m.* coffin, casket.

céréale, *f. & a.* cereal.

céréaline, *f.* cerealin (aleurone); cerealine (a maize preparation).

cérébrine, *f.* cerebrin.

cérémonie, *f.* ceremony.

céréolé, *m.* (*Pharm.*) cerate.

cérésine, cérésite, *f.* ceresin, ceresine.

cerf, *m.* deer, stag, hart.

cérides, *m. pl.* cerium metals.

cérifère, *a.* ceriferous, producing wax.

cérine, *f.* cerin.

cérique, *a.* of or pertaining to cerium, specif., ceric; (*Org. Chem.*) ceric.

cerise, *f.* cherry. — *m.* cherry (color), cerise; cerise (the dye).

cerisette, *f.* dried cherry; a cherry drink.

cerisier, *m.* cherry (tree or wood). — *de Virginie,* (*Bot.*) wild black cherry (*Prunus serotina*).

cérite, *f.* (*Min.*) cerite.

cérium, *m.* cerium.

cerne, *m.* circle, ring, halo.

cerneau, *m.* green walnut kernels.

cerner, *v.t.* encircle, surround, cut round. — *v.r.* become encircled.

céroïde, *a.* resembling wax.

céroléine, *f.* cerolein.

ceromel, *m.* (*Med.*) ceromel.

cérosée, cérosie, cérosine, *f.* cerosin (a waxy substance from sugar cane).

cérotique, *a.* cerotic; cerotyl, ceryl (in the phrase *alcool cérotique,* ceryl alcohol).

céroxyle, *m.* (*Bot.*) Ceroxylon, wax palm.

céroxyline, *f.* palm wax, ceroxyle.

certain, *a.* certain.

certainement, *adv.* certainly.

certes, *adv.* certainly, surely.

certificat, *m.* certificate.

certificateur, *m.* certifier; guarantor.

certifier, *v.t.* certify; guarantee.

certitude, *f.* certitude, certainty; steadiness, sureness (of hand).

cérulé, céruléen, *a.* cerulean, azure, light blue.

céruléine, *f.* (*Dyes*) ceruleïn.

céruléum, *m.* ceruleum (an inorganic pigment).

céruline, *f.* soluble indigo blue.

cérumen, *m.* (*Physiol.*) cerumen.

cérumineux, *a.* ceruminous.

céruse, *f.* white lead, ceruse (the pigment).

cérusier, *m.* white-lead worker.

cérusite, *f.* (*Min.*) cerussite.

cerveau, *m.* brain (the organ).

cervelet, *m.* cerebellum.

cervelle, *f.* brain substance, brains.

cervoise, *f.* cervisia (beer of the ancients).

cervoisier, *m.* a maker of cervisia.

céryle, *m.* ceryl.

cérylique, *a.* ceryl.

ces, *a.* (*pl.* of *ce, cette*) these; those. — *pron.* these, those, they.

césium, *m.* cesium, caesium.

cessant, *p.a.* ceasing, suspended.

cessation, *f.* cessation, ceasing.

cesse, *f.* ceasing, cessation. *sans* —, without ceasing, incessantly.

cesser, *v.t. & i.* cease, stop, discontinue.

cession, *f.* cession, transfer, assignment.

cessionnaire, *n.* assignee, grantee, transferee.

c'est, *abbrev.* (ce est) it is, he is, she is.

c'est-à-dire, that is to say, that is, i.e.

cet, *a.* Used for *ce* before a vowel sound. See *ce.*

cétacé, *a. & m.* (*Zoöl.*) cetacean.

cétazine, *f.* ketazine.

cétimine, *f.* ketimine.

cétine, *f.* cetin.

céto- . keto- .

cétohexose, *f.* ketohexose.

cétone, *f.* ketone.
— *de Michler,* Michler's ketone.

cétonique, *a.* ketonic.

cétopipéridine, *f.* ketopiperidine (piperidone).

cétose, *m.* ketose.

cette, *a. fem.* of *ce.*

ceux, *pron.* they, them, those.

ceux-ci, *pron.* they, them, these; the latter, the last.

ceux-là, *pron.* they, them, those; the former, the first.

cévadille, *f.* (*Bot.*) sabadilla, cevadilla.

cévadine, *f.* cevadine (veratrine of Merck).

Ceylan, *n.* Ceylon.

ch., *abbrev.* (chapitre) chapter.

chablis, *m.* Chablis (a dry white wine).

chabotte, *f.* anvil block, anvil (of a power hammer).

chacrille, *f.* cascarilla.

chacun, *pron.* each, each one; everyone, everybody.

chafée, chaffée, *f.* residue remaining after the extraction of starch from cereals, bran.

chagrin, *n.* shagreen (leather); affliction, grief, trouble; chagrin. — *a.* sad, peevish, morose.

chagriner, *v.t.* shagreen (skins); distress, afflict; vex; chagrin.

chagrineur, chagrinier, *m.* shagreen maker.

chai, *m.* storage room, storage shed (for wines, brandies, etc.).

chaille, *pr. 3 sing. subj.* of chaloir.

chaîne, *f.* chain.
à — ouverte, open-chain.
— *à godets,* chain pump.
— *allemande,* open-link chain.
— *anglaise,* close-link chain.
— *arborescente,* branching chain, branched chain.
— *à rouleaux,* block chain, roller chain.
— *carbonée,* carbon chain, chain of carbon atoms.
— *de transmission,* driving chain, power chain.
— *fermée,* closed chain, ring.
— *hydrocarbonée,* hydrocarbon chain.
— *latérale,* side chain, lateral chain.
— *linéaire,* (*Org. Chem.*) straight chain, unbranched chain.
— *ouverte,* open chain.
— *sans fin,* endless chain.

chaînette, *f.* small chain; (*Math.*) catenary.

chaînon, *m.* link (or, broadly, portion) of a chain (in *Org. Chem.* often best translated as "grouping" or "group.")

chaînon *sulfoné,* sulfonic group, sulfo group.
— *terminal,* end group, terminal group.

chair, *f.* flesh; (*Leather*) flesh side, flesh; (*Metal.*) fiber; chair (of a rail).
— *fossile,* a variety of asbestos.

chaire, *f.* chair, professorship; pulpit.

chais, *m.* = chai.

chaise, *f.* chair; chaise; frame, framework; (*Mach.*) hanger.

chalait, *imp. 3 sing.* of chaloir.

chaland, *m.* customer, patron; barge, lighter.

chalcédoine, *f.* chalcedony.

chalcographie, *f.* chalcography.

chalcopyrite, *f.* (*Min.*) chalcopyrite.

chalcosine, *f.* (*Min.*) chalcocite.

châle, *m.* shawl.

chaleur, *f.* heat.
— *blanche,* white heat.
— *d'échauffement,* the amount of heat required to raise a given body a given temperature interval.
— *de combinaison,* heat of combination.
— *de dissolution,* heat of solution.
— *latente,* latent heat (heat of fusion or of vaporization).
— *rouge,* red heat.
— *soudante* (or *suante*), welding heat.
— *spécifique* (or *spéciale*), specific heat.

chaleureux, *a.* warm, ardent, hot-blooded.

chaloir, *v.i.* matter, signify (used impersonally).

chalkolite, *f.* (*Min.*) chalcolite (torbernite).

chalkopyrite, *f.* (*Min.*) chalcopyrite.

chalkosine, *f.* (*Min.*) chalcocite.

chaloupe, *f.* launch.

chalu, *p.p.* of chaloir.

chalumeau, *m.* blowpipe; tube, pipe; reed; straw; lime twig; (*Music*) wood-wind.
— *à bouche,* mouth blowpipe.
— *à gaz,* gas blowpipe, (gas) blast lamp, (gas) blast burner.
— *à soufflerie,* blowpipe with bellows.
— *oxhydrique,* oxyhydrogen blowpipe.
— *oxyacétylénique,* oxyacetylene blowpipe.

chalybé, *a.* chalybeate.

chambertin, *m.* Chambertin (a red Burgundy wine).

chambourin, *m.* a kind of glass sand; a kind of green glass.

chambranle, *m.* casing, case (as of a door); frame.

chambre, *f.* chamber; room; floor, story; (*Metal.*) cavity, blowhole, honeycomb.
— *à air,* air chamber; (of pneumatic tires) inner tube.
— *à air comprimé,* compressed-air chamber.
— *à gaz,* gas chamber.
— *à poussière,* dust chamber.
— *à vapeur,* steam chamber, steam chest.
— *de Böttcher,* Böttcher's moist chamber.

chambre de combustion, combustion chamber, fire box.
— de plomb, lead chamber.
— des députés, Chamber of Deputies.
— humide, moist chamber.
— noire, (Photog.) camera.
— syndicale, a body of syndics or representatives; a tribunal of a corporation.
chambré, a. chambered; specif., containing cavities, honeycombed.
chambrer, v.r. become honeycombed.
chameau, m. camel.
chamelle, f. female camel.
chamois, m. chamois (animal or leather).
chamoiser, v.t. chamois.
chamoiserie, f. chamois leather factory; chamois leather; chamoising.
chamoiseur, m. chamois dresser.
chamotte, f. (Ceram.) chamotte.
champ, m. field; edge, long narrow side (as of a brick); (pl.) country.
— de force, field of force.
— de vue, — de vision, = champ visuel.
— magnétique, magnetic field.
— tournant, rotating field.
— visuel, visual field, field of vision, field of view.
de —, edgewise, edgeways, on edge.
champagne, m. champagne (wine). — f. In the phrase fine champagne, a superior kind of brandy.
champagniser, v.t. treat (wine) after the method of making champagne.
champêtre, a. rural, country.
champignon, m. fungus; mushroom; mushroom-shaped object, as a rail head.
champlevé, a. & m. (Enameling) champlevé.
chance, f. chance; luck, fortune.
chanceler, v.i. stagger, totter; waver.
chanceux, a. hazardous; lucky.
chanci, p.a. molded, moldy.
chancir, v.i. mold, become moldy.
chancissure, f. mold; moldiness.
chancre, m. canker; (Med.) chancre.
chandelier, m. chandelier, branched fixture; candlestick; an upright support, as the stanchion of a pump.
— d'amphithéâtre, lecture-room fixture (for gas, water, etc.).
— de jauge, (Ceram.) a gage consisting of a vertical rod with horizontal crossrods.
chandelle, f. candle (esp. one containing tallow or other fat, as distinguished from bougie); vertical support, prop, stay, shore, stud; icicle; candle light; artificial light.
— à la baguette, dipped candle, so called from the rod from which the candles are suspended.
— moulée, molded candle.
— romaine. (Pyro.) Roman candle.

chandellerie, f. candle factory or shop, chandlery.
change, m. change; exchange.
changé, p.a. changed; exchanged.
changeant, p.a. changing, changeable, variable.
changement, m. changing, change, alteration, variation; (Railways) switch.
— de marche, (Mach.) reversing gear.
— d'état, change of state.
changer, v.t. change; exchange. — v.i. & r. change.
— de, change.
chant, m. singing; song; melody; chant; canto.
chanter, v.t. & i. sing; chant.
chantier, m. yard (for storage or work); stand for barrels, gantry.
chantignole, f. bracket.
chantonné, a. (of paper) defective.
chantourner, v.t. (Tech.) cut in profile.
chanvre, m. hemp.
— de Calcutta, — de Chine, jute.
— de la Nouvelle-Zélande, New Zealand flax (or hemp), Phormium tenax.
— de l'Inde, Indian hemp (Cannabis sativa, var. indica).
— de Manille, Manila hemp, manila (from Musa textilis).
— du Bengale, jute.
— du Canada, Canadian hemp (Apocynum cannabinum).
— femelle, — porte-graine, female hemp.
— mâle, male hemp.
chanvreux, a. hempen, hemp.
chape, f. cover, lid, cap, capping; (of a still) head; (of a balance) bearing; (Mach.) strap, clasp, catch, loop; (of a mold) cope; molding clay; bed in which tiles are set.
chapeau, m. hat; (Tech.) cap, cover, lid, top, head; (of a still) head; (of beer, etc.) head; (of a chimney) cowl, hood; felt; (Tin) a block of Malacca tin; (Mining) the altered top portion of a vein deposit; (Bot.) pileus.
— de fer, (Mining) gossan, "eisenhut."
chapelet, m. chaplet (wreath, string of beads, or something resembling these); (on liquors) beads, bubbles; chain pump; series.
— électrique, a set of fuses or charges connected so as to be fired simultaneously.
chapelle, f. valve box, valve case; cylinder case; vault, arched part (of a furnace or the like); explosion chamber; chapel.
chapellerie, f. hat making; hat trade; hat shop.
chaperon, m. hood; cover, cap, (of a wall) coping; chaperon.
chapiteau, m. head, cap, cover, hood, dome, top; (of a still) head; (of a column) capital.
chapitre, m. chapter; subject, head.

chapon, *m.* capon; young vine.

chaque, *a.* each, every.

char, *m.* car, chariot; cart, wagon.
— *d'assaut,* (*Mil.*) tank.

charançon, *m.* (*Zoöl.*) weevil.

charbon, *m.* charcoal; char; coal; carbon (as in an electric cell or an arc light); (*Med.*) anthrax, charbon; (*Agric.*) smut.
— *à dessin,* drawing charcoal, charcoal crayon.
— *à filtrer,* filtering charcoal.
— *à lumière,* electric light carbon.
— *à mèche,* cored carbon.
— *animal,* animal charcoal.
— *artificiel,* charcoal prepared by carbonizing a prepared mixture (instead of wood or other natural material), synthetic charcoal (so called).
— *à souder,* soldering carbon.
— *bitumineux,* bituminous coal.
— *collecteur,* (*Elec.*) brush.
— *de bois,* wood charcoal.
— *de cornue,* retort carbon, gas carbon.
— *de houille,* coke; anthracite.
— *de lumière,* lamp carbon, electric light carbon.
— *de Paris,* a synthetic charcoal made from powdered charcoal, coal dust or the like, with pitch as a binder, and used as a briquetted fuel.
— *de pierre,* stone coal (anthracite).
— *de saule,* willow charcoal.
— *de sucre,* a pure charcoal made from sugar.
— *de terre,* (mineral) coal.
— *de tourbe,* peat coal.
— *de vinasse,* vinasse charcoal (esp. that from sugar beets).
— *électrique,* electric carbon (either for arc light or for battery).
— *en poudre,* charcoal dust, powdered charcoal; coal dust.
— *fulminant,* fulminating charcoal (residue obtained by heating tartar emetic to white heat).
— *graphitique,* graphitic carbon.
— *gras,* fat coal, (broadly) soft coal.
— *incombustible,* anthracite.
— *luisant,* anthracite.
— *maigre,* lean coal.
— *minéral,* (mineral) coal.
— *plastique,* plastic coal.
— *pour piles,* battery carbon.
— *roux,* brown (or red) charcoal.
— *végétal,* vegetable charcoal.

charbonnage, *m.* coal mining; coal mine.

charbonnaille, *f.* small coal (or charcoal); (*Metal.*) a mixture of cinders, sand and clay.

charbonné, *p.a.* charred; blackened, marked with black; drawn with charcoal; (*Agric.*) smutty.

charbonnée, *f.* layer of charcoal.

charbonner, *v.t.* char; blacken with charcoal; polish with charcoal. — *v.i. & r.* char; deposit carbon.

charbonnette, *f.* wood for making charcoal.

charbonneux, *a.* of the nature of, or containing, charcoal or carbon, charry; (*Med.*) of or pertaining to anthrax.

charbonnier, *m.* charcoal burner; charcoal (or coal) dealer; coke oven; stoker, fireman; coal cellar; collier. — *a.* pertaining to the charcoal or coal industry.

charbonnière, *f.* charcoal pit or oven.

charcuterie, *f.* pork in various forms, or a place where these products are made or sold.

charcutier, *m.* pork butcher.

chardon, *m.* thistle; teasel.
— *bénit,* blessed thistle (*Cnicus benedictus*).

chardonner, *v.t.* teasel.

charge, *f.* charge (in various senses); load; burden; (*Paper,* etc.) loading; head, pressure (as of water); office, responsibility.
— *admissible,* working load.
— *amorce,* priming charge.
— *brute,* total load.
— *d'amorçage,* priming charge.
— *d'épreuve,* test charge; test load.
— *de rupture,* breaking load.
— *dissimulée,* (*Elec.*) bound charge.
— *explosive,* bursting charge.
— *induite,* induced charge.
— *résiduelle,* residual charge.
— *utile,* useful load.
en —, (*Ceram.*) by piling one upon another the articles to be fired.

chargé, *p.a.* charged, etc. (see charger); (of liquids) turbid; (of the sky) overcast. — *m.* In such phrases as *chargé d'affaires* (chargé d'affaires) and *chargé de cours* (one who conducts a university course altho not a professor).

charge-amorce, *n.* priming charge.

chargement, *m.* charging, etc. (see charger); load, charge; registered letter.

charger, *v.t.* charge; load; exaggerate; register (a letter). — *v.r.* (of liquids) become turbid; (of the sky) become overcast; (with *de*) charge oneself with, undertake.

chargeur, *m.* loader; charger; stoker.

chariot, *m.* wagon; carriage, trolley, car, truck; (*Mach.*) carriage, carrier, slide, slide rest, traveler; chariot.

charité, *f.* charity.

charmant, *p.a.* charming.

charme, *m.* charm; (*Bot.*) hornbeam.

charmer, *v.t.* charm.

charnière, *f.* hinge; (turning) joint.
à —, with hinge(s), hinged.

charnu, *a.* fleshy.

charogne, *f.* carrion.

charpente, *f.* frame, framework; framing; timbering, carpentry.

charpenter, *v.t.* frame, build, plan; cut for framing.

charpenterie, *f.* carpentry.

charpentier, *m.* carpenter.

charpie, *f.* lint.

charquer, *v.t.* dry (meat).

charrée, *f.* ashes, lye ashes; (*Soda*) exhausted black ash, vat waste.

charretée, *f.* cart load, cartful.

charrette, *f.* (two-wheeled) cart.

charriage, *m.* cartage; carrying.

charrier, *v.t.* cart; carry.

charroi, *m.* carriage, drayage.

charroyer, *v.t.* cart, haul.

charrue, *f.* plow.

charte, *f.* charter.

chartré, *a.* chartered.

chartreuse, *f.* chartreuse.

chas, *m.* starch size; eye (of a needle).

chasse, *f.* chase, hunting, pursuit; flooding, flushing; (of machinery) play; driving tool, set, setter, drift, punch, etc.

châsse, *f.* knife edge (of a balance); mounting.

chasser, *v.t.* drive out, expel; drive out, drive away; drive; propel; drive in; chase, pursue, hunt.

chasseur, *m.* hunter; chasseur. — *a.* hunting.

châssis, *m.* frame, framework; chassis; (of a window) sash; (*Founding*) flask; (*Photog.*) plate holder.

chassoir, *m.* driver (any of various tools).

chat, *m.* cat.

châtaigne, *f.* chestnut.

— *de cheval,* — *d'Inde),* horse-chestnut.

— *du Brésil,* brazil nut.

châtaignier, *m.* chestnut tree.

châtain, *a. & m.* chestnut, chestnut brown.

châtaire, *f.* (*Bot.*) catnip.

château, *m.* castle; chateau.

chatoiement, chatoïment, *m.* chatoyancy, chatoyment.

chaton, *m.* kitten; (*Bot.*) catkin; setting (of a ring); bezel.

chatonner, *v.t.* set (a gem).

chatouiller, *v.t.* tickle.

chatoyant, *p.a.* chatoyant.

chatoyement, *m.* = chatoiement.

chatoyer, *v.i.* be chatoyant.

châtrer, *v.t.* castrate; prune.

chaud, *a.* hot; warm; fresh, recent; expensive. — *m.* heat.

à —, with the use of heat, in a heated state, at a high temperature, hot.

chaude, *f.* heat, heating; brisk fire.

— *blanc de lune,* — *blanche,* white heat.

— *grasse,* melting heat; (*Metal.*) welding heat.

chaude *rouge,* red heat.

— *rouge cerise,* cherry-red heat.

— *sombre,* dark red heat.

— *soudante,* — *suante,* welding heat.

chaudière, *f.* boiler; copper, caldron, kettle, pan; (*Sugar,* etc.) evaporator.

— *à bière,* — *à brasser,* (*Brewing*) copper, brewing kettle.

— *à brai,* pitch kettle.

— *à cuire,* (*Sugar*) evaporator.

— *à goudron,* tar pot, tar kettle.

— *à haute pression,* high-pressure boiler.

— *alimentaire,* feed boiler.

— *à retour de flamme,* return-flame boiler.

— *à tubes d'eau,* water-tube boiler.

— *autoclave,* autoclave.

— *à vapeur,* steam boiler, boiler.

— *de fusion,* lead pot.

— *de raffinage,* refining boiler.

— *fixe,* stationary boiler.

— *multitubulaire,* water-tube boiler.

— *sectionnelle,* sectional boiler, water-tube boiler.

chaudra, *fut. 3 sing.* of chaloir.

chaudron, *m.* kettle, (small) boiler, caldron, copper.

chaudronnée, *f.* kettleful, boilerful.

chaudronnerie, *f.* coppersmith's trade; tinman's trade; kettles, pans, etc. or a place where they are made.

chaudronnier, *m.* coppersmith, brazier; dealer in kettles, pans, etc.

chauffable, *a.* capable of being heated, heatable.

chauffage, *m.* heating; warming; firing; fuel.

— *à vapeur,* steam heating.

chauffe, *f.* heating; firing; distillation; fireplace, hearth (of a furnace); stokehole, firehole; furnace (of a foundry); (*Metal.*) heat.

chauffer, *v.t.* heat; warm; fire, fire up; urge forward, push. — *v.i.* heat; warm; be heated.

— *v.r.* get hot or warm; warm oneself.

— *blanc,* heat white-hot.

chaufferette, *f.* a small warming apparatus, as a foot warmer or a chafing dish.

— *à l'acétate,* a warmer containing a supersaturated solution of sodium acetate, which gives off heat on crystallizing.

chaufferie, *f.* stokehole; range of boilers; (*Iron*) chafery.

chauffe-tubes, *m.* an apparatus for heating tubes; specif., an attachment which permits a combustion tube to be supported and heated over an ordinary Bunsen burner.

chauffeur, *m.* fireman, stoker; chauffeur.

chauffe-vin, *m.* a preheater for wine or other alcoholic liquid about to be distilled.

chauffoir, *m.* (*Salt*) one of the shallow basins in which sea water is subjected to the first stage of evaporation.

chauffure, *f.* scaling of iron or steel due to overheating; burnt iron.

chaufour, *m.* limekiln.

chaufournerie, *f.* lime burning.

chaufournier, *m.* lime burner.

chaulage, *m.* liming; whitewashing.

chauler, *v.t.* treat with lime, lime; whitewash.

chaulier, *m.* lime burner.

chaume, *m.* culm, stem (of grasses); stubble; stubble straw; thatch.

chausse, *f.* bag filter, filter bag.

chaussée, *f.* road, roadway; ground floor; dike.

chausser, *v.t.* put on (stockings, etc.); fit, suit; hill, earth up (plants).

chaussette, *f.* sock.

chaussure, *f.* footwear; (of a pile) shoe.

chaut, *pr. 3 sing.* of chaloir.

chauve-souris, *f.* bat.

chaux, *f.* lime; calx.
— *anhydre,* quicklime.
— *azotée,* crude calcium cyanamide (commercial name).
— *calcinée,* quicklime.
— *caustique,* caustic lime, quicklime.
— *commune,* common lime, quicklime.
— *de construction,* building lime.
— *éteinte,* slaked lime.
— *éteinte à sec,* air-slaked lime.
— *fluatée,* calcium fluoride (old name).
— *fusée,* slaked lime.
— *grasse,* fat lime, rich lime.
— *hydratée,* hydrated lime.
— *hydraulique,* hydraulic lime.
— *légères,* hydraulic lime fines, separated by bolting.
— *lourde,* a kind of hydraulic lime which has been subjected to bolting to remove the fines.
— *maigre,* lean lime, poor lime.
— *métallique,* metallic calx; a blue pigment said to be cobaltous arsenate.
— *morte,* dead-burned lime, overburned lime.
— *nitratée,* nitrate of lime (calcium nitrate).
— *phosphatée,* phosphate of lime (calcium phosphate).
— *sodée,* soda lime.
— *sulfatée,* sulfate of lime (calcium sulfate).
— *vive,* quicklime.

chaux-azote, *m.* (commercial) calcium cyanamide, "lime-nitrogen."

chavibétol, *m.* chavibetol.

chavica, *n.* (*Bot.*) any plant of the genus *Piper.*

chavirage, chavirement, *m.* upsetting, turning upside down, capsizing.

chavirer, *v.t. & i.* upset, turn upside down, (*Naut.*) capsize.

cheddite, *f.* (*Expl.*) cheddite.

chef, *m.* head; chief.
— *d'atelier,* works manager.

chef *de file,* leader; first member (of a series); (*Mil.*) file leader.
— *de place,* (*Glass*) foreman glassblower.
— *de travaux,* a person in charge of laboratory work but not lecturing.
— *d'ouvriers,* foreman.
— *du savonnage,* (*Glass*) a workman who has charge of the supplementary grinding of mirror glass, and who marks the spots to be reground.
— *fondeur,* (*Glass*) foreman founder.
en —, in chief; in charge.

chef-ouvrier, *m.* master workman; foreman.

chélidoine, *f.* (*Bot.*) celandine.

chélidonine, *f.* chelidonine.

chélidonique, *a.* chelidonic.

chemin, · m. way; path, track, course; road, passage.
— *de fer,* railway, railroad.
— *faisant,* on the way, by the way.
— *ferré,* metaled road, macadam road.

cheminée, *f.* chimney; flue, stack, shaft, smokepipe, smoke funnel; fireplace, hearth; flame passage (as in a fuse or primer).
— *d'appel,* ventilating flue, ventilating shaft.
— *parajour,* a chimney which is also a shade, as on a magic lantern or a lamp for a microscope.

cheminement, *m.* advance, progress, traveling.

cheminer, *v.i.* go, travel, move on, advance.

chemise, *f.* case, casing, jacket, mantle, shell, wrapper; lining; facing, revetment; shirt; chemise.
— *d'eau,* water jacket.
— *de cylindre,* cylinder jacket.
— *de vapeur,* steam jacket.

chemiser, *v.t.* (put a covering of some kind on), case, jacket, mantle, coat, lute.

chenal, *m.* channel; gate (of a mold); gutter (of a roof).

chenaux, *pl.* of chenal.

chêne, *m.* oak (tree or wood).
— *à écorce,* tanbark oak (any oak yielding tanbark).
— *à feuilles persistantes* (or *vertes*), evergreen oak.
— *blanc,* white oak, Quebec oak (*Quercus alba*); a subspecies of the British oak (it has been called *Q. pedunculata*).
— *de mer,* fucoid seaweed, sea oak.
— *des Indes,* Indian oak, teak.
— *de vie,* live oak.
— *liège,* cork oak, cork (*Quercus suber*).
— *pédonculé,* a variety of the British oak (it has been called *Quercus pedunculata*).
— *rouvre,* British oak (*Quercus robur*).
— *saule,* willow oak (*Quercus phellos*).
— *vert,* evergreen oak, specif. the holm oak (*Quercus ilex*).

chêne *yeuse*, holm oak (*Quercus ilex*).

chènevis, *m.* hemp seed.

chènevotte, *f.* woody part of hemp, boon, shive.

chenille, *f.* caterpillar, larva (of a butterfly or moth); chenille.

chénopode, *m.* (*Bot.*) any member of the genus *Chenopodium* (goosefoot).

— *anthelmintique,* American wormseed (*Chenopodium anthelminticum*).

chénopodée, chénopodiée, *f.* (*Bot.*) any member of a tribe of plants including *Chenopodium* and other genera.

chenu, *a.* hoar, hoary.

chèque, *m.* (*Com.*) check, cheque.

cher, *a.* dear. — *adv.* dear, dearly.

chercher, *v.t.* seek, look for, search for.

aller —, go in quest of, go to bring, go and bring.

— *à,* try to, endeavor to.

chercheur, *m.* finder (optical instrument); research worker; searcher, seeker, hunter. — *a.* searching, exploring.

chère, *a. fem.* of cher. — *f.* cheer.

chèrement, *adv.* dearly, dear.

chérir, *v.t.* cherish.

cherté, *f.* dearness.

chester, *m.* Cheshire cheese.

chétif, *a.* poor, weak, wretched, mean.

cheval, *m.* horse; horse power; donkey engine.

à —, astride, astraddle, on both sides; on horseback.

— *de force,* horse power.

— *de vapeur,* steam horse power; horse power.

— *électrique,* electric horse power.

cheval-an, *m.* horse-power year.

chevalet, *m.* horse (in technical senses), frame, trestle, rack, stand; specif., (*Glass*) a stand on which window-glass cylinders are laid; easel.

— *de rivière,* (*Leather*) beam, tanner's beam.

— *trépied,* three-legged stand, tripod.

cheval-heure, *m.* horse-power hour.

chevalier, *m.* knight.

cheval-jour, *m.* horse-power day.

cheval-vapeur, *m.* horse power.

chevauchement, *m.* lapping, overlapping.

chevé, *p.a.* hollowed out.

chevelu, *a.* hairy, bearded; long-haired. — *m.* hairy growth (as of rootlets), beard.

chevelure, *f.* (head of) hair; (*Astron. & Bot.*) coma; foliage.

chever, *v.t.* hollow out.

cheveu, *m.* hair.

cheveux, *m.pl.* hairs; (collectively) hair.

cheville, *f.* peg; pin, bolt, spike, key, dowel.

cheviller, *v.t.* peg, pin, bolt.

chevillette, *f.* (small) peg, (small) pin.

chèvre, *f.* goat, she-goat; derrick, gin, jack, crane.

chevreau, *m.* kid (animal and leather).

chèvrefeuille, *f.* (*Bot.*) honeysuckle.

chevrette, *f.* (small) gin, jack.

chevreuil, *m.* roe deer.

chevrotain, *m.* musk deer.

chevrotin, *m.* kid leather, kid; fawn; musk deer.

chevrotine, *f.* buckshot.

chez, *prep.* at the house of, at the home of, among, with; (with verbs of motion) to the home of, to.

— *moi,* — *soi,* — *lui,* at home.

de — *moi,* from my home, from home.

chiasse, *f.* (*Metal.*) dross, scum.

chibou, *m.* cachibou, cachibou resin.

chicane, *m.* baffle, baffle plate; obstacle; chicanery.

en —, in the manner of baffle plates, with broken joints, staggered.

chicané, *a.* provided with baffles.

chicorée, *f.* chicory.

chien, *m.* dog (animal or gripping device); a small car running on rails, for use in factories; (of firearms) hammer, cocking piece.

chiendent, *m.* couch grass; (*Pharm.*) triticum.

— *de mer,* — *marin,* wrack, sea wrack.

— *fossile,* amianthus.

— *officinal, Agropyron repens.*

chiffon, *m.* rag; scrap; (*pl.*) dress, fashions.

— *d'essuyage,* wiping cloth, dust cloth.

chiffonier, *m.* rag man, junk man; chiffonier.

— *a.* pertaining to rags or junk.

chiffrage, *m.* numbering, etc. (see chiffrer); valuation.

chiffraison, *f.* graduation figures, graduation.

chiffre, *m.* figure; numeral, digit; cipher.

— *d'affaires,* volume of business.

chiffrer, *v.t.* number; express in figures; write in cipher. — *v.i.* figure, calculate.

chimiatre, *m.* iatrochemist.

chimiatrie, *f.* chemiatry, iatrochemistry.

chimiatrique, *a.* chemiatric, iatrochemical.

chimiatriquement, *adv.* iatrochemically.

chimico-légal, *a.* chemico-legal.

chimico-physique, *a.* chemico-physical.

chimie, *f.* chemistry.

— *agricole,* agricultural chemistry.

— *alimentaire,* food chemistry.

— *analytique,* analytical chemistry.

— *appliquée,* applied chemistry.

— *biologique,* biological chemistry.

— *industrielle,* industrial chemistry.

— *inorganique,* inorganic chemistry.

— *minérale,* inorganic chemistry.

— *organique,* organic chemistry.

— *technique,* technical chemistry.

— *théorique,* theoretical chemistry.

— *végétale,* plant chemistry.

chimie-physique, *f.* physical chemistry.

chimio-. chemo-.

chimiotactique, *a.* chemotactic.

chimiotaxie, *f.* chemotaxis.
chimiothérapie, *f.* chemotherapy.
chimiotropique, *a.* chemotropic.
chimiotropisme, *m.* chemotropism.
chimique, *a.* chemical.
chimiquement, *adv.* chemically.
— *pur,* chemically pure.
chimiqueur, *m.* (*Matches*) composition attendant.
chimisme, *m.* chemism (esp. in the sense of chemical phenomena).
chimiste, *m.* chemist.
— *analyste,* = chimiste-analyste.
— *d'industrie,* industrial chemist.
— *indienneur,* chemist in a calico printery, print chemist.
— *organicien,* organic chemist.
— *spécialiste,* one who is a specialist in some particular branch of chemistry, specializing chemist.
chimiste-analyste, *m.* analytical chemist (esp. one who has completed the regular three-year course and received a diploma).
chimiste-coloriste, *m.* color chemist, dye chemist.
chimitypie, *f.* chemitypy, chemitype.
chimoine, *m.* a kind of imitation marble.
china, *m.* chinaroot (from *Smilax china*); cinchona bark, Peruvian bark.
chinage, *m.* clouding (as of fabrics).
chinagrass, *m.* ramie, China grass.
Chine, *f.* China.
chiner, *v.t.* cloud (as fabrics), render chiné.
chinois, *a.* Chinese.
chinoline, *f.* quinoline.
chinone, *f.* quinone.
chio, *m.* (*Metal.*) taphole.
chipage, *m.* (*Leather*) infiltration.
chiper, *v.t.* (*Leather*) infiltrate with tan liquor.
chiquer, *v.t. & i.* chew.
chirette, *f.* (*Bot. & Pharm.*) chirata.
chirurgical, *a.* surgical.
chirurgie, *f.* surgery.
chirurgien, *m.* surgeon.
chirurgique, *a.* surgical.
chitine, *f.* chitin.
chloracétate, *m.* chloroacetate, chloracetate.
chloracétique, *a.* chloroacetic, chloracetic.
chlorage, *m.* chloring.
chloral, *m.* chloral.
chloralamide, *f.* chloralamide.
chloralimide, *n.* chloralimide.
chloralose, *n.* chloralose.
chloramidure, *m.* chloramide, chloroamide.
— *de mercure,* mercury chloramide, aminomercuric chloride.
chloranile, *m.* chloranil.
chlorate, *m.* chlorate.
— *de potasse,* potassium chlorate, chlorate of potash.

chlorate *de soude,* sodium chlorate.
chloraté, *a.* containing chlorate, chlorate (as, *poudre chloraté,* chlorate powder).
chloration, *f.* chlorination.
chlorazotique, *a.* nitrohydrochloric, nitromuriatic.
chlore, *m.* chlorine.
— *liquide,* liquid chlorine; (*Pharm.*) chlorine water.
chloré, *p.a.* chlorinated, containing chlorine, chloro.
chlorebenzène, *m.* chlorobenzene.
chloréthyle, *m.* ethyl chloride, chloroethane.
chloreux, *a.* chlorous.
chlorhydrate, *m.* hydrochloride, chlorhydrate.
— *d'ammoniaque,* ammonium chloride.
chlorhydrine, *f.* chlorohydrin, chlorhydrin.
— *chromique,* chromyl chloride.
— *sulfurique,* chlorosulfonic acid.
chlorhydrique, *a.* hydrochloric.
chlorique, *a.* chloric.
chlorite, *m.* chlorite.
chloroamidure, chloro-amidure, *m.* = chloramidure.
chloro-antimonié, *a.* containing, or combined with, chlorine and antimony.
chloro-arséniaté, *a.* (*Min.*) containing or combined with chlorine and arsenic acid.
chloro-aurique, *a.* chloroauric.
chlorobenzine, *f.* chlorobenzene.
chlorobromé, *a.* containing, or combined with, chlorine and bromine, bromochloro.
chloro-carbonaté, *a.* chlorocarbonate of.
chlorochromique, *a.* chlorochromic.
chloro-éthyline, *f.* chlorethylin, chloroethylin (compound of the type $ClR''OC_2H_5$).
chloroformation, *f.* chloroforming.
chloroforme, *m.* chloroform.
chloroformer, *v.t.* chloroform.
chloroformiate, *m.* chloroformate.
chloroformique, *a.* chloroformic, of or pertaining to chloroform; chloroformic (acid).
chloroformisation, *f.* chloroforming, chloroformization.
chloroformiser, *v.t.* chloroform, chloroformize.
chloromètre, *m.* chlorometer.
chlorométrie, *f.* chlorometry.
chlorométrique, *a.* chlorometric.
chloronitré, *a.* containing or combined with chlorine and the nitro group, chloronitro.
chlorophylle, *f.* chlorophyll, chlorophyl.
chlorophyllien, *a.* of or pertaining to chlorophyll.
chloropicrine, *f.* chloropicrin.
chloroplatinate, *m.* chloroplatinate.
— *d'ammoniaque,* ammonium chloroplatinate.
chloroplatinique, *a.* chloroplatinic.
chlorose, *f.* chlorosis.

chlorosel, *m.* chloro salt, double chloride.

chlorosubstitué, *a.* chloro-substituted, chloro.

chlorosulfoné, *a.* containing or combined with chlorine and the sulfonic group, chlorosulfo.

chlorosulfure, *m.* chlorosulfide.

chloroxycarbonique, *a.* In the phrase *éther chloroxycarbonique*, ethyl chloroformate.

chlorurage, *m.* chlorination, etc.(see chlorurer); (*Photog.*) specif., intensification with gold chloride.

chlorurant, *p.a.* chlorinating, etc. (see chlorurer). —*m.* chlorinating agent, chloridizing agent.

chloruration, *f.* chlorination, etc. (see chlorurer).

chlorure, *m.* chloride.
 — *acide,* acid chloride.
 — *calcique,* calcium chloride.
 — *d'acétyle,* acetyl chloride.
 — *d'antimoine,* antimony chloride.
 — *d'argent,* silver chloride.
 — *d'azote,* nitrogen chloride.
 — *de baryum,* barium chloride.
 — *de carbone,* carbon chloride, specif. carbon tetrachloride.
 — *de chaux,* chloride of lime, bleaching powder, chlorinated lime.
 — *de chaux liquide,* (*Pharm.*) solution of chlorinated lime.
 — *décolorant,* a chlorine compound that bleaches, as bleaching powder or a hypochlorite.
 — *de cuivre,* copper chloride, specif. cupric chloride.
 — *de fer,* iron chloride, specif. ferric chloride.
 — *de Julin,* hexachlorobenzene.
 — *de mercure,* mercury chloride.
 — *de mercure (bi),* mercuric chloride.
 — *de mercure (proto),* mercurous chloride.
 — *de méthyle,* methyl chloride, chloromethane.
 — *de phosphore,* phosphorus chloride.
 — *de phosphore (per),* phosphorus pentachloride.
 — *de platine,* platinum chloride, specif. platinic chloride.
 — *de plomb,* lead chloride.
 — *de potasse,* eau de Javel, Javel water.
 — *de potassium,* potassium chloride.
 — *de sodium,* sodium chloride.
 — *de soude,* — *de soude liquide,* solution of chlorinated soda, Labarraque's solution.
 — *de soufre,* sulfur chloride, specif. the monochloride, S_2Cl_2.
 — *d'étain,* tin chloride.
 — *d'étain (proto),* stannous chloride.
 — *d'éthyle,* ethyl chloride, chloroethane.
 — *de zinc liquide,* (*Pharm.*) solution of zinc chloride.
 — *d'or,* gold chloride, specif. auric chloride

chlorure *ferreux,* ferrous chloride.
 — *ferrique,* ferric chloride.
 — *ferrique liquide,* (*Pharm.*) solution of ferric chloride.
 — *mercureux,* mercurous chloride.
 — *mercurique,* mercuric chloride.
 — *stanneux,* stannous chloride

chloruré, *p.a.* chlorinated, chloridized, containing chlorine.

chlorurer, *v.t.* chlorinate, chloridize, chloridate. (*Chlorinate* always implies impregnation or combination with chlorine; *chloridize* and *chloridate,* esp. the latter, may also denote impregnation with a chloride. *Chlorurer* is used in both senses.)

choc, *m.* shock, collision, impact, knocking.
 — *direct,* normal impact.
 — *oblique,* oblique impact.

chocolat, *m.* & *a.* chocolate.
 — *au lait,* milk chocolate.
 — *en poudre,* chocolate powder.

chocolaterie, *f.* chocolate manufacturer; chocolate factory.

chocolatier, *m.* chocolate maker or dealer.

choeur, *m.* chorus; choir.

choir, *v.i.* fall.

choisi, *p.a.* chosen, picked, selected; select, choice. — *m.* choice article; chosen person.

choisir, *v.t.* choose, select, pick.

choix, *m.* choice, selection.
 de —, (with a noun) preferred, best.

cholagogue, *m.* & *a.* (*Med.*) cholagog(ue).

cholémie, *f.* (*Med.*) cholemia, cholæmia.

cholérique, *a.* (*Med.*) cholera, choleraic.

cholestérine, *f.* cholesterol, cholesterin.

cholestérique, *a.* cholesteric.

cholihémie, *f.* (*Med.*) cholihemia.

choline, *f.* choline.

cholurie, *f.* (*Med.*) choluria.

chômage, *m.* stoppage, standing idle, inactivity, dead season, slack time.

chômer, *v.i.* suspend work, stand idle; not to work; be out of work; stand in need.

chondrine, *f.* chondrin.

chondrogène, *a.* chondrigenous. — *m.* chondrigen.

choquer, *v.t.* strike against, collide with; attack; shock.

chorée, *f.* (*Med.*) chorea.

choroïde, *f.* (*Anat.*) choroid.

chose, *f.* thing; matter, affair; property.
 quelque —, something, anything.

chou, *m.* cabbage.
 — *palmiste,* palm cabbage; cabbage palm.
 — *potager,* common cabbage.
 — *vert,* kale.
 choux de Bruxelles, Brussels sprouts.

choucroute, *f.* sauerkraut, pickled cabbage.

chouette, *f.* owl.

chou-fleur, *m.* cauliflower.

chou-rave, *m.* kohl-rabi.

choux, *pl.* of chou.

chrême, *m.* chrism, consecrated oil.

chrétien, *a.* Christian.

christe-marine, *f.* (*Bot.*) samphire, specif. sea samphire (*Crithmum maritimum*) and marsh samphire or glasswort (*Salicornia herbacea*).

chromatable, *a.* (*Dyeing*) chromable.

chromatage, *m.* chromating, (*Dyeing*) chroming.

chromate, *m.* chromate.

— (*bi*) *de potasse,* potassium dichromate, potassium bichromate.

— *de baryte,* barium chromate.

— *de plomb,* lead chromate.

— *de potasse,* potassium chromate.

chromaté, *p.a.* chromated, chromed; chromate of.

chromater, *v.t.* chromate, (*Dyeing*) chrome.

chromatie, *f.* chromatism.

chromatine, *f.* (*Biol.*) chromatin.

chromatique, *a.* chromatic.

chromatiquement, *adv.* chromatically.

chromatiser, *v.t.* render chromatic.

chromatisme, *m.* chromatism.

chromatogène, *a.* chromogenic. — *m.* chromogen.

chrome, *m.* chromium, (in certain phrases) chrome.

chromé, *a.* containing or combined with chromium; (*Leather*) chrome-tanned.

chromicyanure, *m.* chromicyanide.

chromique, *a.* chromic.

chromite, *n.* chromite.

— *de fer,* (*Min.*) chromite, chrome iron ore.

chromocre, *m.* chrome ocher (earthy impure Cr_2O_3).

chromocyanure, *m.* chromocyanide.

chromogène, *a.* chromogenic. — *m.* chromogen.

chromogénique, *a.* chromogenic.

chromophile, *a.* (*Biol.*) chromophilous.

chromophotographie, *f.* chromophotography.

chromosphère, *n.* (*Astron.*) chromosphere.

chromosulfurique, *a.* chromosulfuric.

chromotropique, *a.* chromotropic (acid).

chromyle, *m.* chromyl.

chronique, *a.* chronic. — *f.* chronicle.

chronomètre, *m.* chronometer.

chronométrique, *a.* chronometric(al).

chrysalide, *f.* chrysalis.

chrysamine, *f.* (*Dyes*) chrysamine.

chrysaniline, *f.* chrysaniline.

chrysanthème, *m.* chrysanthemum.

chrysarobine, *f.* chrysarobin.

chrysène, *m.* chrysene.

chrysobéryl, chrysobéril, *m.* chrysoberyl.

chrysocale, chrysochalque, *m.* an imitation gold composed of copper, tin and zinc.

chrysocolle, *f.* (*Min.*) chrysocolla.

chrysoïdine, *f.* (*Dyes*) chrysoidine.

chrysolithe, *f.* (*Min.*) chrysolite.

chrysophanique, *a.* chrysophanic.

chrysoprase, *f.* (*Min.*) chrysoprase.

chu, *p.a.* fallen.

chuchoter, *v.i. & t.* whisper.

chute, *f.* fall; falling; failure; (*Metal.*) rejected end of an ingot.

— *d'eau,* waterfall; head of water.

— *de potentiel,* fall of potential, drop of potential.

— *du bas,* (*Metal.*) rejected lower end.

— *du haut,* (*Metal.*) rejected upper end.

chylaire, *a.* chylous, chylaceous.

chyle, *m.* chyle.

chyleux, *a.* chylous, chylaceous.

chylifier, *v.t.* chylify.

chylurie, *f.* (*Med.*) chyluria.

chyme, *m.* (*Physiol.*) chyme.

chymifier, *v.t.* chymify.

Chypre, *f.* Cyprus.

chypre, *m.* Cyprus wine.

ci, *adv.* here; now. (*Ci* may be attached to a word to limit the meaning of ce, cette, celui, etc.; as, *cette homme-ci,* this man; *celui-ci,* this one, this, the latter. Its opposite is *là. Ci* is also used in accounts to indicate a product; as, *6 cornues à 1 franc, ci 6 francs.*)

de —, *de là,* here and there.

par —, *par là,* here and there; now and then; hither and thither.

ci-après, *adv.* below, hereafter.

cible, *f.* target.

ciboulette, *f.* chives.

cicatrice, *f.* scar, cicatrix.

cicatrisant, *p.a.* cicatrisive, cicatrizant. — *m.* cicatrizant.

cicatrisation, *f.* cicatrization.

cicatriser, *v.t.* cicatrize, heal; cicatrize, scar.

ci-contre, *adv.* opposite.

cicutaire, *f.* = ciguë.

cicutine, *f.* cicutine.

ci-dessous, *adv.* below, underneath.

ci-dessus, *adv.* above.

ci-devant, *adv.* formerly, previously. — *a.* former, late, ex-.

cidre, *m.* cider.

— *doux,* sweet cider, new cider.

cidrerie, *f.* cider factory, cider mill.

Cie, *abbrev.* (Compagnie) Company, Co.

ciel, *m.* sky; heaven; roof (as of a mine); crown, top (as of a furnace or boiler); canopy.

à — *couvert,* under cover.

à — *ouvert,* under the open sky, without cover, open.

ciel *couvert,* clouded sky, cloudy weather.

cierge, *m.* candle, wax candle (esp. for ceremonial use); (*Bot.*) cereus; (*Bot.*) mullein.

— *à grandes fleurs,* night-blooming cereus.

cierger, *v.t.* wax.

ciergier, *m.* one who makes or sells wax candles.

cieux, *pl.* of ciel.

cigare, *m.* cigar.

cigogne, *f.* stork; crank, crank lever.

ciguë, *f.* (*Bot.*) hemlock.

— *officinale,* — *ordinaire, Conium maculatum.*

— *vireuse,* water hemlock (*Cicuta virosa*).

ci-joint, *a.* given herewith, annexed.

cil, *m.* eyelash; (*Biol., pl.*) cilia.

ciliaire, *a.* ciliary.

ciller, *v.i.* wink, blink.

cime, *f.* summit, top; (*Bot.*) cyme.

ciment, *m.* cement.

— *à prise lente,* slow-setting cement.

— *à prise rapide,* quick-setting cement.

— *armé,* reinforced cement.

— *artificiel,* artificial cement.

— *de Boulogne,* Roman cement.

— *de fer,* iron-rust cement (Fe, S, NH₄Cl).

— *de grappiers,* see grappiers.

— *de laitier,* — *de scories,* slag cement.

— *fondu,* alumina cement.

— *hydraulique,* hydraulic cement.

— *métallique,* metallic cement.

— *naturel,* natural cement.

— *Portland,* Portland cement.

— *pouzzolane,* pozzuolanic cement, pozzuolana.

— *prompt,* quick-setting cement.

— *romain,* Roman cement.

cimentaire, *a.* pertaining to cement, cement.

cimentation, *f.* cementation.

cimenter, *v.t.* cement; (*Glass*) varnish.

cimentier, *m.* cement maker or user.

cimetière, *m.* cemetery.

cimolée, *f.* cimolite; pipeclay; the ooze from a grindstone.

cinabarin, *a.* vermilion.

cinabre, *m.* cinnabar; (as a color) vermilion.

— *d'antimoine,* antimony cinnabar, antimony vermilion.

— *vert,* green cinnabar (green pigment containing chrome yellow and Prussian blue).

cinchona, *m.* cinchona.

cinchonamine, *f.* cinchonamine.

cinchonicine, *f.* cinchonicine.

cinchonidine, *f.* cinchonidine.

cinchonine, *f.* cinchonine.

cinchoninique, *a.* cinchoninic (4-quinoline-carboxylic).

cinchonique, *a.* cinchonic.

acide —, cinchonic acid (tetrahydro-6-keto-1, 2-pyran-3,4-dicarboxylic acid); less properly, cinchoninic acid.

cinématique, *a.* kinematic(al). — *f.* kinematics.

cinématiquement, *adv.* kinematically.

cinène, *m.* cinene (dipentene, inactive limonene).

cinéol, *m.* cineole, cineol.

cinération, *f.* cineration, incineration.

cinériforme, *a.* in the form of or resembling ashes.

cinérite, *f.* a fine-grained tufa.

cinétique, *a.* kinetic. — *f.* kinetics.

— *chimique,* chemical kinetics.

cinglage, *m.* (*Metal.*) shingling; sailing, course.

cinglard, *m.* shingling hammer.

cingler, *v.t.* (*Metal.*) shingle; lash, switch.

cingleresse, *f.* shingling tongs.

cingleur, *m.* shingler, squeezer. — *a.* shingling.

— *rotatif,* shingling rolls.

cingleuse, *f.* (*Metal.*) shingler, squeezer.

cinnabre, *m.* = cinabre.

cinname, *m.* cinnamon tree (*Cinnamomum* species); cinnamon.

cinnaméine, *f.* cinnameïn.

cinnamène, *m.* cinnamene (styrene).

cinnamique, *a.* cinnamic.

cinnamome, cinnamone, *m.* = cinname.

cinnamyle, *m.* cinnamyl.

cinq, *a. & m.* five; fifth.

cinquantaine, *f.* fifty (as a round number); fiftieth year or anniversary.

cinquante, *a.* fifty; fiftieth. — *m.* fifty.

cinquantième, *a. & m.* fiftieth.

cinquième, *a. & m.* fifth.

cintrage, *m.* bending, etc. (see cintrer); specif., plate bending.

cintre, *m.* bend, curve, bow, arch.

cintré, *p.a.* bent, etc. (see cintrer).

cintrement, *m.* bending, etc. (see cintrer).

cintrer, *v.t.* bend, curve, bow, arch, camber.

cipolin, *m.* cipolin, cipolino (marble).

cirage, *m.* waxing, etc. (see cirer); blacking (for shoes); polishing (of leather); polish (for leather); varnish.

circonférence, *f.* circumference.

circonflexe, *a.* circumflex; crooked, twisted.

circonlocution, *f.* circumlocution.

circonscription, *f.* circumscription; subdivision, division (limited area).

circonscrire, *v.t.* circumscribe.

circonstance, *f.* circumstance.

circuit, *m.* circuit.

— *amortisseur,* (*Elec.*) damping circuit.

— *d'alimentation,* (*Elec.*) supply circuit.

— *de charge,* (*Elec.*) charging circuit.

— *de décharge,* (*Elec.*) discharging circuit.

— *dérivé,* (*Elec.*) derived circuit, shunt circuit

— *primaire,* (*Elec.*) primary circuit.

— *secondaire,* (*Elec.*) secondary circuit.

hors —, (*Elec.*) disconnected.

circulaire, *a.* circular. — *f.* circular; circle.

circulairement, *adv.* circularly.

circulant, *p.a.* circulating, in circulation.

circularité, *f.* circularity.

circulation, *f.* circulation; (*Com.*) traffic.

circulatoire, *a.* circulatory.

circuler, *v.i.* circulate; move round.

　faire —, circulate; cause to move about.

cire, *f.* wax; wax candle.

　— *à cacheter,* sealing wax.

　— *blanche,* white wax.

　— *d'abeille(s),* beeswax.

　— *de Chine,* Chinese wax.

　— *de cirier,* myrtle wax.

　— *de myrica,* myrtle wax.

　— *de palmier,* palm wax.

　— *des doreurs,* gilder's wax.

　— *d'Espagne,* Spanish wax, sealing wax.

　— *du Japon,* Japan wax.

　— *fossile,* fossil wax, ceresin.

　— *jaune,* yellow wax.

　— *minérale,* mineral wax, ozocerite.

　— *vierge,* virgin wax.

ciré, *p.a.* waxed, etc. (see cirer).

cirer, *v.t.* wax; black (shoes); polish (leather); varnish; waterproof.

cireux, *a.* waxy.

cirier, *m.* wax myrtle (*Myrica* sp., esp. *M. cerifera*); wax chandler. — *a.* wax-producing.

ciron, *m.* (*Zoöl.*) mite.

cirrhose, *f.* (*Med.*) cirrhosis.

cirure, *m.* waxing, coating of wax.

cisaillage, *m.* shearing, cutting.

cisaille, *f.* clippings, parings (from coins); (*pl.*) shears (for cutting metal, trimming hedges or the like), nippers (for wire), cutter (for trimming paper).

cisaillement, *m.* shearing; shearing stress; cutting.

cisailler, *v.t.* shear, cut (with shears).

ciseau, *m.* chisel; (*pl.,* **ciseaux**), scissors, (small) shears.

　— *à bois,* wood chisel.

　— *à chaud,* hot chisel.

　— *à froid,* cold chisel.

　— *de calfat,* calking iron.

ciseler, *v.t.* chisel, cut; chase, engrave.

ciselure, *f.* chasing, engraving; chased work.

cisoires, *f.pl.* bench shears.

cissoïde, *f.* (*Math.*) cissoid.

citation, *f.* citation.

cité, *f.* city. — *p.a.* cited.

citer, *v.t.* cite.

citerne, *f.* cistern; water boat.

citerneau, *m.* small cistern.

citoyen, *m.,* **citoyenne,** *f.* citizen.

citraconique, *a.* citraconic.

citragon, *m.* garden balm (*Melissa officinalis*).

citrate, *m.* citrate.

　— *d'ammoniaque,* ammonium citrate.

citrate *d'ammoniaque liquide,* (*Pharm.*) solution of ammonium citrate.

　— *de lithine,* lithium citrate.

　— *de potasse,* potassium citrate.

　— *de sesquioxyde de fer,* ferric citrate.

　— *de soude,* sodium citrate.

citrène, *f.* citrene, *d*-limonene.

citrin, *a.* citrine.

citrine, *f.* lemon oil.

citrique, *a.* citric.

citro-magnésien, *a.* In the phrase *liqueur citro-magnésienne,* an ammoniacal magnesium citrate solution used to precipitate phosphoric acid.

citron, *m.* lemon; any fruit from the genus *Citrus medica* (including lemon, lime and citron). — *a.* lemon-colored, lemon-yellow.

citronnade, *f.* lemonade.

citronné, *a.* like, or smelling of, lemons; flavored with lemon.

citronnellal, *m.* citronellal.

citronnelle, *f.* (*Bot.*) citronella, esp. citronella grass (*Andropogon nardus*); a liqueur flavored with lemon peel.

citronnellique, *a.* citronellic.

citronnellol, *m.* citronellol.

citronnier, *m.* lemon tree (*Citrus medica limon*); any tree of the species *C. medica* (including lemon, lime and citron trees); any tree of the genus *Citrus* (including trees bearing citrous fruits of all kinds).

citrouille, *f.* pumpkin.

civette, *f.* civet (the perfume); civet, civet cat.

civière, *f.* (hand) barrow.

civil, *a.* civil; civilian. — *m.* civilian.

civilisateur, *a.* civilizing.

civilisation, *f.* civilization.

civiliser, *v.t.* civilize.

claie, *f.* screen, sieve; hurdle.

claim, clain, *m.* auriferous ground.

clair, *a.* (of fire, etc.) bright; (of colors, rooms, etc.) light; clear; thin; (of cloth) flimsy; (of money) available. — *m.* light; (of cloth) thin part. — *adv.* clearly, clear.

　— *semé,* thinly sown, scattered.

claircage, *m.* clearing, purging (esp. the freeing of sugar crystals from mother liquor by water, sirup, centrifugalizing or other means).

claircé, *m.* (*Sugar*) clearing sirup, claircé.

claircer, claircir, *v.t.* clear, purge (see claircage).

claire, *f.* bone ash; boiler for sugar refining.

claire-étoffe, *f.* an alloy of tin and lead.

clairement, *adv.* clearly, plainly.

claire-soudure, *f.* = claire-étoffe.

clairet, *a.* rather light (in color); rather thin.

　— *m.* red wine of light color.

clairette, *f.* a white sparkling wine of the Midi (also, the variety of grape from which this wine is made); a disease of silk worms.

claire-voie, *f.* openwork, lattice, grating; skylight.

à —, in openwork, latticed; (of fabrics) open, loose-woven.

clairsemé, *a.* thinly sown, scattered, scarce.

clairvoyant, *a.* clear-sighted.

clamauder, *v.t.* clamp.

clameau, *m.* clamp, cramp, dog; vise.

clapet, *m.* clack valve, clack, flap valve.

— à couronne, cup valve.

clapoter, *v.i.* ripple.

claquer, *v.t. & i.* clap, smack, crack.

clarifiant, *p.a.* clarifying. *— m.* clarifying agent, clarifier, fining.

clarification, *f.* clarification

clarifier, *v.t. & r.* clarify.

clarifleur, *m.* clarifier.

clarté, *f.* light; clearness, specif. transparency, limpidity.

— du jour, daylight.

— du soleil, sunlight.

classe, *f.* class.

classement, *m.* classification, classing.

classer, *v.t.* classify, class.

classeur, *m.* classifier (as for ore); letter file.

classification, *f.* classification.

classifier, *v.t.* classify.

classique, *a.* classic, classical.

clastique, *a.* clastic.

clausthalite, *f.* (*Min.*) clausthalite.

clavalier, *m.* (*Bot.*) prickly ash (*Zanthoxylum*, esp. *Z. americanum*); (*Pharm.*) xanthoxylum.

clavecin, *m.* harpsichord.

fil de —, piano wire.

clavel, *m.* an inferior kind of soda.

clavelée, *f.* sheep pox.

clavetage, *m.* keying.

claveter, *v.t.* key, fasten with keys or cotters.

clavettage, *m.* = clavetage.

clavette, *f.* key, pin, peg, cotter.

— de serrage, tightening key.

clavetter, *v.t.* = claveter.

clayon, *m.* small screen, sieve or grating; small hurdle; wattle.

clef, clé, *f.* key (in various senses); key, plug (of a stopcock); wrench; hook (of a chain); keystone; knot, hitch; (*Glass*) a piece of earthenware fitting over the cracked side of a pot and serving to hold it together.

— à béquille, spanner.

— à écrous, screw wrench, nut wrench, spanner.

— à molette, a wrench with jaws adjusted by turning a small wheel.

— anglaise, monkey wrench

clef *d'appel,* call button.

— de calage, adjusting key, tightening key.

— de tirage, damper.

— d'interruption, (*Elec.*) make-and-break key.

— ouverte, spanner.

clématite, *f.* (*Bot.*) clematis.

clerc, *m.* clergyman; scholar, savant; clerk.

clichage, *m.* stereotyping, stereotypy.

cliché, *m.* engraved plate or block (for printing), cut, engraving; stereotype plate, cliché; (*Photog.*) negative. *— a.* stereotyped.

— à demi-teintes, half-tone cut, half-tone.

— de projection, lantern slide.

— négatif, negative.

clicher, *v.t.* stereotype.

client, *m.* customer; patient; client.

clientèle, *f.* patronage, custom, customers; clientèle.

climat, *m.* climate.

climatique, *a.* climatic.

clin d'œil. twinkling of an eye.

clinique, *a.* clinic, clinical. *— f.* clinic.

clinomètre, *m.* clinometer.

clinorhombique, *a.* clinorhombic (monoclinic).

clinquant, *m.* tinsel, imitation gold or silver foil. *— a.* tinseled, glittering.

cliquet, *m.* click, catch, detent, pawl, ratchet.

cliqueter, *v.i.* click, clank.

cliquetis, *m.* clank, clanking, jingle, jingling.

clisse, *f.* screen, mat (as for draining cheese); wicker cover for a bottle; splint.

clissé, *p.a.* wickered; wicker.

clisser, *v.t.* wicker, cover with wicker; (*Med.*) splint.

clivable, *a.* cleavable.

clivage, *m.* cleavage.

cliver, *v.t.* cleave.

cloche, *f.* bell jar, bell glass; bell (in various senses); cover (for a dish); blister; (*Ceram.*) bubble, bleb; (*Biol.*) a covered glass dish for plate cultures, serum, etc.

— à cultures, culture dish (with cover).

— à douille, open-top bell jar, bell jar having a mouth or tubulation at the top.

— à vide, vacuum bell jar (with lower edge ground to fit air-tight).

clocher, *v.t.* cover with a bell jar.

clochette, *f.* little bell jar; little bell; bell flower.

cloison, *f.* partition, septum.

cloisonnage, *m.* partition work.

cloisonné, *p.a.* divided into compartments; cellular; cloisonné.

cloisonnement, *m.* partitioning; partition.

cloisonner, *v.t.* partition, divide into compartments.

clonique, *a.* (*Med.*) clonic.

clore, *v.t.* close; inclose. *— v.i.* close, shut.

clos, *p.a.* closed, inclosed, shut, ended. — *m.* inclosure, close; specif., vineyard.

— *de vigne,* vineyard.

close, *a. fem.* of clos. — *pr. 1 & 3 sing. subj.* of clore.

closeau, closerie, *m.* small inclosure; small farm.

clôt, *pr. 3 sing.* of clore.

clôture, *f.* close, closing, conclusion; inclosure (something that incloses), fence, wall, etc.; seclusion.

clôturer, *v.t.* close.

clou, *m.* nail; spike, stud, rivet, (shoe) peg.
— *à river,* rivet, riveting nail.
— *aromatique,* clove.
— *à tête perdue,* brad.
— *de bouche,* tack.
— *de girofle,* clove.
— *d'épingle,* brad; wire nail.
— *de soufflet,* tack.
— *en fonte,* cast nail.
— *mécanique,* machine-made nail.
— *zingué,* galvanized nail.

clouage, clouement, *m.* nailing.

clou-épingle, *n.* brad.

clouer, *v.t.* nail; spike, rivet, pin.

clouterie, *f.* nail factory; nail trade.

clupéine, *f.* clupeine.

cm, *abbrev.* (centimètre) centimeter.

cm^c, cm³, *abbrev.* (centimètre cube) cubic centimeter, cu. cm., cc.

cm^q, cm², *abbrev.* (centimètre carré) square centimeter, sq. cm.

coagglutinine, *f.* coagglutinin.

coagulabilité, *f.* coagulability.

coagulable, *a.* coagulable.

coagulant, *p.a.* coagulating.

coagulase, *f.* coagulase.

coagulateur, *a.* coagulatory, coagulating.
— *m.* coagulator, coagulant.

coagulation, *f.* coagulation.
— *sanguine,* coagulation of the blood.

coaguler, *v.t. & r.* coagulate.

coagulum, *m.* coagulum; coagulant.

coaille, *f.* inferior wool (from the tail).

coalescence, *f.* coalescence.

coaliser, *v.t. & r.* unite, combine, league.

coalition, *f.* coalition, combination.

coaltar, *m.* coal tar.

coaltarement, *m.* treating with coal tar, tarring.

coaltarer, *v.t.* treat or impregnate with coal tar.

coaltarisation, *f.* coal tar treatment, tarring.

coaltariser, *v.t.* treat (esp. coat) with coal tar, tar.

coalté, *a.* designating a coal tar preparation used as a disinfectant.

coassocié, *m.* copartner, associate, partner.

coauteur, *m.* co-author, joint author.

cobalt, *m.* cobalt.
— *gris,* cobalt glance, cobaltite.
— *d'outremer,* cobalt ultramarine, cobalt blue.

cobaltage, *m.* coating or plating with cobalt.

cobaltammine, cobaltamine, *f.* cobaltammine, cobaltamine.

cobalteux, *a.* cobaltous.

cobaltico-potassique, *a.* of (trivalent) cobalt and potassium; as, *nitrite cobaltico-potassique,* tripotassium cobaltic nitrite.

cobaltifère, *a.* cobaltiferous.

cobaltine, *f.* (*Min.*) cobaltite, cobaltine.

cobaltique, *a.* cobaltic.

cobaltisage, *m.* = cobaltage.

cobaltiser, *v.t.* cover or plate with cobalt.

cobaye, *m.* guinea pig, cavy.

coboldine, *f.* (*Min.*) linnæite.

cobolt, *m.* powdered metallic arsenic.

coca, *f.* coca.

cocaïne, *f.* cocaine.

cocaïniser, *v.t.* cocainize.

cocalon, *m.* a silkworm cocoon of inferior value.

coche, *f.* notch, nick, cut, slit, groove; sow.

cochenillage, *m.* cochineal dyeing or dye bath.

cochenille, *f.* cochineal; cochineal insect.
— *a.* cochineal.

cocheniller, *v.t.* dye with cochineal.

cochenillier, *m.* cochineal fig, cochineal plant.

cochon, *m.* swine, hog, pig; pork; dross; (*Metal.*) sow, pig.

coco, *m.* coconut; coco, coco palm; coconut shell; licorice water.
noix de —, coconut.

cocon, *m.* cocoon.

coconnage, *m.* formation of cocoons.

cocose, *n.* a butter substitute from coconut oil.

cocotier, *m.* coco, coco palm (*Cocos nucifera*).

coction, *f.* boiling, coction; cooking; (*Physiol.*) digestion.

code, *m.* code; codex, pharmacopeia.

codéine, *f.* codeine.

codex, *m.* codex, pharmacopeia (specif. the French pharmacopeia).

codifier, *v.t.* codify.

coefficient, *m.* coefficient.
— *d'abaissement,* coefficient of (freezing-point) lowering.
— *d'adhérence,* coefficient of adhesion.
— *de dilatation,* expansion coefficient.
— *de sûreté,* — *de sécurité,* coefficient (or factor) of safety.
— *économique,* efficiency.

coenzyme, *f.* coenzyme.

coercible, *a.* coercible, (of gases) compressible.

cœsium, *m.* cesium, cæsium.

cœur, *m.* heart; interior, inner part, middle part, etc.; (in distillation) middle fraction; heartwood; stomach (as in *mal au cœur*).
en —, heart-shaped.

cœurce, *m.* currying knife; planishing knife.

coferment, *m.* coferment (coenzyme).

cofféine, *f.* caffeine.

coffre, *m.* chest; box; trunk; coffer; coffer-dam.

coffre-fort, *m.* strong box, safe.

coffret, *m.* (small) chest, (small) box; muffle.

cognac, *m.* cognac, Cognac brandy.

cognassier, *m.* quince, quince tree (*Cydonia*).

cognée, *f.* ax.

cogner, *v.t.* knock, strike, beat, drive.

cohérence, *f.* coherence, coherency.

cohérent, *a.* coherent.

cohérer, *v.i.* cohere.

cohésion, *f.* cohesion.

cohobateur, *a.* cohobating.

cohobation, *f.* cohobation.

cohober, *v.t.* cohobate.

coiffage, *m.* cap, covering; specif., fuse cap.

coiffe, *f.* cap, cover, hood; mesentery (of slaughtered animals); headdress; lining (of hats).
— *de fusée,* fuse cap.

coiffer, *v.t.* cap, cover the head or top of; dress the hair of.

coin, *m.* corner; wedge; stamp, die; stamp, mark; quoin; corner cupboard.

coinçage, *m.* wedging, fastening with wedges.

coincement, *m.* wedging, state of being wedged, (of machine parts) jamming.

coincer, *v.t.* wedge; jam; drive in (wedges); corner. — *v.r.* (of machine parts) jam.

coïncidence, *f.* coincidence.

coïncider, *v.i.* coincide.

coing, *m.* quince.

coir, *m.* coir (coconut husk fiber).

coke, *m.* coke.
— *de pétrole,* petroleum coke, oil coke.

cokéfaction, cokéification, *f.* coking.

cokéifier, *v.t.* coke.

cokerie, *f.* coking plant.

coketier, *m.* one that makes or sells coke.

cokeur, *m.* coke burner, coke-oven stoker.

col, *m.* neck; collar; (mountain) pass.
à — court, short-necked.
— *de cygne,* swan neck (specif. a curved tube or pipe); gooseneck.
— *droit,* = flacon à large ouverture, under *flacon;* straight neck.

cola, *m.* (*Pharm.*) kola; (*Bot.*) Cola.

colature, *f.* filtration, straining; filtrate, strained liquid. (*Colature* in English is obs. or rare.)

colchicacée, *f.* (*Bot.*) a plant belonging to *Colchicum* or a related genus.

colchicéine, *f.* colchiceine.

colchicine, *f.* colchicine.

colchique, *m.* (*Bot.*) meadow saffron (*Colchicum*).

colcotar, *m.* colcothar.

coléoptère, *m.* (*Zoöl.*) coleopteran; (*pl.*) Coleoptera.

colère, *f.* anger.

colibacille, *m.* (*Bact.*) colon bacillus.

colifichet, *m.* trinket, knicknack; bird cake; (*Ceram.*) small tripod.

colique, *f. & a.* colic.

colis, *m.* package, parcel, bale, case.
— *postal,* parcel (sent by post).
par — postaux, by parcel post.

collaborateur, *m.* collaborator.

collaboration, *f.* collaboration.

collaboratrice, *f.* collaborator.

collaborer, *v.i.* collaborate, work or act jointly.

collage, *m.* gluing, etc. (see coller).

collagène, *a.* collagenous, collagenic. — *m.* collagen.

collapsus, *m.* (*Med.*) collapse.

collationner, *v.t.* compare.

colle, *f.* glue; size, sizing; (adhesive) paste; (adhesive) cement; fining.
— *à bouche,* mouth glue, lip glue.
— *d'amidon,* starch paste.
— *de caoutchouc,* rubber cement.
— *de Flandre purifiée,* gelatin.
— *de pâte,* paste, flour paste.
— *de peau,* skin glue, hide glue.
— *de poisson,* fish glue; isinglass.
— *d'os,* bone glue.
— *du Japon,* agar, Japanese agar.
— *forte,* glue.
— *liquide,* liquid glue.
— *végétale,* vegetable glue, vegetable size.

collecteur, *m.* collector. — *a.* collecting; (of a lens) condensing.

collectif, *a. & m.* collective.

collection, *f.* collection.

collectionner, *v.t.* collect, make a collection of.

collectivement, *adv.* collectively.

collectivité, *m.* collectivity.

collectrice, *a. fem.* of collecteur.

collège, *m.* college.

collégien, *m.* collegian. — *a.* collegiate, college.

collègue, *m.* colleague.

colle-matière, *n.* (also *pl.*) glue stock.

coller, *v.t.* glue, cement, paste, stick; size; fine, clarify (as wine); (*Ceram.*) fasten with slip. — *v.r.* adhere, stick, stick together, cake; be glued, sized, fined, etc. — *v.i.* stick, adhere, cling.

collerette, *f.* flange, rim, collar, ring; (*Bot.*) involucre.

colles-matières, *n.pl.* See colle-matière.

collet, *m.* collar; (*Tech.*) neck, throat, shoulder, shank, collar, collet, etc.; (of a sugar beet or other plant) collar.
— *de jonction,* flange, collar, shoulder.

collette, *f.* a vessel for fining beer, also its contents.

colleur, *m.* gluer, etc. (see coller).

colleuse, *f.* sizing machine, sizer.

collidine, *f.* collidine.

collier, *m.* collar; hoop, ring, belt; necklace.

colliger, *v.t.* collect, gather.

collimateur, *m.* collimator. — *a.* collimating.

collimation, *f.* collimation.

colline, *f.* hill.

collision, *f.* collision.

collodié, *a.* collodionized.

collodion, *m.* collodion.
— *cantharidé,* (*Pharm.*) cantharidal collodion, blistering collodion.
— *élastique,* (*Pharm.*) flexible collodion.
— *styptique,* (*Pharm.*) styptic collodion.
— *vésicant,* (*Pharm.*) blistering collodion, cantharidal collodion.

collodionnage, *m.* collodion varnishing.

collodionné, *p.a.* collodionized.

collodionner, *v.t.* collodionize.

colloïdal (*m.pl.* **colloïdaux**)**,** *a.* colloidal, colloid. — *m.* colloidal substance or preparation.

colloïde, *m.* colloid. — *a.* colloid, colloidal.

collotypie, *f.* collotypy, collotype.

colloxyline, *f.* colloxylin (soluble guncotton).

collutoire, *m.* (*Med.*) collutory, collutorium.

collyre, *m.* (*Med.*) collyrium.

colmater, *v.t.* (*Agric.*) warp.

colocynthine, *f.* colocynthin.

Colombie, *f.* Colombia.

colombier, *m.* dovecote; colombier, columbier (size of paper, about 23 x 34 in.).

colombin, *m.* a long cylinder of ceramic paste used for measuring contraction due to firing or for luting saggers together; (*Metal.*) a kind of lead ore; (*Building*) a mixture of plaster Paris and baryta.

colombine, *f.* dung of domestic fowls.

colombite, *f.* (*Min.*) columbite.

colombium, *m.* columbium.

colombo, *m.* (*Pharm.*) calumba, colombo.
— *d'Amérique,* — *de Mariette,* American columbo, American gentian (*Frasera carolinensis*).

colomine, *f.* (*Ceram.*) talcose clay.

colomnaire, *a.* columnar.

colon, *m.* planter, farmer; colonist, settler; (*Anat.*) colon.

colonial (*m.pl.* **coloniaux**)**,** *a.* colonial.

colonie, *f.* colony.

coloniser, *v.t.* colonize.

colonne, *f.* column; pillar, post; (of a pump) barrel.
— *à rectifier,* rectifying column, fractionating column.
— *à robinet,* upright pipe with stopcock or tap.

colophane, *f.* colophony, rosin.

colophaner, *v.t.* treat with colophony or rosin.

coloquinte, *f.* colocynth (*Citrullus colocynthis*).

colorant, *a.* coloring; specif., dyeing. — *m.* colorant, coloring matter; specif., dye, dyestuff.
— *acridique,* acridine dye.
— *azoaminé,* aminoazo dye.
— *azoïque,* azo dye.
— *azophénolique,* hydroxyazo dye.
— *azoxique,* azoxy dye.
— *basique,* basic dye.
— *bisazoïque,* disazo dye.
— *nitrosé,* nitroso dye.
— *oxazinique,* oxazine dye.
— *thiazinique,* thiazine dye.
— *vital,* (*Biol.*) vital stain.

coloration, *f.* coloration, coloring.

coloré, *p.a.* colored; (of wine) deep-colored; (of tints) bright.

colorement, *m.* coloring, coloration.

colorer, *v.t. & r.* color.

coloriage, *m.* coloring.

colorier, *v.t.* color.

colorieur, *m.* colorer, colorist; (*Calico*) printing roller, print roll.

colorifique, *a.* colorific.

colorigène, *a.* colorigenic, color-producing.

colorimètre, *m.* colorimeter.

colorimétrie, *f.* colorimetry.

colorimétrique, *a.* colorimetric.

colorimétriquement, *adv.* colorimetrically.

coloris, *m.* coloring; hue, tint.

colorisation, *f.* coloration.

coloriste, *m.* colorer; (*Art.*) colorist.

colossal, *a.* colossal.

colostrum, *m.* colostrum.

coltar, colthar, coltarement, etc. See coaltar, coaltarement, etc.

columbium, *m.* columbium.

colza, *m.* colza, specif. rape or rapeseed.
(Distinctions between "huile de colza," "huile de navette" and "huile de rabette" are apparently rather confused. All may be translated "rape oil" or "colza oil," terms which are now practically synonymous in English except as "colza oil" is used of the purer varieties.)

comaniaue, *a.* comanic.

comateux, *a.* (*Med.*) comatose.

combat, *m.* combat, contest, strife. — *pr. 3 sing.* of combattre.

combattre, *v.t.* fight; combat; oppose. — *v.i.* fight, combat, contend, strive.

combien, *adv.* how, how much, how many, how far, how long. — *m.* price.
— *de temps,* how long.

combinabilité, *f.* combinableness.

combinable, *a.* combinable.

combinaison, *f.* combination; compound; scheme, contrivance.
— *binaire,* binary compound.

combinaison *sulfurée,* compound with sulfur, sulfur compound.

— *tertiaire,* tertiary compound.

combinateur, *m.* combiner; *(Elec.)* controller.

— *a.* combining.

combinatoire, *a.* combinatory, combinative.

combiné, *p.a.* combined; concerted. — *m.* compound; set (as of apparatus).

combiner, *v.t.* combine; concert, contrive. — *v.r.* combine, unite.

combineur, *m.* an apparatus for continuous distillation.

comble, *m.* roof; height, summit; overmeasure. — *a.* heaped; quite full.

comblé, *p.a.* filled, etc. (see combler).

comblement, *m.* filling, etc. (see combler).

combler, *v.t.* fill, fill up; supply; fulfill.

comburable, *a.* combustible.

comburant, *a.* supporting combustion. — *m.* supporter of combustion.

comburé, *p.a.* burned; *(Min.)* consisting of oxide or halide.

comburer, *v.t.* burn.

combustibilité, *f.* combustibility.

combustible, *a.* combustible. — *m.* fuel; combustible.

— *aggloméré,* briquetted fuel.

— *ancien,* coal.

— *d'éclairage,* combustible used as an illuminant.

— *gazeux,* gaseous fuel.

— *minéral,* mineral fuel, specif. coal.

— *moderne,* wood and peat.

combustion, *f.* combustion.

— *au libre,* combustion in free air.

— *blanche,* bright combustion (as distinguished from blue flame).

— *bleue,* blue-flame combustion.

— *de la fumée,* smoke consumption.

— *lente,* slow combustion.

— *spontanée,* spontaneous combustion.

— *vive,* quick (or rapid) combustion.

coménique, *a.* comenic.

comestible, *a. & m.* comestible, edible.

comète, *f.* comet.

comique, *a.* comic, comical.

comité, *m.* committee; commission, board, etc.

commandant, *m.* commanding officer, commander.

commande, *f.* order (for goods); driving gear, drive; control, control mechanism; lashing (of ropes).

à — *directe,* direct-acting.

— *desmodromique,* belt drive.

de —, ordered; feigned, sham.

faire une —, give an order.

commandement, *m.* command; commandment.

commander, *v.t.* command, order; *(Com.)* order; *(Mach.)* drive, control, operate. — *v.i.* command, have command. — *v.r.* *(Mach.)* gear together, work together.

commanditaire, *m.* dormant partner. See commandite.

commandite, *f.* commandite, a firm having one or more members (*commanditaires*) who are responsible only for the amount which they have invested, while the rest (*gérants, commandités*) are fully responsible. Called also société en commandite.

commandité, *m.* See commandite.

comme, *adv.* as, as . . . as, like, as if, as much as; as to; how. — *conj.* as.

commemorer, *v.t.* remember, recall to mind.

commencement, *m.* beginning, commencement.

au —, at (or in) the beginning, at first.

commencer, *v.t. & i.* begin, commence.

commensurabilité, *f.* commensurability.

comment, *adv.* how; why. — *m.* how; why, wherefore. — *interj.* what!

commentaire, *m.* commentary; comment.

commenter, *v.t.* comment on. — *v.i.* comment; comment unfavorably.

commerçant, *a.* commercial, mercantile. — *m.* merchant, trader, dealer.

commerce, *m.* commerce; trade, business; tradespeople.

— *de détail,* retail trade.

— *en gros,* wholesale trade.

— *extérieur,* foreign commerce, foreign trade.

— *intérieur,* home (or domestic) trade, inland commerce.

faire le — *de,* trade in, deal in; be in the business of.

commercer, *v.i.* trade, deal, do business; have relations (with).

commercial, *a.* commercial.

commercialement, *adv.* commercially.

commettre, *v.t.* commit; appoint; lay (ropes).

comminuer, *v.t.* comminute.

comminution, *f.* comminution.

commis, *p.a.* committed; appointed; (of ropes) laid. — *m.* clerk.

— *voyageur,* commercial traveler.

commissaire, *m.* commissioner, commissary.

commission, *f.* commission; board; committee; errand; *(Com.)* commission business.

commissionnaire, *m.* commission merchant, agent; errand boy, errand man, porter.

commissionner, *v.t.* commission.

commode, *a.* convenient, commodious, agreeable, comfortable; (of persons) accommodating. — *m.* commode.

commodément, *adv.* conveniently, comfortably.

commodité, *f.* convenience; (*pl.*) water closet.

commotion, *f.* shock; commotion; (*Med.*) concussion.

commun, *a.* common. — *m.* common; generality; commonalty, common people.

communauté, *f.* community.

commune, *f.* commune; town hall; common people.

communément, *adv.* commonly, generally.

communicabilité, *f.* communicability.

communicant, *p.a.* communicating.

communicatif, *a.* communicative; (of inks) copying.

communication, *f.* communication.

communiqué, *p.a.* communicated. — *m.* communication, communiqué.

communiquer, *v.t.* communicate, impart. — *v.i.* communicate; confer. — *v.r.* communicate; be communicated; be communicative.

commutateur, *m.* (electric) switch; commutator.

— *à balais,* brush commutator.

— *à bouchon,* plug switch, pin switch.

— *à bouton,* button switch.

— *à manette,* lever switch, knife switch.

— *disjoncteur,* cut-out, interrupter.

— *inverseur,* current reverser.

— *permutateur,* universal switch.

— *unipolaire,* single-pole switch.

compacité, *f.* compactness.

compact, *a.* compact.

compagnie, *f.* company.

compagnon, *m.* companion; workman.

compagnonnage, *m.* trade union.

comparabilité, *f.* comparability.

comparable, *a.* comparable.

comparablement, *adv.* comparably.

comparaison, *f.* comparison.

à (or *en*) — *de,* in comparison with.

par —, by comparison, comparatively.

sans —, beyond comparison; without making any comparison.

comparaître, *v.i.* appear.

comparateur, *a.* comparing. — *m.* comparator.

comparatif, *a.* comparative.

comparé, *p.a.* compared; comparative.

comparer, *v.t.* compare. — *v.r.* be compared; compare; compare oneself.

compartiment, *m.* compartment; (*fig.*) department, province.

compartimentage, *m.* division into compartments.

compartimenter, *v.t.* divide into compartments.

comparu, *p.p.* of comparaître.

compas, *m.* compass, compasses (the instrument).

— *à coulisse,* slide calipers.

— *à diviser,* dividers.

compas *à pointes sèches,* dividers.

— *de calibre,* calipers.

— *d'épaisseur,* calipers.

— *de route,* ordinary mariner's compass.

— *électrique,* galvanometer.

compassé, *p.a.* measured with (or as with) compasses; formal, precise.

compasser, *v.t.* measure with compasses; proportion; (*fig.*) measure, weigh.

compassier, *m.* compass maker, instrument maker.

compatibilité, *f.* compatibility.

compatible, *a.* compatible.

compatir, *v.i.* be compatible, agree; sympathize.

compendieusement, *adv.* compendiously, briefly.

compendieux, *a.* compendious, abridged.

compendium, *m.* compendium, compend.

compensateur, *m.* compensator. — *a.* compensating, compensation, compensative.

compensatif, *a.* compensating.

compensation, *f.* compensation.

compensatoire, *a.* compensatory.

compensatrice, *fem.* of compensateur.

compenser, *v.t.* compensate; offset. — *v.r.* compensate each other, compensate.

compétemment, *adv.* competently.

compétence, *f.* competence; province, sphere.

compétent, *a.* competent; due; proper.

compétiteur, *m.* competitor, rival. — *a.* competing.

compilateur, *m.* compiler.

compilation, *f.* compilation.

compiler, *v.t.* compile.

complaire, *v.i.* give pleasure. — *v.r.* delight (in), be pleased (at); be delighted.

complaisamment, *adv.* obligingly; willingly.

complaisance, *f.* complaisance; complacency.

complaît, *pr. 3 sing.* of complaire.

complant, *m.* planting, specif. vineyard.

complanter, *v.t.* plant.

complément, *m.* complement.

complémentaire, *a.* complementary, complemental.

complet, *a.* complete. — *m.* complement.

complètement, *adv.* completely.

compléter, *v.t.* complete; make up (as solutions to a given volume); perfect; fill up.

complexe, *a. & m.* complex.

complexité, *f.* complexity.

complication, *f.* complication.

complimenter, *v.t.* compliment.

complimenteur, *a.* complimentary.

compliqué, *p.a.* complicated, complex, intricate.

compliquer, *v.t.* complicate.

complu, *p.p.* of complaire.

comporter, *v.t.* permit, allow, admit of. — *vs* behave, comport oneself.

composant, *a. & m.* component.

composante, *f.* (*Mech.*) component.

composé, *p.a.* compound; composed; composite; formal. — *m.* compound.

— *benzénique,* aromatic compound.

— *carburé,* carbon compound.

— *chloré,* chlorine compound.

— *conjugué,* (*Old Chem.*) conjugate compound.

— *d'addition,* addition compound.

— *défini,* definite compound.

— *de substitution,* substitution product.

— *double,* double compound.

— *hollandais,* Dutch compound (a solution of sodium and potassium silicates used for purifying water).

— *maximum,* a compound in which an element has its highest valence.

— *minimum,* a compound in which an element has its lowest valence.

— *nitré,* nitro compound.

— *non saturé,* unsaturated compound.

— *organo-magnésien,* organic magnesium compound.

— *passager,* transient (or transitory) compound.

— *phosphoré,* phosphorus compound.

— *saturé,* saturated compound.

— *sulfoné,* sulfonated compound, sulfo compound.

composée, *f.* (*Bot.*) composite plant, composite, (*pl.*) Compositæ (called also Asterales).

composer, *v.t.* compose; compound. — *v.i.* compose; come to terms. — *v.r.* be composed; compose oneself.

compositeur, *m.* composer; (*Printing*) compositor.

composition, *f.* composition; compound (in the sense of a composition or mixture).

— *centésimale,* percentage composition.

— *d'amorce,* primer composition.

— *d'artifices,* fireworks composition.

— *fulminante,* fulminating composition, priming composition.

— *fusante,* fuse composition, rocket composition.

— *incendiaire,* incendiary composition, carcass composition.

— *lente,* slow-burning composition.

— *vive,* quick-burning composition.

compost, *m.* compost.

composter, *v.t.* compost, manure with compost.

composto, *m.* a sort of concrete made of stone fragments and cement or pozzuolana mortar.

compound, *a.* compound.

compréhensibilité, *f.* comprehensibility, comprehensibleness.

compréhensif, *a.* comprehensive.

comprenant, *p.a.* comprizing, including; understanding, comprehending.

comprendre, *v.t.* comprize, include, comprehend; understand, comprehend.

comprennent, *pr. 3 pl. indic. & subj.* of comprendre.

compresseur, *m.* compressor; grease cup.

compressibilité, *f.* compressibility, -bleness.

compressible, *a.* compressible.

compressif, *a.* compressive.

compression, *f.* compression.

comprimable, *a.* compressible.

comprimant, *p.a.* compressing; restraining.

comprimé, *p.a.* compressed; restrained, checked. — *m.* compressed tablet, compressed pill.

comprimer, *v.t.* compress; restrain, check.

comprimeur, *m.* compressor.

compris, *p.p.* of comprendre.

non —, not including, not included.

y —, including, included.

comprit, *p.def. 3 sing.* of comprendre.

compromettre, *v.t. & i.* compromise.

— *le rendement,* endanger the yield.

compromis, *p.a.* compromised. — *m.* compromise.

comptabilité, *f.* bookkeeping, accounting, keeping accounts, accounts; accountability, responsibility.

comptable, *a.* accountable, responsible. — *m.* accountant, bookkeeper; accountable person.

comptant, *a.* ready (money). — *m.* ready money, cash. — *adv.* in cash.

compte, *m.* account.

à —, on account.

au bout du —, after all, considering everything.

— *rendu,* report; specif., (*pl.*) the weekly reports of the sessions of the French Academy of Sciences.

compte-fils, *m.* thread counter, weaver's glass.

compte-gouttes, *m.* dropper, drop counter. — *a.* dropping.

compter, *v.t.* count; pay; charge; account, consider. — *v.i.* count, calculate; render an account; be considered; count (on), intend; count (on), rely (on). — *v.r.* be counted; count oneself.

compte-tours, *m.* an instrument for counting revolutions.

compteur, *m.* counter, meter, indicator; reckoner, calculator.

— *à eau,* water meter.

— *à gaz,* gas meter.

— *à seconde,* seconds counter.

— *électrique,* electric meter.

— *sec,* dry meter.

comptoir, *m.* (shop) counter; branch bank, bank.

compulsif, *a.* compulsory.

computation, *f.* computation.

computer, *v.t.* compute, count.

comté, *m.* earldom; (in England) county, shire.

concassage, *m.*, **concassation**, *f.*, **concassement**, *m.* crushing, etc. (see concasser).

concasser, *v.t.* (reduce to fragments) crush, break, comminute, bray, pound.

concasseur, *m.* crusher, etc. (see concasser). — *a.* crushing.

concave, *a.* concave.

concavité, *f.* concavity.

concéder, *v.t.* concede.

concentrable, *a.* capable of being concentrated.

concentrateur, *m.* concentrator.

concentration, *f.* concentration.

concentré, *p.a.* concentrated. — *m.* concentrate.

concentrer, *v.t.* concentrate.

concentrique, *a.* concentric.

concentriquement, *adv.* concentrically.

conception, *f.* conception.

concernant, *p.pr. & prep.* concerning.

concerner, *v.t.* concern, relate to.

concession, *f.* concession.

concessionnaire, *m. & f.* concessionaire, grantee.

concevable, *a.* conceivable.

concevoir, *v.t. & i.* conceive. — *v.r.* be conceived.

conchoïdal, *a.* conchoidal.

conchoïde, *a.* conchoidal. — *f.* conchoid.

conchylien, *a.* containing shells, conchylaceous.

concierge, *m. & f.* doorkeeper, porter.

conciliation, *f.* reconciliation; conciliation.

concilier, *v.t.* reconcile; conciliate.

concis, *a.* concise.

concision, *f.* conciseness.

conclu, *p.p.* of conclure.

concluant, *p.a.* concluding; conclusive.

conclure, *v.t. & i.* conclude. — *v.r.* be concluded.

conclusif, *a.* conclusive.

conclusion, *f.* conclusion. — *adv.* in short.

conclut, *pr. & p.def. 3 sing.* of conclure.

concoction, *f.* concoction.

conçoit, *pr. 3 sing.* of concevoir.

concombre, *m.* cucumber.

concordance, *f.* concordance, agreement.

concordant, *a.* concordant.

concourir, *v.i.* concur, contribute, coöperate; compete; (of time) coincide; (of lines) meet in a common point.

concourme, *f.* a yellow dyestuff.

concours, *m.* concourse; concurrence, coöperation, agreement; competition; (of lines) meeting point.

concourt, *pr. 3 sing.* of concourir.

concréfier, *v.t.* render concrete.

concret, *a.* concrete.

concréter, *v.t. & r.* concrete, solidify, render or become concrete or solid.

concretion, *f.* concretion.

concrétionnaire, **concrétionné**, *a.* concretionary.

concrétionner, *v.r.* assume a concretionary state.

conçu, *p.a.* conceived.

concurrement, *adv.* concurrently, conjointly; in competition.

concurrence, *f.* competition; amount; (*Law*) concurrence.

concurrencer, *v.t.* compete with.

concurrent, *m.* competitor, rival. — *a.* competing; concurrent.

concurrentiel, *a.* competing.

concuter, *v.t.* strike.

concuteur, *m.* striker, plunger.

condamner, *v.t.* condemn; bar, shut up.

condensabilité, *f.* condensability.

condensable, *a.* condensable.

condensant, *p.a.* condensing.

condensateur, *m.* (esp. *Elec.*) condenser. — *a.* condensing.

condensatif, *a.* condensative.

condensation, *f.* condensation.
— *par l' extérieur*, surface condensation.
— *par mélange*, condensation by injection.

condensé, *p.a.* condensed.

condenser, *v.t. & r.* condense.

condenseur, *m.* condenser.
— *à injecteur* (or *à jet*), jet condenser, injection condenser.
— *par surface*, surface condenser.
— *tubulaire*, tubular condenser, surface condenser.

condenseuse, *f.* condensing apparatus, condenser.

condiment, *m.* condiment.

condimentaire, **condimenteux**, *a.* condimental, condimentary.

condit, *m.* (esp. *Pharm.*) comfit, confection.

condition, *f.* condition; service, employment.
à — de, on condition of.
à — que, on condition that, provided that.

conditionné, *p.a.* conditioned.

conditionnel, *a.* conditional.

conditionnellement, *adv.* conditionally.

conditionnement, *m.* conditioning; specif., manner of packing for shipment.

conditionner, *v.t.* condition.

conductance, *f.* (*Elec.*) conductance.

conducteur, *m. & f.* conductor; superintendent, manager, guide, driver, leader, etc.
— *a.* conducting; leading; (*Mach.*) driving, transmitting.

conductibilité, *f.* conductibility, conductivity.

conductible, *a.* conductible, conductive.

conduction, *f.* conduction.

conductivité, *f.* conductivity.

conduire, *v.t.* conduct; convey; carry; drive; (*Mach.*) control. — *v.i.* conduct; lead; drive. — *v.r.* conduct oneself; be conducted.

conduit, *m.* conduit, pipe, tube, duct, channel, passage, canal. — *p.p. & pr. 3 sing.* of conduire.

— *à gaz,* gas pipe.

— *d'échappement,* escape pipe, (*Steam*) exhaust passage.

— *de la vapeur,* steam pipe.

— *de vent,* blast pipe.

conduite, *f.* conduit, pipe, tube, main, hose, channel, etc.; conducting, conduction; direction, superintendence, management; conduct; carrying, conveyance; delivery (of water, etc.).

— *d'amenée,* delivery pipe or tube.

— *d'aspiration,* suction pipe, suction tube.

— *de vent,* ventilation shaft.

cône, *m.* cone.

ajuster —, taper.

— *allumoir,* lighting cone (funnel-shaped chimney to be placed on open furnaces while they are being lighted).

— *de friction,* cone clutch.

— *droit,* right cone.

— *filtrateur,* filtering cone.

— *tronqué,* truncated cone.

donner du —, taper.

en —, conical, cone-shaped.

faire —, make conical, taper.

côné, *p.a.* coned, conical.

côner, *v.t.* cone.

conf., *abbrev.* compare, see, cf.

confection, *f.* making, making up, construction, execution; ready-made clothes; (*Pharm.*) confection.

confectionné, *p.a.* made, etc. (see confectionner); ready-made (clothes). — *m.* manufactured article.

confectionnement, *m.* making, making up, etc. (see confectionner).

confectionner, *v.t.* make, make up, construct, manufacture, execute. — *v.r.* be made, be made up, etc.

confectionneur, *m.* maker, manufacturer; specif., clothes maker; clothier.

conférence, *f.* conference, meeting; lecture; class; comparison.

conférencier, *m.* lecturer. — *v.i.* hold a conference; give an address.

conférer, *v.t.* confer; compare. — *v.i.* confer.

conferve, *f.* conferva, confervoid alga, hair-weed.

confesser, *v.t.* confess.

confiance, *f.* confidence.

confiant, *p.a.* confiding, etc. (see confier); confident.

confidemment, *adv.* confidentially.

confidence, *f.* confidence.

confident, *a.* confidential. — *m.* confidant.

confidentiel, *a.* confidential.

confidentiellement, *adv.* confidentially.

confier, *v.t.* confide; commit. — *v.r.* confide (in); be committed.

configuration, *f.* configuration.

configurer, *v.t.* give configuration or form to.

confin, *m.* confine, border. (Used in the *pl.*)

confiner, *v.t.* confine, — *v.i.* border (on).

confire, *v.t.* preserve (fruits, etc.); pickle (in vinegar); (*Leather*) soak in bate, bate.

confirmatif, *a.* confirmatory, confirmative.

confirmation, *f.* confirmation.

confirmatoire, *a* confirmatory.

confirmer, *v.t.* confirm. — *v.r.* be confirmed.

confisant, *p.a.* preserving, etc. (see confire).

confiserie, *f.* confectionery; confectioner's art, business or shop.

confiseur, *m.* confectioner.

confit, *p.a.* preserved; pickled; (*Leather*) bated. — *m.* preserved meat; mash (for animals); tub, vat (for dressing furs); (*Leather*) bate, puer. — *pr. & p.def. 3 sing.* of confire.

confiture, *f.* preserve(s), jam.

confiturerie, *f.* preserves; preserve making, preserve factory or shop, preserve business.

confiturier, *m.* maker or seller of preserves.

conflit, *m.* conflict.

confondre, *v.t.* confound; mingle, blend; confuse. — *v.r.* be confounded; mingle, blend; be confused; (*Org. Chem.*) be identical or, in some cases, tautomeric (as, la quinone-oxime *se confond* avec le *p*-nitrosophénol).

confondu, *p.a.* confounded, etc.

conformation, *f.* conformation.

conforme, *a.* conformable.

conformément, *adv.* conformably, according.

conformer, *v.t. & i.* conform.

conformité, *f.* conformity.

confortable, *a.* comfortable, comforting. — *m.* comfort.

confronter, *v.t.* confront; compare. — *v.i.* border.

confus, *p.a.* confused.

confusément, *adv.* confusedly.

confusion, *f.* confusion.

congé, *m.* leave; notice (of leaving); discharge.

congédier, *v.t.* discharge, dismiss.

congélabilité, *f.* congealableness.

congelable, *a.* congealable.

congelant, *p.a.* congealing.

congélation, *f.* congelation, congealing, congealment; specif., freezing; cold storage.

congeler, *v.t. & r.* congeal; freeze.

congénère, *n.* congener. — *a.* congeneric.

congestif, *a.* congestive, congested.

congestion, *f.* congestion.

congestionner, *v.t.* congest.

conglomérat, *m.* conglomerate.

conglomération, *f.* conglomeration.

congloméré, *p.a.* conglomerated, conglomerate.

conglomérer, *v.t.* conglomerate.

conglutinant, *a.* agglutinant, agglutinative.

Congo brillant. (*Dyes*) brilliant Congo.

congrégé, *a* congregated.

congrès, *m.* congress.

— *du froid,* a congress of those interested in the refrigerating industry.

congru, *a.* congruous.

congruence, *f.* congruence.

congruité, *f.* congruity.

conhydrine, *f.* conhydrine.

conicine, *f.* conicine (conine).

conicité, *f.* conicalness, conicity.

conifère, *a.* coniferous. — *m.* conifer; (*pl.*) Coniferæ, Pinales.

conifèrine, *f.* coniferin.

coniférylique, *a.* coniferyl (in the phrase *alcool coniférylique,* coniferyl alcohol).

coniforme, *a.* coniform, conical.

conine, *f.* conine.

conique, *a.* conical, conic.

conjecturalement, *adv.* conjecturally.

conjecture, *f.* conjecture.

conjecturer, *v.t.* conjecture.

conjoindre, *v.t.* conjoin, join, unite.

conjoint, *p.a.* conjoined, joined, joint.

conjointement, *adv.* conjointly, jointly.

conjoncteur, *m.* (*Elec.*) circuit closer, key.

conjonctif, *a.* conjunctive, connective.

conjonction, *f.* conjunction.

conjonctive, *f.* conjunctiva.

conjonctivite, *f.* (*Med.*) conjunctivitis.

conjugué, *a.* conjugate; conjugated.

connaissance, *f.* (sometimes *pl.*) knowledge; consciousness, senses; acquaintance.

connaissant, *p.a.* knowing; skilled in; acquainted with.

connaissement, *m.* bill of lading.

connaître, *v.t.* know.

faire —, make known, describe, introduce, give notice of.

se — *à* (or *en*), be versed in, understand.

connectif, *a.* connective.

connexe, *a.* connected; associated.

connexer, *v.t.* connect, join.

connexion, connexité, *f.* connection.

connexion directe, (*Mach.*) direct action.

connu, *p.a.* known; well known. — *m.* known.

connut, *p.def. 3 sing.* of connaître (to know).

conoïde, *a.* & *m.* conoid.

conquérir, *v.t.* & *i.* conquer.

conquête, *f.* conquest.

conquiert, *pr. 3 sing.* of conquérir.

conquis, *p.a.* conquered.

conquit, *p.def. 3 sing.* of conquérir.

consacrer, *v.t.* devote; consecrate; sanction.

conscience, *f.* consciousness; conscience; conscientiousness.

consciencieusement, *adv.* conscientiously.

consciencieux, *a.* conscientious.

consécutif, *a.* consecutive.

consécutivement, *adv.* consecutively.

conseil, *m.* counsel; council; board.

conseiller, *v.t.* counsel, advise. — *v.r.* be advised, seek counsel (from). — *m.* counselor, advisor; councilor.

consentement, *m.* consent.

consentir, *v.t.* consent to. — *v.i.* consent.

conséquemment, *adv.* consistently; consequently.

conséquence, *f.* consequence.

en —, in consequence, consequently.

conséquent, *a.* consistent; consequent. — *m.* consequent.

par —, consequently, therefore.

conservateur, *a.* preservative, conservative; — *m.* preserver; keeper; conservative.

conservatif, *a.* conservative.

conservation, *f.* preservation, conservation; conservatism.

conservatoire, *a.* & *m.* conservatory.

conserve, *f.* conserve (in *Pharm.* now often translated *confection* since the old distinction between conserve and electuary has been abandoned); preserved food; jar; specif., (*Micros.*) staining jar; (*pl.*) colored glasses.

— *alimentaire,* preserved food.

— *d'amandes,* (*Pharm.*) compound powder of almonds.

— *d'écorce d'orange,* (*Pharm.*) confection of orange peel.

— *de fruits,* preserved fruit (dried, pickled, in sirup, etc.).

— *de légumes,* preserved vegetables (dried, canned, etc.).

— *de poissons,* preserved fish, cured fish.

— *de rose,* — *de rose rouge,* (*Pharm.*) confection of rose.

— *de viande,* preserved meat.

conserves alimentaires, preserved food, specif. canned goods.

conserver, *v.t.* preserve. — *v.r.* (of fruits, etc.) keep; be preserved; preserve oneself.

considérable, *a.* considerable; distinguished.

considérablement, *adv.* considerably.

considération, *f.* consideration; esteem, regard.

considéré, *p.a.* considered; esteemed; considerate.

considérer, *v.t.* consider. — *v.r.* be considered; consider oneself.

consignataire, *m.* (*Com.*) consignee.

consignateur, *m.* (*Com.*) consignor, consigner.

consignation, *f.* consignment.

consigner, *v.t.* consign; deposit; order; keep in (soldiers, etc.); refuse admittance to.

consistant, *a.* consistent; (with *en*) consisting.

consistence, *f.* consistency, consistence; credit.

consister, *v.i.* consist.

console, *f.* bracket; wall fixture, hanger.

consolidation, *f.* consolidation.

consolider, *v.t.* consolidate; (*Finance*) fund. — *v.r.* consolidate.

consommateur, *m.* consumer; consummator.

consommation, *f.* consumption; consummation.

consommé, *p.a.* consumed; consummated; consummate.

consommer, *v.t.* consume; consummate.

consomptible, *a.* consumable.

consomption, *f.* consumption.

consortium, *m.* syndicate.

consoude, *f.* (*Bot.*) comfrey (*Symphytum*).
— *officinale, grande* —, common comfrey (*S. officinale*).

constamment, *adv.* constantly; with constancy.

constance, *f.* constant; constancy.

constant, *a.* constant; certain.

constante, *f.* constant.
— *de dissociation*, dissociation constant.
— *empirique*, empirical constant.

constatation, *f.* ascertaining; verification; statement.

constater, *v.t.* ascertain; verify; state.

constipant, *p.a.* constipating.

constiper, *v.t.* constipate.

constituant, *p.a.* constituent; constituting.
— *m.* constituent.

constituer, *v.t.* constitute; put, place, fix.

constitution, *f.* constitution.

constructeur, *m.* constructor, builder.

construction, *f.* construction; building, structure.

construire, *v.t.* construct, build.

consultatif, *a.* consultative, advisory.

consulter, *v.t. & i.* consult. — *v.r.* consider; consult one another; be consulted.

consumant, *p.a.* consuming.

consume, *f.* (*Wine*) loss from evaporation and absorption.

consumer, *v.t.* consume. — *v.r.* be consumed; waste one's money, health, strength or time; waste away.

contact, *m.* contact.

contagieux, *a.* contagious.

contamination, *f.* contamination.

contaminer, *v.t.* contaminate. — *v.r.* be contaminated.

contempler, *v.t.* contemplate.

contemporain, *a.* contemporary, contemporaneous.

contenance, *f.* capacity, contents; extent, area; countenance.

contenant, *p.a.* containing; restraining. — *m.* container.

contenir, *v.t.* contain; restrain.

content, *a.* content, satisfied, pleased, glad.

contentieux, *a.* disputed; contentious.

contention, *f.* containing; intense application, intenseness; contention.

contenu, *p.a.* contained; restrained. — *m.* contents.

conter, *v.t.* tell, relate; report.

conterie, *f.* coarse Venice glass.

contestation, *f.* contestation, dispute.

contester, *v.t.* contest.

contiennent, *pr. 3 pl.* of contenir.

contient, *pr. 3 sing.* of contenir.

contigu, *a.* contiguous, adjacent, adjoining.

contint, *p.def. 3 sing.* of contenir.

continu, *a.* continuous; continued. — *m.* continuum.

continuation, *f.* continuation, continuance.

continuel, *a.* continual.

continuellement, *adv.* continually.

continuer, *v.t. & i.* continue. — *v.r.* be continued; continue.

continuité, *f.* continuity; continuance.

continûment, *adv.* continuously, continually.

contour, *m.* contour, outline; circuit.

contourner, *v.t.* trace the outline of, pass, turn or wind around; twist; distort, deform.

contractant, *p.a.* contracting. — *m.* contracting party, contractor.

contracter, *v.t.* contract. — *v.r.* be contracted, contract.

contractif, *a.* contractive.

contractile, *a.* contractile.

contraction, *f.* contraction.

contracture, *f.* contracture.

contradiction, *f.* contradiction.

contradictoire, *a., m. & f.* contradictory.

contraignant, *p.a.* constraining.

contraignent, *pr. 3 pl.* of contraindre.

contraindre, *v.t.* constrain, compel, force, make.

contraint, *p.p.* of contraindre.

contrainte, *f.* constraint.

contraire, *a.* contrary; injurious. — *m.* contrary.
au —, on the contrary.
au — *de*, contrary to, against.

contrairement, *adv.* contrarily, contrary.

contrariété, *f.* contrariety; difficulty; annoyance.

contrastant, *p.a.* contrasting.

contraste, *m.* contrast.

contraster, *v.t. & i.* contrast.

contrat, *m.* contract.

contrayerva, **contrayerve**, *f.* (*Bot.*) contrayerva.

contre, *prep.* against; close by, near. — *adv.* against, in opposition. — *m.* con, opposing side or point.

— *à* —, side by side.

par —, per contra, as an offset, on the other hand.

tout —, very close; ajar.

contre-, counter-, contra-.

contre-accélération, *f.* negative acceleration, retardation.

contre-balancer, contrebalancer, *v.t.* counterbalance. — *v.r.* counterbalance each other; be counterbalanced.

contrebande, *f.* contraband.

contre-bas, *adv.* downward.

contre-burette, *f.* leveling tube (for Lunge nitrometers and the like).

contre-charge, *f.* counterpoise.

contre-cœur, *m.* back (of a chimney).

à —, reluctantly, unwillingly.

contre-coup, *m.* rebound; counterstroke.

contre-courant, *m.* countercurrent; back draft.

contredire, *v.t. & i.* contradict.

contredisant, *p.a.* contradicting, contradictory.

contredit, *pr. & p.def. 3 sing. & p.p.* of contredire. — *m.* contradiction; rejoinder.

sans —, unquestionably.

contrée, *f.* country, region.

contre-écrou, *m.* set screw (for another screw); lock nut, jam nut, check nut.

contre-électromoteur, -électromotif, *a.* counter-electromotive.

contre-émail, *m.* counterenamel.

contre-émailler, *v.t.* counterenamel.

contre-épreuve, *f.* countertest, check test; counterproof.

contre-essai, *m.* countertest, check test.

contrefaçon, *f.* counterfeiting; counterfeit; (of patents) infringement; forgery; (of books) pirated edition.

contrefacteur, *m.* counterfeiter; infringer; forger; pirate.

contrefaction, *f.* counterfeiting, etc. (see contrefaire).

contrefaire, *v.t.* counterfeit; forge; mimic; feign; disguise.

contrefait, *p.a.* counterfeited, etc. (see contrefaire); counterfeit. — *pr. 3 sing.* of contrefaire.

contreferait, *cond. 3 sing.* of contrefaire.

contre-feu, *m.* back plate (of a hearth).

contre-fiche, *f.* brace, strut, stud, stay.

contrefit, *p.def. 3 sing.* of contrefaire.

contrefont, *pr. 3 pl.* of contrefaire.

contrefort, *m.* buttress, support; reinforcement; (mountain) spur.

contre-hacher, *v.t.* cross-hatch.

contre-hachure, *f.* cross-hatching.

contre-maître, *m.* overseer, foreman, boss.

contremander, *v.t.* countermand.

contremarque, *f.* countermark.

contre-oxydeur, *m.* enameler.

contre-paroi, *f.* casing (of a furnace).

contre-pente, *f.* reverse slope.

contre-peser, *v.t.* counterpoise, counterbalance.

contre-plaque, *f.* back plate; reinforcing plate; cheek plate, guard plate.

contreplaqué, *a.* layered; laminated.

— *en trois épaisseurs,* three-ply.

contrepoids, *m.* counterpoise.

contre-poil, *m.* wrong way (of the hair).

à —, against the grain or the nap, the wrong way.

contrepoison, contre-poison, *m.* antidote.

— *de l'arsenic,* (*Pharm.*) arsenic antidote (ferric hydroxide with magnesium oxide).

contre-pression, *f.* counterpressure, back pressure.

contresens, *m.* wrong sense; (of fabrics) wrong side; wrong way.

contretemps, *m.* mishap, mischance; wrong time, unseasonableness.

à —, inopportunely.

contre-tirage, *m.* back draft.

contre-vapeur, *f.* returning steam, back steam.

contrevenir, *v.i.* offend, transgress.

contrevent, *m.* blast plate; shutter; brace.

contrevient, *pr. 3 sing.* of contrevenir.

contre-vis, *f.* check screw.

contribuant, *p.a.* contributing. — *m.* contributor.

contribuer, *v.i.* contribute.

contribution, *f.* contribution; tax.

contrôlage, *m.* controlling, etc. (see contrôler).

contrôle, *m.* control; checking, check; mark, stamp; register, roll, list; controller's office.

contrôlement, *m.* controlling, etc. (see contrôler).

contrôler, *v.t.* control, check; calibrate; ascertain by checking; stamp, mark (as gold).

contrôleur, *m.* controller, checker, inspector, comptroller.

contrôleur-mesureur, *m.* (*Sugar*) an instrument for recording the amount of beet juice extracted.

controuvé, *a.* fabricated, forged, false.

controverse, *f.* controversy.

controverser, *v.t.* dispute, controvert, debate.

convaincant, *a.* convincing; convicting.

convaincre, *v.t.* convince; convict.

convaincu, *p.a.* convinced; convicted.

convainquant, *p.a.* convincing; convicting.

convallaire, *f.* (*Bot.*) convallaria.

convallarine, *f.* convallarin.

convection, *f.* convection.

convenable, *a.* suitable; expedient; becoming; proper; well-bred. — *m.* propriety.

convenablement, *adv.* suitably.

convenance, *f.* convenience; conformity; agreement; suitableness; propriety.

convenant, *a.* convenient; suitable; proper. — *p.pr.* of convenir.

convenir, *v.i.* admit, acknowledge; agree; suit, be suitable; be expedient; be proper.

convention, *f.* convention.

convenu, *p.p.* of convenir.

convergence, *f.* convergence, convergency.

convergent, *p.a.* converging, convergent.

converger, *v.i.* converge.

converse, *a. & f.* converse.

conversible, *a.* convertible.

conversion, *f.* conversion.

converti, *p.a.* converted. — *m.* convert.

convertibilité, *f.* convertibility.

convertible, *a.* convertible.

convertir, *v.t.* convert. — *v.r.* be converted, change, turn.

convertissable, *a.* convertible.

convertissant, *p.p.* of convertir.

convertissement, *m.* conversion, converting.

convertisseur, *m.* converter.

convexe, *a.* convex.

convexité, *f.* convexity.

convient, *pr. 3 sing.* of convenir.

convier, *v.t.* invite, ask, beg, prompt.

convint, *p. def. 3 sing.* of convenir.

convoi, *m.* (railway) train; convoy; funeral.
— *de marchandises*, freight train, good strain.
— *de petite vitesse*, slow train.
— *direct* (or *exprès*), express train.

convoiteux, *a.* covetous.

convolvuline, *f.* convolvulin.

convoquer, *v.t.* convoke, call.

convulser, *v.t.* convulse.

convulsif, *a.* convulsive.

convulsion, *f.* convulsion.

convulsionner, convulsiver, *v.t.* convulse.

convulsivement, *adv.* convulsively.

cooperatif, *a.* coöperative.

coopérer, *v.i.* coöperate.

coordonnateur, *a.* coördinating.

coordonné, *p.a.* coördinated, coördinate.

coordonnée, *f.* coördinate.
coordonnées polaires, polar coördinates.
coordonnées rectangulaires, rectangular coördinates.

coordonner, *v.t.* coördinate.

copahu, *m.* copaiba.

copaïer, *m.* copaiba tree, any tree of the genus *Copaiva*.

copal, *m.*, **copale**, *f.* copal.

copaline, *f.* (*Min.*) copalite, copaline.

copalme, *m.* copalm.

copayer, *m.* = copaïer.

copeau, *m.* chip, shaving.
— *de sapin*, fir (deal, pine) shaving.
copeaux d'alésage, borings.
copeaux de tour, turnings.

copie, *f.* copy.

copier, *v.t.* copy.

copieusement, *adv.* copiously.

copieux, *a.* copious.

copiste, *m.* copier, copyist.

coprah, copra, *f.*, **copre**, *m.* copra.

coprécipitine, *f.* coprecipitin.

copropriété, *f.* joint property.

copulation, *f.* coupling; copulation.

copuler, *v.t.* couple; (of animals) copulate. — *v.r.* couple.

coq, *m.* cock; (on shipboard) cook.

coque, *f.* shell; cocoon; (of ships) hull; (of a boiler) body; kink; (*Ceram.*) see coque d'œuf, below.
— *d'œuf*, eggshell; (*Ceram.*) a dull porous state of the glaze due to imperfect vitrification.
— *du Levant*, cocculus indicus.

coquelicot, *m.* coquelicot (the corn poppy, *Papaver rhœas*, or its red color).

coqueluche, *f.* (*Med.*) whooping cough.

coqueluchon, *m.* (*Bot.*) aconite.

coqueret, *m.*, **coquerelle**, *f.* (*Bot.*) alkekengi.

coquillage, *m.* shellfish; shell.

coquille, *f.* shell; (*Paper*) a size about 17 × 22 inches; (*Founding*) chill; (*Mach.*) casing, housing; typographical error.

coquiller, *v.t.* (*Founding*) cast in a chill mold.

coquilleux, coquillier, *a.* shelly.

cor, *m.* horn (instrument); corn (callosity).
— *de chasse*, hunting horn, bugle; (*Mach.*) a part looped like a bugle.

corail, *m.* coral.

corailleux, corallien, *a.* coral.

coralline, *f.* corallin.

coraux, *pl.* of corail.

corbeau, *m.* raven; corbel, bracket.

corbeille, *f.* basket.

corbeillée, *f.* basketful, basket.

cordage, *m.* rope, cord, cordage; cording (of wood); measuring (of wood) by cords.
— *goudronné*, tarred rope.

cordat, *m.* a coarse wrapping cloth; coarse serge.

corde, *f.* cord, twine, rope, string, line; thread; (of wood) cord; (*Geom.*) chord; (*Music*) string; (*Glass*) a cordlike defect on blown glass, due to incomplete fusion.
— *à boyau*, catgut.
— *à feu*, fuse, match.
— *à violon*, violin string.
— *à violoncelle*, cello string, violoncello string.
— *de piano*, piano string, piano wire.
— *en fil de fer*, wire rope.

cordé, *p.a.* corded; twisted; (*Bot.*) cordate.

cordeau, *m.* line, string, small cord or rope; fuse.

— *Bickford,* Bickford fuse.

— *d'amorce,* fuse, match.

— *détonant,* detonating fuse.

— *porte-feu,* fuse, match.

cordée, *f.* cord (of wood).

cordeler, *v.t.* twist, twine, lay (into rope).

cordelette, *f.* small cord, string, line.

cordeline, *f.* (*Glass*) ferret; selvedge, list.

corder, *v.t.* cord; twist; measure by the cord.
— *v.r.* become stringy; twist; be corded.

corderie, *f.* cordage making; cordage business; cordage factory, ropewalk.

cordial, *a. & m.* cordial.

cordier, *a.* cordage. — *m.* rope maker or seller.

cordiérite, *f.* (*Min.*) iolite, cordierite.

cordon, *m.* string, cord; strand; band, tape; ribbon; row, string, line, string; cordon; border; milled edge (of a coin).

cordonner, *v.t.* twist; plait, plat, braid.

cordonnerie, *f.* shoe making, trade or shop.

cordonnet, *m.* small cord, string; braid; silk twist; milling (on a coin).

cordonnier, *m.* shoemaker; shoe dealer.

cordouan, *m.* cordovan (leather).

Cordoue, *f.* Cordova.

Corée, *f.* Korea.

coriace, *a.* tough, leatherlike.

coriacé, *a.* coriaceous, leatherlike.

coriaire, *f.* (*Bot.*) tanner's sumac.

coriandre, *f.* (*Bot.*) coriander.

coriandrol, *m.* coriandrol (linaloöl).

corindon, *m.* (*Min.*) corundum.

corindonique, *a.* corundum.

corme, *f.* service berry.

cormier, *m.* service tree (*Sorbus domestica*).

cornaille, *f.* horn raspings.

cornaline, *f.* (*Min.*) carnelian.

cornard, *m.* (*Glass*) a hooked tool for removing crucibles from the furnace.

corne, *f.* horn; prong, point; turned-down corner (of a sheet of paper); fruit of the cornel.

— *de cerf,* hartshorn.

— *d'abondance,* cornucopia, horn of plenty.

corné, *a.* horny, horn, corneous.

cornée, *f.* (*Anat.*)cornea; (*Bot.*) a member of the Cornaceæ.

cornéenne, *f.* (*Petrog.*) aphanite.

corneille, *f.* crow; (*Bot.*) loosestrife.

cornéliane, *f.* (*Min.*) carnelian.

cornéole, *f.* woadwaxen, dyer's weed (*Genista tinctoria*).

corner, *v.i.* sound a horn; (of the ears) ring; smell, stink. — *v.t.* din; horn; turn down the corner of.

cornet, *m.* bag, cornet; little horn; dice box; (*Assaying, Music,* etc.) cornet.

cornetier, *m.* horn worker.

corniche, *f.* cornice; (*Bot.*) water caltrop.

cornier, *a.* corner. — *m.* (*Bot.*) cornel (*Cornus*).

cornière, *f.* angle iron, corner iron.

Cornouailles, *n.* Cornwall.

cornouille, *f.* cornel berry, fruit of the cornel.

cornouiller, *m.* (*Bot.*) any plant of the genus *Cornus* (cornels and dogwoods).

— *à grandes fleurs,* flowering dogwood (*Cornus florida*).

cornu, *a.* horned; angular, cornered; absurd.

cornue, *f.* retort; a large two-handled vessel for carrying liquids; (*Metal.*) converter.

— *à gaz,* gas retort.

— *à goudron,* tar still, tar retort.

— *à lessives,* a vessel for carrying lye.

— *de distillation,* retort.

corollaire, *m.* corollary. — *a.* (*Bot.*) corolline.

corolle, *f.* (*Bot.*) corolla.

corozo, *m.* vegetable ivory (from the corozo palm).

corporation, *f.* guild, union.

corporifier, *v.t.* (*Old Chem.*) corporify, solidify.

corps, *m.* substance, compound; body; (*Mil.*) corps. (Careful modern usage avoids use of the term "body" for substances.)

— *aldéhydique,* aldehydic substance, aldehyde.

— *azoté,* nitrogenous substance, nitrogen compound.

— *comburant,* supporter of combustion.

— *composé,* compound.

— *creux,* hollow body; hollow ware.

— *desséchant,* dehydrating agent, desiccating substance.

— *étranger,* foreign substance; foreign body (e.g. in the eye).

— *gazeux,* gaseous substance, gas; gaseous body, body of gas.

— *gras,* fatty substance (including, in the most general sense, fats, soaps, fatty acids and even lecithin, etc.); specif., fat (*corps gras proprement dit*).

— *isolant,* insulator.

— *muqueux de Malpighi,* Malpighian layer.

— *réducteur,* reducing substance, reducer.

— *simple,* simple substance, element.

corpusculaire, *a.* corpuscular.

corpuscule, *m.* corpuscle.

correct, *a.* correct.

correctement, *adv.* correctly.

correcteur, *m.* corrector; proofreader. — *a.* correcting, corrective.

correctif, *a. & m.* corrective.

correction, *f.* correction; correctness.

corrélatif, *a. & m.* correlative.

corrélation, *f.* correlation.

corrélativement, *adv.* correlatively.

correspondance, *f.* correspondence; communication.

correspondant, *p.a.* corresponding. — *m.* correspondent.

correspondre, *v.i. & r.* correspond.

corrigé, *p.a.* corrected. — *m.* corrected copy.

corriger, *v.t.* correct. — *v.r.* be corrected; correct oneself, reform.

corroborant, *p.a.* corroborating, corroborant. — *m.* (*Med.*) corroborant, tonic.

corroboratif, *a.* corroborative.

corroborer, *v.t.* corroborate.

corrodant, *p.a.* corroding, corrosive; erosive. — *m.* corrosive.

corroder, *v.t.* corrode; erode. — *v.r.* be corroded; be croded.

corroi, *m.* (*Leather*) currying; claying, puddling; puddle.

corroirie, *f.* curriery.

corrompre, *v.t.* corrupt; soften (as leather or iron); deprive of plasticity (as wax). — *v.r.* become corrupted, softened, etc.

corrompu, *p.a.* corrupted, etc. (see corrompre), corrupt; putrid.

corrosif, *a. & m.* corrosive.

corrosion, *f.* corrosion; (*Geol.*) erosion.

corroyage, *m.* currying, etc. (see corroyer).

corroyé-masse, *f.* scrap iron.

corroyer, *v.t.* curry (leather or skins); work, hammer, forge, weld (iron, etc.); puddle, pug (clay or with clay); knead, beat, malax (dough, mortar, etc.); dress down, plane (wood); (*Dyeing*) spread out on the roller.

corroyère, *f.* (*Bot.*) tanner's sumac.

corroyeur, *m.* currier, etc. (see corroyer).

corruption, *f.* corruption; specif., putrefaction.

Corse, *f.* Corsica.

corsé, *p.a.* full-bodied, full-flavored, strong.

corser, *v.t.* give body, consistency, flavor or strength to.

corton, *m.* a kind of Burgundy wine.

corubis, *m.* artificial corundum.

cosecante, *f.* (*Math.*) cosecant.

cosinus, *m.* (*Math.*) cosine.

cosmétique, *a. & m.* cosmetic.

cosmique, *a.* cosmic.

cosmogonie, *f.* cosmogony.

cosse, *f.* pod, husk, hull (of peas, etc.).

cossette, *f.* chip, slice, cossette.

cosseterie, *f.* cutting into chips or slices and the treatment of these (in the case of chicory, drying).

cotangente, *f.* (*Math.*) cotangent.

cotarnine, *f.* cotarnine.

cote, *f.* quota; classification mark; figure indicating a dimension, altitude or the like on a plan or chart; (*Com.*) quotation.

côte, *f.* rib; slope; shore, coast.
    — *à* —, side by side.
    — *de soie,* silk floss, cappadine.
    Côte d'Ivoire, Ivory Coast.

Côte *occidentale,* West Coast (esp. of Africa).

coté, *p.a.* lettered, numbered; (*Com.*) quoted.

côté, *m.* side; (*Printing*) form.
    à —, on one side, sideways, sidewise; obliquely; near, by; close together.
    à — *de,* beside, close to, near.
    — *de la chair,* (*Leather*) flesh side.
    — *poil,* (*Leather*) hair side.
    de —, on one side, sideways; obliquely; awry; aside, apart.
    du —, among; with respect to, regarding.

coteau, *m.* slope, hillside; hill.

Côte-d'Or, *f.* Côte-d'Or (a department of France).

coter, *v.t.* number, letter; (*Com.*) quote.

côtier, *a.* coasting, pertaining to the coast.

côtière, *f.* side stone (of a furnace); (*Founding*) one of the parts of a mold.

cotin, cotinus, *m.* (*Bot.*) fustet (*Cotinus cotinus*).

cotisation, *f.* assessment; clubbing together; share, quota.

cotiser, *v.t.* assess. — *v.r.* club together.

coton, *m.* cotton; down.
    — *absorbant,* absorbent cotton.
    — *azotique,* nitrated cotton, guncotton.
    — *brut,* raw cotton.
    — *caoutchouté,* rubberized cotton.
    — *courte soie,* short-staple cotton.
    — *de verre,* glass wool.
    — *endécanitrique,* Vieille's name for the highest nitration product of cotton.
    — *en laine,* cotton wool, raw cotton.
    — *fulminant,* fulminating cotton, guncotton.
    — *hydrophile,* absorbent cotton.
    — *longue soie,* long-staple cotton.
    — *minéral,* mineral wool, slag wool, mineral cotton.
    — *nitraté,* nitrated cotton.
    — *octonitrique,* a nitrated cotton of the formula $C_{24}H_{26}O_{12}(NO_3)_8$ (according to Vieille).
    — *pyrique,* (*Expl.*) pyrocotton.

cotoniser, *v.r.* become like cotton.

cotonnade, *f.* cotton fabric, cotton goods.

cotonné, *p.a.* cottony, downy; (of hair) woolly.

cotonner, *v.i.* become nappy or downy; (of copper) become covered with light spots. — *v.r.* become nappy or downy.

cotonnerie, *f.* cotton plantation; cotton culture; cotton mill.

cotonneux, *a.* cottony.

cotonnier, *m.* (*Bot.*) cotton plant. — *a.* cotton.

coton-poudre, *m.* guncotton.

côtoyer, *v.t.* go along, coast, skirt.

cotylédon, *m.* cotyledon.

cotylédonaire, *a.* cotyledonary, cotyledonous.

cotylédoné, *a.* cotyledonous.

cou, *m.* neck.
    — *de cygne,* see col de cygne, under *col.*

couchant, *p.p.* of coucher. — *m.* west; decline; setting sun.

couchart, *m.* (*Paper*) coucher.

couche, *f.* layer; bed, stratum; coat, coating; (*Brewing*) couch; bed; hotbed; childbed, parturition.

— *cornée,* horny layer.

— *de Malpighi,* Malpighian layer.

— *d'impression,* (*Painting*) priming coat.

— *isolante,* insulating layer.

coucher, *v.t.* lay, lay flat, lay down, lay on; coat; write down; inscribe, insert; (*Paper*) couch. — *v.i.* lie, lie down; sleep. — *v.r.* go to bed; lie, lie down; (of the sun, etc.) set. — *m.* lying, recumbent position; (of the sun, etc.) setting; bed, bedding.

coucheur, *m.* layer, coater, etc. (see coucher).

couchis, *m.* bed, layer, stratum; (of a boiler) lagging; (*Gardening*) layer.

coucou, *m.* cuckoo; cuckoo clock.

coud, *pr. 3 sing.* of coudre.

coude, *f.* elbow; knee, angle, bend, turn, bow, loop; crank, cranked part.

coudé, *p.a.* bent, kneed, cranked.

couder, *v.t.* bend (in the form of an elbow), knee, crank. — *v.r.* form an elbow.

coudrage, *m.* soaking in tan liquor.

coudran, coudranner, etc. See goudron, goudronner, etc.

coudre, *v.t.* sew; tack on, attach. — *v.i.* sew. — *v.r.* be sewed; be joined.

coudrée, *f.* arid land.

coudrement, *m.* soaking in tan liquor.

coudrer, *v.t.* soak in tan liquor.

coudret, *m.* tanning vat, tan vat.

coudreuse, *f.* an apparatus for stirring tan liquor.

coudrier, *m.* (*Bot.*) hazel, hazel tree.

couenne, *f.* pigskin, (of bacon) rind; any thick skin.

couette, *f.* socket; feather bed.

coulage, *m.* flowing, leaking, casting, etc. (see couler); (loss due to or allowance made for) leakage; waste; scalding, scouring, bucking (in laundries).

— *à noyau* (or *à noyaux*), core casting.

— *en coquille,* chill casting.

— *en sable,* sand casting.

— *en terre,* loam casting.

coulant, *p.a.* flowing, etc. (see couler); fluid; (of machine parts) movable, sliding; (of kilns, etc.) continuous; (of soil) not firm, caving; fluent; accommodating. — *m.* (*Mach.*) slide, slider, runner.

coulé, *p.a.* cast, etc. (see couler, *v.t.*). — *m.* cast, casting.

coulée, *f.* casting; tapping, running off (of metal); discharge; channel from furnace to mold; spray (side channel in a mold);

head, sullage piece (of a casting); (*Geol.*) coulée; running hand.

coulement, *m.* flow, flowing.

couler, *v.i.* flow, run; leak, run out; glide, slide, slip; slip; creep; (of fruits, etc.) drop; (of ships) sink; (*Founding*) run thru the mold. — *v.t.* cast, pour, mold, found; tap (metal or a furnace); filter, strain; cause to flow, pass, run; sink (a ship); pass (time); slip; engrave boldly; lead, run lead into. — *v.r.* slip, glide; ruin oneself.

couleur, *f.* color, colour; specif., dye; a variety of smalt. — *m.* caster (in foundries and ceramic work); scalder, scourer (in laundries).

— *à cuve,* vat dye.

— *à eau,* — *à l'eau,* water color.

— *à l'huile,* oil color.

— *à mordants,* mordant dye, mordant color.

— *au pinceau,* brush color, artist's color.

— *azoïque,* azo dye, azo color.

— *bon teint,* fast color, fast dye.

— *claire,* light color.

— *complémentaire,* complementary color.

— *composée,* secondary color.

— *d'aniline,* aniline color, aniline dye.

— *de chair,* flesh color.

— *de feu de moufle,* (*Ceram.*) overglaze color.

— *de grand feu,* (*Ceram.*) any color applied in or under the glaze.

— *d'imprimerie,* colored printing ink.

— *du recuit,* tempering color.

— *en détrempe,* distemper color.

— *faux teint,* fugitive color, fugitive dye.

— *foncée,* dark color.

— *glacée,* an azo dye produced on the fiber with the use of an iced diazo solution.

— *grand teint,* fast color, fast dye.

— *grattée,* pale color.

— *lavée,* pale color, faint color.

— *matrice,* primary color.

— *minérale,* mineral color.

— *mode,* mode color.

— *pour cuve,* vat color.

— *primitive,* primary color.

— *rabattue,* saddened or darkened color.

— *secondaire,* secondary color.

— *solide,* fast color, fast dye.

— *vitrifiable,* vitrifiable color, enamel color.

mettre en —, color, stain; alter the surface of (a gold alloy) by pickling.

couleuvre, *f.* snake.

couleuvrée, *f.* (*Bot.*) a name for plants which corresponds (but only in part) to snake-root.

— *de Virginie,* Virginia snakeroot (*Aristolochia serpentaria*).

couleuvrine, *f.* (*Bot.*) bistort, snakeweed (*Polygonum bistorta*).

coulis, *m.* grout, thin mortar; molten metal; strained broth.

coulisse, *f.* groove, slot, channel, guide, slide; slotted link; sliding shutter.

à —, sliding, with sliding motion.

coulisseau, *m.* small groove; (piece which slides in a groove) slide block, slider, guide block, guide, tongue.

coulissement, *m.* sliding, sliding motion.

coulisser, *v.i.* slide. *— v.t.* furnish with grooves or guides.

couloir, *m.* strainer, colander; hopper; passage, passageway, lobby.

couloire, *f.* strainer, colander; vessel placed under the spigot of a cask; drawplate.

coulomb, *m.* (*Elec.*) coulomb.

coulombmètre, *m.* coulomb meter.

coulommiers, *m.* a kind of cheese similar to Brie.

coulure, *f.* running, running down (as of glaze), running out (as of cast metal); failure of fruit due to falling off of pollen.

coumarine, *f.* coumarin.

coumarique, *a.* coumaric.

coumarone, *m.* coumarone (benzofuran).

coumarylique, *a.* coumarilic.

coumys, *m.* kumiss.

coup, *m.* stroke; blow, shock; shot, report, discharge (of a firearm); blast; time, turn; draft (of liquor); move, throw (in games); bruise; wound.

à —, all at once.

à — sûr, surely, certainly.

après —, too late, afterward.

— d'arrière, back stroke, return stroke.

— d'avance, forward stroke.

— de dé, — de dés, cast of the dice, (fig.) venture.

— de feu, sudden increase in heat; heating; burning; overheating, overburning; (*Founding*) melting heat.

— d'essai, first attempt; proof shot.

— d'œil, glance, look, sight, view.

— heureux, lucky shot (throw, move), lucky venture.

— sur —, one after another, successively, uninterruptedly.

pour le —, this time, for once.

coupable, *a.* guilty; culpable. *— m.* culprit.

coupage, *m.* cutting, etc. (see couper); mixing (of liquids); specif., addition of fermenting liquid to unfermented to start the fermentation; dilution (e.g. of alcohol); mixture, blend (of wines, etc.).

coupant, *p.a.* cutting, etc. (see couper); sharp. *— m.* (sharp) edge.

coupe, *f.* cutting; cut; section; a shallow cup or glass having a foot and stem.

— en long, longitudinal section.

coupe *en travers,* cross section.

— horizontale, horizontal section, plan.

— transversale, cross section.

— verticale, vertical section.

coupé, *p.a.* cut, etc. (see couper).

coupeau, *m.* cutting, snip, chip. (Cf. copeau.)

coupe-air, *m.* trap.

coupe-betteraves, *m.* sugar-beet slicer, beet-root slicer.

coupe-cannes, *m.* sugar-cane slicer.

coupe-circuit, *m.* (*Elec.*) circuit breaker, cut-out, fuse.

coupellation, *f.* cupellation.

coupelle, *f.* cupel; pan for burning sulfur in a sterilizing apparatus; cup (in certain tech. senses); small cup on a stem; small shovel for gunpowder.

coupeller, *v.t.* cupel.

coupelleur, *m.* cupeller.

coupement, *m.* cutting, etc. (see couper).

coupe-net, *m.* wire cutters, nippers.

couper, *v.t.* cut; cut up, cut off, cut out; (of liquids) mix, blend, dilute (e.g. alcohol), cut. *— v.i.* cut. *— v.r.* cut oneself; be cut; cut; divide, part, cleave; cut each other, cross, intersect; (of fabrics) wear at the folds; contradict oneself.

coupe-racines, *m.* root cutter, specif. beetroot slicer.

couperet, *m.* cutter, (large) knife; cleaver, chopper; enameler's file.

couperose, *f.* copperas, vitriol (*copperas* is now used in English only of the green iron salt, $FeSO_4 \cdot 7 H_2O$); (*Med.*) acne rosacea.

— blanche, white vitriol (zinc sulfate).

— bleue, blue vitriol (copper sulfate).

— verte, green vitriol, copperas.

coupe-tubes, coupe-tuyau, *m.* tube cutter.

coupeur, *m.* cutter. *— a.* cutting.

coupeuse, *f.* cutter.

couplage, *m.* coupling; (*Elec.*) connection.

couple, *f.* couple. *— m.* couple; timber, frame.

couplement, *m.* coupling; (*Elec.*) connection.

coupler, *v.t.* couple; (*Elec.*) connect.

couplet, *m.* couplet; hinge, hinged joint.

coupoir, *m.* cutter.

coupole, *f.* cupola.

coupon, *m.* coupon; fraction (of a share); remnant (of a cloth).

coupure, *f.* cut; ditch, trench; small bank note.

cour, *f.* court.

courade, *f.* a kind of sardine.

courage, *m.* courage; heart; spirit; temper.

couramment, *adv.* commonly, currently; readily.

courant, *p.a.* current, present, ordinary, common; lineal; running. *— m.* current; course, run; current month, inst.; length (as of a building).

courant à haute fréquence, high-frequency current.

— alternatif, alternating current.

— biphasé, two-phase (or diphase) current.

— continu, continuous current.

— d'allumage, lighting current.

— de charge, charging current.

— de retour, return current.

— dérivé, derived current, shunt current.

— direct, direct current.

— discontinu, intermittent current.

— inducteur, inducing current, primary current.

— induit, induced current, secondary current.

— monophasé, single-phase (or monophase) current.

— polyphasé, polyphase current.

— redressé, rectified current.

— tellurique, — terrestre, earth current, ground current.

— triphasé, three-phase (or triphase) current.

— vagabond, stray current.

— vibré, alternating current.

courbable, a. bendable.

courbage, m. bending, curving, bowing.

courbaril, m. courbaril (Hymenæa courbaril).

courbarine, f. courbaril (resin), courbaril copal.

courbe, f. curve; turn, bend; knee, rib; contour. — a. curved.

courbé, p.a. curved, bent, bowed; crooked; bowed down; stooping.

courbement, m. bending, curving, curvature, curvation.

courber, v.t. bend, curve, bow. — v.i. bend, bow, stoop. — v.r. bend, be bent, bow, stoop.

courbure, f. curvature, curve, bend, bending, inflection, flexure.

courent, pr. 3 pl. indic. & subj. of courir.

coureur, m. runner; rover; frequenter.

courge, f. (Bot.) cucurbit, member of the genus Cucurbita (including pumpkin, squash, gourd).

courir, v.i. run; be current; run about, go about; hasten. — v.t. run; run after, follow, frequent; roam over, travel over; overrun; be current in.

couronne, f. crown; ring; halo, corona; rim (of a wheel); hoop.

couronnement, m. crowning; cap, coping, top, top piece; coronation.

— en éventail, wing top (of a burner).

couronner, v.t. crown; cap.

courra, fut. 3 sing. of courir.

courrier, m. mail, post, correspondence; courier, messenger.

courroi, m. (Dyeing) roller.

courroie, f. belt; strap; band.

— de transmission, driving belt, belt.

courroie bandoulière, shoulder strap, shoulder belt.

— sans fin, endless belt.

courroyage, courroyer, etc. See corroyer, corroyage, etc.

cours, m. course; price, rate; market price; length.

avoir —, be current, circulate.

— de chimie, chemistry course (specif., a course of lectures or recitations as distinguished from laboratory work); (as a title) textbook on chemistry.

course, f. course, race, run; journey, trip, walk, ride, drive; distance; stroke (of a piston); throw (of a crank); flight (of a projectile).

coursier, m. mill race; (Calico) a cloth made to travel underneath the cloth to be printed in order to prevent soiling and to improve the printing.

court, a., m. & adv. short. — pr. 3 sing. of courir.

courtage, m. brokerage.

court-circuit, m. (Elec.) short circuit.

court-circuiter, v.t. short-circuit.

courtement, adv. shortly, briefly.

courtier, m. broker.

courtois, a. courteous.

couru, p.p. of courir.

cous, m. whetstone.

cousant, p.a. sewing, etc. (see coudre).

cousent, pr. 3 pl. of coudre.

couseuse, f. sewer; stitcher.

coussin, m. cushion.

coussiner, v.t. cushion, pad.

coussinet, m. pad, cushion; (Mach.) journal bearing, (journal) box, pillow, pillow block, bush, bushing, brass; (of rails) chair; bolster.

— à billes, ball bearing.

— à fileter, — de filière, screw die.

cousso, m. (Pharm.) cusso; (Bot.) any plant of the genus Hagenia.

cousu, p.a. sewed; attached; (of the mouth) closed.

coût, m. cost, price, expense.

coûtant, p.pr. of coûter; (of price) cost.

couteau, m. knife; (of a balance) knife edge; cutter.

— affilé, sharp knife.

— belge, (Expl.) granulating knife.

— émoussé, blunt knife.

— râcloir, scraping knife, scraper.

— rond, round knife.

coutelet, m. small knife.

coutellerie, f. cutlery; cutlery works or shop.

coutelure, f. damage done to leather or skin by the knife used in dressing.

coûter, v.i. cost; be dear (costly); be painful.

coûteux, a. costly, expensive, dear.

coutil, *m.* ticking, tent cloth, drill, duck.

coutume, *f.* custom.

de —, customary, customarily.

coutumier, *a.* accustomed; customary.

coutumièrement, *adv.* customarily, usually.

couture, *f.* seam; sewing, stitching.

couver, *v.t.* brood on, incubate; brood over.

couvercle, *m.* cover, lid; cap, top.

— *à anneau,* ringed cover, ring cover, ringed lid.

couvert, *p.a.* covered, etc. (see couvrir); obscure; secret; deep-colored (wine); cloudy (weather); sheltered (country). — *m.* cover; lodging.

couverte, *f.* covering; facing; (*Ceram.*) glaze, esp. a hard, difficultly fusible aluminous variety.

— *écaille,* (*Ceram.*) a reddish brown glaze resembling tortoise shell.

couverture, *f.* cover, covering; roof, roofing; blanket, coverlet; (*Com.*) security.

couvre, *pr. 3 sing. indic. & subj.* of couvrir.

couvre-amorce, *m.* primer cap, primer cover.

couvre-objet, *m.* (*Micros.*) cover glass.

couvrir, *v.t.* cover; face; screen; outbid. — *v.r.* cover oneself; be covered; be cloudy.

covelline, *f.* (*Min.*) covellite, covelline.

C.r., *abbrev.* Comptes rendus (which see).

crachat, *m.* sputum, spit, spittle.

crachement, *m.* spitting, etc. (see cracher).

cracher, *v.i.* spit; sputter; (of boilers) prime, foam; start at the joints; (*Elec.*) spark; (*Founding*) run over. — *v.t.* spit, spit out, throw out.

Cracovie, *f.* Cracow, Krakow.

cracque, *f.* crack, fissure.

craie, *f.* chalk.

— *de Briançon,* (*Min.*) steatite.

— *lavée,* — *préparée,* (*Pharm.*) prepared chalk.

— *précipitée,* (*Pharm.*) precipitated chalk.

craignant, *p.pr.* of craindre.

craignent, *pr. 3 pl. indic. & subj.* of craindre.

craindre, *v.t.* fear, be afraid of; (of objects) be liable to injury from. — *v.i.* fear, be afraid; hesitate (with *de*).

je crains qu'il ne vienne, I fear he will come.

je crains qu'il ne vienne pas, I fear he will not come.

craint, *p.a.* feared, dreaded.

crainte, *f.* fear, apprehension.

de — *de,* for fear of.

de — *que,* for fear that, lest.

cramoisi, *a. & m.* crimson.

crampe, *f.* cramp.

crampon, *m.* cramp, cramp iron, clamp; staple.

cramponner, *v.t.* cramp, clamp.

cramponnet, *m.* small cramp or clamp; staple.

cran, *m.* notch, nick; defect in welding; (*Bot.*) horseradish.

— *de Bretagne,* horseradish.

cranage, *m.* notching.

crâne, *m.* skull, cranium. — *a.* bold.

crânien, *a.* cranial.

crapaud, *m.* toad; mushroom anchor; armchair.

crapaudine, *f.* bearing, bushing, bush, brass; socket; pivot; escape valve; grating, screen (of a waste pipe).

craquelage, *m.* crackle manufacture.

craquelé, *a.* (*Ceram. & Glass*) crackled. — *m.* crackling.

craquelée, *f.* crackle, crackle ware.

craquèlement, *m.* crackling, crackled state.

craqueler, *v.t.* crackle (as porcelain or glass).

craquelin, *m.* hard biscuit, cracker.

craquellement, *m.* = craquèlement.

craquelure, *f.* cracking (as of varnish or paint).

craquement, *m.* cracking; crackling; creaking.

craquer, *v.i.* crack; crackle; creak.

craquetant, *p.a.* crackling.

craquètement, *m.* crackling, decrepitation.

craqueter, *v.i.* crackle, decrepitate.

craquette, *f.* impurities removed from melted butter.

craqure, *f.* crack, fissure.

crasse, *f.,* **crasses,** *f. pl.* (of metals) dross; (*Leather*) a scum of lime soaps, albuminoid matter, etc.; dirt, filth; fouling.

crasse noire, black dross, specif. a mixture of lead antimonate and antimonite formed in lead refining.

crassement, *m.* fouling.

crasser, *v.t.* foul; dirty.

cratère, *m.* crater.

crayer, *v.t.* chalk, mark with chalk. — *m.* vitrified coal ashes.

crayeux, *a.* chalky.

crayon, *m.* crayon; pencil; (*Elec.*) carbon; sketch, outline.

— *à dessiner,* drawing pencil.

— *à gaine,* incased crayon, pencil.

— *à lumière,* electric-light carbon.

— *d'azotate d'argent,* silver nitrate stick.

— *de carbone,* carbon rod, carbon.

— *de couleur,* colored pencil.

— *de graphite,* — *de mine de plomb,* lead pencil.

— *électrique,* arc-light carbon.

crayonnage, *m.* drawing.

crayonner, *v.t.* draw; mark; sketch.

crayonneux, *a.* chalky.

créance, *f.* credit; credence; claim, debt.

créancier, *m.* creditor.

créateur, *a.* creative. — *m.* creator.

créatine, *f.* creatine.

créatinine, *f.* creatinine.

créatinique, *a.* creatinic.
création, *f.* creation.
créatrice, *a. fem.* of créateur.
crédibilité, *f.* credibility.
crédit, *m.* credit.
créditer, *v.t.* credit.
créer, *v.t.* create.
crémaillère, *f.* rack, toothed bar, rack bar.
crème, *f.* cream; sirupy liqueur.
   — *d'amandes,* almond cream.
   — *de chaux,* cream of lime.
   — *de soufre,* sublimed sulfur, flowers of sulfur.
   — *de tartre,* cream of tartar.
   — *de tartre soluble,* soluble cream of tartar (a
     product of the solution of cream of tartar
     in boric acid or borax).
   — *froide,* cold cream.
crémer, *v.i.* cream. — *v.t.* cremate.
crémeux, *a.* creamy.
crémomètre, *m.* creamometer.
crémone, *f.* a kind of sash latch.
créneler, *v.t.* indent; notch, cog, tooth.
crénelure, *f.* indentation, indenting, notching.
créosol, *m.* creosol.
créosotage, *m.* creosoting.
créosotal, *m.* (*Pharm.*) creosotal.
créosote, *f.* creosote.
   — *du goudron de bois,* wood-tar creosote.
créosoter, *v.t.* creosote.
crêpage, *m.* dressing or glossing of crêpe; crap-
   ing, crimping.
crêpe, *m.* crêpe; specif., crape.
crêper, *v.t.* crape, crimp, crisp.
crépi, *m.* rough coat (of plaster).
crépine, *f.* strainer; fringe; omentum, caul.
crépir, *v.t.* grain (leather); rough-cast, rough-
   coat (walls); crisp (hair).
crépissage, *m.* graining, etc. (see crépir).
crépitant, *p.a.* crepitating, crepitant.
crépitation, *f.,* **crépitement,** *m.* crepitation,
   crackling.
crépiter, *v.i.* crepitate, crackle.
crépon, *m.* crépon (a thick craped fabric).
crépu, *a.* crisped, curled, (of hair) woolly.
crépuscule, *m.* twilight (either dawn or dusk).
crésol, *m.* cresol.
cresson, *m.* (*Bot.*) cress.
   — *de fontaine,* watercress.
   — *Para,* Para cress (*Spilanthes oleracea*).
crésyl, *m.* cresyl (an antiseptic).
crésyle, *m.* cresyl (the radical).
crésylique, *a.* cresylic.
crésylite, *f.* cresylite (trinitro-*m*-cresol).
crésylol, *m.* cresol.
crétacé, *a.* cretaceous.
crête, *f.* crest.
crétin, *m.* (*Med.*) cretin.
cretons, *m.pl.* cracklings, greaves (of tallow).
creusage, creusement, *m.* digging, etc. (see
   creuser).

creuse, *a. fem.* of creux.
creuser, *v.t.* dig, excavate, hollow out; go
   deeply into, study carefully. — *v.r.* become hollow.
creuset, *m.* crucible.
   — *brasqué,* brasqued crucible.
   — *de Gooch,* Gooch crucible.
   — *de Hesse,* Hessian crucible.
   — *de Paris,* French crucible, Paris crucible.
   — *de Rose,* Rose crucible (with tube passing
     thru the cover).
   — *en argent,* silver crucible.
   — *en biscuit,* unglazed porcelain crucible.
   — *en charbon de cornue,* retort-carbon cru-
     cible.
   — *en fer forgé,* wrought-iron crucible.
   — *en fonte,* cast-iron crucible.
   — *en grès,* sand crucible, stoneware crucible.
   — *en platine,* platinum crucible.
   — *en plombagine,* graphite crucible, black-
     lead crucible.
   — *en porcelaine,* porcelain crucible.
   — *en silice,* silica crucible.
   — *en terre,* clay crucible.
   — *en terre de Paris,* = creuset de Paris.
creusetier, creusiste, *m.* crucible maker.
creusure, *f.* hollowing, hollow.
creux, *a.* hollow; (of iron, etc.) tubular; empty;
   deep; (of food) unsubstantial. — *m.*
   cavity, hollow, hole; mold (for casting);
   depth (of a ship or a cog).
   en —, in intaglio.
crevasse, *f.* crevice, fissure, crack; crevasse.
crevasser, *v.t., i. & r.* chink, crack.
crever, *v.i.* burst, break; (of animals) die.
   — *v.t.* burst, break; kill. — *v.r.* burst.
crevette, *f.* shrimp; prawn.
cri, *m.* cry; creaking; rustling (as of silk);
   grating.
   — *de l'étain,* tin cry, creaking of tin.
criblage, *m.* sifting, screening, riddling.
crible, *m.* sieve, screen, riddle.
   — *à secousse,* shaking sieve, swing sieve.
   — *sasseur,* bolter.
cribler, *v.t.* sift, screen, riddle.
cribleur, *m.* sifter.
criblure, *f.* siftings.
cric, *m.* jack, lifting jack.
   — *à vis,* screw jack.
crier, *v.i.* cry, cry out, squeal, bark, etc.; creak;
   rustle; grate; scold. — *v.t.* cry; proclaim;
   offer for sale.
Crimée, *f.* Crimea.
criminel, *a. & m.* criminal.
crin, *m.* (coarse) hair, esp. horsehair; bristles
   — *végétal,* vegetable fiber, esp. palm fiber.
criquet, *m.* locust; cricket.
criqûre, *f.* crack, fissure.
crise, *f.* crisis.
crispation, *f.* shriveling, etc. (see crisper).

crisper, *v.t.* shrivel, contract; wrinkle; irritate.
— *v.r.* shrivel, contract.

crissement, *m.* grating, grinding.

crisser, *v.i.* grate, grind.

crissure, *f.* crease, wrinkle, unevenness.

cristal, *m.* crystal; crystal glass, crystal, flint glass; an object of glass, esp. crystal glass.
— *de roche,* rock crystal.
— *d'Islande,* Iceland spar.
— *mixte,* mixed crystal.

cristaux de soude, soda crystals, sal soda (sodium carbonate decahydrate).

cristaux de tartre, (*Com.*) tartar (recrystallized but not as highly purified as cream of tartar).

cristaux de Vénus, crystals of Venus (old name for crystallized copper acetate).

cristallerie, *f.* art of making crystal glass or objects of it; crystal glass factory.

cristallière, *f.* rock-crystal mine.

cristallifère, *a.* crystalliferous.

cristallin, *a.* crystalline. — *m.* (*Anat.*) crystalline lens.

cristalline, *f.* (*Physiol. Chem.*) crystallin.

cristallinité, *f.* crystallinity.

cristallisabilité, *f.* crystallizability.

cristallisable, *a.* crystallizable; (of acetic acid) glacial.

cristallisant, *p.a.* crystallizing.

cristallisation, *f.* crystallization.
— *fractionnée,* fractional crystallization.

cristallisé, *p.a.* crystallized.

cristalliser, *v.i., t. & r.* crystallize.

cristallisoir, *m.* crystallizing dish; crystallizing pan, crystallizing vessel, crystallizer.

cristallite, *f.* (*Petrog.*) crystallite.

cristallitique, *a.* (*Petrog.*) crystallitic.

cristallogénie, *f.* crystallogeny.

cristallogénique, *a.* crystallogenic.

cristallographe, *m.* crystallographer.

cristallographie, *f.* crystallography.

cristallographique, *a,* crystallographic.

cristallographiquement, *adv.* crystallographically.

cristalloïde, *a. & m.* crystalloid.

cristallométrie, *f.* crystallometry.

cristallométrique, *a.* crystallometric.

cristallotechnie, *f.* crystallotechny, art of producing crystals artificially.

cristaux, *pl.* of cristal.

cristé, *a.* crested, tufted.

criste-marine, *f.* = christe-marine.

critère, critérium, *m.* criterion.

critiquable, *a.* open to criticism.

critique, *a.* critical. — *f.* criticism; critique.
— *m.* critic.

critiquement, *adv.* critically.

critiquer, *v.t.* criticize.

croc, *m.* hook; claw, catch; crunch (the sound).

crocéine, *f.* crocein.

crocéique, *a.* croceic.

crocher, *v.t.* hook.

crochet, *m.* hook, small hook; clasp, catch, pawl, etc.; rabble; bend, turn; steelyard; fang; crochet; (*Glass*) a notched block of iron used as a support for the blowpipe; (*Printing*) bracket, crotchet.
— *de brassage,* rabble.

crocique, *a.* croconic.

crocoïse, *f.* (*Min.*) crocoite, crocoisite.

croconique, *a.* croconic.

crocus, *m.* (*Chem. & Bot.*) crocus.

croire, *v.t.* believe; think. — *v.i.* believe.
— *v.r.* think oneself; depend on oneself; be believed.

croisé, *p.a.* crossed, etc. (see croiser). — *m.* twill; crusader.

croisée, *f.* (window) sash; window; crossing.

croisement, *m.* crossing.

croiser, *v.t.* cross; cross out; lap over; twill; twist slightly. — *v.i.* cross; cruise. — *v.r.* cross, be crossed, cross each other.

croisette, *f.* little cross; (*Min.*) cross-stone.

croiseur, *m.* cruiser; cross lode, cross vein.

croisillon, *m.* crossbar, crossarm, crosspiece.

croissance, *f.* growth.

croissant, *p.a.* growing, etc. (see croître).
— *m.* crescent; crescent-shaped object, e.g. a sickle; tread (of a tire).

croissent, *pr. 3 pl. indic. & subj.* of croître.

croit, *pr. 3 sing.* of croire.

croît, *pr. 3 sing.* of croître.

croître, *v.i.* grow; increase; grow longer; (of a stream) rise. — *v.t.* augment.

croix, *f.* cross; (*Printing*) dagger, †.
en —, crosswise.

cron, *m.* sandy soil.

croquant, *p.a.* crunching, crisp. — *m.* gristle.

croquer, *v.t.* crunch; sketch roughly. — *v.i.* be crisp, give a crunching sound.

croquis, *m.* sketch, outline.

crosne, *m.* Japanese artichoke (*Stachys*). Called also *crosne du Japon.*

crosse, *f.* butt, buttstock, stock; (*Mach.*) crosshead; bend, crook; crosier.
courbé en — de fusil, bent like a gunstock.

crot, *m.* earthenware pot for collecting turpentine.

croton, *m.* (*Bot.*) croton.

crotoné, *a.* (*Pharm.*) containing croton oil.

crotonique, *a.* crotonic.

crotonisation, *f.* crotonization (reaction typically shown in the formation of crotonaldehyde from two molecules of acetaldehyde).

crotte, *f.* dung; dirt, filth, mire.

crotter, *v.t.* dirty, foul.

crottin, *m.* dung (of horses, mules, etc.).
— *de cheval,* horse dung.

crotton, *m.* a lump of sugar retained in sifting.

crouler, *v.i.* fall, fall in, sink.

croupi, *p.a.* stagnant.

croupion, *m.* rump.

croupir, *v.i.* stagnate; decay in water; lie in filth, wallow.

croupissant, *p.a.* stagnating, stagnant.

croupon, *m.* (*Leather*) butt.

croûte, *f.* crust; (*Ceram.*) rough form of what is to be a plate or other flat object; (*Med.*) scab; (*Tanning*) outer, fissured part of bark; (*Founding*) skin.

croyable, *a.* credible, likely.

croyance, *f.* belief, credence.

croyant, *p.a.* believing. — *m.* believer.

cru, *a.* crude, raw; (of water) hard; rude, rough, harsh; indecent. — *m.* growth, (of wine) particular vineyard or region; fabrication. — *p.p.* believed, thought.

crû, *p.p.* of croître.

cruauté, *f.* cruelty.

cruche, *f.* pitcher, jug.

cruchée, *f.* pitcherful, jugful.

cruchette, *f.* small pitcher or jug.

cruchon, *m.* small pitcher or jug; stone bottle.

crucifères, *f.pl.* (*Bot.*) Cruciferæ.

crudité, *f.* crudeness, rawness; (of water) hardness; crudity; (*pl.*) raw vegetables or fruit.

crue, *f.* growth, increase; (of water) rise.

cruel, *a.* cruel; tiresome.

cruenté, *a.* bloody.

cruor, *m.* (*Physiol.*) cruor.

crurent, *p.def. 3 pl.* of croire.

crûrent, *p.def. 3 pl.* of croître.

crusocréatinine, *f.* crusocreatinine.

crustacés, *m.pl.* (*Zoöl.*) Crustacea, crustaceans.

crut, *p.def. 3 sing.* of croire.

crût, *p.def. & imp. subj. 3 sing.* of croître. — *imp. subj. 3 sing.* of croître.

cryogène, *m.* cryogen, freezing mixture.

cryohydrate, *m.* cryohydrate.

cryolithe, cryolite, *f.* (*Min.*) cryolite.

cryomètre, *m.* cryometer.

cryométrie, *f.* cryometry.

cryophore, *m.* (*Physics*) cryophorus.

cryoscopie, *f.* cryoscopy.

cryptogame, *a.* (*Bot.*) cryptogamic, cryptogamous. — *m.* cryptogam.

cryptographique, *a.* cryptographic.

crypton, *m.* krypton, crypton.

cryptopine, *f.* cryptopine.

cubage, *m.* cubic capacity, cubic contents; cubature, determination of cubic contents.

cubature, *f.* cubature.

cube, *m.* cube. — *a.* cubic, cubical, cube.

cubèbe, *m.* cubeb (fruit of *Piper cubeba*).

cubébin, *m.,* **cubébine,** *f.* cubebin.

cuber, *v.t.* cube; measure in cubic units.

cubicite, cubizite, *f.* (*Min.*) cubicite (analcite).

cubilot, *m.* (*Metal.*) cupola, cupola furnace.

cubique, *a.* cubic, cubical, cube. — *f.* cubic.

cuboïde, *a. & m.* cuboid.

cubo-octaèdre, *m.* cuboctahedron, combination of cube and octahedron.

cucurbitacé, *a.* (*Bot.*) cucurbitaceous.

cucurbitacées, *f.pl.* (*Bot.*) Cucurbitaceæ.

cucurbite, *f.* cucurbit.

cudbear, *m.* cudbear.

cueillage, *m.* gathering.

cueillette, *f.* gathering; crop, harvest.

cueillir, *v.t.* gather; coil (rope); nip off (wire).

cuffat, cufat, *m.* (*Mining*) cage.

cuiller, cuillère, *f.* spoon; ladle; (auger) bit.
  cuiller à fondre (or *de coulée*), casting spoon, casting ladle.
  cuillère en corne, horn spoon.
  cuillère en fer, iron spoon, iron ladle.
  cuillère en os, bone spoon.
  cuillère en platine, platinum spoon.
  cuillère en porcelaine, porcelain spoon.
  cuillère en verre, glass spoon.

cuillerée, *f.* spoonful; ladleful.

cuilleron, *m.* bowl (of a spoon).

cuir, *m.* leather; hide; skin.
  — *à empeignes,* upper leather, vamp leather.
  — *à gaz,* leather for gas fittings.
  — *aggloméré,* an imitation leather made from leather scrap, pancake.
  — *à œuvre,* = cuir d'œuvre.
  — *à semelle(s),* sole leather.
  — *bouilli,* waxed leather; molded leather.
  — *brut,* raw hide, green hide.
  — *chagrin,* shagreen.
  — *chamoisé,* chamois leather, chamois.
  — *chevelu,* scalp.
  — *chromé,* chrome-tanned leather.
  — *ciré,* waxed leather, wax leather.
  — *corroyé,* curried leather.
  — *cru,* rough leather.
  — *d'Allemagne,* horse leather.
  — *de bœuf,* ox hide.
  — *de buffle,* buff, buff leather.
  — *de Cordoue,* Cordovan leather, cordovan.
  — *de Hongrie,* Hungary leather.
  — *de molleterie,* soft leather.
  — *de papier,* a paper imitation of leather.
  — *de poule,* fine sheepskin, glove leather.
  — *de Russie,* Russia leather.
  — *de Valachie,* Wallachian leather, leather dressed in scourings.
  — *d'œuvre,* soft or thin leather, such as is used for vamps.
  — *dur,* strong leather.
  — *embouti,* hydraulic leather.
  — *en cire,* = cuir ciré.
  — *en croûte,* leather tanned but not curried.
  — *en huile,* oiled leather.

cuir en plein suif, leather fully curried.
— en poils, green hide, raw hide.
— en suif, tallowed leather.
— étiré, scraped leather.
— factice, artificial leather, imitation leather.
— fort, stout leather, specif. sole leather.
— fossile, a variety of asbestos.
— hongroyé, Hungary leather.
— jusé, ooze leather.
— laqué, enameled leather, patent leather.
— lissé, smooth leather.
— maroquiné, morocco leather, morocco.
— mégissé, tawed hide; tawed leather.
— mou, — molleterie, soft leather.
— moulé, molded leather.
— plaqué, flattened leather, rolled leather.
— plat, leather from a thin hide.
— préparé, dressed leather.
— rond, thick leather.
— salé, salted hide.
— tanné, tanned hide; tanned leather.
— verni, patent leather, japan leather, enamel leather.
— vert, green hide, raw hide.

cuirasse, f. armor, armor plate; cuirass.
cuirassé, p.a. armored, armor-plated; armed, prepared. — m. armored vessel, specif. battleship.
— de station, armored cruiser.
cuirassement, m. armor, armor plate; armoring.
cuirasser, v.t. armor, armor-plate; arm with a cuirass; arm, prepare.
cuire, v.t. burn (brick, cement, lime, etc.); boil (sirup, soap, varnish, silk, etc.); char (wood); cook; bake; ripen. — v.i. cook, be cooked; bake; smart, hurt.
cuirer, v.t. cover with leather.
cuir-toile, n. leather cloth.
cuisage, m. charring.
cuisant, p.a. burning, etc. (see cuire); sharp, keen, biting, smarting; suitable for cooking.
cuisent, pr. 3 pl. indic. & subj. of cuire.
cuiseur, m. burner (of pottery, brick, etc.), kilnman, fireman; boiler (of sirup, silk, etc.); baker (of cheese); cooker; (Paper) cooker, boilerman; (Brewing) kettle man; (Sugar) vacuum panman.
cuisine, f. kitchen; cooking, food; cookery.
cuisse, f. thigh.
cuisson, m. burning, boiling, cooking, etc. (see cuire); burned, boiled or cooked state; smart, smarting.
— à la volée, (Brick) clamp firing.
— en dégourdi, (Ceram.) biscuit baking.
cuit, p.a. burned, boiled, cooked, etc. (see cuire); done; ripe. — pr. 3 sing. of cuire.
— au filet, (of sirup) boiled to thread.
cuite, f. burning, boiling, cooking, etc. (see cuire); batch, heat. — p.a. fem. of cuit.

cuite en grains, (Sugar) boiling to grain.
cuivrage, m. coppering, etc. (see cuivrer); copper sheathing.
— galvanique, copper electroplating.
cuivre, m. copper; (pl.) brasses, brass or copper fittings, instruments or the like.
— ammoniacal, any ammoniated copper compound or solution, specif. cuprammonium (Schweitzer's reagent); (Pharm.) ammoniated copper, copper ammonio-sulfate.
— battu, wrought copper; copper foil.
— blanc, white copper (a white alloy of copper, such as white tombac).
— bleu, (Min.) azurite.
— brut, crude copper, unrefined copper.
— corné, cuprous chloride.
— cru, crude copper, unrefined copper.
— de Chypre, blue vitriol, cupric sulfate.
— électrolytique, electrolytic copper.
— gris, (Min.) gray copper, tetrahedrite.
— jaune, brass.
— laminé, sheet copper.
— natif, native copper.
— noir, (Metal.) black copper, blister copper.
— panaché, (Min.) bornite.
— phosphaté, copper phosphate.
— pyriteux, (Min.) copper pyrites, chalcopyrite.
— rosette, rosette copper.
— rouge, pure (unalloyed) copper; (Min.) red copper, cuprite.
— sulfaté, copper sulfate.
— vierge, virgin copper.
cuivré, p.a. coppered, etc. (see cuivrer); (of color) copper, coppery.
cuivrer, v.t. copper, cover or sheathe with copper, copperplate; treat with copper or a copper salt.
cuivreux, a. cuprous (as, sulfocyanure cuivreux, cuprous sulfocyanide); copper, coppery, cupreous (as, pyrites cuivreuses, copper pyrites).
cuivrique, a. cupric.
cul, m. bottom, breech; (of a ship) poop.
culasse, f. breech, breechblock; (of a cut stone) culet, culasse.
culbuter, v.t. overturn, upset; overthrow, vanquish. — v.i. turn over, somersault, tumble, fall.
culbuteur, m. tripper; rocker, rocking lever.
cul-de-lampe, m. bracket; (of guns) cascabel.
cul-de-poule, m. swelling, protuberance.
culée, f. abutment; part of a hide next the tail.
culilavan, n. culilawan, clove bark.
culinaire, a. culinary.
culminant, p.a. culminating.
culminer, v.i. culminate.

culot, *m.* culot, mass (usually of metal or matte and then also called regulus) at the bottom of a crucible; bottom; base; latest, youngest or smallest member.

culotte, *f.* Y-tube, Y-pipe, Y-joint, Y; breeches.

cultivateur, *m.* farmer, agriculturist; cultivator. — *a.* agricultural, farming.

cultivé, *p.a.* cultivated.

cultiver, *v.t.* cultivate. — *v.r.* be cultivated.

culture, *f.* culture; cultivation; cultivated land; crop.
 — *en gouttelette,* (*Bact.*) droplet culture.
 — *en piqûre,* (*Bact.*) stab culture.
 — *en plaque,* (*Bact.*) plate culture.
 — *en strie,* (*Bact.*) streak culture.

cumène, *m.* cumene (properly used of isopropylbenzene but applied also to a mixture of mesitylene and pseudocumene).

cumin, *m.* (*Bot.*) cumin.
 — *des prés,* caraway.

cuminique, *a.* cumic, cuminic.

cuminol, *m.* cumaldehyde, cuminole.

cumulatif, *a.* cumulative.

cunila, *n.* (*Bot.*) Cunila (a genus of which dittany, *C. origanoides,* is a member).

cupferrone, *f.* cupferron.

cupide, *a.* greedy, covetous.

cupréine, *f.* cupreine.

cupricalcique, *a.* of copper and calcium (as, *acétate cupricalcique,* CuCa(C$_2$H$_3$O$_2$)$_4$ · 6 H$_2$O).

cuprico-uraneux, *a.* of copper and uranium (as, *phosphate cuprico-uraneux,* Cu(UO$_2$)$_2$-(PO$_4$)$_2$ · 8 H$_2$O).

cuprifère, *a.* cupriferous.

cuprique, *a.* of or pertaining to copper (the French word for " cupric " is *cuivrique*).

cupro-ammonique, *a.* cuprammonium.

cupro-potassique, cupropotassique, *a.* of copper and potassium.

cupule, *f.* (small) cup, cupule.

curaçao, *m.* curaçao (a liqueur).

curage, *m.* cleaning, cleansing; (*Bot.*) persicaria.

curain, *m.* saline incrustation; specif., a mixture of sodium and calcium sulfates obtained in the evaporation of sea water.

curare, *m.* curare, curari.

curarine, *f.* curarine.

curatier, *m.* tanner or currier.

curatif, *a.* curative.

curcuma, *m.* turmeric, curcuma; curcuma (the plant).

cure, *f.* cure.

cure-feu, *m.* poker.

curement, *m.* cleaning, cleansing; dredging.

curer, *v.t.* clean, clean out, cleanse.

curette, *f.* cleaner, scraper, spoon (as for exploring for oil or for cleaning a blast hole); (*Med.*) curette.

cureur, *m.* cleaner, cleanser; dredger.

curieusement, *adv.* curiously.

curieux, *a.* curious; careful; interested (in).
 — *m.* curious person or thing; connoisseur.

curin, *m.* = curain.

curiosité, *f.* curiosity.

curseur, *m.* slide, slider; index, pointer. — *a.* sliding.

cursoir, *m.* slide.

curure, *f.* mud, slime (as from a ditch).

curvilatère, *a.* with curved side or sides.

curviligne, *a.* curvilinear, curvilineal.

cuscamine, *f.* cuscamine.

cusconine, *f.* cusconine.

cuscute, *f.* (*Bot.*) dodder (*Cuscuta*).

cusparine, *f.* cusparine.

custode, *m.* custodian, keeper.

cutané, *a.* cutaneous, of the skin.

cuticulaire, *a.* cuticular, of the cuticle.

cuticule, *f.* cuticle.

cutine, *f.* cutin.

cutinisation, *f.* cutinization.

cutiniser, *v.t.* cutinize.

cuvage, *m.* fermentation of grape mash in vats before pressing (in making red wine); tub house.

cuve, *f.* vat; tub, tank, copper, cistern; (pneumatic) trough; jar, cell (as for spectrum analysis); (photographic) tray; (of a blast furnace) interior, esp. the shaft.
 — *à absorption,* absorption cell (for spectral analysis).
 — *à eau,* water trough.
 — *à fermentation,* fermentation vat.
 — *à huile,* (*Metal.*) oil-tempering tank.
 — *à la couperose,* copperas vat.
 — *à lavage,* (*Photog.*) washing tray.
 — *à mercure,* mercury trough.
 — *à moût,* (*Brewing*) mash tun, mash vat.
 — *de fermentation,* fermentation vat.
 — *de précipitation,* precipitation tank.
 — *de trempage,* (*Malt*) steeping cistern.
 — *d'indigo,* indigo vat.
 — *matière,* (*Brewing*) mash vat, mash tun.

cuveau, *m.* small vat, tub or tank.

cuvée, *f.* contents of a vat, tub or tank.

cuvelage, *m.* tubbing, lining (as for wells).

cuveler, *v.t.* tub, line (as a well).

cuve-matière, *m.* (*Brewing*) mash vat, mash tun.

cuve-mouilloire, *n.* steeping cistern.

cuver, *v.t. & i.* ferment (in a vat).

cuverie, *f.* room or building in which fermentation vats are housed.

cuvette, *f.* cistern, reservoir (as of a barometer); jar; tray (as for developing); bulb (of a thermometer); cup (e.g., of a primer); basin; sink; (*Micros.*) staining box or tray; (*Acids*) bottom of a lead chamber, in which the sulfuric acid collects; (*Glass*) a small pot formerly used in refining;

(*Soda*) the mildly heated pan or first hearth of a salt-cake furnace.

cuvette *à photographie*, photographer's tray.
— *à rainures*, (*Micros.*) grooved staining box.
— *d'egouttage*, drip cup.

cuvier, *m.* tub.

cyanacétique, *a.* cyanoacetic, cyanacetic.

cyanamide, *f.* cyanamide.
— *calcique*, calcium cyanamide.

cyanamide-calcium, *f.* calcium cyanamide.

cyanamine. *f.* cyanamine; cyanamide.

cyanate, *m.* cyanate.

cyane, *m.* cyanogen.

cyané, *a.* of or containing cyanogen, cyano.

cyanéthine, *f.* cyanethine (4-amino-2,6-diethyl-5-methylpyrimidine).

cyaneux, *a.* cyanous (*Obs.*).

cyanhydrine, *f.* cyanohydrin, cyanhydrin.

cyanhydrique, *a.* hydrocyanic.

cyanide, *m.* cyanide.

cyanilide, *f.* cyanilide (carbanilonitrile).

cyanine, *f.* cyanin, cyanine (both the plant color and the artificial dye).

cyanique, *a.* cyanic.

cyanite, *f.* (*Min.*) cyanite, kyanite.

cyanofer, *m.* ferrocyanogen.

cyanoferrate, *m.* ferrocyanide, ferrocyanate.

cyanoferre, *m.* ferrocyanogen.

cyanoferrique, *a.* ferrocyanic.

cyanoferrure, *m.* ferrocyanide.

cyanogène, *m.* cyanogen.

cyanogéné, *a.* containing or pertaining to cyanogen.

cyanogénèse, *f.* cyanogenesis.

cyanogénétique, *a.* cyanogenetic.

cyanophycées, *f.pl.* (*Bot.*) Cyanophyceæ (Schizophyceæ).

cyanose, *f.* cyanosis.

cyanosé, cyanotique, *a.* cyanotic, cyanosed.

cyanurate, *m.* cyanurate.

cyanuration, *f.* cyanidation, cyaniding.

cyanure, *m.* cyanide.
— *d'argent*, silver cyanide.
— *de benzyle*, benzyl cyanide. α-tolunitrile.
— *de potasse*, potassium cyanide.
— *jaune*, potassium ferrocyanide.
— *rouge*, potassium ferricyanide.

cyanuré, *a.* cyanided; cyanide; cyanide of.

cyanurer, *v.t.* cyanide.

cyanurique, *a.* cyanuric.

cyclamine, *f.* cyclamin.

cycle, *m.* cycle.
— *fermé*, closed cycle.

cyclique, *a.* cyclic.

cycloheptane, *m.* cycloheptane.

cycloheptène, *m.* cycloheptene.

cyclohexane, *m.* cyclohexane.

cyclohexanol, *m.* cyclohexanol.

cyclohexanone, *f.* cyclohexanone.

cyclohexène, *m.* cyclohexene.

cyclohexylamine, *f.* cyclohexylamine.

cyclohexyle, *m.* cyclohexyl.

cycloïdal, *a.* cycloidal.

cycloïde, *f.* cycloid.

cyclométhylénique, *a.* cyclomethylene (designating the cycloparaffins or polymethylenes).

cyclomètre, *m.* cyclometer.

cyclopentadiène, *m.* cyclopentadiene.

cyclopentane, *m.* cyclopentane.

cyclopentanone, *f.* cyclopentanone.

cyclopentène, *m.* cyclopentene.

cycloptérine, *f.* cyclopterine.

cyclotétranol, *m.* cyclotetranol (cyclobutanol).

cyclotétrylamine, *f.* cyclotetrylamine (cyclobutylamine).

cygne, *m.* swan.

cylindrage, *m.* calendering; rolling; cylindering.

cylindre, *m.* cylinder; roller, roll; arbor, shaft.
— *à chemise*, jacketed cylinder.
— *à cingler*, shingling roll.
— *à fil*, wire roll.
— *à laminer*, roll, roller.
— *à tôle*, plate roll.
— *à vapeur*, steam cylinder.
— *broyeur*, crusher roll, crusher.
— *compresseur*, roller.
— *creux*, hollow cylinder.
— *dégrossisseur*, coarse-crushing roll.
— *droit*, right cylinder.
— *finisseur*, finishing roll.
— *fournisseur*, feeding cylinder, feed roll.
— *presseur*, (*Calico*) pressure cylinder, the cylinder on which the cloth is carried while being printed.

cylindrée, *f.* cylinderful.

cylindrer, *v.t.* calender (paper, cloth, etc.); roll (leather or a road); cylinder.

cylindreur, *m.* calender, calenderer; roller.

cylindrique, *a.* cylindric(al).

cylindriquement, *adv.* cylindrically.

cymène, cymol, *m.* cymene.

cynurine, *f.* kynurine, cynurine.

cyprès, *m.* cypress.

cypripède, *m.* (*Bot.*) cypripedium, lady's-slipper.

cystéine, *f.* cysteïne.

cystine, *f.* cystine.

cystique, *a.* cystic.

cystoïde, *a.* cystoid.

cytisine, *f.* cytisine.

cytologie, *f.* cytology.

cytologique, *a.* cytological, cytologic.

cytolyse, *f.* cytolysis.

cytolytique, *a.* cytolytic.

cytoplasma, cytoplasme, *m.* (*Biol.*) cytoplasm.

cytosine, *f.* cytosine.

D

d'. Contraction of *de.*

d'abord. See abord.

dactylotype, *f.* typewriter.

dactylotypie, *f.* typewriting.

dague, *f.* scraper, scraping knife; dagger, dirk.

daguerréotyper, *v.t.* daguerreotype.

daguerréotypie, *f.* daguerreotypy.

daguerrien, *a.* of Daguerre, daguerreotypic.

dahlia, *m.* (*Dyes*) dahlia; (*Bot.*) dahlia.

dahline, *f.* dahlin (inulin).

daigner, *v.i.* deign; be pleased.

d'ailleurs. See ailleurs.

daim, *m.* buck, deer.

daleau, *m.* = dalot.

dallage, *m.* flagstone pavement, flagging.

dalle, *f.* slab, plate; flagstone, flag; gutter, conduit; (in sugar refining) a pipe conveying the sirup; shelf (of pottery or stone); whetstone; slice (as of fish).

— *rodée,* ground-glass plate.

daller, *v.t.* flag, pave with flagstones.

dalot, *m.* escape hole for liquids, vent.

dam., *abbrev.* (décamètre) decameter.

damage, *m.* ramming, tamping.

damas, *m.* damask (in various senses); specif., damask steel, Damascus steel; damson (plum); (*cap.*) Damascus.

damasquinage, *m.* damascening, damaskeening.

damasquiner, *v.t.* damascene, damaskeen.

damasquinerie, *f.* damascening, damaskeening.

damasquineur, *m.* damascener, damaskeener.

damasquinure, damasquinurie, *f.* damascening.

damassé, *a.* damasked, damask.

dame, *f.* dam; (*Metal.*) dam, damstone; rammer; lady, dame; (at cards) queen; (*pl.*) checkers.

dame-jeanne, *f.* demijohn.

damer, *v.t.* ram, tamp.

dameur, *m.* rammer, tamper.

dammar, *m.* dammar (the resin).

dammara, *m.* (*Bot.*) member of the genus *Dammara.*

damoiselle, *f.* rammer, beetle.

danger, *m.* danger, risk; trouble.

dangereux, *a.* dangerous.

danois, *a.* Danish.

dans, *prep.* in; into; according to.

danser, *v.i.* dance; flicker.

dans-œuvre, *adv.* (of measurements) in the clear.

daphnine, *f.* daphnin.

d'après. See après.

dard, *m.* pointed (dartlike) flame; dart; spindle; (of insects) sting; (of flowers) pistil.

darder, *v.t.* dart; shoot with (or as with) a dart.

dari, *m.* durra, dari (a kind of millet).

dartre, *f.* tetter, skin affection.

date, *f.* date.

daté, *p.a.* dated.

dater, *v.t. & i.* date.

datiscine, *f.* datiscin.

datte, *f.* date (the fruit).

dattier, *m.* date tree, date palm.

daturine, *f.* daturine (hyoscyamine).

dauphin, *m.* dolphin; mouth of a gutter pipe.

Dauphiné, *m.* Dauphiné (old French province).

dauphinelle, *f.* (*Bot.*) larkspur, delphinium.

d'autant. See autant.

davantage, *adv.* more; longer.

de, *prep.* of; from; (of time) in, during; because of; (of manner or means) with; (of an agent) by; (with comparatives) than, by; (with an infinitive) to . . . , of . . . ing (thus, *d'avoir* may usually be translated either "to have" or "of having"); (in partitive expressions) some, any (as, *de l'eau,* some water); (with *traiter*) as.

— *ce que,* because; altho.

dé, *m.* die (*pl.* dice); thimble; block (for propping); gage.

déalbation, *f.* dealbation, bleaching.

déauration, *f.* gilding.

débâcle, *f.* breaking up (of ice); debacle.

débâcler, *v.t.* open; clear.

déballage, *m.* unpacking.

déballer, *v.t.* unpack.

débander, *v.t.* unbind; unbend, relax; disband.

débarquement, *m.* landing, debarkation.

débarquer, *v.t. & i.* land, disembark.

débarrasser, *v.t.* free; clear, clear up, clear away. — *v.r.* free oneself, get rid.

débarrer, *v.t.* unbar; free from inequalities of dyeing.

débarreur, *m.* a mender of defects in dyeing.

débat, *m.* debate; (*Law, pl.*) trial. — *pr. 3 sing.* of débattre.

débâtir, *v.t.* demolish; remove basting from.

débattable, *a.* debatable.

débattre, *v.t. & i.* debate, dispute, discuss.
— *v.r.* struggle; debate; be debated.

débenzolage, *m.* removal of benzole or benzene
(*e.g.* from illuminating gas).

débenzoler, *v.t.* remove benzole or benzene from.

débet, *m.* debit.

débile, *a.* feeble, weak.

débilement, *adv.* feebly, weakly.

débilitant, *p.a.* debilitating.

débilité, *f.* feebleness, debility, weakness.

débit, *m.* delivery, flow, rate of flow; supply;
output; (*Mach.*) feed; debit; sale; (retail)
shop; license (to sell); cutting, sawing (of
wood, stone, etc.); delivery (in speaking).
de —, (*Com.*) in demand.

débitage, *m.* cutting up, sawing up (of wood,
etc.).

débiter, *v.t.* deliver, discharge (liquid, gas, etc.);
retail, sell at retail; cut up, saw up (wood,
stone, etc.); debit; utter, speak. — *v.r.* be
delivered; sell, be sold; (of wood, etc.) cut
up; be told, be reported.

débiteur, *m. & a.* debtor.

débituminisation, *f.* debituminization.

débituminiser, *v.t.* debituminize.

déblai, *m.* excavation, cutting; cut; (*pl.*) ex-
cavated material, rubbish.
sels de —, abraum salts.

déblayement, déblaiement, *m.* cutting, cutting away, excavation; clearing.

déblayer, *v.t.* cut away, clear away; clear.

déboire, *m.* aftertaste; disappointment; dis-
taste.

déboisement, *m.* clearing, deforestation.

déboiser, *v.t.* clear of timber, deforest.

déboiter, *v.t.* disconnect; dislocate.

débonder, *v.t.* unbung; open the vent of.
— *v.i. & r.* escape, run out, burst forth; discharge, be emptied.

débondonner, *v.t.* unbung.

débordement, *m.* overflow; flood, outburst.

déborder, *v.i.* overflow, run over; project; burst
forth. — *v.t.* overflow; project beyond;
outstrip. — *v.r.* overflow; burst forth.

débouché, *m.* outlet; end; expedient; opening,
market.

débouchement, *m.* unstopping, etc. (see
déboucher); outlet.

déboucher, *v.t.* unstop, unstopper, uncork;
open; (*Elec.*) unplug; set (a fuse); punch,
make (a hole). — *v.i.* pass out, debouch.

déboucler, *v.t.* unbuckle; uncurl.

débouillage, *m.* = débouilli.

débouilleur, *m.* (*Dyeing*) boiler; bleacher (of
yarn).

débouilli, *m.* (*Dyeing*) boiling, boil.

débouillir, *v.t.* (*Dyeing*) boil.

débouillissage, *m.* = débouilli.

débourbage, *m.* washing, etc. (see débourber).

débourber, *v.t.* wash, free from mud or slime,
(*Mining*) trunk, sluice; clean, clean out (as
a cistern).

débourbeur, *m.* a worker or apparatus that
cleanses from mud or slime, washer, cleanser. Specif.: a boiler attachment for removing the mud deposited by hard water;
a trommel for removing earthy matter from
ore; an apparatus in which the suspended
matter in gas liquor is deposited.

débourrage, *m.* removal of wadding, etc. (see
débourrer).

débourrer, *v.t.* remove the wadding, padding,
etc. from (see bourre); untamp; (*Leather*)
unhair.

débourser, *v.t.* disburse, expend.

debout, *adv.* upright, standing, erect, up.

débraiser, *v.t.* draw the embers or fire from.

débraser, *v.t.* unsolder, unbraze.

débrayage, *m.* disconnecting, disconnection,
etc. (see débrayer).

débrayer, *v.t.* disconnect, disengage, throw out
of gear, uncouple.

débrider, *v.i.* stop; unbridle one's horse.

débris, *m.* remains, fragments, debris, rubbish.
— *cellulosique*, (in flour analysis) fiber.
— *d'os*, (*Fertilizers*) bone meal.

débrouiller, *v.t.* untangle, unravel, clear up.

débrûler, *v.t.* deoxidize. (*Obs. or Rare.*)

débrutir, *v.t.* rough-polish.

début, *m.* beginning; début; first book; first
throw, first play.

débuter, *v.i.* begin; start; (in games) lead.

déca-. deca-, ten.

deçà, *prep.* on this side of. — *adv.* on this
side.
— *et delà*, here and there, on all sides.
en —, on this side.
par —, on this side, on this side of.

décacheter, *v.t.* unseal, open.

décadence, *f.* decadence, decline.

décagramme, *m.* decagram, 10 grams.

décahydrate, *m.* decahydrate.

décahydrure, *m.* decahydride.

décaisser, *v.t.* unpack, unbox; pay out.

décalage, *m.* unwedging, unkeying; difference,
divergence; (*Elec.*) lead (or lag).

décalcifier, *v.t.* decalcify, delime.

décalitre, *m.* decaliter, 10 liters.

décalquer, *v.t.* transfer (a design).

décamètre, *m.* decameter, 10 meters.

décantage, *m.* decanting, decantation.

décantateur, *m.* decanter.
— *en hélice*, helical decanter (for removing
suspended matter from water).

décantation, *f.* decantation, decanting.

décanter, *v.t.* decant.

décanteur, *m.* decanter. — *a.* decanting.

décapage, décapement, *m.* cleaning, etc. (see
décaper).

décaper, *v.t.* (of metals) clean, scour, pickle; dip; flux (metal before soldering); scrape; uncap.

décapeur, *m.* cleaner, etc. (see décaper).

décarbonater, *v.t.* decarbonate.

décarbonateur, *m.* decarbonator; specif., a tube for removing carbon dioxide from air.

décarbonisation, *f.* decarbonization.

décarboniser, *v.t.* decarbonize.

décarburant, *p.a.* decarbonizing, decarburizing.

décarburateur, *a.* decarbonizing, decarburizing.

décarburation, *f.* decarbonization, decarburation.

décarburer, *v.t.* decarbonize, decarburize.

dècarbureur, *m.* (*Gas*) purifier man.

décarburisation, *f.* decarbonization, decarburization.

décastère, *m.* decastere (10 cubic meters).

décatir, *v.t.* sponge, ungloss (cloth, etc.).

décatissage, *m.* sponging, unglossing.

décatisseur, *m.* sponger (of cloth to remove gloss).

décédé, *p.a. & m.* deceased.

décèlement, *m.* disclosure, disclosing, revealing, detection.

déceler, *v.t.* reveal, disclose, make known, detect.

décembre, *m.* December.

décennal, *a.* decennial, ten-year.

décentrage, *m.*, **décentration,** *f.*, **décentrement,** *m.* decentration, decentering.

décentrer, *v.t.* decenter.

déceptivement, *adv.* deceptively.

décerner, *v.t.* award, confer; decree.

décès, *m.* decease.

décevant, *p.a.* deceiving.

décevoir, *v.t.* deceive.

déchaîner, *v.t.* unchain; release.

décharge, *f.* discharge (in various senses); outlet, vent; place for keeping rubbish.
— *silencieuse*, (*Elec.*) silent discharge.

déchargement, *m.* discharging; unloading.

déchargeoir, *m.* outlet, vent, escape pipe, waste pipe.

décharger, *v.t.* discharge; remove the charge from; unload; empty. — *v.r.* discharge, unload or relieve oneself; be discharged; (of dyes) fade out; (of rivers) empty.

déchargeur, *m.* discharger; unloader; drawer (of coke, etc.).

déchéance, *f.* forfeiture; decadence, decay.

décherra, *fut. 3 sing.* of déchoir.

déchet, *m.* loss, waste; (*pl.*) waste, refuse; diminution.
déchets de verre, broken glass, cullet.

déchiqueter, *v.t.* cut up, cut in pieces.

déchiré, *p.a.* torn, etc. (see déchirer).

déchirement, *m.* tearing, etc. (see déchirer).

déchirer, *v.t.* tear; rend, lacerate, tear open, tear up; torture, distress; slander.

déchirure, *f.* tear, rent; tearing; laceration.

déchloruration, *f.* dechlorination.

déchlorurer, *v.t.* dechlorinate.

déchoir, *v.i.* fall, lose, decline, fall off.
— *d'un brevet,* forfeit a patent.

déchu, *p.p.* of déchoir.

déciare, *m.* deciare, 10 square meters.

décidé, *p.a.* decided.

décidément, *adv.* decidedly, positively.

décider, *v.t.* decide, determine, settle. — *v.i.* decide. — *v.r.* be decided; decide.
— *de,* decide on; decide to.

décidu, *a.* deciduous.

décigramme, *m.* decigram, 0.1 gram.

décilitre, *m.* deciliter, 0.1 liter.

décimal, *a.* decimal.

décimale, *f.* decimal.

décime, *m.* decime, 0.1 franc. —*f.* tenth (part).

décimètre, *m.* decimeter, 0.1 meter.
— *carré,* square decimeter.

décirage, *m.* removal of wax.

décirer, *v.t.* remove wax from.

décisif, *a.* decisive; peremptory.

décision, *f.* decision.

décistère, *m.* decistere, 0.1 cubic meter.

déclanche, *f.* disengaging gear, release.

déclanchement, *m.* = déclenchement.

déclancher, *v.t.* = déclencher.

déclaration, *f.* declaration; verdict.

déclarer, *v.t.* declare; reveal, disclose.
— *v.r.* manifest oneself, appear, occur; declare oneself.

déclenchement, *m.* unloosing, release; (*Mach.*) disengaging, disengagement, also disengaging gear; unlatching.

déclencher, *v.t.* unloose, release; (*Mach.*) disengage, throw out of gear; unlatch.

déclic, *m.* click, catch, pawl, trip; trigger.

déclin, *m.* decline.

déclinaison, *f.* declination; declension.

décliner, *v.t. & i.* decline.

déclive, *a.* sloping, inclined. — *f.* slope.

décliver, *v.i.* slope, be inclined.

déclivité, *f.* declivity, slope.

déclore, *v.t.* unclose, open.

décoaltarisation, *f.* removal of coal tar.

décoaltariser, *v.t.* free from coal tar.

décocté, *m.* decoction (the substance).
— *d'aloès composé,* (*Pharm.*) compound decoction of aloes.
— *de bois de Campêche,* (*Pharm.*) decoction of logwood.

décoction, *f.* decoction (act or substance).

décofféination, *f.* removal of caffeine.

décoiffage, *m.* uncapping (as of a fuse); uncorking.

décoiffer, *v.t.* uncap; uncork.

déçoit, *pr. 3 sing.* of décevoir.

décollage, décollement, *m.* separation (of things which are stuck together), detachment, ungluing.

décoller, *v.t.* separate (things stuck together), detach, unstick, unglue, deglutinate. — *v.r.* come unglued, come off, come apart.

décolletage, *m.* removal of the crown or top of plants (e.g., sugar beets); screw cutting.

décolleter, *v.t.* remove the crown or top of; cut, turn (screws).

décolorant, *p.a.* decolorizing, decolorant.

décoloration, *f.* decolorization, decoloration, decolorizing, decoloring.

décolorer, *v.t.* decolorize, decolor; discolor.

décolorimètre, *m.* decolorimeter.

décoloris, *m.* loss of color.

décoltharisation, *f.* = décoaltarisation.

décombres, *m.pl.* rubbish, débris.

décomposable, *a.* decomposable.

décomposant, *p.a.* decomposing.

décomposé, *p.a.* decomposed; (*Bot.*) decompound.

décomposer, *v.t.* decompose; alter. — *v.r.* decompose; change, be altered.

décomposeur, *m.* decomposer; specif., in the Deacon process, the chamber in which hydrochloric acid is decomposed.

décomposition, *f.* decomposition; alteration.

décompte, *m.* discount, deduction; itemizing, analysis (of an account); disappointment.

décompter, *v.t.* deduct.

déconfiture, *f.* discomfiture; insolvency.

décongélation, *f.* thawing, thawing out.

décongeler, *v.t.* thaw, thaw out.

déconseiller, *v.t.* counsel against.

déconsidérer, *v.t.* discredit.

décor, *m.* decoration.

décorateur, *m.* decorator.

décoratif, *a.* decorative.

décoration, *f.* decoration.

décordonnage, *m.* removal of crust from the stampers of a powder mill.

décorer, *v.t.* decorate.

décorticage, *m.*, **décortication,** *f.* decortication, etc. (see décortiquer).

décortiquer, *v.t.* decorticate, bark, husk, hull, shell, peel, mill.

découdre, *v.t.* unsew, unstitch, rip, rip off.

découler, *v.i.* trickle, flow (gently), drop; (fig.) flow, spring, follow.

découpage, *m.* cutting, etc. (see découper).

découper, *v.t.* cut, cut up, cut off, cut out; punch. — *v.r.* be cut, be cut up; stand out.

découpeur, *m.* cutter.

découpeuse, *f.* cutter; specif., an apparatus or machine used for cutting.

découpoir, *m.* cutter, cutting machine; cutting punch.

découpure, *f.* cutting, cutting out; cut; cutting, part cut out.

décourager, *v.t.* discourage. — *v.r.* be discouraged.

décourber, *v.t.* unbend, straighten.

décousant, *p.pr.* of découdre.

décousu, *p.p.* of découdre.

découvert, *p.a.* discovered; uncovered; exposed. — *m.* deficit.
à —, uncovered; unsheltered; openly; (*Com.*) without security.

découverte, *f.* discovery.

découvrant, *p.pr.* of découvrir.

découvreur, *m.* discoverer.

découvrir, *v.t.* discover; uncover; expose. — *v.r.* be discovered; be visible; reveal oneself; expose oneself; uncover oneself.

décramponer, *v.t.* remove the cramps from.

décrassage, *m.* cleaning, etc. (see décrasser).

décrasser, *v.t.* clean; wash, scour, scrape.

décrayonnage, *m.* cleaning (of a furnace).

décréditer, *v.t.* discredit.

décrépitation, *f.* decrepitation.

décrépiter, *v.i.* decrepitate.

décret, *m.* decree.

décreusage, décreusement, *m.* scouring, etc. (see décreuser).

décreuser, *v.t.* (*Silk*) scour, strip, ungum, degum; scour, cleanse.

décreuseur, *m.* scourer, etc. (see décreuser).

décrire, *v.t.* describe. — *v.r.* be described.

décrit, *p.p. & pr. 3 sing.* of décrire.

décrivant, *p.a.* describing.

décrochage, *m.* unhooking, etc. (see décrocher).

décrocher, *v.t.* unhook, unfasten; disengage.

décroissance, *f.* decrease, diminution, decline.

décroissant, *p.a.* decreasing, diminishing.

décroissement, *m.* decrease, diminution.

décroître, *v.i.* decrease, diminish.

décrotter, *v.t.* clean, clean off, brush.

décrouir, *v.t.* (*Metal.*) anneal.

décrouissage, *m.* (*Metal.*) annealing.

décroûtage, *m.* cleaning (from crust); (*Founding*) dressing down.

décroûter, *v.t.* remove the crust from, clean.

décru, *p.p.* of décroître.

décruage, *m.* = décreusage.

décrue, *f.* decrease, decrement.

décruer, *v.t.* = décreuser.

décrueur, *m.* = décreuseur.

décrûment, décrusage, décrusement, *m.* = décreusage.

décruser, *v.t.* = décreuser.

déçu, *p.p.* deceived.

décuire, *v.t.* thin (sirup). — *v.r.* become thin.

décuivrer, *v.t.* deprive of copper, decopperize.

décuplateur, *a.* multiplying by ten.

décupler, *v.t.* decuple, multiply by ten.

décuvage, *m.*, **décuvaison,** *f.* (*Wine*) racking off from the vat after fermentation, tunning.

décuver, *v.t.* rack off from the vat, tun (wine).

dédaigner, *v.t.* disdain.

dédale, *m.* labyrinth, maze.

dédaléen, dédalien, *a.* inextricable, labyrinthine.

dédaller, *v.t.* remove the flags from, unpave.

dedans, *adv.* inside, within, in. — *m.* inside, interior.

au —, within, inside.

au — *de,* within, inside of, in the interior of.

de —, from within; (of measurements) in the clear.

en —, within, inside, in; (of a burner) at the base; (of measurements) in the clear.

en — *de,* within, inside of.

par —, thru; within.

dédicace, *f.* dedication.

dédicatoire, *a.* dedicatory.

dédier, *v.t.* dedicate.

dédit, *m.* forfeit; bond; retraction.

dédommagement, *m.* damages, compensation.

dédommager, *v.t.* compensate.

dédorage, *m.* ungilding.

dédorer, *v.t.* ungild, remove the gilding from.

dédorure, *f.* ungilding.

dédoublable, *a.* divisible into two, decomposable, resolvable, etc. (see dédoubler).

dédoublage, dédoublement, *m.* division into two, decomposition, resolution, etc. (see dédoubler).

dédoublé, *p.p.* of dédoubler. — *m.* an artificial brandy made by diluting alcohol.

dédoubler, *v.t.* divide into two, decompose, (of racemic compounds, double salts, etc.) resolve; dilute; reduce one half; undouble, unfold; unline, unsheathe. — *v.r.* divide into two, decompose, be resolved; diminish one half; unfold; be unlined.

déductif, *a.* deductive.

déduction, *f.* deduction.

déductivement, *adv.* deductively.

déduire, *v.t.* deduct; deduce; relate.

dédurcir, *v.t.* soften.

défaillance, *f.* failing; fainting; faintness.

défaillir, *v.i.* fail; default; faint.

défaire, *v.t.* unmake, undo, untie, take apart; disengage, ungear; take off; defeat; make way with; rid. — *v.r.* be undone, come loose; weaken, spoil; rid oneself, get rid (of).

défait, *p.p.* & *pr. 3 sing.* of défaire.

défaite, *f.* defeat; subterfuge, evasion.

défalquer, *v.t.* deduct.

défausser, *v.t.* straighten.

défaut, *m.* want, lack, absence, deficiency; defect, flaw; fault; default. — *pr. 3 sing.* of défaillir.

à — *de, au* — *de,* in the absence of, in lieu of; for want of.

faire —, be absent, be missing.

défavorable, *a.* unfavorable.

défavorablement, *adv.* unfavorably.

défécation, *f.* defecation; clarification.

défectible, *a.* imperfect.

défectueusement, *adv.* defectively.

défectueux, *a.* defective.

défendre, *v.t.* defend; prohibit, forbid. — *v.r.* defend oneself; forbear; avoid; decline; deny.

défendu, *p.a.* defended; forbidden, prohibited.

défense, *f.* defense; prohibition; fender; tusk.

défenseur, *m.* defender, champion.

défensif, *a.* defensive.

déféquer, *v.t.* defecate, clarify, clear, purify.

défera, *fut. 3 sing.* of défaire.

déférer, *v.t.* confer, bestow; denounce, accuse. — *v.i.* defer.

déferrage, *m.* removal of iron, deferrization; removal of ironwork; (of a horse) unshoeing.

déferrer, *v.t.* deprive of iron, deferrize; remove ironwork from; unhinge (a door); unshoe (a horse); disconcert.

défervescence, *f.* defervescence, subsidence from a state of effervescence or (*Med.*) of fever.

défi, *m.* defiance.

défiance, *f.* distrust; diffidence.

défibrage, *m.* separation of fibers (see défibrer).

défibrer, *v.t.* remove or separate the fibers of, disintegrate (as wood or sugar cane); grind (wood for making mechanical pulp).

défibreur, *m.* fiber separator; (*Paper*) pulp grinder; (*Sugar*) cane shredder.

défibrination, *f.* defibrination.

défibriner, *v.t.* defibrinate.

déficeler, *v.t.* untie, undo, open (a package).

déficit, *m.* deficit; deficiency.

déficitaire, *a.* deficient.

défier, *v.t.* defy. — *v.r.* be distrustful; be suspicious, have an idea (of, that); defy each other.

défilage, *m.* unthreading; (*Paper*) breaking in of rags, separation of rags into their threads or fibers.

défiler, *v.t.* unthread, unstring; (*Paper*) separate (rags) into threads or fibers, break in.

défileur, *m.* unthreader; (*Paper*) rag-cylinder driver. — *a.* unthreading; (*Paper*) breaking-in.

défileuse, *f.* unthreader; (*Paper*) breaker.

défilochage, *m.* = défilage.

défilocher, *v.t.* = défiler.

défini, *p.a.* definite; defined.

définir, *v.t.* define. — *v.r.* be defined.

définissable, *a.* definable.

définitif, *a.* definitive, final.

définition, *f.* definition.

définitivement, *adv.* definitively, finally.

défit, *p.def. 3 sing.* of défaire.

déflagrant, *p.a.* deflagrating.
déflagrateur, *m.* deflagrator.
déflagration, *f.* deflagration.
déflagrer, *v.i.* deflagrate.
défléchir, *v.t. & i.* deflect.
déflecteur, *m.* deflector.
déflegmateur, *m.* dephlegmator.
déflegmation, *f.* dephlegmation.
déflegmer, *v.t.* dephlegmate, rectify.
déflexion, *f.* deflection.
déflocheuse, *f.* rag-tearing machine, devil.
défont, *pr. 3 pl.* of défaire.
déformateur, *a.* deforming.
déformation, *f.* deformation.
déformer, *v.t.* deform. — *v.r.* become deformed.
défournage, défournement, *m.* taking from an oven, kiln or furnace, drawing.
défourner, *v.t.* take or draw (porcelain, glass, cement, etc.) from an oven, kiln, or furnace.
défourneur, *m.* drawer, kiln (or oven) drawer; (for plate glass) hot-push man.
défourneuse, *f.* (*Glass*) pot wagon.
défraîchir, *v.t. & r.* fade.
défrayer, *v.t.* pay the expense of; entertain.
défricher, *v.t.* clear, make tillable; clear up.
défruiter, *v.t.* strip of fruit; remove the taste of the fruit from (as olive oil).
dégagé, *p.a.* disengaged, etc. (see dégager); free, open; graceful.
dégagement, *m.* disengagement, liberation, emission, evolution; clearing; redemption.
dégager, *v.t.* disengage; liberate, emit, evolve; clear (a way); redeem. — *v.r.* be disengaged; be emitted; be cleared; disengage oneself.
dégaler, *v.t.* clean, comb. (as rabbit skins).
dégarnir, *v.t.* degarnish, strip, untrim.
dégarnissage, *m.* degarnishing; thinning (of sugar beets).
dégât, *m.* damage, destruction, waste.
dégauchir, *v.t.* smooth, dress, plane, polish.
dégélatiner, *v.t.* deprive of gelatin, degelatinize.
dégeler, *v.t., r. & i.* thaw, melt.
dégénération, *f.* degeneration.
dégénéré, *p.a.* degenerated, degenerate.
dégénérer, *v.i.* degenerate.
dégénérescence, *f.* degeneration.
— *graisseuse,* fatty degeneration.
dégénérescent, *a.* degenerative.
dégermer, *v.t.* degerminate, deprive of the germ.
dégermeur, *m.* **dégermeuse,** *f.* degerminator.
déglaçage, déglacement, *m.* removal of luster; removal of ice.
déglacer, *v.t.* deprive of luster (as paper); remove the ice from; thaw, melt.
dégluement, *m.* removal of glue, ungluing.

dégluer, *v.t.* remove the glue from, unglue.
déglycériner, *v.t.* remove the glycerin from (as oils), deglycerin, deglycerinate.
dégommage, *m.* degumming, etc. (see dégommer).
dégommer, *v.t.* degum, ungum, scour (silk).
dégonflement, *m.* deflation, going down, reduction (of swellings).
dégonfler, *v.t.* deflate, reduce (something swollen). — *v.r.* be deflated, be reduced, subside.
dégor, *m.* discharge pipe or tube; specif., (*Distilling*) a pipe carrying spent wash.
dégorgement, dégorgeage, *m.* clearing, etc., (see dégorger).
dégorgeoir, *m.* a mill for scouring cloth; an instrument for cleaning a tube or opening; (of a gun) priming wire; spout where water issues.
dégorger, *v.t.* clear, deabstruct, unstop (as a pipe); cleanse, scour, wash (cloth, leather, etc.); hollow out; disgorge. — *v.i.* discharge, empty; overflow. — *v.r.* discharge; overflow; become unstopped.
dégoudronnage, dégoudronnement, *m.* removal of tar.
dégoudronner, *v.t.* deprive of tar, untar.
dégoudronneur, *m.* tar remover.
dégourdi, *p.a.* slightly warm; (of porcelain) in the state of biscuit, unglazed; reanimated, revived; lively; sharp, shrewd. — *m.* (*Ceram.*) biscuit baking, biscuit fire; also, biscuit, unglazed body.
dégourdir, *v.t.* warm slightly, take the chill off of; cool, allow to cool (to lukewarmness); (*Ceram.*) give the first or biscuit baking to; remove numbness from, revive; (fig.) sharpen, polish. — *v.i.* warm slightly, become lukewarm; cool (to lukewarmness). — *v.r.* revive.
dégoût, *m.* disgust; mortification.
dégoûtant, *p.a.* disgusting.
dégoûter, *v.t.* disgust. — *v.r.* be disgusted.
dégouttement, *m.* dripping, dropping, trickling.
dégoutter, *v.i.* drip, drop, trickle.
dégradation, *f.* degradation; deterioration.
dégrader, *v.t.* degrade; (of colors) degrade, reduce, tone down; deteriorate. — *v.r.* be degraded; degrade oneself.
dégrafer, *v.t.* unhook, unfasten.
dégragène, *m.* degragene, a substance supposed to exist in fish oils.
dégraissage, *m.* scouring, etc. (see dégraisser).
dégraissant, *p.a.* scouring, etc. (see dégraisser). — *m.* scouring agent, cleaner, degreaser; (*Ceram.*) leaning agent, antiplastic or aplastic material.
dégraisse, *m.* fat cut from meat in retailing it (distinguished from leaf fat).

dégraisser, *v.t.* scour, clean; free from fat, degrease, ungrease, unoil; skim, scum (liquids); correct the greasy taste of (wine); fat-extract (filter paper); (*Ceram.*) render less plastic, lean, thin; impoverish (soil); dress down (wood); emaciate. (Cf. dessuintage.)

dégraisseur, *m.* scourer (in various industries); scouring machine (for wool).

dégraissis, *m.* grease removed by scouring, extracting, skimming or the like.

dégraissoir, *m.* a machine or instrument for scouring or degreasing.

dégras, *m.* (*Leather*) dégras, degras.
— *anglais,* sod oil.

degré, *m.* degree; step, stair.
de — en —, step by step.
— *centésimal,* centigrade degree, degree centigrade.
par degrés, by degrees, gradually.

dégreneur, *m.* (*Ceram.*) mill attendant.

dégrever, *v.t.* free (as from tax or duty).

dégrossage, *m.* reduction in size, drawing (of metal).

dégrosser, *v.t.* reduce in size, draw (metals).

dégrossir, *v.t.* (subject to any operation roughly, in a preliminary way) rough down, rough-shape, rough-dress; rough-grind (plate glass); concentrate (ores) roughly; give a first filtration to (water).

dégrossissage, *m.* roughing down, rough-shaping, etc. (see dégrossir).

dégrossisseur, *m.* a workman, machine or tool that roughs down, rough-shapes, etc. (see dégrossir); (*Iron*) roughing roll. — *a.* roughing, rough-shaping, etc., rough, reducing, preliminary.

déguisement, *m.* disguise.

déguiser, *v.t.* disguise.

dégustateur, *m.* taster.

dégustation, *f.* tasting (of wine, etc.).

déguster, *v.t.* taste, test by tasting.

dégut, *m.* daggett (empyreumatic oil of birch).

dehors, *adv.* out, without, outside. — *m.* outside, exterior; foreign countries.
au —, out, outside; abroad; externally.
au — de, outside of.
en —, out, outward, on the outside; frank.
en — de, outside of, out of; apart from.
par —, outside, without.

déhydracétique, *a.* dehydracetic, dehydroacetic.

déhydromucique, *a.* dehydromucic.

déjà, *adv.* already.

déjecteur, *m.* ejector (of boiler scale).

déjection, *f.* (*Geol. & Physiol.*) dejection; (*pl.*) dejecta, excrement.

déjeter, *v.t. & r.* warp.

déjeuner, *m.* breakfast; breakfast service.
— *de soleil,* a fabric that fades rapidly.

delà, *prep.* beyond. (Used chiefly in phrases.)
au —, farther, beyond; more.
au — de, beyond.
de —, from there; beyond; from beyond.
en —, beyond; further on.
par —, beyond; farther; more.

délabrer, *v.t.* shatter, ruin, disorder.

délai, *m.* delay; time allowed for something.
— *de livraison,* time required for delivery.

délaiement, *m.* = délayage.

délainage, *m.* wool pulling.

délainer, *v.t.* strip of wool.

délaineur, *m.* wool puller.

délaissé, *p.a.* abandoned; in disuse.

délaisser, *v.t.* abandon.

délaitage, délaitement, *m.* removal of milk; specif., the working or drying of butter.

délaiter, *v.t.* deprive of milk, work (butter).

délaiteuse, *f.* butter drier.

délassant, *p.a.* refreshing.

délavage, *m.* dilution, weakening (of colors).

délaver, *v.t.* dilute, weaken (colors).

délayage, *m.* mixing, etc. (see délayer); mixture, thinned or tempered material.

délayant, *p.a.* mixing, etc. (see délayer). — *m.* a liquid used for mixing, tempering or thinning.

délayement, *m.* = délayage.

délayer, *v.t.* mix (a powder, paste or the like with a liquid), temper, thin; dilute (colors).

délayeur, *m.* mixer, thinner (see délayer); (*Ceram.*) blunger; (*Cement*) wash mill.

délecter, *v.t.* delight.

délégué, *p.a.* delegated. — *m.* delegate.

délétère, *a.* deleterious, noxious.

délibéré, *p.a.* deliberate; bold.

délibérer, *v.i.* deliberate; decide.

délicat, *a.* delicate.

délicatesse, *f.* delicacy.

délicieux, *a.* delicious; delightful.

délié, *p.a.* untied, loose; fine, thin, slender; shrewd, clever.

délier, *v.t.* untie; loose, free.

délimiter, *v.t.* delimit, demarcate, define.

déliquescence, *f.* deliquescence.

déliquescent, *a.* deliquescent.

déliquium, *m.* deliquium; (*Obs.*), deliquescence, liquid formed by deliquescence.

délirant, *p.a.* delirious.

délire, *m.* delirium.

délissage, *m.* rag cutting (see délisser).

délisser, *v.t.* (*Paper*) cut (rags) to workable size.

délisseuse, *f.* rag cutter.

délit, *m.* (of a stone) edge, side transverse to the grain; offense.

délitage, délitement, *m.* crumbling, etc. (see déliter).

déliter, *v.t. & r.* crumble, disintegrate; specif., effloresce; (of lime) slake; cleave, split, splinter; set (a stone) edgewise.

délitescence, *f.* crumbling, disintegration; specif., efflorescence; (*Med.*) delitescence.

délitescent, *a.* capable of being disintegrated; specif., efflorescent.

délivrance, *f.* delivery; deliverance; (*Physiol.*) expulsion of the after-birth.

délivre, *m.* after-birth.

délivrer, *v.t.* deliver.

delphinine, *f.* delphinine.

délustrer, *v.t.* deprive of luster, ungloss.

délutage, *m.* unluting, removal of the lute; removal of coke from a gas retort.

déluter, *v.t.* unlute, remove the lute from.

démaclage, *m.* stirring of melted glass.

démacler, *v.t.* stir (melted glass).

démagnétiser, *v.t.* demagnetize.

demain, *adv.* & *m.* tomorrow.

demande, *f.* request; application; claim; (*Com.*) order; demand; question.
— *de brevet,* patent application.

demander, *v.t.* ask; want, wish; demand; (*Com.*) order; call for; ask for.

demandeur, *m.* asker; (*Law*) plaintiff.

démangeaison, *f.* itching.

démarcation, *f.* demarcation.

démariage, *m.* thinning (of plants); divorce.

démarrage, *m.* starting.

démarrer, *v.t.* & *i.* start.

démarreur, *m.* (*Mach.*) starter.

démêlage, *m.* disentangling, etc. (see démêler); sorting (of wool); combing (of hair); (*Brewing*) mashing, mash.

démêler, *v.t.* disentangle, reduce to order; distinguish; discover. — *v.r.* be disentangled; be cleared up; stand out.

démence, *f.* dementia.

démentir, *v.t.* belie; contradict; disbelieve.
— *v.r.* contradict oneself or each other, be inconsistent; cease; (of buildings) give way.

démesuré, *a.* immoderate, excessive, enormous.

démesurément, *adv.* immeasurably, excessively, enormously.

déméthyler, *v.t.* deprive of methyl, demethylate.

démettre, *v.t.* dismiss; dislocate. — *v.r.* quit, resign; be dislocated.

demeurant, *p.a.* living, etc. (see demeurer).
— *m.* remainder, rest; survivor.
au —, on the whole, after all.

demeure, *f.* delay; dwelling, residence.
à —, immovable, immovably.

demeurer, *v.i.* live, dwell; stay; remain; stop.

demi, *a.*, *adv.* & *m.* half.
à —, half, by half.

demi-. semi-, half-, demi-, hemi-.

demi-balle, *f.* half-ball, hemisphere.

demi-blanc, *a.* semi-white, nearly white.

demi-brillant, *a.* semi-brilliant, moderately brilliant.

demi-cercle, *m.* semicircle, half circle.

demi-cheval, *m.* half horsepower.

demi-chimique, *a.* (*Paper*) designating a wood pulp made by combined chemical and mechanical treatment, semi-chemical.

demi-circulaire, *a.* semicircular.

demi-conducteur, *m.* (*Elec.*) semi-conductor.

demi-cristal, *m.* a kind of lead glass in the making of which crystal cullet is used.

demi-diamètre, *m.* semidiameter, radius.

demi-disque, *m.* half-disk.

demi-dixième, *m.* half a tenth, twentieth.

demi-doublé, *a.* (*Jewelry*) designating a jewel having the crown of one stone and the lower part of another stone or of glass.

démieller, *v.t.* remove wax from (honey).

demi-ferme, *a.* semi-firm; specif., (*Ceram.*) designating a paste which is rather stiff but can be kneaded by hand.

demi-fin, *a.* (of metals) half-fine. — *m.* twelve-carat gold.

demi-fluide, *a.* semifluid.

demi-fraude, *f.* semi-fraud, partial fraud.

demi-gras, *a.* (of coal) semi-bituminous; (of cheese) made from partially skimmed milk, single.

demi-haut-fourneau, *m.* small blast furnace.

demi-heure, *f.* half-hour.

demi-jour, *m.* twilight, dim light.

demi-journée, *f.* half-day.

demi-litre, *m.* half-liter.

demi-longueur, *f.* half-length.

demi-molécule, *f.* half-molecule.

demi-molle, *a. fem.* (*Ceram.*) designating a paste easily worked by the fingers yet not adhering to them.

demi-moule, *f.* half-mold.

demi-muid, *m.* half hogshead.

déminéralisation, *f.* demineralization, (*Med.*) excessive elimination of mineral matter in the urine.

déminéraliser, *v.t.* demineralize.

demi-onde, *f.* half-wave, half-wave-length.

demi-opale, *f.* semiopal.

demi-porcelaine, *f.* semiporcelain.

demi-précieux, *a.* semiprecious.

demi-profond, *a.* of moderate depth, neither deep nor shallow.

demi-raffiné, *a.* semirefined.

demi-rond, *a.* half-round.

demi-ronde, *f.* half-round file.

démis, *p.p.* & *p. def. 1* & *2 sing.* of démettre.

demi-siècle, *m.* half-century.

demi-sphère, *a.* hemisphere.

demi-sphérique, *a.* hemispheric(al).

démission, *f.* resignation.

demi-teinte, *f.* demitint, half tint.

demi-tour, *m.* half turn.

démocratie, *f.* democracy.

démodé, *a.* out of fashion, old-fashioned.

demoiselle, *f.* beetle, rammer.

démolir, *v.t.* demolish.

démonstrateur, *m.* demonstrator.

démonstration, *f.* demonstration.

démonstrativement, *adv.* convincingly.

démontable, *a.* dismountable, capable of being taken apart.

démontage, *m.* dismounting, etc. (see démonter).

démonter, *v.t.* dismount; take to pieces, take apart, take down, knock down; unset (gems); disconcert. — *v.r.* be dismounted; be dismountable; go to pieces; be disconcerted.

démontrabilité, *f.* demonstrability.

démontrable, *a.* demonstrable.

démontrer, *v.t.* demonstrate; show.

démoulage, *m.* removal from the mold, stripping, lifting.

démouler, *v.t.* remove from the mold, strip, lift.

démouleur, *m.* stripper.

démultiplicateur, *m.* (*Mach.*) a device for reducing velocity, reducer. — *a.* reducing.

démultiplier, *v.t.* (*Mach.*) reduce, gear down.

dénaturant, *p.a.* denaturing. — *m.* denaturant.

dénaturateur, *m.* denaturer.

dénaturation, *f.* denaturing, denaturation; alteration, modification, conversion.

dénaturé, *p.a.* denatured; altered, converted; (of iron) refined; unnatural.

dénaturer, *v.t.* denature; change the nature or properties of, alter, convert; refine (pig iron).

dendrite, *f.* dendrite.

dendritique, *a.* dendritic.

dénégation, *f.,* **déni,** *m.* denial.

dénier, *v.t.* deny.

dénitrant, *p.a.* denitrating.

dénitrer, *v.t.* denitrate.

dénitrification, *f.* denitrification.

dénitrifier, *v.t.* denitrify.

dénombrement, *m.* numbering, enumeration, counting; census; list, catalog.

dénombrer, *v.t.* number, enumerate, count.

dénominateur, *m.* denominator.

dénomination, *f.* denomination.

dénommer, *v.t.* name, denominate. — *v.r.* be named.

dénoncer, *v.t.* denounce; declare; denote.

dénoter, *v.t.* denote.

denrée, *f.* product, produce, good, commodity; specif., food product, foodstuff.
— *alimentaire,* foodstuff, food product, provision.

dense, *a.* dense.

densimètre, *m.* densimeter.

densimétrique, *a.* densimetric.

densité, *f.* density; denseness.

densité *absolue,* absolute density.
— *au mercure,* density by mercury, real density.
— *de vapeur,* vapor density.
— *du courant,* current density.
— *réelle,* real density.

dent, *f.* tooth.
— *de barrage,* catch, detent, pawl, click.
— *de lion,* (*Bot.*) dandelion.
— *de loup,* (*Mach.*) catch, detent, pawl.
— *d'engrenage,* gear tooth; ratchet.
dents de loup, (*Glass*) a long rod with a double hook at the end, for handling the pots in the furnace.

dentaire, dental, *a.* dental.

dent-de-cheval, *f.* a greenish-blue variety of topaz.

dent-de-loup, *f.* see dent de loup, under *dent.*

denté, *a.* toothed, (of wheels, etc.) cogged, (of leaves, etc.) serrated, serrate.

dentelaire, *f.* (*Bot.*) leadwort (*Plumbago,* esp. *P. europæa*).

dentelé, *p.a.* notched, jagged; (*Bot.*) dentate.

denteler, *v.t.* notch, jag, indent, tooth.

dentelle, *f.* lace; lacework.
— *de fil,* — *en fils,* thread lace.

denter, *v.t.* tooth, cog.

dentier, *m.* an instrument for cutting soap; set of teeth (esp. an artificial one), denture.

dentifrice, *m.* dentifrice. — *a.* tooth-cleaning, tooth.

dentine, *f.* dentine.

dentiste, *m.* dentist.

denture, *f.* (*Mach.*) teeth, cogs, gearing, toothing; set of teeth.

dénué, *a.* devoid, destitute.

dénuer, *v.t.* deprive, strip.

dépaqueter, *v.t.* unpack, open.

dépareillé, *a.* incomplete, imperfect; odd.

dépareiller, *v.t.* break, spoil (a set).

déparer, *v.t.* spoil, mar; strip of ornament.

départ, *m.* parting (of metals); separation, division; sorting; departure; place of departure, exit; discharge (of a firearm). — *pr. 3 sing.* of départir.

département, *m.* department.

départir, *v.t.* dispense, bestow. — *v.r.* depart; desist.

dépasser, *v.t.* pass, go beyond; exceed; be higher or longer than; project beyond; surpass; draw out, unwind; (*Metal.*) overpole.

dépêche, *f.* dispatch.

dépêcher, *v.t.* dispatch. — *v.r.* hasten, hurry.

dépeindre, *v.t.* describe, depict.

dépendance, *f.* dependence.

dépendant, *a. & m.* dependent

dépendre, *v.i.* depend; belong. — *v.t.* take down.

dépens, *m.pl.* expense, cost, costs, charges.
aux — *de,* at the expense of.

dépense, *f.* expenditure, expense, outlay; consumption (as of electricity); discharge (as of water); dispensary.

dépenser, *v.t.* expend, spend; consume; waste. — *v.r.* be spent; be consumed.

déperdition, *f.* loss, waste, dissipation; (of gas or liquid) escape.

dépérir, *v.i.* waste away, molder, wither, spoil.

dépeupler, *v.t.* depopulate.

déphlegmateur, etc. See déflegmateur, etc.

déphlogistiquer, *v.t.* (*Old Chem.*) dephlogisticate.

déphosphoration, *f.* dephosphorization.

déphosphorer, *v.t.* dephosphorize.

dépilage, *m.* unhairing; (*Mining*) removal of pillars.

dépilatif, *a.* depilatory.

dépilation, *f.* unhairing, depilation.

dépilatoire, *a. & m.* depilatory.

dépiler, *v.t.* unhair, depilate; clear of pillars.

dépister, *v.t.* discover; mislead.

dépit, *m.* spite, despite.

en — de, in spite of.

déplaçable, *a.* displaceable.

déplacé, *p.a.* displaced; out of place.

déplacement, *m.* displacement, etc. (see déplacer).

déplacer, *v.t.* displace; shift, move. — *v.r.* be displaced; be moved; change one's residence.

déplaire, *v.i.* be displeasing. — *v.r.* not to thrive; not to be pleased.

déplaisant, *a.* unpleasant, disagreeable.

déplâtrage, *m.* removal of plastering.

déplier, *v.t.* unfold.

déplombage, *m.* removal of lead.

déplomber, *v.t.* remove lead from, strip of lead.

déployer, *v.t.* display; deploy.

déplu, *p.p.* of déplaire.

dépolarisant, *p.a.* depolarizing. — *m.* depolarizer.

dépolarisateur, *m.* depolarizer.

dépolarisation, *f.* depolarization.

dépolariser, *v.t.* depolarize.

dépoli, *p.a.* depolished, mat, (of glass) ground, frosted. — *m.* depolishing, depolished surface.

dépolir, *v.t.* depolish, remove the polish of; grind, frost (glass); tarnish. — *v.r.* lose its polish, become dull, tarnish.

dépolissage, dépolissement, *m.* depolishing; grinding, frosting (of glass); tarnishing.

déposage, *m.* deposition, settling.

déposer, *v.t.* deposit; lay aside; depose. — *v.r.* be deposited, deposit; form a deposit, settle. — *v.i.* form a deposit, settle.

dépositaire, *m. & f.* depositary.

dépôt, *m.* deposit; depot; storehouse; deposition; depository.

— *marin,* marine deposit.

dépotage, dépotement, *m.* decanting, decantation.

dépoter, *v.t.* decant; unpot (a plant).

dépouille, *f.* hide, skin; crop (as of grain); spoil, booty; (mortal) remains.

dépouillement, *m.* skinning, etc. (see dépouiller).

dépouiller, *v.t.* skin; strip; deprive; plunder; reap, gather (crops); abstract (accounts); go over, work up (reports, etc.). — *v.r.* strip oneself; be stripped; (of wine) become lighter.

dépourvoir, *v.t.* deprive.

dépourvu, *p.a.* deprived, unprovided (with); destitute, devoid.

au —, unawares.

déprenant, *p.a.* separating, etc. (see déprendre).

déprendre, *v.t.* separate, part. — *v.r.* rid oneself.

dépression, *f.* depression, lowering; reduced pressure, vacuum.

déprimer, *v.t.* depress; disparage.

dépris, *p.a.* separated, parted.

depuis, *prep.* from; since. — *adv.* since. — *que,* since.

dépulpage, *m.* reduction to pulp, pulping.

dépulper, *v.t.* reduce to pulp, pulp.

dépulpeur, *m.* pulper.

dépurateur, dépuratif, *a. & m.* (*Med.*) depurative.

dépuration, *f.* (*Med.*) depuration.

dépuratoire, *a.* depuratory, depurative.

dépurer, *v.t.* (*Med.*) depurate.

député, *p.a.* deputed. — *m.* deputy.

déraciner, *v.t.* uproot; eradicate.

déraisonnable, *a.* unreasonable.

dérangement, *m.* derangement, disorder.

déranger, *v.t.* derange, disarrange, disturb. — *v.r.* be (or get) out of order; trouble oneself; be deranged.

dératisation, *f.* rat extermination.

dérayer, *v.t.* = drayer.

déréglé, *p.a.* out of order; irregular; inordinate; dissolute.

dérégler, *v.t.* put out of order or adjustment. — *v.r.* get out of order.

dérivant, *p.a.* derived, originating, proceeding. — *m.* derivative.

dérivatif, *a. & m.* derivative.

dérivation, *f.* derivation; drift; diversion; deviation; (*Elec.*) shunting, shunt; small tube connected to a larger one.

dérive, *f.* deviation.

dérivé, *p.a.* derived; secondary; derivative; (*Elec.*) shunt, shunted. — *m.* derivative. — *éthéré,* derivative of the nature of an ether (or ester).

— *purique,* purine derivative.

dérivée, *f.* (*Math.*) derivative.

dériver, *v.i.* be derived; deviate; drift. — *v.t.* derive; divert; deflect; unrivet; (*Elec.*) shunt.

dériveter, *v.t.* unrivet.

derle, *f.* kaolin.

dermatite, *f.* dermatitis.

derme, *m.* (*Anat.*) derma, dermis, true skin.

dermique, *a.* dermic, dermal.

dernier, *a. & m.* last.

  au — point, extremely.

dernièrement, *adv.* lately, recently; lastly.

dérobé, *p.a.* shelled; stolen; secret, private.

  à la dérobée, secretly, privately.

dérober, *v.t.* shell, peel; steal; save; hide. — *v.r.* steal away, slip away; be hidden.

dérochage, *m.* (of metals) scouring, pickling, dipping.

dérocher, *v.t.* (of metals) scour, pickle, dip; remove rocks from.

dérompage, *m.* cutting (of rags); removal of excess finish (from cloth).

dérompeuse, *f.* a machine for removing excess finish from cloth.

dérompoir, *m.* rag cutter.

dérompre, *v.t.* cut (rags); remove excess finish from (cloth).

dérouillement, *m.* removal of rust.

dérouiller, *v.t.* remove rust from; rub up.

déroulement, *m.* unrolling.

dérouler, *v.t. & r.* unroll; (fig.) unfold.

dérouter, *v.t.* mislead; disconcert.

derrière, *prep. & adv.* behind. — *m.* back, back part, hind part, rear, (of ships) stern.

  par —, behind, from behind.

des. Contraction of *de les,* of the, from the, etc. (see *de*).

dès, *prep.* from, starting from.

  — après, after.

  — avant, before.

  — là, therefore; from then, ever since.

  — là que, as, since.

  — lors, from that time, ever since; immediately; therefore, accordingly.

  — lors que, since, as.

  — que, as soon as, when; as, since.

dés-, des-. dis-, de-, un-.

désablage, *m.* removal of sand (as from castings).

désabonnement, *m.* discontinuance (of a subscription to a periodical).

désaccord, *m.* disagreement; discord.

désacidification, *f.* deacidification.

désacidifier, *v.t.* deacidify.

désaciération, *f.* unsteeling.

désaciérer, *v.t.* unsteel.

désactiver, *v.t.* render inactive; disarm (mines).

désaffleurer, *v.t.* fail to bring to the proper level (as in pipets); destroy the level of, render uneven.

désagencer, *v.t.* disarrange, disadjust; ungear.

désagrafer, *v.t.* unhook, unclasp; (*Mach.*) release.

désagréable, *a.* disagreeable.

désagréablement, *adv.* disagreeably.

désagrégation, *f.* disintegration, disaggregation.

désagrégeable, *a.* disintegrable.

désagrégeant, *p.a.* disintegrating.

désagréger, *v.t.* disintegrate, disaggregate; disarrange. — *v.r.* disintegrate.

désagrément, *m.* unpleasantness; defect.

désaigrir, *v.t.* remove the sourness or sharpness of, sweeten.

désaimantation, *f.* demagnetization.

désaimanter, *v.t.* demagnetize.

désajuster, *v.t.* put out of order, disarrange.

désallaiter, *v.t.* wean.

désaltérer, *v.t.* quench the thirst of, refresh.

désamidase, *f.* deamidase, desamidase.

désamorçage, *m.* unpriming; exhausting (of an arc).

désamorcer, *v.t.* unprime, remove the priming from (as a siphon or a cartridge case).

désappointer, *v.t.* disappoint.

désapprendre, *v.t.* unlearn, forget.

désappris, *p.a.* unlearned, forgotten.

désargentage, *m.*, **désargentation,** *f.* desilverization, desilvering; disilvering.

désargenter, *v.t.* desilverize, desilver.

désargenture, *f.* = désargentage.

désarmer, *v.t.* disarm.

désarroi, *m.* disarray, disorder, confusion.

désarticuler, *v.t.* disarticulate, disjoint, separate (joint by joint).

désassembler, *v.t.* disassemble.

désassimilateur, *a.* (*Physiol.*) disassimilating, catabolizing.

désassimilation, *f.* disassimilation, catabolism.

désassimiler, *v.t.* disassimilate, catabolize.

désassociation, *f.* disassociation.

désassocier, *v.t.* disassociate.

désassorti, *p.a.* unmatched, odd, ill-sorted.

désastre, *m.* disaster.

désastreusement, *adv.* disastrously.

désastreux, *a.* disastrous.

désaturé, *a.* unsaturated; superheated (steam).

désavantage, *m.* disadvantage.

désavantageusement, *adv.* disadvantageously.

désavantageux, *a.* disadvantageous, unfavorable.

désazotation, *f.* denitrification.

désazoter, *v.t.* denitrify.

descellement, *m.* unsealing, etc. (see desceller).

desceller, *v.t.* unseal, unfasten; (*Glass*) chip off the rough parts of.

descendant, *p.a.* descending, etc. (see descendre). — *m.* descendant.

descenderie, *f.* (*Mining*) a descending passage connecting galleries at different levels.

descendre, *v.i.* descend; go down, subside, fall.
— *v.t.* take down, lower; descend; land.
descendu, *p.a.* descended, etc. (see descendre).
descente, *f.* descent; slope; taking down; pipe; down stroke; (*Med.*) rupture, hernia.
descriptif, *a.* descriptive.
description, *f.* description.
désemballage, *m.* unpacking.
désemballer, *v.t.* unpack.
désembrayage, etc. See débrayage, etc.
désemparer, *v.t.* & *i.* leave, quit.
désemplir, *v.t.* make less full, empty (in part).
— *v.r.* get low, get less full.
désempoisonner, *v.t.* free from poison.
désencroûter, *v.t.* free from crust or scale.
désenflammer, *v.t.* extinguish (a blaze).
désenfler, *v.t.* deflate. — *v.i.* be deflated.
désenflure, *f.* deflation, going down.
désenfourner, *v.t.* = défourner.
désengrener, *v.t.* throw out of gear, ungear, disengage, disconnect.
désenrayer, *v.t.* unlock (a wheel); release (as a brake); uncouple, disconnect.
désentortiller, *v.t.* disentangle.
désenvelopper, *v.t.* open, undo (as a package).
déséquilibre, *m.* absence of equilibrium.
déséquilibré, *a.* disequilibrated, out of equilibrium.
déséquilibrer, *v.t.* throw out of equilibrium.
désert, *a.* & *m.* desert.
déserter, *v.t.* & *i.* desert.
désespérance, *f.* despair.
désespérant, *p.a.* discouraging.
désespéré, *p.a.* desperate; sorry.
désespérément, *adv.* desperately.
désespérer, *v.i.* & *v.* despair. — *v.t.* drive to despair; dishearten, discourage.
désespoir, *m.* despair; concern.
désessencier, *v.t.* deprive of the more volatile part, remove the low-boiling constituents of.
désétamage, *m.* detinning.
désétamer, *v.t.* detin.
désherbant, *p.a.* weeding, eradicating weeds. — *m.* weed eradicator.
déshonneur, *m.* dishonor.
déshuiler, *v.t.* remove the oil from, unoil.
déshydratant, *p.a.* dehydrating. — *m.* dehydrating agent.
déshydratation, *f.* dehydration.
déshydrater, *v.t.* dehydrate. — *v.r.* dehydrate, lose water.
déshydrogénation, *f.* dehydrogenation, dehydrogenization.
déshydrogéner, *v.t.* dehydrogenize, dehydrogenate.
désignation, *f.* designation.
désigner, *v.t.* designate; denote, indicate.
désincrustant, *m.* disincrustant, antiincrustator, boiler compound. — *a.* scale-removing.

désincrustation, *f.* removal of incrustation or scale (as from a boiler).
désincruster, *v.t.* free from incrustation or scale.
désinence, *f.* termination (of a word).
désinfectant, *a.* & *m.* disinfectant.
désinfecter, *v.t.* disinfect.
désinfecteur, *m.* disinfector.
désinfection, *f.* disinfection.
désinfectoire, *m.* disinfecting station.
désintégrateur, *m.* disintegrator.
désintégration, *f.* disintegration.
désintégrer, *v.t.* disintegrate.
désintéressé, *a.* disinterested.
désintéressement, *m.* disinterestedness.
désir, *m.* desire.
désirable, *a.* desirable.
désirer, *v.t.* desire; wish.
désireux, *a.* desirous.
desmine, *f.* (*Min.*) desmine (stilbite).
desmotropie, *f.* desmotropy, desmotropism.
désobéir, *v.i.* disobey.
désobstruant, désobstructif, *a.* & *m.* (*Med.*) deobstruent, aperient.
désodorer, *v.t.* deodorize.
désodorisation, *f.* deodorization.
désodoriser, *v.t.* deodorize.
désœuvrer, *v.t.* (*Paper*) separate (the sheets); render idle.
désolation, *f.* desolation; distress; vexation.
désolé, *p.a.* desolate; distressed; sorry.
désordonner, *v.t.* disorder, disturb.
désordre, *m.* disorder.
désorganisateur, *a.* disorganizing. — *m.* disorganizer.
désorganisation, *f.* disorganization.
désorganiser, *v.t.* disorganize. — *v.r.* become disorganized.
désormais, *adv.* henceforth, hereafter; thenceforth, thereafter.
désosser, *v.t.* bone, remove the bones from.
désoufrage, *m.* desulfurization, desulfuring.
désoufrer, *v.t.* desulfurize, desulfur.
désoxalique, *a.* desoxalic.
désoxydant, *p.a.* deoxidizing. — *m.* deoxidizer.
désoxydation, *f.* deoxidation, deoxidization.
désoxyder, *v.t.* deoxidize. — *v.r.* be deoxidized.
désoxygénation, *f.* deoxygenation, deoxidation.
désoxygéner, *v.t.* deoxygenate, deoxidize.
despumation, *f.* despumation, scumming; formation of froth.
despumer, *v.t.* despumate, skim, clarify (by removing scum).
desquels, desquelles. Contractions of *de lesquels, de lesquelles*, of whom, of which, from whom, from which, etc. (see *de*).
dessabler, *v.t.* remove sand from (as castings).
dessaignage, *m.* freeing from blood.

dessaigner, *v.t.* free from blood, specif., wash (hides or skins) to remove blood and other impurities.

dessalement, dessalage, *m.,* **dessalaison,** *f.* removal of salt, rendering less salt.

dessaler, *v.t.* render less salt, remove salt from (as sea water by distillation).

desséchant, *p.a.* drying, desiccating, desiccative.

dessèchement, *m.* drying, etc. (see dessécher).

dessécher, *v.t. & i.* dry, desiccate; dry up; drain (land); season (wood).

dessécheur, *m.* drier.

dessein, *m.* design.

  à —, by design, on purpose.

  à — *de,* in order to.

  à — *que,* in order that.

  sans —, unintentionally.

desserrage, desserrement, *m.* loosening; slackening; release.

desserrer, *v.t.* loosen; slacken. — *v.r.* be loosened; get loose.

dessert, *pr. 3 sing,* of desservir. — *m.* dessert.

desservir, *v.t.* serve, supply, furnish service to; take away, clear off; disserve, damage.

desseuvrer, *v.t.* separate (sheets of paper).

dessiccant, *p.a.* desiccating, desiccant.

dessiccateur, *m.* desiccator.

  — *à vide,* vacuum desiccator.

dessiccatif, *a. & m.* desiccative.

dessiccation, *f.* desiccation, drying.

dessin, *m.* design; drawing.

dessiner, *v.t.* draw; sketch, outline; set off. — *v.i.* draw. — *v.r.* appear, take shape.

dessoudage, *m.* unsoldering; unwelding.

dessouder, *v.t.* unsolder; unweld. — *v.r.* come unsoldered, or unwelded.

dessoudure, *f.* unsoldering; unwelding.

dessouffler, *v.t.* deflate.

dessoufrage, *m.* desulfurization, desulfuring.

dessoufrer, *v.t.* desulfurize, desulfur.

dessous, *adv. & prep.* below, under, underneath, beneath. — *m.* under side, bottom; wrong side (of cloth); support for a dish.

  de —, under, under-.

  en —, underneath; underhand.

dessuintage, *m.* removal of suint from wool, steeping, scouring. (Suint is preferably removed by steeping in water and is then distinguished from the degreasing or true scouring which follows. However the two operations, singly or combined, are usually included in the term "scouring.")

dessuinter, *v.t.* deprive of suint, steep, scour (see dessuintage).

dessuinteuse, *f.* an apparatus for removing suint from wool.

dessus, *adv.* above, over, on, upon. — *m.* upper side, upper part, top; right side (of cloth); cover (of a book); back (of the hand); advantage.

dessus, *de* —, from off, from; upper, top.

  en —, above, on top, on, upon, over.

destin, *m.* destiny, fate.

destinataire, *m.* addressee, consignee.

destination, *f.* destination.

destiner, *v.t.* destine; resolve.

destructeur, *a.* destructive, destroying.

destructibilité, *f.* destructibility.

destructible, *a.* destructible.

destructif, *a.* destructive.

destruction, *f.* destruction.

désuintage, *m.* = dessuintage.

désulfuration, *f.* desulfuration, desulfurization.

désulfurer, *v.t.* desulfur, desulfurize.

désunir, *v.t.* disunite, disjoin; disjoint.

détacher, *v.t.* detach; unloose, untie, undo, take down; remove spots from, clean. — *v.r.* be detached, get loose, come undone, come out.

détail, *m.* detail; (*Com.*) retail.

  en —, in detail; by retail, retail.

détaillant, *a.* retail. — *m.* retailer.

détailler, *v.t.* detail; cut up; (*Com.*) retail.

détanner, *v.t.* detannate, deprive of tannin.

détartrage, *m.* removal of tartar or scale.

détartrer, *v.t.* remove tartar or scale from, detartarize, scale, fur; clean (casks or boilers); soften (water).

détartreur, *m.* detartarizer, water softener; cleaner (of casks or boilers).

détartrisation, *f.* removal of tartar or scale.

déteignant, *p.a.* decoloring, etc. (see déteindre).

déteignent, *pr. 3 pl.* of déteindre.

déteindre, *v.t.* decolor, bleach. — *v.r.* loss color, fade. — *v.i.* lose color, fade; come off (on).

déteint, *p.p. & pr. 3 sing.* of déteindre.

détendeur, *m.* pressure regulator; expansion valve; reduction valve.

détendre, *v.t.* (of gases) relax the pressure on, expand, allow to expand; unbend, relax, slacken, loosen; take down. — *v.r.* (of gases) expand; unbend, loosen, relax.

détenir, *v.t.* detain, hold, hold back.

détente, *f.* expansion (of gases); expansion valve; (*Steam*) cut-off; detent, catch, trigger; firing mechanism; relaxation.

détenu, *p.p.* of détenir.

détergent, *a. & m.* detergent.

déterger, *v.t.* cleanse.

détériorant, *p.a.* deteriorating.

détérioration, *f.* deterioration.

détériorer, *v.t. & r.* deteriorate.

déterminable, *a.* determinable.

déterminant, *a. & m.* determinant.

détermination, *f.* determination.

déterminé, *p.a.* determined; caused; determinate.

déterminer, *v.t.* determine; cause. — *v.r.* determine, resolve.

déterrer, *v.t.* unearth; dig up, exhume.

détersif, *a. & m.* detersive, detergent.

détient, *pr. 3 sing.* of détenir.

détirer, *v.t.* draw out; stretch.

détonant, *p.a.* detonating, explosive. — *m.* detonating substance, explosive.

détonateur, *m.* detonator.

— *à double effet,* double-action detonator.

— *retardé,* delay-action detonator.

détonation, *f.,* **détonement,** *m.* detonation, explosion.

à détonation, detonating.

détoner, *v.i.* detonate, explode.

détonneler, *v.t.* draw from the tun or cask.

détordre, *v.t.* untwist.

détors, *a.* untwisted.

détortiller, *v.t.* untwist.

détouper, *v.t.* unstop.

détour, *m.* turning, winding; by-way, detour.

détourné, *p.a.* diverted, etc. (see détourner); indirect, roundabout; by-(way).

détourner, *v.t.* divert, deflect, turn aside; deter, dissuade. — *v.i.* turn, turn off. — *v.r.* deviate, turn away, turn off.

détraqué, *p.a.* out of order; deranged.

détraquer, *v.t.* put out of order, disorder, derange; divert. — *v.r.* get out of order; become deranged.

détrempe, *f.* distemper.

détremper, *v.t.* mix (a solid with a liquid), thin; soak, wet; slake (lime); anneal (steel); soften, sweeten, mollify; enervate. — *v.r.* be mixed, be thinned, etc.; (of lime) slake; (of steel) anneal, soften.

détrempeur, *m.* (*Steel*) annealer.

détresse, *f.* distress.

détrichage, *m.* sorting (of wool).

détricher, *v.t.* sort (wool).

détriment, *m.* detriment.

détriplement, *m.* division into three parts.

détripler, *v.t. & r.* divide into three parts.

détritage, *m.* crushing (esp. of olives).

détriter, *v.t.* crush (esp. olives).

détrition, *f.* detrition.

détritique, *a.* (*Geol.*) detrital.

détritoir, *m.* crushing mill (esp. for olives).

détritus, *m.* detritus, débris.

détroit, *m.* strait; pass, defile.

détrôner, *v.t.* dethrone.

détruire, *v.t.* destroy.

détruisant, *p.a.* destroying.

dette, *f.* debt.

deuil, *m.* mourning.

deutane, *m.* ethane.

deutazotate, *m.* (*Old Chem.*) deutonitrate.

— *de mercure liquide,* (*Pharm.*) solution of mercuric nitrate.

deutéroprotéose, *f.* deuteroproteose.

deutiodure, *m.* (*Old Chem.*) deutiodide.

deuto-, deut-. (*Old Chem.*) deuto-, deut-. (It properly designated the second in a series but was sometimes used in the same way as bi- or di-.)

deutocarboné, *a.* (*Old Chem.*) bicarburetted.

deutochlorure, *m.* (*Old Chem.*) deutochloride.

— *de mercure,* mercuric chloride.

deuto-iodure, *m.* deutoiodide, deutiodide.

— *de mercure,* mercuric iodide.

deutosulfure, *m.* (*Old Chem.*) deutosulfide.

deutoxyde, *m.* (*Old Chem.*) deutoxide.

— *de mercure,* mercuric oxide.

— *de plomb,* minium, red lead.

deutylène, *m.* ethylene.

deux, *a.* two; second; a few. — *m.* two; second.

— *fois,* twice.

deuxième, *a.* second. — *m.* second story.

— *jet,* (*Sugar*) second spinning (see sucre de deuxième jet, under *sucre*).

— *titre,* (for gold) a standard of 0.840 fine (about 20 carats).

deuxièmement, *adv.* secondly.

deux-points, *m.* colon (:).

devait, *imp. 3 sing.* of devoir.

dévaler, *v.t.* lower, let down; descend. — *v.r.* go down, come down.

devancer, *v.t.* precede; outstrip, outdo, surpass; anticipate, forestall.

devancier, *m.* predecessor.

devant, *prep. & adv.* before. — *m.* front, fore part; prow (of a boat). — *p.pr.* of devoir.

— *que,* before.

— *que de,* before.

dévaster, *v.t.* devastate.

développable, *a.* developable.

développateur, *m.* (esp. *Photog.*) developer.

développement, *m.* development, etc. (see développer).

développer, *v.t.* develop; unwrap, undo, uncoil, unfold. — *v.r.* develop; be unwrapped; unfold.

développeur, *m.* (esp. *Dyeing*) developer.

devenir, *v.i.* become, grow, get.

devenu, *p.p.* of devenir.

déverdir, *v.i.* lose green color.

déverdissage, *m.* loss of green color.

dévernir, *v.t.* remove the varnish from. — *v.r.* lose the varnish.

dévernissage, *m.* removal or loss of varnish.

devers, *a.* leaning. — *m.* inclination; taper, batter; warping.

par —, before, near, toward.

déversement, *m.* pouring, etc. (see déverser).

déverser, *v.t.* pour; incline, lean; warp. — *v.r.* flow, run; warp. — *v.i.* flow, run over, pour; incline, lean.

déversoir, *m.* overflow, spillway; overfall.

déviateur, *a.* causing deviation, deflecting.

déviation, *f.* deviation.

dévider, *v.t.* wind, reel; unwind.

deviendra, *fut. 3 sing.* of devenir.

deviennent, *pr. 3 pl. indic. & subj.* of devenir.

devient, *pr. 3 sing.* of devenir.

dévier, *v.i. & r.* deviate. — *v.t.* deflect.

deviner, *v.t. & i.* divine.

devint, *p.def. 3 sing.* of devenir.

devis, *m.* estimate.

  — *approximatif,* rough estimate.

  — *estimatif,* estimate.

dévissage, dévissement, *m.* unscrewing.

dévisser, *v.t.* unscrew.

dévitrifiable, *a.* devitrifiable.

dévitrification, *f.* devitrification.

dévitrifier, *v.t.* devitrify. — *v.r.* become devitrified.

dévoiement, *m.* inclination, leaning; diarrhea.

dévoiler, *v.t.* unveil; reveal, disclose.

devoir, *v.t.* owe; should, ought; must, be obliged. — *m.* duty; association of workmen.

dévorer, *v.t.* devour.

dévotion, *f.* devotion.

dévoué, *p.a.* devoted.

dévouer, *v.t.* devote.

dévoyer, *v.t.* turn aside; slope, incline; purge.

devra, *fut. 3 sing.* of devoir.

devrait, *cond. 3 sing.* of devoir.

dextérité, *f.* dexterity, skill.

dextrement, *adv.* skillfully.

dextrine, *f.* dextrin.

dextriner, *v.t.* dextrinate, treat or coat with dextrin.

dextrinique, *a.* dextrinous, pertaining to dextrin.

dextrogyre, *a.* dextrorotatory, dextrogyrate.

dextroracémique, *a.* dextrotartaric, *d*-tartaric.

dextrorsum, *a. & adv.* clockwise, from left to right, (*Bot.*) dextrorse.

dextrose, *f.* dextrose.

dg., *abbrev.* (décigramme) decigram.

diabase, *f.* (*Petrog.*) diabase.

diabète, *m.* (*Med.*) diabetes.

diabétique, *a., m. & f.* (*Med.*) diabetic.

diabétomètre, *m.* diabetometer.

diable, *m.* devil; specif., a machine for tearing rags, for kneading rubber or for opening wool; a barrow or truck, as one for taking glass pots to the furnace.

  à la —, badly, wretchedly.

diacatholicon, *m.* (*Old Pharm.*) diacatholicon.

diacaustique, *a. & m.* diacaustic.

diacétate, *m.* diacetate.

diacétylénique, *a.* diacetylenic, containing two triple bonds.

diachylon, diachylum, *m.* (*Pharm.*) diachylon, diachylum.

diacode, *m.* (*Pharm.*) diacodion, diacodium.

diagnose, *f.* diagnosis.

diagnostique, *a.* diagnostic.

diagnostiquer, *v.t.* diagnose, diagnosticate.

diagomètre, *m.* diagometer.

diagométrie, *f.* diagometry.

diagométrique, *a.* diagometric.

diagonal, *a.* diagonal.

diagonale, *f.* diagonal.

  en —, diagonally.

diagonalement, *adv.* diagonally.

diagramme, *f.* diagram.

diagrède, *m.* (*Old Pharm.*) diagrydium.

dialcool, *m.* dialcohol, diacid alcohol.

dialdéhyde, *m.* dialdehyde.

diallage, *m.* (*Min.*) diallage.

diallagique, *a.* (*Min.*) diallagic.

diallyle, *m.* biallyl, diallyl, 1,5-hexadiene.

dialysable, *a.* dialyzable.

dialyse, *f.* dialysis.

dialyser, *v.t. & i.* dialyze.

dialyseur, *m.* dialyzer. — *a.* dialyzing.

dialytique, *a.* dialytic.

diamagnétique, *a.* diamagnetic.

diamagnétisme, *m.* diamagnetism.

diamagnétomètre, *m.* diamagnetometer.

diamant, *m.* diamond.

  — *à rabot,* glazier's diamond.

  — *brut,* rough diamond.

  — *brut de Ceylan,* zircon, used as a gem.

  — *de vitrier,* glazier's diamond.

  — *taillé,* cut diamond.

diamantaire, *a.* diamond, brilliant. — *m.* diamond cutter.

diamanté, *a.* glittering, glistening, sparkling.

diamantifère, *a.* diamantiferous.

diamantin, *a.* diamantine.

diamantine, *f.* diamantine (a polishing powder).

diamétral, *a.* diametral, diametric(al).

diamétralement, *adv.* diametrically.

diamètre, *m.* diameter.

  — *extérieur,* outside diameter.

  — *intérieur,* inside diameter.

diaminé, *a.* diaminated, diamino.

diamino-acide, *m.* diamino acid.

diamylique, *a.* diamyl.

Diane, *f.* (*Old Chem.*) Diana, silver.

diaphane, *a.* diaphanous, translucent, transparent.

diaphanéité, *f.* diaphaneity, translucency, transparency.

diaphorèse, *f.* diaphoresis.

diaphorétique, *a. & m.* diaphoretic.

diaphragme, *m.* diaphragm.

diaphragmé, *a.* having a diaphragm.

diargentique, *a.* disilver.

diarrhée, *f.* (*Med.*) diarrhea.

diaspore, *m.* (*Min.*) diaspore.

diastase, *f.* diastase; (in general) enzyme.

diastasique, *a.* diastatic, diastasic; enzymic.

diathermane, *a.* diathermic, diathermanous.

diathermanéité, *f.* diathermaneity.

diathermansie, *f.* diathermancy.

diatomée, *f.* (*Bot.*) diatom.

diatomicité, *f.* diatomicity.

diatomique, *a.* diatomic.

diazimide, *f.* diazoimine, diazoimide.

diazine, *f.* diazine.

diazo, *m.* diazo compound.

diazoacétique, *a.* diazoacetic.

diazobenzène, diazobenzol, *m.* diazobenzene.

diazocomposé, *m.* diazo compound.

diazo-dérivé, *m.* diazo derivative.

diazo-imidé, *a.* diazoimino, diazoimido.

diazoïque, *a.* diazo. — *m.* diazo compound.

diazométhane, *m.* diazomethane.

diazonaphtalène, *m.* diazonaphthalene.

diazophénol, *m.* diazophenol.

diazotable, *a.* diazotizable.

diazotant, *p.a.* diazotizing.

diazotation, *f.* diazotization.

diazoter, *v.t.* diazotize.

diazoture, *m.* dinitride.

dibenzyle, *m.* bibenzyl, dibenzyl.

dicalcique, *a.* dicalcium, dicalcic.

dicarbonique, *a.* (*Org. Chem.*) dicarboxylic.

dicarbonylé, *a.* containing two carbonyl groups, dicarbonyl.

dicétone, *f.* diketone.

dicétonique, *a.* diketo, diketonic.

dichloré, *a.* dichloro.

dichroé, *a.* having two colors, bicolor, dichromatic.

dichroïde, *a.* dichroic.

dichroïsme, *m.* dichroism.

dichroïte, *m.* (*Min.*) dichroite (iolite).

dichromatique, *a.* dichromatic.

dicotylédone, dicotylédoné, *a.* dicotyledonous.

dicrésyle, *m.* bicresyl, dicresyl (when used of the compound $CH_3C_6H_4C_6H_4CH_3$ it is better translated *bitolyl*).

dictame, *m.* (*Bot.*) dittany, specif. the fraxinella (*Dictamnus albus*).

— *de Crête,* Cretan dittany (*Amaracus dictamnus*).

dicter, *v.t. & i.* dictate.

dictionnaire, *m.* dictionary.

— *de chimie,* dictionary of chemistry.

didyme, *m.* didymium.

dièdre, *a.* dihedral. — *m.* dihedron.

diélectrique, *a. & m.* dielectric.

diélectrolyse, *f.* (*Med.*) dielectrolysis.

diète, *f.* diet.

— *lactée,* milk diet.

diététique, *a.* dietetic(al). — *f.* dietetics.

diététiquement, *adv.* dietetically.

diéthylé, *a.* diethylated, diethyl.

diéthylénique, *a.* diethylenic, diethylene.

diéthylique, *a.* diethyl.

dieu, *m.* god; (*cap.*) God.

diferreux, *a.* diferrous.

différé, *p.a.* deferred, delayed, postponed.

différemment, *adv.* differently.

différence, *f.* difference.

à la — de, differently from, contrary to.

— *finie,* fundamental difference.

différenciation, *f.* differentiation.

différencier, *v.t.* differentiate, distinguish.

— *v.r.* be differentiated, differ.

différend, *m.* difference, dispute.

différent, *a.* different.

différentiation, *f.* differentiation.

différentiel, *a.* differential.

différentielle, *f.* differential.

différentier, *v.t.* differentiate.

différer, *v.t.* defer, delay, postpone. — *v.r.* be deferred, be put off. — *v.i.* differ; delay.

difficile, *a.* difficult. — *m.* difficulty.

difficilement, *adv.* difficultly, with difficulty.

difficulté, *f.* difficulty; difference, dispute.

sans —, without difficulty, readily; undoubtedly; willingly.

difforme, *a.* deformed.

diffracter, *v.t.* diffract.

diffractif, *a.* diffractive.

diffraction, *f.* diffraction.

diffus, *a.* diffuse.

diffuser, *v.t.* diffuse.

diffuseur, *m.* diffuser.

diffusibilité, *f.* diffusibility.

diffusible, *a.* diffusible.

diffusif, *a.* diffusive.

diffusion, *f.* diffusion.

digérable, *a.* digestible.

digérer, *v.t. & i.* digest. — *v.r.* be digested.

digesté, *m.* a product obtained by digestion with some solvent.

digesteur, *m.* digester. — *a.* digestive.

— *continu,* continuous digester, siphon extractor.

digestibilité, *f.* digestibility.

digestible, *a.* digestible.

digestif, *a. & m.* digestive.

digestion, *f.* digestion.

digitale, *f.* (*Bot. & Pharm.*) digitalis, foxglove.

— *d'Espagne,* Digitalis thapsi.

— *officinale,* — *pourprée,* common or official foxglove (*Digitalis purpurea*).

digitaline, *f.* digitalin.

digitoxine, *f.* digitoxin.

digne, *a.* worthy; dignified.

dignement, *adv.* worthily, deservedly, suitably.

dignité, *f.* dignity.

digresser, *v.i.* digress.

digression, *f.* digression.

digue, *f.* dam, dike; obstacle, barrier.

diguer, *v.t.* dam, dike.

dihalogéné, *a.* dihalogenated, dihalo, dihalo-.

dihexahèdre, *a.* (*Min.*) dihexahedral. — *m.* dihexahedron.

dihydrate, *m.* dihydrate.
— *de térébenthène,* terpinol hydrate, terpin hydrate.
dihydrobenzène, *m.* dihydrobenzene, cyclohexadiene.
dihydrogéné, *a.* dihydrogenated, dihydro.
dihydrorésorcine, *f.* dihydroresorcinol.
dihydrure, *m.* dihydride.
diiodé, *a.* diiodo.
diiodhydrate, *m.* dihydriodide.
dijonnais, *a.* of or pertaining to Dijon, a city of Burgundy.
dika, *m.* (*Bot.*) dika.
dilacérateur, *a.* (*Expl.*) rending, shattering, brisant.
dilacérer, *v.t.* rend, tear, tear up.
dilatabilité, *f.* expansibility; dilatability.
dilatable, *a.* expansible; dilatable.
dilatant, *p.a.* expanding; dilating, distending. — *m.* (*Med.*) dilator.
dilatation, *f.* expansion; dilation.
dilater, *v.t. & r.* expand; dilate, distend.
dilation, *f.* delay, postponement.
dilemme, *m.* dilemma.
diligence, *f.* diligence; promptness, despatch.
diligent, *a.* diligent; prompt.
diligenter, *v.t.* press, push. — *v.i. & r.* hasten.
dilué, *p.a.* diluted, dilute.
diluer, *v.t.* dilute. — *v.r.* be diluted.
diluteur, *a.* diluting, diluent.
dilution, *f.* dilution.
diluvial, diluvien, *a.* diluvial, diluvian.
diluvium, *m.* (*Geol.*) diluvium.
dimanche, *m.* Sunday.
dime, *f.* tithe, tenth; dime (the coin).
dimension, *f.* dimension.
dimensionner, *v.t.* proportion.
dimercure-ammonium, *m.* dimercurammonium; specif. dimercuriammonium (Hg_2N).
dimère, *m.* dimer.
diméthylarsinique, *a.* dimethylarsinic (cacodylic).
diméthylbenzène, *m.* dimethylbenzene, xylene.
diméthylcétone, *f.* dimethyl ketone (methyl ketone, acetone).
diméthylé, *a.* dimethylated, dimethyl.
diméthylique, *a.* dimethyl.
diminuer, *v.t. & i.* diminish.
diminution, *f.* diminution.
dimorphe, *a.* dimorphous, dimorphic.
dimorphisme, *m.,* **dimorphie,** *f.* dimorphism.
dinanderie, *f.* a kind of brassware formerly made in Belgium and elsewhere.
dinaphtyle, *m.* binaphthyl, dinaphthyl.
dindon, *m.* turkey.
diner, *m.* dinner. — *v.i.* dine.
dinitré, *a.* dinitro.
dinitrosé, *a.* dinitroso.
dinitrosorésorcine, *f.* dinitrosoresorcinol.

dioctaèdre, *a.* (*Min.*) dioctahedral.
dioggot, diogot, *m.* (*Leather*) daggett.
dionine, *f.* (*Pharm.*) dionin, dionine.
dioptase, *f.* (*Min.*) dioptase.
dioptrique, *a.* dioptric(al). — *f.* dioptrics.
diorite, *f.* (*Petrog.*) diorite.
dioritique, *a.* dioritic.
dioxindol, *m.* dioxindole (3-hydroxyoxindole).
dioxy-. dioxy-; dihydroxy-. (See note under *oxy-*.)
dioxybutyrique, *a.* dihydroxybutyric.
dioxypurine, *f.* dioxypurine (purinedione).
dioxyquinone, *f.* dihydroxyquinone.
dioxytartrique, *a.* dihydroxytartaric.
dipentène, *m.* dipentene, inactive limonene.
dipeptide, *n.* dipeptide.
diphasé, *a.* diphase, diphasic, two-phase.
diphénol, *m.* diphenol.
diphénolique, *a.* diphenolic.
diphénylacétylène, *m.* diphenylacetylene, tolan.
diphénylamine, *f.* diphenylamine.
diphénylcarbinol, *m.* diphenylcarbinol (benzohydrol).
diphénylcétone, *f.* diphenyl ketone (phenyl ketone, benzophenone).
diphényle, *m.* biphenyl, diphenyl.
diphénylglyoxal, *m.* diphenylglyoxal (benzil).
diphtérie, *f.* diphtheria.
diphtérique, diphtéritique, *a.* diphtheritic, diphtheric.
dipicolique, *a.* dipicolinic.
diploèdre, *m.* (*Cryst.*) diplohedron, diploid.
diploédrique, *a.* (*Cryst.*) diplohedral.
diplombique, *a.* diplumbic.
diplôme, *m.* diploma.
diplômé, *m.* holder of a diploma.
dipropylique, *a.* dipropyl.
dipyridyle, *m.* bipyridyl, dipyridyl, bipyridine.
dire, *v.t.* say; tell; speak; predict. — *v.r.* be said, be told, be spoken; profess to be. — *m.* saying, say, words, statement, opinion.
c'est-à-—, that is to say, that is.
comme qui dirait, as one would say, as much as to say, the equivalent of.
en —, reproach (but, *si le cœur vous en dit,* if you like it).
direct, *a.* direct.
directement, *adv.* directly.
directeur, *m.* director, manager, superintendent. — *a.* directing; (fig.) guiding.
directif, *a.* directive.
direction, *f.* direction; management, administration; branch, department (of administration); directorship, managership, superintendency; steering gear.
directrice, *f.* directrix. — *f. & a., fem.* of directeur.

dirent, *p.def. 3 pl.* of dire.
dirigeable, *a. & m.* dirigible.
diriger, *v.t.* direct; manage, conduct.
dis, *pr. & p.def. 1 & 2 sing.* of dire (to say).
disaccharide, *n.* disaccharide.
disant, *p.pr.* of dire.
disazoïque, *a.* disazo. — *m.* disazo compound.
disbroder, *v.t.* wash (silk) after dyeing.
discernable, *a.* discernible.
discerner, *v.t.* discern. — *v.r.* be discerned.
discoïde, discoïdal, *a.* discoid, discoidal.
discolore, *a.* discolor, bicolor, bicolored.
discontinu, *a.* discontinuous; (*Elec.*) intermittent.
discontinuation, *f.* discontinuance, discontinuation.
discontinuer, *v.t. & i.* discontinue. — *v.r.* be discontinued.
discontinuité, *f.* discontinuity.
disconvenance, *f.* disproportion, incongruity.
disconvenir, *v.i.* be unsuitable; deny.
discordance, *f.* discordance.
discordant, *a.* discordant.
discorder, *v.i.* be discordant.
discourir, *v.i.* discourse; talk.
discours, *m.* discourse, speech.
discrédit, *m.* discredit.
discret, *a.* discreet.
discrétion, *f.* discretion.
  à —, at discretion, at will.
discussion, *f.* discussion.
discutable, *a.* debatable, questionable.
discuter, *v.t.* discuss. — *v.r.* be discussed.
disent, *pr. 3 pl. indic. & subj.* of dire.
disert, *a.* fluent (of speech), eloquent.
disette, *f.* dearth.
disgracieux, *a.* awkward; ungracious.
disjoignant, *p.p.* of disjoindre.
disjoindre, *v.t.* disjoin, separate. — *v.r.* be disjoined, become separated, come apart.
disjoint, *p.a.* disjoined, separated, disjunct.
disjoncteur, *m.* disjunctor; specif., (*Elec.*) circuit breaker, cut-out.
disjonction, *f.* separation, disjunction.
dislocation, *f.* dislocation; dismemberment.
disloquer, *v.t.* dislocate; disjoint, dismember.
disodé, *a.* (*Org. Chem.*) containing two atoms of sodium, disodium, disodio-.
disodique, *a.* disodium.
disomose, *f.* (*Min.*) gersdorffite, nickel glance.
disparaissant, *p.a.* disappearing; transient, ephemeral.
disparaître, *v.i.* disappear.
disparité, *f.* disparity.
disparition, *f.* disappearance.
disparu, *p.p.* of disparaître. — *a.* missing, vanished.
dispendieux, *a.* expensive, costly.
dispensable, *a.* dispensable.
dispensaire, *m.* dispensary.

dispenser, *v.t.* dispense; diffuse (light); excuse, exempt, dispense with. — *v.r.* be dispensed; dispense (with).
dispenseur, *m.* (*Optics*) diffuser.
disperser, *v.t.* disperse. — *v.r.* be dispersed; disperse.
dispersif, *a.* dispersive.
dispersion, *f.* dispersion.
disponibilité, *f.* availability, disposability.
disponible, *a.* available, disposable.
dispos, *a.* active, lively, sprightly.
disposé, *p.a.* disposed.
disposer, *v.t.* dispose; arrange; place in position. — *v.i.* dispose (of). — *v.r.* be disposed or arranged; prepare.
dispositif, *m.* arrangement (as of apparatus); device; disposition; (of a law) enacting part.
disposition, *f.* disposition; arrangement.
disproportionné, *p.a.* disproportionate.
disproportionnel, *a.* disproportional.
disproportionellement, *adv.* disproportionally.
disproportionnément, *adv.* disproportionately.
disproportionner, *v.t.* disproportion.
dispute, *f.* dispute; disputation.
disputer, *v.i.* dispute; contend. — *v.t.* dispute; contend with.
disque, *m.* disk, disc.
disruptif, *a.* disruptive.
dissécable, *a.* dissectible.
dissection, *f.* dissection.
dissemblable, *a.* dissimilar, unlike.
dissemblablement, *adv.* dissimilarly.
dissemblance, *f.* dissimilarity, unlikeness.
dissemblant, *p.a.* dissimilar, unlike.
disséminateur, *a.* disseminating.
dissémination, *f.* dissemination.
disséminer, *v.t.* disseminate.
dissentir, *v.i.* dissent.
disséquer, *v.t.* dissect.
dissertation, *f.* dissertation.
dissimilarité, *f.* dissimilarity.
dissipation, *f.* dissipation.
dissiper, *v.t.* dissipate; divert. — *v.r.* dissipate, be dissipated.
dissociabilité, *f.* dissociability.
dissociable, *a.* dissociable.
dissociation, *f.* dissociation.
dissocier, *v.t. & r.* dissociate.
dissolubilité, *f.* dissolvability, dissolvableness; dissolubility.
dissoluble, *a.* dissolvable; dissoluble.
dissoluté, *m.* solution.
dissolutif, *a.* dissolvent, solvent.
dissolution, *f.* solution; specif., a solution of caoutchouc used as a cement; dissolution.
dissolvant, *a.* dissolving, dissolvent, solvent. — *m.* solvent, (less commonly) dissolvent.
dissolvent, *pr. 3 pl. indic. & subj.* of dissoudre.

dissoudre, *v.t.* dissolve. — *v.r.* dissolve; be dissolved.

dissous, *p.a.* dissolved.

dissout, *pr. 3 sing.* of dissoudre.

dissoute, *a. fem.* of dissous.

dissymétrique, *a.* asymmetric(al), unsymmetrical, (less commonly) dissymetrical.

distance, *f.* distance.
— *explosive,* (*Expl.*) striking distance, radius of effect.

distanhexéthyle, *m.* hexaethyldistannane, $Sn_2(C_2H_5)_6$.

distannique, *a.* distannic.

distant, *a.* distant.

distendre, *v.t. & r.* distend.

distendu, *p.a.* distended.

distension, *f.* distention.

disthène, *m.* (*Min.*) disthene (cyanite).

distillable, *a.* distillable.

distillateur, *m.* distiller.

distillateur-liquoriste, *m.* liqueur maker.

distillateur-parfumeur, *m.* perfume distiller.

distillation, *f.* distillation.
— *fractionnée,* fractional distillation.
— *sèche,* dry distillation.

distillatoire, *a.* distilling, distillatory.

distillatum, *m.* distillate.

distiller, *v.t., i. & r.* distill.

distillerie, *f.* distillery; distilling.

distinct, *a.* distinct; different.

distinctement, *adv.* distinctly.

distinctif, *a.* distinctive.

distinction, *f.* distinction.

distinctivement, *adv.* distinctively.

distingué, *p.a.* distinguished.

distinguer, *v.t. & i.* distinguish. — *v.r.* be distinguished; distinguish oneself.

distordre, *v.t.* distort.

distordu, distors, *a.* distorted.

distorsion, *f.* distortion.

distraire, *v.t.* distract; divert.

distrayant, *p.pr.* of distraire.

distribuer, *v.t.* distribute. — *v.r.* be distributed; (of liquids, etc.) flow.

distributeur, *m.* distributor, distributer; carrier; (of steam) regulator. — *a.* distributing; supply, feed.

distributif, *a.* distributive.

distribution, *f.* distribution; delivery; issue; valve gear (for steam).

distributivement, *adv.* distributively.

disulfoconjugué, *a.* disulfonated, disulfo.

disulfoné, *a.* disulfonated, disulfo.

disulfoner, *v.t.* disulfonate.

dit, *pr. 3 sing., p.def. 3 sing. & p.p.* of dire (to say). — *p.a.* said; told; spoken; called.

ditartrique, *a.* ditartaric.

ditérébénique, *a.* diterpene.

ditertiaire, *a.* ditertiary.

dithionique, *a.* dithionic.

dithymol bilodé. (*Pharm.*) dithymol diiodide, thymol iodide.

dito, *adv.* ditto.

diurèse, *f.* (*Med.*) diuresis.

diurétique, *a. & m.* (*Med.*) diuretic.

diurne, *a.* diurnal, daily.

divaguer, *v.i.* wander, ramble.

divalent, *a.* bivalent, divalent.

divellent, *a.* divellent, drawing asunder (*Obs.*).

divergence, *f.* divergence, divergency.

divergent, *p.a.* diverging, divergent.

diverger, *v.i.* diverge.

divers, *a.* various, different, diverse.

diversement, *adv.* variously, differently, diversely.

diversifier, *v.t.* diversify.

diversion, *f.* diversion.

diversité, *f.* diversity; variety.

divertir, *v.t.* divert. — *v.r.* be diverted.

dividende, *m.* dividend.

dividivi, *m.* divi-divi (*Cæsalpinia* species).

divisé, *p.a.* divided; finely divided, powdered.

divisément, *adv.* dividedly, separately.

diviser, *v.t.* divide; specif., divide finely, powder. — *v.r.* divide, be divided.

diviseur, *m.* divider; (*Math.*) divisor; (*Mach.*) dividing plate. — *a.* dividing.

divisibilité, *f.* divisibility.

divisible, *a.* divisible.

division, *f.* division.

divisionnaire, *a.* divisional, divisionary.

dix, *a. & m.* ten; tenth.

dix-huit, *a. & m.* eighteen; eighteenth.
in —, eighteenmo, 18-mo.

dix-huitième, *a. & m.* eighteenth.

dixième, *a. & m.* tenth.
au —, (of solutions) ten per cent.

dix-neuf, *a. & m.* nineteen; nineteenth.

dix-neuvième, *a. & m.* nineteenth.

dix-sept, *a. & m.* seventeen; seventeenth.

dix-septième, *a. & m.* seventeenth.

dixylyle, *m.* bixylyl, dixylyl.

dizaine, *f.* ten, a total of ten or about ten.
une — *de,* ten, about ten.
une — *d'années,* a decade.

dl., *abbrev.* (décilitre) deciliter(s).

dm., *abbrev.* (décimètre) decimeter(s).

dm^q, dm², *abbrev.* (décimètre carré) square decimeter(s).

docimasie, *f.* docimasy; specif., assaying.

docimasiste, *m.* assayer.

docimastique, *a.* docimastic(al); specif., pertaining to assaying.

docte, *a.* learned. — *m.* learned man.

docteur, *m.* doctor.

doctorat, *m.* doctorate, degree of doctor.

document, *m.* document.

documentaire, *a.* documentary.

documenter, *v.t.* document, base on documentary evidence, support by citations.

dodécaèdre, *m.* (*Cryst.*) dodecahedron.
— *rhomboïdal,* rhombic dodecahedron.
dodécaédrique, *a.* dodecahedral.
doigt, *m.* finger; toe; (*Astron.*) digit.
  à deux doigts de, within an ace of, within a
  hair's breadth of.
— *annulaire,* ring finger.
— *auriculaire,* little finger.
— *du milieu,* middle finger.
— *indicateur,* index finger, forefinger.
— *majeur,* middle finger.
  petit —, little finger.
doigtier, *m.* finger stall, finger cot; (*Bot.*)
  common foxglove (*Digitalis purpurea*).
— *de caoutchouc,* rubber finger stall.
doisil, *m.* spigot hole; spigot.
doit, *pr. 3 sing.* of devoir. — *m.* debtor, Dr.,
  debit side.
doive, *pr. 3 sing. subj.* of devoir; should, ought.
doivent, *pr. 3 pl. indic. & subj.* of devoir.
dolage, *m.* smoothing, etc. (see doler).
doler, *v.t.* smooth (by cutting), shave; buff,
  slick (leather, etc.); dress (wood) with an
  adz.
dolérite, *f.* (*Petrog.*) dolerite.
doléritique, *a.* (*Petrog.*) doleritic.
dolomie, dolomite, *f.* (*Min.*) dolomite.
dolomitique, *a.* dolomitic.
domaine, *m.* domain; property; power.
dôme, *m.* dome.
domestique, *a.* domestic.
dominant, *a.* dominant.
dominer, *v.t.* dominate; be above, overlook.
— *v.i.* dominate, rule, predominate; rise
  (above).
dommage, *m.* damage.
  c'est —, it is a pity.
dommageable, *a.* detrimental, injurious.
dompte-venin, *m.* (*Bot.*) swallowwort, white
  swallowwort (*Cynanchum vincetoxicum*).
don, *m.* gift.
donc, *conj.* therefore, then, so, now.
donnée, *f.* datum, known quantity; information; idea. — *pl.* **données,** data.
donner, *v.t.* give; attribute, ascribe, assign,
  fix; apply; bid. — *v.r.* be given; give oneself up; buy oneself; have; be published;
  be fought. — *v.i.* give; yield, bear; hit,
  strike; run; fall; fight.
— *à la grille,* clean or rake the grate.
— *du cône,* taper.
— *du froid,* (*Metal.*) slacken the fire.
— *du résidu,* leave a residue.
— *la chaude à,* (*Metal.*) heat (esp. to redness).
— *la pression,* apply pressure, turn on pressure.
— *le vent,* turn on the blast, blow.
— *naissance à,* give rise to, produce.
— *sur,* open upon, overlook; censure.
— *vent à,* admit air to; give vent to.
donneur, *m.* giver, donor; guarantor.

dont, *pron.* of whom, of which, whose, whereof;
  from whom, from which; with whom, with
  which, etc. (see *de*).
dorage, *m.* gilding.
doré, *p.a.* gilded, gilt (specif., gold-plated, as
  weights); golden. — *m.* gilding.
dorénavant, *adv.* henceforth, hereafter, in the
  future.
dorer, *v.t.* gild. — *v.r.* be gilded, become golden.
doreur, *m.* gilder.
dormant, *a.* fixed, standing, fast, dead; dormant, sleeping; (of water) stagnant. — *m.*
  sleeper; post; (fixed) frame.
dormir, *v.i.* sleep; (of water) be still, be stagnant; (of fire) smolder.
dormit, *p.def. 3 sing.* of dormir.
dormitif, *a. & m.* (*Med.*) dormitive, soporific.
dort, *pr. 3 sing.* of dormir.
dorure, *f.* gilding.
— *à mordant,* pigment gilding.
— *au feu,* hot gilding, dry gilding.
— *au mercure,* — *au sauté,* amalgam gilding.
— *au trempé,* wet gilding.
— *électrochimique,* electrogilding.
— *galvanique,* electrogilding, electroplating
  with gold.
— *par immersion,* wet gilding.
dos, *m.* back; ridge.
dosable, *a.* determinable; in proper proportions.
dosage, *m.* determination; preparation, mixture (using the proper proportions); proportion, proportions (of ingredients in a
  mixture); (*Wine*) dosage, liqueuring;
  (*Med.*) dosage.
  au —, properly proportioned or constituted.
dose, *f.* proportion, amount; dose.
doser, *v.t.* determine; prepare; mix or proportion properly; fill or furnish with a measured quantity; (*Wine*) dose, liqueur;
  (*Med.*) dose. — *v.r.* be determined, etc.
doseur, *m.* determiner; preparer; regulator;
  doser (of wine or medicine). — *a.* (*Med.*)
  dosing.
dosimétrie, *f.* (*Med.*) dosimetry.
dosimétrique, *a.* dosimetric.
dossier, *m.* file, papers; back (as of a chair).
doter, *v.t.* endow.
douane, *f.* custom, duty; customhouse.
douanier, *m.* customhouse officer. — *a.* of
  customs, custom.
douara, *n.* pearl millet (*Pennisetum typhoideum*).
doublage, *m.* doubling, etc. (see doubler).
double, *a.* double; duplicate. — *m.* double;
  duplicate; copy; thickness (as of paper).
— *adv.* double.
  à — *effect,* double-effect, double-acting.
  à — *fond,* double-bottomed.
  à — *voie,* double-way.
— *liaison,* double bond.

doublé, *p.a.* doubled, etc. (see doubler). — *m.* plated ware, plate.

double-fond, *m.* double bottom.

doublement, *m.* doubling, etc. (see doubler). — *adv.* doubly.

double-pesée, *f.* double weighing.

doubler, *v.t.* double; fold; line, sheath; (*Glass*) flash, coat with a film of colored glass. — *v.r.* be doubled, etc. — *v.i.* double.

doublet, *m.* doublet.

doubleur, *m.* doubler, etc. (see doubler); maker of plated ware; (*Sugar*) part of a cane mill; (in a rolling mill) passer, catcher.

doublier, *m.* (*Calico*) doubler (thick cloth beneath the fabric and the *coursier*).

doublure, *f.* lining; scaling, scale (as a defect in metals); incomplete welding or soldering.

douçâtre, *a.* sweetish.

douce, *a. fem.* of doux.

douce-amère, *f.* bittersweet (*Solanum dulcamara*).

douceâtre, *a.* sweetish.

doucement, *adv.* sweetly, softly, etc. (see doux).

doucereux, *a.* sweetish, insipidly sweet. — *m.* sweetishness, insipidity.

doucette, *f.* an inferior quality of soda ash; (*Soap*) a solution of potassium carbonate; (*Bot.*) corn salad (*Valerianella*).

douceur, *f.* sweetness, softness, etc. (see doux). *en —,* gently, mildly; by degrees.

douche, *f.* douche.

douci, *m.* grinding (as of glass or a gem). — *p.p.* of doucir.

doucir, *v.t.* grind, smooth by grinding (as glass).

doucissage, *m.* fine grinding, grinding smooth.

doué, *p.a.* endowed (with), possessed (of).

douer, *v.t.* endow.

douille, *f.* socket; stem (of a funnel); neck (of a bell jar with top opening); sleeve; eye; (*Mach.*) box, bushing, bush; case.

douillet, *a.* soft, tender, delicate.

douillon, *m.* wool of inferior quality.

douleur, *m.* pain.

douloureusement, *adv.* painfully.

douloureux, *a.* painful; sore; sorrowful.

dourine, *f.* (*Med.*) dourine (disease of horses).

dousil, *m.* spigot hole; spigot.

doute, *m.* doubt. *hors de —,* beyond doubt, undoubtedly. *sans —,* without doubt, undoubtedly, no doubt, probably.

douter, *v.i.* doubt. — *v.r.* suspect, have a suspicion. — *de,* doubt, distrust.

douteusement, *adv.* doubtfully.

douteux, *a.* doubtful.

douve, *f.* stave; ditch; fluke, flukeworm; (*Bot.*) spearwort (*Ranunculus* species).

Douvres, *n.* Dover.

doux, *a.* sweet; soft; gentle; mild; smooth; easy; (of wine, etc.) unfermented; (of water) fresh; (of steel, etc.) mild. — *adv.* sweetly, softly, gently, etc. — *m.* sweet, soft, gentle, etc.

douzaine, *f.* dozen.

douze, *a. & m.* twelve; twelfth.

douzième, *a. & m.* twelfth.

douzil, *m.* spigot hole; spigot.

doyen, *m.* dean; elder, senior.

D^r, *abbrev.* (Docteur) Doctor, Dr.

drachme, *f.* dram (the weight); drachma.

dragage, *m.* dredging; dragging.

drage, *f.* brewer's grains.

dragée, *f.* bonbon; sugarplum; small shot. — *pharmaceutique,* pharmaceutical lozenge.

drageoire, *f.* groove.

dragon, *m.* dragon; spot in a diamond; dragoon.

dragonnier, *m.* dragon tree (*Dracæna draco*).

drague, *f.* dredge, dredger; drag; brewer's grains.

draguer, *v.t.* dredge; drag.

dragueur, *m.* dredger.

drain, *m.* drain; drain tile.

drainage, *m.* drainage, draining.

drap, *m.* cloth; sheet (for a bed). — *maléfique* = malfil.

drapant, *m.* (*Paper*) pressing board.

drapeau, *m.* flag.

draper, *v.t.* make (a fabric) into finished cloth by fulling, dressing, etc.; drape; censure.

draperie, *f.* cloth manufacture; drapery, cloth (esp. woolen stuffs); drapery, cloth selling.

drapier, *m.* draper, cloth seller. — *a.* cloth.

drastique, *a.* drastic. — *m.* (*Med.*) drastic, strong purgative.

drayage, *m.* (*Leather*) shaving, buffing.

drayer, *v.t.* (*Leather*) shave, buff (with a knife or emery wheel to equalize the thickness).

drayoire, *f.* shaving knife, currier's knife.

drayure, *f.* shaving (of hide).

drêche, *f.* (often *pl.,* **drèches**) spent wash, slops, the residue left after distilling starchy materials (as grain, potatoes; cf. vinasse); spent malt, brewer's grains; (less often) malt.

Dresde, *m.* Dresden.

dressage, *m.* dressing, etc. (see dresser).

dressé, *p.a.* dressed, etc. (see dresser).

dressement, *m.* dressing, etc. (see dresser).

dresser, *v.t.* dress; straighten, level, smooth; set, set up, arrange; erect, raise; train; draw up.

dressoir, *m.* something for dressing or straightening or leveling, dresser, straightening rod, etc.; table, sideboard, dresser.

drille, *f.* drill, bit, borer; (*Paper*) rag.

drogue, *f.* drug.

droguer, *v.t.* drug.

droguerie, *f.* drugs; drug trade; drug shop.

droguiste, *m. & f.* druggist, dealer in drugs. — *a.* drug, dealing in drugs.

droit, *m.* right; law; duty, toll; fee. — *a.* right; straight; upright; (*Org. Chem.*) right-handed, dextro. — *adv.* right; upright, perpendicularly.

à bon —, with good reason, justly.

de —, rightful; rightfully.

— *de douane,* customs duty.

— *d'entrée,* import duty.

— *protecteur,* protective duty, protective tariff.

droite, *f.* right line, straight line; right hand, right.

droitement, *adv.* rightly, justly, judiciously.

drôle, *a.* amusing, funny, droll.

dropax, *m.* (*Pharm.*) dropax.

drousser, *v.t.* scribble (wool).

dru, *a.* thick, crowded; brisk, lively, vigorous. — *adv.* thickly.

druse, *f.* (*Min.*) druse.

du. Contraction of *de le,* of the, from the, in the, some, etc. (see *de*).

dû, *p.p. & p.a.* (from devoir), due, owing, owed, ought. — *m.* due.

il a — *aller,* he has had to go; he must have gone.

il aurait — *savoir,* he ought to have known.

dualine, *f.* dualin (a kind of dynamite).

dualisme, *m.* dualism.

dualiste, *a.* dualistic. — *m.* dualist.

dualistique, *a.* dualistic.

duboisine, *f.* duboisine.

duc, *m.* duke; owl (of certain kinds).

duché, *m.* duchy, dukedom.

ductile, ductible, *a.* ductile.

ductilimètre, *m.* ductilimeter.

ductilité, *f.* ductility.

duitage, *m.* weft, woof.

duite, *f.* weft thread, weft yarn.

dulcamarine, *f.* dulcamarin.

dulcifiant, *p.a.* dulcifying. — *m.* dulcifier.

dulcification, *f.* (*Old Chem.*) dulcification.

dulcifier, *v.t.* (*Old Chem.*) dulcify (sweeten, free from acidity, bitterness, or the like).

dulcine, *f.* dulcin.

dulcitane, *f.* dulcitan.

dulcite, *f.* dulcitol, dulcite.

dûment, *adv.* duly.

Dunkerque, *n.* Dunkirk.

duplicata, *m.* duplicate.

duplication, *f.* duplication.

duplicité, *f.* doubleness; duplicity.

duquel. Contraction of *de lequel,* of whom, of which, of the which, from whom, from which, etc. (see *de*).

dur, *a.* hard; harsh, rough; stiff; hardened, inured. — *adv.* hard; stoutly. — *m.* hard.

durabilité, *f.* durability, durableness.

durable, *a.* durable.

durablement, *adv.* durably.

duralumine, *f.* duralumin.

duramen, *m.* (*Bot.*) duramen, heartwood.

durant, *prep.* during. — *a.* lasting, long.

— *que,* during . . . that, while.

durcir, *v.t., i. & r.* harden.

— *au feu,* fire-harden.

durcissage, *m.* hardening.

durcissant, *p.a.* hardening.

durcissement, *m.* hardening.

durcisseur, *a.* hardening. — *m.* hardener.

durée, *f.* duration, time, time interval; life.

durelin, *m.* British oak (*Quercus robur*).

durement, *adv.* hardly, hard; harshly.

dure-mère, *f.* (*Anat.*) dura mater.

durent, *pr. 3 pl. indic. & subj.* of durer. — *p. def. 3 pl.* of devoir.

durer, *v.i.* last, endure, continue, remain.

dureté, *f.* hardness, etc. (see dur); callosity.

durillon, *m.* callosity; specif., corn.

dus, *a. pl.* of dû.

dut, *p. def. 3 sing.* of devoir.

duvet, *m.* down; (of cloth) nap.

duveteux, duveté, *a.* downy.

dyke, *m.* (*Geol.*) dike, dyke.

dynamagnite, *f.* dynamagnite, Hercules powder.

dyname, dynamie, *m.* a unit of work, 1000 kilogrammeters.

dynamique, *a.* dynamic(al). — *f.* dynamics.

dynamitage, *m.* dynamiting.

dynamite, *f.* dynamite.

dynamiterie, *f.* dynamite factory.

dynamiteur, *m.* dynamite maker; dynamite handler; dynamite.

dynamitière, *f.* dynamite magazine.

dynamo, *f.* (*Elec.*) dynamo.

dynamogène, *m.* (*Expl.*) dynamogen.

dynamophore, *a.* energy-producing (said of foods such as starch, sugar, fats).

dyne, *f.* (*Physics*) dyne.

dysenterie, *f.* (*Med.*) dysentery.

dyslysine, *f.* dyslysin.

dyspnée, *f.* (*Med.*) dyspnea, dyspnœa.

dysprosium, *m.* dysprosium.

dyssymétrie, *f.* dissymmetry, asymmetry.

dyssymétrique, *a.* asymmetrical, unsymmetrical.

E

eau, *f.* water; (*pl.*, eaux) clear liquid, filtrate (as distinguished from precipitate); (often *pl.*, eaux) liquor, liquid; aqueous extract; gloss; sweat; tears.

— à grande —, with excess of water.
— à blanchir, whitewash.
— acidulée, carbonated water, aërated water, charged water.
— africaine = eau d'Égypte.
— alcaline, alkaline water, alkali water.
— amidonnée, starch water, water containing starch.
— ardente, ardent spirit, distilled liquor (esp. brandy), eau de vie.
— à refroidir, cooling water.
— aromatique, an aromatic distillate made from flowers, berries, leaves or the like macerated in water.
— blanche, (*Pharm.*) lead water, diluted solution of lead subacetate; bran mash.
— bromée, bromine water.
— calcaire, calcareous water, limewater (water containing calcium salts).
— camphrée, (*Pharm.*) camphor water.
— céleste, eau céleste (solution of cupric ammonium sulfate).
— cémentatoire, (*Mining*) cement water (water containing metallic salts).
— chaude, hot water.
— chlorée, chlorine water.
— claire, clear water.
— courante, running water.
— créosotée, (*Pharm.*) creosote water.
— crue, hard water.
— d'alimentation, feed water (as for boilers); drinking water.
— d'amandes amères, (*Pharm.*) bitter almond water.
— d'ammoniaque, ammonia water, aqua ammoniæ.
— d'ammoniaque forte, (*Pharm.*) stronger ammonia water, aqua ammoniæ fortior.
— d'aneth, (*Pharm.*) dill water.
— d'ange, angel water (an old perfume).
— d'anis, (*Pharm.*) anise water.
— de baryte, baryta water.
— de brome, bromine water.
— de carottes, carrot water.
— de chaux, limewater (solution of calcium hydroxide).

eau de Chine = eau d'Égypte.
— de chlore, chlorine water.
— de chloroforme, (*Pharm.*) chloroform water.
— de ciel, rain water.
— de Cologne, cologne, Cologne water, eau de Cologne.
— de combinaison, combined water, water of constitution or combination (usually applied to water not removed by ordinary drying at 100–120°).
— de constitution, water of constitution.
— de cristallisation, water of crystallization, crystal water.
— de façonnage, (*Ceram.*) water which evaporates from the paste on exposure to dry air at ordinary temperature.
— de fenouil, (*Pharm.*) fennel water.
— de fleur d'oranger, orange-flower water.
— de fontaine, well water, spring water.
— de Goulard, (*Pharm.*) Goulard water.
— d'Égypte, a solution of silver nitrate used for turning hair black.
— de Javel, eau de Javel, Javel water. (The spelling *Javelle* is also used in French and English.)
— de Labarraque, Labarraque's solution (or fluid or liquor), eau de Labarraque.
— de laitue, water distilled from lettuce, used as a mild sedative.
— de la reine de Hongrie, (*Pharm.*) Hungary water.
— de lavage, wash water.
— de levure, yeast extract.
— de Luce, (*Old Pharm.*) eau de luce (mixture of oil of amber, alcohol and ammonia, sometimes with balm of Gilead added).
— de malt, malt extract, malt water.
— de Mendererus = esprit de Mendererus, under *esprit*.
— de menthe verte, (*Pharm.*) spearmint water.
— de mer, sea water.
— de mine, mine water.
— de naphe(?), orange flower water.
— dentifrice, tooth lotion, liquid dentifrice.
— de Perse = eau d'Égypte.
— de piment de la Jamaïque, (*Pharm.*) pimento water.
— de pluie, rain water.
— de puits, well water.
— de rase, — de raze, oil of turpentine.

118

eau *de riz*, arrack (from rice); rice water.
— *de rose*, rose water.
— *de savon*, soapsuds.
— *de Seltz*, Seltzer water; (in general) carbonated water, aërated water, soda water.
— *de source*, spring water.
— *de strontiane*, strontia water.
— *de toilette*, toilet water.
— *de trempe*, (*Metal.*) hardening water.
— *de vie*, = eau-de-vie.
— *de ville*, city water, tap water.
— *d'interposition*, included water.
— *d'iode*, iodine water, an aqueous solution of iodine.
— *distillée*, distilled water.
— *distillée de cannelle*, (*Pharm.*) cinnamon water.
— *distillée de carvi*, (*Pharm.*) caraway water.
— *distillée de fleurs d'oranger*, (*Pharm.*) orange flower water.
— *distillée de hamamelis*, (*Pharm.*) hamamelis water, witchhazel extract.
— *distillée de laurier-cerise*, (*Pharm.*) cherry-laurel water.
— *distillée de menthe poivrée*, (*Pharm.*) peppermint water.
— *distillée de rose*, (*Pharm.*) rose water.
— *distillée de sureau*, (*Pharm.*) elder-flower water.
— *divine de Fernel*, (*Pharm.*) yellow mercurial lotion.
— *dormante*, stagnant water, still water.
— *douce*, fresh (not salt) water; soft water.
— *dure*, hard water.
— *ferrée*, iron water, chalybeate water; water in which red-hot iron has been quenched.
— *ferrugineuse*, iron water, chalybeate water.
— *fraîche*, fresh (not stale) water; cool water.
— *froide*, cold water.
— *gazeuse*, water containing dissolved gases; specif., carbonated water, charged water, aërated water.
— *gazeuse simple*, carbonic acid water, "plain soda."
— *glacée*, ice water.
— *hygrométrique*, moisture, hygrometric water (sometimes, specif., water remaining in a substance under ordinary dry conditions but expelled by heating to 120° or less).
— *incrustante*, hard water (producing an incrustation in boilers).
— *magnésienne*, (*Pharm.*) solution of magnesium carbonate, fluid magnesia.
— *médicinale*, (*Pharm.*) water, medicated water.
— *mère*, mother liquor.
— *minérale*, mineral water.
— *oxygénée*, hydrogen peroxide, oxygenated water.

eau *parfumée*, scented water, perfumed water.
— *phagédénique*, (*Pharm.*) yellow mercurial lotion, yellow wash.
— *phagédénique noire*, (*Pharm.*) black mercurial lotion, black wash.
— *phéniquée*, a dilute aqueous solution of phenol, weak carbolic acid.
— *potable*, potable water, drinking water.
— *régale*, aqua regia, nitrohydrochloric acid.
— *rouillée*, water in which red-hot iron has been quenched.
— *salée*, salt water, specif. an aqueous solution of sodium chloride.
— *saline*, saline water.
— *saumâtre*, brackish water.
— *seconde*, weak nitric acid; lye water.
— *sédative de Raspail*, (*Pharm.*) sedative water.
— *séléniteuse*, a natural water containing calcium sulfate.
— *souterraine*, underground water.
— *sulfureuse*, sulfur water.
— *tellurique*, ground water, water which has been in contact with the earth as distinguished from meteoric water (rain water, snow water).
— *thermale*, thermal water, hot-spring water.
— *tombée*, rainfall.
— *trouble*, turbid water.
— *usée*, waste water.
— *végéto-minérale*, (*Pharm.*) Goulard water.
— *vitriolique*, (*Mining*) water containing metallic sulfates, esp. copper sulfate.
— *vive*, spring water.
eaux ammoniacales, ammoniacal liquor (from distillation of coal, wood, etc.); ammoniacal solution or filtrate.
eaux d'égout, sewage.
eaux des villes, refuse water from cities, sewage.
eaux faibles, (*Tech.*) a weak solution, e.g. one made by leaching sodium aluminate with water.
eaux folles = eaux sauvages (below).
eaux fortes, (*Tech.*) a strong solution, strong liquor.
eaux grasses, water containing grease.
eaux résiduaires, residual liquor (or liquid); waste waters (from mines or factories).
eaux rouges, (*Soda*) red liquor.
eaux sauvages, water that has run over the surface without penetrating the earth; surface water (in the narrow sense), flood water.
eaux sures, (*Starch*) sour liquor.
eaux vannes, liquid manure obtained from urinals, privy vaults, etc.; specif., stale urine used as a source of ammonium salts.
faire —, leak.

eau-de-vie, *f.* brandy (or, in general, distilled liquor), eau de vie, aqua vitæ.
— *de grain,* whisky.
— *de marc,* marc brandy.
— *de vin,* brandy (in the strict sense).
eau-forte, *f.* aqua fortis (nitric acid, esp. a medium strength used in etching).
eau-mère, *f.* mother liquor.
eaux, *pl.* of eau.
eaux-mères, *f.pl.* mother liquors.
eaux-vannes, *f.pl.* = eaux vannes, under *eau.*
ébarbage, ébarbement, *m.* trimming, etc. (see ébarber).
ébarber, *v.t.* (remove the edges, ends or roughnesses of) trim, clip, pare, nip, chip, edge, scrape, clean.
ébarbure, *f.* trimming, etc. (see ébarber).
ébauchage, *m.* rough-shaping, etc. (see ébaucher).
ébauche, *f.* rough-shaping (as of a gem); sketch, rough sketch; rough draft.
ébaucher, *v.t.* rough-shape, roughcast; sketch.
ébène, *f.* ebony (wood).
ébéner, *v.t.* ebonize.
ébénier, *m.* ebony (tree).
ébeurrer, *v.t.* remove the butter from.
ébiseler, *v.t.* bevel, chamfer.
éblouir, *v.t.* dazzle. — *v.r.* be dazzled.
éblouissant, *p.a.* dazzling.
éblouissement, *m.* dazzling.
ébonite, *f.* ebonite.
ébouillant, *p.pr.* of ébouillir.
ébouillantage, *m.* scalding.
ébouillanter, *v.t.* scald.
ébouillanteuse, *f.* scalder.
ébouillir, *v.i.* boil down; boil away.
éboulement, *m.* falling in, falling down, crumbling; collapse; landslide.
ébouler, *v.r. & i.* fall in, fall down, crumble.
— *v.t.* cause to fall down, throw down.
éboulis, *m.* fallen material, débris, rubbish.
ébourrer, *v.t.* = débourrer.
ébout, *pr. 3 sing.* of ébouillir.
ébranlement, *m.* shaking, agitation, disturbance.
ébranler, *v.t.* shake, agitate. — *v.r.* shake, be shaken, be agitated.
ébrécher, *v.t.* notch (as a knife); break; damage.
ébronder, *v.t.* deoxidize (iron wire).
ébrouage, *m.* (*Dyeing*) washing (of wool).
ébroudage, *m.* drawing (of wire; see ébroudir).
ébroudeur, *m.* wire drawer.
ébroudir, *v.t.* draw (wire), esp. thru a fine gage.
ébrouer, *v.t.* (*Dyeing*) wash (wool); hull, husk.
ébrouissage, *m.* (*Dyeing*) washing (of wool).
ébruiter, *v.t.* make known, report, rumor.
ébrutage, *m.* grinding (of diamonds).
ébulliomètre, *m.* ebulliometer, ebullioscope.

ébullioscope, *m.* ebullioscope.
ébullioscopie, *f.* ebullioscopy, observation of boiling points.
ébullition, *f.* boiling, (less often) ebullition.
éburge, *n.* (*Glass*) an iron shovel used for introducing the pots into the furnace.
éburine, *f.* eburin, eburine, eburite.
éburnation, *f.* = éburnification; (*Med.*) eburnation.
éburnification, *f.* conversion of powdered bone and ivory into eburin by heat and pressure.
ébutter, *v.t.* flesh, slate (hides).
écachement, *m.* crushing, etc. (see écacher).
écacher, *v.t.* crush, squeeze, squash, bruise; flatten (wire); (*Paper*) squeeze moisture from.
écaillage, *m.* scaling, etc. (see écailler).
écaille, *f.* scale; flake, chip; shell.
écaillement, *m.* scaling, etc. (see écailler).
écailler, *v.t.* scale; chip off, flake off; shell.
— *v.r.* scale, scale off, chip off; shell.
écaillette, *f.* little scale; little shell.
écailleux, *a.* scaly, squamous.
écale, *f.* shell; hull, husk.
écaler, *v.t.* shell; hull, husk.
écarlate, *f. & a.* scarlet.
— *de Biebrich,* Biebrich scarlet.
— *de Hollande,* natural cochineal scarlet.
— *de Venise,* Venetian scarlet (an old scarlet dye made from kermes).
écarner, *v.t.* round off, remove the angles of.
écart, *m.* variation, deviation, difference; error; (*Optics*) dispersion; scarf joint; stepping aside; digression; strain.
à l'—, aside, apart.
— *absolu,* absolute error.
— *moyen,* mean error.
écartement, *m.* removal, etc. (see écarter); distance (apart), space, spacing.
écarter, *v.t.* remove; disperse, scatter; spread open, separate; turn aside, divert; keep away. — *v.r.* deviate, depart; diverge, open; disperse, be dispersed.
écatir, écatissage, = catir, catissage.
ecbolique, *a. & m.* (*Med.*) ecbolic.
ecboline, *f.* ecboline (ergotinine).
ecgonine, *f.* ecgonine.
échafaud, *m.* (also *Metal.*) scaffold; platform.
échafaudage, *m.* (also *Metal.*) scaffolding; mass.
échalote, *f.* shallot (*Allium ascalonicum*).
échancrer, *v.t.* channel, groove; hollow out, indent, notch.
échancrure, *f.* channel, groove; hollow, indentation, notch.
échange, *m.* exchange.
en — de, in exchange for, in return for.
échangeable, *a.* exchangeable.
échanger, *v.t.* exchange.
échantignole, *f.* bracket.

échantillon, *m.* sample; specimen; pattern; model; gage; mold, matrix; (*Founding*) modeling board.

échantillonage, *m.* sampling, etc. (see next def.).

échantilloner, *v.t.* sample, take samples of; gage, test (as weights); mix different samples of (e.g. sugar) in order to obtain a standard product, blend; make up (e.g. iron) into given patterns or shapes; sort.

échappade, *f.* (*Ceram.*) a system of refractory partitions in a kiln. (Certain articles are fired in the compartments thus formed instead of in saggers.)

échappée, *f.* interval; clear space; width (of an opening); pitch (of a screw); escapade.

échappement, *m.* escape; escapement; outlet; flue (of a reverberatory furnace); (*Steam*) exhaust.

échapper, *v.i. & t.* escape. — *v.r.* escape; vanish; forget oneself; come undone.
— *à,* — *de,* escape from, escape.
laisser —, let off (steam, etc.); let escape, let go, let slip, pass over.

écharde, *f.* splinter, fragment; prickle.

écharnage, écharnement, *m.* fleshing (of hides).

écharner, *v.t.* flesh (hides).

écharnoir, *m.* fleshing knife.

écharpe, *f.* brace, tie; outrigger; scarf; sash.

écharper, *v.t.* card; pass a rope around; slash.

échauboulure, *f.* pimple.

échaudage, *m.* scalding; whitewash; whitewashing.

échaude, *f.* welding heat.

échauder, *v.t.* scald; whitewash.

échaudillon, *m.* piece of iron at welding heat.

échaudoir, *m.* scalding vat; scalding house.

échaudure, *f.* scald.

échauffe, *m.* (*Leather*) sweating.

échauffé, *p.a.* heated; (of wood, etc.) rotten.

échauffement, *m.* heating, etc. (see échauffer).

échauffer, *v.t.* heat; overheat; warm; vex, anger. — *v.r.* heat, become heated or hot; become overheated; warm, become warm; (of wood) dry-rot; (of yarn) become rotten.

échéance, *f.* expiration; (*Com.*) maturity.

échéant, *p.pr.* of échoir.

échec, *m.* check, defeat, failure, reverse.

échelle, *f.* scale; ladder.
— *d'eau,* water gage.
— *de proportion,* diagonal scale.
— *graduée,* graduated scale.
— *mobile,* sliding scale.
— *proportionnelle,* diagonal scale.
— *réduite,* reduced scale.

échelon, *m.* step; (of a ladder) rung, round.

échelonnement, *m.* arrangement, disposition, distribution (at intervals, in a series).

échelonner, *v.t.* arrange (in a series), distribute, place at intervals; proportion.

échenal, écheneau, échenet, *m.* gutter, channel; (*Founding*) gate.

écherra, *fut. 3 sing.* of échoir.

échet, *pr. 3 sing.* of échoir.

écheveau, *m.* skein, hank.

échevette, *f.* skein (in France 100 meters).

échine, *f.* spine; echinus (a rounded molding).

échinocoque, *m.* (*Zoöl.*) echinococcus.

échinoderme, *m.* (*Zoöl.*) echinoderm.

échinon, *m.* cheese mold.

échiquier, *m.* checkerwork; checkerboard; exchequer.

écho, *m.* echo.

échoir, *v.i.* fall due; expire; happen, occur.

échouer, *v.i.* strand, run aground; fail.

échoyait, *imp. 3 sing.* of échoir.

échu, *p.p.* of échoir.

éclabousser, *v.t.* splash, spatter.

éclair, *m.* lightning; flash, gleam; (*Assaying*) fulguration, blick.

éclairage, *m.* lighting, illuminating, illumination.
— *au gaz,* gas lighting.
— *du fond,* (in spectra) continuous illumination, continuous light.
— *électrique,* electric lighting.

éclairant, *p.a.* illuminating, lighting.

éclaircie, *f.* clear spot, clear space; clearing.

éclaircir, *v.t.* clear, clear up; clarify; thin; brighten, polish; inform. — *v.r.* clear, clear up, become clear; become thin; become bright.

éclaircissage, éclaircissement, *m.* clearing, etc. (see éclaircir).

éclairé, *p.a.* lighted, illuminated; well informed, experienced, enlightened; open.

éclairement, *m.* lighting, etc. (see éclairer); clearness, clarity.

éclairer, *v.t.* light, illuminate; enlighten; watch; reconnoiter. — *v.i.* give light, shine, gleam; lighten, flash. — *v.r.* be lighted, be illuminated; inform oneself.

éclampsie, *f.* (*Med.*) eclampsia.

éclat, *m.* burst, bursting; loud and sudden noise, report, crash; fragment, splinter, shiver, chip; luster (as of minerals); brightness, brilliancy; flash; crack (in wood); splendor.
— *métallique,* metallic luster; metallic fragment or splinter.

éclatant, *p.a.* bursting, etc. (see éclater); brilliant, bright; loud; shrill.

éclatement, *m.* bursting, etc. (see éclater).

éclater, *v.i.* burst, shiver, split, splinter; make a loud and sudden noise, crack, thunder; shine, be brilliant, flash, glitter; (*Elec.*) spark; break out; be revealed. — *v.t.* burst, shiver, split, rend; chip. — *v.r.* burst, shiver, split, splinter, split.

éclateur, *m.* (*Elec.*) spark gap.
— *réglable*, adjustable spark gap.

éclipser, *v.t.* eclipse. — *v.r.* be eclipsed.

éclisse, *f.* splinter, splint; fish, fishplate.

éclore, *v.i.* dawn, appear; open, bloom; hatch.

éclosion, *f.* dawning, appearance; blossoming, bloom; hatching, hatch.

écluse, *f.* sluice; floodgate; dam; lock.

école, *f.* school; schoolhouse; blunder.

écolier, *m.* scholar, pupil. — *a.* (of paper) white and of medium quality.

écollage, *m.* fleshing (of hides).

économe, *a.* economical. — *m.* steward, fiscal agent, bursar, etc.

économie, *f.* economy; organism; savings.

économique, *a.* economic(al). — *f.* economics.

économiquement, *adv.* economically.

économiser, *v.t.* economize; save. — *v.i.* economize. — *v.r.* be economized.

écope, *f.* ladle; scoop.

écorçage, *m.* stripping, barking, peeling.

écorce, *f.* bark; rind, peel; (earth's) crust; (*Bot. & Anat.*) cortex; (fig.) surface.
— *à tan*, tanbark.
— *d'azédarach*, (*Pharm.*) margosa bark, azadirachta, azedarach.
— *de bigarade*, bitter orange peel.
— *de bourdaine*, (*Pharm.*) frangula.
— *de cerisier de Virginie*, (*Pharm.*) wild cherry.
— *de chêne*, oak bark; (*Pharm.*) white oak, quercus.
— *de citron*, lemon peel.
— *de géoffrée*, (*Pharm.*) cabbage-tree bark.
— *de grenade*, (*Pharm.*) pomegranate, granatum.
— *de la racine de cotonnier*, (*Pharm.*) cotton root bark.
— *de la racine de grenadier*, (*Pharm.*) pomegranate.
— *de limon*, lemon peel.
— *de magnolier*, (*Pharm.*) magnolia bark, magnolia.
— *de mancône*, see mancône.
— *de margousier*, = écorce d'azédarach.
— *de quillaja*, (*Pharm.*) soap bark, quillaja.
— *de tan*, tanbark.
— *d'evonymus*, (*Pharm.*) euonymus, wahoo.
— *d'oranges amères*, bitter orange peel.
— *d'oranges douces*, sweet orange peel.
— *d'orme*, elm bark, specif. (*Pharm.*) slippery elm.
— *de racine de berbérides*, berberis root bark.
— *de ronce noire*, (*Pharm.*) rubus, blackberry.
— *de Winter*, (*Pharm.*) Winter's bark.
— *éleuthérienne*, (*Pharm.*) cascarilla, cascarilla bark, eleuthera bark.

écorcement, *m.* stripping, barking, peeling.

écorcer, *v.t.* strip (of bark or rind), bark, peel.

écorcher, *v.t.* skin; taste rough to; grate on; fleece; murder (a language).

écorcier, *m.* place where tanbark is kept.

écorner, *v.t.* break off the corner of, chip; (fig.) cut into, impair.

écornure, *f.* chip, splinter, broken-off corner.

écossais, *a.* Scotch, Scottish.

Écosse, *f.* Scotland.

écosser, *v.t.* husk, hull, shell.

écoulement, *m.* flowing off, running out, outflow, efflux, draining, discharge; vent; (*Com.*) sale.

écouler, *v.i. & r.* flow out, run out, flow away; (of people) disperse; (*Com.*) sell; (of time) pass. — *v.t.* sell, sell off.

écourter, *v.t.* curtail, shorten; dock.

écoussage, *m.* (*Ceram.*) stain, dark spot.

écoutant, *p.a.* listening. — *m.* hearer, auditor.

écouter, *v.t.* listen to.

écraminage, *m.* (*Leather*) removal of the fatty matter from the flesh side of the hide.

écran, *m.* screen; shield; baffle plate.

écrasage, *m.* crushing, etc. (see écraser).

écrasé, *p.a.* crushed, etc. (see écraser); flat.

écrasement, *m.* crushing, etc. (see écraser).

écraser, *v.t.* crush; squeeze; flatten; bruise; masticate (rubber).

écrasite, *f.* (*Expl.*) ecrasite.

écrémage, *m.*, **écrémaison**, **écrémation**, *f.* skimming; separating (as of milk by machine).

écrémé, *p.a.* skimmed; separated; skim (milk).

écrémer, *v.t.* skim; separate (as milk or soapy water, in a centrifugal).

écrémette, *f.* skimmer, skimming ladle.

écrémeuse, *f.* separator (for liquids, esp. milk); milk pan, milk crock.

écrémoir, *m.*, **écrémoire**, *f.* skimmer.

écrémures, *f.* skimmings, skim, scum; (*Glass*) glass gall, sandiver.

écrevisse, *f.* crawfish; large tongs for handling heavy weights; (*Astron.*) Cancer.

écrier, *v.t.* scour (iron wire). — *v.r.* cry out.

écrin, *m.* case; jewel case, casket.

écrire, *v.t.* write; spell.

écrit, *pr. 3 sing. & p.p.* of écrire. — *m.* writing; pamphlet.

écriteau, *m.* bill, placard.

écriture, *f.* writing; (*Com., pl.*) accounts, books; Scripture.

écrivain, *m.* writer; clerk.

écrivant, *p.pr.* of écrire.

écrivent, *pr. 3 pl.* of écrire.

écriveur, *m.* writer.

écrou, *m.* screw nut, (female or internal) screw.
— *ailé*, — *à oreilles*, wing nut, wing screw, thumb nut, fly nut.
— *de serrage*, clamp nut, set nut.

écrouir, *v.t.* (*Metal.*) cold-hammer, hammer harden, cold-harden. — *v.r.* cold-harden.

écrouissage, écrouissement, *m.* cold-hammering, cold-hardening, hammer-hardening.

écroûter, *v.t.* remove the crust from.

écru, *a.* (of cloth, etc.) raw, unbleached, unwashed; (of iron) badly worked.

éctogan, *m.* a trade name for zinc peroxide.

écu, *m.* shield; crown (the coin, at present in France the 5-franc piece); (*pl.*) money.

écueil, *m.* rock, reef.

écuelle, *f.* porringer, bowl, saucer; (*Oils*) ecuelle, a saucer-shaped metal vessel covered on the inside with short spikes and used in obtaining essential oil from fruits (*e.g.* oranges).

écumage, *m.* skimming, scumming; ridding.

écumant, *p.a.* foaming, foamy, frothing, frothy; scummy.

écume, *f.* foam, froth; scum; dross, scoria.
— *de fer,* (*Min.*) a porous form of hematite.
— *de fonte,* (*Metal.*) kish.
— *de mer,* meerschaum; sea foam.

écumer, *v.t.* skim, scum, clear of froth or scum; skim off, remove by skimming; clear, rid. — *v.i.* foam, froth; become covered with scum or dross.

écumeresse, *f.* large skimmer.

écumette, *f.* small skimmer.

écumeur, *m.* skimmer, scum remover.

écumeux, *a.* foamy, foaming, frothy.

écumoire, *f.* skimmer, scummer.

écurage, écurement, *m.* cleaning, scouring.

écurer, *v.t.* clean, scour.

écurier, *f.* stable.

édicule, *m.* small edifice or structure.

édification, *f.* building, building up; edification.

édifice, *m.* edifice, structure.

édifier, *v.t.* build, build up; satisfy; edify.

Édimbourg, *m.* Edinburgh.

éditer, *v.t.* publish; (less commonly) edit.

éditeur, *m.* publisher; (less commonly) editor.

édition, *f.* edition.

éducateur, *m.* educator; raiser, breeder. — *a.* educational.

éducation, *f.* education; breeding, raising.

éducatrice, *fem.* of éducateur.

éduction, *f.* eduction, exhaustion (of steam).

édulcorant, *a. & m.* edulcorant.

édulcoration, *f.* edulcorating, edulcoration.

édulcorer, *v.t.* edulcorate.

effacer, *v.t.* efface, obliterate, erase.

effaçure, *f.* effacement, erasure, obliteration.

effectif, *a.* effective; actual, real.

effectivement, *adv.* in fact, in reality, indeed; effectively.

effectuer, *v.t.* effect, perform, do, make, accomplish, execute, effectuate.

effervescence, *f.* effervescence.

effervescent, *a.* effervescent.

effet, *m.* effect; (*Com.*) bill; (public) funds.
à — *retardé,* delay-action.
— *brisant,* rending effect, shattering effect.
— *mécanique,* mechanical effect.
— *utile,* useful effect.
en —, in fact, in reality, in effect, indeed.

effeuiller, *v.t.* strip of leaves.

efficace, *a.* efficacious, effective.

efficacement, *adv.* efficaciously, effectively.

efficacité, *f.* efficacy, efficaciousness.

efficient, *a.* efficient.

effilage, *m.* drawing out, etc. (see effiler).

effilé, *p.a.* drawn out (as a glass tube), tapered, thinned; slender; raveled; fringed.

effiler, *v.t.* draw out (as a glass tube), taper, thin; ravel, unravel. — *v.r.* be drawn out, taper; ravel, ravel out.

effilocher, *v.t.* ravel out, pick to pieces, disintegrate.

effleurage, effleurement, *m.* removal of the surface, etc. (see effleurer).

effleurer, *v.t.* remove the surface of, scrape, shave; (*Leather*) scrape or shave the hair side of; graze; touch; touch on; skim along, skim over; strip of flowers. — *v.r.* be scraped, etc.; fade, wither.

effleurir, *v.i. & r.* effloresce.

efflorescence, *f.* efflorescence.

efflorescent, *a.* efflorescent.

effluent, *a.* effluent.

effluve, *m.* (nondisruptive) electric discharge; effluvium.

effondrer, *v.t.* break in, cause to give way; dig up, plow deep. — *v.r.* give way, fall.

effondrilles, *f.pl.* sediment, dregs, lees.

effort, *m.* effort; force, stress; strain.
— *de cisaillement,* shearing stress.
— *de rupture,* breaking stress.
— *de tension,* tensile stress.

effrayant, *p.a.* frightful, terrible, appalling.

effriter, *v.t.* render friable, cause to crumble; exhaust (soil). — *v.r.* crumble.

effroyable, *a.* frightful, fearful.

effroyablement, *adv.* frightfully, dreadfully.

effulguration, *f.* flashing, glowing.

effusion, *f.* effusion.

égal, *a.* equal; even; same, alike. — *m.* equal.
à l'— *de,* equal to, like; in comparison with.

également, *adv.* equally, alike, likewise. — *m.* equalization.

égaler, *v.t.* equal; equalize; even, level.

égalisation, *f.* equalization; evening.

égaliser, *v.t.* equalize; even, level, smooth.

égaliseur, *m.* equalizer; regulator.

égalisoir, *m.* separating sieve (for gunpowder).

égalitaire, *a.* equalizing, leveling.

égalité, *f.* equation; equality; evenness; uniformity.
à — *de,* supposing the equality of, supposing , , . to be equal (or, to be the same).

égard, m. regard, respect.

à l'— de, with regard to, with respect to, regarding, as to, in comparison with.

à tous égards, in all respects.

égaré, p.a. led astray, stray, wandering, lost; scattered; mislaid; misguided; disordered.

égarement, m. wandering; error; wildness; mislaying.

égarer, v.t. lead astray; let wander; mislay; disorder, derange. — v.r. go astray, stray, wander; be mislaid, be lost.

égaux, m.pl. of égal.

égayer, v.t. enliven; rinse (linen); prune.

égermage, m. degerming.

égermer, v.t. remove the germ of, degerm.

égide, f. protection; ægis.

église, f. church.

égout, m. dripping, drip; sewer; drain, gutter, channel; (Glass) dropping board; (Founding) vent.

— de deuxième jet, (Sugar) sirup from which two crops of crystals have been removed in the centrifugals.

— de premier jet, (Sugar) sirup from which one crop of crystals has been removed in the centrifugals.

égouttage, égouttement, m. draining, drainage; dripping, drip.

égoutter, v.t. drain; let drip. — v.i. drain; drip. — v.r. drain.

égoutteur, m. drainer; drain. — a. draining, drain, dripping.

égouttoir, m. drainer, drain board, draining board, draining rack.

égoutture, f. drainings.

égrainage, etc. See égrenage, etc.

égrappage, m. picking (esp. of grapes) from the bunch; cleaning (of ore).

égrapper, v.t. pick from the bunch (as grapes); clean (ore).

égratigner, v.t. scratch.

égratignure, f. scratch.

égrenage, égrènement, m. shelling; picking; ginning (of cotton).

égrener, v.t. shell (as corn); pick (as grapes from the bunch); gin (cotton).

égrisage, m. grinding (as of glass or diamond).

égrisé, m., **égrisée,** f. diamond powder.

égriser, v.t. grind (glass, marble, etc.).

égrugeoir, m. mortar; (Gunpowder) rubber, mealer.

égruger, v.t. grind, pulverize; meal (gunpowder).

égrugeur, a. grinding, pulverizing.

égrugeure, f. grindings.

Égypte, f. Egypt.

eisen-ram, m. (Min.) a porous form of hematite.

éjecter, v.t. eject.

éjecteur, m. ejector. — a. ejecting.

éjection, f. ejection.

élaborant, élaborateur, a. elaborating.

élaboration, f. elaboration.

élaborer, v.t. elaborate.

élagage, m. pruning.

élaguer, v.t. prune.

élaïdine, f. elaïdin.

élaïdique, a. elaïdic.

élaïne, f. olein, elaïn.

élaïomètre, m. eleometer, elæometer.

élaldéhyde, m. paraldehyde.

élan, m. start, spring, jump, sally, impulse; ardor, enthusiasm; elk.

élancé, a. slender, slim, tall.

élancer, v.r. spring, rush, dash, fly, soar.

élargir, v.t. & r. enlarge, extend, widen.

élargissement, m. enlargement, extension, widening.

élasticité, f. elasticity.

élastification, f. elasticizing.

élastine, f. elastin.

élastique, a. & m. elastic.

élatérine, f. elaterin.

élatérite, f. (Min.) elaterite.

élatérium, m. (Pharm.) elaterium; (Bot.) wild cucumber (Ecballium elaterium).

élatéromètre, m. elatrometer, elaterometer.

élavage, m. washing out, washing.

élaver, v.t. wash out, wash (as rags).

élaveuse, f. washer. — a. washing.

électif, a. elective; selective.

élection, f. election, choice.

d'—, preferred, best, choice.

électivement, adv. electively; selectively.

électivité, f. electivity; selectivity.

électre, m. electrum.

électricien, m. electrician.

électricisme, m. electric phenomena in general.

électricité, f. electricity.

électrique, a. electric, electrical.

électriquement, adv. electrically.

électrisable, a. electrifiable.

électrisant, p.a. electrifying.

électrisation, f. electrification.

électriser, v.t. electrify, electrize. — v.r. become electrified.

électriseur, m. electrifier.

électro-affinité, f. electro-affinity.

électro-aggloméré, m. battery carbon.

électro-aimant, m. electromagnet.

électroanalyse, f. electroanalysis.

électrocapillaire, a. electrocapillary.

électrocapillarité, f. electrocapillarity.

électrochimie, f. electrochemistry.

électrochimique, a. electrochemical.

électrochimiste, m. electrochemist.

électrode, f. electrode.

électrodynamique, f. electrodynamics. — a. electrodynamic.

électrolysable, a. electrolyzable.

électrolysation, *f.* electrolyzing, electrolyzation.

électrolyse, *f.* electrolysis.

électrolyser, *v.t.* electrolyze.

électrolyseur, *m.* electrolyzer.

électrolyte, *m.* electrolyte.

électrolytique, *a.* electrolytic.

électrolytiquement, *adv.* electrolytically.

électromagnétique, *a.* electromagnetic.

électromagnétisme, *m.* electromagnetism.

électrométallurgie, *f.* electrometallurgy.

électrométallurgique, *a.* electrometallurgical.

électromètre, *m.* electrometer.

électrométrique, *a.* electrometric(al).

électromoteur, *a.* electromotive. — *m.* electromotor.

électromotrice, *fem.* of électromoteur.

électron, *m.* electron.

électronégatif, *a.* electronegative

électronique, *a.* electronic.

électrophore, *m.* electrophorus.

électropolaire, *a.* electropolar.

électropositif, *a.* electropositive.

électroscope, *m.* electroscope.

électroscopique, *a.* electroscopic.

électrostatique, *a.* electrostatic. —*f.* electrostatics.

électrotechnique, *a.* electrotechnic(al). —*f.* electrotechnics.

électrotrieuse, *f.* magnetic ore separator.

électrotype, *m.* electrotype.

électrotypie, *f.* electrotypy, electrotype.

électrotypique, *a.* electrotype, electrotypic.

électrum, *m.* electrum.

électuaire, *m.* (*Pharm.*) electuary (now often translated *confection*, the old distinction between electuary and conserve having been abandoned).

— *aromatique,* aromatic confection.

— *de poivre,* confection of pepper.

— *de séné composé,* confection of senna.

— *de soufre,* confection of sulfur.

— *lenitif,* confection of senna.

élégamment, *adv.* elegantly.

élégant, *a.* elegant.

élément, *m.* element.

élémentaire, *a.* elementary.

élémi, *m.* elemi.

éléolat, éléolate, *m.* (*Pharm.*) a preparation having an essential oil as a base.

éléolé, *m.* (*Pharm.*) a preparation having an oil as base.

éléolithe, *m.* (*Min.*) eleolite, elæolite.

éléoptène, *m.* eleoptene, elæoptene.

éléosaccharum, *m.* (*Pharm.*) elæosaccharum.

éleuthérien, *a.* (*Pharm.*) eleuthera, cascarilla.

élevage, *m.* raising (of animals), stock raising.

élévateur, *m.* elevator. — *a.* elevating, elevatory.

élévateur *à godets,* bucket elevator.

élévation, *f.* elevation; increase (of price); (*Math.*) involution.

élévatoire, *a.* elevatory, elevating.

élève, *m. & f.* student, scholar, pupil; young animal or plant. — *f.* raising (of animals).

— *bénévole,* special student, special.

— *chimiste,* chemistry student.

— *physicien,* physics student.

élevé, *p.a.* raised, elevated, etc. (see élever); high.

élèvement, *m.* raising, etc. (see élever).

élever, *v.t.* raise; hoist; elevate; erect; produce; rear; exalt. — *v.r.* rise; be raised, be erected, be produced, etc.

élevure, *f.* pimple, pustule.

éliminateur, *a.* eliminating, eliminative.

élimination, *f.* elimination.

éliminatoire, *a.* eliminative, eliminatory.

éliminatrice, *fem.* of éliminateur.

éliminer, *v.t.* eliminate.

élire, *v.t.* elect.

élixation, *f.* (*Old Pharm.*) elixation, seething.

élixir, *m.* elixir.

— *de propriété,* (*Pharm.*) tincture of aloes and myrrh.

— *de salut,* (*Pharm.*) compound tincture of senna, elixir salutis.

— *fébrifuge d'Huxam,* (*Pharm.*) Huxham's tincture of bark (compound tincture of cinchona).

— *parégorique,* (*Pharm.*) paregoric, camphorated tincture of opium.

— *vitriolique,* (*Pharm.*) aromatic sulfuric acid, elixir of vitriol.

ellagique, *a.* ellagic.

ellago-tannique, *a.* ellagotannic.

elle, *fem. pron.* she, her, it.

ellébore, *m.* (*Bot.*) hellebore.

elléborine, *f.* helleborin; (*Bot.*) helleborine.

elles, *fem. pron.* they, them.

ellipse, *f.* ellipse; (*Grammar*) ellipsis.

ellipsoïdal, *a.* ellipsoidal.

ellipsoïde, *m. & a.* ellipsoid.

elliptique, *a.* elliptic(al).

éloge, *m.* eulogy.

faire l'— de, sing the praises of, praise.

élogieux, *a.* eulogistic, commendatory.

éloigné, *p.a.* removed, etc. (see éloigner); distant, remote; averse.

éloignement, *m.* removal, etc. (see éloigner); distance; absence; neglect.

éloigner, *v.t.* remove; defer; banish; avert; estrange. — *v.r.* go away, withdraw, depart, be distant, be different, be averse.

élu, *p.a.* elected, elect.

élucider, *v.t.* elucidate.

éluder, *v.t.* elude, evade. — *v.r.* be evaded.

émacié, *a.* emaciated, very thin.

émail, *m.* enamel; (*Ceram.*) glaze (tho restricted by some to the thicker and less fusible glazes, whether transparent or opaque, and by others to the opaque glazes which are called enamels in English).
— *de cobalt,* smalt.

émaillage, *m.* enameling; glazing (see émail).

émailler, *v.t.* enamel; glaze (see émail).

émaillerie, *f.* enameling; glazing (see émail).

émailleur, *m.* enameler; glazer (see émail).

émaillure, *f.* enameling; enamel work.

émanatif, *a.* emanative.

émanation, *f.* emanation; (fig.) offspring.

émaner, *v.i.* emanate.

émaux, *pl.* of émail.

emballage, *m.* packing; packing case.

emballer, *v.t.* pack, pack up. — *v.r.* hurry off, run away; (of a reaction) become violent; (of machinery) race.

emballeur, *m.* packer.

emballotter, *v.t.* pack, bale.

embargo, *m.* embargo.

embarillage, *m.* barreling.

embariller, *v.t.* barrel.

embarquer, *v.t.* ship; embark. — *v.r.* embark.

embarras, *m.* obstruction; encumbrance; embarrassment.

embarrassant, *p.a.* embarrassing.

embarrasser, *v.t.* obstruct; embarrass; entangle. — *v.r.* be embarrassed; be entangled; (with *de*) trouble oneself (about).

embase, *f.* (*Tech.*) shoulder, seat, base.

embassure, *f.* walls (of a glass furnace).

embaumer, *v.t.* embalm.

embellir, *v.t.* embellish, improve (in appearance), ornament, adorn.

embellissement, *m.* embellishment, improvement, ornamentation, adornment.

emblée. In the phrase *d'emblée,* at once, at the first onset, with ease.

emblème, *m.* emblem.

emboire, *v.t.* cover or coat with oil or wax. — *v.r.* soak in.

emboîtement, *m.* fitting, etc. (see emboîter).

emboîter, *v.t.* fit, fit in, nest, join, joint, engage, put together, dovetail, mortise, clamp. — *v.r.* fit together or into each other, fit, go together, be engaged.

emboîture, *f.* joint, joining, juncture, socket, clamp, collar, box, frame.

embouchure, *f.* mouthpiece; mouth.

embout, *m.* tip, ferrule.

embouteillage, *m.* bottling.

embouteiller, *v.t.* bottle.

embouti, *p.a.* stamped; covered with metal. — *m.* stamped work, stamped piece.

emboutir, *v.t.* stamp, shape by stamping; cover with metal.

emboutissage, *m.* stamping; sheathing with metal.

embranchement, *m.* branching; junction; branch.

embrancher, *v.t.* join, connect. — *v.r.* join, join each other; branch.

embrasé, *p.a.* burning, on fire; fiery.

embrasement, *m.* conflagration, fire, burning.

embraser, *v.t.* set on fire, fire, kindle. — *v.r.* take fire, fire, kindle.

embrasser, *v.t.* embrace.

embrayage, *m.* engaging, engagement, etc. (see embrayer); engaging gear, clutch, coupling.

embrayer, *v.t.* engage, throw into gear, connect.

embrochage, *m.* insertion, etc. (see embrocher).
système par —, spindle system, skewer system (of cop dyeing).

embrocher, *v.t.* (*Elec.*) (1) insert, cut in, (2) arrange in series; put on a spindle; spit.

embrouillement, *m.* clouding, etc. (see next def.).

embrouiller, *v.t.* cloud, fog, obscure; embroil.

embrunir, *v.t.* brown; make too brown.

embryon, *m.* embryo.

embryonnaire, *a.* embryonic, embryo.

embu, *p.p.* of emboire.

embuvant, *p.pr.* of emboire.

émeraude, *f.* & *a.* emerald.

émergence, *f.* emergence, emerging; emergency.

émerger, *v.i.* emerge.

émeri, *m.* emery.
à l'—, coated with emery, emery; ground with emery, ground; provided with a ground stopper; with emery grinding.

émeriser, *v.t.* coat with emery, emery.

émérite, *a.* emeritus; skilled by long practice.

émersion, *f.* emerging, emergence, emersion.

émerveillable, *a.* astonishing.

émerveiller, *v.t.* astonish. — *v.r.* wonder.

émet, *pr. 3 sing.* of émettre.

émétine, *f.* emetine.

émétique, *a.* & *m.* emetic; specif., tartar emetic.
— *de bore,* soluble cream of tartar, a substance prepared from a solution of cream of tartar in boric acid or borax.
— *de fer,* ferric potassium tartrate, (*Pharm.*) ferrum tartaratum.

émettre, *v.t.* emit; express, utter.

émeut, *pr. 3 sing.* of émouvoir.

émier, *v.t.* crumble, crum.

émiettement, *m.* crumbling, fragmentation.

émietter, *v.t.* & *r.* crumble, break into pieces.

émigrer, *v.i.* emigrate.

émincer, *v.t.* mince.

éminemment, *adv.* eminently.

éminence, *f.* eminence, eminency.
en —, *par* —, eminently.

éminent, *a.* eminent; high, elevated.

émis, *p.a.* emitted; expressed, uttered.

émissif, *a.* emissive.

émission, *f.* emission.

émit, *p.def. 3 sing.* of émettre.

emmagasinage, emmagasinement, *m.* storing, storage; accumulation.

emmagasiner, *v.t.* store; amass, accumulate.

emmanché, *p.a.* having a handle, helved; fastened, attached, joined; begun.

emmanchement, *m.* providing with a handle; fastening, attaching, attachment, joint; beginning.

— *à baïonnette,* bayonet joint, bayonet attachment.

emmancher, *v.t.* provide with a handle, helve; fasten, attach, join; begin. — *v.r.* fit on; be done.

emmêler, *v.t.* entangle; complicate.

emménagogue, *a. & m.* (*Med.*) emmenagog(ue).

emmener, *v.t.* take away, carry away.

emmieller, *v.t.* honey, sweeten with honey.

emmouflement, *m.* putting in the muffle furnace.

emmoufler, *v.t.* (*Ceram.*) put in the muffle furnace.

émodine, *f.* emodin.

émollient, *a. & m.* (*Med.*) emollient.

émonctoire, *m.* (*Physiol.*) emunctory.

émonder, *v.t.* prune, trim.

émotion, *f.* emotion; disturbance, agitation.

émoudre, *v.t.* grind, sharpen (edged tools).

émoulage, *m.* grinding (of edged tools).

émoulant, *p.pr.* of émoudre.

émoulu, *p.a.* ground, sharpened.

émousser, *v.t.* free from foam; dull, blunt; bevel, chamfer; clear of moss.

émousseur, *m.* foam remover (e.g. a perforated tube thru which steam is blown).

émouvoir, *v.t.* move, rouse, raise, stir up. — *v.r.* be moved, be roused, rise, be anxious.

empailler, *v.t.* pack or wrap with straw; stuff.

empaquetage, *m.* packing, packing up.

système par —, compact system (of cop dyeing).

empaqueter, *v.t.* pack, pack up, wrap up.

emparer, *v.r.* (with *de*) take up, absorb, assimilate, dissolve; take possession of, seize.

empasme, *m.* (*Pharm.*) empasm, empasma.

empasteler, *v.t.* dye blue with woad.

empâtage, *m.* embedding, etc. (see empâter); (*Soap*) the process of making soap paste, either by the cold process or by boiling; (*Brewing*) the moistening and softening of malt as a preliminary to mashing.

empâtement, *m.* embedding, etc. (see empâter).

empâter, *v.t.* embed; make into paste, make pasty, impaste; lay on thickly, impaste (as colors); fill with paste; clog (as a file); stuff (fowls).

empattement, *m.* base, foundation, footing, foot.

empêché, *p.a.* prevented; hindered; perplexed; occupied; preoccupied.

empêchement, *m.* impediment, hindrance, obstacle.

empêcher, *v.t.* prevent; impede, hinder, oppose. — *v.r.* refrain, help.

empeigne, *f.* (*Leather*) vamp, upper.

empereur, *m.* emperor.

empesage, *m.* starching.

empeser, *v.t.* starch, stiffen with starch.

empester, *v.t.* infect, taint, contaminate. — *v.i.* stink, have a foul odor.

empêtrer, *v.t.* entangle, embarrass.

emphase, *f.* emphasis.

emphatique, *a.* emphatic(al).

emphatiquement, *adv.* emphatically.

emphysémateux, *a.* (*Med.*) emphysematous.

emphysème, *m.* (*Med.*) emphysema.

empierrement, *m.* macadamizing; broken stone (for roads), road metal, macadam, stone ballast (for railways).

empierrer, *v.t.* macadamize, metal.

empiétement, *m.* encroachment, intrusion.

empiéter, *v.t.* encroach on, invade, infringe on. — *v.i.* encroach, intrude.

empilage, empilement, *m.* piling, piling up, stacking; pile, stack.

empiler, *v.t.* pile, pile up, stack.

empire, *m.* empire.

empirer, *v.t.* render worse. — *v.i.* grow worse.

empirie, *f.* empiricism, experimenting.

empirique, *a.* empirical, empiric.

empiriquement, *adv.* empirically.

empirisme, *m.* empiricism.

empiriste, *m.* empiricist; empiric.

empiristique, *a.* empiristic, empirical.

emplacement, *m.* site, location, position, ground; emplacement.

emplastique, *a.* emplastic, adhesive.

emplâtre, *m.* (*Pharm.*) plaster.

— *adhésif,* adhesive plaster.

— *calmant,* — *céphalique,* opium plaster.

— *de belladone,* belladonna plaster.

— *de cantharides,* cantharides cerate, cantharides plaster.

— *de gomme ammoniac mercuriel,* ammoniacum and mercury plaster.

— *de litharge,* lead plaster, litharge plaster.

— *de menthol,* menthol plaster.

— *de plomb,* lead plaster, diachylon plaster.

— *de poix cantharidé,* warming plaster, cantharidal pitch plaster.

— *de poix de Bourgogne,* pitch plaster, Burgundy pitch plaster.

— *de savon,* soap plaster.

— *diachylon,* diachylon, (in modern usage) lead plaster.

— *d'iodure de plomb,* lead iodide plaster.

emplâtre *d'opium*, opium plaster.
— *mercuriel*, mercurial plaster.
— *odontalgique*, opium plaster.
— *résineux*, adhesive plaster.
— *simple*, lead plaster.
— *temporal*, opium plaster.
— *vésicatoire*, cantharides cerate, blistering plaster.

emplette, *f.* purchase.

empli, *p.a.* filled, full. — *m.* (*Sugar*) a structure in which crystallizers for low-grade beet sugar are placed to be heated.

emplir, *v.t. & r.* fill, fill up.

emplissage, *m.* filling.

emploi, *m.* use; employment; office, position.

emploie, *pr. 3 sing.* of employer.

emploiement, *m.* employment.

emplombage, *m.* leading; specif., lining or coating with lead; (*Metal.*) smelting of silver or gold ore with lead ore as a means to separating the precious metal.

emplomber, *v.t.* lead (see emplombage).

employé, *p.a.* employed. — *m.* clerk; employee.

employer, *v.t.* employ. — *v.r.* be employed; busy oneself.

employeur, *m.* employer.

empoigner, *v.t.* seize, grasp.

empois, *m.* starch paste.
— *d'amidon*, starch paste.
— *de cobalt*, smalt.

empoise, *f.* (*Mach.*) bearing.

empoisonnant, *p.a.* poisoning, poisonous.

empoisonnement, *m.* poisoning.

empoisonner, *v.t.* poison.

empoisser, *v.t.* pitch, cover with pitch.

empontillage, *m.* (*Glass*) manipulation with the punty, or pontil.

empontiller, *v.t.* (*Glass*) manipulate or shape with the punty or pontil.

emporte-pièce, *m.* punch, puncher.

emporter, *v.t.* carry away, carry off, remove; carry. — *v.r.* be carried off, be removed; fly into a rage.
l'— sur, outweigh; surpass, prevail over.

empourpré, *p.a.* purple; purpled.

empourprer, *v.t.* purple, color purple.

empreignent, *pr. 3 pl.* of empreindre.

empreindre, *v.t.* imprint, stamp, impress.
— *v.r.* be imprinted; leave a print.

empreint, *p.a.* imprinted, stamped, impressed.
— *pr. 3 sing.* of empreindre.

empreinte, *f.* impression, imprint, stamp; (*Painting*) first coat, priming.

empressement, *m.* haste, eagerness; assiduity.

empresser, *v.r.* hasten; be eager; crowd.

emprimerie, *f.* tanning vat, tan vat.

emprise, *f.* encroachment.

emprisonner, *v.t.* imprison.

emprunt, *m.* borrowing; loan.
d'—, borrowed; assumed; artificial.

emprunté, *p.a.* borrowed; assumed; artificial; embarrassed.

emprunter, *v.t.* borrow.

empuantir, *v.r.* become putrid.

empyreumatique, *a.* empyreumatic.

empyreume, *m.* empyreuma.

ému, *p.p.* of émouvoir.

émule, *m.* rival, competitor.

émulsif, *a.* emulsive. — *m.* emulsive substance.

émulsine, *f.* emulsin.

émulsion, *f.* emulsion.
— *d'amande*, (*Pharm.*) emulsion of almond, milk of almond.
— *de chloroforme*, (*Pharm.*) chloroform emulsion, chloroform mixture.
— *d'essence de térébenthine*, (*Pharm.*) emulsion of oil of turpentine.
— *d'huile de foie de morue*, (*Pharm.*) emulsion of cod liver oil.
— *d'huile de ricin*, (*Pharm.*) castor oil mixture, emulsion of castor oil.

émulsionnant, *p.a.* emulsifying.

émulsionnement, *m.* emulsification.

émulsionner, *v.t.* emulsify.

en, *prep.* in; at, to, for, like; (with *p.pr.*) in, by, while (or not translated). — *pron.* (instead of *de lui, d'elle*, etc.) of him, of her, of it, of them, from him (her, it, them), etc. (see *de*); his, hers, its, theirs; some, any.
— *ce que*, from the fact that, in the respect that, because.

enantiomorphe, *a.* enantiomorphous, enantiomorphic.

encadrement, *m.* framing; frame.

encadrer, *v.r.* frame; surround; insert.

encaissage, *m.* incasing, etc. (see encaisser).

encaissant, *p.a.* incasing, etc. (see encaisser).

encaisse, *f.* cash on hand.

encaissement, *m.* incasing, etc. (see encaisser); embankment.

encaisser, *v.t.* incase, inclose; pack (in a case or box); collect, receive (money); bed (a road).

encan, *m.* auction.

encaquer, *v.t.* barrel; (fig.) pack.

encartouchage, *m.* cartridge filling.

encartoucher, *v.t.* put or make into cartridges.

en-cas, encas, en cas, *m.* anything for use in an emergency.

encastage, *m.* (*Ceram.*) placing in saggers.

encaster, *v.t.* (*Ceram.*) place in saggers.

encasteur, *m.* (*Ceram.*) placer, sagger stacker, setter hand.

encastrement, *m.* fitting in, etc. (see encastrer); groove, bed, seat, mortise, recess.

encastrer, *v.t.* fit in, let in, set in, bed (in a groove or other recess). — *v.r.* fit together.

encaustique, *m.* polishing wax (any of various preparations containing wax and used on wood, leather, etc.); an impregnating material for preserving sculptures; encaustic painting, encaustic. — *a.* pertaining to or designating the above preparations; encaustic.

encaver, *v.t.* put or store in a cellar.

enceignant, *p.pr.* of enceindre.

enceindre, *v.t.* surround, enclose.

enceint, *pr. 3 sing. & p.p.* of enceindre.

enceinte, *f.* inclosure, enclosure, (inclosed) space, place, chamber, hall; circumference, circuit. — *a.* pregnant.

encens, *m.* incense.
— *femelle,* incense from a kind of juniper (*Juniperus lycia*).
— *indien,* — *mâle,* frankincense.

enchaînement, *m.* linking, linkage; connection; arrangement; chaining; captivation.

enchaîner, *v.t.* link (e.g. atoms); connect; arrange, coördinate; chain; bind; cap-. tivate. — *v.r.* be linked, be connected, etc.

enchanter, *v.t.* enchant; delight.

enchanteur, *a.* enchanting, delightful.

enchâsser, *v.t.* insert, set, set in; enshrine.

enchâssure, *f.* insertion, etc. (see enchâsser).

enchaussenage, *m.* (*Leather*) liming.

enchaussener, enchaussumer, *v.t.* lime (hides).

enchaussenoir, enchaussumoir, *m.* (*Leather*) lime pit, liming vat.

enchaux, *m.* (*Leather*) lime liquor.

enchère, *f.* bid, bidding, auction.

enchérir, *v.t.* raise the price of; bid for. — *v.i.* bid; become dearer.
— *sur,* outbid; surpass.

enchérissement, *m.* rise, increase (in price).

enchevaucher, *v.t.* rabbet.

enchevêtrement, *m.* entangling; interlacement.

enchevêtrer, *v.t.* entangle, tangle, (of crystals) interlace; halter. — *v.r.* become entangled or interlaced.

encirer, *v.t.* wax; oil (cloth).

enclanchement, etc. = enclenchement, etc.

enclave, *f.* boundary; enclosure; recess.

enclaver, *v.t.* inclose, enclose; fit, lock, let in, wedge, key.

enclenchement, *m.* throwing into gear, engaging, locking; engaging gear, locking gear.

enclencher, *v.t.* throw into gear, engage, lock.

enclin, *a.* inclined.

encliquetage, *m.* ratchet mechanism or gear.
— *à frottement,* friction catch.

enclore, *v.t.* inclose, enclose, include.

enclos, *p.a.* inclosed, included. — *m.* inclosure.

enclume, *f.* anvil.

enclumeau, enclumot, *m.,* **enclumette,** *f.* small anvil.

encoche, *f.* notch.

encochement, *m.* notching.

encocher, *v.t.* notch.

encoffrer, *v.t.* inclose; hoard.

encoignure, encognure, *f.* corner, angle; elbow.

encollage, *m.* sizing; size; gumming.

encoller, *v.t.* size; gum.

encolleur, *m.* sizer.

encolleuse, *f.* sizing machine, sizer.

encollure, *f.* (*Metal.*) welding point.

encombrant, *p.a.* obstructing, encumbering; cumbersome, bulky.

encombre, *m.* obstacle, impediment.

encombrement, *m.* obstruction; encumbrance; congestion; space occupied or required.

encombrer, *v.t.* obstruct; encumber.

encontre. In the phrases: *à l'encontre,* to the contrary, in the opposite sense; *à l'encontre de,* counter to, contrary to, against.

encore, encor, *adv.* still, yet, besides, also, again.
— *que,* tho, altho.

encouder, *v.t.* bend into an elbow.

encourageant, *p.a.* encouraging.

encouragement, *m.* encouragement.

encourager, *v.t.* encourage.

encourir, *v.t.* incur.

encouru, *p.a.* incurred.

encrage, *m.* inking.

encrassement, *m.* fouling, etc. (see encrasser).

encrasser, *v.t.* foul, make foul or dirty, soil, choke (as a filter or a grate). — *v.r.* become fouled, dirty or choked.

encre, *f.* ink.
— *à copier,* copying ink.
— *à écrire,* writing ink.
— *à marquer,* marking ink.
— *argentée,* silver ink.
— *à tampon,* ink for pads, stamping ink.
— *copiante,* copying ink.
— *de Chine,* India ink, China ink.
— *d'imprimerie,* — *d'impression,* printing ink, printer's ink.
— *dorée,* gold ink.
— *dragon,* indelible ink.
— *indélébile,* indelible ink.
— *noire,* black ink.
— *rouge,* red ink.
— *sympathique,* sympathetic ink.

encré, *p.a.* inked.

encrené, *a.* (*Metal.*) twice heated and hammered.

encrenée, *f.* iron twice heated and hammered.

encrer, *v.t.* ink.

encreur, *a.* inking.

encrier, *m.* inkstand; ink block.

encrivore, *a.* ink-removing, ink-destroying.

encroûtement, *m.* incrusting; incrustation, crust.

encroûter, *v.t.* incrust; plaster. — *v.r.* become incrusted.

encuivrage, *m.* (of firearms) metal fouling.

encuvage, *m.* putting into a vat, tub or tank.

encuvement, *m.* = encuvage; hollow, depression.

encuver, *v.t.* put into a vat, tub or tank.

encyclie, *f.* one of a series of concentric rings.

encyclopédie, *f.* encyclopedia.

encyclopédique, *a.* encyclopedic.

endaubage, *m.* preserved meat, tinned beef.

endauber, *v.t.* preserve, tin, can (meat).

endécanitrique, *a.* hendecanitro.

endenter, *v.t.* provide with teeth, tooth, cog; scarf; dovetail.

endiguer, *v.t.* dam, dam up, dam in.

endocellulaire, *a.* endocellular.

endochrôme, *m.* (*Bot.*) endochrome.

endogène, *a.* (*Biol.*) endogenous. — *m.* endogen.

endommager, *v.t.* damage.

endoplasme, *m.* (*Biol.*) endoplasm.

endormir, *v.t.* put to sleep; lull, calm.

endosmomètre, *m.* endosmometer.

endosmose, *f.* endosmosis, endosmose.

endosmotique, *a.* endosmotic.

endosperme, *m.* (*Bot.*) endosperm.

endospermique, *a.* (*Bot.*) endospermic.

endosser, *v.t.* indorse; put on.

endothermique, *a.* endothermic.

endotoxine, *f.* endotoxin.

endotryptase, *f.* endotryptase.

endroit, *m.* place; locality; side, point; right side (of a fabric); passage (in a book).

   à l'— de, with respect to, with regard to.

   à son —, with regard to it (him, her).

enduire, *v.t.* coat, cover, treat, overlay, dope.

enduisant, *p.pr.* of enduire.

enduit, *m.* coat, coating, layer; (something used for coating) wash, paint, plaster, etc.; (for airplanes) dope; (*Ceram.*) glaze, glazing. — *pr. 3 sing. & p.p.* of enduire.

   — de noir, blacking; (*Founding*) black wash.

endurcir, *v.t.* harden. — *v.r.* become hardened, harden.

endurcissement, *m.* hardening.

endurer, *v.t.* endure; allow, permit.

énergétique, *a.* energetic; energy-producing.

énergie, *f.* energy.

   — chimique, chemical energy.

   — cinétique, kinetic energy.

   — libre, free energy.

   — mécanique, mechanical energy.

   — thermique, heat energy, thermal energy.

   — utilisable, available energy.

énergique, *a.* energetic.

énergiquement, *adv.* energetically.

enfaiter, *v.t.* fill (a glass pot) properly so that the batch in it has a rounded top; cover the ridge of (a roof).

enfance, *f.* infancy, childhood.

enfant, *m. & f.* child; infant.

enfantement, *m.* childbirth.

enfanter, *v.t.* bring forth, give birth to.

enfer, *m.* hell.

   — de Boyle, a vessel for heating mercury, devised by Boyle.

enfermer, *v.t.* confine; inclose; include, contain; lock up, shut up, shut in.

enfieller, *v.t.* render bitter; embitter.

enfiévré, *a.* feverish.

enfilade, *f.* row, series, sequence; suite.

enfiler, *v.t.* thread; string; thread on; go along (a road); begin on; head (pins).

enfilure, *f.* row, series, string; stringing.

enfin, *adv.* in short, at last, finally.

enflammé, *p.a.* ignited, set on fire; burning, on fire; inflamed.

enflammer, *v.t.* set fire to, ignite, kindle — *v.r.* take fire, ignite.

enflé, *p.a.* inflated, swollen, swelled, puffed up, turgid.

enflement, *m.* inflation, etc. (see enfler).

enfler, *v.t.* inflate, swell, blow up, puff out; bloat; exaggerate. — *v.r.* swell, swell up, be inflated, (of waters) rise.

enfleurage, *m.* (*Perfumery*) enfleurage.

enfleurer, *v.t.* impregnate (fatty substances) with the odorous principles of flowers.

enflure, *f.* inflation, swelling, swell; tumor; pride.

enfoncé, *p.a.* sunk, sunken, etc. (see enfoncer); deep.

enfoncement, *m.* sinking, etc. (see enfoncer); recess; hollow; depth.

enfoncer, *v.t.* sink, lower, plunge; drive (in); break in, break open; deepen; bottom (casks). — *v.r.* sink; plunge.

enfonçure, *f.* hollow, hole, cavity, depression; bottom (of a cask).

enformer, *v.t.* shape, fashion.

enfouir, *v.t.* bury.

enfourchement, *m.,* **enfourchure,** *f.* fork, crotch.

enfournement, enfournage, *m.,* **enfournée,** *f.* putting in a furnace or oven.

enfourner, *v.t.* put in the furnace or oven; start, begin. — *v.r.* start, enter (into).

   — en échappade, (*Ceram.*) put into compartments in the kiln (cf. échappade).

enfourneur, *m.* one who puts something into a furnace or kiln; (*Glass*) pot filler; (*Ceram.*) oven man, kiln man.

enfreignant, *p.pr.* of enfreindre.

enfreindre, *v.t.* infringe, violate.

enfui, *p.p.* of enfuire.

enfuir, *v.r.* leak, escape, run out; run away; disappear.

enfumage, *m.* smoking, etc. (see enfumer); (*Ceram.*) heating to about 100° to expel moisture.

enfumer, *v.t.* smoke; fill with smoke; smoke out; intoxicate. — *v.r.* be smoked.

enfutailler, enfûter, *v.t.* cask, barrel, tun.

enfuyant, *p.pr.* of enfuire.

engagement, *m.* engagement, etc. (see engager).

engager, *v.t.* engage; insert; begin; invite; urge; induce; pawn. — *v.r.* engage oneself; begin; get entangled; enter; enlist.

engallage, *m.* galling (see engaller).

engaller, *v.t.* gall, impregnate with gallnut decoction.

engeance, *f.* breed.

engendrement, *m.* production, formation; breeding.

engendrer, *v.t.* produce, form, generate; breed, beget; cause. — *v.r.* be produced; be bred.

engin, *m.* device, appliance, engine, instrument, gear; hoist, windlass.
— *de lancement,* — *de mise de feu,* firing device.

englober, *v.t.* include; envelop; unite; blend.

engloutir, *v.t.* swallow up.

engluage, *m.* = engluement.

engluanter, *v.t.* = engluer.

engluement, *m.* coating with glue or glutinous material, liming.

engluer, *v.t.* coat with glue or glutinous material (as tar), lime.

engobage, *m.* (*Ceram.*) coating with engobe; specif., slip painting.

engobe, *f.* (*Ceram.*) engobe (slip or paste applied in a thin layer over a body of different composition).

engober, *v.t.* (*Ceram.*) coat with engobe.

engommage, *m.* gumming; (*Ceram.*) glazing.

engommer, *v.t.* gum; (*Ceram.*) glaze.

engorgement, *m.* stopping up, etc. (see engorger).

engorger, *v.t.* stop up, choke, choke up, obstruct, congest. — *v.r.* be stopped up, etc.

engouffrer, *v.t.* ingulf. — *v.r.* be ingulfed; (of water, wind, etc.) rush (into).

engourdir, *v.t.* numb, benumb; dull, deaden.

engourdissement, *m.* torpor; numbness; dullness.

engrais, *m.* fertilizer, manure; fattening feed or pasture.
— *azoté,* nitrogenous fertilizer.
— *chimique,* chemical fertilizer.
— *Chosko,* a fertilizer made by precipitating liquid manure (*eaux vannes*) with magnesium and iron sulfates.
— *complet,* complete fertilizer.
— *composé,* mixed fertilizer.
— *d'os,* bone fertilizer, bone meal.
— *flamand,* liquid manure made from human excrement.
— *phosphaté,* phosphate fertilizer.
— *potassique,* potash fertilizer.

engraissage, *m.* fertilizing, etc. (see engraisser).

engraissé, *p.a.* fertilized; fattened, fat; greasy; enriched.

engraissement, *m.* = engraissage; fertilizer.

engraisser, *v.t.* fertilize, manure; fatten; grease, oil; soil with grease. — *v.r.* fatten, grow fat; (of liquids) thicken. — *v.i.* fatten; thrive.

engrenage, *m.* gear, gearing, train of wheels, wheel work; stowage (of casks); complication.
à —, geared.
— *à vis sans fin,* worm gear.
— *conique,* bevel gear.
— *cylindrique,* — *droit,* spur gear.

engrènement, *m.* engaging, etc. (see engrener).

engrener, *v.t.* engage, gear, put into gear; tooth, indent; feed with grain; begin. — *v.r.* gear, be in gear, catch, be put in gear. — *v.i.* be in gear; begin.

engrenure, *f.* gearing, catching; toothing; cogging.

engrumeler, *v.t., r. & i.* clot, coagulate.

enhydre, *a.* (*Min.*) enhydrous, containing water in cavities.

énigme, *f.* enigma.

enivrant, *p.a.* intoxicating.

enivrement, *m.* intoxication.

enivrer, *v.t.* intoxicate.

enkysté, *a.* encysted.

enlacer, *v.t.* lace, enlace, interlace, entwine.

enlevage, *m.* (*Calico*) discharge, discharging.

enlèvement, *m.* removing, removal, etc. (see enlever).

enlever, *v.t.* remove, take away, carry away, take off, take out; (*Calico*) discharge; lift up; buy up; take. — *v.r.* be removed, etc.; come out, come off; rise, be lifted; be sold.

enluminer, *v.t.* color; illuminate; flush, redden.

enluminure, *f.* coloring; colored print.

ennemi, *m.* enemy. — *a.* hostile.

ennoblir, *v.t.* ennoble.

ennui, *m.* ennui, tedium; grief; care.

ennuyant, *a.* tiresome, tedious.

ennuyer, *v.t.* tire, weary, annoy. — *v.r.* grow weary. — *v.i.* be annoying.

énoncé, *p.a.* stated, etc. (see énoncer). — *m.* statement, enunciation, declaration, expression.

énoncer, *v.t.* state, enunciate, declare, express.

énonciation, *f.* statement, enunciation, expression; proposition.

énorme, *a.* enormous.

érormément, *adv.* enormously.

énormité, *f.* enormousness; absurdity; enormity.

énouer, *v.t.* burl (cloth).

enquérir, *v.r.* inquire.

enquête, *f.* inquiry, investigation.

enquiert, *pr. 3 sing.* of enquérir.

enquis, *p.p.* of enquérir.

enraciner, *v.t.* root. — *v.r.* root, be rooted.

enragé, *a.* mad; raging, furious. — *m.* madman.

enrayage, enrayement, *m.* checking, etc. (see enrayer).

enrayer, *v.t.* check, restrain; stop, jam (machinery); brake; lock (as a wheel); spoke (a wheel). — *v.r.* (of machinery) stop, fail to work, jam.

enregistrement, *m.* registration, registering, recording, record; registry.

enregistrer, *v.t.* register, record.

enregistreur, *m.* register, registering device. — *a.* registering.

enrichir, *v.t.* enrich. — *v.r.* be enriched.

enrichissement, *m.* enrichment.

enrobage, *m.* covering, coating; (*Micros.*) embedding.

enrober, *v.t.* cover, coat; specif., cover with a protecting envelope (as foods with gelatin or paraffin); (*Micros.*) embed.

enrochage, *m.* caking, hardening (of gunpowder or the like).

enrochement, *m.* caking, hardening (as of gunpowder); enrockment, riprap.

enrocher, *v.t. & r.* cake, harden.

enrouillé, *p.a.* rusted, rusty.

enrouiller, *v.t. & r.* rust.

enroulage, enroulement, *m.* winding (as of a dynamo or of thread); rolling, rolling up.
— *en dérivation,* (*Elec.*) shunt winding.
— *en série,* (*Elec.*) series winding.

enrouler, *v.t.* wind (wire, etc.); roll, roll up.
— *v.r.* be wound; be rolled.

ensabler, *v.t.* sand.

ensachage, *m.* bagging, sacking.

ensacher, *v.t.* bag, sack, put into sacks.

ensafraner, *v.t.* dye with saffron, ensaffron.

ensanglanter, *v.t.* stain with blood.

enseigne, *f.* sign; flag; ensign.

enseignement, *m.* instruction, teaching.

enseigner, *v.t.* teach, instruct; show.

enseigneur, *m.* teacher, instructor.

ensemble, *adv.* together, at the same time.
— *m.* ensemble, whole, aggregate, general effect or compass; unity.

ensemencement, *m.* sowing, seeding; scattering; (*Med.*) inoculation.

ensemencer, *v.t.* sow, seed; scatter (like seed).

enserrer, *v.t.* contain; inclose; squeeze.

ensilage, *m.* storing in a silo.

ensiler, *v.t.* store in a silo.

ensilotage, ensiloter. See ensilage, ensiler.

ensimage, *m.* oiling (of wool, cloth, etc.).

ensimer, *v.t.* oil, grease (wool, cloth, etc.).

ensoufrer, *v.t.* sulfur, sulfurize, treat with sulfur.

ensoufroir, *m.* sulfuring stove; place in which articles are bleached with the vapor of burning sulfur.

ensucrer, *v.t.* sugar.

ensuifer, *v.t.* tallow, treat with tallow.

ensuit, *pr. 3 sing.* of ensuivre.

ensuite, *adv.* then, next, afterward.
— *de,* after.

ensuivre, *v.r.* follow, ensue.

entacher, *v.t.* vitiate; taint, contaminate.

entaille, *f.* mark, graduation mark; cut, groove, score, notch, nick, jag, slash.
— *repère,* guide mark, guide notch.

entailler, *v.t.* cut, score, groove, notch, gash.

entaillure, *f.* = entaille.

entame, *f.* = entamure.

entamer, *v.t.* begin on, open up; begin cutting cut into, cut the first piece from; touch; cut (slightly), incise; injure; encroach on; prevail on or over.

entamure, *f.* cut, incision; first piece or quantity cut.

entartrage, *m.* incrustation, scale.

entartré, *a.* covered with scale.

entartrer, *v.r.* become covered with scale.

entassement, *m.* heap, pile, mass; heaping, piling, massing.

entasser, *v.t.* heap, pile, pile up, mass. — *v.r.* be heaped, etc.; accumulate; crowd.

entendre, *v.t.* hear; hear of; mean; understand. — *v.r.* be heard; be understood; agree; understand. — *v.i.* hear.
— *à,* hear to, consent to.

entendu, *p.a.* heard, understood, etc. (see entendre); capable, intelligent; ordered, contrived.
bien —, see under *bien.*

entente, *f.* understanding, agreement; meaning; intelligence, ability; entente.

entérokinase, *f.* enterokinase.

enterrement, *m.* burying, burial.

enterrer, *v.t.* bury.

en-tête, *m.* heading (of reading matter).

entêté, *a.* stubborn; affected with headache; intoxicated; infatuated.

enthousiasme, *m.* enthusiasm.

enthousiasmer, *v.t.* fill with enthusiasm.

enthousiaste, *m. & f.* enthusiast. — *a.* enthusiastic.

entier, *a.* entire; whole, integral (numbers); self-willed, obstinate. — *m.* whole, entirety, total, totality; (*Arith.*) integer.
en —, whole, entire; wholly, entirely, fully, in full.

entièrement, *adv.* entirely, wholly, fully.

entoilé, *a.* cloth-covered; mounted on cloth.

entonnage, entonnement, *m.,* **entonnaison,** *f.* barreling, tunning, putting in casks.

entonner, *v.t.* barrel, tun, put in casks; swallow; intone, sing.

entonnerie, *f.* (*Brewing*) place where casks or kegs are filled.

entonnoir, *m.* funnel, (*Mil.*) crater.

en —, funnel-shaped, funnel-like, infundibular.

— *à boule,* bulbed funnel; bulb-shaped funnel, separating funnel.

— *à brome,* bromine funnel.

— *à filtrer à chaud,* — *à filtration chaude,* funnel for hot filtration, hot water (hot air, steam) funnel.

— *ajouré,* openwork funnel, filtering basket (funnel with large holes in its sides).

— *à longue douille,* long-stemmed funnel.

— *à mercure,* mercury funnel (bulb-shaped, with stem drawn out).

— *à robinet,* funnel with stopcock.

— *à séparation,* separating funnel, separatory funnel.

— *à tamis,* sieve funnel, funnel containing a screen.

— *cannelé,* channeled funnel, ribbed funnel.

— *en grès,* stoneware funnel.

— *en porcelaine,* porcelain funnel.

— *en verre,* glass funnel.

— *filtrateur,* filtering funnel.

— *percé de trous,* perforated funnel.

entorse, *f.* strain, sprain; shock.

entortiller, *v.t.* twist; wrap up, wrap; entangle. — *v.r.* twist, twine, wind; get entangled.

entour, *m.* environ; associate.

à l'—, in the vicinity.

entourage, *m.* (something that surrounds) case, casing, border, frame, mounting, setting; surroundings, entourage.

entourer, *v.t.* surround, encircle.

en-tout-cas, *m.* = en-cas.

entrailles, *f.pl.* entrails, intestines, bowels; womb; heart, core.

entraînable, *a.* capable of being carried away.

entraînement, *m.* carrying away, etc. (see entraîner); (*Mach.*) communicating gear; enthusiasm; impulse.

— *à la vapeur,* distillation with steam.

entraîner, *v.t.* carry away, carry along, carry over (in distillation), carry down (in precipitation), sweep away; involve; win, captivate; train.

entrave, *f.* hindrance, obstacle, trammel.

entraver, *v.t.* hinder, impede, fetter.

entre, *prep.* between; among.

par —, between.

entre-. inter, reciprocally, mutually, each other, one another.

entre-bâiller, *v.t.* open partly, set ajar.

entre-choquer, *v.r.* collide, encounter one another, clash, conflict.

entrecouper, *v.t.* intersect; interrupt; intersperse.

entre-croiser, entrecroiser, *v.t. & r.* intercross, intersect, interweave.

entre-deux, entredeux, *m.* part or space between, interval, middle, partition. — *adv.* indifferently, so-so.

entrée, *f.* entrance, entry, inlet, admission, ingress, opening (into); mouth (of a stream); entrée.

entrefer, entre-fer, *m.* (*Elec.*) air gap.

entrelacement, *m.* interlacing, interlacement, interweaving, intertwining.

entrelacer, *v.t.* interlace, interweave, intertwine, twine, braid.

entre-luire, entreluire, *v.i.* shine feebly.

entremêlement, *m.* intermixing, intermixture.

entremêler, *v.t.* intermix, intermingle, mix.

entremise, *f.* intervention, interposition, mediation, medium, agency.

entreposage, *m.* storage; placing in bond.

entreposer, *v.t.* store, deposit, warehouse; place in bond.

entreposeur, *m.* warehouse keeper; bonded warehouse keeper.

entrepositaire, *m.* bonder; depositary.

entrepôt, *m.* warehouse; bonded warehouse; mart.

entreprenant, *p.a.* enterprising; venturesome.

entreprendre, *v.t.* undertake, attempt; attack. — *v.i.* encroach; make an attack (on).

entrepreneur, *m.* contractor, employer, one who engages to carry out an enterprise.

entrepris, *p.p.* of entreprendre.

entreprise, *f.* undertaking, project, enterprise; contract; establishment; encroachment.

entrer, *v.i.* enter; be admitted. — *v.t.* enter, bring in, put in, insert; import.

— *en ligne de compte,* enter into account, be worth considering.

entre-temps, *m.* mean time, interval. — *adv.* (also **entretemps**) meantime, meanwhile.

entretenir, *v.t.* maintain, keep, keep up; talk with, converse with. — *v.r.* be maintained, be kept, be kept up; maintain oneself, support oneself; support each other; talk, converse.

entretien, *m.* maintenance; upkeep; repair; care; support; supply; clothing, clothes; talk, conversation.

entretient, *pr. 3 sing.* of entretenir.

entretoise, *f.* crosspiece, crossbar, stay, tie, rib, girder, bracing; stay bolt.

entrevit, *p.def. 3 sing.* of entrevoir.

entrevoir, *v.t.* suspect (as a new element); see or foresee dimly, catch a glimpse of.

entrevoyant, *p.pr.* of entrevoir.

entrevu, *p.p.* of entrevoir.

entropie, *f.* entropy.

entr'ouvert, *p.a.* opened, open; half-open.

entr'ouvrir, *v.t.* open, spread open; open partly. — *v.r.* open, spread open; half-open.

énumérer, *v.t.* enumerate.

envahir, *v.t.* invade, overrun.

envahissement, *m.* invasion.

envaser, *v.t.* fill up with mud or silt.

enveloppant, *p.a.* enveloping.

enveloppe, *f.* envelope; cover, covering, wrapping; casing, case, jacket; appearance.

enveloppement, *m.* envelopment, enveloping, wrapping, etc. (see envelopper).

envelopper, *v.t.* envelop; wrap up, inclose, surround, cover; involve.

envergure, *f.* extent, spread (as of wings).

enverra, *fut. 3 sing.* of envoyer.

enverrage, *m.* (*Glass*) addition of cullet to the pots in order to facilitate fusion.

enverrer, *v.t.* add cullet to (see enverrage).

envers, *m.* reverse, back side, wrong side. — *prep.* toward, to, with regard to.

à l'—, wrong side out, upside down, wrong.

envie, *f.* envy; longing, desire.

envier, *v.t.* envy; long for, desire.

envieux, *a.* envious; desirous.

environ, *adv.* about.

environner, *v.t.* surround, encompass.

environs, *m.pl.* environs, suburbs.

envisager, *v.t.* regard, consider; face.

envoi, *m.* sending, shipment; parcel or goods sent, shipment; envoy.

— *contre remboursement*, C.O.D. shipment.

— *franco*, sent free.

envoie, *pr. 3 sing.* of envoyer.

envoiler, *v.r.* warp, bend (in tempering).

envoler, *v.r.* fly, fly away, fly off.

envoyer, *v.t.* send; send forth.

— *chercher*, send for.

envoyeur, *m.* sender.

enzymatique, *a.* enzymic, enzymatic.

enzyme, *f.* enzyme.

éocène, *a. & m.* (*Geol.*) Eocene.

éolipile, éolipyle, *m.* eolipile, æolipile.

éosine, *f.* eosin.

éosinophile, *a.* (*Biol.*) eosinophil(e).

épaillage, *m.* removal of straw and other vegetable material from wool or wool cloth, either mechanically or chemically (in the latter case called carbonizing); stripping (sugar cane) of the lower leaves; cleaning (of gold).

épailler, *v.t.* deprive (wool, or wool cloth) of vegetable matter (see épaillage); strip (sugar cane) of the lower leaves; clean (gold).

épais, *a.* thick. — *m.* thickness; (*Elec.*) density.

épaisseur, *f.* thickness.

épaissir, *v.t., r. & i.* thicken.

épaississant, *p.a.* thickening. — *m.* thickener, thickening agent.

épaississement, *m.* thickening; thickness.

épanchement, *m.* effusion; outpouring.

épancher, *v.t.* pour out, effuse. — *v.r.* be effused, (*Med.*) be extravasated; overflow.

épandage, *m.* spreading; scattering; distribution; shedding; (*Sewage*) disposal of sewage by flowing it upon land, land treatment, irrigation.

épandre, *v.t.* spread; scatter, distribute; shed. — *v.r.* spread.

épanouir, *v.t.* open, expand; cause to bloom. — *v.r.* expand, spread; open, bloom.

épanouissement, *m.* expansion, spread, splay; opening (of flowers); (*Elec.*) pole piece.

épargne, *f.* economy, thrift; savings.

épargner, *v.t.* spare; save, economize.

éparpillement, *m.* scattering.

éparpiller, *v.t.* scatter.

épars, *a.* scattered; sparse. — *m.* bar, crossbar.

épaule, *f.* shoulder.

épaulement, *m.* shoulder; (*Mil.*) epaulment.

épauler, *v.t.* support, aid; shoulder.

épeautre, *m.* (*Bot.*) spelt.

épée, *f.* sword.

épeler, *v.t.* spell.

épellation, *f.* spelling.

éperon, *m.* spur; buttress; breakwater.

éperonner, *v.t.* spur, spur on.

épervière, *f.* (*Bot.*) hieracium, hawkweed; specif., (*Pharm.*) rattlesnake weed (*Hieracium venosum*).

éphémère, *a.* ephemeral.

épi, *m.* ear, head (of grain); spike; tuft; jetty; dike.

épiage, épiation, *f.* earing, heading (of grain).

épice, *f.* spice.

épicé, *p.a.* spiced; pungent, biting.

épicer, *v.t.* spice.

épicerie, *f.* groceries (spices, tea, coffee, sugar, etc.); grocery (shop); grocery business.

épichlorhydrine, *f.* epichlorohydrin.

épicier, *m.* grocer (dealer in spices, sugar, coffee, etc.).

épidémie, *f.* epidemic.

épidémique, *a.* epidemic.

épiderme, *m.* (*Anat.*) epidermis.

épidote, *m.* (*Min.*) epidote.

épierrer, *v.t.* remove stones from.

épierreur, *m.*, **épierreuse**, *f.* stone remover.

épiguanine, *f.* epiguanine.

épilage, *m.*, **épilation**, *f.* epilation, depilation, unhairing.

épilatoire, *a. & m.* epilatory, depilatory.

épiler, *v.t.* epilate, eradicate (hair).

épilobe, *m.* (*Bot.*) willow-herb.

épinage, *m.* (*Soap*) drawing off of the aqueous layer after the soap has been salted out.

épinard, *m.* spinach.

épinceler, épinceter, épincer, *v.t.* burl (cloth).

épine, *f.* thorn; spine; (*Soap*) a tube used to draw off the aqueous layer from beneath the soap; (*Metal.*) a sharp point on the surface of metallic copper.

épinéphrine, *f.* epinephrine (adrenaline).
épineux, *a.* thorny, spiny, prickly.
épine-vinette, *f.* (*Bot.*) barberry.
épingle, *f.* pin.
épingler, *v.t.* pin; prick.
épinglette, *f.* priming wire; pricker.
épinière, *a.* spinal.
épipastique, *a.* (*Pharm.*) epipastic.
épisarcine, *f.* episarcine.
épispastique, *a. & m.* (*Pharm.*) epispastic.
épisser, *v.t.* splice.
épister, *v.t.* reduce to paste (by grinding).
épithéliome, *m.* (*Med.*) epithelioma.
épluchage, *m.* picking; scrutiny.
éplucher, *v.t.* pick, clean; scrutinize; weed.
épluchure, *f.* pickings, leavings, parings, etc.
épointer, *v.t.* break the point of; point.
éponge, *f.* sponge.
— **de platine,** platinum sponge (specif., purified material as distinguished from *mousse de platine*).
épongeage, *m.* sponging.
éponger, *v.t.* sponge.
époque, *f.* epoch.
épouser, *v.t.* espouse.
époussetage, *m.* dusting.
épousseter, *v.t.* dust.
épousseteux, *a.* dusting, dust-removing.
époussetoir, *m.* small dusting brush.
époussette, *f.* duster, dustbrush, dustcloth.
époûti, *m.* foreign matter in wool cloth.
époutier, époutir, *v.t.* burl (cloth).
épouvantable, *a.* frightful, dreadful.
épouvanter, *v.t.* frighten, terrify, dismay.
épreignant, *p.pr.* of épreindre.
épreindre, *v.t.* express, press, squeeze.
épreint, *p.a.* expressed, pressed, squeezed.
— *pr.* 3 *sing.* of épreindre.
épreuve, *f.* test; trial; experiment; proof; assay.
à *l'*—, on trial.
à *l'*— **de,** proof against.
à *l'*— **de** *l'eau,* waterproof.
à *l'*— **du** *feu,* fireproof.
— *à chaud,* hot test.
— *à froid,* cold test.
— *à outrance,* resistance test, breaking test, bursting test.
— *de chaleur,* heat test.
— *de durée,* endurance test.
— *de pression,* pressure test.
— *d'essai,* test.
éprouver, *v.t.* test; try, prove; experience, undergo; feel.
éprouvette, *f.* cylinder (narrow upright vessel, usually of glass); gage; test bar, test piece; (*Assaying*) éprouvette, small spoon for fluxes; probe.
— *à cuvette supérieure,* cylinder with enlarged top, hydrometer jar.

éprouvette *à dessécher les gaz,* drying cylinder, drying tower, calcium chloride jar.
— *à pied,* cylinder with foot, cylinder.
— *à rainure,* cylinder with a vertical channel for a thermometer.
— *divisée,* — *graduée,* graduated cylinder.
épuisable, *a.* exhaustible.
épuisant, *p.a.* exhausting.
épuisé, *p.a.* exhausted, spent; out of print.
épuisement, *m.* exhaustion.
épuiser, *v.t.* exhaust; extract; spend; drain.
épulpeur, *m.* (*Sugar*) an apparatus formerly used for extracting beet juice from the pulp.
épurateur, *m.* purifier; specif., scrubber; refiner (as of oils); cleaner. — *a.* purifying, refining, cleaning.
épuratif, *a.* purifying, purificative.
épuration, *f.* purification, purifying; refining; scrubbing (of gases); cleaning; refinement.
épuratoire, *a.* purifying, purificatory.
épuré, *p.a.* purified; refined; pure.
épurement, *m.* purification, refinement.
épurer, *v.t.* purify; refine; scrub (gases); clean, cleanse. — *v.r.* be purified, etc.
éq., *abbrev.* (équivalent) equivalent.
équarrir, *v.t.* square; cut up (animals).
équarrissage, *m.* squaring; cutting up of animals (esp. old horses, mules, etc.), knacker's trade.
équateur, *m.* equator; (*cap.*) Ecuador.
équation, *f.* equation.
— *chimique,* chemical equation.
équerrage, *m.* bevel, beveling, dihedral angle.
équerre, *f.* square; right-angled iron or brace; right-angled elbow.
à —, square, at right angles.
équerrer, *v.t.* square; bevel.
équiangle, *a.* equiangular.
équidistant, *a.* equidistant.
équilatéral, équilatère, *a.* equilateral.
équilibrage, *m.* equilibration, balancing.
équilibre, *m.* equilibrium.
— *chimique,* chemical equilibrium.
— *mécanique,* mechanical equilibrium.
équilibrer, *v.t.* equilibrate, balance.
équimoléculaire, *a.* equimolecular.
équipage, *m.* equipment; outfit; gear, gearing; equipage; crew.
équipe, *f.* set, gang, crew, party (of workmen).
équipé, *p.a.* equipped.
équipement, *m.* equipment.
équiper, *v.t.* equip.
équipotentiel, *a.* equipotential.
équitable, *a.* equitable.
équivalant, *p.pr.* of équivaloir.
équivalemment, *adv.* in an equivalent manner.
équivalence, *f.* equivalence, equivalency.
équivalent, *a. & m.* equivalent.

équivalentiste, *m.* an adherent of the doctrine of equivalents, equivalentist.

équivaleur, *f.* equivalence.

équivaloir, *v.i.* be equivalent.

équivaut, *pr. 3 sing.* of équivaloir.

équivoque, *a.* equivocal. — *f.* equivocalness; equivocation.

érable, *m.* maple (tree or wood).

erbine, *f.* erbia, erbium oxide.

erbium, *m.* erbium.

erbue, *f.* clay used as a flux.

ère, *f.* era.

érection, *f.* erection.

érepsine, *f.* erepsin.

ergot, *m.* ergot; spur.
— *de seigle*, ergot of rye.

ergoté, *a.* ergoted, spurred (as rye); spurred.

ergotine, *f.* ergotinine; (*Pharm.*) ergotin.

ergotinine, *f.* ergotinine.

ergotique, *a.* ergotic.

éricacé, *a.* (*Bot.*) ericaceous.

ériger, *v.t.* erect.

érigéron, *m.* (*Bot.*) erigeron.

ernolithe, *m.* ernolith (a plastic composition).

éroder, *v.t.* erode.

érosif, *a.* erosive.

érosion, *f.* erosion.

errant, *a.* wandering, errant.

erratique, *a.* erratic.

errer, *v.i.* err; wander.

erreur, *f.* error.
— *de lecture*, error in reading.

erroné, *a.* erroneous.

erronément, *adv.* erroneously.

érugineux, *a.* eruginous, æruginous.

éruptif, *a.* eruptive.

éruption, *f.* eruption.

érythème, *m.* (*Med.*) erythema.

érythrine, *f.* erythrin.

érythrite, *f.* erythrol, erythritol, erythrite.

érythrodextrine, *f.* erythrodextrin.

érythrophléine, *f.* erythrophleine.

érythrophylle, *f.* (*Bot.*) erythrophyll(in).

érythroxyle, *m.* (*Bot.*) any member of the genus *Erythroxylum*.

es, *pr. 2 sing.* of être.

ès, *prep.* in the.

esc., *abbrev.* (escompte) discount.

escalier, *m.* staircase, stairs.

escarbille, *f.* (coal) cinder.

escarboucle, *f.* (*Min.*) carbuncle.

escargot, *m.* snail; spiral; spiral staircase.
en —, spiral.

escarner, *v.t.* flesh (hides).

escarpé, *a.* steep; escarped.

Escaut, *m.* Scheldt (river).

eschare, *f.* (*Med.*) eschar, scab.

eschel, *n.* eschel (variety of smalt).

esclavage, *m.* slavery.

esclave, *m. & f.* slave. — *a.* slave; slavish.

escompte, *m.* (*Com.*) discount.

escompter, *v.t.* discount. — *v.r.* be discounted.

escope, *f.* scoop.

escoupe, *f.* round shovel, scoop.

esculine, *f.* esculin, æsculin.

ésérine, *f.* eserine (physostigmine).

espace, *m.* space.
— *clos*, closed (or inclosed) space.
— *mort*, dead space.

espacement, *m.* spacing, distance, interval.

espacer, *v.t.* space, place at a distance apart, separate.

Espagne, *f.* Spain.

espagnol, *a.* Spanish. — *m.* Spanish (language); (*cap.*) Spaniard.

espèce, *f.* species; kind, sort; (*pl.*) specie.
— *chimique*, chemical species.

espérance, *f.* hope.

espérer, *v.t.* hope, hope for, expect. — *v.i.* hope.

espoir, *m.* hope, expectation.

esprit, *m.* spirit; mind; wit.
— *de bois*, wood spirit.
— *de camphre*, (*Pharm.*) spirit of camphor.
— *de cannelle*, (*Pharm.*) spirit of cinnamon.
— *de Mindérérus*, (*Pharm.*) spirit of Mindererus, solution of ammonium acetate.
— *de nitre*, spirit of niter.
— *de pétrole*, petroleum spirit.
— *de raifort composé*, (*Pharm.*) compound spirit of horseradish.
— *de sel*, spirit of salt (old name for hydrochloric acid).
— *de sel ammoniac*, spirit of sal ammoniac (old name for aqueous ammonia).
— *de vin*, spirit of wine, (ethyl) alcohol.
— *de vinaigre*, glacial acetic acid.
— *de vitriol*, spirit of vitriol (old name for sulfuric acid, esp. when somewhat diluted).
— *d'orange composé*, (*Pharm.*) compound spirit of orange.
— *pyroacétique*, pyroacetic spirit (old name for acetone).
— *pyroligneux*, — *pyroxylique*, pyroligneous spirit, pyroxylic spirit (wood spirit, wood alcohol).

esprit-de-vin, *m.* spirit of wine, alcohol.

esquille, *f.* splinter (of bone).

esquilleux, *a.* splintery.

esquine, *f.* chinaroot (rootstock of *Smilax china*).

esquisse, *f.* sketch, outline.

esquisser, *v.t.* sketch, outline.

esquiver, *v.t.* evade.

essai, *m.* test; testing; analysis; sample, portion; assay, assaying; trial, proof; essay; foretaste.
— *à blanc*, blank test.
— *à chaud*, hot test.
— *à froid*, cold test.

essai *à la touche*, drop test, spot test, touch test.

— *de durée*, endurance test.

— *de pression*, pressure test.

— *par voie humide*, wet analysis, analysis by the wet method, wet test, wet assay.

— *par voie sèche*, dry analysis, dry test, dry assay.

essayage, *m.* testing, etc. (see essayer).

essayer, *v.t.* test; assay; analyze; try.

essayerie, *f.* assay laboratory, assay office.

essayeur, *m.* assayer.

esse, *f.* the letter S; S-shaped object; wire gage.

esséminer, *v.t.* disseminate, scatter.

essence, *f.* (essential) oil; specif., oil of turpentine; essence; any volatile liquid; specif., gasoline, petrol; (of trees) species, variety. (English usage tends to restrict the word "essence" to alcoholic solutions of essential oils, altho it is sometimes used of the oils themselves and of other volatile liquids.)

— *anglaise*, an American oil of turpentine sold in England and Antwerp.

— *antihystérique*, (*Pharm.*) fetid spirit of ammonia.

— *artificielle*, any artificial substitute for essential oils.

— *d'absinthe*, oil of wormwood.

— *d'amandes amères*, oil of bitter almond.

— *d'aneth*, oil of dill.

— *d'anis*, oil of anise.

— *d'automobile*, gasoline, petrol.

— *de bigarade*, oil of bitter-orange peel.

— *de bouleau*, — *de bétula*, oil of sweet birch, oil of betula.

— *de cajeput*, oil of cajuput.

— *de camomille romaine*, (Roman) camomile oil, (*Pharm.*) oil of chamomile.

— *de cannelle de Chine*, oil of cinnamon, oil of cassia, oil of Chinese cinnamon.

— *de carvi*, oil of caraway.

— *de chénopode anthelmintique*, oil of chenopodium, oil of American wormseed.

— *de citron*, oil of lemon.

— *de Cognac*, an artificial mixture of esters of fatty acids, used to adulterate brandy.

— *de copahu*, oil of copaiba.

— *de coriandre*, oil of coriander.

— *de cubèbe*, oil of cubeb.

— *de eucalyptus*, oil of eucalyptus.

— *de fenouil*, oil of fennel.

— *de gaulthérie*, oil of gaultheria, oil of wintergreen.

— *de genièvre*, oil of juniper.

— *de hédéoma*, oil of hedeoma, oil of pennyroyal.

— *de lavande*, oil of lavender.

— *de malt*, extract of malt.

essence *de menthe poivrée*, oil of peppermint.

— *de menthe verte*, oil of spearmint.

— *de miaouli*, oil of miaouli (from *Melaleuca* species).

— *de mirbane*, oil (or essence) of mirbane, nitrobenzene.

— *de moutarde*, volatile oil of mustard.

— *de muscade*, (volatile) oil of nutmeg, oil of myristica.

— *de petit-grain*, see under *petit-grain*.

— *de pétrole*, petroleum spirit (a rather loose term applied to volatile petroleum distillates).

— *de Portugal*, oil of Portugal (sweet) orange peel.

— *de pouliot américaine*, oil of (American) pennyroyal, oil of hedeoma.

— *d'érigéron*, oil of erigeron, oil of fleabane.

— *de romarin*, oil of rosemary.

— *de rose*, oil of rose, attar of roses, essence of roses.

— *de sabine*, oil of savin.

— *de santal*, oil of santal, oil of sandalwood.

— *de savon*, an alcoholic soap solution.

— *de templine*, oil from the cones of the silver fir (*Abies picea*).

— *de térébenthine*, oil (or spirit, or essence) of turpentine.

— *de térébenthine purifiée*, (*Pharm.*) rectified oil of turpentine.

— *de thym*, oil of thyme.

— *de valériane*, oil of valerian.

— *grasse*, resinified oil of turpentine, used in porcelain painting.

— *indienne*, an oil of camphor formerly used in painting.

— *minérale*, gasoline, petrol.

— *naturelle*, essential oil (volatile oil of natural origin).

— *rectifiée*, (*Painting*) rectified oil of turpentine.

(For the meaning of other phrases with plant names, as *essence d'ail*, *essence d'aspic*, etc. in which *essence* means " (essential or volatile) oil," see the plant name in question.)

essencifier, *v.t.* (*Alchemy*) essentificate.

essentiel, *a. & m.* essential.

essentiellement, *adv.* essentially.

esser, *v.t.* gage (wire).

essieu, *m.* axle, axletree, spindle, pin.

essor, *m.* flight; progress, advance.

essorage, *m.* drying (see essorer); seasoning.

essorer, *v.t.* dry (without heating, as by exposure to air, wringing or centrifugalizing); season.

essoreuse, *f.* centrifugal machine or apparatus, centrifuge, centrifugal drier, centrifugal filter.

essuie-mains, *m.* hand towel, towel.

essuie-plume(s), *m.* penwiper.

essuyage, *m.* wiping.

essuyant, *p.a.* wiping; drying; enduring.
— *m.* (*Soap*) drier.

essuyer, *v.t.* wipe, wipe off; dry; endure, stand.

est, *m.* east. — *pr. 3 sing.* of être.

estacade, *f.* stockade; framework; platform.

estagnon, *m.* a copper or tinplate container used for transporting oils, etc.

estampage, *m.* stamping.

estampe, *f.* stamp.

estamper, *v.t.* stamp.

estampeuse, *f.* stamper, stamping machine.

estampillage, *m.* stamping.

estampille, *f.* stamp; mark; trademark.

estampiller, *v.t.* stamp.

estampilleuse, *f.* stamping machine, stamp.

estampoir, *m.* stamp, punch.

est-ce que. Used in asking questions: *est-ce que le corps fond?* does the substance melt?

estimation, *f.* estimation; estimate.

estime, *f.* estimation; esteem.

estimer, *v.t.* estimate; value; esteem.

estival, *a.* summer, estival, æstival.

estoc, *m.* stock; tuck (a sword).

estomac, *m.* stomach.

estragol, *n.* estragole.

estragon, *m.* (*Bot.*) tarragon, less commonly called estragon (*Artemisia dracunculus*).

estraquelle, *f.* (*Glass*) ladle.

estrasse, *f.* floss silk.

estrigue, *m.* annealing oven, leer.

estropier, *v.t.* mutilate, mangle, cripple.

estuaire, *m.* estuary; strand.

esturgeon, *m.* (*Zoöl.*) sturgeon.

esule, *f.* (*Bot.*) spurge.

et, *conj.* and.

étable, *f.* stable.

établi, *p.a.* established, etc. (see établir). — *m.* bench, workbench; frame (as of a lathe).

établir, *v.t.* establish; fix, set, lay, place; erect, set up; settle.

établissement, *m.* establishment; settlement.

étage, *m.* story, floor; layer; row, tier; degree; step; (*Mining*) level.

étager, *v.t.* arrange in rows or tiers.

étagère, *f.* shelf; set of shelves.

étai, *m.* stay, prop.

étaient, *imp. 3 pl.* of être.

étaim, *m.* fine carded wool.

étain, *m.* tin; pewter vessel; pewter.
— *battu,* tinfoil.
— *de glace,* bismuth.
— *en blocs,* block tin.
— *en feuilles,* sheet tin.
— *en grenaille,* granulated tin.
— *en verges,* bar tin.
— *gris,* gray tin; (*Old Chem.*) bismuth.

étain-grain, *m.* grain tin.

était, *imp. 3 sing.* of être.

étal, *m.* butcher stall (or shop).
— *de boucherie,* butcher stall (or shop).

étalage, *m.* display; (of a furnace, *pl.*) boshes.

étaler, *v.t.* spread, spread out; display, show.
— *v.r.* spread, spread out; smear; (of ink) blot.

étalon, *m.* standard; stallion. — *a.* standard; (of glassware, etc.) standard, normal.

étalonnage, étalonnement, *m.* standardization, standardizing; calibration; gaging; stamping, sealing (of weights, etc.).

étalonner, *v.t.* standardize; calibrate; gage; stamp, seal (weights, etc.).

étamage, *m.* tinning, etc. (see étamer); tinned work.
— *au zinc,* zincking, galvanizing.
— *des glaces,* silvering of mirrors.
— *galvanique,* galvanizing.

étamer, *v.t.* tin; (in general) coat with a metal, plate, silver, galvanize, zinc, etc.

étamine, *f.* bolting cloth, tammy; (*Bot.*) stamen.

étampage, *m.* stamping; swaging.

étampe, *f.* stamp; swage; punch; die.

étamper, *v.t.* stamp; swage; punch.

étamure, *f.* an alloy used for tinning; the layer produced by tinning.

étanche, *a.* tight.
à —, tight.
— *à la vapeur,* — *de vapeur,* steam-tight.
— *à l'eau,* — *d'eau,* water-tight.

étanchéité, *f.* tightness.

étancher, *v.t.* stop, stanch; make tight; calk; quench (thirst).

étançon, *m.* prop, stay, stud, shore.

étang, *m.* pond.

étant, *p.pr.* of être.
— *donné,* given.

étape, *f.* stage; halting place.

état, *m.* state; calling, profession, trade; list; register; account.
à l'— *libre,* in the free state.
en l'—, this being the case.
— *critique,* critical state.
— *gazeux,* gaseous state.
— *naissant,* nascent state.
faire — *de,* depend on; intend; value.
faire — *que,* think, believe.

état-major, *m.* staff; staff office.

États-Unis, *m.pl.* United States.

étau, *m.* vise.
— *à main,* — *portatif,* hand vise.

étaux, *m.pl.* of étau; *m.pl.* of étal.

étayer, *v.t.* prop, stay, support, shore up.

été, *m.* summer. — *p.p.* of être (to be).

éteignant, *p.pr.* of éteindre.

éteindre, *v.t.* extinguish; put out (fire, lights); quench (flame, metals, etc.); slake (lime); cancel; dim; allay. — *v.r.* be extinguished, go out, be quenched; become extinct; die; die away; be dimmed.

éteint, *p.a.* extinguished, slaked, etc. (see éteindre); out; faint, dying; extinct. — *pr. 3 sing.* of éteindre.

étendage, *m.* dilution, diluting; stretching, etc. (see étendre); *(Glass)* flattening; *(Calico)* steaming; (for clothes, etc.) drying, drying rack or drying room.

étendre, *v.t.* dilute; stretch, stretch out, spread; spread out, extend; flatten (window glass). — *v.r.* be diluted; spread, stretch, extend, reach; dwell (upon).

étendu, *p.a.* dilute, diluted; stretched, etc. (see étendre); extensive.

étendue, *f.* extension; extent, compass, length.

éternel, *a.* eternal.

éternuement, éternûment, *m.* sneezing, sneeze.

éternuer, *v.i.* sneeze.

étes, *pr. 2 pl.* of être (to be).

étêter, *v.t.* remove the head of; top (trees).

éthal, *m.* ethal (cetyl alcohol).

éthalique, *a.* palmitic (acid); cetyl (alcohol).

éthanal, *m.* ethanal (acetaldehyde).

éthane, *m.* ethane.

éthane-diol, *m.* ethanediol.

éthanol, *m.* ethanol, ethyl alcohol.

éther, *m.* ether; ester. (Some French organic writers prefer to apply the term "ether" to esters and "oxide" to ethers.)

— *acétique,* acetic ester (or, according to old usage, acetic ether), specif. ethyl acetate. (Other phrases of similar translation need not be given here; see the adjectives.)

— *acide,* ester; acid ester (corresponding to acid salt).

— *amylazoteux,* — *amylnitreux,* amyl nitrite.

— *anesthésique,* anesthetic ether (specially purified ethyl ether).

— *azoteux,* nitrous ester, (old usage) nitrous ether; specif., ethyl nitrite.

— *azoteux alcoolisé,* *(Pharm.)* spirit of nitrous ether, sweet spirit of niter.

— *azotique,* nitric ester, specif. ethyl nitrate.

— *chlorhydrique,* hydrochloric ester, (organic) chloride; specif., ethyl chloride, chloroethane.

— *de pétrole,* petroleum ether.

— *éthylique,* ethyl ether; ethyl ester.

— *hydriodique,* ethyl iodide, iodoethane.

— *hydrique,* hydric ether (ethyl ether).

— *hydrochlorique,* ethyl chloride, chloroethane.

— *méthyliodhydrique,* methyl iodide, iodomethane.

— *méthylique,* methyl ether; methyl ester.

— *méthylsilicique,* methyl silicate.

— *monochlorhydrique,* (organic) monochloride; specif., ethyl chloride, chloroethane.

éther *neutre,* neutral ester (as contrasted with acid ester); neutral ether (ether having a neutral reaction).

— *nitreux,* nitrous ester (or ether), specif. ethyl nitrite.

— *nitrique,* = éther azotique.

— *officinal,* *(Pharm.)* purified ether, ether.

— *officinal alcoolisé,* *(Pharm.)* spirit of ether.

— *ordinaire,* ordinary ether, ethyl ether.

— *pur,* *(Pharm.)* = éther officinal.

— *pyroacétique,* pyroacetic ether (old name for acetone).

— *sel,* ethereal salt, ester.

— *sulfuré,* thio ether, sulfur ether, (organic) sulfide.

— *sulfurique,* sulfuric ether (ordinary or ethyl ether); sulfuric ester, specif. ethyl sulfate.

— *sulfurique alcoolisé,* *(Pharm.)* spirit of ether.

— *vinique,* old name for ethyl ether.

éthéré, *a.* ethereal; specif., pertaining to ethers (or to esters), as, *dérivé éthéré,* ethereal derivative.

éthérène, *m.* an old name for ethylene.

éthérificateur, *m.* etherifying (or esterifying) apparatus, etherifier (or esterifier).

éthérification, *f.* esterification; etherification. (See the note under *éther*).

éthérifier, *v.t.* esterify; etherify. (See *éther*.) — *v.r.* be esterified, esterify; be etherified.

éthérique, *a.* pertaining to ordinary ether, ethereal, ether; pertaining to ethers (or to esters).

éthérisation, *f.* etherification (or esterification; see *éther*); *(Med.)* etherization.

éthériser, *v.t.* etherify; esterify; etherize.

éthéro-alcoolique, *a.* ether-alcohol, of or pertaining to ether and alcohol.

éthérolat, *m.* a product obtained by distilling ether from aromatic material.

éthérolature, *f.* ethereal tincture.

éthérolé, *m.* an ethereal solution of medicinal principles.

éthérolique, *a.* (of medicines) having ether as the excipient.

éther-oxyde, *m.* ether, (organic) oxide.

éther-sel, *m.* ethereal salt, ester.

éthionique, *a.* ethionic.

éthiops, *m.* *(Old Chem.)* ethiops.

— *martial,* ethiops martial (black oxide of iron).

— *minérale,* ethiops mineral (black mercuric sulfide).

— *per se,* black oxide of mercury, mercurous oxide.

éthoxy-. ethoxy-.

éthoxyle, *m.* ethoxyl.

éthylamide, *f.* ethylamine.

éthylamine, *f.* ethylamine.

éthylaniline, *f.* ethylaniline.

éthylbenzène, éthylbenzine, *f.* ethylbenzene.

éthyle, *m.* ethyl.

éthylé, *p.a.* ethyl, ethylated.

éthylène, *m.* ethylene.

éthylène-diamine, *f.* ethylenediamine.

éthylène-disulfoné, *a.* ethylenedisulfonic; *s*-ethanedisulfonic (incorrect use).

éthylène-dithiol, *m.* ethylenedithiol(*s*-ethane-dimercaptan).

éthylénique, *a.* ethylene, (less often) ethylenic.

éthylénolactique, *a.* ethylene lactic (hydra-crylic).

éthyler, *v.t.* ethylate, introduce ethyl into.

éthyliaque, *f.* ethylamine. (*Obs. or Rare*).

éthylidénique, *a.* ethylidene.

éthylidénolactique, *a.* ethylidene lactic (desig-nating lactic acid proper).

éthylique, *a.* ethyl, ethylic.

éthylmalonique, *a.* ethylmalonic.

éthylphénylcétone, *f.* ethyl phenyl ketone (propiophenone).

éthylsulfurique, *a.* ethylsulfuric.

éthylure, *m.* ethide (binary compound of ethyl).

étincelant, *p.a.* sparkling, sparking.

étinceler, *v.i.* sparkle, spark.

étincelle, *f.* spark.

— *électrique,* electric spark.

étincellement, *m.* sparkling, sparking.

étiolé, *p.a.* etiolated.

étioler, *v.t. & r.* etiolate.

étique, *a.* emaciated.

étiqueter, *v.t.* label; ticket; docket.

étiquette, *f.* label; etiquette.

étirable, *a.* capable of being drawn or stretched out, (of metals) ductile, (of rubber, etc.) elastic.

étirage, *m.* drawing, etc. (see étirer); (*Ceram.*) compression (thru a template).

étiré, *p.a.* drawn, etc. (see étirer); long (in proportion to width).

étirer, *v.t.* draw, draw out (as metals or glass tubing); beat out, roll (metal); stretch (as rubber); stretch out, lengthen. — *v.r.* be drawn out; be stretched; stretch.

étisie, *f.* emaciation.

etnite, *f.* (*Expl.*) etnite.

étoffe, *f.* fabric, cloth, stuff; material, stuff; (*Leather*) a tawing liquor consisting of a solution of alum and salt.

étoffé, *p.a.* stuffed; padded: ample; comfort-able; stout, stocky.

étoile, *f.* star; star-shaped crack; (of a centri-fuge) head, star-shaped part bearing the tubes; (*Mach.*) star wheel.

étoiler, *v.t.* crack (radially); star.

étonnament, *adv.* astonishingly.

étonnant, *p.a.* astonishing.

étonnement, *m.* astonishment, wonder.

étonner, *v.t.* astonish; crack, rend (as minerals by heating and quenching).

étouffant, *p.a.* suffocating, stifling.

étouffée, *f.* stewing, stew; (*Charcoal*) the second stage in burning, during which the principal decomposition occurs.

étouffement, *m.* suffocation, stifling.

étouffer, *v.t.* suffocate, stifle; smother; sup-press. — *v.i.* be suffocating or stifling. — *v.r.* be suffocating; be suffocated; choke.

étouffoir, *m.* vessel into which new charcoal is put to prevent oxidation; extinguisher (for charcoal fires); damper.

étoupage, *m.* (*Mach.*) packing, stuffing.

étoupas, *m.* a coarse cloth made from tow.

étoupe, *f.* tow; oakum; (*Mach.*) packing, stuffing; waste (as of cotton).

étoupement, *m.* stopping, packing.

étouper, *v.t.* stop (with tow, oakum, cotton wax, etc.); (*Mach.*) pack, stuff; calk.

étouperie, *f.* tow packing cloth.

étoupille, *f.* primer, fuse, match (the last term is now used chiefly in pyrotechnics).

— *à friction,* friction primer, friction fuse.

— *à percussion,* percussion primer, percussion fuse.

— *de Bickford,* Bickford fuse.

— *de mise de feu,* primer, fuse, match.

— *de sûreté,* safety fuse.

— *électrique,* electric primer, electric fuse.

— *fulminante,* friction primer.

— *percutante,* percussion primer, percussion fuse.

étoupiller, *v.t.* provide (pyrotechnic devices) with quick match.

étourdi, *p.a.* stunned; numb; slightly warm; parboiled; allayed; giddy, heedless.

étourdir, *v.t.* stun; benumb; warm slightly (as water); parboil; allay (as pain).

étourdissement, *m.* stunning, etc. (see étour-dir); dizziness; astonishment.

étrange, *a.* strange.

étrangement, *adv.* strangely.

étranger, *a.* foreign; strange, unacquainted, unknown. — *m.* foreigner; stranger; for-eignness, strangeness; foreign land, foreign parts.

à l'—, abroad.

étrangeté, *f.* strangeness; strange thing.

étranglement, *m.* constriction (as in a tube); (*Pyro.*, etc.) choke; wiredrawing; throt-tling; strangling; condensing.

étrangler, *v.t.* constrict, narrow; (*Pyro.*, etc.) choke; wiredraw (metal); throttle (steam); strangle; condense (a subject).

étrangleur, *m.* throttling device, throttle valve.

étrangloir, *m.* (*Pyro.*) choking frame.

être, *v.i.* be; have (in certain auxiliary uses, as *il est venu,* he has come). — *m.* being, existence.

en —, be, be at, be arrived, be of it, be in it. be in for it.

être (continued).

est-ce que. See this in the vocabulary.

— *à,* — *de,* etc. See the meanings of the prepositions.

— *vivant,* living being.

il est, there is, there are, he is, it is.

étrécir, *v.t.* narrow; — *v.r.* narrow, become narrower; shrink, contract.

étrécissement, *m.* narrowing; shrinking, shrinkage, contraction; narrowness.

étreignant, *p.pr.* of étreindre.

étreindelle, *f.* a bag of horsehair or other coarse material in which are enclosed the woolen press bags used in pressing oil meal or fatty acids.

étreindre, *v.t.* bind, tie, tie up, clasp.

étreint, *p.a.* bound, etc. (see étreindre).

étreinte, *f.* constraint; binding, tie; embrace.

étrésillon, *m.* prop, stay, shore.

étrier, *m.* stirrup (specif., in a balance the stirrup-shaped support for the pan); support (of various kinds), strap, hoop, band, etc.

étroit, *a.* narrow; tight; strict; close, intimate.

étroitement, *adv.* narrowly, etc. (see étroit).

étroitesse, *f.* narrowness, etc. (see étroit).

étude, *f.* study; survey (as of a railway); office (as of a lawyer).

à l'—, under investigation.

étudiant, *m.* student. — *p.pr.* of étudier.

étudier, *v.t. & i.* study. — *v.r.* be studied; study.

étui, *m.* case; box, sheath, cover, envelope.

étuvage, *m.* stoving, etc. (see étuver); specif.: (*Ceram.*) enfumage; (*Leather*) a heating of the oiled skins in steam ovens in the process of making chamois.

étuve, *f.* oven, closet, bath (for drying, for bacterial cultures, etc.); drying room, drying stove, stove; sweating room, (hot-air or vapor) bath.

— *à air,* air bath, hot-air oven, hot-air closet.

— *à cultures,* (*Bact.*) incubator.

— *à eau,* hot-water oven (not water bath, tho the latter may be combined with it).

— *à dessécher,* drying oven, drying closet.

— *à dessécher dans le vide,* vacuum drying oven.

— *à vapeur,* steam oven.

— *électrique,* electric oven.

— *froide,* — *glacière,* ice closet, refrigerator (specif. a double-walled case cooled by a current of ice water and used for maintaining substances at zero).

— *stérilisatrice,* sterilizer, sterilizing oven.

étuvement, *m.* stoving, etc. (see étuver).

étuver, *v.t.* stove, heat (moderately) or dry in an oven, dry; (*Med.*) foment; (*Cooking*) stew.

eu, *p.p.* of avoir (to have).

eucalypte, eucalyptus, *m.* eucalyptus.

eucalyptol, *m.* eucalyptole (cineole).

euchlorine, *f.* euchlorine (mixture of chlorine dioxide and chlorine).

euclase, *f.* (*Min.*) euclase.

euclidienne, *a. fem.* Euclidean.

eudialite, *f.* (*Min.*) eudialyte.

eudiomètre, *m.* eudiometer.

eudiométrie, *f.* eudiometry.

eudiométrique, *a.* eudiometric(al).

eugénine, *f.* eugenin.

eugénol, *m.* eugenol.

euglobuline, *f.* euglobulin.

eupatoire, *f.* (*Bot.*) eupatory (*Eupatorium*).

— *des Grecs,* common agrimony (*A. eupatoria*).

euphorbe, *f.* (*Bot.*) euphorbia; (*Pharm.*) euphorbium.

euphorbiacée, *f.* (*Bot.*) euphorbiaceous plant; (*pl.*) Euphorbiaceæ.

euphotide, *f.* (*Petrog.*) euphotide.

euphraise, euphrasie, *f.* (*Bot.*) Euphrasia, specif. euphrasy, eyebright (*E. officinalis*).

eupione, *f.* eupione, eupion (an empyreumatic oil).

eurent, *p.def. 3 pl.* of avoir (to have).

eurhodine, *f.* eurhodine.

eurhodol, *m.* eurhodol.

européen, *a.* European. — *m.* (*cap.*) European.

europhène, *m.* (*Pharm.*) europhen.

europium, *m.* europium.

eut, *p.def. 3 sing.* of avoir (to have).

eût, *imp. subj. 3 sing.* of avoir (to have).

eutectique, *a. & m.* eutectic.

eutexie, *f.* eutexia.

point d'—, eutectic point.

eux, *pron.* they, them.

eux-mêmes, *pron.* themselves.

évacuant, *a. & m.* (*Med.*) evacuant.

évacuation, *f.* evacuation; emptying; (of goods) exportation, shipping; (*Steam*) exhaust, eduction; (*Mach.*) evacuation pipe.

évacuer, *v.t. & i.* evacuate; empty; export, ship. — *v.r.* be evacuated, etc.

évader, *v.r.* evade, escape.

évaluable, *a.* estimable, appreciable, evaluable.

évaluation, *f.* estimation, estimate, evaluation, valuation.

évaluer, *v.t.* estimate, evaluate, value. — *v.r.* be estimated, be valued; be worth.

évanescent, *a.* evanescent.

évanouir, *v.r.* vanish, disappear; faint. *faire* —, (*Math.*) eliminate.

évanouissement, *m.* disappearance; fainting.

évaporable, *a.* evaporable.

évaporateur, *m.* evaporator.

évaporatif, *a.* evaporating, evaporative.

évaporation, *f.* evaporation.

évapportoire, *a.* evaporating.

évaporé, *p.a.* evaporated; giddy, heedless.

évaporer, *v.t.* evaporate; (fig.) vent. — *v.r.* evaporate; (fig.) find vent; be heedless, be wild.

— *à sec*, evaporate to dryness.

évaporeuse, *f.* evaporator, evaporating apparatus.

évaporisation, *f.* evaporation.

évaporo-distillatoire, *a.* for or pertaining to both evaporation and distillation.

évaporomètre, *m.* evaporometer, atmometer.

évasé, *p.a.* widened, enlarged, flaring, bell-shaped, funnel-shaped, bell-mouthed, splayed.

évasement, *m.* widening, enlargement, spreading, splay; width (at the opening).

évaser, *v.t.* widen, enlarge, make flaring. — *v.r.* widen, become flaring.

évasion, *f.* escape; evasion.

évasure, *f.* widening, (flaring) opening.

éveil, *m.* warning, alarm.

éveiller, *v.t.* wake, awake; call forth. — *v.r.* wake, awake.

événement, *m.* event; emergency.

évent, *m.* (open) air; alteration due to exposure to air; (of wine, etc.) mustiness, flatness; air hole; escape hole, gas port; (of a fuse) flash hole (also, punching hole); cavity, honeycomb (in metals); crack.

éventable, *a.* capable of being aired; alterable in air; etc. (see éventer).

éventage, *m.* airing, etc. (see éventer).

éventail, *m.* fan; batswing burner, batswing; (enameler's) screen.

en —, fan-shaped.

éventer, *v.t.* air, expose to the air; alter by exposure to air, specif. make (wine, etc.) musty or flat; discover, scent, find out; fan (a person). — *v.r.* be altered by exposure to air, (of wine) become flat; be discovered; fan oneself.

éventrer, *v.t.* rip open, eviscerate, gut.

éventualité, *f.* eventuality, contingency.

éventuel, *a.* contingent; eventual.

éventuellement, *adv.* contingently, depending on circumstances; eventually.

éventure, *f.* crack.

évêque, *m.* bishop.

éverdumer, *v.t.* bleach, blanch (green things).

évidage, *m.* hollowing, etc. (see évider).

évidé, *p.a.* hollowed, hollowed out, hollow; grooved.

évidement, *m.* hollow, cavity, recess, aperture, cut, groove; hollowing, grooving.

évidemment, *adv.* evidently; certainly.

évidence, *f.* evidence; obviousness, clearness.

évident, *a.* evident.

évider, *v.t.* hollow, hollow out, scoop out; groove.

évidure, *f.* = évidement.

évier, *m.* sink; stone gutter.

éviscérer, *v.t.* eviscerate.

évitable, *a.* avoidable.

évitement, *m.* avoiding, avoidance; (railway) siding, switch.

éviter, *v.t.* avoid.

évolué, *a.* evolved, developed.

évoluer, *v.i.* evolve; perform evolutions.

évolutif, *a.* evolutional, evolutionary.

évolution, *f.* evolution.

évoquer, *v.t.* evoke.

exact, *a.* exact; precise, strict, punctual.

exactement, *adv.* exactly.

exactitude, *f.* exactness, exactitude.

exagération, *f.* exaggeration.

exagérer, *v.t.* exaggerate.

exalbuminé, *a.* (*Bot.*) exalbuminous.

exalgine, *f.* (*Pharm.*) exalgin, exalgine.

exaltation, *f.* exaltation; excitement.

exalter, *v.t.* exalt; excite.

examen, *m.* examination.

en —, under examination.

— *d'admission*, entrance examination.

— *écrit*, written examination.

examinateur, *m.* examiner.

examiner, *v.t.* examine.

excavateur, *m.* excavator.

excavation, *f.* excavation, excavating.

excaver, *v.t.* excavate.

excédant, *a.* exceeding, excess, in excess; surplus; tiresome.

excédent, *m.* excess, surplus. — *a.* surplus, excess, in excess.

excéder, *v.t.* exceed, go beyond; tire, weary.

excellement, *adv.* excellently.

excellence, *f.* excellence, excellency.

par —, preëminent, preëminently, par excellence.

excellent, *a.* excellent.

exceller, *v.i.* excel.

— *sur*, excel.

excentrer, *v.t.* decenter, make eccentric.

excentricité, *f.* eccentricity.

excentrique, *a. & m.* eccentric.

excentriquement, *adv.* eccentrically.

excepté, *prep.* except, excepting. — *p.a.* excepted.

— *que*, except that, unless.

excepter, *v.t.* except.

exception, *f.* exception.

par —, by way of exception, as an exception.

exceptionnel, *a.* exceptional.

exceptionnellement, *adv.* exceptionally.

excès, *m.* excess.

à l' —, *jusqu'à l'*—, to excess; immoderately.

dans l'—, in excess; excessively.

excessif, *a.* excessive.

excessivement, *adv.* excessively, exceedingly.

excipient, *m.* (*Pharm.*) excipient.

excise, *f.* excise.

excitant, *a. & m.* excitant.
excitateur, *a.* exciting. — *m.* exciter; (*Elec.*) discharger, excitator.
excitatif, *a.* excitative.
excitation, *f.* excitation; excitement.
excitatrice, *f.* (*Elec.*) exciter. — *fem.* of excitateur.
exciter, *v.t.* excite.
exclure, *v.t.* exclude.
exclusif, *a.* exclusive.
exclusion, *f.* exclusion; (*Steam*) cut-off. *à l'— de,* exclusive of, excepting.
exclusivement, *adv.* exclusively.
exclut, *pr. & p.def. 3 sing.* of exclure.
excorier, *v.t.* excoriate.
excortiquer, *v.t.* excorticate, decorticate.
excrément, *m.* excrement.
excrémenteux, excrémentiel, excrémentitiel, *a.* excremental, excrementitious.
excreta, *m.pl.* excreta.
excréter, *v.t.* excrete.
excréteur, *a.* excretory.
excrétion, *f.* excretion.
excrétoire, *a.* excretory.
excroissance, *f.* excrescence.
excursion, *f.* excursion.
excusablement, *adv.* excusably.
excuser, *v.t.* excuse.
exécutable, *a.* executable, practicable.
exécuter, *v.t.* execute.
exécuteur, exécutif, *a.* executive.
exécution, *f.* execution.
exemplaire, *m.* copy; model. — *a.* exemplary.
exemple, *m.* example. *par —,* for example, for instance; indeed!
exempt, *a.* free; exempt.
exempter, *v.t.* exempt.
exercer, *v.t.* exercise; practice.
exercice, *m.* exercise; practice; financial period, fiscal year.
exfoliation, *f.* exfoliation.
exfolier, *v.t. & r.* exfoliate.
exhalaison, *m.* exhalation (something exhaled).
exhalation, *f.* exhalation (the act).
exhalatoire, *f.* a kind of salt evaporator.
exhaler, *v.t.* exhale. — *v.r.* be exhaled; evaporate.
exhaussement, *m.* raising; erection; height.
exhausser, *v.t.* raise; erect.
exhaustion, *f.* exhaustion.
exhibition, *f.* exhibition.
exhilarant, *a.* exhilarant. *gaz —,* laughing gas.
exigeant, *a.* exacting.
exigence, *f.* need, necessity, exigency; exactingness; unreasonableness.
exiger, *v.t.* require, necessitate, need, demand, exact. — *v.r.* be required, etc.
exigu, *a.* small, narrow, scanty, petty.
existant, *p.a.* existing, existent; on hand.

existence, *f.* existence; living being; (*Com.*) stock on hand.
exister, *v.i.* exist.
exitèle, *f.* (*Min.*) valentinite.
exocellulaire, *a.* exocellular.
exogène, *a.* (*Bot.*) exogenous. — *m.* exogen.
exorbitamment, *adv.* exorbitantly, excessively.
exosmose, *f.* exosmosis, exosmose.
exosmotique, *a.* exosmotic.
exothermique, exotherme, *a.* exothermic.
exotique, *a.* exotic, foreign.
exotiquement, *adv.* exotically.
expansibilité, *f.* expansibility.
expansible, *a.* expansible.
expansif, *a.* expansive.
expansion, *f.* expansion.
expansivité, *f.* expansiveness.
expectation, *f.* expectation.
expectorant, *a. & m.* (*Med.*) expectorant.
expédient, *a. & m.* expedient.
expédier, *v.t.* send, forward, ship; despatch; expedite; draw up (documents).
expéditeur, *m.* sender, shipper. — *a.* sending, forwarding.
expéditif, *a.* expeditious, speedy; prompt.
expédition, *f.* sending, forwarding, shipment, transmission; copy, duplicate; expedition; despatch.
expéditionnaire, *m.* sender; shipper; commission agent; copying clerk. — *a.* shipping, forwarding; copying; expeditionary.
expéditivement, *adv.* expeditiously, speedily.
expérience, *f.* experiment; test; experience. — *à blanc,* blank experiment, blank test.
expérimental, *a.* experimental.
expérimentalement, *adv.* experimentally.
expérimentaliste, *m.* experimentalist.
expérimentateur, *m.* experimenter.
expérimentation, *f.* experimentation.
expérimentatrice, *f.* experimenter.
expérimenté, *a.* experienced; tried, tested.
expérimenter, *v.t.* try, test, make experiment of; experience. — *v.i.* experiment.
expert, *a. & m.* expert.
expertement, *adv.* expertly.
expertise, *f.* (expert) investigation, survey; (expert) report, valuation.
expertiser, *v.t.* examine, investigate, make a survey of.
expiration, *f.* expiration.
expirer, *v.t. & i.* expire.
explicable, *a.* explicable, explainable.
explicatif, *a.* explanatory, explicative.
explication, *f.* explanation, explication.
explicite, *a.* explicit.
explicité, *f.* explicitness.
explicitement, *adv.* explicitly.
expliquer, *v.t.* explain. — *v.r.* be explained, give an explanation.

exploitant, *m.* exploiter, operator (of mines, etc.). — *a.* exploiting.

exploitation, *f.* exploitation; working, operating (of mines, factories, etc.); business, works, mine, farm, etc.

exploiter, *v.t.* exploit, work, operate.
— *à la poudre,* blast.

exploiteur, *m.* exploiter; operator, manager, farmer, grower, etc.

explorateur, *m.* explorer. — *a.* exploring, exploratory.

exploration, *f.* exploration.

explorer, *v.t.* explore.

exploser, *v.i.* explode.

exploseur, *m.* exploder.

explosible, *a.* explosive.

explosif, *a. & m.* explosive.
— *à la cellulose nitrée,* nitrocellulose explosive.
— *à la nitroglycérine,* nitroglycerin explosive, dynamite.
— *binaire,* binary explosive (one containing two ingredients).
— *chimique,* a chemical compound used as an explosive, explosive compound.
— *de sûreté,* safety explosive.
— *Favier,* Favier explosive.
— *mécanique,* a mechanical mixture capable of explosive interaction, explosive mixture.
— *mixte,* mixed explosive (as one containing an inert base).
— *réglé,* an explosive the action or sensitiveness of which has been regulated by admixture or other treatment.
— *simple,* an explosive consisting of a single chemical compound.

explosion, *f.* explosion.
— *de deuxième ordre,* explosion of the second order.
— *de premier ordre,* explosion of the first order.
— *sympathique,* sympathetic explosion, explosion by influence.
faire —, explode.

explosivité, *f.* explosiveness.

exponentiel, *a.* exponential.

exportateur, *m.* exporter.

exportation, *f.* exportation; export.

exportatrice, *f.* exporter.

exporter, *v.t.* export. — *v.r.* be exported.

exporteur, *m.* exporter.

exposant, *m.* exponent; exhibitor; petitioner.
— *p.pr.* of exposer.

exposé, *p.a.* exposed; explained. — *m.* exposition, statement, account; exposé.

exposer, *v.t.* expose; explain, expound. — *v.r.* expose oneself; be exposed.

expositif, *a.* explanatory, descriptive.

exposition, *f.* exposition; exposure; exhibition; statement, account.

exprès, *a.* express. — *m.* messenger. — *adv.* purposely; expressly.

expressément, *adv.* expressly; purposely.

expressif, *a.* expressive.

expression, *f.* expression.

exprimable, *a.* expressible.

exprimer, *v.t.* express. — *v.r.* be expressed; express oneself.

expulser, *v.t.* expel.

expulseur, expulsif, *a.* expulsive.

expulsion, *f.* expulsion.

expultrice, *fem.* of expulseur.

exquis, *a.* exquisite.

exquisement, *adv.* exquisitely.

exsiccateur, *m.* desiccator.

exsiccation, *f.* desiccation, drying.

exsudant, *a.* exuding; (*Med.*) sudorific.

exsudat, *m.* exudate.

exsudation, *f.* exudation.

exsuder, *v.t. & i.* exude.

extensibilité, *f.* extensibility.

extensif, *a.* extensive; expansive.

extension, *f.* extension; extent.

extérieur, *a.* exterior, external, outer, outside, foreign. — *m.* exterior; outside; foreign parts.
à l'—, abroad; outside.

extérieurement, *adv.* on the outside, externally, outwardly.

exterminateur, *m.* exterminator. — *a.* exterminating.

exterminer, *v.t.* exterminate.

externe, *a.* external; exterior (angle).

extincteur, *m.* extinguisher. — *a.* extinguishing.

extinction, *f.* extinction; slaking (of lime); suppression; loss.
— *spontanée,* air slaking.

extinctrice, *fem.* of extincteur.

extinguible, *a.* extinguishable.

extirpation, *f.* extirpation, eradication.

extirper, *v.t.* extirpate, eradicate.

extra, *a. & m.* extra.

extra-blanc, *a.* extra white.

extra-calciné, *a.* supercalcined, calcined beyond the usual degree.

extra-courant, *m.* (*Elec.*) extra current, self-induced current.

extracteur, *m.* extractor; (*Gas*) exhauster.

extractif, *a. & m.* extractive.

extractiforme, *a.* extractiform.

extraction, *f.* extraction; (*Mach.*) discharge.

extra-fin, *a.* extra fine.

extraire, *v.t.* extract.

extrait, *m.* extract. — *p.p. & pr. 3 sing.* of extraire.
— *d'actée à grappes,* (*Pharm.*) extract of cimicifuga, extract of black cohosh.
— *de colchique acétique,* (*Pharm.*) extract of colchicum corm, acetic extract of colchicum.
— *de fiel de bœuf,* (*Pharm.*) purified oxgall.
— *de garance,* madder extract.

extrait *de Goulard,* (*Pharm.*) Goulard's extract, solution of lead subacetate.

— *de Javel,* a concentrated hypochlorite solution used for sterilizing water, etc., Javel extract.

— *de saturne,* (*Pharm.*) Goulard's extract.

— *de viande,* meat extract.

— *éthéré,* ethereal extract, ether extract; (*Pharm.*) oleoresin.

— *éthéré de capsique,* (*Pharm.*) oleoresin of capsicum.

— *liquide,* (*Pharm.*) fluidextract, fluid extract.

extrait *liquide aromatique,* (*Pharm.*) aromatic fluidextract.

— *tannant,* tanning extract.

— *thébaïque,* (*Pharm.*) extract of opium.

extraordinaire, *a.* extraordinary.

extrapolation, *f.* extrapolation.

extra-pur, *a.* extra pure.

extrayant, *p.a.* extracting.

extrême, *a. & m.* extreme.

extrêmement, *adv.* extremely.

extrémité, *f.* extremity.

F

f., *abbrev.* franc; (frère) brother.
fabricant, *m.* manufacturer.
fabricant-liquoriste, *m.* liqueur maker.
fabrication, *f.* manufacture, manufacturing, making, make; fabrication.
fabrique, *f.* factory, manufactory, works, mill, plant; manufacture, making, make; fabric, structure.
— *de matières colorantes,* dye factory.
fabriquer, *v.t.* make, manufacture; fabricate.
— *v.r.* be made, be manufactured.
fabuleux, *a.* fabulous.
face, *f.* face.
en — de, opposite; in the front of; in the face of.
faire — à, be opposite; face.
facette, *f.* facet, (small) face.
facetter, *v.t.* facet, cut facets upon.
fâcheusement, *adv.* sadly, etc. (see fâcheux).
fâcheux, *a.* sad, regrettable, unfavorable, troublesome, disagreeable.
facile, *a.* easy; facile, ready, quick.
facilement, *adv.* easily, readily.
facilité, *f.* facility; ease, readiness.
faciliter, *v.t.* facilitate.
façon, *f.* fashion; manner; sort; way; making, make, workmanship; appearance; care, attention; operation; (*Agric.*) culture, dressing; ceremony, ado.
de — à, de — que, de telle — que, so that, so as, in such a way as (or that).
façonnage, façonnement, *m.,* **façonnerie,** *f.* fashioning, etc. (see façonner).
façonner, *v.t.* fashion, shape, form, model; make, make up, work; finish off, finish; adorn, (of fabrics) figure; accustom.
fac-similaire, *a.* facsimile.
fac-similé, *m.* facsimile.
fac-similer, *v.t.* facsimile.
factage, *m.* delivery (as of letters); porterage.
facteur, *m.* factor; maker, manufacturer; clerk; carrier (as of letters); porter.
— *de charge,* load factor.
— *de puissance,* power factor.
— *de sécurité,* factor of safety.
factice, *a.* artificial, factitious; sham. —*m.* something artificial; specif., artificial rubber.
facticement, *adv.* artificially, factitiously.
faction, *f.* faction; (*Mil.*) sentry.
factorielle, *f.* (*Math.*) factorial.

facture, *f.* invoice, bill; composition.
facturer, *v.t.* invoice, bill.
facturier, *m.* invoice book; invoice clerk.
facule, *f.* (*Astron.*) facula.
facultatif, *a.* optional.
facultativement, *adv.* optionally.
faculté, *f.* faculty; (*pl.*) means, ability.
fadasse, *a.* very insipid; very dull.
fade, *a.* insipid; flat (as wine); dull (colors).
fadement, *adv.* insipidly.
fadet, *a.* somewhat insipid; rather dull.
fadeur, *f.* insipidity, insipidness; flatness; dullness.
fagot, *m.* fagot; bundle.
fagoue, *f.* sweetbread.
fahlunite, *f.* (*Min.*) fahlunite.
faible, *a.* weak; feeble; slight; light (in weight). — *m.* weak point, weak part; weakness; weak person.
faiblement, *adv.* weakly, feebly, slightly.
faiblesse, *f.* weakness; feebleness; slightness; lightness; faint.
faiblir, *v.i.* weaken, grow weak, lose strength; yield, relax, slacken, fail.
faïence, *f.* glazed earthenware, faïence, delftware (in English *faïence* is used esp. of the finer, decorative kinds).
faïencerie, *f.* glazed earthenware factory, manufacture or business.
faïencier, *m.* maker or seller of glazed earthenware.
faillant, *p.a.* failing, etc. (see faillir).
faille, *f.* fissure; (*Geol.*) dike; grogram (cloth). — *pr. subj.* of falloir. — *pr. 3 sing. subj.* of faillir.
failli, *a. & m.* bankrupt. — *p.p.* of faillir.
faillibilité, *f.* fallibility.
faillible, *a.* fallible.
faillir, *v.i.* fail; err; be on the point (of), be near, be nearly.
faillite, *f.* bankruptcy, failure.
faim, *f.* hunger.
faîne, *f.* beech mast, beech nuts.
faire, *v.t.* make; do; cause; perform; produce; form; study; play; say; accustom; charge (for goods); matter, signify. — *v.i.* do; matter, signify, be. — *v.r.* be made, be done, etc., (of a precipitate or the like) be formed, form; become; make oneself; be. — *m.* making, doing, execution.

146

faire *que*, be the reason that, cause that; see that.

ne — que, make only, do only, do nothing but.

ne — que de venir, to have just come.

(For other phrases commencing with *faire* see the next important word in the phrase).

faisable, *a.* feasible.

faisait, *imp. 3 sing.* of faire.

faisant, *p.pr.* of faire.

faisceau, *m.* bundle; group, lot, number; (*Optics*) pencil; (*Anat.*) fasciculus; sheaf.

faiseur, *m.* maker, doer, etc. (see faire); author.

faisselle, *f.* a basket or vessel in which cheese is drained.

fait, *p.a.* made, done, etc. (see faire); fit; finished; grown; ripe; (of prices) fixed; (of words) common. — *m.* fact; deed, act; matter, affair, point, question; portion, share. — *pr. 3 sing.* of faire.

dans le —, in fact.

de —, in fact, indeed; actual, de facto.

en — de, in point of.

être — de, become, suit.

par le —, in fact; by the fact.

faite, *m.* top, summit; ridge; watershed.

faîtière, *a.* ridge. — *f.* ridge tile; skylight.

faix, *m.* burden, load.

falaise, *f.* cliff.

fallacieusement, *adv.* fallaciously.

fallacieux, *a.* fallacious.

falloir, *v.i.* be wanting, be needing; be necessary, be needful, be requisite, must; ought, should. — *v.r.* (with *en*) be wanting, be nearly.

comme il faut, properly, as should be; well bred.

il a fallu un litre, one liter was needed.

il fallait venir, you (he, they, etc.) should have come (or ought to have come).

il faudra qu'il vienne, he will have to come.

il me faut un entonnoir, I need a funnel.

fallu, *p.p.* of falloir.

falsificateur, *m.* adulterator; falsifier.

falsification, *f.* adulteration, sophistication; falsification.

falsifier, *v.t.* adulterate, sophisticate; debase (coin); falsify, counterfeit, forge.

falun, *m.* shell marl.

falunage, *m.* application of shell marl to land; shell marl production.

faluner, *v.t.* treat (land) with shell marl.

falunière, *f.* shell-marl pit.

fameux, *a.* famous.

familiariser, *v.t.* familiarize. — *v.r.* become familiar.

familiarité, *f.* familiarity.

familier, *a.* familiar.

familièrement, *adv.* familiarly.

famille, *f.* family.

fanage, *m.* fading; wilting; tedding; foliage, leaves.

fane, *f.* dry leaf.

faner, *v.t.* fade; wilt; turn, ted (hay). — *v.r.* fade; wilt.

fange, *f.* mud, mire.

fangeux, *a.* muddy, miry, dirty.

fantaisie, *f.* fancy; floss silk (thread).

farcir, *v.t.* stuff.

fard, *m.* (face) paint; false show, disguise.

fardeau, *m.* burden, load.

farder, *v.t.* paint (the face); disguise. — *v.i.* weigh heavily; sink, settle.

farinacé, *a.* farinaceous.

farine, *f.* flour; meal; farina.

— *d'avoine*, oatmeal.

— *de blé*, wheat flour.

— *de bois*, (fine) sawdust, wood flour.

— *de graine de lin*, — *de lin*, linseed meal, flaxseed meal.

— *de maïs*, corn meal, maize flour.

— *de montagne*, mountain flour (= farine fossile, which see).

— *de moutarde*, flour of mustard.

— *de première*, flour of the best grades.

— *de seigle*, rye flour.

— *d'orge*, barley flour.

— *dure*, hard-wheat flour.

— *fossile*, fossil farina (usually diatomaceous earth but sometimes a powdery variety of calcite).

— *jaune*, yellow corn meal.

— *lactée*, a preparation of condensed milk and flour (such as Nestlé's food).

— *tendre*, soft-wheat flour.

fariner, *v.t.* flour, convert into or dust with flour.

farineux, *a.* floury; powdery; mealy; farinaceous. — *m.* starchy food; legume (including peas, beans, lentils, etc.).

farinuleux, *a.* containing flour deprived of gluten.

faro, *m.* a kind of beer made in Belgium.

fascicule, *m.* fascicle, small bundle, (of a publication) part.

fasciner, *v.t.* fascinate.

fasse, *pr. 1 & 3 sing. subj.* of faire.

fastidieux, *a.* tedious, tiresome.

fatal, *a.* inevitable; fatal.

fatalement, *adv.* inevitably; unfortunately; fatally.

fatigant, *p.a.* fatiguing, wearisome, tedious.

fatigue, *f.* fatigue.

fatiguer, *v.t.* fatigue, tire, weary; strain, impair. — *v.r.* be fatigued; be strained, be impaired. — *v.i.* work hard; be fatiguing.

fau, *m.* beech, beech tree.

faubourg, *m.* suburb; faubourg.

faucher, *v.t.* mow, cut down, cut.

faude, *m.* place where charcoal is burned; place where ores are calcined.

fauder, *v.t.* fold (double).

faudra, *fut. 3 sing.* of falloir and of faillir.

faudrait, *cond. 3 sing.* of falloir and faillir.

faulde, *m.* = faude.

faulx, *f.* scythe; currying knife; *(Anat.)* falx.

faune, *f.* fauna.

fausse, *a. fem.* of faux.

faussement, *adv.* falsely. — *m.* bending, etc. (see fausser).

fausser, *v.t.* bend, warp, spring; vitiate, affect adversely (as an analytical result); strain, distort; break (one's word, etc.). — *v.r.* be bent, be warped, etc.; bend, warp, spring.

fausset, *m.* spigot, peg, pin.

fausseté, *f.* falsity, falseness; falsehood.

faut, *pr.* of falloir. — *pr. 3 sing.* of faillir.

faute, *f.* fault; default, lack, want.

  faire —, fail.

  — *de*, for lack of, for want of.

  sans —, without fail; without fault, faultless.

fautif, *a.* faulty, imperfect.

fautivement, *adv.* erroneously, wrongly.

fautre, *m.* *(Paper)* felt.

fauve, *a.* fallow, fawn, tawny, yellowish brown, (of leather) fair. — *m.* fawn color, yellowish brown.

faux, *a.* false; imitation, counterfeit, mock. — *m.* falseness; imitation; forgery; falsehood; scythe. — *adv.* falsely, erroneously. — *pr. 1 & 2 sing.* of faillir.

  à —, falsely, wrongly; unevenly; out of perpendicular.

  fausse améthyste, a purple variety of fluorspar.

  fausse oronge, fly agaric, fly amanita (*Amanita muscaria*).

  — *argent*, mica.

  — *cumin*, nutmeg flower (*Nigella sativa*).

  — *diamant*, zircon.

  — *fenouil*, deadly carrot (*Thapsia garganica*).

  — *gambier*, the root of *Dioscorea rhipogonoides*, used in dyeing.

  — *jour*, false light; *(Micros.)* any light that interferes with observation.

  — *teint*, fugitive (color or dye).

faux-frais, *m.pl.* incidental expenses.

faux-fuyant, *m.* shift, evasion; by-path.

faux-jour, *m.* See faux jour (under *faux*).

faveur, *m.* favor.

  à la — *de*, by means of, by help of.

  en — *de*, for the sake of, in consideration of, in behalf of.

favorable, *a.* favorable.

favorablement, *adv.* favorably.

favori, *a. & m.* favorite.

favoriser, *v.t.* favor; assist.

favorite, *fem.* of favori.

fayard, *m.* beech, beech tree.

fayence, fayencerie, etc. See faïence, etc.

fébrifuge, *a. & m.* *(Med.)* febrifuge.

fécal, *a.* fecal.

féces, *m.pl.* feces, fæces.

fécond, *a.* fertile, fruitful; fecund, prolific; abundant; genial.

fécondant, *p.a.* fertilizing, rendering fertile; genial.

fécondateur, *a.* fecundating.

fécondation, *f.* fecundation.

féconder, *v.t.* fecundate; fertilize.

fécondité, *f.* fecundity; fertility.

fécule, *f.* starch (especially in its domestic and medicinal uses; cf. amidon), fecula; specif., potato starch.

  — *de blé*, wheat starch.

  — *de maïs*, corn starch, maize starch.

  — *de pommes de terre*, potato starch.

féculence, *f.* feculence.

féculent, *a.* starchy, containing starch; feculent, muddy, turbid. — *m.* starchy material; starchy vegetable.

féculer, *v.t.* convert into starch or fecula.

féculerie, *f.* starch factory; starch manufacture.

féculeux, *a.* starchy, containing starch.

féculier, *m.* starch maker, starch manufacturer.

féculomètre, *m.* feculometer (instrument for determining starch).

feignant, *p.pr.* of feindre.

feindre, *v.t.* feign; scruple.

feint, *p.a.* feigned; imitation, counterfeit.

feldspath, *m.* *(Min.)* feldspar.

  — *adulaire*, adularia.

  — *argiliforme*, kaolin.

  — *aventuriné*, aventurine feldspar.

  — *nacré*, pearly feldspar, moonstone.

  — *opalin*, opaline feldspar, labradorite.

feldspathique, *a.* feldspathic.

fêle, *f.* *(Glass)* blowpipe.

fêler, *v.t. & r.* crack.

félin, *a.* feline. — *m.* cat (one of the Felidæ).

felle, *f.* = fêle.

fêlure, *f.* crack, chink.

F.E.M., *abbrev.* (force électromotrice) electromotive force, E.M.F.

femelle, *f. & a.* female.

femme, *f.* woman; wife.

fenchène, *m.* fenchene.

fenchol, *m.* fenchol.

fenchone, *f.* fenchone.

fenchyle, *m.* fenchyl.

fendage, *m.* splitting, etc. (see fendre).

fenderie, *f.* slitting, cutting (of iron) into narrow pieces; slitting machine; slitting mill.

fendeur, *m.* splitter, etc. (see fendre).

fendille, *f.* little crack.

fendillement, *m.* cracking, splitting, chinking.

fendiller, *v.t. & r.* crack, split, chink, chap.

fendre, *v.t.* split; cleave; rend; slit; crack, chink, chap; cut thru. — *v.r.* split, be split, part, be rent; crack, chink, chap, splinter.

fendu, *p.a.* split, cleft, etc. (see fendre).

fenêtre, *f.* window; opening; (*Mach.*) port.

fenêtrer, *v.t.* perforate; provide with windows.

fénolone, fénone, *f.* fenchone.

fénolyle, *m.* fenchyl.

fenouil, *m.* (*Bot.*) fennel.

— *amer*, bitter fennel, wild fennel.

— *bâtard*, dill.

— *d'eau*, water fennel (*Œnanthe phellan-drium*).

— *d'Italie*, Roman fennel, sweet fennel.

— *doux*, sweet fennel.

— *marin*, sea fennel, samphire (*Crithmum maritimum*).

— *puant*, dill.

fenouillette, *f.* fennel water, fennel brandy.

fente, *f.* slit, slot; split, crack, chink, cleft, fissure, crevice; cut, groove.

fenugrec, *m.* fenugreek (*Trigonella fœnum-grœcum*).

fer, *m.* iron; (also *pl.*) a kind of spring pincers used in making crystal glass; (iron) tool; horseshoe.

— *aciéré*, steeled iron.

— *aciéreux*, steely iron.

— *affiné*, refined iron.

— *à grain*, granular iron.

— *à gros grains*, coarse-grained iron.

— *aigre*, brittle iron, cold-short iron.

— *à la catalane*, Catalan iron (wrought iron made from the ore).

— *à la française*, iron made by a method similar to the Catalan process.

— *à loupe*, bloom iron.

— *à massiau*, see massiau.

— *à nerf*, fibrous iron.

— *argileux*, (*Min.*) iron clay, clay ironstone.

— *à souder*, soldering iron.

— *au bois*, — *au charbon de bois*, charcoal iron.

— *battu*, forged iron, wrought iron.

— *blanc*, white cast iron, white iron (see also fer-blanc).

— *Bravais*, a colloidal solution of ferric hydroxide obtained by dialyzing a solution of it in ferric chloride, Bravais iron.

— *brûlé*, burnt iron, overburnt iron.

— *brut*, crude iron; specif., puddled bar.

— *carbonaté*, iron carbonate.

— *carbonaté argileux*, (*Min.*) siderite mixed with clay, clay ironstone.

— *carbonaté spathique*, (*Min.*) spathic iron ore, siderite.

— *carbonaté terreux*, (*Min.*) clay ironstone.

— *carbonyle*, iron carbonyl.

— *carburé*, iron containing carbon; iron carbide.

— *cassant*, brittle iron, short iron.

— *cassant à chaud*, hot-short iron, red-short iron.

— *cassant à chaud noir*, black-short iron.

fer *cassant à froid*, cold-short iron.

— *cémenté*, cement iron, cement steel.

— *cendreux*, iron that is flawy from ash spots.

— *chromaté*, iron chromate.

— *chromé*, (*Min.*) chrome iron ore, chrome iron, chromite.

— *corroyé*, wrought iron.

— *coulé*, cast iron, pig iron.

— *creux*, tubular iron.

— *cru*, crude iron.

— *cylindré*, rolled iron.

— *de ferraille*, scrap iron, old iron.

— *de fonte*, cast iron.

— *de forge*, wrought iron.

— *de lance*, spearhead, lance head.

— *de lune*, white-hot iron.

— *de masse*, scrap iron.

— *de riblons*, scrap iron.

— *doux*, soft iron.

— *écroui*, hammer-hardened iron.

— *émaillé*, enameled iron.

— *en bandes*, hoop iron.

— *en barres*, bar iron.

— *encrené*, iron twice heated and hammered.

— *en feuilles*, sheet iron.

— *en gueuse*, pig iron.

— *en lames*, sheet iron.

— *en limaille*, iron filings.

— *en loupes*, bloom iron.

— *en plaques*, iron plate, boiler plate.

— *en rubans*, hoop iron.

— *en saumon*, pig iron.

— *en tournure*, iron turnings.

— *étamé*, tinned iron; plated iron.

— *étiré*, rolled iron.

— *fendu*, slit iron.

— *fibreux*, fibrous iron.

— *fin*, refined iron.

— *fin au bois*, charcoal iron.

— *fondu*, ingot iron; cast iron (usually called *fonte*); molten iron.

— *fondu à air chaud*, hot-blast iron.

— *fondu à air froid*, cold-blast iron.

— *forgé*, wrought iron.

— *fort*, strong iron, forge iron.

— *galvanisé*, galvanized iron.

— *gratté*, filed iron, brushed iron.

— *gris*, gray iron.

— *hydraté limoneux*, (*Min.*) bog iron ore (porous limonite).

— *laminé*, rolled iron, laminated iron.

— *limoneux*, (*Min.*) limonite.

— *magnétique*, (*Min.*) magnetite, magnetic iron.

— *malléable*, malleable iron.

— *manganésé*, (*Min.*) triplite.

— *marchand*, merchant iron, merchant bar.

— *martelé*, hammered iron.

— *mêlé*, mottled iron (cast iron between white and gray).

fer *métallique*, metallic iron.
— *météorique*, meteoric iron.
— *métis*, red-short iron.
— *micacé*, (*Min.*) micaceous iron ore, scaly hematite.
— *mitis*, mitis metal.
— *mou*, soft iron.
— *natif*, native iron.
— *nerveux*, fibrous iron.
— *noir*, black iron, esp. black sheet iron.
— *oligiste*, (*Min.*) hematite, oligist iron (the latter name is infrequent).
— *oölithique*, (*Min.*) oölitic iron ore.
— *oxydé*, iron oxide, oxide of iron.
— *oxydé rouge*, (*Min.*) red iron ore, ordinary hematite.
— *oxydulé*, (*Min.*) magnetite.
— *pailleux*, flawy iron.
— *pisiforme*, (*Min.*) pea ore, pisiform iron ore (granular form of limonite).
— *plat*, hoop iron.
— *platiné*, platinized iron, specif. platinum-plated iron; rolled iron.
— *puddlé*, puddled iron.
— *raffiné*, refined iron.
— *réduit*, (*Pharm.*) reduced iron, powder of iron.
— *roulé*, rolled iron.
— *rouverin*, hot-short (or red-short) iron.
— *soudé*, weld iron.
— *spathique*, (*Min.*) spathic iron, siderite.
— *spéculaire*, (*Min.*) specular iron, hematite (exhibiting metallic luster).
— *tenace*, tough iron.
— *tendre*, cold-short iron.
— *titané*, (*Min.*) titanic iron ore, ilmenite.
— *zingué*, galvanized iron.
fera, *fut. 3 sing.* of faire.
ferait, *cond. 3 sing.* of faire.
féramine, *f.* (*Min.*) pyrite, iron pyrites. *Obs.*
fer-à-nerf, *m.* fibrous iron.
fer-béton, *m.* ferro-concrete.
fer-blanc, *m.* tin plate.
— *brillant*, tin plate proper (as distinguished from terneplate).
— *terne*, terneplate.
ferblanterie, *f.* tinwork; tinware; tinware business.
ferblantier, *m.* one who makes or sells tinware.
féret, *m.* = ferret.
fermail, *m.* clasp.
fermant, *p.pr.* of fermer.
ferme, *a.* firm; binding (sale). — *adv.* firmly.
— *m.* farm; farmhouse; lease; girder, truss, rib, frame; roof.
fermé, *p.a.* closed, etc. (see fermer); close, tight.
ferme-circuit, *m.* (*Elec.*) circuit closer.
fermement, *adv.* firmly.
ferment, *m.* ferment. — *a.* fermentative.
— *figuré*, — *organisé*, organized ferment.

fermentable, *a.* fermentable.
fermentaire, fermentatif, *a.* fermentative, fermenting.
fermentation, *f.* fermentation; (fig.) ferment.
— *à chapeau*, top fermentation in which the head is agitated by the escaping gas.
— *alcoolique*, alcoholic fermentation.
— *ascendante et descendante*, fermentation in which the head alternately rises and falls.
— *basse*, low (or bottom) fermentation.
— *circulaire*, fermentation characterized by a vertical circulation, the solid matter rising in one part of the vat and falling in another.
— *diastasique*, diastatic fermentation; enzymic fermentation.
— *en bulles*, fermentation characterized by the formation of bubbles of good size in the head.
— *haute*, high (or top) fermentation.
— *mousseuse*, fermentation characterized by excessive foaming.
— *nitreuse*, nitrous fermentation.
— *panaire*, fermentation of bread dough, leavening.
— *par dépôt*, bottom fermentation.
— *sous le chapeau*, top fermentation which proceeds quietly without agitation of the head.
— *superficielle*, top fermentation.
— *tumultueuse*, violent fermentation, specif. the first fermentation of wine.
fermenter, *v.i. & t.* ferment.
fermentescibilité, *f.* fermentability.
fermentescible, *a.* fermentable.
fermer, *v.t.* close; shut; fasten, bolt, pin, lock; stop, stop up. — *v.r. & i.* close, shut, be closed.
fermeté, *f.* firmness.
fermeture, *f.* closing, closure, etc. (see fermer); seal; fastening (of any kind); shutter, blind; (*Steam*) cut-off; (*Ordnance*) fermeture.
— *à rodage*, ground joint, ground seal.
— *hermétique*, hermetic closing, hermetic seal.
fermier, *m.* farmer.
fermoir, *m.* clasp, catch, snap; chisel.
ferraille, *f.* (also *pl.*) old iron, scrap iron.
ferrasse, *f.* an iron vessel used in fusing metals; (*Glass*) an iron plate, as one on which the glass rests in a flattening furnace, one on which glass is cast, or one on which the pots are set.
ferrate, *m.* ferrate.
ferre, *f.* fine iron grindings; (*Glass*) bottle pincers.
ferré, *a.* impregnated with iron, chalybeate; fitted or bound with iron; ironed; (of roads) metaled, macadamized; (of horses) shod; versed (in); harsh, hard.

ferrer, *v.t.* fit or mount with iron; iron, smooth; metal, macadamize (a road); shoe (a horse).
ferrement, *m.* ironwork; iron tool or piece.
ferrerie, *f.* ironwork; iron business, iron trade.
ferret, *m.* hard kernel in stones; (*Glass*) ferret; tag (as of a shoe lace).
— *d'Espagne*, (*Min.*) red iron ore, hematite.
ferreux, *a.* ferrous.
ferrico-. ferric (in double salts).
ferrico-ammonique, *a.* ferric ammonium.
ferrico-potassique, *a.* ferric potassium.
ferricyanhydrique, *a.* ferricyanic.
ferricyanogène, *m.* ferricyanogen.
ferricyanure, *m.* ferricyanide.
ferrifère, *a.* ferriferous, iron-bearing.
ferrique, *a.* ferric.
ferrite, *n.* ferrite.
ferro, *m.* iron alloy (as ferromanganese, etc.); (*Photog.*) blue print.
ferro-aluminium, *m.* ferro-aluminium, ferro-aluminum.
ferro-chrome, *m.* ferrochrome, ferrochromium.
ferro-ciment, *m.* ferro-concrete.
ferrocyanate, *m.* ferrocyanate, ferrocyanide.
ferrocyanhydrique, *a.* ferrocyanic.
ferrocyanique, *a.* ferrocyanic.
ferrocyanogène, *m.* ferrocyanogen.
ferrocyanure, *m.* ferrocyanide.
ferro-manganèse, *m.* ferromanganese.
ferromolybdène, *m.* ferromolybdenum.
ferronerie, *f.* iron foundry; ironworks; hardware.
ferro-nickel, *m.* ferronickel; specif., nickel steel (when the percent of nickel is low).
ferroprussiate, *m.* ferroprussiate (ferrocyanide).
ferro-silicium, *m.* ferrosilicon.
ferroso-ferrique, *a.* ferrosoferric.
ferrotier, *m.* glass worker.
ferro-titane, ferrotitan, *m.* ferrotitanium.
ferro-tungstène, *m.* ferrotungsten.
ferrotypie, *f.* (*Photog.*) ferrotype, tintype.
ferrugineux, *a.* ferruginous, chalybeate. — *m.* ferruginous medicine, chalybeate.
ferrugino-calcaire, *a.* pertaining to or containing iron and lime, ferrugino-calcareous.
ferruginosité, *f.* ferruginous quality.
ferrure, *f.* ironwork; shoeing, shoes (of horses).
fertile, *a.* fertile.
fertilement, *adv.* fertilely, fruitfully.
fertilisable, *a.* fertilizable.
fertilisant, *p.a.* fertilizing.
fertilisation, *f.* fertilization.
fertiliser, *v.t.* fertilize.
fertilité, *f.* fertility, fertileness.
férule, *f.* ferule; (*Bot.*) Ferula.
férulique, *a.* ferulic.
feston, *m.* festoon.
festonner, *v.t.* festoon.
fête, *f.* feast, festival, fête; birthday.

fétide, *a.* fetid.
fétidité, *f.* fetidness, fetidity.
fétu, *m.* straw.
fétus, *m.* fetus, fœtus.
feu, *m.* fire; flame; light; fireplace, hearth; household, home; heat. — *a.* flame (color); late, deceased.
à — *nu*, with the naked or free flame (not using a sand bath, oil bath or other protection).
en —, on fire; (of a furnace) in operation, in blast.
— *Catalan*, (*Metal.*) Catalan forge.
— *d'affinerie*, (*Metal.*) fining forge.
— *d'artifice*, firework, fireworks.
— *de Bengale*, Bengal light.
— *d'éclairage*, light.
— *de couleur*, colored light.
— *de raffinerie*, (*Metal.*) refining hearth.
— *de réduction*, reducing flame.
— *de signaux*, signal light, signal firework.
— *d'oxydation*, oxidizing flame.
— *follet*, ignis fatuus.
— *grégeois*, Greek fire.
— *grisou*, fire damp.
— *nu*, naked fire, naked flame, free flame.
mettre à —, fire, start up (a furnace); put in blast, blow in (a blast furnace).
mettre en —, burn; = mettre à feu.
mettre hors de —, *mettre hors* —, (*Metal.*) put out of blast.
mettre le — à, set on fire, touch off.
feuillage, *m.* foliage.
feuille, *f.* leaf; foil; sheet; plate; scale, flake; list; bill; newspaper.
— *anglaise*, (*Rubber*) cut sheet.
— *d'argent*, silver leaf, silver foil.
— *de cuivre*, sheet of copper.
— *de fer*, plate or sheet of iron.
— *de fer-blanc*, sheet of tin (plate).
— *d'étain*, tinfoil.
— *de ventre*, the middle layer in pasteboard.
— *de verre*, sheet of glass.
— *d'or*, gold leaf, gold foil.
— *laminée*, rolled sheet.
— *sciée*, (*Rubber*) cut sheet.
— *volante*, loose sheet.
feuille-morte, *a. & m.* dull yellow (resembling the color of dead leaves).
feuillet, *m.* leaf; layer; thin plate; (of crystals) leaf, plate; folio.
feuilleté, *a.* foliated. — *p.p.* of feuilleter.
feuilleter, *v.t.* turn the leaves of, peruse.
feuilletis, *m.* edge (of a cut stone); soft place in slate.
feuillette, *f.*, small leaf, leaflet; cask (containing about 125 liters).
feuillir, *v.i.* put forth leaves, leaf, leave.
feuillu, *a.* leafy; specif., having broad leaves as distinguished from needles.

feuillure, *f.* rabbet, groove, slit, recess.

feutrable, *a.* capable of being felted.

feutrage, *m.* felting.

feutrant, *p.a.* felting.

feutre, *m.* felt; padding, packing.

feutré, *p.a.* felted; felt; padded, packed.

feutrement, *m.* felting.

feutrer, *v.t.* felt; pad, pack (as with felt).

fève, *f.* bean (refers specif. to the genus *Faba;* cf. haricot).

— *de cacao,* cocoa bean.

— *de Calabar,* Calabar bean (from *Physostigma venenosum*).

— *de haricot,* kidney bean.

— *de Saint-Ignace,* Saint Ignatius's bean (from *Strychnos ignatia*).

— *de soja,* soy bean, soja bean.

— *igasurique,* = fève de Saint-Ignace.

— *pichurim,* pichurim bean (from *Nectandra puchury*).

— *tonka,* — *Tonka,* tonka bean.

féverole, *f.* (dried) bean; horse bean.

févier, *m.* (*Bot.*) locust tree (*Gleditsia*).

février, *m.* February.

fibre, *f.* fiber, fibre; grain (of wood).

— *de bois,* wood fiber; grain of wood.

— *libérienne,* (*Bot.*) bast fiber, bast.

— *ligneuse,* wood fiber.

— *textile,* textile fiber.

— *vulcanisée,* vulcanized fiber.

fibré, *a.* fibered, fibrous.

fibreux, *a.* fibrous.

fibrillaire, *a.* fibrillar, fibrillary.

fibrille, *f.* small fiber, fibril, fibrilla.

fibrine, *f.* fibrin.

fibrineux, *a.* fibrinous.

fibrinogène, *m.* fibrinogen.

fibrinolyse, *f.* fibrinolysis.

fibrocartilage, *m.* (*Anat.*) fibrocartilage.

fibrocartilagineux, *a.* fibrocartilaginous.

fibrocellulaire, *a.* fibrocellular, fibroareolar.

fibro-ciment, *m.* a building material consisting of cement mixed with asbestos fiber.

fibrolite, fibrolithe, *f.* (*Min.*) fibrolite.

ficeler, *v.t.* tie, tie up (with string).

ficelle, *f.* string, twine, packthread; trick.
— *pr. 1 & 3 sing.* of ficeler.

fiche, *f.* (something to be thrust or driven in) pin, peg, bolt, plug, stake, etc.; ticket, card; memorandum (on a card); knife or plug (of an electric switch).

ficher, *v.t.* thrust, stick in, plunge in, drive in, insert; point (masonry).

fidèle, *a.* faithful.

fidèlement, *adv.* faithfully.

fidélité, *f.* fidelity, faithfulness.

fiel, *m.* gall.

— *de bœuf,* ox gall.

fielleux, *a.* like gall, bitter as gall.

fiente, *f.* dung.

fienter, *v.t.* dung, manure.

fier, *a.* proud; fierce; fiery; great. — *v.r.* trust, confide (in).

fièvre, *f.* fever.

fiévreux, *a.* feverish. — *m.* fever patient.

figement, *m.* congealment, congealing, setting, solidifying, solidification.

figer, *v.t. & r.* congeal, set, solidify.

figue, *f.* fig.

— *de Barbarie,* Barbary fig, prickly pear (*Opuntia opuntia*).

figuier, *m.* fig tree, fig.

figuratif, *a.* figurative; representing or symbolizing something. — *m.* representation, diagram.

figure, *f.* figure; face, countenance.

figurer, *v.t.* represent; figure. — *v.r.* be represented or figured. — *v.i.* appear, figure; look.

fil, *m.* wire; thread; filament, fiber; string; yarn; edge (of sharp tools); grain (of wood); vein (in stones); course (as of a stream).

— *à plomb,* plumb line.

— *conducteur,* (*Elec.*) conducting wire, conductor.

— *couvert,* (*Elec.*) insulated wire.

— *d'acier,* steel wire.

— *d'aller,* (*Elec.*) lead wire.

— *d'amorce,* priming wire.

— *d'araignée,* (*Optics*) spider line, cross wire.

— *d'archal,* brass wire.

— *de chauffage,* (*Elec.*) resistance wire.

— *de clavecin,* piano wire.

— *de cuivre,* copper wire.

— *de fer,* iron wire.

— *de laiton,* brass wire.

— *de ligne,* (*Elec.*) line wire.

— *d'emballage,* pack thread, wrapping twine.

— *de platine,* platinum wire.

— *de retour,* (*Elec.*) return wire.

— *de soie,* silk thread.

— *fusible,* (*Elec.*) fuse wire.

— *métallique,* wire.

— *nu,* (*Elec.*) uninsulated wire.

— *recouvert,* (*Elec.*) insulated wire.

— *retors,* twine; double yarn.

— *thermique,* (*Elec.*) heating wire.

— *tranchant,* cutting edge, sharp edge.

filage, *m.* spinning, etc. (see filer); (*Metal.*) addition of fresh lead in the cupellation of argentiferous lead; (*Bot.*) Filago.

filament, *m.* filament.

— *de charbon,* carbon filament.

— *de maïs,* corn silk, (*Pharm.*) zea.

filamenteux, *a.* filamentous.

filandre, *f.* thread, string, filament, specif.: a defect in glass; gossamer; a soft vein in marble.

filandreux, *a.* stringy, thready, filamentous; streaky, veiny.

filant, *p.pr.* of filer; (of liquids) ropy.

filardeux, *a.* (of stones) veiny, containing veins of softer material, streaky.

filasse, *f.* tow; (in general) unspun fibers.

filature, *f.* spinning; spinning mill; twine mill.

file, *f.* file; row.

filé, *p.p.* of filer. — *m.* thread.

filer, *v.t.* spin; wiredraw, draw; run, draw (lines); spin out; conduct. — *v.i.* (of liquids) flow in a viscous stream, be ropy, rope; spin; (of stars) shoot; (of lamps) smoke; slip away; purr.

filet, *m.* thread, fiber, filament; (of a screw) thread; (of a liquid) fine stream; net, netting, network; dash, stroke; (*Printing*) rule.

— *à droite,* right-handed thread (of screws).

— *à gauche,* left-handed thread.

— *de vis,* screw thread.

filetage, *m.* threading (of screws); wire-drawing.

fileter, *v.t.* thread, cut a thread on; wiredraw, draw.

filière, *f.* drawplate; screw plate; a perforated plate for molding plastic substances; wire gage.

filiforme, *a.* filiform.

filigrane, *m.* filigree; (*Paper*) watermark. — *a.* watermarked.

filigrané, *p.a.* filigreed, filigree; (*Paper*) watermarked.

filin, *m.* rope, cordage.

fille, *f.* daughter; girl; maid; nun.

filon, *m.* (*Geol. & Mining*) vein, lode.

filon-couche, *n.* vein-bearing rock.

filonien, *a.* pertaining to or containing veins or lodes (as, *gîte filonien,* vein deposit).

filoselle, *f.* filoselle (coarse silk thread).

fils, *m.* son; boy. — *pl.* of fil.

filtrage, *m.* filtering, filtration.

filtrant, *p.a.* filtering, filter.

filtrateur, *a.* filtering. — *m.* filterer.

filtration, *f.* filtration, filtering.

filtratum, *m.* filtrate.

filtre, *m.* filter; strainer. — *a.* filter, filtering.

— *à analyse,* analytical filter, (circle of) filter paper.

— *à aspiration,* suction filter.

— *à bougie,* candle filter, cylinder filter.

— *à eau,* water filter.

— *à l'amiante,* asbestos filter.

— *à plis,* plaited filter, folded filter.

— *à pression,* pressure filter.

— *à sable,* — *au sable,* sand filter.

— *à vide,* vacuum filter.

— *Berkefeld,* Berkefeld filter.

— *centrifuge,* centrifugal filter.

— *Chamberland,* Chamberland filter.

— *clos,* closed (or inclosed) filter.

— *dégrossisseur,* preliminary filter.

filtre *en charbon,* charcoal filter.

— *en papier,* paper filter.

— *finisseur,* final filter.

— *mécanique,* mechanical filter.

— *plissé,* plaited filter, folded filter.

— *sans pli,* unplaited filter.

— *sans cendres,* ashless filter.

filtre-fontaine, *m.* a large vessel containing a filter and having a cock at the bottom.

filtre-presse, *m.* filter press.

filtrer, *v.t. & i.* filter. — *v.r.* be filtered, filter.

— *dans le vide,* filter in vacuo, filter with suction.

filtreur, *m.* filterer, strainer.

fin, *a.* fine; (of a cutting edge or an optical image) sharp; slender; acute. — *m.* end; fine metal; fine coal, coal dust; fine point, essence; fine linen. — *adv.* quite.

à la —, at last; in the end; after all.

en — *de compte,* in the end, finally, after all.

fine champagne, a superior variety of brandy.

finage, *m.* (*Metal., etc.*) fining.

final, *a.* final.

finalement, *adv.* finally.

financiel, *a.* financial.

financier, *a.* financial. — *m.* financier.

financièrement, *adv.* financially.

fine, *f.* = fine champagne (under *fin*); (*pl.*) fines (specif., coal dust). — *a. fem.* of fin.

finement, *adv.* finely.

fine-métal, *m.* (*Iron*) white pig.

finerie, *f.* (*Iron*) finery, refinery.

finesse, *f.* fineness; sharpness (as of an optical image); keenness; nicety; delicacy, finesse, subtlety.

fin-fin, *a.* very fine, extra fine.

fini, *p.a.* finished; finite. — *m.* finish; finite.

finir, *v.t.* finish; end, complete. — *v.i.* end; cease, terminate. — *v.r.* end, cease; be finished, be completed.

finissage, *m.* finishing.

finisseur, *m.* finisher. — *m.* finishing, final.

fiole, *f.* flask; (less often) bottle; phial, vial.

— *à attaques,* a pear-shaped flat-bottomed flask used in analysis for decompositions.

— *à gouttes,* dropping bottle.

— *à jet,* washing bottle, wash bottle.

— *à matières grasses,* fat-extraction flask.

— *à vide,* filtering flask.

— *ballon,* round flask (any flask with more or less nearly spherical body), balloon flask (when short-necked).

— *ballon à fond plat,* flat-bottomed flask, Florence flask.

— *bouchée à l'émeri,* flask with ground stopper, glass-stoppered flask.

— *conique,* conical flask, specif. Erlenmeyer flask.

— *d'Erlenmeyer,* Erlenmeyer flask.

— *d'extraction,* extraction flask.

fiole *en Bohême,* flask of Bohemian glass.
— *en Iéna,* flask of Jena glass.
— *étalon,* standard flask, normal flask.
— *forme bouteille,* bottle-shaped flask.
— *forme poire,* pear-shaped flask.
— *jaugée,* graduated flask, volumetric flask.
— *tubulée,* tubulated flask.
fiole-ballon, *n.* = fiole ballon, under *fiole.*
fiorite, *f.* (*Min.*) fiorite.
firent, *p.def. 3 pl.* of faire.
firme, *f.* (*Com.*) firm.
fis, *p.def. 1 & 2 sing.* of faire.
fisc, *m.* State treasury, exchequer, fisc.
fiscal, *a.* fiscal.
fissent, *imp. 3 pl. subj.* of faire.
fissile, *a.* fissile.
fissilité, *f.* fissility.
fissuration, *f.* fissuration.
fissure, *f.* fissure.
fissurer, *v.t. & r.* fissure.
fistule, *f.* fistula.
fistuleux, *a.* fistulous, fistular.
fit, *p.def. 3 sing.* of faire.
fît, *imp. 3 sing. subj.* of faire.
fixage, *m.* fixing.
fixateur, *m.* fixing agent (as *Photog.*); fixative; stabilizer; fixer. — *a.* fixing; stabilizing.
fixe, *a.* fixed (specif., not volatilizable). — *m.* fixed substance; fixed star; fixed salary.
fixé, *p.a.* fixed.
fixement, *adv.* fixedly.
fixer, *v.t.* fix. — *v.r.* fix, become fixed.
fixité, *f.* fixity, fixedness; stability.
Fl. Symbol for fluorine.
flache, *f.* flaw; crack; depression; inequality.
flacheux, *a.* flawy.
flacon, *m.* bottle.
— *à baume du Canada,* balsam bottle.
— *à bec,* lipped bottle.
— *absorbeur,* absorption bottle.
— *à capsule à vis,* bottle with screw cap.
— *à collections,* specimen bottle, museum bottle.
— *à colorants,* staining bottle.
— *à cuvette,* a bottle having a depression around the base of the neck to catch liquid that runs down.
— *à densité,* specific gravity bottle, pycnometer.
— *à double bouchage,* a bottle with a cap fitting on over the stopper.
— *à étroite ouverture,* narrow-mouthed bottle.
— *à lait,* milk bottle.
— *à large ouverture,* a (comparatively) wide-mouthed bottle (a very wide-mouthed bottle, as one for salts, is called *bocal*).
— *à l'émeri,* = flacon bouché à l'émeri.
— *à pied,* a bottle with a special base (as for museum use).
— *à pression,* pressure bottle.

flacon *à réactifs,* reagent bottle.
— *à robinet,* bottle with stopcock (usually at the bottom).
— *à saccharification,* saccharification bottle, pressure bottle.
— *à sirop,* sirup bottle (esp. a wide-mouthed lipped bottle with sloping shoulder).
— *à tare,* weighing bottle, weighing tube.
— *bouché à l'émeri,* bottle with ground stopper, glass-stoppered bottle.
— *camphored,* a sort of wide-mouthed bottle.
— *compte-gouttes,* dropping bottle.
— *conique,* conical bottle.
— *de Woulfe,* — *de Woolf,* Woulfe bottle.
— *gramme,* gram bottle, specific gravity bottle.
— *laveur,* washing bottle (for gases).
— *marin,* a stocky square bottle.
— *pour réactifs,* reagent bottle.
— *prismatique,* a wedge-shaped bottle used in color tests.
— *récepteur,* receiving bottle.
— *tubulé,* tubulated bottle.
flair, *m.* smell, scent; sense of smell.
flairer, *v.t.* smell; smell out.
flamand, *a.* Flemish. — *m.* Flemish (language); (*cap.*) Fleming.
flambage, *m.* flaming, etc. (see flamber).
flambart, *m.* grease skimmed from the surface of water in which meat (esp. pork) is cooked; half-consumed charcoal.
flambé, *p.a.* flamed, etc. (see flamber); (*Ceram.*) flambé. — *m.* (*Ceram.*) flambé glaze.
flambeau, *m.* torch; candle; candlestick; light.
flamber, *v.t.* flame, pass thru or before fire; sterilize (instruments, etc. with a flame); singe (cloth, etc.). — *v.i.* flame, blaze; yield (to pressure).
flambeur, *m.* (*Textiles*) singer.
flamboiement, *m.* flaming, etc. (see flamboyer).
flamboyer, *v.i.* flame, blaze, flash, glitter.
flambure, *f.* (*Dyeing*) unevenness of coloring.
flamme, *f.* flame; streamer, pennant.
— *de Bengale,* Bengal light.
— *éclairante,* luminous flame.
— *nue,* naked flame, free flame.
— *oxydante,* — *d'oxydation,* oxidizing flame.
— *réductrice,* — *de réduction,* reducing flame.
flammé, *a.* flamelike, wavy; (*Ceram.*) flambé. — *m.* (*Ceram.*) flambé glaze.
flammèche, *f.* flaming or burning particle, spark; (in the Leblanc soda process) candle.
flammette, *f.* small flame.
flan, *m.* disk, plate, planchet.
flanc, *m.* flank, side.
Flandre, *f.* Flanders.
flanelle, *f.* flannel.

flanquer, *v.t.* flank; throw, toss.

flasque, *a.* flabby, flaccid, soft, weak. — *f.* flask (as for mercury). — *m.* plate, side plate, cheek plate; support.

flatter, *v.t.* flatter; soothe; caress.

flatteur, *a.* flattering; caressing.

flatueux, *a.* flatulent.

flatuosité, *f.* flatulence, flatulency.

flavaniline, *f.* flavaniline.

flavone, *f.* flavone.

flavopurpurine, *f.* flavopurpurin.

fléau, *m.* beam (e.g., of a balance); bar; flail; (fig.) scourge.

flèche, *f.* arrow; sprout, shoot; (of sugar cane) top, head; deflection, bend, sag; (of an arch or trajectory) height; spire.

fléchir, *v.t.* bend; (fig.) move. — *v.i.* bend; yield, give way; (of prices) fall. — *v.r.* bend; yield.

fléchissement, *m.* bending, etc. (see fléchir); deflection; (of an arch) height.

flegme, *m.* phlegm; (*Distilling*) low wines.

flétrir, *v.t.* wither, fade; brand. — *v.r.* wither, fade.

flétrissure, *f.* withering, fading; brand; stigma.

fletter, *v.t.* smooth, dress, esp. on the edges (as glass or pottery).

fleur, *f.* flower; surface; bloom (on fruits, etc.); hair side (of a hide or skin); mold (on wine, etc.); inner bark, part of the bark richest in tannin; fine flour.
à — de, level with, even with, flush with; near.
— d'alun, (*Min.*) feather alum, alunogen.
— d'antimoine, flower of antimony, antimonious acid.
— de chaux, (*Min.*) a powdery form of calcite.
— de cinabre, a powdery form of cinnabar (native mercuric sulfide).
— de farine, fine flour.
— de muscade, mace.
— (or fleurs) de soufre, flowers of sulfur.
— (or fleurs) de vin, wine mold (*Mycoderma vini*).
fleurs d'arnique, (*Pharm.*) arnica flowers, arnica.
fleurs d'arsenic, flowers of arsenic, sublimed arsenic trioxide.
fleurs de benjoin, flowers of benzoin (benzoic acid).
fleurs de bismuth, flowers of bismuth, sublimed bismuth trioxide.
fleurs de carthame, (*Pharm.*) safflower.
fleurs de cobalt, (*Min.*) cobalt bloom, erythrite.
fleurs de tous les mois, (*Pharm.*) calendula, marigold.
fleurs de zinc, flowers of zinc, zinc oxide (sublimed).

fleurage, *m.* an opaqueness produced on glass by hydrofluoric acid; grits, groats.

fleuraison, *f.* (*Bot.*) flowering, efflorescence.

fleurant, *a.* fragrant.

fleurée, *f.* foam which forms in dyeing with indigo.

fleurer, *v.i.* smell, give forth an odor.

fleuret, *m.* floss silk; choice; borer, drill, bit.
— de laine, choice wool.

fleuri, *a.* flowery, flowered; florid.

fleurir, *v.i.* flower, bloom; flourish.

fleuron, *m.* flower, ornament.

fleuve, *m.* river.

flexibilité, *f.* flexibility.

flexible, *a.* flexible. — *m.* flexible tube or pipe.

flexion, *f.* bending, flection, flexion; inflection.

flint, flint-glass, *m.* flint glass.

flocon, *m.* flock, floccule, flake; (of smoke) puff.

floconnement, *m.* flocculation.

floconner, *v.i.* flocculate.

floconneux, *a.* flocculent, flocky, flaky.

floculant, *p.a.* flocculating. — *m.* flocculating agent.

floculation, *f.* flocculation.

floculer, *v.t. & r.* flocculate.

floraison, *m.* (*Bot.*) flowering, efflorescence.

floran, florau, *m.* (*Paper*) beating trough, beater.

flore, *f.* flora.

florentin, *a.* Florentine.

Floride, *f.* Florida.

florissait, *imp. 3 sing.* of fleurir.

florissant, *a.* flourishing.

floss, *m.* (*Metal.*) floss: (a) vitrified material floating in a puddling furnace, (b) white cast iron for conversion into steel.

flot, *m.* wave, billow; flood; sea; floating; float.

flotomètre, *m.* an instrument for measuring the height of liquids by means of a float.

flôtre, flotre, *m.* felt.

flottant, *p.a.* floating; flowing, waving; wavering, uncertain.

flotte, *f.* float; buoy; fleet; navy.

flottement, *m.* floating; wavering.

flotter, *v.i.* float; waver, fluctuate. — *v.t.* float.

flotteur, *m.* float. — *a.* floating.
— index, index float.

fluate, *m.* fluate (old name for fluoride).

fluaté, *a.* fluate (fluoride) of. (*Obs.*)

fluctuant, *a.* fluctuating.

fluctuation, *f.* fluctuation.

fluctuer, *v.i.* fluctuate; waver.

fluctueux, *a.* fluctuating.

fluide, *a. & m.* fluid.

fluidement, *adv.* fluidly, with fluidity.

fluidifiant, *a.* fluidifying, fluidizing.

fluidificateur, *a.* fluidifying, fluidizing.

fluidification, *f.* fluidification, fluidizing.
fluidifier, *v.t.* fluidify, fluidize, render fluid.
fluidique, *a.* fluidic, fluid.
fluidité, *f.* fluidity, fluidness; fluency.
fluoborate, *m.* fluoborate.
fluoborique, *a.* fluoboric.
fluor, *m.* fluorine; (*Min.*) fluorite. — *a.* In the phrase *spath fluor,* fluor spar.
fluoranthène, *m.* fluoranthene.
fluoré, *a.* containing fluorine, fluoric.
fluorène, *m.* fluorene.
fluorescéine, *f.* fluorescein.
fluorescence, *f.* fluorescence.
fluorescent, *a.* fluorescent.
fluorescine, *f.* fluorescin.
fluorhydrate, *m.* hydrofluoride.
fluorhydrique, *a.* hydrofluoric.
fluorifique, *a.* containing fluorine, fluoric.
fluorine, *f.* (*Min.*) fluorite, fluor spar.
fluorique, *a.* fluoric (hydrofluoric). (*Obs.*)
fluoroforme, *m.* fluoroform.
fluoromètre, *m.* fluorometer, fluormeter.
fluorure, *m.* fluoride.
 — *de carbone,* carbon fluoride, specif. the tetrafluoride.
fluosel, *m.* a salt which may be regarded as an oxysalt in which fluorine has replaced oxygen partially or completely.
fluosilicate, *m.* fluosilicate.
fluosilicique, *a.* fluosilicic.
flux, *m.* flux; flow.
 — *blanc,* white flux.
 — *noir,* black flux.
 — *réducteur,* reducing flux.
fluxion, *f.* fluxion; inflammation.
F.O., *abbrev.* (flamme oxydante) oxidizing flame.
focal, *a.* focal.
fœtal, *a.* fetal, fœtal.
fœtus, *m.* fetus, fœtus.
foi, *f.* faith; proof.
foie, *m.* liver; (*Old Chem.*) liver, hepar.
 — *d'antimoine,* liver of antimony, hepar anti- monii.
 — *de morue,* cod liver.
 — *de soufre,* liver of sulfur, hepar.
foin, *m.* hay.
fois, *m.* time.
 à la —, at once, at the same time, together.
 de — *à autre,* from time to time.
 deux —, twice.
 une —, once.
 une — *que,* when once; once as.
foison, *f.* abundance. — *adv.* abundantly.
foisonnant, *p.a.* swelling, etc. (see foisonner); abundant, extensive.
foisonner, *v.i.* swell; increase, multiply; abound.
fol. Alternative masc. form of fou.
foliacé, *a.* foliaceous.
foliation, *f.* foliation.

folie, *f.* madness; insanity; folly; passion.
folié, *a.* foliated; (*Bot.*) foliate, leafy.
foliole, *f.* leaflet, foliole.
folle, *a. fem.* of fou.
follicule, *f.* follicle.
fomentation, *f.* fomentation.
fomenter, *v.t.* foment.
fonçage, *m.* deepening, etc. (see foncer).
foncé, *a.* (of colors) deep, dark; deepened, etc. (see foncer).
fonceau, *m.* a table or slab on which glass pots are molded.
foncement, *m.* deepening, etc. (see foncer); esp., the sinking of wells.
foncer, *v.t.* deepen, darken, sadden (a color); dig, sink (a hole or shaft); lower; bottom (as casks); dry and solidify (sugar loaves); prime, size, put the ground coat on.
foncièrement, *adv.* thoroly; at bottom.
fonction, *f.* function; office.
 en — *de,* as a function of.
 — *linéaire,* linear function.
fonctionnaire, *m.* functionary, official.
fonctionnel, *a.* functional.
fonctionnement, *m.* functioning; operation, action.
fonctionner, *v.i.* function; work, operate, act.
fond, *m.* bottom; foundation, base; ground, background; ground coat; further end, back; depth. — *pr. 3 sing.* of fondre. — See also fonds.
 à —, completely, thoroly.
 à — *plat,* flat-bottomed.
 à — *rond,* round-bottomed.
 au —, at bottom, in reality.
 de — *en comble,* completely, entirely.
 faire —, rely, depend.
fondage, *m.* founding.
fondamental, *a.* fundamental; foundation.
fondamentalement, *adv.* fundamentally; completely.
fondant, *m.* flux; (*Med.*) dissolvent; a soft bonbon. — *p.pr.* of fondre. — *a.* (*Med.*) dissolvent.
 — *de couverture,* a covering of flux used as a protection from the air.
 — *réducteur,* reducing flux.
fondation, *f.* foundation.
fondé, *p.a.* founded; well founded; author- ized. — *m.* authorized person.
 — *de pouvoir,* attorney.
fondement, *m.* foundation; reliance.
fonder, *v.t.* found, base. — *v.r.* be founded; rely.
fonderie, *f.* foundry; founding.
 — *de bronze,* bronze foundry (or founding), brass foundry (or founding).
 — *de cuivre jaune,* — *de laiton,* brass foundry, brass founding.
 — *de fer,* iron foundry, iron founding.
fondeur, *m.* founder; melter; smelter; caster.

fondoir, *m.* melting house.
— *de suif,* tallow melting house.
fondre, *v.t.* melt, fuse; smelt; found; cast;
dissolve; (*Med.*) dissolve, resolve; blend.
— *v.i.* melt, fuse; melt away; dissolve;
sink, fall in; fall (upon), rush (upon),
burst. — *v.r.* melt, fuse; melt away;
blend, be blended.
fondrier, *m.* (*Salt*) the end wall of a furnace.
fondrilles, *f.pl.* lees, dregs, sediment.
fonds, *m.* fund; ground, land; soil; (fig.) field;
landed estate; business. — *pl.* of fond.
— *courants,* current funds.
— *de roulement,* working capital.
— *social,* capital (of a company).
fondu, *p.a.* melted, fused, etc. (see fondre).
fondure, *f.* melted grease.
fonger, *v.i.* (of paper) blot.
fongine, *f.* fungin.
fongoïde, *a.* fungoid.
fongueux, *a.* fungous.
fongus, *m.* fungus.
font, *pr. 3 pl.* of faire.
fontaine, *f.* fountain; spring; receptacle or
fixture for delivering liquids, esp. water.
— *à robinet,* water fixture; water reservoir
with cock.
fonte, *f.* cast iron, pig iron, pig; (in general)
cast metal, crude metal; melting, fusion;
melt, fused product; founding; casting;
smelting; blend, blending (as of wools);
lot of skins curried at one time; (of type)
font; (formerly) brass, bronze.
en —, cast-iron.
— *à affiner,* forge pig.
— *à canon,* gun pig.
— *à canon en bronze,* gun metal.
— *à canon en fer,* gun pig.
— *aciérée,* steeled iron, steeled pig.
— *à l'acide,* acid fusion; (*Fats*) rendering in
the presence of acid.
— *à l'air chaud,* hot-blast pig.
— *à l'air froid,* cold-blast pig.
— *à l'alkali,* alkali fusion.
— *à noyau,* cored casting.
— *au bois,— au charbon de bois,* charcoal iron,
charcoal pig.
— *au coke,* coke pig.
— *aux cretons,* rendering of fats by simple
heating, the cracklings being strained off.
— *bigarrée,* mottled pig.
— *blanche,* white cast iron, white pig.
— *blanche miroitante,* specular cast iron.
— *blanchie,* fine iron.
— *caverneuse,* porous cast iron, porous pig.
— *crue,* crude cast metal; specif. pig iron.
— *cuivreuse,* (*Metal.*) a process for obtaining
silver or gold by production of a copper
matte and subsequent separation of the
precious metal.

fonte *d'acier,* cast steel.
— *d'affinage,* forge pig.
— *d'art,* a quality of zinc used in making
imitation bronze; artistic casting.
— *de bois,* charcoal iron, charcoal pig.
— *de chrome,* cast chromium, crude chromium.
— *de fer,* cast iron, pig iron, pig.
— *de manganèse,* cast manganese, crude
manganese.
— *de moulage,* foundry iron, foundry pig.
— *dure,* hard cast iron, chilled cast iron.
— *électrique,* electric smelting.
— *en creux,* hollow casting.
— *en gueuse,* pig iron.
— *en pain de sucre,* (*Glass*) fusion which is
proceeding satisfactorily, as shown by the
resemblance of the batch to a sugar loaf.
— *en saumon,* pig iron.
— *épurée,* refined cast iron.
— *graphiteuse,* graphitic pig.
— *grise,* gray cast iron, gray pig.
— *limailleuse,* very graphitic iron, black
(dark gray) cast iron.
— *maculée,* mottled cast iron, mottled pig.
— *malléable,* malleable cast iron.
— *mêlée,* cast iron of mixed gray and white
color, mottled pig.
— *mitis,* mitis casting; mitis metal.
— *moulée,* cast iron, iron castings.
— *noire,* black (very graphitic) cast iron.
— *plombeuse,* (*Metal.*) = emplombage.
— *pour acier,* cast iron for conversion into steel.
— *raffinée,* refined cast iron, refined pig.
— *rubanée,* mottled cast iron.
— *silicieuse,* siliceous cast iron.
— *spiegel,* spiegeleisen.
— *tendre,* soft (gray) cast iron.
— *truitée,* mottled cast iron.
forage, *m.* boring, drilling.
force, *f.* force; power; strength; much, many;
(*pl.*) shears.
à —, hard, extremely.
à — *de,* by the strength of, by dint of, by
means of.
de —, by force, forcibly.
— *centrifuge,* centrifugal force.
— *centripète,* centripetal force.
— *de cheval,* horsepower.
— *d'impulsion,* momentum.
— *d'inertie,* inertia, vis inertiæ.
— *en chevaux,* horsepower.
— *électromotrice,* electromotive force.
— *morte,* (*Mech.*) vis mortua, dead force.
— *motrice,* motive power.
— *vive,* (*Mech.*) vis viva, living force.
forcement, *m.* forcing.
forcément, *adv.* forcibly; necessarily.
forcer, *v.t.* force.
forcettes, *f.pl.* small shears.

forcite, *f.* (*Expl.*) forcite.
— *antigrisouteuse,* firedamp forcite.
forer, *v.t.* bore, drill.
forerie, *f.* boring, drilling; boring machine, drill; boring house.
forestier, *a.* forest. — *m.* forester.
foret, *m.* drill, borer, boring bit, boring tool, gimlet, auger.
forêt, *f.* forest.
foret-râpe, *m.* a combined drill and rasp for sampling sugar beets.
foreuse, *f.* boring machine, drilling machine.
forfait, *p.a.* forfeited. — *m.* contract; crime.
forge, *f.* forge.
— *à l'anglaise,* rolling mill.
— *d'affinerie,* (*Metal.*) refinery.
forgé, *p.a.* forged; wrought (as iron).
forgeable, *a.* forgeable, malleable.
forgeage, forgement, *m.* forging.
forger, *v.t.* forge, work, hammer.
forgeron, *m.* smith, blacksmith.
forgis, *m.* wire iron.
formaldéhyde, *m.* formaldehyde.
formanilide, *f.* formanilide.
format, *m.* size.
formateur, formatif, *a.* formative; creative.
formation, *f.* formation.
forme, *f.* form; mold; shape; (*pl.*) manners.
formel, *a.* formal.
formellement, *adv.* formally.
formène, *m.* methane.
— *perchloré,* carbon tetrachloride, tetrachloromethane.
forménique, *a.* methane, pertaining to methane.
former, *v.t.* form; make, shape, mold. — *v.r.* form, be formed.
formiamide, *f.* formamide.
formiate, *m.* formate.
— *d'ammoniaque,* ammonium formate.
— *de potasse,* potassium formate.
— *de soude,* sodium formate.
formidable, *a.* formidable; fearful; enormous.
formique, *a.* formic.
formobenzoïlique, *a.* A name used to designate mandelic or phenylglycolic acid.
formochloride, *m.* chloroform.
formogène, *a.* generating formaldehyde.
formol, *m.* formol; (loosely) formaldehyde.
formoler, *v.t.* treat with formol or formaldehyde.
formonitrile, *m.* formonitrile (hydrocyanic acid).
formulable, *a.* capable of being formulated.
formulaire, *m.* formulary.
formulation, *f.* formulation.
formule, *f.* formula.
— *brute,* empirical formula (indicating the relative or, often, the absolute number of atoms of different kinds but not their mode of union).

formule *chimique,* chemical formula.
— *de constitution,* constitutional formula, structural formula.
— *développée,* structural formula.
— *empirique,* empirical formula.
— *globale,* summary formula.
— *rationnelle,* rational formula.
formuler, *v.t.* formulate, express in a formula.
— *v.r.* be formulated, be stated; have the formula.
formyle, *m.* formyl.
fort, *a.* strong; thick; hard, difficult, severe; skillful, clever; (of soil) heavy. — *m.* strong part, thick part, best part, main part, strength, depth, height; forte; strong person; fort. — *adv.* strongly, hard, fast; greatly, extremely, very.
— *peu soluble,* very slightly soluble, almost insoluble.
fortement, *adv.* strongly, forcibly, firmly, hard.
fortifiant, *p. a.* strengthening, fortifying. — *m.* (*Med.*) strength-giving medicine, tonic.
fortifier, *v.t.* strengthen, fortify.
fortuit, *a.* fortuitous, accidental.
fortuitement, *adv.* fortuitously, by chance, as it happens.
fortune, *f.* fortune; fortunate thing.
de —, by chance, accidentally.
fortuné, *a.* fortunate.
forure, *f.* drilled hole, bored hole, bore.
fosse, *f.* pit, hole, excavation, cavity; ditch, trench; grave; (*Anat.*) fossa.
— *à mouler,* molding pit.
— *d'aisances,* cesspool.
— *de fonderie,* foundry pit.
— *nasale,* (*Anat.*) nasal fossa.
— *septique,* (*Sewage*) septic tank.
fossé, *m.* ditch, drain.
fossile, *a. & m.* fossil.
fou (or *fol*), *a.* crazy; (of things) unsteady; irregular; (*Mach.*) loose, idle; (*Bot.*) wild, foolish; prodigious; wanton.
foudre, *m.* lightning, thunderbolt, thunder; tun, vat, large cask (containing as much as 300 hl.), specif. one sometimes used for the first fermentation of wine.
foudroiement, *m.* striking with lightning, etc. (see foudroyer).
foudroyante, *f.* (*Pyro.*) a rocket making a noise in imitation of thunder.
foudroyer, *v.t.* strike by lightning or by any powerful electric discharge; thunderstrike; destroy, overwhelm. — *v.i.* thunder.
fouet, *m.* whip; whipcord; lash; whipping; (*Glass*) a workman who anneals bottles.
fouetter, *v.t.* whip; lash, beat, pelt; streak. — *v.i.* beat, lash, pelt.
fougère, *f.* fern.
— *mâle,* male fern, (*Pharm.*) aspidium.
fouille, *f.* excavation, digging; cavity.

fouiller, *v.t.* excavate; dig; search, explore.
— *v.i.* dig; search, explore.
fouir, *v.t.* dig; sink (a well); burrow.
foulage, *m.* pressing, etc. (see fouler).
foule, *f.* multitude, great number; crowd; fulling.
en —, in crowds, in great quantity.
foulée, *f.* pile (of skins); blast (of a bellows); tread (of a step).
foulement, *m.* pressing, etc. (see fouler).
fouler, *v.t.* press, crush, squeeze (as grapes); compress (as air); full, mill (cloth or leather); tread, tramp, trample; oppress; sprain.
foulerie, *f.* fulling mill.
fouleur, *m.* (*Textiles,* etc.) fuller, mill man.
fouloir, *m.* crusher (as for grapes); beater; rammer, tamper; fulling mill.
— *trémie,* hopper crusher.
— *pressoir,* a combination crusher and press for fruits.
foulon, *m.* fuller (of cloth); (*Leather*) a vat in which leather is given a supplementary tanning.
foulonner, *v.t.* full, mill.
foulonnier, *m.* fuller, mill man.
foulure, *f.* fulling, milling; treading; sprain.
four, *m.* furnace; kiln; oven; stove.
— *à acier,* steel furnace.
— *à aludels,* aludel furnace, Bustamente furnace.
— *à bleuir,* bluing kiln (for tile).
— *à boulanger,* baker's oven; (*Coke*) beehive oven.
— *à briques,* brick kiln.
— *à carbonater,* (*Soda*) black-ash furnace.
— *à cémenter,* cementation furnace.
— *à chambre,* chambered furnace or kiln, specif. a gas-retort furnace.
— *à chaux,* lime kiln.
— *acide,* acid-lined furnace.
— *à coke,* coke oven.
— *à coke à récupération,* by-product coke oven.
— *à coupeller,* cupellation furnace, muffle furnace.
— *à creusets,* crucible furnace.
— *à cuire,* (*Glass*) annealing oven, leer.
— *à cuve,* shaft furnace.
— *à cuvette,* (*Soda*) a salt-cake furnace using pots, as distinguished from a mechanical furnace.
— *à dalles,* shelf burner (as for burning pyrites smalls).
— *à émailler,* enameling furnace (a kind of muffle furnace).
— *à étendre,* (*Glass*) flattening furnace or oven.
— *à fondre,* melting furnace (as for glass).
— *à galères,* gallery furnace, galley.

four *à gaz,* gas furnace; gas kiln; gas oven.
— *à incinérer,* incinerator.
— *à manche,* small shaft furnace, cupola.
— *à moufle,* muffle furnace.
— *à pain,* baking oven, bread oven.
— *à pétrole,* petrol furnace, gasoline furnace.
— *à plâtre,* plaster kiln, specif. one for burning gypsum to plaster of Paris.
— *à poix,* pitch oven.
— *à puddler,* puddling furnace.
— *à réchauffer,* reheating furnace, heating furnace.
— *à recuire,* annealing furnace or oven.
— *à refroidir,* (*Glass*) a furnace (or compartment of a furnace) in which window glass is cooled gradually after being flattened.
— *à résine,* resin oven.
— *à réverbère,* reverberatory furnace.
— *à ruche,* (*Coke*) beehive oven.
— *à sel de soude,* soda-ash furnace.
— *à sole,* hearth with bed.
— *à soude,* soda furnace.
— *à souffler.* = four-ouvreau.
— *à soufre,* brimstone burner, sulfur furnace.
— *à tubes,* tube furnace.
— *à tuiles,* tile kiln.
— *basique,* basic-lined furnace.
— *belge,* (*Zinc*) Belgian furnace.
— *belge-silésien,* (*Zinc*) a combined Belgian and Silesian furnace.
— *d'affinage,* — *d'affinerie,* (*Iron*) refinery, finery.
— *continu,* — *coulant,* continuous furnace or kiln
— *d'attrempage,* (*Glass*) a furnace in which the pots are given a preliminary heating, pot arch.
— *de fusion,* melting furnace.
— *de galère,* gallery furnace, galley.
— *de recuisson,* (*Glass*) annealing oven, leer.
— *de travail.* = four-ouvreau.
— *électrique,* electric furnace.
— *liégeois,* (*Zinc*) Liége furnace, Belgian furnace.
— *Martin,* — *Martin-Siemens,* Siemens-Martin furnace, open-hearth furnace.
— *mécanique,* mechanical furnace.
— *Perrot,* Perrot furnace (a gas furnace for heating crucibles or one for cupellation).
— *rotatif,* (*Cement,* etc.) rotary kiln, rotatory kiln, rotating kiln; (*Metal.*) rotating hearth.
— *silésien,* (*Zinc*) Silesian furnace.
— *soufflé,* blast furnace.
— *tubulaire,* tube furnace.
— *tuyère,* blast furnace.
(For other phrases see the corresponding phrases under *fourneau.*)
fourbir, *v.t.* furbish, clean, burnish.
fourche, *f.* fork.

fourché, *p.a.* forked, branched.

fourcher, *v.i. & r.* fork, branch.

fourchette, *f.* (small) fork; table fork.

fourchon, *m.* prong (of a fork).

fourchu, *a.* forked, branching.

fourchure, *f.* fork, forking, branching.

fourgon, *m.* wagon, car, van; poker, stirrer.

fourmi, *f.* ant.

fourmiller, *v.i.* swarm, abound; tingle.

fournaise, *f.* furnace.

fourneau, *m.* furnace; kiln; oven; stove; chamber (of a mine); (military) mine.
— *à bassine,* a stove or brazier consisting of a deep vessel with a circular top.
— *à calciner,* calcining furnace, calciner.
— *à charbon,* charcoal stove or furnace; charcoal pit.
— *à coupelle,* cupellation furnace.
— *à flammes,* (*Mercury*) a double reverberatory furnace used at Idria.
— *à gaz,* gas stove; gas furnace.
— *à main,* portable stove or furnace.
— *à rôtir,* roasting furnace.
— *à sécher,* drying stove, drying furnace.
— *à tubes,* a furnace or stove provided with inlet and outlet tubes for the use of gaseous fuel; tube furnace (for heating tubes).
— *à vases clos,* closed-vessel furnace (including chamber, muffle, crucible and retort furnaces).
— *à vent,* blast furnace.
— *catalan,* Catalan forge (or furnace).
— *d'artifice,* pyrotechnic furnace.
— *de calcinage,* calcining furnace, calciner.
— *de cuisine,* kitchen stove, kitchen range.
— *de grillage,* roasting furnace.
— *de laboratoire,* laboratory furnace, chemical furnace.
— *d'émailleur,* enameler's furnace (a kind of muffle furnace).
— *de raffinage,* refining furnace.
— *d'essai,* assay furnace.
— *de verrerie,* glass furnace.
— *générateur,* generator (for fuel gas).
— *rotatoire,* rotary furnace, revolving furnace.
haut —, blast furnace.
(For other phrases see the corresponding phrases under *four.*)

fournée, *f.* charge (in a furnace or kiln), batch, baking, ovenful, kilnful.

fournette, *f.* small reverberatory furnace.

fourni, *p.a.* furnished, provided; thick.

fournir, *v.t.* furnish, provide, supply. — *v.i.* (with *à*) provide (for), pay (for).

fournissement, *m.* furnishing, etc. (see fournir); share (of capital).

fournisseur, *m.* furnisher, supplier, purveyor, seller, source; (*Founding*) ingate.

fourniture, *f.* furnishing, etc. (see fournir); provision, supply; supplies; specif., bedding.

four-ouvreau, *m.* (*Glass*) a furnace with large openings at which the workmen heat the glass while blowing it, " blowing hole."

fourrage, *m.* forage, fodder, food for animals (esp. hay, straw, etc.); foraging.

fourrager, *v.t. & i.* forage; pillage; rummage.
— *a.* forage, fit for forage.

fourré, *p a.* thrust, inserted, etc. (see fourrer); bushy, woody (country); furry, furred.

fourreau, *m.* cover, covering, case, casing, sheath, scabbard.

fourrer, *v.t.* thrust, insert, put; cram, stuff; line with fur; line with wood; serve (rope).

fourreur, *m.* furrier.

fourrier, *m.* quartermaster.

fourrure, *f.* fur; furring, wood lining; nest (of tubs, coppers or the like).

four-séchoir, *m.* drying oven, drying furnace.

fourvoyer, *v.t.* mislead. — *v.r.* err, be wrong.

fouteau, foyard, *m.* beech, beech tree.

foyer, *m.* hearth; fire; light, source of light focus; furnace; fire box; foyer, lobby.
— *à grille,* grate hearth, grate furnace.
— *alandier,* = alandier.
— *à réchauffer,* reheating hearth.
— *d'affinage,* — *d'affinerie,* (*Metal.*) refinery, finery hearth.
— *de raffinerie,* (*Metal.*) refinery, fining hearth.
— *écossais,* (*Lead*) Scotch hearth.
— *gazogène,* gas producer, producer.
— *réel,* real focus.
— *supérieur,* (of a blast furnace) part above the boshes, main body, belly.
— *virtuel,* virtual focus.

F.R., *abbrev.* (flamme réductrice) reducing flame.

fracas, *m.* crash, crashing, noise; tumult.

fracasser, *v.t.* shatter, break.

fraction, *f.* fraction.

fractionnaire, *a.* fractional.

fractionné, *p.a.* fractional; fractionated.

fractionnement, *m.* fractioning, fractionation.

fractionner, *v.t.* fractionate, fraction.

fracture, *f.* fracture; breaking.

fracturer, *v.t.* break, fracture.

fragile, *a.* fragile, brittle, frail.

fragilité, *f.* fragility, fragileness, frailness.

fragment, *m.* fragment.

fragmentaire, *a.* fragmentary.

fragmentation, *f.* fragmentation.

fragmenter, *v.t.* break into fragments.

fragrance, *f.* fragrance.

fragrant, *a.* fragrant.

frai, *m.* spawning; spawn; small fry; fraying, etc. (see frayer). See also frais.
— *de grenouille,* frog spawn.

fraîche, *a. fem.* of frais.

fraîchement, *adv.* freshly, recently; coolly.

fraîcheur, *m.* coolness; freshness; chill; gentle breeze.

fraîchir, *v.i.* become cooler; (of wind) freshen.

frais, *a.* fresh; recent; cool. — *m.* freshness; coolness; fresh air; cool place. — *adv.* freshly; recently; coolly. — *m.pl.* expense, expenses, cost, costs, charges. — *pl.* of frai.

— *scolaires*, tuition, tuition fee.

fraise, *f.* strawberry; cutting file, cutter, milling tool; reamer; countersink; mesentery, crow (of a calf, lamb or the like); ruffle, frill.

fraiser, *v.t.* bore; ream; mill; plait.

fraiseuse, *f.* milling machine.

fraisier, *m.* strawberry (plant).

fraisil, *m.* coal cinders.

framboise, *f.* raspberry.

framboisé, *a.* having a raspberrylike surface, mammillated; having a raspberry odor or flavor.

franc, *a.* pure, unadulterated; true, real; full, complete, whole; free; frank. — *adv.* fully, completely, quite; frankly. — *m.* franc (19.3 cents or 9.4 d.).

français, *a.* French. — *m.* French (language); (*cap.*) Frenchman.

Francfort, *m.* Frankfort, Frankfurt.

Francfort-sur-Mein, *m.* Frankfort on the Main, Frankfurt am Main.

franche, *a. fem.* of franc.

franchement, *adv.* frankly; purely; free.

franchir, *v.t.* pass; pass over, cross; exceed; surmount; jump over; clear.

franchise, *f.* exemption (as from duty or tax); immunity; franking privilege; frankness; freedom.

franchissement, *m.* passing, etc.(see franchir).

franco, *adv.* free, carriage paid, postpaid.

frange, *f.* fringe.

franger, *v.t.* fringe.

frangibilité, *f.* frangibility, frangibleness.

frangible, *a.* frangible, breakable, fragile.

franguline, *f.* frangulin.

frappage, *m.* striking, etc. (see frapper); stamp, impression.

frappé, *p.a.* struck; stamped; iced; (of cloth) strong, close; forcible, spirited.

être — de, be affected with, have, exhibit; be struck with.

frappement, *m.* striking, etc. (see frapper).

frapper, *v.t.* strike; hit, knock, beat, stamp; press into cakes (as soap); ice (wines, etc.); affect. — *v.i.* strike; hit, knock, beat.

frappeur, *m.* striker, beater, knocker, stamper. — *a.* striking, beating, etc.

fraude, *f.* fraud.

frauder, *v.t.* adulterate; defraud; smuggle in.

frauduleusement, *adv.* fraudulently.

frauduleux, *a.* fraudulent.

frayer, *v.t.* fray, rub, graze, wear; open (a way). — *v.i.* associate; wear, wear away.

frein, *m.* brake; check, curb; bit, bridle.

freindre, *v.i.* shrink.

freiner, *v.t.* brake.

freinte, *f.* shrinkage, loss (as of a substance during manufacture).

frelatage, frelatement, *m.*, **frelatation, frelaterie**, *f.* adulteration, sophistication.

frelater, *v.t.* adulterate, sophisticate.

frelateur, *m.* adulterator, sophisticator.

frêle, *a.* frail, fragile.

frelon, *m.* hornet.

frémir, *v.i.* be agitated, vibrate, tremble, shake, quiver, (of water) simmer; rustle, murmur, roar.

frémissement, *m.* agitation, vibration, etc. (see frémir).

frêne, *m.* ash (tree or wood).

— *épineux*, prickly ash (*Zanthoxylum*).

fréquemment, *adv.* frequently.

fréquence, *f.* frequency.

fréquent, *a.* frequent.

fréquentation, *f.* attendance (on courses); frequenting, frequentation.

fréquenter, *v.t.* attend (classes); frequent. — *v.i.* resort (to); associate (with).

frequin, *m.* a cask for sugar, sirup, etc.

frère, *m.* brother.

fresque, *f.* fresco.

fret, *m.* freight; chartering (of a vessel).

fréter, *v.t.* freight, load; charter.

frette, *f.* hoop, ring, band, tire.

fretter, *v.t.* hoop; flange; build up (a gun).

friabilité, *f.* friability.

friable, *a.* friable.

friche, *f.* fallow, fallow land.

friction, *f.* (*Med.*) rubbing, friction.

frictionner, *v.t.* rub.

frigidité, *f.* frigidity, coldness, frigidness.

frigorifère, *m.* cold chamber, refrigerating chamber, refrigerator, refrigerating apparatus.

frigorification, *f.* refrigeration, production of cold. (See frigorifier).

frigorifier, *v.i.* produce cold. — *v.t.* refrigerate, esp. at low temperature, freeze (as meat).

frigorifique, *a.* refrigerating, refrigeration, cold-producing, frigorific; (of mixtures) freezing.

frire, *v.t.*, *i.* & *r.* fry.

Frise, *f.* Friesland.

friser, *v.t.* curl, crisp; graze; be near. — *v.i.* curl, crisp.

frissonner, *v.i.* shiver, quiver, shake.

frit, *p.p.* & *pr. 3 sing.* of frire.

frittage, *m.* fritting.

fritte, *f.* frit; fritting.

fritter, *v.t.* & *i.* frit, fuse partially.

fritteur, *m.* fritter.

fritteux, *a.* of the nature of a frit, fritty; suitable for making frit.

frittier, *m.* fritter.

friture, *f.* frying; frying oil, frying fat; fried material.

froid, *a.* cold. — *m.* cold; coldness.
à —, cold, in the cold.
froidement, *adv.* coldly, frigidly, coolly.
froideur, *f.* coldness, chill.
froidure, *f.* cold, coldness, cold weather.
froidureux, *a.* cold.
froisser, *v.t.* rub; rumple; bruise; offend.
froissure, *f.* bruise; rumple, wrinkle.
frôlement, *m.* grazing, brushing, contact.
frôler, *v.t.* graze, brush, pass over.
fromage, *m.* cheese; a cheese-shaped support
for crucibles, crucible stand.
— à la crème, cream cheese.
— à pâte ferme (or résistante), hard cheese.
— à pâte molle, soft cheese.
— cru, uncooked cheese.
— cuit, cooked cheese.
— de Hollande, Dutch cheese.
— d'Eidam, Edam cheese.
— demi-gras, cheese from partially skimmed
milk, single cheese.
— fermenté, fermented cheese.
— frais, fresh or unfermented cheese, cottage
cheese.
— gras, rich cheese (either cream cheese,
whole-milk cheese, or filled cheese containing an equivalent amount of fat).
— maigre, skim-milk cheese.
— mou, soft cheese.
fromager, *m.* cheese mold; cheese maker;
cheese dealer; (*Bot.*) silk-cotton tree.
fromagerie, *f.* cheese factory; cheese room;
cheese shop; cheese trade.
fromageux, *a.* cheesy.
froment, *m.* wheat.
— aubaine, hard wheat, durum wheat.
— de Turquie (or des Indes, or d'Espagne),
(Indian) corn, maize.
— épeautre, spelt.
— touselle, a kind of beardless wheat.
fromentacé, *a.* frumentaceous.
fromentacée, *f.* frumentaceous plant.
fromental, *m.* oat grass; rye grass.
fromenteux, *a.* wheat-producing, wheat.
frondaison, *f.* foliage; leafing time.
fronde, *f.* sling; frond (of ferns); foliage.
front, *m.* front; face; brow, forehead; head
de —, in front; simultaneously, at the same
time; abreast.
— de taille, (*Mining*) working face.
frontière, *f. & a.* frontier, boundary, border.
frontispice, *m.* frontispiece.
frottage, *m.* rubbing, etc. (see frotter).
frottement, *m.* friction; rubbing, etc. (see
frotter).
— de glissement, sliding friction.
— de roulement, rolling friction.
frotter, *v.t.* rub; grind (as glass); scour;
polish. — *v.i.* rub each other; rub up
(with), associate; meddle (with). — *m.*
rubbing, friction.

frotteur, *m.* rubber; friction piece; (*Elec.*)
rubbing contact.
frottis, *m.* thin coat, wash (as of a transparent
color); polish.
frottoir, *m.* rubber; polisher.
frou-frou, froufrou, *m.* rustling, froufrou.
fructifier, *v.i.* fructify; be fruitful.
fructose, *f.* fructose.
fructueusement, *adv.* fruitfully, profitably.
fructueux, *a.* fruitful; profitable.
fruit, *m.* fruit.
fruits de fenouil, (*Pharm.*) fennel, fennel fruit,
fennel seed.
fruité, *a.* fruity.
fruiter, *v.i.* fruit, bear fruit.
fruitier, *a.* fruit, fruit-bearing. — *m.* fruiterer;
fruit room; fruit rack.
frumentacé, *a.* frumentaceous, cereal.
frustrer, *v.t.* deprive; frustrate.
fuchsine, *f.* (*Dyes*) fuchsine, fuchsin.
fucose, *n.* fucose.
fugace, *a.* fugitive; fleeting.
fugacité, *f.* fugitiveness; transiency.
fugitif, *a.* fugitive.
fui, *p.p.* of fuir.
fuir, *v.i.* escape, leak; (of colors) fade; fly,
flee; retreat, recede; evade, avoid. — *v.t.*
flee from, avoid, escape.
fuite, *f.* leak, leakage, leaking, escape; fading;
flight; evasion, shift; avoidance.
fulguration, *f.* (*Assaying*) fulguration, brightening; flashing.
fulgurite, *m.* fulgurite.
fuligineux, *a.* sooty, smoky, fuliginous.
fuliginosité, *f.* fuliginosity, sootiness, soot.
fulmicoton, fulmi-coton, *m.* guncotton.
— soluble, soluble guncotton, pyroxylin.
fulminant, *p.a.* fulminating.
fulminate, *m.* fulminate.
— d'argent, silver fulminate.
— de mercure, mercury fulminate.
fulminaterie, *f.* fulminate factory.
fulmination, *f.* fulmination, detonation.
fulminer, *v.i.* detonate; fulminate, rage.
fulminique, *a.* fulminic.
fulminurique, *a.* fulminuric.
fulmi-paille, *f.* (*Expl.*) nitrated straw.
fulmison, *m.* (*Expl.*) nitrated bran.
fumage, *m.* fuming; smoking; (*Agric.*) manuring.
fumaison, *m.* (*Agric.*) manuring, dunging.
fumant, *p.a.* fuming; smoking; steaming;
reeking.
fumarate, *m.* fumarate.
fumarine, *f.* fumarine.
fumarique, *a.* fumaric.
fumarolle, *f.* (*Geol.*) fumarole.
fumé, *p.a.* smoked, smoky; fumed; manured.
fumée, *f.* fume; smoke; smoke gas, flue gas.
— de Londres, London smoke (denoting the
color of a dark glass made for optical use).

fumée, *sans* —, smokeless.

fumer, *v.i.* fume; smoke; steam; reek. — *v.t.* fume; smoke; manure.

fumerolle, fumerole, *f.* (*Geol.*) fumarole.

fumeron, *m.* half-burnt (or smoky) charcoal.

fumet, *m.* aroma, odor.

fumeterre, *m.* (*Bot.*) fumitory (*Fumaria*).

fumeux, *a.* smoky; (of wine) fumy, heady.

fumier, *m.* manure, dung; dunghill.

— *de ferme,* farm manure.

fumigateur, *m.* fumigator.

fumigation, *f.* fumigation.

fumigatoire, *a.* fumigating.

fumigène, *a.* smoke-producing, fumigenic. — *m.* smoke-producing substance, fumigen.

fumiger, *v.t.* fumigate.

fumiste, *m.* builder or repairer of chimneys, furnaces, ovens, etc.; stove setter.

fumisterie, *f.* building or repair of chimneys, ovens, furnaces, etc.

fumivore, *a.* smoke-consuming. — *m.* smoke consumer.

fumure, *f.* manuring, fertilizing; manure, fertilizer.

funeste, *a.* fatal, disastrous; sinister.

fungine, *f.* fungin.

funiculaire, *a.* funicular. — *m.* funicular curve, catenary; funicular (or cable) railway; funicular machine.

fur, *m.* In the phrases *à fur et mesure, au fur et à mesure,* in proportion.

furanique, *a.* of or pertaining to furan.

α-furazol, *m.* α-furazole (isoxazole).

β-furazol, *m.* β-furazole (oxazole).

furent, *p.def. 3 sing.* of être (to be).

fureur, *m.* fury, rage, passion.

furfuracé, *a.* furfuraceous, branny, scurfy.

furfurane, *m.* furan (or furane), furfuran (or furfurane).

furfurane-carbonique, *a.* furancarboxylic.

furfuranique, *a.* of or pertaining to furan (or furfuran).

furfurine, *f.* furfurine.

furfurol, *m.* furfurole (better called furaldehyde or furfural).

furfurolique, *a.* In the phrase *alcool furfurolique,* furfuryl alcohol (2-furancarbinol).

furieux, *a.* furious; tremendous, extraordinary.

furoncle, *m.* (*Med.*) furuncle, boil.

furonculose, *f.* (*Med.*) furunculosis.

fusain, *m.* spindle tree (*Evonymus*); charcoal for drawing; charcoal crayon; charcoal drawing.

fusant, *p.a.* fusing, etc. (see fuser); (of fuses, etc.) time; (of compositions) fuse.

fuseau, *m.* spindle (in various senses); pin, axle, arbor.

fusée, *f.* fuse, fuze; rocket; fusee, fuzee; (*Glass*) a tool for enlarging openings; (*Mach.*) axle arm, barrel, journal, spindle; (*Spinning*) spindleful, cop.

fusée *à baguette,* rocket and stick.

— *à concussion,* concussion fuse.

— *à double effet,* double-action fuse, combination fuse.

— *à durée,* time fuse.

— *à effet retardé,* delay-action fuse.

— *à la Congreve,* Congreve rocket.

— *à obus,* shell fuse; shell rocket.

— *à temps,* time fuse.

— *de Bickford,* Bickford fuse.

— *d'éclair,* flash rocket.

— *d'éclairage,* light rocket.

— *de sûreté,* safety fuse.

— *d'honneur,* (*Pyro.*) signal rocket.

— *éclairante,* rocket flare.

— *éclatante,* star rocket.

— *fusante,* time fuse.

— *instantanée,* instantaneous fuse, quick match.

— *lente,* slow match.

— *lumineuse,* light rocket.

— *mixte,* combination fuse.

— *percutante,* percussion fuse.

— *réglée,* time fuse.

— *retardée,* delay-action fuse.

— *vive,* quick match.

— *volante,* sky rocket.

fusée-détonateur, *n.* detonating fuse.

fusée-signal, *f.* signal rocket.

fuselé, *a.* spindle-shaped, fusiform.

fuser, *v.i.* fuse, melt; (of salts on glowing charcoal) react with crackling or sputtering; (of pyrotechnic compositions) burn without exploding; (of colors) spread.

fusibilité, *f.* fusibility.

fusible, *a.* fusible.

fusil, *m.* (small) gun, rifle, (formerly) musket; steel (for striking fire or sharpening).

fusillette, *f.* small rocket.

fusion, *f.* fusion.

— *à la soufflerie,* sealing with the blast.

— *aqueuse,* aqueous fusion.

— *ignée,* igneous fusion.

fusionner, *v.t., r. & i.* fuse, unite, blend.

fussent, *imp. 3 pl. subj.* of être (to be).

fustet, fustel, *m.* fustet (*Cotinus cotinus* or its wood).

fustoc, fustok, *m.* fustic.

fut, *p.def. 3 sing.* of être (to be).

fût, *m.* cask, barrel; shaft, trunk, pole, post; stock, frame, brace, handle. — *imp. 3 sing. subj.* of être (to be).

futaie, *f.* timber forest; large tree.

futaille, *f.* cask, barrel; casks, barrels.

futaillerie, *f.* wood for casks.

futée, *f.* a cement composed of glue and sawdust.

futile, *a.* futile.

futur, *a. & m.* future.

fuyant, *p.a.* escaping, etc. (see fui.); leaky; (of colors) fugitive.

G

gâchage, *m.* mixing, etc. (see gâcher).

gâche, *f.* staple.

gâchée, *f.* batch, mixing (of cement, etc.).

gâcher, *v.t.* mix, temper (cement, mortar, etc.); slake (lime); bungle, botch.
— *clair,* — *lâche,* mix thin.
— *serré,* mix stiff.

gâchis, *m.* mortar; mud, mire; slush; (fig.) mess.

gadolinite, *f.* (*Min.*) gadolinite.

gadolinium, *m.* gadolinium.

gadoue, *f.* night soil.

gagner, *v.t.* gain; get, obtain, earn, reach.
— *v.i.* gain; improve.

gai, *a.* gay; brisk, spirited, lively; (of colors) bright; (of mechanical parts) loose, having play.

gaïac, *m.* guaiacum (tree or wood).

gaïac-, (in names of compounds) guaiac-.

gaïacol, *m.* guaiacol.

gailleterie, *f.* coal in small lumps; small lump of coal.

gailleteux, *a.* lumpy, lump, rubbly (as coal).

gailletin, *m.* coal in small pieces; small piece of coal.

gaillette, *f.* coal in lumps, rubbly coal; lump or piece of coal.

gain, *m.* gain.

gaine, *f.* sheath, case; shaft; flue; (of a fuse or shell) gaine, priming tube.

gainer, *v.t.* sheath, incase.

gainerie, *f.* making or selling of sheaths or cases; case maker's art or business.

gaize, *f.* (*Petrog.*) gaize.

galactagogue, *a. & m.* (*Med.*) galactagog(ue).

galactane, *m.* galactan.

galactate, *m.* lactate.

galactidensimètre, *m.* galactodensimeter, galactometer.

galactique, *a.* lactic; (*Astron.*) galactic.

galactomètre, *m.* galactometer.

galactométrie, *f.* galactometry.

galactonique, *a.* galactonic.

galactose, *f.* galactose; (*Physiol.*) galactosis.

galactotimètre, *m.* galactotimeter (a form of lactometer).

galaheptonique, *a.* galaheptonic.

galalithe, *m.* galalith.

galanga, *m.* (*Pharm.*) galangal, galanga; (*Bot.*) any member of the genus *Alpina.*

galazyme, *m.* a fermented beverage made from milk.

galbanifère, *a.* producing galbanum.

galbanum, *m.* galbanum.

galbe, *m.* contour; swell, entasis; (of a blast furnace) body.

gale, *f.* (*Glass*) gall, scum; (*Med.*) itch; (*Bot.*) scurf; mange.

galène, *f.* (*Min.*) galena, galenite.

galénique, *a.* (*Min.*) galenic(al); (*Pharm. & Med.*) Galenic(al).

galère, *f.* gallery furnace, galley.

galerie, *f.* gallery; shelf, frame, platform (attached to the side of something); flue (of a boiler); (*Mining*) drift, adit, gallery.

galet, *m.* pebble; gravel, shingle; disk (placed over the column of liquid in a polarimeter); friction roller (or disk); roller, roll.
— *de roulement,* friction roller.

galetage, *m.* (*Gunpowder*) making press cake.

galeter, *v.t.* (*Gunpowder*) press into press cake.

galet-guide, *m.* roller guide.

galette, *f.* (flat) cake; specif., (*Gunpowder*) press cake; disk; ship biscuit.

galeux, *a.* (of glass) covered with gall or scum; itchy, mangy, scurfy.

Galice, *f.* Galicia (Spain).

Galicie, *f.* Galicia (Austria).

galicien, *a.* Galician.

galipot, *m.* galipot, gallipot.

galipoter, *v.t.* coat with galipot.

gallate, *m.* gallate.
— *basique de bismuth,* (*Pharm.*) bismuth subgallate.

galle, *f.* gall (of plants); specif., nutgall, gallnut; gallfly; knot, lump, concretion (as in founding sand).
— *d'Alep,* Aleppo gall.
— *de chêne,* oak gall.
— *de Chine,* Chinese gall (from sumach).

galléine, *f.* gallein.

Galles, *f.* Wales.

gallifère, *a.* containing or yielding gallium.

gallinace, *f.* (*Min.*) obsidian.

gallique, *a.* gallic; Gallic.

gallium, *m.* gallium.

gallocyanine, *f.* (*Dyes*) gallocyanin(e).

gallois, *a.* Welsh. — *m.* (*cap.*) Welshman.

gallon, *m.* an excrescence similar to the nutgall but found on or replacing acorns; gallon.

gallotannique, *a.* gallotannic.

galvanique, *a.* galvanic.

galvaniquement, *adv.* galvanically.

galvanisation, *f.* galvanization, galvanizing.

galvaniser, *v.t.* galvanize.

galvano, *m.* (*Printing*) electrotype, electro.

galvanocérame, *m.* a piece of electroplated ceramic ware.

galvanomètre, *m.* galvanometer.
— *à miroir,* — *à réflexion,* mirror galvanometer, reflecting galvanometer.
— *enrégistreur,* recording galvanometer.

galvanoplastie, *f.* galvanoplastics, -plasty.
galvanoplastique, *a.* galvanoplastic.
galvanotyper, *v.t.* electrotype.
gambir, *m.* gambier, gambir.
— *cubique,* gambier in cubes.
gamin, *m.* gamin, boy; (*Glass*) assistant, apprentice.
gamme, *f.* gamut, scale.
Gand, *n.* Ghent.
gangreneux, gangréneux, *a.* (*Med.*) gangrenous.
gangue, *f.* (*Geol. & Mining*) gangue, gang.
ganil, *m.* (*Petrog.*) ganil (a kind of limestone).
ganse, *f.* loop; cord.
gant, *m.* glove; gauntlet.
— *de laboratoire,* laboratory glove.
ganterie, *f.* glove making; glove trade.
garançage, *m.* madder dyeing.
garance, *f.* madder. — *a.* madder-colored, scarlet.
garancer, *v.t.* dye with madder.
garancerie, *f.* madder dyehouse; madder dyeing.
garanceur, *m.* madder dyer.
garanceux, *m.* a coloring matter extracted from the residues from madder dyeing.
garancière, *f.* madder field; madder dyehouse.
garancine, *f.* garancin, garancine.
garant, *m. & f.* guarantor, guaranty, surety; voucher, warrant.
garantie, *f.* guarantee; guaranteeing; security; pledge; protection; assay.
garantir, *v.t.* guarantee, warrant; protect, shelter, preserve.
garçon, *m.* boy; (of workmen) assistant, journeyman, boy, man; bachelor.
— *brasseur,* brewer.
garde, *f.* guard; protection, custody. — *m.* guard; keeper, warden.
garde-corps, *m.* railing, rail, guard rail.
garde-feu, *m.* fire guard; (*Metal.*) furnace foreman.
garde-fou, *m.* = garde-corps.
garde-four, garde-fourneau, *m.* furnace man.
garder, *v.t.* keep; guard. — *v.r.* keep, remain unspoiled; beware (of); keep (from). — *v.i.* beware.
garde-temps, *m.* chronometer.
garde-vue, *m.* eye shade; lamp shade.
gare, *f.* (railway) station; switch, shunt.
gargariser, *v.t. & r.* gargle.
gargarisme, *m.* gargle, (*Med.*) gargarism.
garnir, *v.t.* furnish, supply, provide, stock, fit, equip; fill; line, face; trim; mount; pack, stuff; garnish, decorate.
garnissage, garnissement, *m.* furnishing, etc. (see garnir).
garnissage basique, (*Metal.*) basic lining.

garniture, *f.* filling; lining; mounting; fitting(s); trimming(s); covering; furnishing, furniture; edge, border; set, assortment; outfit; ornament; (*Mach.*) packing, stuffing; bearing box.
— *basique,* (*Metal.*) basic lining.
garou, *m.* spurge flax (*Daphne gnidium*); mezereon (*Daphne mezereum*).
garouille, *f.* kermes oak (*Quercus coccifera*) or its bark.
garus, *m.* (*Pharm.*) elixir of Garus (a stomachic containing cinnamon, saffron, nutmeg, etc.).
Gascogne, *f.* Gascony.
gaspillage, *m.* wasting, squandering.
gaspiller, *v.t.* waste, squander.
gastrique, *a.* gastric.
gâte, *m.* spoiled part.
gâteau, *m.* cake.
gâter, *v.t.* spoil; waste. — *v.r.* spoil, be spoiled, decay.
gauche, *a.* left; left-handed; levorotatory, levo; crooked; awkward. — *f.* left hand, left side, left wing.
à —, to the left; on the left; amiss.
gauchir, *v.t. & i.* warp, put or get out of true.
gauchissement, *m.* warping.
gaudage, *m.* dyeing with weld.
gaude, *f.* weld, yellowweed (*Reseda luteola*); weld (the yellow dye from weld).
gauder, *v.t.* dye with weld.
gaufre, *f.* honeycomb; waffle.
gaufrer, *v.t.* emboss; goffer, plait, crimp.
gaule, *f.* pole, rod, staff, stick.
gaulthérie, *f.* (*Bot.*) Gaultheria.
— *couchée,* wintergreen (*G. procumbens*).
gayac, *m.* guaiacum (tree or wood).
gaylussite, *f.* (*Min.*) gaylussite.
gaz, *m.* gas.
— *acide carbonique,* carbonic acid gas (carbon dioxide).
— *aérogène,* air carburetted by passage thru light hydrocarbons, air gas.
— *à l'air,* producer gas (made with dry air).
— *à l'eau,* water gas.
— *à l'huile,* oil gas.
— *ammoniac, — ammoniacal,* ammonia gas, gaseous ammonia.
— *asphyxiant,* asphyxiating gas.
— *azote,* nitrogen gas.
— *brûlés,* burnt gases, flue gases, exhaust gases.
— *carbonique,* carbon dioxide gas.
— *carburé,* carburetted gas.
— *d'échappement,* (of explosion engines) exhaust gases.
— *de chauffage,* fuel gas.
— *d'éclairage,* illuminating gas.
— *de combustion,* gases of combustion.
— *de générateur,* generator gas, producer gas.

gaz *de houille,* coal gas.
— *des marais,* marsh gas.
— *détonant,* detonating gas.
— *de ville,* city gas.
— *exhilarant,* — *hilarant,* laughing gas.
— *mixte,* mixed gas, specif. semi-water gas.
— *naturel,* natural gas.
— *nobles,* noble gases, inert gases.
— *parfait,* perfect gas, ideal gas.
— *pauvre,* poor gas, specif. producer gas (made with dry air).
— *perdu,* waste gas.
— *rare,* rare gas, noble gas.
— *riche,* rich gas.
— *Riché,* Riché gas (a fuel gas made from wood or other organic matter).
— *Siemens,* Siemens gas, producer gas.
— *suffocant,* asphyxiating gas.
— *tonnant,* detonating gas, explosive gas mixture.
gazage, *m.* gassing, etc. (see gazer).
gaze, *f.* gauze.
— *métallique,* wire gauze, metallic gauze.
gazéifiable, *a.* gasifiable.
gazéificateur, *m.* gasifier, specif. a form of oil burner.
gazéification, *f.* gasification; carbonating.
gazéifier, *v.t.* gasify; carbonate, charge (as water).
gazéiforme, *a.* gasiform, gaseous.
gazéité, *f.* gaseousness, gaseous nature or state.
gazer, *v.t.* gas; specif., (Metal.) reduce with gaseous fuel; cover with gauze; veil.
gazette, *f.* newspaper, gazette; (Ceram.) sagger.
gazeux, *a.* gaseous.
gazier, *a.* gas, of or pertaining to gas; gas man; gauze maker.
gazifère, *a.* gas-producing, yielding gas.
gazochimie, *f.* the chemistry of gases.
gazochimique, *a.* pertaining to the chemistry of gases.
gazofacteur, *m.* gasifier, specif. an apparatus for gasifying coal.
gazogène, *m.* gas generator; specif., producer, generator (for fuel gas); gas-impregnating apparatus, specif. one for carbonating liquids. — *a.* carbonating.
gazolène, *m.,* **gazoléine, gazoline,** *f.* gasoline, gasolene.
gazolyte, *a.* gasifiable, tending to gasify.
gazomètre, *m.* gasometer, gas holder.
— *à cloche,* a gas holder with a bell which can be raised and lowered.
gazométrie, *f.* gasometry.
gazométrique, *a.* gasometric(al).
gazon, *m.* grass; turf, sod.
gaz-platine, *m.* an early incandescent lamp consisting of a platinum basket hung in a flame.
géant, *m.* giant. — *a.* giant, gigantic, huge.

géine, *f.* geïn (humic acid).
gel, *m.* gel; frost.
gelable, *a.* congealable.
gélatine, *f.* gelatin.
— *animale,* gelatin.
— *de guerre,* explosive gelatin.
— *détonante,* explosive gelatin.
— *végétale,* vegetable gelatin (gliadin).
gélatiné, *a.* gelatined, coated with gelatin, treated with gelatin.
gélatine-dynamite, *n.* gelatin dynamite.
gélatineur, *m.* gelatin maker.
gélatineux, *a.* gelatinous.
gélatiniflable, *a.* gelatinizable.
gélatinifier, *v.t.* convert into gelatin or jelly, gelatinate, gelatinize.
gélatiniforme, *a.* gelatiniform.
gélatinisant, *p.a.* gelatinizing. — *m.* gelatinizing agent, gelatinizer.
gélatinisation, *f.* gelatinization, gelatinizing.
gélatiniser, *v.t.* gelatinize.
gélatino-bromure, *f.* gelatino-bromide.
gélatino-chlorure, *f.* gelatino-chloride.
gélation, *f.* congelation, gelation.
gélatose, *f.* gelatose.
gelée, *f.* jelly; frost, freeze, freezing.
— *glycérinée,* (Pharm.) glycerinated gelatin.
geler, *v.t., i. & r.* freeze, congeal.
gélif, *a.* congealing easily, having a relatively high freezing point; (of stones, etc.) susceptible to rending by frost.
gélification, *f.* jellification, jellying.
gélifier, *v.t. & r.* jellify, jelly.
géliforme, *a.* of the nature of a jelly.
gélignite, *f.* (Expl.) gelignite.
gélose, *f.* gelose; (loosely) agar.
gelsémine, *f.* gelsemine; (Bot.) gelsemium.
Gémeaux, *m.pl.* (Astron.) Gemini.
gémir, *v.i.* groan, moan.
gemmage, *m.* incising, tapping (of trees for turpentine).
— *à mort,* severe tapping resulting in the death of the tree.
— *à vie,* tapping in such a way as not to kill the tree.
gemme, *f.* gem; pine oleoresin, turpentine; (Bot.) bud. — *a.* gem, precious; gemmy, glittering.
sel —, rock salt.
gemmer, *v.t.* incise, tap (pine trees). — *v.i.* bud.
gencive, *f.* (Anat.) gum.
gêne, *f.* difficulty, trouble, inconvenience.
gêner, *v.t.* impede, hinder, embarrass; inconvenience, annoy, trouble.
général, *a. & m.* general.
en —, in general, generally.
généralement, *adv.* generally, in general.
généralisable, *a.* generalizable.
généralisateur, *a.* generalizing.
généralisation, *f.* generalization.

généraliser, *v.t.* generalize, render general.
— *v.r.* become generalized; become common.

généralité, *f.* generality.

générateur, *m.* generator; specif., a substance from which another substance is formed, original substance. — *a.* generating, generative; specif., designating substances from which others are formed (as, *une alliage et ses métaux générateurs,* an alloy and its component metals; *le mélange générateur d'un composé,* the mixture from which a compound was formed).
— *à vapeur,* — *de vapeur,* steam generator, boiler.
— *du gaz,* gas generator.

génératif, *a.* generative.

génération, *f.* generation.

génératrice, *f.* generatrix. — *a. fem.* of générateur.

généreusement, *adv.* generously.

généreux, *a.* generous; (of land) fertile.

générique, *a.* generic.

génériquement, *adv.* generically.

genèse, *f.* genesis, formation.

génésique, *a.* genetic.

génésiquement, *adv.* genetically.

genestrelle, genestrolle, *f.* (*Bot.*) woadwaxen, greenweed, dyer's weed (*Genista tinctoria*).

genêt, *m.* (*Bot.*) broom.
— *à balai(s),* — *commun,* common broom (*Cytisus scoparius*).
— *des teinturiers,* dyer's broom, woadwaxen (*Genista tinctoria*).
— *d'or,* common broom.

génétique, *a.* genetic.

Genève, *f.* Geneva.

genévrette, *f.* a drink made from wild fruits and flavored with gin.

genévrier, *m.* (*Bot.*) juniper (*Juniperus*).

génie, *m.* engineering; engineers; genius.

genièvre, *m.* gin (the liquor); juniper; juniper berry.

genièvrerie, *f.* gin distillery.

génisse, *f.* heifer.

genou, *m.* knee; (*Tech.*) elbow; (*Mach.*) joint, specif. ball-and-socket joint.

genouillère, *f.* joint, connection, specif. elbow joint; toggle joint; kneepiece.

genre, *m.* kind, sort, type; genus; style, mode; gender; genre (painting).
— *Saxe,* Saxon type (said of a shallow type of small evaporating dish).

gens, *m.pl.* people, folks, persons, men; nations; servants, attendants.

gent, *f.* tribe, race.

gentiane, *f.* (*Bot.*) gentian.
— *jaune,* yellow gentian, bitterwort (*Gentiana lutea*).

gentianine, *f.* gentianin.

gentianose, *f.* gentianose.

gentil, *a.* pretty; delicate; gracious.

gentilhomme, *m.* nobleman; gentleman.

génuine, *a.* genuine.

génuinité, *f.* genuineness.

géode, *f.* (*Geol.*) geode.

géodésique, *a.* geodetic.

géodique, *a.* geodic.

géoffrée, *f.* (*Bot.*) Geoffræa.

géographie, *f.* geography.

géographique, *a.* geographic(al).

géographiquement, *adv.* geographically.

géologie, *f.* geology.

géologique, *a.* geologic(al).

géologiquement, *adv.* geologically.

géologue, *m.* geologist.

géométral, *a.* geometrical.

géométrie, *f.* geometry.

géométrique, *a.* geometric(al).

géométriquement, *adv.* geometrically.

gérance, *f.* management, managership.

géraniacées, *f.pl.* (*Bot.*) Geraniaceæ.

géranial, *m.* geranial (citral).

géraniol, *m.* geraniol.

géranique, *a.* geranic.

géranium, *m.* (*Bot.*) geranium.
— *rosat,* rose geranium (*Pelargonium capitatum*).

gérant, *m.* manager, director, agent. — *a.* managing, directing.

gerbe, *f.* sheaf; small heap (of sea salt).

gerber, *v.t.* pile; bind in sheaves.

gerbiforme, *a.* sheaflike.

gercer, *v.t., r. & i.* crack; chap.

gerçure, *f.* crack, fissure, cleft, chink; chap.
— *de trempe,* (*Metal.*) temper crack.

gercuré, *a.* cracked, fissured; (of wood) shaky.

gérer, *v.t.* manage, conduct, carry on.

gergelin, *m.* oil of sesame.

germanique, *a.* Germanic, German.

germaniser, *v.t.* Germanize.

germanium, *m.* germanium.

germant, *p.a.* germinating.

germe, *m.* germ.

germer, *v.i. & t.* germinate.

germinal, *a.* germinal.

germination, *f.* germination, germinating.

germoir, *m.* malt floor.

gésier, *m.* gizzard.

gésir, *v.i.* lie, rest, be.

gesse, *f.* (*Bot.*) vetchling (*Lathyrus*), pea.

gestation, *f.* gestation.

gestion, *f.* management, administration, direction.

gestionnaire, *a.* managerial. — *m.* manager.

gibelet, *m.* gimlet.

gibier, *m.* game.

gible, *m.* set of bricks in the kiln.

giclée, *f.* spray.

gicler, *v.i.* squirt, spurt, spray.

gicleur, *m.* sprayer, atomizer, nozzle. — *a.* squirting, spraying, atomizing.

gigantesque, *a.* gigantic, huge, colossal.

gillénie, *f.* (*Bot.*) Gillenia, Indian physic (*Porteranthus*).

gillon, *m.* (*Bot.*) mistletoe (*Viscum album*).

gin, *m.* whisky; (abusively) gin.

gingas, *m.* ticking (the cloth).

gingembre, *m.* ginger.
— *blanc,* white ginger, Jamaica ginger, peeled ginger.
— *officinal,* common ginger (*Zinziber officinale*).

gingibrine, *f.* powdered ginger.

ginguet, *a.* of little value, poor.

ginseng, *m.* (*Bot. & Pharm.*) ginseng.

giobertite, *f.* (*Min.*) magnesite (magnesium carbonate). Cf. magnésite.

gipon, *m.* wool cloth used for tallowing hides.

girandole, *f.* chandelier; (*Pyro.*, etc.) girandole.

girasol, *m.* (*Min.*) fire opal, girasol; (*Bot.*) heliotrope, girasol.

giration, *f.* gyration.

giratoire, *a.* gyratory, gyrational.

girofle, *m.* clove.

giroflée, *a.* In the phrase *cannelle giroflée,* clove bark.

giroflier, *m* (*Bot.*) clove tree.

gisant, *p.pr.* of *gésir.*

gisement, *m.* (mineral) deposit; position of minerals in the earth; bearing.
— *en amas,* massive deposit.
— *en couche,* stratified deposit.
— *en filons,* vein deposit.

gisent, *pr. 3 pl.* of *gésir.*

git, *pr. 3 sing.* of *gésir.*

gite, *m.* (mineral) deposit; bed; lodging; home; lower millstone; leg (of beef).
— *filonien,* vein deposit.
— *sédimentaire,* sedimentary deposit, stratified deposit.

givre, *m.* white frost, hoar frost; crystallization resembling frost, specif. that of vanillin (or benzoic acid) on the surface of vanilla.

givré, *a.* frosty, frosted.

Gl. Symbol of *glucinium* (beryllium).

glaçage, *m.* icing, etc. (see glacer).

glace, *f.* ice; freezing point; plate glass; looking glass, mirror; glass, lens; glass window; flaw (in a precious stone); ice cream, ice; glazing (for pastry); icing, sugar crust.
— *fondante,* melting ice.
— *pilée,* crushed ice.

glacé, *p.a.* iced, etc. (see glacer); glacé; icy, ice, cold.

glacer, *v.t.* ice; freeze; chill; glaze; calender; vitrify. — *v.r.* freeze; vitrify. — *v.i.* freeze.

glacerie, *f.* plate-glass works or business; ice cream business, making of ices.

glaceur, *m.* glazer; calenderer.

glaceux, *a.* (of gems) flawed, flawy; icy, cold.

glaciaire, *a.* (*Geol.*) glacial.

glacial, *a.* glacial; icy, freezing, cold.

glaciale, *f.* (*Bot.*) ice plant (*Mesembryanthemum crystallinum*).

glacier, *m.* plate-glass maker; maker of, or dealer in, ices; (*Geol.*) glacier.

glacière, *f.* ice chamber, refrigerator; apparatus for making ice or ices, freezer; ice pail; ice house; ice cellar.
— *à bascule,* a small rocking apparatus for making ice or ices by the use of freezing mixtures.

glacis, *m.* gentle slope, sloping surface, glacis; (*Painting*) glaze, coat of transparent pigment.

glaçon, *m.* piece of ice, cake of ice; icicle.

glaçure, *f.* (*Ceram.*) glaze, glazing.
sous la —, under the glaze, underglaze.
sur la —, on the glaze, on-glaze.

glaie, *f.* vault of a glass furnace.

glairage, *m.* glairing, coating with glair.

glaire, *f.* glair (white of egg or any similar viscous substance).

glairer, *v.t.* glair, coat with glair.

glaireux, *a.* glairy, glaireous, viscous.

glairine, *f.* glairin.

glaise, *f.* clay; vault of a glass furnace. — *a.* In the phrase *terre glaise,* clay soil, clay.

glaiser, *v.t.* clay, treat with clay.

glaiseux, *a.* clayey.

glaisière, *f.* clay pit.

gland, *m.* acorn; tassel; (*Anat.*) glans.

glande, *f.* gland.
— *surrénale,* suprarenal gland, adrenal gland.

glandulaire, glanduleux, *a.* glandular, -ulous.

glaner, *v.t. & i.* glean.

glaubérite, *f.* (*Min.*) glauberite.

glauconie, glauconite, *f.* (*Min.*) glauconite.

glauque, *a.* glaucous, bluish green.

glette, *f.* litharge.

gleucomètre, *m.* glucometer, gleucometer.

gliadine, *f.* gliadin.

glissant, *p.a.* sliding, slipping; slippery.

glissement, *m.* sliding, slipping, gliding.

glisser, *v.t.* slide, slip, glide. — *v.i.* slip, slide. — *v.r.* slip, glide, steal.

glisseur, *m.* slider; (*Mach.*) sliding piece, slide block.

glissière, *f.* slide bar, slide, slide face, guide bar, guide.

glissoir, *m.* slider, slide block, slide.

glissoire, *f.* slider, sliding piece; slide, slide block, guide.

global, *a.* lump, total, entire, gross, summary.

globaux, *masc. pl.* of global.

globe, *m.* globe; ball; glass cover (as for a clock).

globeux, *a.* globose, globular.

globiforme, *a.* globular, globe-shaped.

globine, *f.* globin.

globulaire, *a.* globular, spherical.

globule, *m.* globule; (*Biol.*) corpuscle; pellet.
— *blanc*, white corpuscle (of the blood).
— *du sang*, blood corpuscle.
— *rouge*, red corpuscle (of the blood).
— *sanguin*, blood corpuscle.

globuleux, *a.* globulous, globular.

globulicide, *a.* destroying blood corpuscles.

globulin, *m.* (*Physiol.*) leucocyte.

globuline, *f.* globulin.

globulinurie, *f.* (*Med.*) globulinuria.

globulolyse, *f.* (*Physiol.*) hemocytolysis.

globulose, *f.* globulose.

gloire, *f.* glory, honor, splendor; (*Pyro.*) fixed sun.

glomérule, *f.* small mass; (*Bot.*, etc.) glomerule, glomerulus.

glonoïne, *f.* (*Pharm.*) glonoïn, glonoïne (nitroglycerin solution).

glorieux, *a.* glorious, splendid.

glossaire, *m.* glossary; vocabulary.

glouglou, *m.* gurgle, gurgling.

glouteron, *m.* common burdock (*Arctium lappa*); goose grass, cleavers (*Gallium*).

glover, *m.* (*Acids*) Glover tower.

glu, *f.* birdlime; cement, glue.
— *marine*, marine glue.

gluant, *a.* glutinous, sticky, adhesive, viscous; (fig.) tenacious.

glucine, *f.* glucina (beryllia).

glucinium, *m.* glucinum, glucinium (beryllium).

glucique, *a.* glucic.

glucoheptonique, *a.* glucoheptonic.

glucomètre, *m.* glucometer.

gluconique, *a.* gluconic.

glucosane, *f.* glucosan.

glucose, *f.* glucose.

glucoser, *v.t.* treat with glucose.

glucoserie, *f.* glucose factory; glucose manufacture.

glucoside, *m.* glucoside.

glucosier, *m.* glucose maker.

gluer, *v.t.* smear with birdlime or other viscous substance, lime, make sticky.

glutamine, *f.* glutamine.

glutamique, *a.* glutamic.

glutarique, *a.* glutaric.

gluten, *m.* gluten; glue, cement.

glutinatoire, *a.* agglutinating, agglutinative.

glutine, *f.* glutin (old name for gliadin and for gelatin).

glutineux, *a.* glutinous.

glycémie, *f.* (*Med.*) glycemia.

glycéré, glycérat, *m.* (*Pharm.*) glycerite, glycerin (solution of a medicinal substance in glycerin).
glycérate simple, = glycéré d'amidon.
— *d'alun*, glycerin of alum.
— *d'amidon*, glycerite of starch, glycerin of starch.
— *de borax*, glycerin of borax.
— *de boroglycéride*, glycerite of boroglycerin, solution of boroglyceride.
— *de pepsine*, glycerin of pepsin.
— *de phénol*, glycerite of phenol, glycerite of carbolic acid.
— *de sousacétate de plomb*, glycerin of subacetate of lead.
— *de tannin*, glycerite (or glycerin) of tannic acid or of tannin.
— *d'hydrastis du Canada*, glycerite of hydrastis.

glycéride, *f.* glyceride.

glycérine, *f.* glycerin (in *Org. Chem.* better called glycerol).
— *phénique*, (*Pharm.*) glycerite of phenol.
— *tannique*, (*Pharm.*) glycerite (or glycerin) of tannic acid.

glycériner, *v.t.* glycerinate, glycerinize.

glycérineux, *a.* containing or resembling glycerin.

glycérique, *a.* glyceric; of glycerin.

glycéro-acétique, *a.* designating the esters of glycerol with acetic acid.

glycéro-arséniate, *m.* glycero-arsenate.

glycéro-arsénique, *a.* glycero-arsenic.

glycéroborique, *a.* glyceroboric.

glycérolé, *m.* = glycéré.
— *d'acide phénique*, glycerite (or glycerin) of phenol.

glycérophosphate, *m.* glycerophosphate.

glycérophosphorique, *a.* glycerophosphoric.

glycérose, *f.* glycerose.

glycéryle, glycile, *m.* glyceryl.

glycide, *m.* glycide.

glycin, *m.* glycin (a photographic developer).

glycination, *f.* formation of or combination with glycine, glycination.

glycine, *f.* glycine; glucina (beryllia); (*Bot.*) Glycine, or a plant of this genus.

glycinium, *m.* glucinum (beryllium).

glycique, *a.* glucic.

glycocholique, *a.* glycocholic.

glycocolle, *m.* glycocoll (glycine).

glycogénase, *f.* glycogenase.

glycogène, *m.* glycogen. — *a.* glycogenous, glycogenic.

glycogénèse, glycogénie, *f.* glycogenesis, glycogeny.

glycogénique, *a.* glycogenic.

glycol, *m.* glycol.

glycolamide, *f.* glycolamide.

glycolique, *a.* glycolic.

glycolyse, *f.* glycolysis.

glycolytique, *a.* glycolytic.

glycomètre, *m.* glucometer.

glycoprotéide, *n.* glucoprotein, glycoprotein.

glycosamine, *f.* glucosamine.

glycosazone, *f.* glucosazone.

glycose, *f.* glucose, (rarely) glycose.

glycoside, *n.* glucoside.

glycosurie, *f.* glycosuria, glucosuria.

glycosurique, *a.* glycosuric.

glycuronique, *a.* glucuronic, glycuronic.

glycyrrhizine, *f.* glycyrrhizin.

glyoxal, *m.* glyoxal.

glyoxalidine, *f.* glyoxalidine (imidazoline).

glyoxaline, *f.* glyoxaline (imidazole).

glyoxylique, *a.* glyoxylic, glyoxalic.

gneiss, *m.* (*Petrog.*) gneiss.

gneisseux, gneissique, *a.* gneissic, gneissose.

go, *adv.* In the phrase *tout de go*, readily, immediately.

gobelet, *m.* cup (taller than *tasse* and frequently without a handle); goblet; (*Pyro.*) case.

gobeleterie, *f.* manufacture of drinking glasses or, more broadly, of glass vessels; hollow-glass works.

gobetis, *m.* (*Masonry*) pointing.

goder, *v.i.* bag, pucker.

godet, *m.* cup; shallow dish (as one for stains); cell; bucket (of an elevator, waterwheel or the like); bagging, puckering.

— *à graisse,* grease cup.

— *à huile,* oil cup.

— *de fusée,* fuse cup.

goémon, *m.* seaweed; specif., fucoid.

goitre, *m.* (*Med.*) goiter, goitre.

golfe, *m.* gulf.

gommage, *m.* gumming.

gomme, *f.* gum; (*Med.*) gumma; rubber.

— *adragant(e),* — *adraganthe,* tragacanth, gum tragacanth.

— *ammoniaque,* ammoniac, gum ammoniac.

— *animé,* animé, gum animé (or animi).

— *arabique,* gum arabic.

— *d'acajou,* cashew gum.

— *de Bassora,* Bassora gum.

— *de l'or,* = gomme des sages.

— *des funérailles,* bitumen.

— *des sages,* (*Old Chem.*) mercury.

— *du Sénégal,* Senegal gum, gum Senegal.

— *élastique,* gum elastic, caoutchouc, rubber.

— *explosive,* explosive gelatin.

— *haut-du-fleuve,* Galam gum (a kind of gum arabic).

— *laque,* lac, gum lac.

— *pseudo-adragant(e),* sassa gum.

— *rouge,* red gum; (*Old Chem.*) sulfur.

— *sauvage,* (*Rubber*) wild rubber, as contrasted with plantation rubber.

gommé, *a.* gummed, gum, gummy.

gomme-gutte, *f.* gamboge, gum guttæ.

gomme-laque, *f.* lac, gum lac.

gommelaquer, *v.t.* treat with gum lac, shellac.

gommeline, *f.* gommelin (dextrin).

gommement, *m.* gumming.

gommer, *v.t.* gum; mix with gum. — *v.r.* gum, stick.

gomme-résine, *f.* gum resin.

— *ammoniaque,* ammoniac, gum ammoniac.

— *d'euphorbe,* euphorbium.

gommeux, *a.* gummy, gummous; (of trees) gummiferous, gum-yielding.

gommier, *m.* gum-bearing tree, gum tree, specif. the acacia.

gommifère, *a.* gummiferous, gum-yielding.

gommo-résineux, *a.* gum-resinous.

gommose, *f.* gummose; (*Bot.*) gummosis.

gond, *m.* hinge.

gonder, *v t.* hinge.

gondoler, *v.r. & i.* (of plates, etc.) buckle.

gonflage, *m.* inflation.

gonflé, *p.a.* swelled, swollen, distended, inflated.

gonflement, *m.* swelling, swell, distention, inflation.

gonfler, *v.t.* swell, swell out, distend, inflate. — *v.i. & r.* swell, swell up.

goniomètre, *m.* goniometer.

gorge, *f.* throat; opening, orifice, mouth, neck; groove, channel, furrow; hollow, recess; choke (of a gun or rocket); gorge.

gorge-de-pigeon, *a.* of the iridescent color of a pigeon's breast.

gorger, *v.t.* (*Pyro.*) fill; gorge.

gorgonzola, *m.* Gorgonzola cheese, Gorgonzola.

gosier, *m.* throat, gullet.

gosiller, *v.i.* be carried over (said of the liquid which is being distilled for brandy).

gouache, *f.* water colors; water-color painting.

gouaché, *a.* water-color.

goudron, *m.* tar.

— *à calfater,* navy pitch.

— *de bois,* wood tar.

— *de houille,* coal tar.

— *de lignite,* brown-coal tar, lignite tar.

— *de Norvège,* Norway tar.

— *de pétrole,* petroleum tar.

— *des gazogènes,* producer tar.

— *des hauts fourneaux,* blast-furnace tar.

— *de tourbe,* peat tar.

— *minéral,* mineral tar, maltha; coal tar.

— *végétal,* vegetable tar (usually meaning wood tar).

goudronnage, *m.* tarring.

— *gras,* (*Pyro.*) a mixture of pitch and tallow.

— *sec,* (*Pyro.*) a mixture of pitch and rosin.

goudronner, *v.t.* tar.

goudronnerie, *f.* tar works, place where tar is prepared; place where tar is stored.

goudronneux, *a.* tarry.

gouet, *m.* arum (plant of the genus *Arum* or some related genus).
— *à trois feuilles*, jack-in-the-pulpit, Indian turnip (*Arisæma triphyllum*).

gouffre, *m.* gulf, abyss.

gouger, *v.t.* gouge.

goujon, *m.* gudgeon, pin, peg, dowel.

goulot, *m.* neck (of a bottle or other vessel).

goupille, *f.* pin; peg, key, splint.

goupiller, *v.t.* pin, key, peg, dowel.

goupillon, *m.* test tube brush (or one for bottles, for jars, for lamp chimneys, or the like); sprinkler (as for a forge).

goupillonner, *v.t.* cleanse with a test tube brush, bottle brush or the like.

gourde, *f.* gourd; flask, canteen.

goure, *f.* adulterated drug.

gourer, *v.t.* adulterate (drugs).

goureur, *m.* adulterator.

gousse, *f.* pod; (of garlic) clove.
— *de vanille*, vanilla bean.

gousset, *m.* bracket, brace; armpit; gusset.

goût, *m.* taste; (less often) odor.

goûter, *v.t.* taste; (less often) smell; relish, like. — *v.r.* taste; be tasted; be liked; like each other. — *v.i.* taste; smell; lunch.
— *m.* lunch.

goutte, *f.* drop; (*Pharm.*) minim; (*Assaying*) bead; blister (in a casting); globule of oxidized metal (in an ingot); (*Med.*) gout.
— *à* —, drop by drop, dropwise.

gouttelette, *f.* droplet, little drop.

goutter, *v.i.* drop, drip.

goutteux, *a.* gouty. — *m.* gouty person.

gouttière, *f.* gutter, trough, spout; channel, groove, furrow; drip, eaves; roof.

gouvernail, *m.* rudder; helm; steering apparatus.

gouverneur, *m.* (*Mach.*) governor, regulator.

gouverne, *f.* guidance, guide.

gouvernement, *m.* government; management, direction; steering.

gouverner, *v.t.* govern; steer; manage.

gouverneur, *m.* governor; helmsman.

gouvion, *m.* large bolt.

goyave, *f.* guava (the fruit).

goyavier, *m.* guava tree, guava.

grabeau, *m.* (*Pharm.*) fragment, refuse, garble.

grabelage, *m.* (*Pharm.*) sorting to remove refuse.

grabeler, *v.t.* (*Pharm.*) cull the refuse from, garble.

grâce, *f.* grace; thanks.
— *à*, thanks to, owing to, on account of.

gracieux, *a.* graceful; gracious.

gradatif, *a.* gradational.

gradation, *f.* gradation.

grade, *m.* rank, grade; degree; (*Math.*) grad.

gradin, *m.* step; shelf; bench.

graduation, *f.* graduation.

gradué, *p.a.* graduated. — *m.* graduate.

graduel, *a.* gradual.

graduellement, *adv.* gradually.

graduer, *v.t.* graduate.

grain, *m.* grain (in various senses); berry; bead; (*Pharm.*) pellet; (*Mach.*) bushing, step; (*Glass*) opaque inclusion, stone; (*Med.*) pustule, pock; shower; squall.
à grains fins, fine-grained.
à grains serrés, close-grained.
à gros grains, large-grained, coarse-grained.
— *à facettes*, grain showing facets of crystals, faceted grain.
— *aiguillé*, grain showing acicular crystals.
— *d'amorce*, (*Pyro.*) primer.
— *de bœuf*, buffalo berry, bullberry (*Lepargyrea* or its fruit).
— *de raisin*, grape.
— *d'orge*, barleycorn.
grains de lin, linseed, flaxseed.
— *truité*, mottled grain.

graine, *f.* seed; berry; eggs (of silkworms).
— *à vers*, wormseed.
— *d'ambrette*, amber seed, musk seed.
— *d'Avignon*, Avignon berry, French berry.
— *d'écarlate*, kermes, kermes berry.
— *de girofle*, cardamom.
— *de lin*, flaxseed, linseed.
— *de paradis*, grain of paradise.
— *de Perse*, Persian berry.
— *de Turquie*, (Indian) corn, maize.
— *jaune*, yellow berry.
— *joyeuse*, fenugreek seed.
— *musquée*, musk seed, amber seed.
graines d'anis, aniseed.
graines de coton (or *de cotonnier*), cottonseed.
graines de lin, linseed, flaxseed.
graines des Moluques, Molucca grains, tilley seed.
graines de Tilly, tilley, tilley seed (from *Croton tiglium*).
— *tinctoriale*, kermes, kermes berry.

graineler, grainer, etc. See greneler, etc.

graissage, *m.* greasing; lubricating; grease; becoming ropy, roping (as of wine).

graisse, *f.* fat; grease; (*Wines*) viscous disease, alteration causing ropiness; (*Glass*) a milky and intractable state due to excess of sulfate in proportion to lime.
— *à canon*, gun wax.
— *alimentaire*, grease (or fat) for food purposes.
— *animale*, animal fat.
— *balsamique*, — *benzoinée*, (*Pharm.*) benzoinated lard.
— *de bitume*, purified bitumen.
— *de laine*, wool grease, (when purified) lanolin.
— *de machines*, lubricating grease.

graisse *de ménage,* grease for household use, specif. oleomargarin(e).

— *de mouton,* mutton suet, mutton tallow.

— *de réserve,* reserve fat.

— *du beurre,* butter fat.

— *du porc,* lard.

— *industrielle,* industrial grease.

— *minérale,* crude paraffin.

— *végétale,* vegetable fat.

— *verte,* grease from restaurants and kitchens, dripping, kitchen stuff.

graisser, *v.t.* grease; lubricate; clog (a file). — *v.i.* (of wine) become ropy or oily.

graisserie, *f.* grease business; grease shop.

graisseur, *m.* lubricator, greaser, oiler. — *a.* lubricating, greasing.

— *à godet,* grease cup.

— *mécanique,* self-acting lubricator.

graisseux, *a.* greasy; fatty.

graissier, *m.* grease dealer. — *a.* grease, dealing in grease.

gram, *m.* (*Bact.*) Gram solution, Gram stain.

gramen, *m.* (*Bot.*) grass.

graminée, *f.* (*Bot.*) grass, gramineous plant, any member of the Poaceæ.

grammaire, *f.* grammar.

grammatite, *f.* (*Min.*) grammatite (tremolite).

gramme, *m.* gram, gramme (15.433 grains).

grand, *a.* great; large; long, wide, high, tall, full; main, chief; grand. — *m.* adult; maximum; great man; dignitary; (in schools) senior; grandee; greatness, grandeur.

à grande eau, in excess of water.

en —, on a large scale; (of pictures, etc.) full-size; at large; grandly.

— *air,* open air.

— *autel,* fire bridge (in a reverberatory furnace).

— *basilic,* (*Bot.*) common basil, sweet basil (*Ocimum basilicum*).

— *boucage,* (*Bot.*) burnet saxifrage (*Pimpinella saxifraga*).

grande ciguë, (*Bot.*) poison hemlock (*Conium maculatum*).

grande digitale, (*Bot.*) common foxglove.

grande joubarbe, (*Bot.*) common houseleek.

grande masse, main body (of a blast furnace).

grande mauve, (*Bot.*) wild mallow (*Malva sylvestris*).

grande vitesse, great velocity; full speed; (*Railways*) express, dispatch by passenger train, quick dispatch.

— *feu,* (*Ceram.,* etc.) heating to the highest temperature desired, full fire.

— *foyer,* boshes (of a blast furnace).

— *gamin,* — *garçon,* a glassblower's chief assistant, called variously first footman, ball maker, ball holder, servitor, etc.

grands blés, wheat and rye.

grand *soleil,* (*Bot.*) common sunflower.

— *teint,* fast dye or color; fast.

le — Océan, the Pacific Ocean.

grand-duc, *m.* grand-duke.

Grande-Bretagne, *f.* Great Britain.

grandement, *adv.* greatly; grandly.

grandeur, *f.* greatness, largeness, etc. (see *grand*); size; quantity; magnitude; height, stature.

grandir, *v.i.* grow; increase. — *v.t.* magnify; make great; cause to grow. — *v.r.* make oneself greater or taller.

grandissement, *m.* enlargement; magnification; growth.

granit, *m.* granite.

granitelle, *f.* (*Petrog.*) granitell, aplite.

graniteux, granitique, *a.* granitic.

granitoïde, *a.* granitoid.

granulage, *m.* granulation.

granulaire, *a.* granular.

granulateur, *m.* granulator.

granulation, *f.* granulation.

granulatoire, *m.* granulator (for metals).

granule, *m.* granule.

granulé, *p.a.* granulated. — *m.* granulated sugar.

granuler, *v.t.* granulate, grain.

granuleux, *a.* granular, granulated.

granuliforme, *a.* granuliform, granular.

granulose, *f.* granulose.

graphique, *a.* graphic, graphical. — *m.* diagram, drawing, graph.

graphiquement, *adv.* graphically.

graphitation, *f.* graphitization; manufacture of graphite.

graphite, *m.* graphite, plumbago, black lead.

grappe, *f.* cluster, bunch, (*Bot.*) raceme; agglomeration, mass, lump; grapeshot, grape; cluster of bullets; powdered madder.

— *de raisin,* bunch of grapes; grapeshot.

grappiers, *m.pl.* the nonpulverulent residue obtained when hydraulic lime is slaked. A slow-setting cement called *ciment de grappiers* is prepared from it.

grappillon, *m.* small cluster.

gras, *a.* fat; fatty; greasy, oily, unctuous; rich; thick; (*Org. Chem.*) aliphatic; (of wine) ropy, oily; (of glass) milky and hard to work (see *graisse*); (of metals) soft; (of casting sand) loamy; (of soil) heavy, clayey; (of plants) fleshy; moist; slippery; meat, flesh. — *m.* fat; flesh, meat; (*Soap*) an impure precipitate formed in making white soap. — *adv.* thick, thickly.

grateron, *m.* goose grass, cleavers (*Galium*); burdock (*Arctium lappa*); sweet woodruff (*Asperula odorata*); etc.

gratification, *f.* gratuity; allowance; reward.

gratiole, f. (*Bot.*) Gratiola.
— *officinale,* hedge hyssop (*G. officinalis*).
gratioline, f. gratiolin.
gratis, adv. & a. gratis, free.
grattage, m. scraping, etc. (see gratter).
gratte, f. scraper.
gratte-boësse, f. wire brush, scratch brush.
gratter, v.t. scrape; scratch; erase; tease (cloth); flatter. — v.i. scrape; scratch.
grattoir, m. scraper; rake; eraser.
gratuit, a. gratuitous.
gratuitement, adv. gratuitously.
grauwacke, f. (*Petrog.*) graywacke, greywacke.
gravats, m.pl. = gravois.
grave, a. heavy; grave; (of sounds) deep.
graveler, v.t. gravel.
graveleux, a. gravelly.
gravelle, f. (*Med.*) gravel.
gravement, adv. gravely, seriously.
graver, v.t. engrave; etch. — v.r. be engraved; be etched; (*Pyro.*) split, crack.
graveur, m. engraver.
gravier, m. gravel; (*Salt*) a low mound on which sea salt is heaped.
gravillon, m. coarse gravel.
gravimètre, m. gravimeter (a kind of hydrometer).
gravimétrique, a. gravimetric.
gravir, v.i. & t. climb, climb up, ascend.
gravitation, f. gravitation.
gravité, f. gravity.
— *spécifique,* specific gravity.
graviter, v.i. gravitate.
gravivolumètre, m. an instrument for measuring out an exact weight of liquid.
gravois, m.pl. screenings (of plaster); rubbish.
gravure, f. engraving; etching.
gré, m. will, inclination; liking.
de — à —, by mutual agreement.
en —, in good part, kindly.
grec, a. & m. Greek.
Grèce, f. Greece.
grecque, a.fem. of grec. — f. fretwork, fret.
greffe, f. graft; grafting; record office.
greffer, v.t. graft; attach; implant. — v.r. be grafted; be attached.
grège, a. (of silk) raw.
grêle, a. slender, thin, small. — m. hail.
grêler, v.t. pass (wax) thru the *grêloir.* — v.i. hail.
grêlet, a. rather slender or thin.
grêloir, m., **grêloire,** f. a perforated vessel for converting wax into ribbons, threads or grains before bleaching.
grêlon, m. hailstone.
grenade, f. pomegranate; grenade; (*cap.*) Grenada.
— *à fusil,* rifle grenade.
— *à main,* hand grenade.
— *fumigène,* smoke grenade.

grenade *fusante,* time grenade.
— *incendiaire,* incendiary grenade.
— *suffocante,* asphyxiating grenade.
grenadier, m. pomegranate tree; grenadier.
grenadin, a. In the phrase *sirop grenadin,* sirup of pomegranate.
grenadine, f. sirup of pomegranate, grenadine.
grenage, m. granulation, graining, corning.
grenaille, f. granulated metal; granulated form (of metal); granular carbon; grain; small shot; refuse grain, tailings.
en —, granulated.
— *de plomb,* granulated lead.
— *d'étain,* granulated tin.
— *de zinc,* granulated zinc.
grenaillement, m. granulation.
grenailler, v.t. granulate.
grenat, m. & a. garnet.
grenatique, a. of or containing garnets.
greneler, v.t. grain (leather, paper, etc.).
grener, v.t. grain; granulate, corn; stipple. — v.i. seed; (of silkworms) lay eggs.
greneter, v.t. grain (leather, etc.).
grénetine, f. grenetine (a pure form of gelatin).
grenette, f. little grain; (*Expl.*) powder grains remaining on the sieve; Avignon berries or pellets made from their powder.
grenier, m. loft; granary; attic, garret.
en —, in bulk.
grenoir, m. corning house (for gunpowder); granulator, granulating machine, (for gunpowder) corning machine.
grenouille, f. frog; triturator (for antimony sulfide).
grenu, a. grained, grainy, granular. — m. grain, granular structure.
grès, m. sandstone, grit, gritstone; paving stone; scouring sand; sandy loam; (*Ceram.*) stoneware.
— *à aiguiser,* grindstone.
— *à meule,* millstone grit.
— *cérame,* stoneware.
— *de Hesse,* the refractory material of Hessian crucibles, made from fireclay and sand.
— *dur,* gritstone.
— *fin,* (*Ceram.*) fine stoneware.
— *houiller,* carboniferous sandstone.
— *ordinaire,* (*Ceram.*) common stoneware.
grès-cérame, m. stoneware.
grésil, m. broken or powdered glass; sleet.
grésillé, p.a. shriveled; (*Ceram.*) having rough, pitted glaze. — m. (*Ceram.*) roughness or pitting of the glaze, due to reducing vapors in the kiln.
grésillement, m. shriveling, etc. (see grésiller); chirping (of crickets); sound like that of sleet.
grésiller, v.t. shrivel, shrivel up; trim (the edges of a piece of glass); (*Ceram.*) give a rough, pitted surface to. — v.i. sleet.

gresserie, *f.* sandstone quarry; sandstone work; (*Ceram.*) stoneware.

grève, *f.* mortar sand, coarse sand; sand bank; strand; strike (of workers).

faire —, *se mettre en* —, strike.

grever, *v.t.* burden, encumber.

gréviste, *m.* striker. — *a.* striking.

grief, *m.* grievance.

grieu, *m.* fire damp.

griffage, *m.* seizing, gripping; scratching.

griffe, *f.* claw; catch, pawl, prong, hook, catch; stamped signature; stamp (for signatures); root, rootlet (of certain plants).

griffes de girofle, clove stems.

griffer, *v.t.* seize with, or as with, claws, grip; scratch, claw.

grignard, *m.* hard stoneware, used in construction; confusedly crystallized gypsum, found in plaster stone.

grignons, *m.pl.* olive husks, marc of olives; crusts.

grillage, *m.* roasting, etc. (see griller).

— *au four à cuve,* roasting in kilns.

— *en stalles,* (*Metal.*) roasting in stalls.

— *en tas,* roasting in heaps.

grillager, *v.t.* lattice, grate.

grille, *f.* grate, grating; grid; furnace (as for combustions, for Kjeldahl flasks, for evaporations); (*Metal.*) roasting furnace, calcining furnace, roaster, calciner; fire bar; lattice; grille; grill.

— *à analyse(s),* — *à combustion,* combustion furnace.

— *à attaque,* furnace for decompositions (as one for Kjeldahl flasks).

— *à feu,* fire grate.

— *à manche,* strainer.

— *d'analyse,* combustion furnace.

— *de foyer,* fire grate.

— *en escalier,* step grate.

griller, *v.t.* roast, calcine; singe (cloth); scorch, parch, burn; broil, grill.

grilloir, *m.* roaster, roasting kiln; singeing machine, singer.

grimpage, *m.* climbing.

grimper, *v.i. & r.* climb (up); (of salts) creep.

grincer, *v.i.* grind, grate; gnash. — *v.t.* gnash.

grindélie, *f.* (*Bot.*) gum plant (*Grindelia*); (*Pharm.*) grindelia.

griotte, *f.* a sour cherry; a spotted marble.

griottier, *m.* a kind of cherry tree.

grippage, *m.* seizing, etc. (see gripper).

grippeler, *v.r.* shrink.

grippement, *m.* seizing, etc. (see gripper).

gripper, *v.t.* seize, grip, catch; snatch; shrink, contract. — *v.r.* shrink, shrivel. — *v.i.* (*Mach.*) grip, bind, wedge, grind.

grippure, *f.* (*Mach.*) worn part.

gris, *a.* gray, grey; (of paper) unbleached, brown; (of wine) pale; tipsy. — *m.* gray, grey.

gris *acier,* steel gray.

— *ardoise,* slate gray.

— *argenté,* silver gray.

— *clair,* light gray.

— *d'acier,* steel gray.

— *de fer,* iron gray; (*Dyeing*) iron liquor.

— *de perle,* pearl gray.

— *de zinc,* zinc dust.

— *foncé,* dark gray.

— *perle,* pearl gray.

grisaille, *f.* grizzled color, gray, mixed white and black. — *a.* grizzled, gray.

grisard, *a.* dark gray, or grayish brown.

grisâtre, *a.* grayish, greyish.

griser, *v.t.* gray, make gray; shade (with lines); intoxicate. — *v.i.* become gray. — *v.r.* become intoxicated.

grisou, *m.* fire damp.

grisoumètre, *m.* an apparatus for analyzing air for fire damp.

grisouteux, *a.* containing fire damp.

Groenland, *m.* Greenland.

grogner, *v.i.* grunt; grumble.

groisil, *m.* broken glass, cullet.

gronder, *v.i.* growl, roar, rumble; grumble.

gros, *a.* big, large, great; coarse, gross, thick; heavy; (of colors) dark, deep; (of persons) rich. — *m.* bulk, mass, body, main part; large-sized material, specif. lump coal; (*Com.*) wholesale; revenue; grogram (a fabric). — *adv.* much, largely, high.

en —, wholesale; in general.

— *bleu,* a coarse variety of smalt.

gros-blanc, *m.* a cement composed of whiting and glue.

groseille, *f.* currant. — *a.* currant-colored, currant-red, currant.

— *à maquereau,* gooseberry.

groseillier, *m.* currant bush, currant; (*à maquereau*) gooseberry bush, gooseberry.

grosse, *a.fem.* of gros. — *f.* gross; large piece.

grossement, *adv.* in general; grossly.

grosserie, *f.* wholesale trade.

grossesse, *f.* pregnancy.

grosseur, *m.* size; swelling, tumor.

grossier, *a.* coarse; rough, rude, gross; imperfect. — *m.* coarse material.

grossièrement, *adv.* coarsely; roughly, rudely, grossly; imperfectly.

grossièreté, *f.* coarseness; coarse language.

grossir, *v.i.* grow large or larger, increase, swell. — *v.t.* magnify; enlarge, make larger, increase, swell. — *v.r.* grow larger, increase; be magnified; be enlarged.

grossissant, *p.a.* magnifying; enlarging; increasing.

grossissement, *m.* magnification, magnifying; magnifying power; enlargement; increasing, swelling.

grossulaire, *a.* grossular.

grotte, *f.* grotto.

groupe, *m.* group; set.

— *directeur,* directing group, (in dyes) chromophore.

— *haptophore,* haptophorous group, haptophore group.

— *imidé,* imido (or imino) group.

— *sulfonique,* sulfonic group, sulfo group.

— *toxophore,* toxophorous group, toxophore group.

— *yttrique,* yttrium group.

groupement, *m.* grouping; group; gathering; (*Elec.*) connecting up, connection.

— *prosthétique,* prosthetic group.

grouper, *v.t.* group; (*Elec.*) connect up, join up. — *v.i.* group. — *v.r.* group; gather.

— *en dérivation,* — *en quantité,* (*Elec.*) connect in parallel.

— *en série,* — *en tension,* (*Elec.*) connect in series.

gruau, *m.* meal, coarse flour, specif. wheat meal; grits, groats; flour; gruel; small crane.

gruauter, *v.t.* grind, reduce to meal.

grue, *f.* crane.

— *à vapeur,* steam crane.

gruer, *v.t.* grind, reduce to meal.

grugeon, *m.* a large piece of sugar.

gruger, *v.t.* crunch.

grume, *f.* bark.

grumeau, *m.* clot; flock; lump, particle.

grumeleux, *a.* clotted, clotty, lumpy.

grumelure, *f.* small hole (in cast metal).

grumillon, *m.* piece of forge scale.

gruyère, *m.* Gruyère cheese, Schweitzer cheese.

guaco, *m.* (*Bot. & Pharm.*) guaco.

guaéthol, *m.* guaiethol (2-ethoxyphenol).

guanase, *f.* guanase.

guaner, *v.t.* manure with guano.

guanidine, *f.* guanidine.

guanier, *a.* of or yielding guano, guaniferous.

guanine, *f.* guanine.

guano, *m.* guano.

— *de chauves-souris,* bat guano.

— *de poissons,* fish guano, fish manure.

— *de viande,* meat guano, meat meal.

— *marin,* — *polaire,* a fertilizer made from whale refuse.

— *phosphaté,* phosphatic guano.

guède, *f.* woad, pastel (either dye or plant).

guéder, *v.t.* dye with woad; cram, stuff.

guéderon, guédron, guédon, *m.* woad dyer.

guêpe, *f.* wasp.

guère, *adv.* (with *ne*) scarcely, hardly, little, not long, not much, few.

guérir, *v.t.* cure. — *v.i. & r.* recover; heal; be cured (of).

guérison, *m.* cure, recovery, healing.

guérite, *f.* a small chamber or shelter, as a condensation chamber in the manufacture of zinc oxide, a sentry box, a lookout, a conning tower, a locomotive cab or a telephone booth.

— *respiratoire,* respiration chamber.

guerre, *f.* war; warfare; war department.

guerrier, *a.* warlike, martial. — *m.* warrior.

guesdre, *f.* woad, pastel.

guesdron, *m.* woad dyer.

guet, *m.* watch.

guetter, *v.t.* watch; lie in wait for.

gueulard, *m.* throat, mouth (of a blast furnace); mouth (of a sewer).

gueule, *f.* mouth; (of a gun) muzzle, mouth; (*Founding*) gate.

gueuse, *f.* (*Iron*) pig.

gueuse-mère, *f.* (*Iron*) sow.

gueuset, *m.* (*Iron*) small pig.

gui, *m.* (*Bot.*) mistletoe (*Viscum album*).

guichet, *m.* wicket; little window; shutter.

guidage, *m.* guiding, etc. (see guider).

guide, *m.* guide. — *f.* rein.

guider, *v.t.* guide; conduct, direct; steer.

guildive, *f.* a kind of rum.

guillage, *m.* (*Brewing*) working, fermenting.

guillemet, *m.* quotation mark, quote.

guillemeter, *v.t.* put in quotation marks.

guiller, *v.i.* (*Brewing*) work, ferment.

guilloire, *f.* fermentation vat, gyle vat.

guimauve, *f.* (*Bot.*) marshmallow (*Althœa*).

guinand, *m.* (*Glass*) a hollow cylinder used in stirring optical glass.

guinder, *v.t.* raise, hoist, lift.

guinée, *f.* guinea; (*cap.*) Guinea.

guipage, *m.* wrapping, winding; insulation.

guiper, *v.t.* wind, wrap (with tape or the like).

guise, *f.* manner, fashion, way, wise.

en — *de,* instead of, by way of.

gummifère, *a.* gummiferous, gum-bearing.

gurgu, gurjum, gurjun, *m.* gurjun, gurjun balsam (from *Dipterocarpus* species).

gutta, *f.* Short for gutta-percha.

gutta-percha, gutta-perca, *f.* gutta-percha.

gutte, *f.* gamboge, gum guttæ.

guttifère, *a.* (*Bot.*) guttiferous; (*Min.*) in the form of drops or tears.

gymnosperme, *a.* (*Bot.*) gymnospermous. — *m.* gymnosperm.

gypse, *m.* gypsum.

gypseux, *a.* gypseous.

gypsifère, *a.* gypsiferous.

gypsomètre, *m.* gypsometer (apparatus for determining sulfates in wine).

gypsométrique, *a.* gypsometric, pertaining to the determination of sulfates in wine.

gyratoire, *a.* gyratory.

H

habile, *a.* capable, competent, skillful, clever.

habilement, *adv.* ably, skillfully, cleverly.

habileté, *f.* ability, skill, cleverness.

habillage, *m.* dressing, etc. (see habiller).

habillement, *m.* clothing.

habiller, *v.t.* dress; cover, wrap; clothe; fit, become. — *v.i.* fit, be becoming.

habit, *m.* attire, clothes, garment, suit, habit, coat, dress coat.

habitant, *m.* inhabitant, resident.

habitation, *f.* dwelling, habitation; haunt; settlement; plantation.

habiter, *v.t.* inhabit, live in. — *v.i.* dwell.

habitude, *f.* habit.

   d'—, ordinarily.

habituel, *a.* usual, customary; habitual.

habituellement, *adv.* usually, customarily; habitually.

habituer, *v.t.* habituate, accustom.

hache, *f.* ax.

hacher, *v.t.* chop, hack, cut (in pieces), hash, mince, shred; hatch, shade.

hachette, *f.* hatchet.

   — à marteau, = hachette-marteau.

hachette-marteau, *n.* hatchet with hammer head.

hachisch, hachich, *m.* hashish.

hachoir, *m.* cutter, chopper; chopping board.

hachure, *f.* hatching, shading.

hæm-. See hém-.

haï, *p.p.* of haïr.

haie, *f.* hedge.

haine, *f.* hatred, hate, enmity.

haïr, *v.t.* hate, abhor, detest.

haleine, *f.* breath.

   en —, in working trim, in practice.

haler, *v.t. & i.* haul, pull.

haleter, *v.i.* pant, breathe quickly.

halimètre, *m.* = halomètre (first def.).

halle, *f.* market, market place.

halo, *m.* halo.

halochimie, *f.* the chemistry of salts.

halochimique, *a.* pertaining to the chemistry of salts.

halogénalkyl-magnésien, *a.* In the phrase *composé halogénalkyl-magnésien,* alkylmagnesium halide.

halogène, *m.* halogen. — *a.* halogen, halogenous.

halogéné, *p.a.* halogenated, halogen.

halogénique, *a.* designating the residues of oxacids (as NO_3, SO_4, etc.).

halographe, *m.* one who writes about salts.

halographie, *f.* halography, description of salts.

haloïde, *a.* haloid, halide. — *m.* halide.

halologie, *f.* a treatise on salts.

halologique, *a.* pertaining to the knowledge or description of salts.

halomètre, *m.* an apparatus for determining the amount of salts present (as in sirups); an instrument for measuring the angles of salt crystals, goniometer.

halométrie, *f.* determination of salts.

halométrique, *a.* pertaining to the determination of salts.

halotechnie, *f.* the art of preparing salts.

halotechnique, *a.* pertaining to the preparation or manufacture of salts.

haltère, *m.* dumb-bell.

halurgie, *f.* halurgy. (*Obs.*)

halurgique, *a.* relating to salt making. (*Obs.*)

hambourg, *m.* small cask, keg; (*cap.*) Hamburg.

hameau, *m.* hamlet.

hameçon, *m.* fish hook.

hampe, *f.* staff, shank, handle; (*Bot.*) stem, stalk.

hanche, *f.* hip, haunch; (*Tech.*) leg.

hanebane, *f.* (*Bot.*) henbane (*Hyoscyamus niger*).

hangar, *m.* open shed.

happant, *p.a.* adhering, adherent (as clay to the tongue).

happelourde, *m.* imitation gem.

happement, *m.* adhering, etc. (see happer).

happer, *v.i.* adhere, stick. — *v.t.* snap up, snatch, seize.

harasse, *f.* crate (as for glass or porcelain).

harderie, *f.* iron sulfate (*Obs.?*).

hardi, *a.* bold; daring, hardy, venturesome.

hardiesse, *f.* boldness.

hareng, *m.* herring.

haricot, *m.* bean (refers specif. to the genus *Phaseolus;* cf. fève).

harmaline, *f.* harmaline.

harmonie, *f.* harmony.

harmonieusement, *adv.* harmoniously.

harmonieux, *a.* harmonious.

harmonique, *a.* harmonic, harmonical.

176

harmoniquement, *adv.* harmonically.

harmoniser, *v.t. & r.* harmonize.

harmotome, *m.* (*Min.*) harmotome.

harnais, *m.* harness; equipment; gear, gearing.

harveyiser, *v.t.* (*Metal.*) harveyize, Harveyize.

hasard, *m.* chance; accident; hazard, risk; bargain.

à tout —, at all events, in any case.

au —, at random.

au — *de*, at the risk of.

par —, by chance, by accident, chance.

hasarder, *v.t., r. & i.* hazard, risk, venture.

hasardeux, *a.* hazardous; venturesome.

haschisch, *m.* hashish.

hatchetine, *f.* (*Min.*) hatchettite, hatchettine.

hâte, *f.* haste, hurry, hastiness.

à la —, *en* —, in haste, hastily, hurriedly.

hâter, *v.t. & r.* hasten, hurry, quicken.

hâtif, *a.* hasty; premature; (of fruits) early.

hâtivement, *adv.* hastily; early.

haufwerk, *m.* run of mine, mine run.

hausse, *f.* rise (in price), advance; something which serves to raise, lift, support; second course (of a reverberatory furnace); rear sight (of small arms).

hausser, *v.t.* raise. — *v.i.* rise. — *v.r.* rise; be raised; (of weather) clear.

haut, *a.* high; deep; upper; (of colors) bright. — *m.* height; top. — *adv.* highly; high; above; back; loudly, aloud.

de — *en bas,* from above; from top to bottom.

d'en —, from above.

en —, above; up; upstairs.

en — *de,* at the top of.

haute pression, high pressure.

— *fourneau,* blast furnace.

le — *Rhin,* the Upper Rhine.

par en —, upward; by the upper way; at the top; up above.

par —, upward.

hautement, *adv.* highly; openly; loudly, aloud.

hauteur, *f.* height; altitude; depth; loftiness.

à la — *de,* on a level with, abreast with, equal to.

haut-fourneau, *m.* blast furnace.

haüyne, *f.* (*Min.*) haüyne, haüynite.

havane, *a.* light chestnut. — *m.* Havana cigar or tobacco. — *f.* (*cap.,* with *La*) Havana.

havre, *m.* harbor; (*cap.,* with *Le*) Havre.

Haye (La), *f.* The Hague.

hebdomadaire, *a.* weekly.

hebdomadairement, *adv.* weekly, once a week.

hébéter, *v.t.* stupefy.

hébétude, *f.* (*Med.*) hebetude, dullness.

hébraïque, *a.* Hebrew, Hebraic.

hébreu, *a. & m.* Hebrew.

hect., *abbrev.* (hectolitre) hectoliter.

hectare, *m.* hectare (10,000 sq. m.; 2.471 acres).

hectogramme, hecto, *m.* hectogram (100 grams).

hectolitre, hecto, *m.* hectoliter, hectolitre (100 liters; 26.417 U.S. gals., 22.0097 Brit.).

hectostère, *m.* hectostere (100 cu. m., 130.8 cu. yds.).

hédéoma, *m.* (*Bot.*) hedeoma, (American) pennyroyal.

hédonal, *m.* (*Pharm.*) hedonal.

hélianthe, *m.* (*Bot.*) sunflower, helianthus.

hélianthine, *f.* helianthin (methyl orange).

hélice, *f.* helix; (*Mech.*) screw.

hélicine, *f.* helicin.

hélicoïdal, *a.* helicoidal.

hélicoïde, *a.* helicoid, helical. — *m.* helicoid.

héliochromie, *f.* heliochromy, color photography.

héliochromique, *a.* heliochromic.

héliographie, *f.* heliography.

héliographique, *a.* heliographic.

héliogravure, *f.* heliogravure.

hélioplastie, *f.* a process of photo-engraving.

héliotropine, *f.* heliotropin (piperonal).

héliotypie, *f.* heliotypy.

hélium, *m.* helium.

hellébore, *m.* (*Bot.*) hellebore.

helminthe, *m.* intestinal worm, helminth.

helminthique, *a. & m.* (*Med.*) helminthic.

helvétique, *a.* Swiss.

hématéine, *f.* hematein, hæmatein.

hématie, *f.* red blood corpuscle, erythrocyte.

hématimètre, *m.* hematimeter, hæmatimeter.

hématine, *f.* hematin, hæmatin.

hématique, *a.* hematic, hæmatic.

hématite, *f.* (*Min.*) hematite. — *a.* hematitic, blood-red, blood.

— *brune,* brown hematite, limonite.

— *rouge,* red hematite, ordinary hematite.

hématocrite, *m.* hematocrit, hæmatocrit.

hématogène, *m.* hematogen, hæmatogen.

hématoïdine, *f.* hematoidin, hæmatoidin.

hématolyse, *f.* hematolysis, hemolysis.

hématoporphyrine, *f.* hematoporphyrin.

hématoxyle, *m.* logwood tree (*Hæmatoxylon*).

hématoxyline, *f.* hematoxylin, hæmatoxylin.

hématurie, *f.* (*Med.*) hematuria, hæmaturia.

hémialbumose, *f.* hemialbumose.

hémicirculaire, *a.* semicircular.

hémicylindrique, *a.* hemicylindrical.

hémièdre, *a.* (*Cryst.*) hemihedral.

hémiédrie, *f.* (*Cryst.*) hemihedrism, hemihedry.

— *plagièdre,* plagihedral hemihedrism.

hémiédrique, *a.* hemihedral, (rarely) hemihedric.

hémimellique, *a.* hemimellitic.

hémine, *f.* hemin, hæmin.

hémiperméable, *a.* semipermeable.

hémipinique, *a.* hemipic, hemipinic.

hémiprismatique, *a.* (*Cryst.*) hemiprismatic.

hémisphère, *m.* hemisphere.

hémisphérique, *a.* hemispheric(al).

hémitrope, *a.* (*Cryst.*) hemitrope, hemitropic.

hémitropie, *f.* (*Cryst.*) hemitropism, hemitropy.

hémochromogène, *m.* hemochromogen.

hémocytomètre, *m.* hemacytometer, hemocytometer.

hémoglobine, *f.* hemoglobin, hæmoglobin.

hémoglobinomètre, *m.* hemoglobinometer.

hémoglobinurie, *f.* hemoglobinuria.

hémolyse, *f.* hematolysis, hemolysis.

hémolysine, *f.* hemolysin, hæmolysin.

hémolytique, *a.* hematolytic, hemolytic.

hémomètre, *m.* hemometer, hemadynometer.

hémopyrrol, *m.* hemopyrrole.

hémorragie, hémorrhagie, *f.* hemorrhage.

hémorragique, hémorrhagique, *a.* hemorrhagic.

hémorroïde, hémorrhoïde, *f.* hemorrhoid, pile.

hémostatique, *a.* & *m.* hemostatic, hæmostatic.

henequen, *m.* henequen, henequin, sisal hemp.

henné, henneh, *m.* henna.

hépar, *m.* (*Old Chem.*) hepar.

hépatique, *a.* hepatic. — *f.* (*Bot.*) hepatica.

hépatisation, *f.* hepatization.

hépatiser, *v.t.* hepatize.

hépatite, *f.* (*Min.*) hepatite; (*Med.*) hepatitis.

hepsomètre, *m.* an apparatus for regulating the evaporation of sugar solutions.

hepsométrie, *f.* measurement of the degree of evaporation of sugar solutions.

hepsométrique, *a.* pertaining to *hepsométrie*.

heptadécanoïque, *a.* heptadecoic, heptadecylic.

heptaméthylène, *m.* heptamethylene (cycloheptane).

heptane, *m.* heptane.

heptose, *f.* heptose.

heptylène, *m.* heptylene.

heptylique, *a.* heptyl.

héraut, *m.* herald.

herbacé, *a.* herbaceous.

herbe, *f.* herb; grass.

  en —, green, unripe; prospective; in embryo.

  — *à fièvre,* = herbe parfaite.

  — *à jaunir,* yellowweed (*Reseda luteola*); woadwaxen (*Genista tinctoria*).

  — *à loup,* wolfsbane, monkshood (*Aconitum*).

  — *à rayons,* madder.

  — *à Robert,* herb Robert.

  — *à savon,* soapwort.

  — *au scorbut,* scurvy grass (*Cochlearia officinalis*).

  — *au soleil,* sunflower.

  — *au verre,* saltwort, glasswort (*Salsola kali*).

  — *aux chantres,* hedge mustard (*Sisymbrium officinale*).

herbe *aux chats,* catnip; cat thyme.

  — *aux tanneurs,* tanner's sumac.

  — *aux teinturiers,* woadwaxen, dyer's greenweed.

  — *de Saint Antoine,* willow herb (*Chamænerion angustifolium*).

  — *de Saint-Philippe,* woad.

  — *des Juifs,* yellowweed (*Reseda luteola*).

  — *dragon,* — *dragonne,* tarragon.

  — *empoisonnée,* alkekengi; belladonna, nightshade.

  — *jaune,* yellowweed (*Reseda luteola*).

  — *marine,* seaweed.

  — *mauvaise,* weed.

  — *parfaite,* boneset (*Eupatorium perfoliatum*).

herber, *v.t.* grass, bleach on the grass.

herbivore, *a.* herbivorous. — *m.* herbivore.

herbue, *f.* pasture land; (*Metal.*) clay flux.

herculéen, *a.* Herculean.

héréditaire, *a.* hereditary.

hérédité, *f.* heredity; heirship, inheritance.

hérissé, *a.* bristling, bristly; covered; rough; prickly.

hérisser, *v.t.* bristle; cover; (*Masonry*) roughcast. — *v.i.* bristle; become rough.

hérisson, *m.* hedgehog; (*Mach.*) cog wheel, spur wheel.

héritage, *m.* heritage, inheritance.

hériter, *v.t.* & *i.* inherit.

héritier, *m.* heir.

herméticité, *f.* state of being hermetic, airtightness.

hermétique, *a.* hermetic(al); Hermetic, Hermetical (pertaining to Hermes Trismegistus, or to alchemy).

hermétiquement, *adv.* hermetically.

hermétisme, *m.* Hermetism (alchemical doctrine or cult).

Hérodote, *m.* Herodotus.

héroïque, *a.* heroic.

héros, *m.* hero.

herse, *f.* harrow.

herser, *v.t.* harrow.

hésitation, *f.* hesitation.

hésiter, *v.i.* hesitate.

hespéridine, *f.* hesperidin.

hétérocline, *f.* (*Min.*) a ferruginous braunite.

hétérocyclique, *a.* heterocyclic.

hétérogène, *a.* heterogeneous.

hétérogénéité, *f.* heterogeneity, heterogeneousness.

hétéromorphisme, *m.,* **hétéromorphie,** *f.* (*Cryst.*) heteromorphism, heteromorphy.

hétéroprotéose, *m.* heteroproteose.

hétéroxanthine, *f.* heteroxanthine.

hêtre, *m.* beech (either tree or wood).

heuchère, *f.* (*Bot.*) heuchera.

  — *d'Amérique,* alumroot (*Heuchera americana*).

heure, *f.* hour; time.

  à cette —, at this time, now.

  à l'— *qu'il est,* at the present time.

  pour l'—, for the present.

  sur l'—, on the instant, immediately.

  (For other phrases see under *bon* and *tout*).

heureusement, *adv.* happily, advantageously, fortunately.

heureux, *a.* happy; fortunate; pleasing.

heurt, *m.* shock, knock; bruise; crown (as of a bridge).

heurter, *v.t.* hit, strike, knock, run against, collide with; shock. — *v.r. & i.* knock, hit, strike, collide.

heurtoir, *m.* catch, stop; buffer, hurter; sill.

hexachloré, *a.* hexachloro.

hexachlorure, *m.* hexachloride.

hexaèdre, *a.* hexahedral. — *m.* hexahedron.

hexaédrique, *a.* hexahedral.

hexagonal, *a.* hexagonal.

hexagone, *m.* hexagon.

hexahydraté, *a.* hexahydrated.

hexahydrobenzénique, *a.* of or pertaining to hexahydrobenzene (cyclohexane).

hexahydrobenzoïque, *a.* hexahydrobenzoic (cyclohexanecarboxylic).

hexahydrophénol, *m.* hexahydrophenol (cyclohexanol).

hexahydropyridique, *a.* of or pertaining to hexahydropyridine (piperidine).

hexahydrure, *m.* hexahydride.

hexaméthylène, *m.* hexamethylene (radical and compound; the latter is better called cyclohexane).

hexaméthylénique, *a.* of or pertaining to hexamethylene (cyclohexane).

hexanitré, *a.* hexanitrated, hexanitro.

hexaoxybenzène, *m.* hexahydroxybenzene.

hexatomique, *a.* hexatomic.

hexite, *f.* hexitol, hexite.

hexonique, *a.* (of acids) hexonic; (of bases) hexone.

hexosane, *m.* hexosan.

hexose, *f.* hexose.

hexyle, *m.* hexyl.

hexylène, *m.* hexylene.

hexylénique, *a.* hexylene, of hexylene.

hexylique, *a.* hexyl, hexylic.

hibou, *m.* owl.

hidrotique, *a. & m.* (*Med.*) hidrotic, sudorific.

hie, *f.* ram, beetle, pile driver.

hier, *adv.* yesterday. — *v.t.* ram, drive. — *v.i.* creak.

hilarant, *a.* exhilarating.

  gaz —, laughing gas (nitrous oxide).

hile, *m.* (*Bot.*) hilum.

hippurate, *m.* hippurate.

hippurique, *a.* hippuric.

hirudine, *f.* hirudin.

histidine, *f.* histidine.

histochimie, *f.* histochemistry.

histochimique, *a.* histochemic(al).

histogène, *a.* histogenetic, histogenic.

histogénèse, *f.* histogenesis.

histogénie, *f.* histogeny.

histogénique, *a.* histogenetic, histogenic.

histoire, *f.* history; story.

histologie, *f.* histology.

histologique, *a.* histologic, histological.

histologiquement, *adv.* histologically.

histologiste, *m.* histologist.

histolyse, *f.* histolysis.

histone, *f.* histone.

historien, *m.* historian.

historique, *a.* historic, historical.

historiquement, *adv.* historically.

hiver, *m.* winter.

hivernal, *a.* winter; wintry.

hiverner, *v.i.* hibernate, winter.

hl., *abbrev.* hectoliter(s), hectolitre(s).

hoche, *f.* notch, nick.

hocher, *v.t.* shake; notch, nick.

hollandais, *a.* Dutch. — *m.* Dutch (language); (*cap.*) Dutchman.

hollande, *f.* Dutch porcelain; holland (the fabric); (*cap.*) Holland. — *m.* Dutch cheese, esp. Edam cheese.

hollander, *m.* hollander, pulper.

holmium, *m.* holmium.

holocristallin, *a.* (*Petrog.*) holocrystalline.

holoèdre, *m.* (*Cryst.*) holohedron. — *a.* holohedral.

holoédrie, *f.* (*Cryst.*) holohedrism.

holoédrique, *a.* (*Cryst.*) holohedral.

homard, *m.* lobster.

homéopathie, *f.* (*Med.*) homeopathy.

homéotropie, *f.* homeotropy (analysis of liquids by behavior of their drops).

homme, *m.* man.

homocaféique, *a.* homocaffeic.

homofluorescéine, *f.* homofluorescein.

homofocal, *a.* homofocal, confocal.

homogène, *a.* homogeneous.

homogénéifier, *v.t.* homogenize.

homogénéisateur, *a.* homogenizing. — *m.* homogenizer, homogenizing apparatus.

homogénéisation, *f.* homogenization.

homogénéiser, *v.t.* homogenize.

homogénéité, *f.* homogeneity, homogeneousness.

homogènement, *adv.* homogeneously.

homogentinisique, homogentisique, *a.* homogentisic.

homologable, *a.* capable of being homologized.

homologie, *f.* homology.

homologique, *a.* homologic(al), homologous.

homologiquement, *adv.* homologically.

homologue, *a.* homologous.

homologuer, *v.t.* homologize; (*Law*) homologate.

homophtalimide, *f.* homophthalimide.

homopyrocatéchine, *f.* homopyrocatechol.

homopyrrol, *m.* homopyrrole.

homoquinine, *f.* homoquinine.

Hongrie, *f.* Hungary.

hongrieur, *m.* = hongroyeur.

hongroierie, *f.* = hongroyage.

hongrois, *a.* Hungarian. — *m.* Magyar (language); (*cap.*) Hungarian.

hongroyage, *m.* preparation of Hungary leather (tanning with alum and salt followed by dressing with tallow or other fatty material).

hongroyer, *v.t.* convert into Hungary leather, subject to *hongroyage* (which see).

hongroyeur, *m.* maker of Hungary leather.

honnête, *a.* honest; proper; decent; reasonable, moderate; civil.

honnêtement, *adv.* honestly, etc. (see honnête).

honnêteté, *f.* honesty; propriety; civility; kindness; present.

honneur, *m.* honor, honour.

en —, in favor.

honorable, *a.* honorable.

honorablement, *adv.* honorably.

honoraire, *a.* honorary. — *m.* (also *pl.*) honorarium.

honorer, *v.t.* honor; be an honor to.

honte, *f.* shame.

honteusement, *adv.* shamefully.

honteux, *a.* shameful; ashamed; timid.

hôpital, *m.* hospital.

horaire, *a.* hourly; horary. — *m.* time-table.

hordéine, *f.* hordein.

horizon, *m.* horizon.

horizontal, *a.* horizontal.

horizontalement, *adv.* horizontally.

horloge, *f.* clock, timepiece.

hormis, *prep.* except, excepting, but.

— *que,* except that.

hormonal, *a.* of or pertaining to hormones.

hormone, *f.* hormone.

hornblende, *f.* (*Min.*) hornblende.

horreur, *m.* horror.

hors, *prep.* outside, without, out of; except, but. — *adv.* outside, out.

— *de,* out of.

horticole, *a.* horticultural.

hospitalier, *a.* hospitable.

hostile, *a.* hostile.

hostilité, *f.* hostility.

hôte, *m.* host; guest.

hôtel, *m.* mansion, house, hall, building; hotel.

— *de ville,* city hall, city building.

hotte, *f.* hood; basket(for ore, etc.); hod; scuttle.

— *à bon* (or *fort*) *tirage,* hood with a good (or strong) draft.

hottée, *f.* basketful; hodful; scuttleful.

hotter, *v.t.* carry in a basket, hod or scuttle.

houblon, *m.* hop (the plant or its flower).

houblonnage, *m.* hopping, addition of hops.

houblonner, *v.t.* hop, impregnate with hops.

houblonnier, *a.* pertaining to hops, hop. — *m.* hop raiser.

houblonnière, *f.* hop field, hop garden.

houe, *f.* hoe.

— *à cheval,* horse hoe, cultivator.

houement, *m.* hoeing.

houer, *v.t. & i.* hoe.

houille, *f.* (mineral) coal.

— *à courte flamme,* short-flame coal.

— *à longue flamme,* long-flame coal.

— *anthraciteuse,* anthracite coal.

— *blanche,* water power, specif. that of springs and waterfalls (said to be so called in allusion to the mountain snows which feed the streams).

— *bleue,* wind power (from the blue of the sky?).

houille *brune,* brown coal, lignite.

— *carbonisée,* coke.

— *crue,* coal (as distinguished from coke).

— *de forge,* forge coal.

— *demi-grasse,* a short-flame coking coal.

— *éclatante,* anthracite coal, glance coal.

— *flambante,* flaming coal.

— *grasse,* fat coal, coking coal.

— *luisante,* anthracite coal, glance coal.

— *maigre,* lean coal, nonbituminous coal; specif., semi-anthracite.

— *maréchale,* forge coal.

— *noire,* black coal, coal (as distinguished from lignite or from water power and wind power).

— *non collante,* non-caking coal.

— *proprement dite,* coal in the proper sense (by some applied to bituminous coal).

— *sans flamme,* non-flaming coal.

— *sèche,* dry coal; non-coking coal.

— *tendre,* soft coal.

— *terreuse,* earthy coal, lignite.

— *tout-venante,* rough coal, run of mine.

— *verte,* water power from ordinary watercourses (said to be so called in allusion to the greenness of the banks).

houiller, *a.* coal, of or containing coal.

houillère, *f.* coal mine, colliery.

houilleur, *m.* coal miner.

houilleux, *a.* coal, coaly, containing coal.

houillite, *f.* anthracite (old name).

houle, *f.* swell, billow.

houlette, *f.* shepherd's crook; (*Tech.*) any of several tools, as a glassmaker's shovel or spatula, a founder's scoop or a garden trowel.

houppe, *f.* tuft.

houppé, *a.* tufted.

houppette, *f.* small tuft.

houssage, *m.* dusting, sweeping.

housse, *f.* housing, cover; (*Ceram.*) rough form of what is to be a vase or other hollow object.

housser, *v.t.* dust, sweep.

houssoir, *m.* duster, whisk, broom (as of feathers or hair).

houx, *m.* (*Bot.*) holly, holly tree.

huche, *f.* bin, chest; trough, tub.

huilage, *m.* oiling.

huile, *f.* oil.

— à l'—, (*Painting*) in oil, oil.

— à brûler, burning oil, specif. kerosene.

— absolue de Braconnot, olein.

— à graisser, lubricating oil.

— alimentaire, oil for food, edible oil.

— à manger, edible oil; specif., table oil, salad oil.

— à mécanisme, — à mouvement, machine oil.

— animale, animal oil; specif., bone oil.

— à usine, lubricating oil.

— blanche, white oil, specif., (*Painting*) poppy oil.

— brute, crude oil.

— camphrée, (*Pharm.*) camphor liniment.

— comestible, edible oil.

— copal à tableaux, (*Painting*) poppy or linseed oil to which copal has been added.

— crue, raw oil.

— cuite, boiled oil; linseed oil varnish.

— d'amandes douces, sweet-almond oil, expressed oil of almond.

— d'arachide, peanut oil, groundnut oil.

— d'arsenic, (*Old Chem.*) butter of arsenic, caustic oil of arsenic (arsenic trichloride).

— de baleine, whale oil, train oil.

— de blanc de baleine, sperm oil, spermaceti oil.

— de bouche, edible oil.

— de Brésil, copaiba.

— de cacao, cacao butter, cocoa butter.

— de cade, oil of cade, cade oil.

— de cantharides térébenthinée, (*Pharm.*) blistering liquid.

— de cétacés, whale oil.

— de cheval, caballine oil, melted horse grease.

— de coco, coconut oil.

— de colza, colza oil (a variety of rapeseed oil).

— de coprah, copra oil, coconut oil.

— de coton, cottonseed oil.

— de couperose, oil of vitriol, concentrated sulfuric acid; specif., Nordhausen acid, fuming sulfuric acid.

— de créosote, creosote oil.

— de croton, croton oil.

— de cuisine, cooking oil, edible oil.

— de foie de morue, codliver oil.

— de fusel, fusel oil.

— de goudron, tar oil.

— de grain, fusel oil.

huile de graissage, lubricating oil.

— de graisse, lard oil.

— de houille, coal-tar oil.

— de lampe, lamp oil, burning oil.

— de lin, linseed oil.

— de lin crue, raw linseed oil.

— de lin cuite, boiled linseed oil.

— de machine, machine oil, engine oil.

— de marc, oil extracted from marc (as of grapes or olives).

— de morue, cod oil, codliver oil.

— de naphte, a petroleum distillate used as a solvent, for preserving sodium, etc.

— de navette, rape oil, rapeseed oil, colza oil.

— d'enfer, worst grade of oil.

— de noix, nut oil (specif., in painting, walnut oil).

— d'ensimage, oil for oiling wool, etc., textile oil.

— de palma-christi, castor oil.

— de palme, palm oil.

— de paraffine, paraffin oil; (*Pharm.*) liquid petrolatum, liquid paraffin.

— de pétrole, petroleum; any oil obtained from petroleum, specif. kerosene.

— de pied(s) de bœuf, neat's-foot oil.

— de pied de mouton, sheep's-foot oil.

— de pierre, rock oil, petroleum.

— de poisson, fish oil.

— de poix, cade oil.

— de pomme de terre, (literally, potato oil) fusel oil or, specif., fermentation amyl alcohol.

— de ravison, ravison oil.

— de recense, an inferior olive oil extracted from the marc.

— de résine, resin oil; rosin oil.

— de ressence, = huile de recense.

— de ricin, castor oil.

— de rorqual rostré, dœgling oil.

— de roses de l'Orient, attar of roses.

— de schiste, schist oil.

— de semence de lin, linseed oil.

— de semences de cotonnier, cottonseed oil.

— de Sénégale, Senegal oil, African palm oil.

— de sésame, sesame oil, benne oil, gingili seed oil.

— des Hollandais, oil of the Dutch chemists (ethylene chloride).

— des pieds du gros bétail, neat's-foot oil.

— de succin, oil of amber, amber oil.

— de suif, tallow oil, olein.

— de table, table oil, salad oil.

— de tartre, (*Old Chem.*) salt of tartar (a pure potassium carbonate), esp. when in a deliquesced state.

— de térébenthine, oil of turpentine, spirits of turpentine.

— de terre, earth oil, petroleum.

huile *d'éther*, (*Pharm.*) ethereal oil, heavy oil of wine.
— *détonante*, explosive oil, specif. nitroglycerin.
— *de veau marin*, seal oil.
— *de vin pesante*, heavy oil of wine.
— *de vitriol*, oil of vitriol.
— *distillée*, distilled oil, volatile oil.
— *d'os*, bone oil, Dippel's oil.
— *douce*, sweet oil, specif., olive oil.
— *douce de vin*, sweet oil of wine (a complex residue from the manufacture of ether); oil of wine, œnanthic ether.
— *du styrax*, liquid storax.
— *empyreumatique*, empyreumatic oil.
— *essentielle*, essential oil.
— *essentielle de pétrole*, a heavy petroleum spirit used in painting.
— *éthérée*, ethereal oil.
— *explosive* = huile détonante.
— *fixe*, fixed oil.
— *froide*, an oil that has a low freezing point.
— *grasse*, fatty oil.
— *grasse à tableaux*, (*Painting*) boiled linseed oil.
— *grasse blanche*, (*Painting*) a bleached and boiled poppy oil.
— *industrielle*, industrial oil.
— *lampante*, purified oil; lamp oil, burning oil, specif. kerosene.
— *légère*, light oil.
— *lithargée*, — *lithargyrée*, oil that has been treated with litharge, boiled oil.
— *lourde*, heavy oil.
— *lubrifiante*, lubricating oil.
— *mangeable*, edible oil.
— *minérale*, mineral oil, specif. petroleum.
— *moyenne*, middle oil.
— *non-siccative*, non-drying oil.
— *phénolique*, carbolic oil.
— *phosphorée*, (*Pharm.*) phosphorated oil.
— *pour rouge turc*, Turkey-red oil.
— *pyrogénée*, empyreumatic oil.
— *rance*, rancid oil.
— *siccative*, drying oil.
— *tournante*, rank olive oil.
— *végétale*, vegetable oil.
— *verte*, green oil, specif. anthracene oil.
— *vierge*, virgin oil, oil from the first light pressing.
— *volatile*, volatile oil.
— *volatile de goudron*, oil of tar, tar oil.
— *volatile de térébenthine*, oil of turpentine, spirits of turpentine.
— *volatile éthérée*, ethereal oil, heavy oil of wine.
mettre en —, (*Leather*) dress with oil, stuff, oil.
(For the meaning of any phrase not included here, as *huile de cèdre*, *huile de chènevis*, etc. see the other important word in the phrase.)

huilement, *m.* oiling.
huiler, *v.t.* oil. — *v.i.* exude oil.
huilerie, *f.* oil works; oil mill; oil shop.
huileuse, *f.* machine for making oil; oiler
— *a. fem.* of huileux.
huileux, *a.* oily.
huilier, *m.* oiler, oil can; oil manufacturer; oil merchant; oil cruet; oil stand, castor.
— *a.* of or pertaining to oil.
huis clos. closed door, private.
huit, *a. & m.* eight; eighth.
huitaine, *f.* eight days, week; eight, about eight.
huitième, *a. & m.* eighth.
huître, *f.* oyster.
humain, *a.* human; humane. — *m.* human being, man.
humainement, *adv.* humanly; humanely.
humanitaire, *a.* humanitarian.
humanité, *f.* humanity.
humant, *p.pr.* of humer.
humate, *m.* humate.
humectage, *m.* moistening, dampening.
humectant, *p.a.* moistening; (*Med.*) humectant, diluent. — *m.* (*Med.*) diluent.
humectation, *f.* moistening, dampening, humectation.
humecter, *v.t.* moisten, dampen, render humid.
— *v.r.* be moistened, become moist or damp.
humer, *v.t.* suck, suck up, suck in; inhale, sniff, snuff.
humescent, *a.* becoming humid.
humeur, *m.* humor, humour; fluid; ill humor; sucker, inhaler, sniffer.
— *amniotique*, amniotic fluid.
— *aqueuse*, (*Anat.*) aqueous humor.
humeux, *a.* of or pertaining to humus.
humide, *a.* humid, moist, damp, wet.
humidement, *adv.* humidly, moistly, damply.
humidier, *v.t.* moisten, dampen, humidify.
humidification, *f.* humidification, moistening, dampening.
humidifier, *v.t.* humidify, render humid, moisten, dampen.
humidifuge, *a.* repelling moisture.
humidité, *f.* humidity; humidness, moisture, moistness, dampness.
humique, *a.* humic.
humuline, *f.* humulin (lupulin).
humus, *m.* humus.
hurler, *v.i.* howl.
hyacinthe, *f. & a.* hyacinth.
hyacinthine, *f.* (*Min.*) hyacinth.
hyalin, *a.* hyaline. — *f.* hyalin.
hyalite, *f.* (*Min.*) hyalite; (*Med.*) hyalitis.
hyalogène, *m.* hyalogen. — *a.* hyalogen, hyalogenic.
hyaloïde, hyaloïdien, *a.* hyaloid.
hyalosidérite, *f.* (*Min.*) hyalosidérite.

hyalotechnie, *f.* art of making and working glass.

hyalotechnique, *a.* pertaining to *hyalotechnie.*

hyalurgie, *f.* art of making glass.

hyalurgique, *a.* pertaining to glass making.

hybride, *m. & a.* hybrid.

hyckori, *m.* hickory (wood and tree).

hydantoïne, *f.* hydantoin.

hydracide, *m.* hydracid.

hydracrylique, *a.* hydracrylic.

hydragogue, *a. & m.* (*Med.*) hydragog, hydragogue.

hydralcool, *m.* alcohol diluted with water, esp. under-proof spirit.

hydralcoolique, *a.* = hydro-alcoolique.

hydramide, *m.* hydramide.

hydranthracène, *m.* any hydrogen addition product of anthracene.

hydrargyre, *m.* hydrargyrum, mercury.

hydrargyride, *a.* pertaining to mercury.

hydrargyrie, *f.* (*Med.*) hydrargyriasis, mercurialism.

hydrargyrique, *a.* hydrargyric, mercurial.

hydrargyrisme, *m.* hydrargyrism, mercurialism.

hydrargyrure, *m.* alloy with mercury, amalgam.

hydraste, *m.* (*Bot.*) hydrastis.
— *du Canada,* goldenseal (*Hydrastis canadensis*).

hydrastine, *f.* hydrastine.

hydrastinine, *f.* hydrastinine.

hydratable, *a.* capable of being hydrated.

hydratant, *p.a.* hydrating.

hydratation, *f.* hydration.

hydrate, *m.* hydrate; hydroxide.
— *d'alumine,* aluminium hydroxide.
— *de carbone,* carbohydrate.
— *de chaux,* hydrated lime, calcium hydroxide.
— *de chloral,* chloral hydrate.
— *de chloral butylique,* (*Pharm.*) butyl-chloral hydrate (2,2,3-trichloro-1,1-butanediol).
— *de peroxyde de fer,* ferric hydroxide.
— *de phényle,* phenol.
— *de potasse,* caustic potash.
— *de soude,* caustic soda.

hydraté, *p.a.* hydrated; hydrate of; hydroxide of.

hydrater, *v.t.* hydrate. — *v.r.* combine with water, become hydrated; (loosely) absorb water, deliquesce.

hydratique, *a.* hydrate, pertaining to hydrates.

hydratropique, *a.* hydratropic.

hydraulicien, *m.* hydraulic engineer, hydraulician.

hydraulicité, *f.* hydraulicity.

hydraulique, *a.* hydraulic. — *f.* hydraulics.

hydrazine, *f.* hydrazine.

hydrazinique, *a.* hydrazine, of hydrazine.

hydrazobenzène, *m.* hydrazobenzene.

hydrazoïque, *a.* hydrazo; hydrazoic.

hydrazone, *f.* hydrazone.

hydréléon, *m.* (*Med.*) hydrelæon, hydrelæum (*Obs.*).

hydrémie, *f.* (*Med.*) hydremia, hydræmia.

hydrer, *v.t.* hydrogenize, hydrogenate.

hydriodate, *m.* hydriodide.

hydriodique, *a.* hydriodic.

hydrique, *a.* aqueous, of water, water.

hydro-alcoolique, *a.* dilute alcoholic, water-alcohol.

hydro-alumineux, *a.* (*Min.*) hydroaluminous, containing alumina and the elements of water.

hydroaromatique, *a.* hydroaromatic.

hydrobenzénique, *a.* hydroaromatic.

hydrobenzoïne, *f.* hydrobenzoïn.

hydrobilirubine, *f.* hydrobilirubin.

hydroboracite, *f.* (*Min.*) hydroboracite.

hydrobromate, *m.* hydrobromide.

hydrobromique, *a.* hydrobromic.

hydrocarbonate, *m.* hydrocarbonate, hydrous (or, sometimes, basic) carbonate.
— *de zinc,* (*Pharm.*) precipitated zinc carbonate.

hydrocarbone, *m.* hydrocarbon; carbohydrate.

hydrocarboné, *a.* hydrocarbon, hydrocarbonic, hydrocarbonaceous; carbohydrate.

hydrocarburateur, *m.* carburettor.

hydrocarbure, *m.* hydrocarbon.
— *acétylénique,* acetylene hydrocarbon.
— *allénique,* allene hydrocarbon.
— *éthylénique,* ethylene hydrocarbon.
— *forménique,* methane hydrocarbon, paraffin.
— *non saturé,* unsaturated hydrocarbon.
— *saturé,* saturated hydrocarbon.

hydrocèle, *f.* (*Med.*) hydrocele.

hydrocellulose, *f.* hydrocellulose.

hydrocérame, *m.* a water cooler of porous ware.

hydrocéramique, *a.* hydroceramic, composed of porous clay.

hydrochlorate, *m.* hydrochloride.
— *d'ammoniaque,* ammonium chloride.
— *de chaux,* calcium chloride.
— *de soude,* sodium chloride.

hydrochlorique, *a.* hydrochloric.

hydrocinnamique, *a.* hydrocinnamic.

hydrocotarnine, *f.* hydrocotarnine.

hydrocoumarique, *a.* hydrocoumaric.

hydrocyanate, *m.* hydrocyanide.

hydrocyanique, *a.* hydrocyanic.

hydrodynamique, *f.* hydrodynamics. — *a.* hydrodynamic, hydrodynamical.

hydro-électrique, *a.* hydro-electric.

hydro-extracteur, *m.* hydro-extractor.

hydroferrocyanique, *a.* hydroferrocyanic (ferrocyanic).

hydrofluate, *m.* hydrofluoride.
hydrofluorique, *a.* hydrofluoric.
hydrofluosilicique, *a.* hydrofluosilicic, fluosilicic.
hydrofuge, *a.* repelling water, protecting against moisture; waterproof, watertight.
hydrogénable, *a.* hydrogenizable, hydrogenatable.
hydrogénation, *f.* hydrogenization, hydrogenation.
hydrogénant, *p.a.* hydrogenizing, hydrogenating. — *m.* hydrogenizer, hydrogenizing agent, reducing agent.
hydrogénase, *f.* hydrogenase.
hydrogénation, *f.* hydrogenation.
hydrogène, *m.* hydrogen.
— *antimonié*, antimoniuretted hydrogen, stibine.
— *arsénié*, arseniuretted hydrogen, arsine.
— *bicarboné*, bicarburetted hydrogen (ethylene).
— *carboné*, hydrocarbon.
— *disponible*, (*Coal*) disposable hydrogen (the excess of hydrogen in a coal over that necessary to form water with the oxygen present).
— *naissant*, nascent hydrogen.
— *phosphoré*, phosphuretted hydrogen (hydrogen phosphide, specif. phosphine).
— *sélénié*, seleniuretted hydrogen (hydrogen selenide).
— *silicié*, siliciuretted hydrogen (silicon hydride, hydrogen silicide).
— *sulfuré*, sulfuretted hydrogen, hydrogen sulfide.
hydrogéner, *v.t.* hydrogenize, hydrogenate. — *v.r.* be hydrogenized, be hydrogenated.
hydrogénique, *a.* of hydrogen, of hydrogen.
hydrogéniser, *v.t.* hydrogenize.
hydrogénure, *m.* hydride.
hydroiodique, *a.* hydriodic.
hydrolat, *m.* (*Pharm.*) water, medicated water.
— *de fleurs d'oranger*, orange flower water.
hydrolysable, *a.* hydrolyzable.
hydrolysation, *f.* hydrolyzation.
hydrolyse, *f.* hydrolysis.
hydrolyser, *v.t.* hydrolyze.
hydromel, *m.* hydromel.
hydrométallurgie, *f.* hydrometallurgy.
hydrométallurgique, *a.* hydrometallurgical.
hydromètre, *m.* hydrometer.
hydrométrie, *f.* hydrometry.
hydrométrique, *a.* hydrometric, hydrometrical.
hydrominéral, *a.* relating to mineral waters.
hydronaphtalène, *m.* hydronaphthalene (any hydrogen addition product of naphthalene).
hydronaphtol, *m.* hydronaphthol (any hydrogen addition product of naphthol).

hydroombellique, *a.* hydroumbellic.
hydrophane, *f.* (*Min.*) hydrophane. — *a.* hydrophanous.
hydrophile, *a.* (of cotton, etc.) absorbent; (*Bot.*) hydrophilous.
hydrophosphate, *m.* (*Min.*) hydrated phosphate.
hydrophthorique, *a.* hydrofluoric (*Obs.*).
hydropique, *a.* (*Med.*) dropsical, hydropic. — *m.* dropsical person.
hydropisie, *f.* (*Med.*) dropsy.
hydroplastie, *f.* metal plating in a solution without the electric current.
hydropneumatique, *a.* hydropneumatic.
hydropyridique, *a.* hydropyridine.
hydroquinone, *f.* hydroquinol, hydroquinone.
hydrorrhée, *f.* (*Med.*) hydrorrhea.
hydrosilicate, *m.* (*Min.*) hydrous silicate.
hydrosiliceux, *a.* (*Min.*) containing silica and the elements of water.
hydrosol, *m.* hydrosol.
hydrostatique, *a.* hydrostatic(al). — *f.* hydrostatics.
hydrosulfate, *m.* hydrosulfide.
hydrosulfite, *m.* hydrosulfite, hyposulfite.
hydrosulfure, *m.* hydrosulfide; (formerly) sulfide.
hydrosulfureux, *a.* hydrosulfurous, hyposulfurous.
hydrosulfurique, *a.* hydrosulfuric.
hydrothermal, *a.* hydrothermal.
hydrotimètre, *m.* hydrotimeter.
hydrotimétrie, *f.* hydrotimetry.
hydrotimétrique, *a.* hydrotimetric.
hydrotypie, *f.* a variety of collotype process.
hydroxéthylénamine, *f.* β-hydroxyethylamine (2-aminoethanol).
hydroxyde, *m.* hydroxide.
hydroxydé, *a.* combined with hydrogen and oxygen, hydroxide of.
hydroxylamine, *f.* hydroxylamine.
hydroxyle, *m.* hydroxyl.
hydroxylé, *a.* hydroxylated, hydroxyl.
hydroxyl-imidé, *a.* hydroxyimido.
hydroxyurée, *f.* hydroxyurea.
hydruration, *f.* hydrogenization, hydrogenation.
hydrure, *m.* hydride.
— *d'éthyle*, ethyl hydride (ethane).
hygiène, *f.* hygiene.
hygiénique, *a.* hygienic.
hygiéniquement, *adv.* hygienically.
hygrine, *f.* hygrine.
hygrique, *a.* hygric.
hygromètre, *m.* hygrometer.
hygrométricité, *f.* hygrometricity; hygroscopicity.
hygrométrie, *f.* hygrometry.
hygrométrique, *a.* hygrometric(al); hygroscopic.

hygrométriquement, *adv.* hygrometrically.
hygroscopicité, *f.* hygroscopicity.
hygroscopie, *f.* hygroscopy.
hygroscopique, *a.* hygroscopic, hygroscopical.
hygroscopiquement, *adv.* hygroscopically.
hyocholalique, *a.* hyocholic.
hyoscine, *f.* hyoscine.
hyoscyamine, *f.* hyoscyamine.
hyperbole, *f.* hyperbole.
— *équilatère,* equilateral hyperbola.
hyperbolique, *a.* hyperbolic, hyperbolical.
hyperboliquement, *adv.* hyperbolically.
hyperboloïde, *m.* & *a.* (*Geom.*) hyperboloid.
hyperchlorhydrie, *f.* (*Med.*) hyperchlorhydria.
hyperchlorique, *a.* hyperchloric (perchloric).
hyperémie, hyperhémie, *f.* (*Med.*) hyperemia, hyperæmia.
hyperglycémie, *f.* hyperglycemia, hyperglycæmia.
hypériodique, *a.* hyperiodic (periodic).
hyperoxyde, *m.* hyperoxide (peroxide).
hypertonique, *a.* hypertonic.
hypertrophie, *f.* (*Med.* & *Biol.*) hypertrophy.
hypertrophié, *a.* hypertrophied.
hyphe, *f.* (*Bot.*) hypha.
hypnal, *m.* (*Pharm.*) hypnal.
hypoazoteux, *a.* hyponitrous.
hypoazotique, *a.* hyponitric (" hyponitric acid " is an old name for nitrogen peroxide).
hypoazotite, *m.* hyponitrite.
hypochloreux, *a.* hypochlorous.
hypochlorhydrie, *f.* (*Med.*) hypochlorhydria.
hypochlorique, *a.* hypochloric (" hypochloric acid " is an old name for chlorine dioxide).
hypochlorite, *m.* hypochlorite.
— *de chaux,* hypochlorite of lime, bleaching powder.
— *de potasse,* hypochlorite of potash, eau de Javel.
— *de soude,* hypochlorite of soda, eau de Labarraque.
hypoderme, *m.* (*Bot.*) hypoderma; (*Zoöl.*) hypodermis, hypoderm.

hypodermique, *a.* hypodermic.
hypodermiquement, *adv.* hypodermically.
hypoglycémie, *f.* hypoglycemia.
hypo-iodeux, *a.* hypo-iodous.
hypophosphate, *m.* hypophosphate.
hypophosphite, *m.* hypophosphite.
— *de chaux,* calcium hypophosphite.
— *de fer,* iron (specif. ferric) hypophosphite.
— *de potasse,* potassium hypophosphite.
— *de soude,* sodium hypophosphite.
— *manganeux,* manganous (or manganese) hypophosphite.
hypophosphoreux, *a.* hypophosphorous.
hypophosphorique, *a.* hypophosphoric.
hypophyse, *f.* hypophysis.
hyposulfate, *m.* hyposulfate (dithionate).
hyposulfite, *m.* hyposulfite, hyposulphite (still often used to designate salts of $H_2S_2O_3$ and then better translated *thiosulfate*).
— *de potasse,* hyposulfite of potash (potassium thiosulfate).
— *de soude,* hyposulfite of soda, " hypo " (sodium thiosulfate).
hyposulfureux, *a.* hyposulfurous, hyposulphurous (still often used to designate the acid $H_2S_2O_3$ and then better translated *thiosulfuric*).
hyposulfurique, *a.* hyposulfuric (old synonym of *dithionic*).
hypoténuse, *f.* (*Geom.*) hypotenuse.
hypothèque, *f.* mortgage.
hypothéquer, *v.t.* mortgage, hypothecate.
hypothèse, *f.* hypothesis.
hypothétique, *a.* hypothetic, hypothetical.
hypothétiquement, *adv.* hypothetically.
hypothionique, *a.* dithionic. *Obs.*
hypotonique, *a.* hypotonic.
hypoxanthine, *f.* hypoxanthine.
hypoxyde, *m.* suboxide.
hysope, hyssope, *f.* (*Bot.*) hyssop.
hystérésie, hystérésis, *f.* hysteresis.
hystérie, *f.* (*Med.*) hysteria.

I

iatrochimie, *f.* iatrochemistry.
iatrochimique, *a.* iatrochemical.
iatrochimiste, *m.* iatrochemist.
ibérien, *a.* Iberian.
ichor, *m.* (*Med.*) ichor.
ichtyocolle, ichthyocolle, *f.* ichthyocolla, isinglass.
ichtyol, *m.* (*Pharm.*) ichthyol.
ici, *adv.* here; now.
    par —, this way, by here, thru here.
icica, *f.* conima, Brazilian elemi.
ici-même, *adv.* in this very place, in this work.
iconogène, *m.* (*Photog.*) eikonogen.
ictère, *m.* (*Med.*) icterus, jaundice.
idéal, *a.* & *m.* ideal.
idéaliser, *v.t.* idealize.
idéaliste, *m.* idealist. — *a.* idealistic.
idée, *f.* idea; bit, small amount; sketch, outline.
identification, *f.* identification.
identifier, *v.t.* identify.
identique, *a.* identical, same.
identiquement, *adv.* identically.
identité, *f.* identity.
idiotique, *a.* idiomatic.
idiotisme, *m.* idiom.
idite, *f.* iditol, idite.
idocrase, *f.* (*Min.*) idocrase (vesuvianite).
idolâtre, *a.* idolatrous. — *m.* idolater.
idole, *f.* idol.
idrialite, idrialine, *f.* (*Min.*) idrialite, idrialine.
idryle, *m.* idryl (fluoranthene).
Iéna, *n.* Jena; Jena glass.
if, *m.* (*Bot.*) yew, yew tree; draining rack (for bottles).
igname, *f.* yam.
igné, *a.* igneous.
ignescence, *f.* ignescence.
ignescent, *a.* ignescent.
ignicolore, *a.* fire-colored, flame-colored.
ignifère, *a.* producing fire, igniferous.
ignifugation, *f.* fireproofing; fireproof (that is, more or less difficultly inflammable) state.
ignifuge, *a.* rendering noninflammable or difficultly inflammable, fireproofing; fire-extinguishing (as a grenade). — *m.* fireproofing agent; fire-extinguishing agent.
ignifugeage, *m.* = ignifugation.
ignifuger, *v.t.* render difficultly inflammable, fireproof.
ignigène, *a.* produced by or with the aid of fire (see sel ignigène, under *sel*).

ignition, *f.* ignition.
ignorance, *f.* ignorance.
ignorant, *a.* ignorant.
ignorer, *v.t.* not know, be ignorant of.
il, *pron.* he; it; there.
    — *n'y a pas,* there is not, there are not
    — *n'y a pas de,* there is no.
    — *y a,* there is, there are.
île, *f.* island, isle.
ilicine, *f.* ilicin.
ilipé, *n.* (*Bot.*) illipe.
illégal, *a.* illegal, unlawful.
illégitime, *a.* illegitimate.
illimitable, *a.* illimitable.
illimité, *a.* unlimited.
illisible, *a.* illegible.
illogique, *a.* illogical.
illogiquement, *adv.* illogically.
illuminant, *p.a.* illuminating.
illumination, *f.* illumination.
illuminer, *v.t.* illuminate.
illusion, *f.* illusion.
    — *d'optique,* optical illusion.
illusionner, *v.t.* deceive, delude.
illusoire, *a.* illusory, illusive, deceptive.
illustration, *f.* illustration; illustriousness; celebrity.
illustre, *a.* illustrious. — *m.* illustrious man.
illustré, *p.a.* illustrated.
illustrer, *v.t.* illustrate.
îlot, *m.* little island, islet.
ils, *pron.* they.
ilvaïte, *f.* (*Min.*) ilvaite.
image, *f.* image.
    — *réelle,* real image.
    — *réfléchie,* reflected image.
    — *renversée,* inverted image.
    — *virtuelle,* virtual image.
imagerie, *f.* chromo printing, color printing; picture trade.
imaginable, *a.* imaginable.
imaginaire, *a.* & *f.* imaginary.
imagination, *f.* imagination.
imaginer, *v.t.* imagine; conceive; devise.
imbiber, *v.t.* soak; imbibe; imbue. — *v.r.* be soaked; be imbued; soak, soak in.
imbibition, *f.* soaking; imbibition; imbuement; (*Silver*) addition of argentiferous material to molten lead, which absorbs the silver.
imboire, *v.t.* moisten, wet; = emboire; imbue; imbibe. — *v.r.* be imbued.

186

imbrifuge, *a.* rain-proof, waterproof.
imbrûlable, *a.* unburnable, incombustible.
imbu, *p.p.* of imboire.
imbuvable, *a.* undrinkable, not fit to drink.
imbuvant, *p.pr.* of imboire.
imidazol, *m.* imidazole.
imidazoline, *f.* imidazoline.
imidazolone, *f.* imidazolone.
imide, *f. & m.* imide.
imidé, *a.* imido; (when the substance is of the nature of an imine and not an imide) imino.
imido-éther, *m.* imido ester, (less correctly) imido ether.
imidogène, *m.* imidogen.
iminé, *a.* imino.
imitateur, *m.* imitator. — *a.* imitative, imitating.
imitatif, *a.* imitative.
imitation, *f.* imitation.
imiter, *v.t.* imitate.
immangeable, *a.* uneatable, not fit to eat.
immaniable, *a.* unmanageable.
immatériel, *a.* immaterial.
immaturité, *f.* immaturity.
immédiat, *a.* immediate; (of analysis) proximate.
immédiatement, *adv.* immediately.
immense, *a.* immense.
immensément, *adv.* immensely.
immensité, *f.* immensity, immenseness.
immerger, *v.t.* immerse.
immersif, *a.* immersion.
immersion, *f.* immersion.
immesurable, *a.* immeasurable.
immesuré, *a.* unmeasured.
immeuble, *a.* immovable, real (estate). — *m.* real estate; landed estate; house.
immiscibilité, *f.* immiscibility.
immiscible, *a.* immiscible.
immobile, *a.* immovable, immobile; motionless.
immobilisation, *f.* rendering immovable, etc. (see immobiliser).
immobiliser, *v.t.* render immovable, fasten, fix; stop (as a machine); throw out of action, disconnect; immobilize.
immobilité, *f.* immobility, fixedness.
immodéré, *a.* immoderate, excessive.
immortel, *a.* immortal.
immortelle, *f.* everlasting, immortelle, specif. (*Pharm.*) cudweed, life everlasting.
immuable, *a.* immutable.
immunisant, *p.a.* immunizing.
immunisation, *f.* immunization.
immuniser, *v.t.* immunize.
immunité, *f.* immunity.
impact, *m.* impact.
impair, *a.* odd, uneven.
impalpable, *a.* impalpable.
imparfait, *a.* imperfect.

imparfaitement, *adv.* imperfectly.
impastation, *f.* impastation.
impatientant, *p.a.* tiring, provoking.
impayable, *a.* invaluable; ridiculous, queer.
impayé, *a.* unpaid.
impénétrabilité, *f.* impenetrability.
impénétrable, *a.* impenetrable.
impératif, *a.* imperative.
impérativement, *adv.* imperatively.
impératoire, *f.* imperatoria, (specif.) masterwort.
imperceptibilité, *f.* imperceptibility.
imperceptible, *a.* imperceptible.
imperceptiblement, *adv.* imperceptibly.
imperfection, *f.* imperfection.
impérissabilité, *f.* imperishability.
impérissable, *a.* imperishable.
impermanence, *f.* impermanence, impermanency.
impermanent, *a.* not permanent, impermanent.
imperméabilisateur, *a.* rendering impermeable; specif., waterproofing, waterproof.
imperméabilisation, *f.* rendering impermeable; specif., waterproofing.
imperméabiliser, *v.t.* render impermeable; specif., waterproof.
imperméabilité, *f.* impermeability, impermeableness; specif., waterproof quality.
imperméable, *a.* impermeable; specif., waterproof. — *m.* waterproof fabric.
— *à l'air,* air-tight.
— *à l'eau,* waterproof, watertight.
imperméablement, *adv.* impermeably.
impersonnel, *a.* impersonal.
impitoyable, *a.* unpitying, pitiless, merciless.
impitoyablement, *adv.* pitilessly, mercilessly.
implicite, *a.* implicit.
impliquer, *v.t.* imply; implicate.
impondérabilité, *f.* imponderability.
impondérable, *a.* imponderable.
importable, *a.* importable.
importance, *f.* importance.
important, *a.* important. — *m.* important point, main point; important person.
importateur, *m.* importer. — *a.* importing.
importation, *f.* importation.
importer, *v.i.* be important, matter. — *v.t.* import. — *v.r.* be imported.
n'importe, no matter.
imposant, *p.a.* imposing.
imposer, *v.t.* impose; levy (a tax); tax; awe. — *v.i.* impose.
— *au poids,* tax by weight.
— *au volume,* tax by volume.
impossibilité, *f.* impossibility.
impossible, *a.* impossible.
impôt, *m.* tax, impost, duty.
impracticabilité, *f.* impracticability.
impracticable, *a.* impracticable.

imprégnation, *f.* impregnation.

imprégner, *v.t.* impregnate.

impressif, *a.* impressive.

impression, *f.* impression (in various senses); printing (specif., textile printing); effect; (*Painting*) priming, ground.

impression *des tissus,* — *sur étoffes,* textile printing.

impressionnant, *p.a.* impressive, impressing.

impressionner, *v.t.* impress, make an impression on; (*Photog.*) produce an image on.

imprévoyance, *f.* improvidence, lack of foresight.

imprévu, *p.a.* unforeseen, unexpected. — *m.* something unforeseen, unexpected circumstance.

imprimé, *p.a.* printed, etc. (see imprimer). — *m.* print, printed paper or book, (*pl.*) printed matter.

imprimer, *v.t.* print; imprint, impress; give, communicate; (*Painting*) prime.

imprimerie, *f.* printing; printing house, printing establishment, press.

imprimeur, *m.* printer; pressman.

imprimeuse, *f.* printing machine.

imprimure, *f.* priming, ground coat, ground.

improbabilité, *f.* improbability.

improbable, *a.* improbable.

improbablement, *adv.* improbably.

improductif, *a.* unproductive.

improductivement, *adv.* unproductively.

improductivité, *f.* unproductiveness, -tivity.

imprononçable, *a.* unpronounceable.

impropre, *a.* improper, unfit, unsuited, inaccurate, incorrect.

— *au service,* unserviceable.

improprement, *a.* improperly, etc. (see impropre).

improviser, *v.t.* improvise.

improviste. In the phrase *à l'improviste,* suddenly, unexpectedly, unawares.

impuissance, *f.* impotence.

impuissant, *a.* impotent.

impulser, *v.t.* impel.

impulsif, *a.* impulsive, impelling.

impulsion, *f.* impulse; impulsion.

impulvérisé, *a.* unpulverized, unpowdered.

impur, *a.* impure.

impurement, *adv.* impurely.

impureté, *f.* impurity.

impurifié, *a.* unpurified.

imputer, *v.t.* impute; reckon in (on an account).

imputrescibiliser, *v.t.* render imputrescible.

imputrescibilité, *f.* imputrescibility.

imputrescible, *a.* imputrescible.

inacceptable, *a.* unacceptable.

inaccommodable, *a.* irreconcilable.

inaccompagné, *a.* unaccompanied.

inaccompli, *a.* unaccomplished, incomplete.

inaccordable, *a.* irreconcilable.

inaccoutumé, *a.* unaccustomed.

inactif, *a.* inactive.

inaction, *f.* inaction.

inactivement, *adv.* inactively.

inactivité, *f.* inactivity.

inadéquat, *a.* inadequate.

inadhérent, *a.* unadhesive, not adherent.

inalliabilité, *f.* incapability of being alloyed; incompatibility.

inalliable, *a.* not capable of being alloyed, non-alloying; incompatible.

inallié, *a.* unalloyed.

inaltérabilité, *f.* inalterability.

inaltérable, *a.* inalterable, unalterable.

inaltéré, *a.* unaltered.

inanalysable, *a.* unanalyzable.

inanalysé, *a.* unanalyzed.

inanimé, *a.* inanimate; unanimated.

inaperçu, *a.* unperceived, unobserved.

inappétence, *f.* inappetence, want of appetite.

inapplicabilité, *f.* inapplicability.

inapplicable, *a.* inapplicable.

inappréciable, *a.* inappreciable, unappreciable.

inappréciablement, *adv.* inappreciably.

inapprêté, *a.* unfinished, undressed.

inapte, *a.* inapt, unsuited, unfit, incapable.

inaptitude, *f.* inaptitude, unfitness.

inattaquable, *a.* unattackable.

inattaqué, *a.* unattacked.

inattendu, *a.* unexpected, unlooked for.

inaugurer, *v.t.* inaugurate.

inautorisé, *a.* unauthorized.

incalcinable, *a.* incapable of being calcined.

incalciné, *a.* uncalcined.

incalculable, *a.* incalculable.

incandescence, *f.* incandescence.

incandescent, *a.* incandescent.

incapable, *a.* incapable.

incapacité, *f.* incapacity.

incarnadin, *a.* incarnadine, pale red, flesh-colored. — *m.* pink, flesh color.

incarnat, *a.* flesh-red, rosy, bright pink.

incassable, *a.* unbreakable.

incendiaire, *a., m. & f.* incendiary.

incendie, *m.* fire, burning, conflagration.

incendier, *v.t.* set fire to; burn; provoke.

incération, *f.* mixing or covering with wax, inceration.

incérer, *v.t.* mix or cover with wax, incerate; render waxy.

incertain, *a.* uncertain.

incertainement, *adv.* uncertainly.

incertitude, *f.* uncertainty.

incessamment, *adv.* immediately, at once; incessantly, unceasingly, continually.

incessant, *a.* incessant, unceasing.

incidemment, *adv.* incidentally.

incidence, *f.* incidence.

incident, *a. & m.* incident.

incidentel, *a.* incidental.
incinérateur, *m.* incinerator.
incinération, *f.* incineration.
incinérer, *v.t.* incinerate, burn to ashes.
inciser, *v.t.* incise.
incision, *f.* incision.
incivilisé, *a.* uncivilized.
inclinaison, *f.* inclination.
inclination, *f.* inclination; affection.
incliné, *p.a.* inclined.
incliner, *v.t. & i.* incline. — *v.r.* incline; bend; bow.
inclus, *p.a.* inclosed, enclosed.
inclusif, *a.* inclusive.
inclusion, *f.* inclusion; inclosure.
inclusivement, *adv.* inclusively, inclusive.
incoagulable, *a.* uncoagulable, incoagulable.
incoercibilité, *f.* incoercibility.
incoercible, *a.* incoercible.
incohérence, *f.* incoherence, incoherency.
incohérent, *a.* incoherent.
incohésion, *f.* incoherence, (rarely) incohesion.
incoloration, *f.* colorlessness.
incolore, *a.* colorless, uncolored.
incomber, *v.i.* be incumbent (on).
incomburant, *p.a.* not supporting combustion.
incombustibilité, *f.* incombustibility.
incombustible, *a.* incombustible.
incommensurabilité, *f.* incommensurability.
incommensurable, *a.* incommensurable.
incommensurablement, *adv.* incommensurably.
incommodant, *a.* disagreeable, annoying.
incommode, *a.* inconvenient; disagreeable.
incommoder, *v.t.* inconvenience, trouble, disturb; hinder; indispose; impair.
incommodité, *f.* inconvenience; indisposition.
incomparablement, *adv.* incomparably.
incompatibilité, *f.* incompatibility.
incompatible, *a.* incompatible.
incompatiblement, *adv.* incompatibly.
incomplet, *a.* incomplete.
incomplètement, *adv.* incompletely.
incompressibilité, *f.* incompressibility.
incompressible, *a.* incompressible.
incomprimé, *a.* uncompresssed.
inconcevable, *a.* inconceivable.
inconciliabilité, *f.* irreconcilability.
inconciliable, *a.* irreconcilable.
inconciliablement, *adv.* irreconcilably.
inconducteur, *a.* nonconducting.
inconfiance, *f.* distrust.
inconformité, *f.* unconformity.
incongelable, *a.* uncongealable, nonfreezing.
incongelé, *a.* uncongealed, unfrozen.
incongru, *a.* incongruous.
incongruité, *f.* incongruity.
inconnu, *a. & m.* unknown.
inconnue, *f.* unknown quantity, unknown.
inconsciemment, *adv.* unconsciously.

inconscience, *f.* unconsciousness.
inconscient, *a.* unconscious.
inconséquence, *f.* inconsistency.
inconséquent, *a.* inconsistent; inconsiderate.
inconstant, *a.* inconstant.
incontaminé, *a.* uncontaminated.
incontestablement, *adv.* incontestably.
incontinu, *a.* not continuous.
inconvénient, *m.* inconvenience; disadvantage.
inconverti, *a.* unconverted.
inconvertible, inconvertissable, *a.* inconvertible, unconvertible.
incorporer, *v.t.* incorporate.
incorrect, *a.* incorrect.
incorrectement, *adv.* incorrectly.
incorrection, *f.* incorrectness; inaccuracy.
incorrigé, *a.* uncorrected.
incrément, *m.* increment.
incristallisable, *a.* uncrystallizable.
incroyable, *a.* incredible.
incroyablement, *adv.* incredibly.
incrustation, *f.* incrustation.
incruster, *v.t.* incrust; inlay.
incuit, *a.* unburned, unboiled, etc. (see cuire). — *m.* (also *pl.*) unburned or underburned material.
inculte, *a.* uncultivated; uncultured.
incurver, *v.t.* incurve, curve inward.
indamine, *f.* indamine.
indaniline, *f.* indaniline.
inde, *m.* indigo, indigo blue; logwood. — *f.* (*cap.*) India.
 Inde anglaise, Indes anglaises, Indes brittaniques, British India (usually understood as including Burma).
 Inde française, Indes françaises, French India. — *plate,* indigo in flat cakes, Dutch indigo.
 Indes néerlandaises, Dutch East Indies.
 Indes occidentales, West Indies.
 Indes orientales, East Indies.
indébrouillable, *a.* inextricable.
indécis, *a.* undecided; uncertain.
indécomposable, *a.* indecomposable, undecomposable.
indécomposé, *a.* undecomposed.
indécrit, *a.* undescribed.
indédoublable, *a.* indivisible, undecomposable, unresolvable, etc. (see dédoubler).
indéfini, *a.* indefinite; undefined; unlimited.
indéfiniment, *adv.* indefinitely.
indéfinissable, *a.* indefinable, undefinable.
indéfinité, *f.* indefiniteness.
indéfrichable, *a.* uncultivable.
indélébile, *a.* indelible.
indélébilement, *adv.* indelibly.
indemniser, *v.t.* indemnify.
indemnité, *f.* allowance; indemnity.
indémontrable, *a.* indemonstrable, undemonstrable.

indène, *m.* indene.

indenter, *v.t.* indent.

indépendamment, *adv.* independently.

indépendance, *f.* independence.

indépendant, *a.* independent.

indéréglable, *a.* not liable to get out of order, foolproof.

Indes. See inde.

indescriptible, *a.* indescribable, undescribable.

indescriptiblement, *adv.* indescribably.

indétermination, *f.* indeterminiteness; indetermination.

indéterminé, *a.* indeterminate.

index, *m.* index; index finger, forefinger. — *a.* index.

— *flotteur*, floating index, index float.

indican, *m.* indican.

indicanurie, *f.* (*Med.*) indicanuria.

indicateur, *m.* indicator; gage; index finger, forefinger; guide. — *a.* indicatory, indicating.

— *coloré*, colored indicator.

— *de la vitesse*, speed indicator.

— *de marche*, (*Elec.*) meter.

— *de niveau d'eau*, water gage.

— *de pression*, pressure gage, pressure indicator.

— *de vide*, vacuum gage.

indicateur-robinet, *m.* gage-cock.

indicatif, *a.* indicative, indicatory.

indication, *f.* indication.

indicatrice, *f.* (*Math.*, etc.) indicatrix. — *a. fem.* of indicateur.

indice, *m.* sign, mark, indication; index; (*Analysis*) (1) value, number, (2) trace.

— *d'acétyle*, acetyl value, acetyl number.

— *de Hehner*, Hehner value, Hehner number.

— *de réfraction*, index of refraction, refractive index.

— *de saponification*, saponification value (or number).

— *d'iode*, iodine value, iodine number.

— *opsonique*, opsonic, index.

indicible, *a.* unspeakable, inexpressible.

indien, *a.* Indian. — *m.* (*cap.*) Indian.

indienne, *f.* printed (or painted) cotton cloth, cotton print, printed calico; (*cap.*) Indian.

indiennerie, *f.* cotton printery, cotton print works, calico printery.

indienneur, *m.* cotton printer, calico printer.

indifféremment, *adv.* indifferently.

indifférent, *a.* indifferent.

indigène, *a.* indigenous, native; (*Com.*) domestic, home. — *m. & f.* native.

indigent, *a.* indigent, poor.

indigeste, *a.* indigestible; undigested.

indigne, *a.* unworthy.

indigo, *m.* indigo.

— *blanc*, indigo white.

— *bleu*, indigo blue.

indigo *sauvage*, wild indigo (*Baptisia*, esp. *B. tinctoria*).

indigocarmine, *f.* indigo carmine.

indigomètre, *m.* indigometer.

indigo-purpurine, *f.* indigopurpurin.

indigoterie, *f.* indigo plantation; indigo factory.

indigotier, *m.* indigo plant (specif. *Indigofera tinctoria*); indigo manufacturer.

indigotine, *f.* indigotin.

indigotique, *a.* indigotic.

indiquer, *v.t.* indicate; sketch. — *v.r.* be indicated.

indirect, *a.* indirect.

indirectement, *adv.* indirectly.

indirubine, *f.* indirubin.

indiscret, *a.* indiscreet.

indiscutable, *a.* indisputable, unquestionable.

indispensable, *a.* indispensable.

indisputablement, *adv.* indisputably.

indissolubilité, *f.* indissolubility.

indissoluble, *a.* undissolvable, indissoluble.

indissous, *a.* undissolved.

indistinct, *a.* indistinct.

indistinctement, *adv.* indistinctly.

indistinguible, *a.* indistinguishable.

indium, *m.* indium.

individu, *m.* individual.

individualité, *f.* individuality.

individuel, *a.* individual.

individuellement, *adv.* individually.

indivisé, *a.* undivided.

indivisibilité, *f.* indivisibility.

indivisible, *a.* indivisible.

indivisiblement, *adv.* indivisibly.

in-dix-huit, *a. & m.* octodecimo, 18mo.

Indo-Chine, *f.* Indo-China.

indochinois, *a.* Indo-Chinese.

indogène, *m.* indogen.

indol, *m.* indole.

indolcarbonique, *a.* indolecarboxylic.

indolore, *a.* painless.

indomptable, *a.* indomitable; untamable.

indophénine, *f.* indophenin.

indophénol, *m.* indophenol.

indosable, *a.* indeterminable.

indosé, *a.* undetermined, etc. (see doser). — *m.* undetermined matter or quantity.

in-douze, *a. & m.* duodecimo, 12mo.

indoxyle, *m.* indoxyl.

indoxylique, *a.* indoxyl; (of the acid) indoxylic.

indu, *a.* undue; not due.

indubitablement, *adv.* indubitably.

inducteur, *a.* (*Elec.*) inducing. — *m.* inductor.

inductif, *a.* inductive.

inductile, *a.* not ductile, inductile.

inductilité, *f.* lack of ductility, inductility.

induction, *f.* induction; inducement.

— *propre*, (*Elec.*) self-induction.

induire, *v.t.* induce; infer; lead.

induisant, *p.pr.* of induire.

induit, *p.a.* induced; inferred; led. — *m.* induced circuit, secondary circuit; armature (of a dynamo).

induline, *f.* induline.

indûment, *adv.* unduly.

indurer, *v.t.* indurate.

indurite, *f.* (*Expl.*) indurite.

industrie, *f.* industry; occupation, calling; skill, dexterity.

— *agricole,* farming industry, agriculture.

— *commerciale,* commerce, trade.

— *d'art,* manufacturing, manufacture.

— *des transports,* transportation.

— *du livre,* book industry.

— *extractive,* extractive industry.

— *frigorifique,* refrigeration industry.

— *manufacturière,* manufacturing, manufacturing industry.

— *privée,* private industry.

— *sucrière,* sugar industry.

— *tinctoriale,* dyeing industry.

— *vinicole,* wine-growing industry.

industriel, *a.* industrial. — *m.* one engaged in industry, industrial (specif., manufacturer).

industriellement, *adv.* industrially.

industrieusement, *adv.* skillfully.

industrieux, *a.* skillful, dexterous.

inébriant, inébriatif, *a.* intoxicating.

inédit, *a.* unpublished.

ineffectif, *a.* ineffective.

inefficace, *a.* inefficacious, ineffectual.

inégal, *a.* inequal; uneven.

inégalement, *adv.* unequally; unevenly.

inégaliser, *v.t.* render unequal or uneven.

inégalité, *f.* inequality.

inégaux, *masc. pl.* of inégal.

inélastique, *a.* inelastic, unelastic.

inéluctable, *a.* irresistible, ineluctable.

inencrassable, *a.* nonfouling.

ineptie, *f.* inaptness, inaptitude, ineptitude.

inépuisable, *a.* inexhaustible.

inépuisé, *a.* unexhausted.

inépuré, *a.* unpurified, unrefined.

inéquitable, *a.* inequitable.

inerte, *a.* inert.

inertie, *f.* inertia.

inespéré, *a.* unhoped for, unexpected.

inestimable, *a.* inestimable.

inétirable, *a.* incapable of being drawn out, (of metals) not ductile.

inévitabilité, *f.* inevitability.

inévitable, *a.* inevitable.

inévitablement, *adv.* inevitably.

inexact, *a.* inexact, inaccurate; not punctual.

inexactement, *adv.* inexactly, inaccurately.

inexactitude, *f.* inexactness, inaccuracy.

inexcusablement, *adv.* inexcusably.

inexistant, *a.* nonexistent.

inexistence, *f.* nonexistence.

inexorablement, *adv.* inexorably.

inexpédient, *a.* inexpedient.

inexpérience, *f.* inexperience.

inexpérimenté, *a.* inexperienced; untried.

inexplicable, *a.* inexplicable.

inexplicablement, *adv.* inexplicably.

inexpliqué, *a.* unexplained.

inexploitable, *a.* unexploitable, incapable of being developed.

inexploité, *a.* unexploited, undeveloped.

inexploré, *a.* unexplored.

inexplosible, inexplosif, *a.* nonexplosive, inexplosive, unexplosive.

inexprimable, *a.* inexpressible.

inexprimablement, *adv.* inexpressibly.

inexprimé, *a.* unexpressed.

inextensible, *a.* inextensible, not extensible.

inextinguible, *a.* inextinguishable.

inextricable, *a.* inextricable.

inextricablement, *adv.* inextricably.

infaillibilité, *f.* infallibility.

infaillible, *a.* infallible.

infailliblement, *adv.* infallibly.

infaisable, *a.* unfeasible, impracticable.

infâme, *a.* infamous; filthy.

infanterie, *f.* (*Mil.*) infantry.

infatigable, *a.* indefatigable.

infatigablement, *adv.* indefatigably.

infavorable, *a.* unfavorable, unfavourable.

infécond, *a.* infertile, infecund.

infécondité, *f.* infertility, infecundity.

infect, *a.* foul, unwholesome, infectious.

infecter, *v.t.* infect.

infectieux, *a.* infectious.

infection, *f.* infection.

inférence, *f.* inference.

inférer, *v.t.* infer. — *v.r.* be inferred.

inférieur, *a.* inferior; lower. — *m.* inferior.

inférieurement, *adv.* below; inferiorly.

infériorité, *f.* inferiority.

infermenté, *a.* unfermented.

infermentescible, *a.* unfermentable, nonfermentable.

infernal, *a.* infernal.

infertile, *a.* infertile, unfertile.

infidèle, *a.* treacherous, unreliable; unfaithful; infidel.

infiltration, *f.* infiltration.

infiltrer, *v.t. & r.* infiltrate.

infime, *a.* lowest, very low.

infini, *a. & m.* infinite.

à *l'*—, infinitely, ad infinitum.

infiniment, *adv.* infinitely.

— *petit,* infinitely small; minute particle; (*Math.*) infinitesimal.

infinité, *f.* infinity.

infinitésimal, *a.* infinitesimal.

infinitésimalement, *adv.* infinitesimally.

infinitésime, *a. & f.* infinitesimal.

infirmation, *f.* nullification; weakening.

infirmer, *v.t.* declare void, nullify; weaken.
inflammabilité, *f.* inflammability.
inflammable, *a.* inflammable.
inflammateur, *m.* igniter, firing device; igniting charge.
inflammation, *f.* inflammation; ignition; (*Expl.*) firing, fire.
— *spontanée,* spontaneous ignition.
infléchir, *v.t.* inflect, bend.
inflexibilité, *f.* inflexibility, inflexibleness.
inflexible, *a.* inflexible, rigid, unyielding.
inflexion, *f.* inflection.
influence, *f.* influence.
influencer, *v.t.* influence.
influent, *a.* influential.
influer, *v.i.* have influence.
in-folio, *m.* & *a.* folio.
information, *f.,* **informations,** *f.pl.* inquiry, investigation; information, intelligence.
informe, *a.* unformed, shapeless; informal.
informé, *p.a.* informed. — *m.* inquiry.
informer, *v.t.* inform. — *v.r.* inquire, ask.
informulé, *a.* unformulated.
infortune, *f.* misfortune.
infortuné, *a.* unfortunate.
infra-rouge, *a.* infra-red.
infrastructure, *f.* substructure, groundwork.
infréquemment, *adv.* infrequently, seldom.
infréquence, *f.* infrequence, infrequency.
infréquent, *a.* infrequent, uncommon.
infriable, *a.* not friable.
infructueusement, *adv.* unprofitably.
infructueux, *a.* unfruitful, unprofitable, unsuccessful.
infus, *a.* intuitive, native.
infusé, *p.a.* infused. — *m.* (*Pharm.*) infusion.
infuser, *v.t.* infuse. — *v.r.* be infused.
infusibilité, *f.* infusibility.
infusible, *a.* infusible.
infusion, *f.* infusion.
infusoire, *a.* infusorial.
infusoires, *m.pl.* Infusoria.
infusum, *m.* infusion.
ingaranti, *a.* unguaranteed, unwarranted.
ingélif, *a.* not freezing readily; (of stones, etc.) not rent by frost.
ingénieur, *m.* engineer.
— *chimiste,* chemical engineer.
— *civil,* civil engineer.
— *électricien,* electrical engineer.
— *des mines,* mining engineer.
— *des poudres et salpêtres,* explosives engineer.
— *gazier,* gas engineer.
— *mécanicien,* mechanical engineer.
— *opticien,* maker of optical instruments.
— *sanitaire,* sanitary engineer.
ingénieusement, *adv.* ingeniously.
ingénieux, *a.* ingenious.
ingéniosité, *f.* ingeniousness, ingenuity.
ingénu, *a.* ingenuous.

ingénuité, *f.* ingenuousness.
ingérer, *v.t.* ingest. — *v.r.* interfere.
ingestion, *f.* ingestion.
ingrain, *m.* (*Bot.*) spelt.
ingrat, *a.* ungrateful, unpleasing, unpromising.
ingrédient, *m.* ingredient.
inguérissable, *a.* incurable.
inhabile, *a.* unskillful, inapt; incapable.
inhabilement, *adv.* unskillfully, inexpertly.
inhabileté, *f.* unskillfulness, inexpertness.
inhabitable, *a.* uninhabitable.
inhabitué, *a.* unaccustomed, unhabituated.
inhalateur, *m.* inhaler. — *a.* inhaling.
inhalation, *f.* inhalation.
inhaler, *v.t.* inhale.
inhérent, *a.* inherent.
inhibiteur, *a.* inhibiting, inhibitory, inhibitive.
inhibition, *f.* inhibition.
inhumectation, *f.* lack of moisture, dryness.
inhumecté, *a.* unmoistened.
inimaginable, *a.* unimaginable.
inimitié, *f.* enmity, hostility.
ininflammable, *a.* uninflammable, noninflammable.
ininflammation, *f.* absence of inflammation.
inintelligible, *a.* unintelligible.
inintentionnel, *a.* unintentional.
inintentionellement, *adv.* unintentionally.
ininterrompu, *a.* uninterrupted, continuous.
ininterruption, *f.* noninterruption, continuity.
initial, *a.* initial.
initialement, *adv.* initially.
initier, *v.t.* initiate.
injectable, *a.* injectable; intended for injection.
injecter, *v.t.* inject; impregnate (by injection). — *v.r.* be injected; be flushed, be congested.
injecteur, *m.* injector. — *a.* injecting.
— *aspirant,* suction injector.
injection, *f.* injection.
injudicieusement, *adv.* injudiciously.
injudicieux, *a.* injudicious.
injure, *f.* injury; abuse.
injurieusement, *adv.* injuriously.
injurieux, *a.* injurious.
injuste, *a.* unjust.
injustement, *adv.* unjustly.
injustice, *f.* injustice.
injustifiable, *a.* unjustifiable.
innocuité, *f.* innocuousness, harmlessness.
innombrable, *a.* innumerable, numberless.
innombrablement, *adv.* without number.
innovation, *f.* innovation.
innover, *v.t.* innovate, introduce. — *v.i.* innovate.
innumérable, *a.* innumerable.
innumérablement, *adv.* without number.
inobscurci, *a.* unobscured.
inobservable, *a.* not observable, unobservable.
inobservé, *a.* unobserved.

inobstruable, *a.* incapable of being obstructed.

inobstrué, *a* unobstructed.

inoccupé, *a.* unoccupied.

in-octavo, *m. & a.* octavo, 8vo.

inoculabilité, *f* inoculability.

inoculable, *a.* inoculable.

inoculation, *f.* inoculation.

inoculer, *v.t.* inoculate.

inodore, inodorant, *a.* odorless, inodorous.

inodulaire, *a.* (*Med.*) inodular.

inoffensif, *a.* inoffensive; (*Mach.*) out of action.

inoffensivement, *adv.* inoffensively.

inofficiel, *a.* unofficial, nonofficial.

inofficiellement, *adv.* unofficially.

inondation, *f.* inundation, flood, deluge.

inonder, *v.t.* inundate, overflow, flood.

inopiné, *a.* unexpected, unforeseen.

inopinément, *adv.* unexpectedly.

inopportun, *a.* inopportune.

inopportunément, *adv.* inopportunely.

inorganique, *a.* inorganic.

inosique, *a.* inosic, inosinic.

inosite, *f.* inositol, inosite.

inositurie, inosurie, *f.* (*Med.*) inosituria, inosuria.

inoubliable, *a.* unforgettable.

inouï, *a.* unheard, unheard of.

inoxydabilité, *f.* unoxidizability, inoxidability, unoxidability.

inoxydable, *a.* unoxidizable, inoxidable.

inquart, inquartation, *f.* (*Assaying*) quartation, inquartation.

inquarter, *v.t.* subject to quartation.

in-quarto, *m. & a.* quarto.

inquiet, *a.* unquiet, uneasy, anxious.

insaisissable, *a.* imperceptible; not seizable.

insaisissablement, *adv.* imperceptibly.

insalubre, *a.* unhealthy, unhealthful.

insalubrité, *f.* unhealthfulness.

insapide, *a.* tasteless, insipid.

insapidité, *f.* tastelessness, insipidness.

insaponifiable, *a.* unsaponifiable.

insatisfait, *a.* unsatisfied.

insaturable, *a.* insaturable, not saturable, unsaturable.

insaturé, *a.* unsaturated.

insciemment, *adv.* unknowingly, unconsciously.

inscriptible, *a.* inscribable.

inscription, *f.* inscription; registration.

inscrire, *v.t.* inscribe; write down, record, register, enter. — *v.r.* be inscribed; register.

inscrit, *p.a.* inscribed, etc. (see inscrire).

inscrivant, *p.pr.* of inscrire.

insécable, *a.* indivisible.

insecte, *m.* insect.

insecticide, *m.* insecticide. — *a.* insecticidal.

insectivores, *m.pl.* (*Zoöl.*) Insectivora.

insécurité, *f.* insecurity.

in-seize, *m. & a.* sixteenmo, 16mo.

insensibilisateur, *a.* producing insensibility, anesthetizing; (*Expl.*) desensitizing. — *m.* anesthetic; anesthetizing apparatus; (*Expl.*) desensitizer.

insensibiliser, *v.t.* render insensible, anesthetize; (*Expl.*) desensitize.

insensible, *a.* insensible; insensitive.

insensiblement, *adv.* insensibly; insensitively.

insensitif, *a.* insensitive.

inséparable, *a.* inseparable.

inséparablement, *adv.* inseparably.

insérable, *a.* insertable.

insérer, *v.t.* insert. — *v.r.* be inserted.

insertion, *f.* insertion.

insidieux, *a.* insidious.

insigne, *a.* remarkable, signal.

insignifiance, *f.* insignificance.

insignifiant, *a.* insignificant.

insipide, *a.* tasteless, insipid.

insipidement, *adv.* insipidly.

insipidité, *f.* tastelessness, insipidity.

insister, *v.i.* insist.

insolation, *f.* insolation, exposure to the sun.

insoler, *v.t.* expose to the rays of the sun.

insolide, *a.* unsolid, not solid.

insolidement, *adv.* unsolidly.

insolidité, *f.* want of solidity.

insolite, *a.* unusual, unwonted.

insolitement, *adv.* unusually.

insolubilisation, *f.* rendering insoluble.

insolubiliser, *v.t.* render insoluble.

insolubilité, *f.* insolubility.

insoluble, *a.* insoluble.

insolublement, *adv.* insolubly.

insolution, *f.* nonsolution.

insolvabilité, *f.* (*Com.*) insolvency.

insolvable, *a.* (*Com.*) insolvent.

insomnie, *f.* insomnia.

insondable, *a.* unfathomable.

insonore, *a.* not sonorous, nonsonorous.

insoudable, *a.* unweldable; unsolderable.

insoupçonné, *a.* unsuspected.

insoutenable, *a.* untenable; insupportable.

inspecter, *v.t.* inspect.

inspecteur, *m.* inspector.

inspection, *f.* inspection; inspectorship.

inspirant, *p.a.* inspiring.

inspiration, *f.* inspiration.

inspirer, *v.t.* inspire.

instabilité, *f.* instability, unstableness.

instable, *a.* unstable.

installateur, *m.* installer.

installation, *f.* installation.

installer, *v.t.* install.

instamment, *adv.* urgently, earnestly.

instance, *f.* instance; insistence.

instant, *m. & a.* instant.

    à l'—, instantly, immediately.

instantané, *a.* instantaneous. — *m.* instantaneous photograph, snapshot.

instantanéité, *f.* instantaneousness.

instantanément, *adv.* instantaneously.

instar. In the phrase *à l'instar de,* after the manner of, after, like.

instaurer, *v.t.* establish, found.

instiller, *v.t.* instill. — *v.r.* be instilled.

instinct, *m.* instinct.

instinctif, *a.* instinctive.

instinctivement, *adv.* instinctively.

instituer, *v.t.* institute.

institut, *m.* institute; institution.
— *agronomique,* agricultural college or institute.
— *chimique,* chemical institute.

instituteur, *m.* institutor, founder; teacher, instructor.

institution, *f.* institution.

instructeur, *m.* instructor, teacher.

instructif, *a.* instructive.

instruction, *f.* instruction; (*Law*) inquiry.

instruire, *v.t.* instruct; teach; train; inform; (*Law*) examine, investigate.

instruit, *p.a.* instructed, etc. (see instruire); learned, well-educated; aware. — *pr. 3 sing.* of instruire.

instrument, *m.* instrument; implement, tool; utensil; apparatus.
— *de ménage,* household utensil.
instruments de chimie, chemical instruments (or utensils or apparatus).
instruments d'optique, optical instruments, optical apparatus.

instrumental, *a.* instrumental.

insu, *m.* ignorance.
à l'— *de,* unknown to.

insuccès, *m.* failure, lack of success.

insuffisamment, *adv.* insufficiently.

insuffisance, *f.* insufficiency; incapacity.

insuffisant, *a.* insufficient; incompetent.

insufflation, *f.* blowing in or on, insufflation.

insuffler, *v.t.* breathe or blow in or on; breathe or blow into, insufflate; inflate.

insupportable, *a.* insupportable, unbearable.

insurmontable, *a.* insurmountable.

intact, *a.* intact, untouched.

intaille, *m.* intaglio.

intarissable, *a.* inexhaustible.

intégrable, *a.* (*Math.*) integrable.

intégral, *a.* complete, entire; integral.

intégrale, *f.* (*Math.*) integral.

intégralement, *adv.* completely, entirely, wholly; integrally.

intégrant, *a.* integral, integrant.

intégration, *f.* integration.

intégrer, *v.t.* integrate.

intégrité, *f.* integrity.

intellectuel, *a.* intellectual.

intelligemment, *adv.* intelligently.

intelligence, *f.* intelligence.

intelligent, *a.* intelligent.

intelligibilité, *f.* intelligibility.

intelligible, *a.* intelligible; intellectual.

intempérant, intempéré, *a.* intemperate.

intempérie, *f.* inclemency, severity (of weather).

intempestif, *a.* inopportune, unseasonable.

intenable, *a.* untenable.

intendance, *f.* direction, management, superintendence, intendancy.

intendant, *m.* director, manager, superintendent, intendant; steward.

intense, *a.* intense.

intensif, *a.* intensive.

intensifier, *v.t.* intensify.

intensité, *f.* intensity, intenseness.
— *du courant,* (*Elec.*) current intensity, current strength.

intensivement, *adv.* intensively.

intenter, *v.t.* (*Law*) bring, enter.

intention, *f.* intention.
à l'— *de,* for the sake of, on account of.

intentionnel, *a.* intentional.

intentionnellement, *adv.* intentionally.

interatomique, *a.* interatomic.

intercalation, *f.* intercalation; insertion.

intercaler, *v.t.* intercalate; insert; interpose.

intercellulaire, *a.* intercellular.

intercepter, *v.t.* intercept; cut off, stop.

interchangeable, *a.* interchangeable.

interdiction, *f.* prohibition, interdiction.

interdire, *v.t.* forbid, prohibit, interdict; suspend; dumfound.

intéressant, *p.a.* interesting.

intéresser, *v.t.* interest; injure, affect. — *v.r.* be interested; take an interest.

intérêt, *m.* interest.

interférence, *f.* interference.

interférent, *a.* interfering.

interférer, *v.i.* interfere.

interfolier, *v.t.* interleave.

intérieur, *a.* interior; internal; inner, inward.
— *m.* interior; inside; home; private life, family life.
à l'—, (*Med.*) internally.

intérieurement, *adv.* interiorly, inside, on the inside, within, internally, inwardly.

interjeter, *v.t.* introduce, interject.

interligne, *m.* space between lines; interlineation. — *f.* (*Printing*) lead.

interligner, *v.t.* interline; (*Printing*) lead.

intermède, *m.* intermediate; interlude.

intermédiaire, *a.* intermediate, intermediary.
— *m.* medium, agency; intermediate, intermediary, intermedium; (*Com.*) middleman.

intermédiairement, *adv.* intermediately.

intermédiat, *a.* intermediate.

intermicellaire, *a.* intermicellar.

interminable, *a.* interminable, endless.
interminablement, *adv.* interminably.
intermittent, *a.* intermittent.
international, *a.* international.
internationalement, *adv.* internationally.
interne, *a.* internal. — *m.* (*Med.*) interne.
interpolaire, *a.* interpolar.
interpolateur, *a.* interpolating. — *m.* interpolator.
interpolation, *f.* interpolation.
interpoler, *v.t.* interpolate.
interposer, *v.t.* interpose. — *v.r.* be interposed; interpose.
interposition, *f.* interposition.
interprétation, *f.* interpretation.
interpréter, *v.t.* interpret.
interroger, *v.t.* interrogate, examine, consult.
interrompre, *v.t.* interrupt.
interrupteur, *m.* interrupter; specif. (*Elec.*) circuit breaker. — *a.* interrupting, interruption.
intersection, *f.* intersection.
interstellaire, *a.* interstellar.
interstice, *m.* interstice.
interstitiel, *a.* interstitial.
intertubulaire, *a.* intertubular.
intervalle, *m.* interval.
intervenir, *v.i.* intervene; interpose.
interversibilité, *f.* invertibility.
interversible, *a.* invertible.
interversion, *f.* inversion.
interverti, *p.a.* inverted, invert.
intervertir, *v.t.* invert; interchange.
intervertissement, *m.* inverting, inversion.
intervient, *pr. 3 sing.* of intervenir.
intestin, *m. & a.* intestine.
intestinal, *a.* intestinal.
intime, *a.* intimate.
intimement, *adv.* intimately.
intimer, *v.t.* notify, give notice to.
intimité, *f.* close relation; inmost part; intimacy.
intitulé, *m.* title. — *p.a.* entitled.
intoxicant, *p.a.* poisoning, poisonous, poison.
intoxication, *f.* poisoning, (*Med.*) intoxication. — *saturnine,* lead poisoning.
intoxiquer, *v.t.* poison.
intra-atomique, *a.* intra-atomic.
intracellulaire, *a.* intracellular.
intraduisible, *a.* untranslatable.
intraitable, *a.* intractable.
intramoléculaire, *a.* intramolecular.
intramusculaire, *a.* intramuscular.
intransférable, *a.* nontransferable.
intransmissible, *a.* nontransmissible.
intransmuable, *a.* nontransmutable, intransmutable.
intransparence, *f.* nontransparence, opacity (or translucency).
intransparent, *a.* nontransparent.

intransportable, *a.* untransportable.
intravasculaire, *a.* intravascular.
intraveineux, *a.* intravenous.
intrinsèque, *a.* intrinsic.
intrinsèquement, *adv.* intrinsically.
introduction, *f.* introduction.
introductoire, *a.* introductory.
introduire, *v.t.* introduce. — *v.r.* be introduced; enter, get in.
introduit, *p.p. & pr. 3 sing.* of introduire.
introuvable, *a.* undiscoverable, not to be found.
intrus, *m.* intruder. — *p.a.* intruded.
intrusif, *a.* intrusive.
intrusion, *f.* intrusion.
intuitif, *a.* intuitive.
intuitivement, *adv.* intuitively.
intumescence, *f.* intumescence.
intumescent, *a.* intumescent.
inulase, *f.* inulase.
inuline, *f.* inulin.
inusable, *a.* that cannot be worn out, durable.
inusé, *p.a.* unused, unworn.
inusité, *a.* not in use; unusual; obsolete, antiquated.
inutile, *a.* useless.
inutilement, *adv.* uselessly.
inutilisable, *a.* unutilizable.
inutilisé, *a.* unutilized; rendered useless.
inutiliser, *v.t.* render useless.
inutilité, *f.* inutility, uselessness.
invalide, *a. & m.* invalid.
invalidement, *adv.* invalidly.
invalider, *v.t.* invalidate.
invalidité, *f.* invalidity.
invar, *m.* invar (nickel-steel alloy).
invariabilité, *f.* invariability, invariableness.
invariable, *a.* invariable.
invariablement, *adv.* invariably.
invariant, *m.* (*Math.*) invariant.
invendable, *a.* unsalable, unsaleable.
invendu, *a.* unsold.
inventaire, *m.* inventory.
inventer, *v.t.* invent. — *v.r.* be invented.
inventeur, *m.* inventor.
inventif, *a.* inventive.
invention, *f.* invention.
inventorier, *v.t.* inventory.
invérifiable, *a.* unverifiable.
invérifié, *a.* unverified.
inverse, *a.* inverse. — *m.* contrary, reverse.
inversement, *adv.* inversely.
inverseur, *m.* reverser; inverter.
inversion, *f.* reversal; inversion.
invertase, *f.* invertase.
invertébré, *a. & m.* invertebrate.
inverti, *p.a.* reversed; (of sugars, etc.) inverted.
invertine, *f.* invertin (invertase).
invertir, *v.t.* reverse; invert (sugar, etc.).
investigateur, *m.* investigator. — *a.* investigating, inquiring.

investigation, *f.* investigation.

inviscant, *p.pr.* of invisquer.

inviscation, *f.* coating with a viscous material, smearing, daubing.

invisibilité, *f.* invisibility.

invisible, *a.* invisible.

invisiblement, *adv.* invisibly.

invisquer, *v.t.* coat with viscous material, smear, daub.

invitant, *p.a.* inviting.

inviter, *v.t.* invite.

involontaire, *a.* involuntary.

involontairement, *adv.* involuntarily.

involution, *f.* involution.

invoquer, *v.t.* invoke, appeal to.

invraisemblable, *a.* improbable, unlikely.

invraisemblablement, *adv.* improbably.

invraisemblance, *f.* improbability, unlikelihood.

iodate, *m.* iodate.

— *de potasse,* potassium iodate.

iode, *m.* iodine.

— *sublimé,* sublimed iodine.

iodé, *p.a.* iodinated, containing iodine, iodo, iodine; coated or impregnated with iodine, iodized.

ioder, *v.t.* combine with iodine, iodinate; impregnate with iodine, iodize.

iodeux, *a.* iodous.

iodhydrate, *m.* hydriodide.

— *d'ammoniaque,* ammonium iodide.

iodhydrique, *a.* hydriodic.

iodifère, *a.* iodiferous, containing iodine.

iodique, *a.* iodic.

iodisme, *m.* (*Med.*) iodism.

iodoamidonné, *a.* starch-iodide.

iodo-aurate, *f.* iodaurate, iodoaurate.

iodo-aurique, *a.* iodauric, iodoauric.

iodobenzine, *f.* iodobenzene.

iodoforme, *m.* iodoform.

iodoformé, *a.* iodoformized, iodoform.

iodo-iodure, *m.* iodine-iodide (see next term).

iodo-ioduré, *a.* iodine-iodide (designating an iodide solution in which iodine is dissolved).

iodol, *m.* (*Pharm.*) iodol.

iodonitré, *a.* (*Org. Chem.*) iodonitro.

iodothyrine, *f.* iodothyrin.

ioduration, *f.* iodination; iodization; iodation.

iodure, *m.* iodide.

— *d'alcoyle,* alkyl iodide.

— *d'amidon,* iodide of starch, iodized starch.

— *d'argent,* silver iodide.

— *de fer,* iron iodide.

— *de formyle,* iodoform (old name).

— *de méthyle,* methyl iodide, iodomethane.

— *de méthylène,* methylene iodide, diiodomethane.

— *de plomb,* lead iodide.

iodure *de potassium ioduré,* iodized potassium iodide, a solution of potassium iodide in which iodine has been dissolved.

— *de soufre,* sulfur iodide.

— *d'éthyle,* ethyl iodide, iodoethane.

— *mercureux,* mercurous iodide.

— *mercurique,* mercuric iodide.

ioduré, *p.a.* combined with iodine, iodinated; treated with or containing iodine, iodized, iodated; treated with or containing an iodide, iodized.

iodurer, *v.t.* iodinate; iodize; iodate.

iolite, iolithe, *f.* (*Min.*) iolite.

ion, *m.* ion.

ionisation, *f.* ionization.

ioniser, *v.t.* ionize.

ionium, *m.* ionium.

ionone, *f.* ionone.

ipéca, *m.* ipecac.

ipécacuana, ipécacuanha, *m.* ipecacuanha.

ipécacuanate, *m.* ipecacuanhate.

ipécacuanique, *a.* ipecacuanhic.

ira, *fut. 3 sing.* of aller (to go).

irait, *cond. 3 sing.* of aller (to go).

irichromatine, *f.* a benzene solution of asphalt used in producing iridescence.

iridescent, *a.* iridescent.

iridié, *a.* alloyed with or containing iridium.

iridien, *a.* (*Anat.*) iridian.

iridine, *f.* iridin.

iridique, *a.* iridic; (*Anat.*) iridian, iridal.

iridium, *m.* iridium.

iridosmine, *f.* iridosmium.

iris, *m.* iris.

— *de Florence,* Florentine iris (*Iris florentina,* the rhizome of which is called orris root).

irisable, *a.* capable of iridescence.

irisage, *m.* production of iridescence.

irisation, *f.* irisation, iridescence.

irisé, *a.* irised, iridescent, irisated.

iriser, *v.t.* iris, make iridescent. — *v.r.* become iridescent.

irlandais, *a.* Irish. — *m.* Irish; (*cap.*) Irishman.

Irlande, *f.* Ireland.

irone, *f.* irone.

irradiation, *f.* irradiation.

irradier, *v.i.* irradiate, emit rays.

irrationnel, *a.* irrational.

irrationnellement, *adv.* irrationally.

irréalisable, *a.* unrealizable, unattainable; unfeasible, impracticable.

irrecevable, *a.* inadmissible.

irréconciliabilité, *f.* irreconcilability.

irréconciliable, *a.* irreconcilable.

irréconciliablement, *adv.* irreconcilably.

irrécouvrable, *a.* irrecoverable.

irrécupérable, *a.* irrecoverable, irreparable.

irrécusable, *a.* unexceptionable.

irréductibilité, *f.* irreducibility, -cibleness.
irréductible, *a.* irreducible.
irréduit, *a.* unreduced.
irréel, *a.* unreal.
irréfrangible, *a.* irrefrangible.
irréfutable, *a.* irrefutable.
irrégularité, *f.* irregularity.
irrégulier, *a.* irregular.
irrégulièrement, *adv.* irregularly.
irréparablement, *adv.* irreparably.
irréprochable, *a.* irreproachable.
irrésistible, *a.* irresistible.
irrésistiblement, *adv.* irresistibly.
irrésolu, *a.* unsolved; irresolute.
irrésoluble, *a.* irresolvable, unresolvable.
irrespirabilité, *f.* irrespirability.
irrespirable, *a.* irrespirable.
irréussite, *f.* lack of success, failure.
irréversible, *a.* irreversible.
irriguer, *v.t.* irrigate.
irritant, *a.* irritant, irritating; annulling.
— *m.* irritant.
irritation, *f.* irritation.
irriter, *v.t.* irritate.
isatine, *f.* isatin.
isatinique, *a.* isatic, isatinic.
isatropique, *a.* isatropic.
isatyde, *m.* isatide.
ischurétique, *a.* (*Med.*) ischuretic.
iséthionique, *a.* isethionic.
isinglass, *m.* isinglass.
— *végétal*, agar.
islandais, *a.* Iceland, Icelandic.
Islande, *f.* Iceland.
isoamyle, *m.* isoamyl.
isoamylique, *a.* isoamyl.
isoapiol, *m.* isoapiole.
isobutane, *m.* isobutane.
isobutyrique, *a.* isobutyric.
isocèle, *a.* (*Geom.*) isosceles.
isochromatique, *a.* isochromatic.
isochrone, isochronique, *a.* isochronal, isochronous, isochronic.
isochroniquement, *adv.* isochronally.
isocinnamique, *a.* isocinnamic.
isocréatinine, *f.* isocreatinine.
isocyanate, *m.* isocyanate.
— *de phényle*, phenyl isocyanate.
isocyanique, *a.* isocyanic.
isocyanurique, *a.* isocyanuric.
isodimorphe, *a.* isodimorphous, isodimorphic.
isodulcite, *f.* isodulcitol, isodulcite.
isodynamique, isodyname, *a.* isodynamic.
isolable, *a.* isolable.
isolant, *a.* isolating; insulating. — *m.* insulating material, insulator.
isolateur, *m.* insulating device, insulator. — *a.* isolating; insulating.
isolatif, *a.* isolating; insulating.
isolation, *f.* isolation; insulation.

isolatrice, *a.fem.* of isolateur.
isolé, *a.* isolated; insulated; detached.
isolement, *m.* isolation; insulation.
isolément, *adv.* in an isolated manner, separately, individually.
isoler, *v.t.* isolate; insulate. — *v.r.* be isolated; be insulated; isolate oneself.
isoleucine, *f.* isoleucine.
isologue, *a.* isologous. — *m.* isolog(ue).
isoloir, *m.* insulator, specif. insulating stool.
isomaltose, *f.* isomaltose.
isomère, *a.* isomeric. — *m.* isomer.
— *optique*, optical isomer.
isomérie, *f.* isomerism.
— *de position*, position isomerism.
— *géométrique*, geometrical isomerism.
— *physique*, physical isomerism.
— *stéréochimique*, stereochemical isomerism.
isomérique, *a.* isomeric.
isomériser, *v.t.* isomerize, convert into an isomer.
isomérisme, *m.* isomerism.
isométrique, *a.* isometric, isometrical.
isomorphe, *a.* isomorphous, isomorphic.
isomorphisme, *m.*, **isomorphie**, *f.* isomorphism.
isonicotianique, *a.* isonicotinic.
isonitrosé, *a.* isonitroso.
isopelletiérine, *f.* isopelletierine.
isophtalique, *a.* isophthalic.
isopropylique, *a.* isopropyl.
isopyromucique, *a.* isopyromucic.
isoquinoléine, *f.* isoquinoline.
isoquinoléique, *a.* isoquinoline.
isosaccharique, *a.* isosaccharic.
isosafrol, *m.* isosafrole.
isoscèle, *a.* (*Geom.*) isosceles.
isostrychnine, *f.* isostrychnine.
isosuccinique, *a.* isosuccinic.
isosulfocyanate, *m.* isosulfocyanate (isothiocyanate).
— *d'allyle*, allyl isothiocyanate, allyl mustard oil.
isotétramorphe, *a.* isotetramorphous.
isotherme, *a.* isothermal. — *f.* isotherm.
isothermique, *a.* isothermic, isothermal.
isothermiquement, *adv.* isothermally.
isothio-urée, *f.* isothiourea, thiopseudourea.
isotonie, *f.* isotonicity, isotonic state.
isotonique, *a.* isotonic.
isotope, *m.* isotope.
isotopie, *f.* isotopy.
isotrimorphe, *a.* isotrimorphous, -phic.
isotrope, *a.* isotropic, isotropous.
isotropie, *f.* isotropy, isotropism.
iso-urée, *f.* isourea, pseudourea.
isoxanthine, *f.* isoxanthine.
isoxazol, *m.* isoxazole.
issu, *a.* issued, issuing; proceeding (from), the result (of); descended.

issue, *f.* issue; (*pl.*) offal, refuse, waste. — *a.* *fem.* of issu.

à l'— de, at the end of, on leaving.

issues de la boucherie, waste parts of slaughtered animals, offal.

issues de la meunerie, by-products of flour (screenings, bran, shorts).

issues des ordinaires, garbage, slop.

isthme, *m.* isthmus.

itaconique, *a.* itaconic.

Italie, *f.* Italy.

italien, *a.* Italian. — *m.* Italian.

italique, *a. & m.* (*Printing*) italic.

italiqué, *a.* italicized.

ivoire, *m.* ivory; (of the teeth) dentine.

— *artificiel,* artificial ivory.

— *végétal,* vegetable ivory (from *Phytelephas macrocarpa*).

ivoirine, *f.* ivorine (imitation ivory).

ivraie, *f.* (*Bot.*) rye grass (*Lolium*).

ivresse, *f.* intoxication, drunkenness.

ixomètre, *m.* a viscosimeter.

J

j'. Contraction of *je*, I.

jable, *m.* that part of a glass pot where the walls and bottom join; chime (of a cask); groove (of a cask stave).

jaborandi, *m.* (*Pharm. & Bot.*) jaborandi.

jachère, *f.* fallow, fallowness; fallow land.

jacinthe, *f.* hyacinth.

jadaïque, *a.* relating to or resembling jade.

jade, *m.* jade.

jadéite, *f.* (*Min.*) jadeite.

jadien, *a.* of or containing jade.

jadis, *adv.* of old, formerly. — *a.* old, former.

jaguar, *m.* (*Zoöl.*) jaguar (*Felis onca*).

jaïet, *m.* jet.

jaillir, *v.i.* spout, gush, spurt; burst forth, (of light) flash; (of the electric spark) pass, jump.

    faire —, pass (an electric spark); spout, flash, etc.

jaillissement, *m.* spouting, gushing, spurting, etc. (see jaillir).

jais, *m. & a.* jet.

jalap, *m.* (*Pharm. & Bot.*) jalap.

jalapine, *f.* jalapin.

jalapique, *a.* jalapic.

jallot, *m.* = jalot.

jalon, *m.* stake, pole (for surveying, etc.).

jalonner, *v.t.* stake out, mark out.

jalot, *m.* a vessel in which cakes of tallow are molded.

jalouse, *a.fem.* of jaloux.

jalousement, *adv.* jealously.

jalousie, *f.* jealousy; grating; jalousie, blind.

    à —, like a grating.

jaloux, *a.* jealous.

Jamaïque, *f.* Jamaica.

jamais, *adv.* ever; (with *ne* expressed or understood) never.

    à —, *pour* —, forever, for ever.

jambage, *m.* jamb; down stroke (of a letter).

jambe, *f.* leg; shank.

jambière, *f.* gaiter; (*Mach.*) sleeve. — *a.* leg.

jambon, *m.* ham.

jante, *f.* rim (of a wheel).

janvier, *m.* January.

Japon, *m.* Japan.

japon, *m.* sapan wood; Japanese porcelain.

japonais, *a.* Japanese. — *m.* Japanese (language); (*cap.*) Japanese (man).

japonique, *a.* japonic (acid); Japonic, Japan.

japonner, *v.t.* (*Ceram.*) fire so as to give the appearance of Japanese porcelain.

jardin, *m.* garden.

jardinage, *m.* gardening; garden produce; garden ground; dark spot (in a gem).

jardineux, *a.* (of gems) cloudy, having dark spots.

jardinier, *a.* garden. — *m.* gardener.

jargon, *m.* (*Min.*) jargon.

jarre, *f.* jar.

jarret, *m.* (of pipes, etc.) elbow; hock, ham.

jasmin, *m.* (*Bot.*) jasmine, jasmin, jessamine.

    — *sauvage*, false jasmine, yellow jasmine (*Gelsemium sempervirens*).

jasmol, *m.* jasmole.

jaspagate, *f.* (*Min.*) agate jasper.

jaspage, *m.* imitation of jasper.

jaspe, *m.* (*Min.*) jasper.

jasper, *v.t.* jasperize, marble, mottle, cloud.

jaspure, *f.* jasperizing, marbling, mottling.

jatte, *f.* bowl, porringer; bowlful.

jattée, *f.* bowlful.

jauge, *f.* gage, gauge; standard.

    — *à tréfiler*, wire gage.

    — *de vapeur*, steam gage.

    — *du vide*, vacuum gage.

jaugeage, *m.* gaging, gauging, calibration; tonnage (of a ship).

    — *par écoulement*, calibration for delivery.

    — *par emplissage*, calibration for contents.

jauge-carcasse, *a.* (of wires) very fine (less than half a millimeter in diameter).

jauger, *v.t.* gage, gauge, calibrate.

jaugeur, *m.* gager, gauger, calibrator. — *a.* gaging, gauging, measuring, calibrating.

jaunâtre, *a.* yellowish.

jaune, *a. & m.* yellow.

    — *acide*, acid yellow.

    — *anglais*, Victoria yellow.

    — *brillant*, brilliant yellow.

    — *citron*, lemon yellow.

    — *clair*, light yellow.

    — *congo*, Congo yellow.

    — *d'acridine*, acridine yellow.

    — *d'alizarine*, alizarin yellow.

    — *d'antimoine*, antimony yellow.

    — *de cadmium*, cadmium yellow.

    — *de Cassel*, Cassel yellow (a lead oxychloride).

    — *de chrome*, chrome yellow.

jaune *de fer*, Mars yellow.
  — *de Hesse*, Hessian yellow.
  — *de Mars*, Mars yellow.
  — *de Naples*, Naples yellow.
  — *de Paris*, Paris yellow.
  — *de quinoléine*, quinoline yellow.
  — *de Vérone*, Verona yellow.
  — *de zinc*, zinc yellow.
  — *d'ocre*, yellow ocher.
  — *d'œuf*, yellow of egg, egg yolk.
  — *d'or*, golden yellow; (*Dyes*) gold yellow.
  — *d'urane*, uranium yellow.
  — *foncé*, dark yellow.
  — *indien*, Indian yellow.
  — *métanile*, metanil yellow.
  — *minéral*, mineral yellow (a lead oxychloride).
  — *N*, yellow N.
  — *pâle*, pale yellow.
  — *serin*, canary yellow.
  — *sidérin*, a yellow pigment prepared by boiling together solutions of ferric chloride and potassium dichromate.
  — *soleil*, sun yellow.
  — *solide*, fast yellow.
  — *soufre*, sulfur yellow.
  — *Victoria*, Victoria yellow.
jaune-capucine, *m.* nasturtium yellow.
jaune-citron, *m.* lemon yellow.
jaune-paille, *m.* straw yellow.
jaune-pâle, *a.* pale yellow.
jaunet, *a.* yellowish.
jaunir, *v.t. & i.* yellow, turn yellow.
jaunissage, *m.* = jaunissement.
jaunisse, *f.* (*Med.*) jaundice.
jaunissement, *m.* yellowing, turning or coloring yellow.
Java, *f.* Java.
javanais, *a.* Javanese.
javelle, *f.* small heap (of sea salt); (*cap.*) incorrect spelling of Javel (see eau de Javel under *eau*); handful, swath (of cut grain); bundle, faggot.
javellisation, *f.* treatment with Javel water or Javel extract (as for sterilization).
jayet, *m.* jet.
je, *pron.* I.
jervine, *f.* jervine.
jésus, *m.* a size of paper (about 55 x 72 cm.).
jet, *m.* throw, throwing, cast, toss; jet, stream, gush, spouting, spout, spurt; flash; outline, sketch; sprout, shoot; (*Founding*) (*1*) casting, pouring, (*2*) jet.
  — *à la mer*, jettison.
  — *à moule*, (*Founding*) gate, runner, channel, jet.
  — *d'acier*, casting of steel.
  — *d'eau*, jet of water; fountain jet, fountain; rain strip.
  — *de condensation*, condensing jet.

jet *de coulée*, (*Founding*) feedhead, feeding head.
  — *de feu*, (*Pyro.*) fire sheaf.
  — *de flamme*, jet of flame.
  — *de fonte*, (*Founding*) deadhead.
  — *de lumière*, flash of light.
jet *de pierre*, stone's throw.
  — *de sable*, sand blast.
  — *de terre*, earth thrown out in excavating.
  — *de vapeur*, steam jet.
jetage, *m.* throwing, etc. (see jeter).
jeter, *v.t.* throw; cast; hurl, toss; sprinkle, scatter; put, place; emit; discharge; throw out; utter; throw away; lay (foundations); calculate. — *v.i.* shoot, sprout; run, discharge. — *v.r.* be thrown, be cast, etc.; throw oneself; fall, flow.
  — *en moule*, cast; pour (concrete) into the form.
jeu, *m.* play; clearance; backlash; action, working, functioning; set; explosion (as of a mine); game; stake; trick; freak.
  en —, (*Mach.*) in gear, in action.
  — *d'orgues*, organ.
  — *perdu*, (*Mach.*) lost motion.
jeudi, *m.* Thursday.
jeun. In the phrase *à jeun*, fasting, without food, without food and drink.
jeune, *a.* young; younger, junior. — *m.* young person; young (of an animal).
jeûne, *m.* fast, fasting; privation, dearth.
jeûner, *v.i.* fast.
jeunesse, *f.* youth.
jeûneur, *m.* faster.
joaillerie, *f.* jewelry; jeweler's trade.
joaillier, *m.* jeweler, jeweller.
joie, *f.* joy.
joignant, *p.a.* joining, etc. (see joigner); adjoining, adjacent, next.
joindre, *v.t.* join; unite; adjoin, be adjacent to; assemble; add. — *v.i.* join, be in contact, fit. — *v.r.* join, be joined, unite, meet.
  — *à chaud*, weld.
joint, *p.a.* joined, united, etc. (see joindre); joint. — *m.* joint; seam. — *pr. 3 sing.* of joindre.
  — *à boulet*, ball-and-socket joint.
  — *à l'émeri*, ground joint.
  — *articulé*, (*Mach.*) knuckle joint.
  — *en about*, butt joint.
  — *que*, — *à ce que*, beside which, and in addition.
jointée, *f.* double handful.
jointif, *a.* joined, touching, in contact.
jointoyer, *v.t.* (*Masonry*) point.
jointure, *f.* joint, joining, junction.
joli, *a.* pretty, handsome, fine; good, ample.
joliment, *adv.* well, agreeably, finely; very.
jonc, *m.* cane, reed, rush; specif., (*Bot.*) a member of the family Juncaceæ, (true) rush.

jonc *odorant,* lemon grass.

joncacées, *f.pl.* (*Bot.*) Juncaceæ.

jonchée, *f.* a little cheese made in a wicker basket; heap; strewed material.

joncher, *v.t.* strew, cover, litter.

jonction, *f.* junction; joining, joint.

jongleur, *m.* juggler; trickster.

jonquille, *f.* (*Bot.*) jonquil.

joseph, *a.* see papier joseph, under *papier;* (of cotton) spun.

joubarbe, *f.* houseleek (*Sempervivum*); stonecrop (*Sedum*).

— *âcre,* common stonecrop (*Sedum acre*).

joue, *f.* cheek (in various senses); side, sidepiece, jaw, flange, etc.

jouer, *v.i.* play; have play, have clearance; be loose, work loose; work, act; go off, explode, be discharged, be sprung; start, spring, warp. — *v.t.* play; use, work; imitate; deceive, make game of. — *v.r.* play, amuse oneself; deceive oneself.

— *le rôle,* play the role, play the part.

jouet, *m.* clamp, fishplate; plaything, sport.

joug, *m.* yoke; (*Mach.*) crosshead.

jouir, *v.i.* (with *de*) enjoy.

jouissance, *f.* enjoyment; the right to interest on capital.

joule, *m.* (*Physics*) joule.

jour, *m.* day; daylight; light; opening, aperture; space, interval; way.

à —, open, openwork; (*Com.*) up to date.

dans le —, (of measurements) in the clear.

mettre au —, bring to light.

journal, *m.* (news)paper; journal; diary; daybook; ship's log.

journalier, *a.* daily, diurnal. — *m.* day laborer.

journalisme, *m.* journalism.

journée, *f.* day, daytime; day's pay; day's work; day work; day's journey.

journellement, *adv.* daily, every day.

joyau, *m.* jewel, gem.

judiciaire, *a.* judicial. — *f.* judgment.

judiciairement, *adv.* judicially.

judicieusement, *adv.* judiciously.

judicieux, *a.* judicious.

juge, *m.* judge.

jugement, *m.* judgment.

juger, *v.t. & i.* judge.

juglandine, *f.* juglandin.

juglon, *m.* juglone.

jugulaire, *a. & f.* jugular.

juif, *m.* Jew. — *a.* Jewish.

juillet, *m.* July.

juin, *m.* June.

jujube, *f.* jujube (the fruit). — *m.* jujube (the paste).

jujubier, *m.* jujube tree, jujube (*Zizyphus*).

julep, *m.* (*Pharm.*) julep.

jumeau, *a. & m.* twin; double.

jumelle, *a. & f. fem.* of jumeau. — *f.* binocular; cheek, sidepiece; (*Pyro.*) double rocket.

— *de campagne,* field glasses.

— *de théâtre,* opera glass.

jumenteux, *a.* (*Med.*) designating urine that has a thick, colored deposit.

jurassique, *a. & m.* (*Geol.*) Jurassic.

juré, *p.a.* sworn. — *m.* juryman, juror.

jurer, *v.t.* swear. — *v.i.* swear; clash, jar squeak, screech, grate.

juridiction, *f.* jurisdiction; body of magistrates.

juridique, *a.* juridical, judicial.

juridiquement, *adv.* juridically, judicially.

jus, *m.* juice.

— *cru,* raw juice, crude juice.

— *de betteraves,* sugarbeet juice.

— *de Brésil,* brazilwood liquor (a decoction of brazilwood or redwood).

— *de cannes,* cane juice.

— *de réglisse,* licorice (or liquorice) juice.

— *de tannée, —* tannant, tanning liquor, tan liquor.

jusant, *m.* ebb, ebb tide.

jusée, *f.* tanning liquor, tan liquor.

jusque, *prep.* (often with *à*) to, till, until, up to, as far as, even to, even.

jusqu'à ce que, till, until.

jusqu'à refus, to excess.

jusqu'à trouble, to cloudiness, till clouding occurs.

jusqu'ici, to this place, thus far, till now.

jusque-là, to that place, so far, till then.

jusque-là que, to such a degree that.

jusqu'où, how far.

jusquiame, *f.* (*Bot.*) henbane (*Hyoscyamus*).

— *noire,* black henbane (*Hyoscyamus niger*).

juste, *a.* just; exact, accurate, correct, right, true; tight, close-fitting. — *adv.* just; rightly, right, well; tightly, tight. — *m.* just; just man.

à — titre, justly, rightly, with reason.

au —, just, exactly.

justement, *adv.* just, exactly; justly.

justesse, *f.* accuracy, exactness, precision, correctness; justness; fitness.

justice, *f.* justice; court of justice.

justifiable, *a.* justifiable.

justifiablement, *adv.* justifiably.

justifier, *v.t.* justify.

jute, *m.* jute.

juter, *v.i.* exude juice.

juteux, *a.* juicy.

juxtaposé, *p.a.* juxtaposed, in juxtaposition.

juxtaposer, *v.t.* juxtapose, place side by side. — *v.r.* be juxtaposed.

juxtaposition, *f.* juxtaposition.

K

kaieput, *m.* cajuput.
kaïnite, *f.* (*Min.*) kainite.
kaki, *a.* khaki. — *m.* Japanese persimmon.
kakodyle, *m.* cacodyl.
kakodylique, *a.* cacodylic; cacodyl.
kaléidoscopique, *a.* kaleidoscopic(al).
kali, *m.* kali, glasswort (*Salsola kali*); potash.
kalisme, *m.* poisoning caused by potash.
kalium, *m.* potassium. (*Obs.*)
kallitypie, *f.* (*Photog.*) kallitype.
kaolin, *m.* kaolin.
kaolinique, *a.* kaolinic.
kaolinisation, *f.* kaolinization.
kaoliniser, *v.t.* kaolinize.
kaoutchouc, *m.* caoutchouc, rubber.
kapok, *m.* kapok, Java cotton.
karabé, *m.* amber. (*Obs.*)
karabique, *a.* karabic (succinic).
karat, *m.* carat.
karature, *m.* = carature.
karyokinèse, *f.* (*Biol.*) karyokinesis, mitosis.
kassu, *m.* kassu (betel nut extract).
kathode, *m.* cathode.
kava, kawa, *f.* kava (the plant or the drink).
kéfir, képhir, *f.* kefir, kephir.
kératine, *f.* keratin.
kératinisation, *f.* keratinization.
kératiniser, *v.t.* keratinize.
kératogène, *a.* keratogenous.
kératoïde, *a.* keratoid.
kératoplastique, *a.* hardening the skin; keratoplastic.
kératose, *f.* (*Med.*) keratosis.
kérite, *f.* kerite (an insulating material).
kermès, *m. & a.* kermes.
— minéral, — médicinal, kermes mineral.
kérosène, *m.* kerosene.
kérosolène, *m.* kerosolene, rhigolene.
ketmie, *f.* (*Bot.*) hibiscus.
khantchin, *m.* a Chinese liquor made from millet.
khôl, *m.* kohl, kohol.
kieselguhr, kieselgur, *m.* diatomaceous earth, kieselguhr.
kilo, *m.* kilo (kilogram).

kilogramme, *m.* kilogram (1000 grams, 2.2046 lbs.).
kilogrammètre, *m.* kilogrammeter (about 7¼ foot pounds).
kilomètre, *m.* kilometer (0.62137 mile).
kilométrique, *a.* kilometric, kilometrical.
kilométriquement, *adv.* by kilometers.
kilowatt, *m.* kilowatt.
kilowatt-heure, *m.* kilowatt hour.
kina, *m.* cinchona bark; quinine.
kinase, *f.* kinase.
kinate, *m.* quinate.
kinétite, *f.* (*Expl.*) kinetite.
kinine, *f.* quinine.
kinique, *a.* quinic.
kino, *m.* kino.
— de l'Inde, East India kino.
kinovine, *f.* quinovin.
kir, *m.* (*Geol.*) bituminous earth.
kirsch, kirschwasser, *m.* kirsch, kirschwasser.
klaubage, *m.* picking, sorting.
knopper, Knopper, *f.* (nut) gall, gallnut, specif. that produced by *Cynips quercus calicis* on certain European oaks.
kohol, koheul, *m.* kohl.
kola, *m.* (*Pharm.*) kola; (*Bot.*) Cola.
kolatier, *m.* kola tree (*Cola acuminata*).
koumis, koumys, *m.* kumiss, koumiss.
koussine, *f.* kosin, kousin, koosin (from cusso).
kousso, *m.* = cousso.
krabb, *m.* a long iron rod formerly used in flattening window glass.
krypton, *m.* krypton.
kulhaven, *n.* an earthenware cylinder formerly used in annealing objects of crystal glass.
kummel, kümmel, *m.* kümmel (the liqueur); caraway.
kupfernickel, *m.* (*Min.*) niccolite, kupfernickel.
kvas, *m.* kvass, kvas (Russian beer).
kw.-an, *abbrev.* (kilowatt-an) kilowatt year.
kwas, *m.* kvass, kvas (Russian beer).
kyanisation, *f.* kyanization.
kyaniser, *v.t.* kyanize.
kyste, *m.* cyst.
kystique, kysteux, *a.* cystic.

L

L. A symbol sometimes used for lithium.

l'. Contraction of *le* or *la*.

la, *article, fem. sing.* the. — *pron.* her, she, it.

là, *adv.* there; then.

  ce . . . -là, cette . . . -là, that . .

  ces . . . -là, those . . .

  de là, thence, from there, from then, from that, after that.

  là-bas, là-dedans, etc., see below.

  là ou, where.

  par là, that way; thereby, by that.

lab, *m.* rennet, lab.

là-bas, *adv.* down there, down yonder.

labdanum, *m.* labdanum.

label, *m.* label, mark.

labenzyme, *m.* the enzyme of rennet, rennin.

labeur, *m.* labor, toil; (*Agric.*) cultivation.

labferment, *m.* = labenzyme.

labié, *a.* (*Bot.*) labiate. — *f.* a member of the Menthaceæ (Labiatæ).

labile, *a.* labile.

laboratoire, *m.* laboratory; (in gas analysis) absorption chamber or pipet.

  — *absorbeur,* absorption chamber or pipet.

  — *d'analyse,* analytical laboratory.

  — *de pyrotechnie,* fireworks laboratory, pyrotechnic works.

  — *de recherches chimiques,* chemical research laboratory, chemical laboratory.

laborieusement, *adv.* laboriously.

laborieux, *a.* laborious.

laboriosité, *f.* laboriousness.

labour, *m.* tillage, plowing; tilled land, plowed land.

labourable, *a.* tillable, arable.

labourage, *m.* tillage; farming.

labourer, *v.t.* till, plow; plow up. — *v.i.* till, plow; drudge.

labradorite, *f.,* **labrador,** *m.* (*Min.*) labradorite.

labyrinthe, *m.* labyrinth.

labyrinthique, *a.* labyrinthine, labyrinthic.

lac, *m.* lake.

  — *salant,* salt lake.

laçage, *m.* lacing.

laccase, *f.* laccase.

laccine, *f.* laccin.

laccique, *a.* laccaic, laccic.

lac-dye, *m.* lac dye.

lacer, *v.t.* lace.

lacérer, *v.t.* lacerate.

lacet, *m.* lace; loop; zigzag, winding; snare.

lâche, *a.* loose; slack; lax; indolent; mean, base, cowardly. — *adv.* loosely, loose.

lâchefer, *m.* (*Metal.*) tap bar.

lâchement, *adv.* loosely, feebly, sluggishly. — *m.* loosening, etc. (see lâcher).

lâcher, *v.t.* loosen, loose, relax, slacken; turn on (cocks, etc.), open; let go, release; fire, discharge, let off; let out, let slip. — *v.i. & r.* loosen, slacken, get loose; escape; (of a gun) go off.

lacis, *m.* network, plexus.

lac-laque, *f.* lac lake.

lacmoïde, *m.* lacmoid.

là-contre, *adv.* against that, to the contrary.

lacryma-christi, *m.* Lachryma Christi (a wine).

lacrymal, *a.* lachrymal.

lacrymogène, *a.* tear-producing, lachrymatory. — *m.* (*Mil.*) lachrymator, tear gas.

lacrymule, *f.* small tear.

lactaire, *a.* lactary.

lactalbumine, *f.* lactalbumin.

lactame, *n.* lactam.

lactamide, *f.* lactamide.

lactamique, *a.* of or pertaining to a lactam; lactamic (designating the acid usually known as alanine).

lactase, *f.* lactase.

lactate, *m.* lactate.

  — *de fer,* lactate of iron, iron lactate, specif (as in *Pharm.*) ferrous lactate.

  — *de magnésie,* magnesium lactate.

lactation, *f.* (*Physiol.*) lactation.

lacté, *a.* (of, containing or resembling milk) milky, milk, lacteal; (*Anat.*) lacteal; (*Bot.*) lactescent.

lacter, *v.t.* treat or combine (as flour) with milk.

lactéiforme, *a.* resembling milk.

lactéine, lactéoline, *f.* lactein, condensed milk.

lactescence, *f.* milkiness, lactescence.

lactescent, *a.* lactescent (milky, or yielding a milky juice).

lacticémie, *f.* autointoxication due to lactic acid in the blood.

lactide, *f.* lactide.

lactique, *a.* lactic.

lactobacilline, *f.* culture of lactic-acid bacteria.

lacto-butyromètre, *m.* lactobutyrometer.

lactodensimètre, *m.* lactodensimeter.

lactoglobuline, *f.* lactoglobulin.

lactoline, *f.* lactolin (potassium lactate and lactic acid, used as a mordant); = lactéine.

lactomètre, *m.* lactometer.

lactone, *f.* lactone.

lacto-nécessaire, *m.* an outfit for the rapid analysis of milk.

lactoprotéine, *f.* lactoprotein.

lactoscope, *m.* lactoscope.

lactose, *f.* lactose, milk sugar.

lactosérum, *m.* milk serum.

lactoviscosimètre, *m.* lactoviscosimeter.

lactucarium, *m.* lactucarium.

lacune, *f.* lacuna, empty or blank space, gap.

lacuneux, *a.* lacunose, full of lacunas.

lacustre, *a.* lacustrine.

ladanifère, *a.* bearing labdanum.

ladanum, *m.* labdanum, ladanum.

là-dedans, *adv.* in there, in, within; therein.

là-dehors, *adv.* out there, out.

là-dessous, *adv.* under there, underneath, below.

là-dessus, *adv.* on that, over that, about that, thereupon, thereafter.

là-devant, *adv.* yonder, on ahead.

ladite, *fem.* of ledit (which see).

lagre, *m.* a sheet of glass on which window glass is flattened.

là-haut, *adv.* up there, above, upward.

laiche, *f.* (*Bot.*) sedge.

laid, *a.* ugly; bad (weather); naughty.

lainage, *m.* woolen stuffs, woolens; fleece; teaseling.

laine, *f.* wool.

— *beige,* natural wool.

— *de bois,* wood wool; wood fiber; pine wool.

— *de Cachemire,* Kashmir (or Cashmere) wool (from the Kashmir goat).

— *de fer,* iron wool.

— *de laitier,* slag wool, mineral wool.

— *de mouton,* sheep's wool.

— *de pin,* pine wool, vegetable wool.

— *de scories,* slag wool, mineral wool.

— *de tourbe,* peat wool.

— *de verre,* glass wool.

— *en suint,* wool in the grease.

— *en toison,* fleece wool.

— *philosophique,* philosopher's wool (sublimed zinc oxide).

laine-renaissance, *f.* wool regeneration, wool recovery.

lainer, *v.t.* teasel, tease (fabrics).

lainerie, *f.* woolen goods, woolens; manufacture of woolens; woolen shop or business; teaseling machine.

laineux, *a.* woolly.

lainier, *a.* wool, pertaining to wool.

laisser, *v.t.* leave; let, allow, permit; leave out, omit; let go; (with *de* and infinitive) cease, leave off (as, *la chose ne laisse pas d'être vraie,* the thing is still true, is nevertheless true, is certainly true).

(Phrases in which *laisser* is followed by another verb are entered under those verbs.)

laisser-passer, *m.* pass.

lait, *m.* milk.

— *caillé,* curdled milk.

— *clair,* whey.

— *concentré,* condensed milk (or evaporated milk).

— *condensé,* condensed milk.

— *cru,* raw milk.

— *d'amande(s),* milk of almond(s), emulsion of almond.

— *d'ammoniaque,* (*Pharm.*) ammoniacum mixture.

— *d'ânesse,* ass's milk.

— *d'asafœtida,* milk (or emulsion) of asafetida.

— *de beurre,* buttermilk.

— *de brebis,* ewe's milk.

— *de chaux,* milk of lime.

— *de chèvre,* goat's milk.

— *de femme,* woman's milk, human milk.

— *de gaïac,* lait de gayac, (*Pharm.*) guaiacum mixture.

— *de jument,* mare's milk.

— *de malt,* (*Brewing*) a milky liquid obtained by grinding malt with water.

— *de soufre,* milk of sulfur.

— *de vache,* cow's milk.

— *écrémé,* skimmed milk, skim milk.

— *en poudre,* milk powder, powdered milk.

— *homogénéisé,* homogenized milk.

— *maternisé,* milk of animals (esp. cow's milk) modified to resemble human milk.

— *mercuriel,* (*Pharm.*) white precipitate, ammoniated mercury.

— *végétal,* (*Bot.*) latex.

laitance, laite, *f.* milt (secretion of fishes); (*Cement*) laitance (a gelatinous exudation from concrete).

laiterie, *f.* dairy; dairying; dairy shop.

laiterol, *m.* (*Metal.*) flosshole plate.

laiteux, *a.* milky; milk; lacteal.

laitier, *m.* slag, scoria, (*Iron*) cinder; (*Bot.*) milkwort (*Polygala*); milkman. — *a.* milk. (A distinction is made by some between *laitier,* slag, and *scorie,* scoria; the former resulting from the gangue and flux, the latter from the metal and the furnace; but both words are most commonly translated *slag.*)

laiton, *m.* brass.

— *de fonte,* cast brass.

— *en feuilles,* sheet brass.

laitonnage, *m.* brass plating, brassing.

laitonner, *v.t.* brass-plate, brass.

laitue, *f.* lettuce.

laize, *f.* width, breadth (of cloth).

lama, *m.* (*Zoöl.*) llama, lama.

— *alpaga,* alpaca.

— *vigogne,* vicuña.

lambeau, *m.* shred, scrap; rag; fragment.

lambris, *m.* paneling, wainscoting; (fig.)canopy.

lame, *f.* (a thin flat piece, esp. of metal) foil, plate, sheet; lamina, scale; blade; slice; strip; slat; (*Micros.*, etc.) slide; band; wave, billow.

— *d'eau,* sheet of underground water; (*Steam*) water space.

— *de gaz,* tongue of gas.

— *de platine,* (piece of) platinum foil.

— *de poudre,* cake of powder, milled cake.

— *porte-objet,* (*Micros.*) slide.

lamellaire, *a.* lamellar.

lamellation, *f.* lamellar arrangement.

lamelle, *f.* lamella, platelet, leaflet, scale; (*Expl.*) grain, small disk; (*Micros.*) cover glass.

— *couvre-objet,* cover glass.

lamellé, *a.* lamellate, lamellated, foliated.

lamelleux, *a.* lamellar, lamellose.

lamellifère, *a.* lamelliferous.

lamelliforme, *a.* lamelliform.

lamentablement, *adv.* lamentably.

lamenter, *v.t., i. & r.* lament.

lamette, *f.* small plate, platelet, lamella, small leaf, leaflet; clasp.

laminage, *m.* rolling, etc. (see laminer).

laminaire, *a.* laminated, laminar.

lamine, *f.* lamina, thin plate.

laminé, *p.a.* rolled; sheet; calendered; flattened.

laminer, *v.t.* roll (as metal into plates or sheets), laminate; calender (paper, cloth, etc.); flatten.

laminerie, *f.* rolling mill.

lamineur, *m.* roller (workman); = laminoir.

— *a.* rolling.

lamineux, *a.* laminar, laminate, laminose.

laminoir, *m.* rolling machine, roller, rolls, roll; rolling mill, flatting mill.

lampant, *a.* (of oil) suitable for burning in a lamp; (of olive oil) purified, good, bright, clear.

lampe, *f.* lamp.

— *à acétylène,* acetylene lamp.

— *à alcool,* alcohol lamp, spirit lamp.

— *à arc,* arc lamp, arc light.

— *à braser,* brazing lamp.

— *à esprit de vin,* spirit lamp.

— *à essence,* benzine lamp, gasoline lamp, gasoline torch.

— *à gaz,* gas lamp, gas burner.

— *à huile,* oil lamp (one using vegetable or animal oil).

lampe *à incandescence,* incandescent lamp.

— *à pétrole,* kerosene lamp, oil lamp.

— *à souder,* soldering lamp.

— *à vapeur de mercure,* mercury vapor lamp.

— *d'émailleur,* enameler's lamp.

— *de poche,* pocket lamp, flash light.

— *de sûreté,* safety lamp.

— *électrique,* electric lamp, specif. a flash light.

— *témoin,* indicator lamp.

lampe-heure, *f.* lamp hour.

lamper, *v.i.* (of the sea) phosphoresce. Cf. lampant.

lamperon, *m.* wick holder; bowl (of a lamp).

lampisterie, *f.* lamp manufacture.

lançage, *m.* launching, etc. (see lancer).

lance, *f.* lance; nozzle; rod, pole, staff.

lance-bombes, *m.* bomb-throwing device, trench mortar, minenwerfer.

lance-flammes, *m.* flame projector.

lancement, *m.* launching, etc. (see lancer).

lance-poudre, *m.* an apparatus for throwing powder on articles to be enameled or soldered.

lancer, *v.t.* launch; throw, dart, shoot, cast, emit; issue, send out, put out; start.

— *v.r.* spring, fly, rush; launch out.

lancette, *f.* lancet; (in general) tool with a lance-like blade.

langage, *m.* language.

langue, *f.* tongue; language; (of a balance) pointer, needle.

— *de chien,* (*Bot.*) hound's-tongue (*Cynoglossum officinale*).

languette, *f.* tongue (in technical senses); small tongue; (of a balance) pointer, needle; (of paper) small strip.

langueur, *m.* languor, languidness.

languir, *v.i.* languish; long; flag, drag.

languissant, *p.a.* languishing; slow, dull.

lanière, *f.* strap, thong; strip.

lanoline, *f.* lanolin, lanoline.

lanterne, *f.* lantern (in various senses); a glass window thru which the color of the gases is observed in the chamber process for sulfuric acid; a measure for loading shells; (*Founding*) core barrel (or core rod).

lanthane, *m.* lanthanum.

lapidaire, *m. & a.* lapidary.

lapin, *m.* rabbit.

— *de garenne,* wild rabbit.

— *du Brésil,* guinea pig.

lapis, lapis-lazuli, *m.* (*Min.*) lapis lazuli.

lapon, *a.* Lappic. — *m.* Lappic; (*cap.*) Laplander.

Laponie, *f.* Lapland.

laps, *m.* lapse.

laquage, *m.* lacquering; laking.

laque, *f.* lake; lac, gum lac. — *m.* lacquer, lacker. — *a.* lac.

laque *anglaise*, a cochineal lake of intense color, the finest quality of which is called crimson lake.

— *brûlée*, a cochineal lake containing a little black.

— *carminée*, lâke (proper), cochineal lake (of which carmine and crimson lake are varieties).

— *de bois rouge*, redwood lake.

— *de garance*, madder lake.

— *de gaude*, weld lake.

— *de Smyrne*, Smyrna lake (from madder).

— *en bâton*, stick-lac.

— *en boules de Venise*, a variety of redwood (or brazilwood) lake.

— *en écailles*, — *en feuilles*, shellac, shell lac.

— *en fils*, lac melted and drawn into threads.

— *en grains*, seed lac.

— *en plaques*, shellac.

— *fine*, a variety of cochineal lake.

— *minérale*, a green pigment made by precipitating a solution of zinc and copper sulfates with sodium carbonate.

— *ordinaire*, a variety of cochineal lake.

— *plate*, shellac.

— *plate d'Italie*, a variety of redwood lake.

laqué, *a.* lacquered, lackered; (of blood) laked.

laquebleu, *m.* litmus.

laquelle, *pron. fem.* who, whom, which, that.

laquer, *v.t.* lacquer, lacker; lake.

laqueux, *a.* of or pertaining to lac.

lard, *m.* bacon.

    gros —, — *gras*, fat bacon.

    petit —, — *maigre*, lean bacon.

lardacé, *a.* lardaceous.

larder, *v.t.* interlard; stab, pierce; lard.

lardeux, *a.* lardaceous, fat, lardy.

large, *a.* broad, wide; large, great, ample; liberal, generous; lax, unscrupulous. — *m.* breadth, width; open sea, high sea. — *adv.* largely; grandly.

    au —, off, away; well off; comfortably, spaciously.

    au long et au —, far and wide.

    de —, in width, wide.

largement, *adv.* widely, wide; largely, liberally, copiously.

largeur, *f.* breadth, width, wideness; (*Railways*) gage; (of a ship) beam.

    — *dans œuvre*, width in the clear.

    — *du jour*, width of opening.

    — *en fond*, width at the bottom.

largue, *a.* slack, loose; (*Mach.*) started.

larguer, *v.t.* let go; slacken (a rope); let off (steam).

larix, *m.* (*Bot.*) larch, larix.

larme, *f.* tear; drop; specif., (*Glass*) a drop of fused material from the vault or crown of the furnace.

larmeux, *a.* (of resins, etc.) in tears.

larmoiement, *m.* lachrymation, watering of the eyes, weeping.

larmoyer, *v.i.* weep, shed tears.

larve, *f.* larva.

las, *a.* weary, fatigued, tired.

laser, *m.* laser (the gum resin); laserwort.

lasser, *v.t.* fatigue, weary, tire. — *v.r.* tire, become fatigued.

lasting, *m.* lasting (the cloth).

latemment, *adv.* latently.

latent, *a.* latent.

latéral, *a.* side, lateral.

latéralement, *adv.* laterally, on the side, sideways, sidewise.

latex, *m.* (*Bot.*) latex.

laticifère, *a.* (*Bot.*) laticiferous.

latin, *a.* Latin; lateen (sail). — *m.* Latin.

latitude, *f.* latitude.

latrine, *f.* latrine, privy.

latte, *f.* lath; strip (as of iron); (*Ceram.*) a paddle or shovel for handling the moist clay; straight saber.

lattis, *m.* lathing, lath work.

laudanine, *f.* laudanine.

laudanisé, *a.* containing laudanum.

laudanosine, *f.* laudanosine.

laudanum, *m.* (*Pharm.*) laudanum.

    — *de Sydenham*, Sydenham's laudanum, wine of opium.

laumonite, *f.* (*Min.*) laumontite, laumonite.

lauracées, *f.pl.* (*Bot.*) Lauraceæ.

lauréole, *f.* (*Bot.*) daphne.

    — *femelle*, mezereon (*Daphne mezereum*).

    — *mâle*, spurge laurel (*Daphne laureola*).

laurier, *m.* laurel.

    — *benzoin*, spicebush (*Benzoin benzoin*).

    — *commun*, — *noble*, true laurel, bay laurel, bay tree (*Laurus nobilis*).

laurier-cerise, *m.* cherry laurel (*Laurocerasus laurocerasus*).

laurier-rose, *m.* oleander (*Nerium oleander*).

laurier-sauce, *m.* true laurel (*Laurus nobilis*).

laurine, *f.* laurin.

laurinées, *f.pl.* (*Bot.*) Lauraceæ.

lauryle, *m.* lauryl.

lavable, *a.* washable.

lavabo, *m.* washstand, lavabo.

lavage, *m.* washing, etc. (see laver); spilled liquid, slop.

    — *à grand eau*, (*Ores*) sluicing.

    — *à la cuve*, (*Ores*) tossing.

    — *au crible*, (*Ores*) riddling.

lavande, *f.* (*Bot.*) lavender.

    — *aspic*, — *en épi*, — *mâle*, aspic, French lavender (*Lavandula spica*).

    — *commune*, — *femelle*, — *vraie*, true or garden lavender (*Lavandula vera*).

    — *triste*, sea lavender (*Limonium*, esp. *L. carolinianum*).

lavandière, *f.* laundress, washerwoman; washing machine.

lavasse, *f.* a hard siliceous stone; downpour.

lave, *f.* lava.

— *de Volvic*, an Auvergne lava much used as a material resistant to chemicals.

— *émaillée*, enameled lava, glazed lava.

— *fusible*, artificial mastic.

lavé, *p.a.* washed; (of colors) thin, faint, pale, light.

lavée, *f.* quantity of material (e.g. wool) washed at one time.

lavement, *m.* washing; (*Med.*) enema.

laver, *v.t.* wash; wash out; wash off; cleanse, purify; buddle (ores).

— *le four*, (*Ceram.*) cleanse the kiln of deposited carbon by means of an oxidizing atmosphere.

laverie, *f.* lavatory (as one consisting of a sink and drainboard); (operation of) washing; place where ores or other materials are washed, washing works.

laveur, *m.* washer; washing apparatus (e.g. a gas-washing bottle); washing machine; (*Gas*) purifier. — *a.* washing, wash.

— *de coupes*, (*Micros.*) section washer.

laveuse, *fem.* of laveur and similarly used.

lavique, *a.* lava, lavatic, lavic.

lavis, *m.* wash; wash drawing, wash design.

lavoir, *m.* washer, washing apparatus, washing machine; washhouse, laundry; washing place; cleaning rod (for guns); buddle (for ores).

lavure, *f.* washings (as precious metal obtained from waste by washing); dishwater; swill, slop; (*Med.*) mucous and bloody supersecretions from the intestines.

laxatif, *a. & m.* (*Med.*) laxative.

laxité, *f.* looseness, slackness; laxity.

layette, *f.* drawer; tray (for powder); box; trunk; baby linen, layette.

lazulite, *m.* (*Min.*) lazulite.

le, *article, masc. sing.* the. (In many cases it should be omitted in translating: as, *le platine est coûteux*, platinum is expensive; *si l'on exclut*, if one excludes.) — *pron.* him, he, it.

lé, *m.* breadth, width (of fabrics).

lebererz, *m.* hepatic (or liver-brown) cinnabar.

lécanorine, *f.* lecanorin (lecanoric acid).

léchage, *m.* licking, etc. (see lécher).

lécher, *v.t.* lick; touch, pass over lightly (as with tongues); finish, finish off.

lécithine, *f.* lecithin.

leçon, *f.* lesson; lecture; reading.

lecteur, *m.* reader; proofreader.

lecture, *f.* reading; proofreading.

ledit, *a. & m.* the said (person or thing), the above, the aforementioned, the same.

lédon, *m.*, **lède**, *f.* (*Bot.*) Labrador tea (*Ledum*).

légal, *a.* legal.

légalement, *adv.* legally, lawfully.

légaliser, *v.t.* authenticate; legalize.

légalité, *f.* legality, lawfulness; law.

légendaire, *a.* legendary.

légende, *f.* legend.

— *explicative*, explanatory legend, explanation (of the letters on a drawing, or the like).

léger, *a.* light; slight. — *m.* lightness.

à la légère, lightly; light.

légèrement, *adv.* lightly; slightly.

légèreté, *f.* lightness; slightness.

légiférer, *v.i.* legislate.

législateur, *m.* legislator. — *a.* legislative.

législatif, *a.* legislative.

législation, *f.* legislation.

légitime, *a.* legitimate. — *f.* legal share.

légitimement, *adv.* legitimately.

légitimer, *v.t.* justify; legitimate.

léguer, *v.t.* bequeath.

légume, *m.* vegetable (specif., leguminous vegetable); legume.

légumine, *f.* legumin.

légumineuses, *f.pl.* (*Bot.*) Leguminosæ, leguminous plants.

légumineux, *a.* leguminous.

lehm, *m.* loam.

leibnizien, *a.* Leibnitzian.

léiocome, leiogomme, *m.* leiocome (dextrin).

lemnien, *a.* Lemnian.

lendemain, *m.* next day, following day, morrow.

le — de, the day after.

sans —, without continuation, without a sequel.

lénitif, *a. & m.* lenitive.

lent, *a.* slow.

lentement, *adv.* slowly, slow.

lenteur, *f.* slowness.

avec —, slowly.

lenticulaire, lenticulé, *a.* lenticular.

lentille, *f.* lens; lentil; bob (of a pendulum).

— *collectrice*, condensing lens, condenser.

— *convergente*, converging lens.

— *de rétrogradation*, a lens-shaped dephlegmator.

— *divergente*, diverging lens.

lentisque, *m.* mastic tree (*Pistacia lentiscus*).

lépidine, *f.* lepidine.

lépidolithe, *m.* (*Min.*) lepidolite.

lequel, *pron. masc.* who, whom, which, that.

les, *article pl.* the. (See *le*.) — *pron.* them, they.

lesdites, lesdits, *pl.* of ladite, ledit. See ledit.

lésion, *f.* injury, damage; (*Med.*) lesion.

lesquelles, *pron. pl.* of laquelle.

lesquels, *pron. pl.* of lequel.

lessivage, *m.*, **lessivation**, *f.* lixiviation, etc. (see lessiver).

lessivage à la vapeur, extraction with steam, steaming.

lessive, f. lixivium; lye; washing; bucking; wash.
— brute, (Soda) vat liquor (lixivium from black ash).
— caustique, caustic alkali solution, specif. (as in Pharm.) solution of potassium hydroxide.
— de recuit, (Soap) lye already used in boiling or salting out, with or without the addition of fresh alkali.
— des savonniers, caustic soda solution.
— de soude alcalino-salée, (Soap) a lye prepared by adding slaked lime to a solution of sodium carbonate and salt.
— de soude brute (or douce), (Soap) a lye made by lixiviating a mixture of sodium carbonate (as black ash) and slaked lime.
— résiduaire, residual lye, waste lye (e.g. sulfite liquor).

lessiver, v.t. lixiviate, leach; extract (with a liquid); wash with alkali or other detergent), scour, buck.

lessiveur, m. lixiviator, leacher, etc. (see lessiver). — a. lixiviating, leaching, etc.
— de chiffons, (Paper) rag boiler.

lessiveuse, f. an apparatus for lixiviating, washing or scouring, as a rag boiler, a pulp digester or a bucking kier; washerwoman, laundress.

lest, m. ballast.

leste, a. light, active, smart, free.

lester, v.t. ballast.

léthargie, a. lethargy.

lettrage, m. lettering.

lettre, f. letter.
— chargée, registered letter (containing money).
— de voiture, waybill; bill of lading.
— morte, of no effect.
— recommandée, registered letter.
lettres grasses, black letters, heavy-faced type.
lettres maigres, light letters, ordinary or light-faced type.

leucéine, f. leucine.

leucémie, f. (Med.) leucemia, leucocythemia.

leucémique, a. (Med.) leucemic, leucocythemic.

leucine, f. leucine.

leucique, a. leucic.

leucite, f. (Min.) leucite; (Med.) sclerotitis.

leucobase, f. leuco base.

leucocidine, f. leucocidin.

leucocytaire, a. leucocytic.

leucocyte, m. leucocyte, white blood corpuscle.

leucocythémie, f. leucocythemia, leucemia.

leucocytolyse, f. leucocytolysis.

leucocytose, f. leucocytosis.

leucodérivé, m. leuco derivative.

leucolyse, f. leucolysis.

leucomaïne, f. leucomaine.

leuconique, a. leuconic.

leur, pron. & a. to them, them, their, theirs.

levage, m. raising, etc. (see lever).

levain, m. leaven, leavening agent.
— artificiel, artificial yeast, cultivated yeast.
— de chef, leaven consisting of dough which has fermented until it is very light.
— de pâte, dough saved out for use as leaven.

levant, m. east; (cap.) Levant. — p.pr. of lever.

levé, p.a. raised, etc. (see lever); erect; up.
— m. surveying, plotting, drawing.

levée, f. raising; removal, removing; crop; stroke (of a piston); surveying, plotting; levee; levy.
— de cuite, in the process of making Marseilles soap, the removal of the curd from the lye and placing of it in vessels where it is left to become marbled.

lever, v.t. raise; lift, lift up; hoist; set up; remove; revoke; take up; take; survey, plot, sketch. — v.i. rise; (of plants) come up. — v.r. rise; be raised; (of weather) clear; (of window glass) warp. — m. rising; surveying, plotting; survey, plot, plan, design, sketch.
— sur lessive, separate (soap) from the lye.

levier, m. lever; handle, arm, bar.
— coudé, crank lever, bent lever.
— de soupape, valve handle, valve lever.

levier-clef, n. (Elec.) make-and-break key.

lévigateur, m. levigator (e.g. an apparatus for levigating sugar-beet pulp).

lévigation, f. levigation, levigating.

léviger, v.t. levigate.

lévo-. levo-, lævo-.

lévogyre, a. levorotatory, levogyre, lævogyre.

lèvre, f. lip; edge, border, rim.

lévuline, f. levulin.

lévulinique, lévulique, a. levulinic.

lévulosane, m. levulosan.

lévulose, f. levulose.

levure, f. yeast.
— basse, bottom yeast, low-fermentation yeast.
— de bière, — de brasserie, beer yeast, brewer's yeast, brewery yeast, barm.
— de culture, cultivated yeast.
— de dépôt, bottom yeast.
— de vin, wine yeast.
— durable, permanent yeast, zymin.
— ferment, yeast enzyme.
— grise, the gray upper layer of yeast which is rejected in making compressed yeast.
— haute, top yeast, high-fermentation yeast.
— morte, dead yeast (specif. Buchner's evaporated yeast extract).
— pressée, compressed yeast, pressed yeast.
— sauvage, wild yeast.

levure *superficielle*, top yeast, surface yeast.
— *végétale*, vegetating yeast.
levure-mère, *f.* mother yeast, parent yeast.
levurier, *m.* maker or seller of yeast.
levuromètre, *m.* an apparatus for measuring the fermentative activity of yeast.
lexique, *m.* lexicon, dictionary. — *a.* lexical.
lézard, *m.* lizard.
lézarde, *f.* crevice, crack.
lézarder, *v.t. & r.* crack, crevice, chink.
liage, *m.* binding, tying, etc. (see lier).
liais, *m.* a very hard, close-grained limestone.
liaison, *f.* connection; bond, linkage; union, junction, joining, joint; tie; mortar.
— *éthylénique*, ethylene linkage, double bond.
liane, *f.* (*Bot.*) liana.
liant, *a.* pliant, pliable, soft, easily worked; flexible; elastic (as a spring); cohesive; (of metals) tenacious, tough, malleable; supple; gentle, compliant. — *m.* pliancy, pliability, etc. (see above); binding agent, binder.
lias, *m.* (*Geol.*) Lias.
liasique, liassique, *a.* (*Geol.*) Liassic.
liasse, *f.* package, packet, bundle, file; cord, tape, string.
libelle, *f.* bubble (in a crystal inclusion); libel.
libellé, *p.a.* drawn up, worded. — *m.* drawing up, wording.
libeller, *v.t.* draw up, word.
liber, *m.* (*Bot.*) bast, phloëm, inner bark.
libéral, *a.* liberal.
libéralement, *adv.* liberally.
libération, *f.* liberation, setting free; discharge, release.
libérer, *v.t.* liberate, set free; discharge.
libérien, *a.* (*Bot.*) relating to bast or phloëm.
liberté, *f.* liberty, freedom; (*Mach.*) play, clearance.
en —, free; freely.
— *du commerce*, freedom of trade.
mettre en —, set free, liberate.
mise en —, setting free, liberation.
libidibi, *n.* (*Bot.*) divi-divi.
libraire, *m.* bookseller.
libraire-éditeur, *m.* bookseller and publisher.
librairie, *f.* bookstore, bookseller's shop; book trade.
libre, *a.* free; (of paper) unstamped; (*Mach.*) out of gear, disengaged.
à l'air —, au —, in the open air, in free air.
libre-échange, *m.* free trade.
librement, *adv.* freely.
lice, *f.* lists, field; warp (of fabrics); list, border; rail.
licence, *f.* license; licentiate's degree.
licencié, *m.* licensee; licentiate (holder of a degree intermediate between bachelor and doctor).
licet, *m.* permission, permit.

lichen, *m.* (*Bot.*) lichen.
— *Carragaheen*, carrageen, Irish moss.
— *des rochers*, (*Bot.*) archil.
— *d'Islande*, Iceland moss, Iceland lichen.
— *tartareux*, *Lecanora tartarea* (from which one kind of archil is prepared).
lichénine, *f.* lichenin.
lichénique, *a.* lichenic (fumaric).
licite, *a.* licit, lawful, allowable.
lie, *f.* lees, dregs, sediment, grounds.
— *de vin*, wine lees; dark red or purple color.
lié, *p.a.* bound, tied, etc. (see lier).
liège, *m.* cork.
liégeois, *a.* of or pertaining to Liége.
liégeux, *a.* corky, of the nature of cork.
liement, *m.* binding, tying, etc. (see lier).
lien, *m.* bond, tie, link; band, strap, brace; ligature.
lier, *v.t.* bind, tie, join, attach, connect, fasten, fasten together, link, knit; relate; enter into, engage in; thicken (as a sauce); (*Soap*) emulsify. — *v.r.* be bound, be tied, etc.; combine, unite; become intimate; thicken.
lierre, *m.* ivy.
— *terrestre*, ground ivy (*Glecoma hederacea*).
lieu, *m.* place; turn, order; estate, rank; (*Geom.*) locus.
au — de, au — et place de, instead of, in place of, as a substitute for.
au — que, whereas, when in fact.
il y a — de, it is well to, it is necessary to.
lieue, *f.* league.
lièvre, *m.* hare.
ligament, *m.* ligament.
ligature, *f.* ligature; tie, joint, union.
ligaturer, *v.t.* tie, bind, fasten.
ligérien, *a.* relating to the Loire or its basin.
ligne, *f.* line (in various senses); order, rank.
en — de compte, into account.
hors de —, out of line, out of the way.
hors —, exceptional, extraordinary.
— *à plomb*, plumb line, vertical line.
— *blanche*, white line; untarred cord.
— *continue*, continuous line.
— *courbe*, curved line.
— *d'eau*, level line; water line.
— *de chaînette*, catenary.
— *de chemin de fer*, railway line.
— *de feu*, firing line.
— *de foi*, fiducial line.
— *de mire*, line of sight.
— *de repère*, reference line.
— *de terre*, ground line.
— *de tir*, line of fire.
— *de visée*, line of sight.
— *droite*, straight line, right line.
— *ferrée*, railway.
— *goudronnée*, tarred line.
— *infléchie*, bent line, inflected line.

ligne *interrompue*, broken line.
— *limite*, limiting line, limit line.
— *pleine*, solid line, continuous line.
— *pointillée*, — *ponctuée*, dotted line.
mettre en — *de compte*, take into account.
ligner, *v.t.* mark with parallel lines, rule.
ligneux, *a.* ligneous, woody. — *m.* lignin.
lignification, *f.* lignification.
lignifier, *v.t. & r.* lignify.
lignine, *f.* lignin.
lignite, *m.* (*Min.*) lignite.
lignocellulose, *f.* lignocellulose.
lignose, *f.* lignose.
lignosité, *f.* ligneous quality, woodiness.
ligroïne, *f.* ligroïn.
ligue, *f.* league.
lilacine, *f.* lilacin (syringin).
lilas, *m. & a.* lilac.
liliacé, *a.* (*Bot.*) liliaceous.
liliacées, *f.pl.* (*Bot.*) Liliaceæ.
lilial, *a.* lily, of or like lilies.
limace, *f.* Archimedean screw; (*Zoöl.*) slug.
limaçon, *m.* (*Zoöl.*) snail; (*Anat.*) cochlea; (*Math.*) limaçon.
limage, *m.* filing.
limaille, *f.* filings, file dust.
— *de fer*, iron filings.
— *de fonte*, (*Iron*) kish.
limailleux, *a.* (*Iron*) kishy.
limas, *m.* = limace.
limature, *f.* filing; filings.
limbe, *m.* limb, border, edge.
lime, *f.* file; lime (the fruit).
— *bâtarde*, bastard file.
— *carrée*, square file.
— *demi-douce*, second-cut file.
— *demi-ronde*, half-round file (a file with one flat and one curved surface).
— *douce*, smooth file.
— *en queue de rat*, rat-tail file.
— *grosse*, coarse file.
— *plate*, flat file (file with two flat surfaces).
— *ronde*, round file.
— *triangulaire*, triangular file, three-cornered file, three-square file.
limer, *v.t.* file; (fig.) polish.
limette, *f.* lime (the fruit).
limettier, *m.* any of several plants of the genus *Citrus*.
limeur, *m.* filer; filing machine. — *a.* filing.
limitatif, *a.* limiting.
limitation, *f.* limitation.
limitativement, *adv.* in a limited manner.
limite, *f. & a.* limit.
limité, *p.a.* limited.
limiter, *v.t.* limit.
limon, *m.* mud, slime; lemon; shaft, thill.
limonade, *f.* lemonade.
— *purgative au citrate de magnésie*, (*Pharm.*) solution of magnesium citrate.

limonade *sèche au citrate de lithine*, (*Pharm.*) effervescent lithium citrate.
limonage, *m.* (*Agric.*) application of mud or silt to land to improve it.
limone, *f.* limonin.
limonène, *m.* limonene.
limoneux, *a.* muddy, slimy, oozy.
limonier, *m.* lemon tree (*Citrus medica limon*).
limonine, *f.* limonin.
limonite, *f.* (*Min.*) limonite.
limoniteux, *a.* (*Min.*) limonitic.
limpide, *a.* limpid, clear.
limpidité, *f.* limpidity, limpidness, clearness.
lin, *m.* flax; linen.
huile de —, linseed oil.
— *minéral*, — *incombustible*, — *vif*, amianthus, fibrous asbestos.
— *sauvage*, (*Bot.*) toadflax.
linaire, *f.* (*Bot.*) toadflax (*Linaria*).
— *commune*, common toadflax (*L. linaria*).
linaloé, *f.* (*Bot.*) linaloa.
linalol, *m.* linaloöl.
linarine, *f.* linarin.
linéaire, *a.* linear.
linette, *f.* linseed, flaxseed.
linge, *m.* cloth, rag; linen.
— *américain*, cloth coated with celluloid or the like.
— *en caoutchouc*, cloth coated with rubber, or (by a misnomer) with celluloid or the like.
lingerie, *f.* linen trade; linen warehouse; linen.
lingot, *m.* ingot; slug (of metal).
lingotière, *f.* ingot mold.
— *à eau*, water-cooled ingot mold.
lingotiforme, *a.* ingot-shaped.
linguet, *m.* pawl, catch, stop.
linier, *a.* flax, of flax.
linière, *f.* flax field.
liniment, *m.* (*Pharm.*) liniment.
— *ammoniacal*, ammonia liniment.
— *calcaire*, lime liniment, carron oil.
— *crotoné*, liniment of croton oil.
— *opiacé*, liniment of opium.
— *phosphoré*, phosphorated oil.
— *savonneux camphré*, soap liniment.
— *savonneux ioduré*, liniment of potassium iodide with soap.
— *sinapisé composé*, liniment of mustard.
— *térébenthiné*, turpentine liniment.
— *térébenthiné acétique*, liniment of turpentine and acetic acid.
— *volatil*, ammonia liniment, volatile liniment.
linition, *f.* coating, anointing.
linoléique, *a.* linoleic.
linoléum, *m.* linoleum.
linoxine, *f.* linoxin, linoxyn.
linteau, *m.* lintel.
linters, *n.pl.* linters (short cotton fiber).

lion, *m.* lion.
liparolé, *m.* (*Pharm.*) ointment, unguent.
lipase, *f.* lipase.
lipasique, *a.* lipase, of lipase.
lipémie, *f.* (*Med.*) lipemia, lipæmia.
lipochrome, *m.* lipochrome.
lipogénèse, *f.* (*Physiol.*) lipogenesis.
lipoïde, *a. & m.* lipoid.
lipolyse, *f.* (*Physiol.*) lipolysis.
lipolytique, *a.* lipolytic.
lipomateux, *a.* (*Med.*) lipomatous.
lipome, *m.* (*Med.*) lipoma.
lipurie, *f.* (*Med.*) lipuria.
liquater, *v.t.* liquate, subject to liquation.
 — *v.r.* be liquated, undergo liquation.
liquation, *f.* (*Metal.*) liquation, eliquation.
liquéfaction, *f.* liquefaction.
liquéfiable, *a.* liquefiable.
liquéfié, *p.a.* liquefied.
liquéfier, *v.t. & r.* liquefy.
liquescence, *f.* liquescence.
liqueur, *f.* liquor; liquid, fluid; solution;
 liqueur.
 — *amniotique,* amniotic fluid.
 — *anodine nitreuse,* (*Pharm.*) spirit of nitrous
 ether.
 — *antiseptique,* antiseptic solution.
 — *aqueuse,* aqueous liquid; aqueous solution.
 — *arsénicale de Fowler,* (*Pharm.*) Fowler's
 solution, solution of potassium arsenite.
 — *arsénicale hydrochlorique,* (*Pharm.*) solution of arsenous acid.
 — *cupro-potassique,* — *de Fehling,* Fehling's
 solution.
 — *d'acétate de fer,* (*Pharm.*) solution of ferric
 acetate.
 — *d'ammoniaque,* ammonia water.
 — *d'ammoniaque vineuse,* (*Pharm.*) spirit of
 ammonia.
 — *de Boudin,* (*Pharm.*) solution of arsenous
 acid.
 — *de Donovan,* (*Pharm.*) Donovan's solution,
 solution of arsenous and mercuric iodides.
 — *de Fehling,* Fehling('s) solution.
 — *de Fowler,* (*Pharm.*) Fowler's solution.
 — *de Labarraque,* = eau de Labarraque,
 under *eau.*
 — *des cailloux,* (*Old Chem.*) liquor of flints
 (water-glass solution).
 — *de Schweitzer,* Schweitzer's reagent.
 — *de Van Swieten,* Van Swieten's liquor or
 solution (an alcoholic solution of mercuric
 chloride).
 — *de virage* = liqueur témoin.
 — *d'Hoffman,* (*Pharm.*) spirit of ether,
 anodyne liquor of Hoffman.
 — *épreuve,* testing solution.
 — *fumante de Boyle,* Boyle's fuming liquid
 (ammonium sulfide solution).

liqueur *fumante de Cadet,* Cadet's fluid (or
 liquid), Cadet's fuming liquid.
 — *fumante de Libavius,* fuming liquor of
 Libavius (stannic chloride).
 — *hémostatique de Monsel,* (*Pharm.*) solution
 of ferric subsulfate, Monsel's solution.
 — *iodo-iodurée,* iodine-iodide solution.
 — *nervine de Bang,* (*Pharm.*) compound
 spirit of ether.
 — *surnageante,* supernatant liquid.
 — *témoin,* indicator liquid.
 — *titrée,* titrated solution, standardized solution.
 — *type,* standard solution.
 — *vésicant,* (*Pharm.*) blistering liquid.
liquidambar, *m.* liquidambar.
liquidambaré, *a.* of liquidambar.
liquidation, *f.* liquidation; (*Soap*) treatment
 of soap with weak lye or water to remove
 impurities.
liquide, *a. & m.* liquid.
 — *conservateur,* preserving liquid.
 — *d'ascite.* ascitic fluid.
 — *excitateur,* exciting liquid.
 — *témoin,* test liquid, liquid used for comparison.
liquider, *v.t.* liquidate. — *v.r.* pay off, settle.
liquidité, *f.* liquidity, liquidness.
liquoreux, *a.* (of wine) liqueur-like, combining
 strength with sweetness.
liquoriste, *m.* a maker or seller of liqueurs.
lire, *v.t.* read. — *v.r.* be read. — *f.* lira (an
 Italian coin equivalent to the franc).
lis, *m.* (*Bot.*) lily.
 — *des teinturiers,* dyer's weed, yellow weed
 (*Reseda luteola*).
lisage, *m.* (*Dyeing*) agitation of hanks of silk.
lisait, *imp. 3 sing.* of lire.
lisant, *p.pr.* of lire and of liser.
Lisbonne, *f.* Lisbon.
lise, *m.* (*Dyeing*) a rod or pole for supporting
 hanks of silk in the vat.
lisent, *pr. 3 pl. indic. & subj.* of lire and of liser.
liser, *v.t.* (*Dyeing*) agitate (hanks of silk) in
 the vat.
liséré, *m.* border, edge; edging, binding. — *p.a.*
 bordered, edged.
lisérer, *v.t.* border, edge, bind.
lisibilité, *f.* legibility.
lisible, *a.* legible; readable.
lisiblement, *adv.* legibly.
lisière, *f.* selvage, selvedge, list; border.
lissage, *m.* smoothing, etc. (see lisser); (*Founding*) black wash.
lisse, *a.* smooth; sleek; glossy. — *m.* smoothness. — *f.* a cylinder for polishing leather;
 warp (of cloth); string, cord; rail.
lissé, *m.* smoothness, glossiness, gloss, glaze,
 polish. — *p.a.* smoothed, smooth, glossed,
 etc. (see lisser).

lissée, *f.* glazing (of powder).

lisser, *v.t.* smooth, gloss, glaze (as gunpowder), calender, sleek; polish, burnish; (*Founding*) black wash.

lisseur, *m.* smoother, glosser, etc. (see lisser).

lisseuse, *fem.* of lisseur. (Used both of persons and of machines.)

lissoir, *m.* (of tools) smoother, glosser, etc. (see lisser); glazing barrel (for powder); glazing room or house (for powder); smoothing room (for paper). — *a.* smoothing, glossing, etc.

lissure, *f.* smoothing, glossing, etc. (see lisser); smoothness, glossiness, gloss, glaze, polish.

liste, *f.* list.

lit, *m.* bed; layer; course, direction. — *pr. 3 sing.* of lire.

— *à fusion,* (*Metal.*) mixing bed.

— *de carrière,* natural bed (of a stone).

— *de fusion,* (*Metal.*) charge (for smelting).

— *de gueuse,* (*Metal.*) sow channel.

— *de grillage,* (*Metal.*) a bed of fuel used in roasting or calcining.

— *de pains,* (*Sugar*) a stage or rack having holes in which sugar loaves are set to drain.

— *filtrant,* filter bed, filtering bed.

liteau, *m.* strip, band; stripe.

liter, *v.t.* place in beds or layers; cover the selvage of (fabrics) before dyeing.

literie, *f.* bedding.

litharge, *f.* litharge.

lithargé, lithargyré, *a.* containing litharge.

lithiase, lithiasie, *f.* (*Med.*) lithiasis.

lithine, *f.* lithia (lithium oxide); lithium hydroxide.

lithiné, *a.* combined with or containing lithia.

lithinifère, *a.* containing lithia or lithium.

lithique, *a.* lithic.

acide —, lithic acid (uric acid).

lithium, *m.* lithium.

lithocolle, *f.* a lapidary cement.

lithofellique, lithofellinique, *a.* lithofellic, lithofellinic.

lithofracteur, *m.* (*Expl.*) lithofracteur.

lithographie, *f.* lithography; lithograph; lithographing establishment.

lithographier, *v.t.* lithograph.

lithographique, *a.* lithographic.

lithoïde, *a.* lithoid, lithoidal.

litigieux, *a.* disputable, contested; litigious.

litre, *m.* liter, litre (61.022 cu. in.).

littéraire, *a.* literary.

littéral, *a.* literal.

littéralement, *adv.* literally.

littérature, *f.* literature; learning.

livarot, *m.* Livarot cheese.

livèche, *f.* (*Bot.*) lovage (esp. *Levisticum levisticum*).

livide, *a.* livid. — *m.* livid color.

lividité, *f.* lividness, lividity.

livrable, *a.* deliverable.

livraison, *f.* delivery; part, fascicle.

livrancier, *m.* deliverer, shipper; contractor. — *a.* delivering.

livre, *m.* book. — *f.* pound (as now used in France the word refers to the half kilogram); franc (which has replaced the livre, an old monetary unit of varying value); pound (as in *livre sterling,* pound sterling). — *anglaise,* English pound, pound sterling. — *métrique,* half kilogram.

livrer, *v.t.* deliver; yield up; abandon.

livret, *m.* small book, booklet; notebook; book (for gold leaf or the like); catalog.

lixiviateur, *m.,* **lixiviateuse,** *f.* lixiviator.

lixiviation, *f.* lixiviation, leaching.

lixiviel, *a.* lixivial.

lixivier, *v.t.* lixiviate, leach.

lobe, *m.* lobe.

lobélie, *f.* (*Bot.*) lobelia.

— *enflée,* Indian tobacco (*Lobelia inflata*).

lobéline, *f.* lobeline (an alkaloid); lobelin (a resin used in pharmacy).

local, *a.* local. — *m.* place, locality, premises, room, building.

localement, *adv.* locally.

localisation, *f.* localization.

localiser, *v.t.* localize; locate.

localité, *f.* locality.

locataire, *m. & f.* tenant, renter.

locatif, *a.* pertaining to a tenant or to rent.

location, *f.* renting; rent.

locaux, *pl. masc.* of local.

locher, *v.t.* shake, shake loose; (*Sugar*) detach (a sugar loaf) from the mold by jarring.

locomotive, *f.* locomotive, engine.

loess, *m.* (*Geol.*) loess.

logarithme, *m.* logarithm.

— *vulgaire,* common logarithm (to the base 10).

logarithmique, *a.* logarithmic(al). — *f.* logarithmic curve.

loge, *f.* lodge, cell, booth, (theater) box.

logé, *p.a.* put up, etc. (see loger); (of wine) in casks.

logement, *m.* bed, groove, seat; letting in; lodging, lodgings, quarters, room.

loger, *v.t.* put up, pack; put, keep; lodge; quarter, billet; (*Tech.*) let in, set in. — *v.i.* lodge, live, stay.

logique, *a.* logical. — *f.* logic.

logiquement, *adv.* logically.

logis, *m.* house, dwelling, home; building; inn. opening (of a glass furnace).

loi, *f.* law; standard (of coinage).

— *de Mariotte,* Mariotte's law (usually called Boyle's law in English).

— *des brevets,* patent law.

— *naturelle,* natural law.

— *périodique,* periodic law.

loin, *adv.* far, to a great distance, (of time) long.

  au —, at a distance, to a distance, far off.

  de —, from afar, from a distance, at a distance, long ago.

  de — *en* —, at long intervals.

  — *de,* far from.

  — *que* (with *subj.*), far from (with *p.pr.*; as, *loin qu'il soit malade,* far from being sick).

lointain, *a.* far, distant, remote. — *m.* distance.

loisible, *a.* allowable, permissible.

loisiblement, *adv.* allowably, permissibly.

loisir, *m.* leisure.

lombaire, *a.* (*Anat.*) lumbar.

Lombardie, *f.* Lombardy.

londonien, *a.* London, of London.

Londres, *f.* London.

long, *a.* long; slow. — *m.* length. — *adv.* long; much.

  à la longue, in the long run, in time.

  au —, at length, at great length.

  au — *de,* along.

  de —, *en* —, lengthwise.

  de — *en large, en* — *et en large,* back and forth, up and down.

  le — *de,* along.

longer, *v.t.* go along, run along, follow.

longitudinal, *a.* longitudinal.

longitudinalement, *adv.* longitudinally.

longtemps, *adv.* a long time, long.

longue, *a. fem.* of long.

longuement, *adv.* a long time, at length.

longuet, *a.* a little long, rather long.

longueur, *f.* length; slowness.

  en —, lengthwise; at length.

  — *d'onde,* wave length.

longue-vue, *f.* spyglass.

looch, *m.* (*Pharm.*) lincture, loch, electuary.

lophine, *f.* lophine.

lopin, *m.* bit, piece, morsel; (*Metal.*) bloom.

loquet, *m.* latch; catch, clasp, pawl.

lorétine, *f.* loretin.

lorgnette, *f.* binocular.

  — *de spectacle,* opera glass.

lorgnon, *m.* eyeglass, eyeglasses.

lors, *adv.* then (now chiefly in phrases).

  dès —, from that time, since then; consequently, then.

  dès — *que,* since, seeing that.

  — *de,* at the time of, at the moment of, during, in.

  — *même que,* altho, even tho.

  pour —, in that case; at that time.

lorsque, *conj.* when.

losange, *f.* lozenge, rhomb, diamond.

lot, *m.* lot; prize (in a lottery).

loterie, *f.* lottery.

lotier, *m.* (*Bot.*) lotus.

lotion, *f.* lotion.

  — *à l'acétate de plomb,* (*Pharm.*) lead water, diluted solution of lead subacetate.

lotionner, *v.t.* wash, bathe (with lotions).

lotir, *v.t.* sample (ores); sort, grade (as grain); lot, divide into or arrange in lots, portion off; allot, assign; put in possession (of).

lotissage, lotissement, *m.* sampling, etc. (see lotir).

louable, *a.* praiseworthy, commendable.

louage, *m.* leasing, renting, hire, hiring.

louange, *f.* praise.

louanger, *v.t.* praise, commend.

louche, *a.* turbid, cloudy, muddy; ambiguous; squinting. — *m.* turbidity, cloudiness, muddiness. — *f.* ladle; broach, reamer.

louchement, *m.* becoming turbid, clouding.

louchet, *m.* narrow spade.

louchir, *v.i.* become turbid or cloudy, cloud.

louchissement, *m.* becoming turbid, clouding.

louer, *v.t.* lease, rent, hire; praise.

louis, *m.* louis (now, the 20-franc piece).

loup, *m.* wolf; (*Metal.*) salamander, bear; (*Paper,* etc.) willow, devil; defect, fault; mask; facepiece (of a gas mask).

loupe, *f.* magnifying glass, reading glass, lens; (*Metal.*) bloom; (*Med.*) wen; (*Bot.*) wart.

  — *à main,* reading glass.

  — *Coddington,* Coddington lens.

  — *microscope,* magnifying glass (esp. one of relatively high power).

lourd, *a.* heavy; slow, dull; great, grave; expensive; (of weather) close.

lourdement, *adv.* heavily; grossly, greatly.

lourdeur, *f.* heaviness, etc. (see lourd).

loutre, *f.* (*Zoöl.*) otter.

loyal, *a.* sincere, honest; loyal.

loyalement, *adv.* sincerely, faithfully.

loyauté, *f.* honesty, sincerity; loyalty.

loyer, *m.* rent; salary, wages; (fig.) reward; (fig.) price.

lu, *p.a.* read.

lubréfaction, etc. See lubrifaction, etc.

lubrifacteur. = lubrificateur.

lubrifaction, *f.* lubrication.

lubrifiage, *m.* lubrication.

lubrifiant, *p.a.* lubricating. — *m.* lubricant.

lubrificateur, *a.* lubricating. — *m.* lubricator.

lubrification, *f.* lubrication.

lubrifier, *v.t.* lubricate.

lubrifieur, *m.* lubricator (specif., oil cup).

lucide, *a.* lucid, clear.

lucidement, *adv.* lucidly, clearly.

lucidité, *f.* lucidity, lucidness, clearness.

luciférase, *f.* luciferase.

luciole, *f.* firefly, lightning bug.

Lucques, *f.* Lucca.

lucratif, *a.* lucrative.

lueur, *f.* glimmer, gleam, glow, flash, light.

lui, *pron.* him, he, it; (as indirect object of a verb) to him, to her, him, her. — *p.p.* of luire.

lui-même, *pron.* himself, (less often) itself.

luire, *v.i.* shine, emit light; gleam, glitter, glisten, sparkle, glow.

luisance, *f.* shining, etc. (see luire).

luisant, *p.a.* shining, gleaming, etc. (see luire); glossy, lustrous. — *m.* gloss, luster, sheen.

luisante, *f.* (*Astron.*) bright star.

luisard, *m.* (*Min.*) micaceous hematite.

lumière, *f.* light; opening, orifice, mouth, hole, vent (as of a gun or fuse), port (as for steam), slit (as for sighting); sight, sight hole (of instruments). — *a.* bright, brilliant.

  — *à arc,* arc light.

  — *à gaz,* gaslight.

  — *à incandescence,* incandescent light.

  — *à vapeur,* steam port.

  — *d'échappement,* (*Steam,* etc.) exhaust port.

  — *du jour,* daylight.

  — *électrique,* electric light.

  — *froide,* light without heat, luminescence.

  — *noire,* (*Physics*) dark light (applied to certain invisible radiations, esp. ultraviolet rays).

  — *oxhydrique,* oxyhydrogen light, limelight.

  — *polarisée,* polarized light.

  — *solaire,* sunlight.

lumignon, *m.* wick end; candle end.

luminaire, *m.* lighting; luminary.

luminescence, *f.* luminescence.

luminescent, *a.* luminescent.

lumineusement, *adv.* luminously.

lumineux, *a.* luminous.

luminifère, *a.* luminiferous.

luminosité, *f.* luminosity, luminousness.

lunaire, *a.* lunar. — *f.* (*Bot.*) satinpod (*Lunaria*).

lundi, *m.* Monday.

lune, *f.* moon; (*Alchemy*) luna, silver; caprice.

  — *cornée,* silver chloride.

lunel, *m.* Lunel wine, Lunel.

lunetier, *m.* spectacle maker.

lunette, *f.* telescope; spyglass, glass; (*pl.*) spectacles; (*pl.*) goggles; peephole, sighthole (also, the glass covering such a hole, sight glass); eyepiece (of a gas mask); (*Glass*) linnet hole; (*Tech.*) annular object, as a ring gage or a shell chuck; lunette.

  — *à viseur,* sighting telescope.

  — *polarimétrique,* polariscope.

lupin, *m.* (*Bot.*) lupine.

lupinine, *f.* lupinin (the glucoside); lupinine (the alkaloid).

lupulin, *m.* lupulin.

lupuline, *f.* lupulin; (*Bot.*) black medic.

lurent, *p.def. 3 pl.* of lire.

lusol, *m.* crude benzene used as an illuminant.

lussent, *imp. subj. 3 pl.* of lire.

lustrage, *m.* lustering, lustring, glossing, etc. (see lustrer).

lustre, *m.* luster, lustre; chandelier.

  — *cantharide,* (*Ceram.*) silver luster (*Obs.*).

lustré, *p.a.* lustered, etc. (see lustrer); lustrous, glossy.

lustrer, *v.t.* luster, gloss, glaze, polish, calender, stake (leather), dress.

lustroir, *m.* polisher, specif. polishing cloth.

lut, *m.* lute, luting. — *p.def. 3 sing.* of lire.

lutation, *f.* luting.

lutécium, *m.* lutecium.

lutéine, *f.* lutein.

lutéoline, *f.* luteolin.

luter, *v.t.* lute.

lutidine, *f.* lutidine.

lutidique, *a.* lutidinic, lutidic.

lutidone, *f.* lutidone.

lutte, *f.* struggle, contest; wrestling.

lutter, *v.i.* strive, struggle; compete; wrestle.

luxe, *m.* luxury; luxuriance; richness.

Luxembourg, *m.* Luxemburg.

luzerne, *f.* (*Bot.*) lucern, lucerne, alfalfa.

lycée, *f.* lycée, secondary school; lyceum.

lycéen, *m.* lycée pupil. — *a.* of a lycée.

lychnide, *f.*, **lychnid,** *m.* (*Bot.*) lychnis.

lycope, *m.* (*Bot.*) bugleweed (*Lycopus*).

  — *de Virginie,* *Lycopus virginicus.*

lycopode, *m.* lycopodium.

lyddite, *f.* (*Expl.*) lyddite.

lydien, *a.* Lydian.

lydienne, *f.* (*Min.*) touchstone, Lydian stone.

lymphatique, *a.* (*Anat.*) lymphatic.

lymphe, *f.* (*Physiol.*) lymph; (*Bot.*) sap.

Lyon, *f.* Lyons, Lyon.

lyonnais, *a.* of or pertaining to Lyons (or Lyon).

lys, *m.* lily. (See lis.)

lysatine, *f.* lysatine.

lysidine, *f.* lysidine.

lysimètre, *m.* lysimeter.

lysine, *f.* lysine (the amino acid); lysin (the class name for bacteriolysins, etc.).

lysol, *m.* (*Pharm.*) lysol.

M

m., *abbrev.* (mètre) meter; (méta) meta; (minute) minute; (mon) my; (midi) south; (*cap.*, Monsieur) Mr., Sir; (*cap.*, Majesté) Majesté.

m², *abbrev.* square meter.

m³, *abbrev.* cubic meter.

m'. Contraction of *me*.

ma, *a. fem.* my.

macadam, *m.* macadam; macadam road.

macadamisage, macadamisation, *f.* macadamizing, macadamization.

macadamiser, *v.t.* macadamize.

macaroni, *m.* macaroni.

macédoine, *f.* medley; (*cap.*) Macedonia.

macératé, *m.* the liquid product of maceration.

macérateur, *m.* macerater, macerator, specif. (*Brewing*) mash tun. — *a.* macerating.

macération, *f.* maceration, macerating.

macératum, *m.* = macératé.

macéré, *p.a.* macerated. — *m.* = macératé.

macérer, *v.t.*, *r. & i.* macerate.

mâché, *p.a.* chewed, etc. (see mâcher).

mâche-bouchon, mâche-bouchons, *m.* cork press.

mâchefer, *m.* cinders, ashes, scoria, dross; specif., coal cinders; forge scale.

mâchement, *m.* chewing, etc. (see mâcher).

mâcher, *v.t.* chew; disintegrate (paper) with moistening so as to form a pulp; (of a saw or the like) cut roughly; (fig.) mince.

mâchicatoire, *m.* masticatory.

machinal, *a.* of or like a machine, mechanical.

machinalement, *adv.* mechanically.

machine, *f.* machine, engine; (*Elec.*) specif., dynamo; mechanism.
— *à calandrer,* calender, calendering machine.
— *à chandelles,* candle machine.
— *à cingler,* shingling machine, shingler.
— *à détente,* expansion engine.
— *à diviser,* dividing engine.
— *à écrire,* typewriter.
— *à feu,* heat engine.
— *à force centrifuge,* centrifugal machine.
— *à gaz,* gas engine; refrigerating machine using ammonia or other gas.
— *à glace,* ice machine.
— *à huile minérale,* oil engine.
— *alimentaire,* feeding engine.
— *alternative,* (*Elec.*) alternating-current dynamo.

machine *à pétrole,* gasoline engine; oil engine.
— *à pression,* press.
— *à triturer,* triturating machine or apparatus.
— *à vapeur,* steam engine.
— *à vent,* = machine soufflante.
— *bétonnière,* concrete mixer.
— *centrifuge,* centrifugal machine, centrifuge.
— *dynamo,* dynamo.
— *électrique,* electric machine.
— *fixe,* stationary engine.
— *frigorifique,* refrigerating machine.
— *motrice,* motor, prime mover, engine.
— *rotative,* rotary engine or machine.
— *soufflante,* blower, blowing engine; fan.

machine-outil, *n.* machine tool.

machinerie, *f.* machinery; machine room, engine room; construction of machines.

machiniste, *m.* machinist; engineman, engineer.

mâchoire, *f.* jaw; clamp, clip; (*Anat.*) maxilla; dolt, blockhead.

mâchurer, *v.t.* smear, soil, daub, spot, mark.

macis, *m.* mace.

macle, *f.* (*Min.*) macle (chiastolite, also a crossed twin crystal).

maclé, *p.a.* (of glass) stirred; (of crystals) of the nature of crossed twins.

macler, *v.t.* stir (melted glass). — *v.r.* (*Min.*) crystallize in crossed twins.

maclurine, *f.* maclurin.

mâcon, *m.* Mâcon wine, Mâcon.

maçon, *m.* mason.

maçonnage, *m.* masonry.

maçonner, *v.t.* build, repair or line (in masonry).

maçonnerie, *f.* masonry.

macquevin, *m.* a kind of liquor composed of brandy and boiled wine.

macrophage, *m.* (*Physiol.*) macrophage.

macroscopique, *a.* macroscopic.

maculage, *m.,* **maculation,** *f.* spotting, maculation.

macule, *f.* spot, stain, macula, macule.

maculer, *v.t.* spot, soil, maculate.

madame, *f.* madam, madame, Mrs.

madéfaction, *f.* (*Pharm.*) moistening, wetting.

madéfier, *v.t.* (*Pharm.*) moisten, wet.

mademoiselle, *f.* young lady, Miss.

madère, *m.* madeira, Madeira wine; (*cap.*) Madeira.

madériser, *v.t.* make (wine) like madeira.

madrage, *m.* (*Soap*) the production of mottling or marbling (as by stirring in soaps of aluminium and iron).

madré, *a.* spotted, speckled, mottled, marbled.

madrier, *m.* plank, thick board.

madrure, *f.* spot, speckle, mark.

magasin, *m.* storehouse, storeroom, warehouse, magazine; shop, store; supply chamber, magazine.

en —, in stock.

magasinage, *m.* storage.

magasinier, *m.* warehouseman, storekeeper.

magdaléon, *m.* roll, stick (as of sulfur); (*Pharm.*) magdaleon (*Obs.*).

magenta, *m.* magenta.

magie, *f.* magic.

— *blanche,* white magic, natural magic.

— *noire,* black magic, black art.

magique, *a.* magic, magical.

magistère, *m.* magistery.

— *de bismuth,* magistery of bismuth (precipitated basic bismuth nitrate).

— *de coquilles d'huîtres,* (*Pharm.*) prepared oyster shell.

— *de soufre,* magistery of sulfur (precipitated sulfur).

magistral, *a.* masterly; authoritative; magisterial; (*Pharm.*) magistral. — *m.* (*Metal.*) magistral, roasted copper pyrites.

magma, *m.* magma.

magnanimement, *adv.* magnanimously.

magnésie, *f.* magnesia.

— *blanche,* — *anglaise,* (*Pharm.*) magnesium carbonate, magnesia alba.

— *calcinée,* calcined magnesia, magnesium oxide, magnesia.

— *calcinée pesante,* (*Pharm.*) heavy magnesium oxide, heavy magnesia, magnesia ponderosa.

— *légère,* (*Pharm.*) light magnesia.

— *liquide,* (*Pharm.*) solution of magnesium carbonate, fluid magnesia.

— *lourde,* (*Pharm.*) heavy magnesia.

— *noire,* black manganese, pyrolusite (*Obs.*).

magnésié, *a.* combined with or containing magnesia.

magnésien, *a.* magnesian, magnesia, magnesium, of or pertaining to magnesia or magnesium. — *m.* magnesium derivative.

magnésique, *a.* magnesium, magnesic.

magnésite, *f.* (*Min.*) sepiolite, hydrous magnesium silicate. (By some the word is used in the same sense as the English *magnesite,* magnesium carbonate, but this is generally considered incorrect.)

magnésium, *m.* magnesium.

magnésothermie, *f.* magnesothermy (production of heat by the chemical combination of magnesium).

magnétique, *a.* magnetic.

magnétiquement, *adv.* magnetically.

magnétisation, *f.* magnetization.

magnétiser, *v.t.* magnetize.

magnétisme, *m.* magnetism.

magnétite, *f.* (*Min.*) magnetite.

magnéto, *f.* (*Elec.*) magneto.

magnétochimie, *f.* magnetochemistry.

magnéto-électrique, *a.* magneto-electric(al).

magnifique, *a.* magnificent.

magnolier, *m.* (*Bot.*) magnolia tree, magnolia.

magnum, *m.* magnum (two-liter bottle).

mai, *m.* May.

maie, *f.* trough (as one for kneading clay); bin.

maigre, *a.* lean, thin, poor, meager, slender; (of coal) lean, nonbituminous; (of lime) lean, not slaking freely; (of type) light-faced; (of cheese) skim-milk. — *m.* lean meat, lean.

maigreur, *f.* leanness, thinness, etc. (see maigre).

mail, *m.* mallet, maul.

maille, *f.* mesh, opening; link; eye, loop; stitch; spot, speck; mallet.

maillechort, *m.* maillechort, German silver.

maillet, *m.* mallet, maul.

mailloche, *f.* beetle, maul, mallet.

maillon, *m.* link; stitch; noose.

main, *f.* hand; scoop; handle; handful; (of paper) 25 sheets; (of silk) skein; hook.

à la —, with the hand; by hand; in hand; at hand; handy; by guess; obliging.

— *coulante,* — *courante,* handrail.

sous la —, at hand; in the hands (of).

sous —, underhand, secretly.

main-brune, *a.* designating a coarse gray paper.

main-d'œuvre, *f.* manual labor, hand labor; price for labor.

maine, *f.* handful.

main-forte, *f.* assistance, help, aid.

mainmise, *f.* seizure.

maint, *a.* many a, many.

maintenant, *adv.* now, at present. — *p.pr.* of maintenir.

— *que,* now that.

maintenir, *v.t.* keep; maintain; keep up, support, hold fast. — *v.r.* keep, remain, be maintained, stand one's ground, hold one's own, keep up.

maintien, *m.* maintenance, preservation; attitude, bearing.

maintiendra, *fut. 3 sing.* of maintenir.

maintient, *pr. 3 sing.* of maintenir.

maïolique, *f.* (*Ceram.*) majolica.

mairain, *m.* = merrain.

maire, *m.* mayor.

mais, *conj.* but. — *adv.* more; indeed.

maïs, *m.* maize, (Indian) corn.

maison, *f.* house; home; household.
— *de commerce,* business house, commercial firm.
— *de santé,* private hospital, sanitarium.
— *de ville,* — *commune,* town house, mayor's offices.

maître, *m.* master; owner, proprietor; head, chief; (sometimes) master workman, foreman. — *a.* master; clever; main, principal, head; arrant.
— *de conférences,* (in French universities) lecturer.
— *de verrerie,* glass manufacturer.
— *ès arts,* master of arts.
— *souffleur,* (*Glass*) head blower.
— *verrier,* master glassmaker, head or employing glass worker.

maîtresse, *f.* mistress. — *a.* notable, superior, masterly; chief, main, principal.

maîtrise, *f.* mastery, mastership.

maîtriser, *v.t.* master, control.

majestueux, *a.* majestic.

majeur, *a.* greater, major; great, important; of age. — *m.* middle finger.

majolique, *f.* (*Ceram.*) majolica.

majoration, *f.* increase (in price), additional cost; allowance, excess.

majorer, *v.t.* increase the cost or price of; overvalue; add an allowance to (to meet reduction of any kind).

majorité, *f.* majority.

Majorque, *f.* Majorca.

majuscule, *a. & f.* (of letters) capital.

mal, *m.* evil; trouble; injury; pain; illness; ill. — *adv.* ill, badly, wrong, badly off, on bad terms. — *a.* bad.
— *chimique,* (*Med.*) necrosis of the lower jaw.

malachite, *f.* (*Min.*) malachite.

malade, *a.* ill, sick, diseased, in bad state. — *m. & f.* sick person.

maladie, *f.* disease; malady, illness, sickness, disorder; mania, passion.
— *cutanée,* skin disease.

maladif, *a.* sickly, unhealthy.

maladresse, *f.* awkwardness, clumsiness.

maladroit, *a.* awkward, unskillful, clumsy.

maladroitement, *adv.* awkwardly, unskillfully.

malaga, *m.* Malaga wine, Malaga.

malaguette, *f.* melegueta pepper, grains of Paradise, Guinea grains.

malais, *a.* Malay, Malayan.

malaise, *m.* discomfort, uneasiness.

malaisé, *a.* difficult, hard.

malaisément, *adv.* with difficulty, hardly.

Malaisie, *f.* Malaysia, Malay Archipelago.

malambo, *m.* malambo (bark of *Croton malambo*).

malate, *m.* malate.

malavisé, *a.* imprudent, rash.

malaxage, *m.,* **malaxation,** *f.* malaxation, malaxing, malaxage.

malaxer, *v.t.* malax, malaxate, soften by kneading or rubbing or mixing; massage.

malaxeur, *m.* malaxator, worker, mixer, (for clay) pugger. — *a.* malaxing, malaxating, mixing.

mâle, *a. & m.* male.

maléique, *a.* maleic.

malfaisant, *a.* injurious, harmful; malicious.

malfil, *m.* a woolen cloth used for making press bags for oil meal, etc.

malgré, *prep.* in spite of, notwithstanding.
— *qu'on en ait,* in spite of oneself.

malhabile, *a.* unskillful, clumsy.

malheur, *m.* misfortune; unhappiness.
par —, unfortunately.

malheureusement, *adv.* unfortunately, unluckily, unhappily.

malheureux, *a.* unfortunate, unhappy, unlucky, wretched, pitiful.

maligne, *fem.* of malin.

malin, *a.* malignant; roguish; cunning, sly.

malique, *a.* malic.

malle, *f.* trunk; mail.

malléabiliser, *v.t.* malleableize.

malléabilité, *f.* malleability, malleableness.

malléable, *a.* malleable.

malléer, *v.t.* beat, hammer (metal into sheets).

malléine, *f.* mallein.

malon, *m.* brick.

malonique, *a.* malonic.

malonitrile, *m.* malonitrile; malononitrile (incorrect use).

malpighien, *a.* (*Anat.,* etc.) Malpighian.

malpropre, *a.* dirty, unclean; improper.

malpropreté, *f.* dirt, dirtiness; indecency.

malsain, *a.* unhealthy, unwholesome.

malt, *m.* malt.
— *brun,* brown malt.
— *d'orge,* barley malt.
— *feutré,* felted malt.
— *pelleté,* turned malt, plowed malt.
— *touraillé,* kiln-dried malt.
— *vert,* green malt (undried malt).

maltage, *m.* malting.

maltais, *a.* Maltese.

maltase, *f.* maltase.

Malte, *f.* Malta.

malter, *v.t.* malt.

malterie, *f.* malt house, malt factory; malting.

malteur, *m.* maltster, maltman, malt maker.

malthe, *f.* maltha, mineral tar.

maltose, *f.* maltose.

malvacées, *f.pl.* (*Bot.*) Malvaceæ.

malvoisie, *f.* malmsey (the wine and grape).

mamelle, *f.* breast; udder; bag; (*Anat.*) mamma.

mamelon, *m.* rounded or mammillary mass or projection; nipple, teat, mammilla; hummock, mamelon; (*Mach.*) gudgeon.

mamelonné, *a.* mammillary; rounded.

mamillaire, *a.* mammillary.

mammaire, *a.* (*Anat.*) mammary.

mammifère, *a.* mammiferous, mammalian. — *m.* mammal.

mammiforme, *a.* mammiform, mammillary.

manche, *m.* handle; holder, helve, haft. — *f.* hose, pipe, tube; shaft, chute; sleeve; channel, strait.

la Manche, the (British) Channel.

— *à bouton,* handle in the form of or terminating in a button.

— *de cuir,* leather hose.

— *de toile,* canvas hose.

— *d' Hippocrate,* Hippocrates' bag (or sleeve), straining bag.

manchette, *f.* cuff.

manchon, *m.* sleeve; collar, hoop, ring, flange, jacket, casing, case, box, housing, etc.; (of a stopcock) shell, socket; (incandescent) mantle; clutch, coupling; (*Glass*) cylinder, muff; (*Paper*) blanket; muff.

— *d'accouplement,* clutch, coupling.

— *de refroidissement,* cooling jacket.

— *fileté,* screw sleeve.

— *pour éclairage,* incandescent mantle.

manchonnier, *m.* (*Glass*) cylinder blower.

mancône, *m.* sassy bark, mancona bark (also the tree, *Erythrophlœum guineense*).

mandarine, *f.* mandarin, mandarin orange.

mandarinier, *m.* mandarin tree (*Citrus nobilis*).

mandat, *m.* mandate; order; check, draft; warrant.

— *de poste,* post-office order.

mandataire, *m.* agent, proxy, attorney.

mandélique, *a.* mandelic.

mander, *v.t.* send for, order.

mandragore, *f.* (*Bot.*) Mandragora.

— *officinale,* mandrake (*Mandragora officinarum*).

mandrin, *m.* mandrel, mandril; shaper, former; form; punch; drift; swage; chuck; gage (for hollow objects).

mandriner, *v.t.* work, shape or form upon or with a mandrel; punch, gage, etc. (see mandrin).

manette, *f.* (small) handle; specif., hand lever or key (as on an electric switch).

manganate, *m.* manganate.

mangane, *m.* manganese (*Obs.*).

manganèse, *m.* manganese.

manganésé, *a.* containing manganese.

manganésien, *a.* of or containing manganese, manganesian.

manganésifère, *a.* manganiferous.

manganésique, *a.* manganic (*Obs.*).

manganeux, *a.* manganous.

manganique, *a.* manganic.

manganite, *f.* manganite.

manganium, *m.* manganese (*Obs.*).

manganoferrosilicium, *m.* an alloy of manganese, iron and silicon.

manganoso-, manganous (as, *sel manganosoammonique,* manganous ammonium salt).

mangeable, *a.* eatable, edible.

mangeaille, *f.* feed (of certain animals).

manger, *v.t.* eat; eat up, consume; eat away, wear away, corrode; destroy; neglect, slight. — *v.i.* eat, (of animals) feed; (of lye used in making soap) lose strength, be neutralized. — *v.r.* be eaten, be consumed, etc. — *m.* food.

manglier, *m.* (*Bot.*) mangrove (*Rhizophora*).

mangue, *f.* (*Bot.*) mango.

manguier, *m.* mango, mango tree (*Mangifera indica*).

mani, *m.* a resin from the Guiana candlewood tree.

maniable, *a.* manageable; easy to handle or work, handy, workable.

maniage, *m.* handling, etc. (see manier).

manie, *f.* mania.

maniement, *m.* handling, etc. (see manier); rustle (of silk on handling).

manier, *v.t.* handle; manipulate; touch, feel; work; manage. — *m.* handling; touch, feel, handle.

manière, *f.* manner, way, method, fashion.

de — à, so as.

de — que, in such a way that, so that.

— *de voir,* opinion, point of view.

par — de, by way of.

manifestation, *f.* manifestation.

manifeste, *a.* manifest, evident.

manifestement, *adv.* manifestly.

manifester, *v.t.* manifest, display, show.

manigaux, *m.pl.* lever (of a bellows).

maniguette, *f.* = malaguette.

manilla, *m.* Manila hemp, manila, manilla.

manille, *f.* shackle; manila rope; Manila hemp, manila; (*cap.*) Manila.

maniment, *m.* = maniement.

manioc, *m.* (*Bot.*) manioc, cassava.

manipulateur, *m.* manipulator; (telegraph) key.

manipulation, *f.* manipulation.

manipuler, *v.t. & i.* manipulate.

manipuleur, *m.* manipulator.

manivelle, *f.* crank, crank arm, winch.

manne, *f.* manna; basket, hamper.

— *en sorte,* (*Pharm.*) common manna.

mannequin, *m.* mannikin; basket, hamper.

mannitane, *f.* mannitan.

mannite, *f.* mannitol, mannite.

mannité, *a.* (of wine, etc.) altered as a result of mannitic fermentation.

mannitique, *a.* mannitic.
mannoheptite, *f.* mannoheptitol, mannoheptite.
mannoheptonique, *a.* mannoheptonic.
mannonique, *a.* mannonic.
mannosaccharique, *a.* mannosaccharic.
mannose, *f.* mannose.
manœuvre, *f.* working, operation; maneuver, manœuvre; rope, tackle. — *m.* workman, laborer (*grand manœuvre* being a higher, and *petit manœuvre* a lower, grade).
manœuvrer, *v.t.* operate, work, manipulate, manage, maneuver. — *v.i.* work; maneuver.
manomètre, *m.* manometer, pressure (or vacuum) gage.
— *à mercure,* mercury manometer, mercurial gage.
— *du vide,* vacuum gage.
— *enregistreur,* recording manometer, recording gage.
manométrie, *f.* manometry.
manométrique, *a.* manometric, manometrical.
manouvrier, *m.* workman, laborer.
manquant, *p.a.* missing, wanting. — *m.* missing person or thing.
manque, *m.* want, lack, deficiency; failure.
manqué, *a.* defective; unsuccessful; missed.
manquement, *m.* want, lack; failure; fault.
manquer, *v.i.* be lacking, be wanting; be short, be in want; fail; miss. — *v.t.* miss; pass by. — *v.r.* fail; be missed.
— *de,* lack, want, be wanting in; (with infinitive) just escape, come near.
s'en — de, be far from it, be missing.
manteau, *m.* mantle (in various senses, e.g. the outer wall and casing of a blast furnace); shell (of a mold); hood; mantel, mantelpiece.
mantisse, *f.* (*Math.*) mantissa.
manuel, *a.* & *m.* manual.
manuellement, *adv.* manually, with the hand.
manufacture, *f.* factory, mill, manufactory; the workmen of a factory.
manufacturer, *v.t.* manufacture.
manufacturier, *m.* manufacturer. — *a.* manufacturing.
manuscrit, *m.* & *a.* manuscript.
manutention, *f.* management, administration; handling (of goods); (*Mil.*) bakery.
manutentionnaire, *m.* manager.
manutentionner, *v.t.* handle (merchandise); make, prepare; (*Mil.*) bake (bread).
manzanilla, *m.* manzanilla (a light, dry sherry).
mappe, *f.* map.
mappemonde, *f.* map of the world.
maquereau, *m.* mackerel.
maquiller, *v.t.* paint (the face).
maraicher, *m.* market gardener. — *a.* market.

marais, *m.* marsh, moor, bog, swamp; market garden, truck farm.
— *salant,* salt garden, saline (for making sea salt).
marasque, *f.* (*Bot.*) marasca (a kind of cherry).
marasquin, *m.* maraschino (a liqueur).
marbre, *m.* marble; marbling; slab, stone, block, plate (often of cast iron); specif., (*Glass*) marver, marble.
— *dur,* granite.
— *factice,* artificial (or imitation) marble.
marbré, *a.* marbled, variegated.
marbrer, *v.t.* marble (as soap or paper); (*Glass*) marver.
marbrière, *f.* marble quarry.
marbrure, *f.* marbling.
marc, *m.* marc, residue, residuum.
— *de soude,* (*Soda*) exhausted black ash, vat waste.
— *de vendange,* marc of grapes.
marcasite, marcassite, *f.* (*Min.*) marcasite.
marceline, *f.* (*Min.*) a variety of braunite; marceline (a thin silk fabric).
marchand, *m.* merchant, dealer, trader; buyer. — *a.* commercial, mercantile, market, merchant; marketable, merchantable; wholesale (price).
marchandage, *m.* bargaining.
marchandise, *f.* merchandise, goods, wares.
marche, *f.* course, procedure; progress; operation, functioning; movement, motion; rate, speed; walk, walking; gait; march; step, stair; treadle, pedal; (*Metal.*) fire bridge.
— *à suivre,* procedure.
— *continue,* continuous operation.
mettre en —, start, set going.
marché, *m.* market; price, rate; bargain, sale, contract.
— *indigène,* domestic market, home market.
bon —, cheap, cheaply.
— *mondial,* world market.
marchepied, *m.* step; footboard; stepping-stone.
marcher, *v.i.* go, move, progress, travel, run; function, work, go; walk; march; behave. — *v.t.* tread; trample. — *m.* walking, walk.
faire —, make go, run; tread (as clay).
marcheur, *m.* treader; walker, pedestrian.
marcheux, marchoir, *m.* kneading trough (for clay); kneading place; kneader; treader.
marcotte, *f.* (*Horticulture*) layer.
mardi, *m.* Tuesday.
mare, *f.* pool, pond.
marécage, *m.* marsh, swamp, bog.
marécageux, *a.* marshy, swampy, boggy.
maréchal, *m.* marshal; (*maréchal ferrant*) horseshoer, farrier, blacksmith. — *a.* (of coal) forge.

maréchalerie, *f.* blacksmith shop, farriery.
marée, *f.* tide; fresh (sea) fish.
— *descendante,* falling tide, ebb tide.
— *montante,* rising tide, flood tide.
odeur de —, odor of fresh fish, fishy odor.
margarate, *m.* margarate.
margarine, *f* margarine (butter substitute);
margarin (glyceryl margarate).
margarinerie, *f.* margarine manufacture;
margarine factory.
margarinier, *m.* margarine maker.
margarique, *a.* margaric.
margarite, *f.* (*Min.*) margarite.
margaroïde, *a.* resembling margarine.
marge, *f.* margin; time, spare time.
margeoir, *m.* (*Glass*) a plate used for closing
the mouth of a covered pot or the opening
of a furnace.
marger, *v.t.* (*Glass*) close, stopper (a pot or
furnace).
marginal, *a.* marginal.
marginer, *v.t.* margin.
margousier, *m.* (*Bot.*) margosa, China tree.
mari, *m.* husband.
mariage, *m.* marriage.
marier, *v.t.* marry; unite; adapt. — *v.r.*
marry; blend, harmonize.
marin, *a.* marine, sea. — *m.* seaman, mariner.
marinade, *f.* pickle, brine; pickled meat;
French dressing.
marine, *f.* navy; marine; navigation.
mariné, *a.* pickled; sea-damaged.
mariner, *v.t. & i.* pickle.
marjolaine, *f.* (*Bot.*) marjoram.
— *commune,* sweet marjoram (*Origanum
majorana*).
— *sauvage,* wild marjoram (*Origanum vulgare*).
marmelade, *f.* marmalade.
marmite, *f.* pot; pan, kettle, boiler, copper,
etc.; retort (as for nitric acid); potful.
— *de Papin,* Papin digester.
marmitée, *f.* potful, kettleful.
marmorisation, *f.* marmarization, marmoriza-
tion.
marmoriser, *v.t.* marmarize, marmorize.
marmorite, *f.* marmorite (a veined opaque
glass).
marnage, *m.* marling (of land).
marne, *f.* marl.
marner, *v.t.* marl.
marneux, *a.* marly, marlaceous.
marnière, *f.* marl pit.
Maroc, *m.* Morocco.
marocain, *a.* Moroccan. — *m.* (*cap.*) Moroccan.
maroquin, *m.* morocco, morocco leather. — *a.*
imitating morocco.
maroquinage, *m.* morocco tanning (or dressing).

maroquiné, *p.a.* morocco-tanned, morocco-
dressed; (of paper) morocco.
maroquiner, *v.t.* morocco-tan, morocco-
dress.
maroquinerie, *f.* morocco tanning (or dressing); morocco factory; morocco trade;
morocco article.
maroquineur, maroquinier, *m.* morocco tanner, morocco dresser.
marouette, *f.* = maroute.
marouflage, *m.* glued-on lining (as of canvas).
maroufle, *f.* a kind of strong glue.
maroufler, *v.t.* line (with canvas or the like,
glued on).
maroute, *f.* (*Bot.*) mayweed (*Anthemis cotula*).
marquage, *m.* marking, etc. (see marquer).
marquant, *a.* striking, conspicuous, remarkable.
marque, *f.* mark; brand; trade-mark; sign;
stamp; marker; tally; buoy.
— *d'eau,* water gage.
— *de fabrique,* trade-mark.
mille marques, the number of one thousand,
the thousand mark.
marquer, *v.t.* mark; mark out; brand; stamp;
note; indicate; appoint, assign. — *v.i.* be
noted, be remarkable or distinguished;
appear.
marqueter, *v.t.* mark with spots; inlay.
marqueterie, *f.* marquetry; (fig.) patchwork.
marquette, *f.* a cake of virgin wax.
marqueur, *m.,* **marqueuse,** *f.* marker,
stamper.
marquise, *f.* (*Pyro.*) a kind of rocket.
marron, *m.* chestnut (specif. the large sweet
European chestnut); chestnut (color),
maroon; (*Pyro.*) maroon; marker, tag,
ticket, check; mold core; stencil. — *a.*
unlicensed; runaway; run wild.
— *d'Inde,* horse-chestnut.
marronier, *m.* chestnut tree.
faux —, — *d'Inde,* horse-chestnut tree.
marrube, *m.* (*Bot.*) horehound (*Marrubium*).
— *blanc,* — *commun,* white horehound, common horehound (*M. vulgare*).
mars, *m.* March; (*Old Chem.*) Mars, iron.
marsala, *m.* Marsala, Marsala wine.
marseillais, *a.* of Marseilles.
marsouin, *m.* (*Zoöl.*) porpoise.
marteau, *m.* hammer; striker, knocker,
clapper.
— *à glace,* hammer for crushing ice.
— *à queue,* tilt hammer.
— *à vapeur,* steam hammer.
— *cingleur* (*Metal.*) shingling hammer.
— *de géologue,* geologist's hammer.
— *d'essayeur,* assayer's hammer.
— *pioche,* a combined hammer and pick.
marteau-foulon, *m.* a heavy hammer for
softening hides.

marteau-pilon, *m.* power hammer, drop hammer (e.g. a steam hammer).

martelage, *m.* hammering.
— *au martinet,* tilting.

marteler, *v.t.* hammer; hammer out; worry.

martelet, *m.* small hammer.

martial, *a.* martial, (*Old Chem.*) of iron.

martiner, *v.t.* hammer, tilt.

martinet, *m.* tilt hammer; martin (the bird).

martre, *f.* (*Zoöl.*) marten.

marum, *m.* (*Bot.*) cat thyme (*Teucrium marum*).

marute, *f.* (*Bot.*) a plant of the genus Anthemis; specif., mayweed (*A. cotula*).

mascaron, *m.* mask; specif., an ornamented piece of cast iron used as a support for one of the leveling screws of a balance or other instrument.

masque, *m.* mask; screen, shield, cover.
— *A.R.S.,* see A.R.S.
— *respirateur,* respirator, gas mask.

masquer, *v.t.* mask; screen, cover, conceal.

masse, *f.* mass; fund, funds, capital; lot, quantity (varying with the article); share, quota; scrap iron; bed, block (of stone); body (as of a furnace); sledge hammer; maul, beetle, rammer; area (on a map); stake (at play); (*Elec.*) ground.
en —, in a mass, en masse, in a body, together.
— *active,* (*Com.*) assets.
— *cuite,* (*Sugar*) massecuite.
— *demi-affinée,* white pig iron.
— *d'équilibrage,* counterpoise, counterweight.
— *passive,* (*Com.*) liabilities.
— *pilulaire de Vallet,* (*Pharm.*) mass of ferrous carbonate, Vallet's mass.
par —, by the mass, by the lump.

massé, *m.* (*Metal.*) bloom. — *a.* massed.

masseau, *m.* (*Metal.*) shingled bloom.

masselet, *m.* (*Metal.*) small bloom.

masselotte, *f.* (*Founding*) feedhead, feeding head (also, a similar excess of material in candle molding); weight; plunger (of a fuse).

masser, *v.t.* mass; massage. — *v.r.* form masses.

massiau, *m.* (*Metal.*) shingled bloom.

massicot, *m.* massicot; (*Mach.*) paper cutter.

massif, *a.* massive.· — *m.* solid mass; main part, solid part; shell (of a furnace); (*Geol.*) massif, massive; mass of masonry (e.g. a pier); block, group (as of furnaces); (*Pyro.*) driít.

massivage, *m.* tamping, packing (as of concrete).

massivement, *adv.* massively.

massiver, *v.t.* tamp, pack (as concrete).

massiveté, *f.* massiveness.

massoque, *f.* (*Metal.*) bloom, slab.

massoquette, *f.* small bloom.

massoquin, *m.* (*Metal.*) bloom, slab.

massou, *m.* (*Salt*) a plank table or mold.

massoy, *n.* massoy, massoy bark.

massue, *f.* club.

mastic, *m.* mastic; cement; putty.
— *artificiel,* artificial mastic.
— *asphaltique,* asphaltic cement, asphaltic mastic, asphalt.
— *bitumineux,* bituminous cement.
— *d'asphalte,* = mastic asphaltique.
— *de fer,* — *de fonte,* — *de limaille de fer,* iron cement.
— *de vitrier,* glazier's putty.
— *naturel,* = mastic asphaltique.

masticage, *m.* cementing; puttying.

masticatoire, *a. & m.* masticatory.

masticine, *f.* masticin.

mastiquer, *v.t.* cement; putty; masticate.

mat, *a.* mat, dead, dull, unpolished; heavy (as bread). — *m.* mat surface, dull surface, mat.

mât, *m.* mast, pole.

matasiette, mataziette, *f.* a kind of dynamite.

maté, *m.* (*Bot.*) maté, mate, Paraguay tea.

matelas, *m.* layer; lining; backing; cushion; mattress.
— *à air,* air cushion.

matelasser, *v.t.* pad, stuff.

matelassure, *f.* padding, lining; backing.

matelot, *m.* sailor, seaman.

mater, *v.t.* mat, render mat, dull, deaden; compact, make heavy (as dough); flatten, beat down, clinch, smooth (metals); overlay with metal.

matérialiser, *v.t.* materialize.

matériaux, *m.pl.* materials, material.
— *bruts,* raw material(s).
— *de construction,* building materials.

matériel, *a.* material. — *m.* material; stock, stores; plant; materiality.
— *de guerre,* war material, military stores.
— *roulant,* rolling stock.

matériellement, *adv.* materially; positively.

materniser, *v.t.* modify (milk) so as to imitate human milk.

maternité, *f.* maternity; lying-in hospital.

mathématicien, *m.* mathematician.

mathématique, *a.* mathematical. — *f.* mathematics.

mathématiquement, *adv.* mathematically.

matico, *m.* matico (esp. *Piper angustifolium*).

matière, *f.* matter; material; (*pl.*) bullion; (*pl.*) contents (of a book); (*Pyro.*) composition; (*Law*) case.
en — *de,* in point of, with respect to.
— *amylacée,* starchy material, amylaceous material.
— *antiplastique,* antiplastic material.
— *azotée,* nitrogenous matter or substance.

matière *brute*, raw material.
— *carbonée*, carbonaceous material or matter.
— *caséeuse*, cheesy or caseous matter.
— *colorante*, coloring matter, specif. (1) dye, dyestuff, (2) pigment.
— *de réserve*, reserve material.
— *détonante*, explosive material, explosive; specif., detonator.
— *étrangère*, foreign matter or material.
— *intermédiaire*, intermediate.
— *isolante*, insulating material, insulator.
— *lubréfiante*, lubricating material, lubricant.
— *médicale*, materia medica.
— *organique*, organic matter.
— *plastique*, plastic material.
— *première*, — *primitive*, raw material.
— *protéique*, protein substance, protein.
— *réfractaire*, refractory material.
— *résiduaire*, residual product, by-product.
matières d'argent, silver bullion.
matières d'or, gold bullion.
— *sucrée*, saccharine matter or material, sugar material.
— *tannante*, tanning material.
matin, *m.* morning. — *adv.* early.
matinée, *f.* morning, forenoon; matinée.
matir, *v.t.* mat, render mat, dull, deaden.
matoir, *m.* stamp; riveting hammer; deadening tool.
matras, *m.* matrass, bolthead; matrass, mattrass (hard glass tube).
— *d'essayeur*, assayer's matrass (hard glass tube).
matricaire, *f.* (*Bot.*) Matricaria.
matrice, *f.* matrix. — *a.* mother, parent; (of colors) simple, primary.
mattage, *m.* matting, dulling, etc. (see mater).
matte, *f.* (*Metal.*) mat, matte, matt.
— *blanche*, (*Copper*, etc.) white metal.
— *bleue*, (*Copper*) blue metal.
— *bronze*, — *brute*, (*Copper*) coarse metal.
— *concentrée*, (*Copper*) fine metal.
— *crue*, crude mat, rough mat.
— *de cuivre*, copper mat.
— *de plomb*, lead mat, hard lead.
matteau, *m.* bundle (of hanks).
matter, *v.t.* = mater.
mattoir, *m.* = matoir.
maturatif, *a.* (*Med.*) maturative. — *m.* maturant, maturative.
maturation, *f.* maturing; ripening; (*Med. & Biol.*) maturation; (*Metal.*) refining.
maturer, *v.t.* mature, ripen; (*Metal.*) refine.
maturité, *f.* maturity; ripeness.
maudit, *a.* cursed, wretched, abominable.
mauresque, *a.* Moorish.
mauvais, *a.* bad; evil, ill. — *adv.* bad, badly; wrong, amiss.
faire —, be hard; be trying; (of the weather) be disagreeable.

mauvais *goût*, — *goûts*, (*Alcohol*) a fraction of disagreeable taste obtained in rectifying. The first fraction (*mauvais goûts de tête*) contains aldehyde, the last fraction (*mauvais goûts de queue*) contains amyl alcohol.
mauve, *f.* mauve; (*Bot.*) mallow. — *a.* mauve.
— *sauvage*, wild mallow (*Malva sylvestris*).
mauvéine, *f.* mauveine, mauvein, mauvine (a dye).
maux, *m.pl.* of mal.
maxima, *a. fem. & m. pl.* maximum.
maxime, *f.* maxim.
maximum, *m. & a.* maximum.
au —, at the maximum; in the highest (or higher) state of valence.
maye, *f.* = maie.
Mayence, *n.* Mainz, Mayence.
mayonnaise, *f.* mayonnaise.
mazéage, mazage, *m.* fining (or refining) of cast iron, esp. to remove silicon.
mazée, *f.* (*Iron*) fine metal, refined metal.
mazéer, mazer, *v.t.* (*Iron*) fine, refine.
mazerie, *f.* (*Iron*) fining hearth, refinery.
me, *pron.* me, to me.
mécanicien, *m.* mechanician, machinist, engineer. — *a.* mechanical, machine.
mécanique, *a.* mechanical. — *f.* mechanics; machine, mechanism; machination.
— *chimique*, chemical mechanics.
mécaniquement, *adv.* mechanically.
mécaniser, *v.t.* mechanize, make mechanical.
mécanisme, *m.* mechanism.
— *chimique*, chemical mechanism.
méchant, *a.* bad, wrong, wicked, unkind.
mèche, *f.* wick; match (for use as a fuse, for fumigation, etc.); tinder; bit, drill; heart, core.
— *d'amorce*, quick match.
— *incendiaire*, slow match.
mécher, *v.t.* match (as casks); (*Sugar*) mix the contents of (a diffuser) by introducing liquid at the bottom.
mechoacan, *m.* (*Bot. & Pharm.*) mechoacan.
mécompte, *m.* miscount; mistake; misconception.
méconate, *m.* meconate.
méconine, *f.* meconin.
méconinique, *a.* meconinic.
méconique, *a.* meconic.
méconium, *m.* (*Med.*) meconium.
méconnaissable, *a.* unrecognizable.
méconnaître, *v.t.* not recognize; disregard, slight; disown.
méconnu, *p.p.* of méconnaître.
mécontent, *a.* dissatisfied; discontented.
médaille, *f.* medal; medallion.
médaillé, *a.* medaled. — *m.* medalist.
médaillon, *m.* medallion.
médecin, *m.* physician, doctor.
médecine, *f.* medicine.

médian, *a.* median.

médical, *a.* medical.

médicament, *m.* medicament, medicine.

médicamentaire, *a.* medicinal, medicamental.

médicamentation, *f.* medicamentation, treatment with medicine.

médicamenter, *v.t.* treat with medicine, medicate.

médicamenteux, *a.* medicinal, medicamental.

médicateur, *a.* medicinal, remedial.

médication, *f.* medication.

médicinal, *a.* medicinal.

médicinalement, *adv.* medicinally.

médicinier, *m.* physic nut (*Jatropha curcas*).

médiocre, *a.* moderate; mediocre.

médiocrement, *adv.* moderately; indifferently.

méditer, *v.t.* meditate, meditate upon, contemplate. — *v.i.* meditate.

méditerrané, *a.* mediterranean, interior.

Méditerranée, *f.* Mediterranean (Sea).

méditerranéen, *a.* of the Mediterranean.

médium, *m.* medium.

médius, *m.* middle finger, medius.

médoc, *m.* Médoc, Médoc wine; a gem stone.

médullaire, *a.* medullary, medullar.

médulle, *f.* (*Biol.*) medulla.

méfier, *v.r.* be distrustful or suspicious.

mégie, *f.* (*Leather*) tawing.

mégir, mégisser, *v.t.* taw.

mégis, *m.* tawing paste. — *a.* tawed.

mégisserie, *f.* tawing; tawing establishment.

mégissier, *m.* tawer.

meilleur, *a.* better; best. — *m.* best. — *adv.* better.

   faire —, be better.

méjuger, *v.t.* misjudge.

mélam, *m.* melam.

mélamine, *f.* melamine.

mélange, *m.* mixture; mixing, blending, mingling; (*Brewing*) mashing; crossing (as of breeds or wires); medley; jumble.

   — *anglais,* a mixture of two thirds Para and one third ordinary rubber.

   — *de Laming,* (*Gas*) Laming's mass.

   — *détonant,* explosive mixture.

   — *frigorifique,* freezing mixture.

   — *magnésien,* magnesia mixture.

   — *réfrigérant,* freezing mixture.

   — *tonnant,* explosive mixture, detonating mixture.

mélangeoir, *m.* mixer; (*Gunpowder*) mixing barrel.

mélanger, *v.t.* mix; blend; mingle; intermix. — *de,* mix with.

mélangeur, *m.* mixer; (*Brewing*) masher. — *a.* mixing.

mélanine, *f.* melanin.

mélanite, *f.* (*Min.*) melanite.

mélaphyre, *m.* (*Petrog.*) melaphyre.

mélasse, *f.* molasses.

mélassigène, *a.* melassigenic.

mélassique, *a.* molasses, relating to molasses; (of an acid) melassic.

mélasso-calcique, *a.* of or relating to molasses and calcium or lime, molasses-lime.

mêler, *v.t.* mix, mingle, blend; entangle; mix up. — *v.r.* mix, be mixed; meddle.

mélèze, *m.* (*Bot.*) larch (*Larix*).

mélézitose, *f.* melezitose.

mélibiose, *f.* melibiose (melebiose).

mélilot, *m.* (*Bot.*) melilot (*Melilotus*).

   — *officinal,* yellow melilot (*M. officinalis*).

mélilotique, *a.* melilotic.

mélinite, *f.* (*Expl.*) melinite.

mélisone, *n.* (*Min.*) wulfenite.

mélisse, *f.* (*Bot.*) balm, balm mint (*Melissa*).

   — *officinale,* garden balm (*M. officinalis*).

mélissipalmitique, *a.* In the phrase *éther mélissipalmitique,* myricyl (or melissyl) palmitate.

mélissique, *a.* (of an acid) melissic; (of an alcohol) myricyl, melissyl.

mélitose, *f.* melitose (raffinose).

mélizitose, *f.* melizitose (melezitose).

mellate, *m.* mellitate.

melléolé, *m.* (*Pharm.*) electuary.

melléolique, *a.* designating a medicine composed of honey and a powder.

mellique, *a.* mellitic.

mellite, *m.* (*Min.*) mellite; (*Pharm.*) honey-mellite.

   — *de borax,* (*Pharm.*) borax honey.

   — *de rose,* (*Pharm.*) honey of rose.

   — *de vinaigre,* (*Pharm.*) oxymel.

   — *de vinaigre scillitique,* (*Pharm.*) oxymel of squill.

   — *simple,* (*Pharm.*) clarified honey.

mellithate, *m.* mellitate.

mellithe, *m.* (*Min.*) mellite.

mellitique, mellithique, *a.* mellitic.

mellon, *m.* mellon, (less desirably) mellone.

mellophanique, *a.* mellophanic.

mellyle, *m.* mellityl.

mélodieux, *a.* melodious.

méloé, *m.* (*Zoöl.*) oil beetle (*Meloe*).

melon, *m.* (*Bot.*) melon.

   — *d'eau,* watermelon.

membrane, *f.* membrane.

   — *du corps vitré,* hyaloid membrane.

   — *muqueuse,* mucous membrane.

membrané, membraneux, *a.* membranous.

membre, *m.* member.

membrure, *f.* frame, framing, framework.

même, *a.* same; (with pronouns) self. — *m.* same thing, same. — *adv.* even, also, more.

   à — de, able to, in a position to.

   au — temps, at the same time.

   au — temps que, at the same time with, simultaneously with.

   dans le — sens que, directly as.

même, continued.

  dans le — temps, at once.

  de —, the same, in the same way.

  de — que, in the same way as, the same as, just as, as well as.

  de — que . . . de —, just as . . . so.

  en — temps, at the same time, together.

  un — échantillon, the same sample, a single sample.

mémoire, *f.* memory. — *m.* memorandum, note; report; bill, account; memoir; memorial (the paper).

mémorandum, *m.* memorandum; memorandum book.

mémorial, *m.* memorandum book; memoirs; memorial (the document); (*Com.*) day-book.

menacer, *v.t.* threaten, menace.

ménage, *m.* housekeeping; household, house, home; household goods; management; economy, thrift.

ménagé, *p.a.* managed, etc. (see ménager); cautious, careful, prudent.

ménagement, *m.* caution, circumspection, care.

ménager, *v.t.* manage; manage carefully, take care of, make the most of, economize; spare; obtain, secure. — *v.r.* obtain, secure; be managed, be cared for, etc.; take care of oneself; be careful; get on, get on well; develop. — *a.* household; (of water) waste; careful, saving. — *m.* economizer; housekeeper.

mène, *f.* (*Soap*) a series of lye tanks (*barquieux*).

mener, *v.t.* lead, conduct, take, carry, convey; drive; direct; carry on; manage; draw (a line); steer (a boat); treat.

meneur, *m.* leader, conductor, etc. (see mener).

méningite, *f.* (*Med.*) meningitis.

ménispermacées, ménispermées, *f.pl.* (*Bot.*) Menispermaceæ.

ménisque, *m.* meniscus.

mensonger, *a.* untrue, false; deceitful.

menstrue, *m.* menstruum, solvent.

menstrues, *f.pl.* (*Physiol.*) menses.

mensuel, *a.* monthly.

mensuellement, *adv.* monthly; by the month.

mensurabilité, *f.* measurability, mensurability.

mensurable, *a.* measurable, mensurable.

mensurateur, *a.* measuring.

mensuration, *f.* mensuration.

ment, *pr. 3 sing.* of mentir.

mental, *a.* mental.

mentalement, *adv.* mentally.

menteur, *a.* deceitful, false; lying. — *m.* liar.

menthe, *f.* (*Bot.*) mint; a liqueur flavored with mint.

  — à épi, spearmint.

  — à feuilles rondes, round-leaved mint (*Mentha rotundifolia*).

menthe *anglaise,* peppermint.

  — aquatique, water mint (*Mentha aquatica*).

  — crépue, curled mint (*Mentha silvestris crispa*).

  — de chat, catnip, catmint (*Nepeta cataria*).

  — de cheval, horsemint (*Monarda punctata*).

  — des champs, corn mint (*Mentha arvensis*).

  — poivrée, peppermint (*Mentha piperita*).

  — pouliot, pennyroyal (*Mentha pulegium*).

  — romaine, — verte, spearmint (*Mentha spicata,* or *viridis*).

  — sauvage, horsemint (*Mentha longifolia*).

menthène, *m.* menthene.

menthol, *m.* menthol.

mentholé, mentholique, *a.* of or containing menthol.

menthone, *f.* menthone.

mention, *f.* mention.

mentionner, *v.t.* mention.

mentir, *v.i.* lie, tell a falsehood, deceive.

menton, *m.* chin.

mentonnet, *m* cam, tappet, wiper; catch, lug, ear; cleat; flange.

mentonnière, *f.* (*Assaying*) muffle plate.

menu, *a.* small; fine, minute; minor, petty, inconsiderable. — *m.* material in small pieces or particles (all below a certain size), smalls, fines; detailed account, detail; bill of fare, menu; lower classes. — *adv.* small, fine.

  — charbon, small coal (esp. fine charcoal).

  menues houilles, small coal, smalls, slack, coal dust.

  — plomb, small shot, bird shot.

  menus sels, (any) salt in small pieces or particles, fines.

menuiserie, *f.* joinery, carpentry; joiner's or carpenter's shop; making of small objects of metal, also the objects themselves.

méphitique, *a.* mephitic, noxious.

  air — (*Old Chem.*) mephitic air (carbon dioxide).

méphitiser, *v.t.* render poisonous or offensive, vitiate.

méphitisme, *m.* vitiation of the air by foul or poisonous gases.

méplat, *a.* flat.

méprendre, *v.r.* be mistaken.

mépris, *m.* contempt. — *p.p.* of méprendre.

  au — de, without regard to, in defiance of.

méprise, *f.* mistake, error, oversight, blunder.

mépriser, *v.t.* scorn, disdain, despise.

mer, *f.* sea.

  Mer Azov, mer d'Azov, Sea of Azov.

  — Baltique, Baltic Sea.

  — Blanche, White Sea.

  — Caspienne, Caspian Sea.

  — de Chine méridionale, China Sea.

  — des Antilles, Caribbean Sea.

  — d'Oman, Arabian Sea.

mer *du Japon,* Sea of Japan.
— *du Nord,* North Sea.
— *Égée,* Ægean Sea.
— *Jaune,* Yellow Sea.
— *Méditerranée,* Mediterranean Sea.
— *Noire,* Black Sea.
— *Rouge,* Red Sea.
mercaptal, *m.* mercaptal.
mercaptan, *m.* mercaptan.
mercenaire, *a. & m.* mercenary.
mercerisage, *m.* mercerizing, mercerization.
merceriser, *v.t.* mercerize.
merci, *f.* mercy. — *m.* thanks.
mercredi, *m.* Wednesday.
mercure, *m.* mercury; (*cap.*) Mercury.
— *argental,* (*Min.*) argental mercury, amalgam.
— *doux à la vapeur,* (*Pharm.*) calomel obtained as an impalpable powder by bringing the vapor in contact with steam (sometimes called hydrosublimate of mercury).
— *fulminant,* fulminating mercury (mercuric fulminate).
— *précipité blanc,* (*Pharm.*) white precipitate, ammoniated mercury.
mercuréthyle, *m.* mercury ethyl, $Hg(C_2H_5)_2$.
mercureux, *a.* mercurous.
mercuriale, *f.* (*Bot.*) mercury (*Mercurialis*); price list, schedule, tariff; reprimand.
mercurialiser, *v.t.* mercurialize.
mercurialisme, *m.* (*Med.*) mercurialism.
mercuriaux, *m.pl.* (*Pharm.*) mercurials.
mercuriel, *a.* mercurial, mercury.
mercurifère, *a.* containing mercury.
mercurification, *f.* mercurification.
mercurique, *a.* mercuric.
mercuroso-mercurique, *a.* mercuroso-mercuric.
merde, *f.* excrement.
— *d'oie,* a greenish yellow color.
mère, *f.* mother; matrix, mold. — *a.* mother; principal; fine, pure.
eau —, mother liquor.
— *de vinaigre,* mother of vinegar.
— *goutte,* unpressed must (which runs from the grapes before pressing begins), first runnings.
— *perle,* mother of pearl.
méridien, *m. & a.* meridian.
méridienne, *f.* meridian, meridian line.
méridional, *a.* southern, south; meridional.
— *m.* southerner, one living in the south.
mérinos, *m.* merino.
merisier, *m.* (*Bot.*) wild cherry (*Prunus avium*).
méristème, *m.* (*Bot.*) meristem.
mérite, *m.* merit; person of merit.
mériter, *v.t.* merit, deserve; need, require.
— *v.i.* be worthy, deserve. — *v.r.* be merited.

méritoire, *a.* meritorious.
merluche, *f.* stockfish (esp. dried codfish); (*Zoöl.*) hake, merluce (*Merluccius*).
merlus, *m.* = merluche.
peau en —, see under *peau.*
merrain, *m.* stavewood, staves; clapboard.
merveille, *f.* wonder, marvel.
à —, wonderfully, marvelously.
merveilleusement, *adv.* wonderfully, marvelously.
merveilleux, *a.* wonderful, marvelous; excellent; strange, odd. — *m.* marvelous; marvelous part.
mes, *a. pl.* my.
més-. mis-; mes-.
mésaconique, *a.* mesaconic.
mésaventure, *f.* mishap, accident.
mescal, *m.* mescal.
mésestimer, *v.t.* underrate, undervalue.
mésitine, *f.* (*Min.*) mesitite, mesitine.
mésitylène, *m.* mesitylene.
mésitylénique, *a.* mesitylenic.
mésotartrique, *a.* mesotartaric.
mésothorium, *m.* mesothorium.
mésoxalique, *a.* mesoxalic.
mésozoïque, *a.* (*Geol.*) Mesozoic.
mesquin, *a.* mean, poor, narrow, stingy.
message, *m.* message.
messager, *m.* messenger; carrier.
messagerie, *f.* rapid transport, conveyance (by rail, boat or otherwise); goods conveyed, shipment, freight; (formerly) coach or coach office.
messieurs, *m.pl.* gentlemen, sirs, Messrs.
mesurable, *a.* measurable.
mesurage, *m.* measuring, measurement; surveying; weighing, considering; proportioning.
mesure, *f.* measure; measurement; position (to do something); proportion.
à —, in proportion, accordingly.
à — *de,* in proportion to, according to.
à — *que,* in proportion as, according as, as soon as, as.
— *à pied,* a measure (specif. a graduated cup or graduate) standing on a foot or base.
— *divisée,* graduate.
— *en ruban,* tape measure.
outre —, beyond measure, excessively.
sans —, without measure, immeasurably, unmeasured.
mesurer, *v.t.* measure; weigh, consider; proportion. — *v.r.* be measured; measure oneself.
mesureur, *m.* measurer. — *a.* measuring.
mésuser, *v.t.* misuse.
met, *pr. 3 sing.* of mettre. (See also mets.)
méta-. meta-, -m-.
méta-antimonique, *a.* metantimonic.
métabisulfite, *m.* metabisulfite, pyrosulfite.

métabolique, a. (Biol.) metabolic.

métabolisme, m. (Biol.) metabolism.

métacellulose, f. metacellulose (fungin).

métachromasie, f. metachromatism.

métachromatine, f. metachromatin.

métachromatique, a. metachromatic.

métadérivé, m. meta derivative.

métadiazine, f. metadiazine.

métahémipinique, a.m-hemipic, metahemipic.

métal, m. metal.

— à canon, gun metal.

— à collets, bearing metal.

— alcalin, alkali metal.

— alcalino-terreux, alkaline-earth metal.

— à miroirs, speculum metal.

— anglais, — argentin, Britannia metal, britannia.

— blanc, white metal.

— britannique, Britannia metal, britannia.

— brut, raw metal, coarse metal.

— console, an alloy of copper, zinc and tin.

— d'Aich, Aich metal (alloy of copper, zinc and iron).

— d'Alger, a white metal composed of tin 60, lead 34.6, antimony 5.4.

— de Bath, Bath metal (a variety of brass).

— de canon, gun metal.

— décapé, pickled metal, scoured metal.

— de cloche, bell metal.

— de Darcet, Darcet's metal (bismuth, lead and tin with sometimes mercury).

— de fonte, cast metal.

— delta, delta metal (essentially copper, zinc and iron).

— de Muntz, Muntz metal (copper 60–64, zinc 40–36).

— de prince, super-refined copper.

— de Rose, Rose's metal (bismuth 50, lead 25, tin 25).

— doux, soft metal; antifriction metal.

— du prince Robert, Prince Rupert's metal, Prince's metal.

— écumé, scum of metal, specif. kish.

— natif, native metal.

— ondulé, corrugated metal.

— précieux, precious metal.

— rebelle, refractory metal.

— Roma, a phosphor bronze containing manganese.

— sterro, sterro metal, delta metal.

— utile, useful metal (specif., Metal., the quantity of workable metal obtained from an ingot or charge).

— vierge, virgin metal.

métalbumine, f. metalbumin, pseudomucin.

métaldéhyde, f. metaldehyde.

métalléité, f. metallicity, metalleity.

métallescence, f. metallic luster.

métallescent, a. of or designating metallic luster.

métallifère, a. metalliferous.

métalliforme, a. metalliform.

métallin, a. metalline, metallic.

métallique, a. metallic.

métalliquement, adv. in metal, in specie.

métallisage, m., **métallisation,** f. metallization, metallizing.

métalliser, v.t. metallize.

métallochimie, f. the chemistry of metals.

métallochimique, a. relating to the chemistry of metals.

métallochromie, f. metallochromy.

métallogénie, f. the science of the origin of metals, the science of metalliferous deposits.

métallographe, m. metallographist.

métallographie, f. metallography.

métallographique, a. metallographic.

métalloïde, m. metalloid. — a. metalloid(al).

métalloïdique, a. metalloid, metalloidal.

métallo-organique, a. metalorganic, organometallic.

métallurgie, f. metallurgy.

métallurgique, a. metallurgical, metallurgic.

métallurgiquement, adv. metallurgically.

métallurgiste, m. metallurgist.

métamère, a. metameric. — m. metamer.

métamérie, f. metamerism, (rarely) metamery.

métamorphique, a. metamorphic.

métamorphisme, m. metamorphism.

métamorphose, f. metamorphosis.

métamorphoser, v.t. metamorphose.

métanile, a. metanil.

métaphosphate, m. metaphosphate.

métaphosphorique, a. metaphosphoric.

métaphtalique, a. metaphthalic (isophthalic).

métaphysique, f. metaphysics. — a. metaphysical.

métaphysiquement, adv. metaphysically.

métaphyte, n. (Bot.) metaphyte.

métargon, m. metargon (a supposed element since shown to be carbon monoxide).

métasérie, f. meta series.

métastannique, a. metastannic.

métatitanique, a. metatitanic.

métaux, pl. of métal.

métavanadique, a. metavanadic.

méteil, m. mixed wheat and rye, meslin.

météorifuge, a. (Pharm.) antiflatulent.

météorique, a. meteoric.

météorite, m. meteorite.

météoritique, a. meteoritic.

météorologie, f. meteorology.

météorologique, a. meteorologic(al).

méthane, m. methane.

méthanique, a. methane, pertaining to methane or the methane (paraffin) series.

méthanol, m. methanol, methyl alcohol.

méthanomètre, m. methanometer.

méthanure, *m.* any compound which may be regarded as formed by the replacement, in methane, of hydrogen by metal.

méthémoglobine, *f.* methemoglobin, methæmoglobin.

méthode, *f.* method; process; elementary treatise.

— *à la catalane,* (*Metal.*) Catalan process.

— *des moindres carrés,* (*Math.*) method of least squares.

— *du touchau,* (*Assaying*) touch-needle method, touchstone method.

— *marseillaise,* Marseilles method (esp. of making soap).

— *pondérale,* (*Analysis*) gravimetric method.

méthodique, *a.* methodical.

méthodiquement, *adv.* methodically.

méthylal, *m.* methylal.

méthylamine, *f.* methylamine.

méthylaniline, *f.* methylaniline.

méthylarsinique, *a.* methanearsonic, methylarsonic.

méthylate, *m.* methylate.

méthyle, *m.* methyl.

méthylène, *m.* (*Org. Chem.*) methylene; (*Com.*) wood alcohol, wood spirit, methyl alcohol, methanol.

— *de bois,* wood spirit, wood alcohol.

méthylène-protocatéchique, *a.* methylene-protocatechuic (piperonylic).

méthylénique, *a.* methylene, of methylene.

méthylfurfurane, *m.* methylfuran.

méthylfurfurol, *m.* methylfuraldehyde.

méthyliodhydrique, *a.* In the phrase *éther méthyliodhydrique,* methyl iodide.

méthylique, *a.* methyl, of methyl, methylic.

méthylisation, *f.* methylation.

méthyliser, *v.t.* methylate, (rarely) methylize.

méthylorange, *m.* methyl orange.

méthylorcine, *f.* methylorcinol.

méthylpentaméthylène, *m.* methylpentamethylene (methylcyclopentane).

méthylphénylcétone, *f.* methyl phenyl ketone (acetophenone).

méthylquinoléine, *f.* methylquinoline.

méthylsilicique, *a.* In the phrase *éther méthylsilicique,* methyl silicate.

méthylthiophène, *m.* methylthiophene.

méthylure, *m.* methide.

méthylxanthine, *f.* methylxanthine.

méthysticine, *f.* methysticin.

métier, *m.* trade; profession; occupation; loom, frame.

— *à bras,* hand loom, hand frame.

— *à filer,* spinning frame.

— *à tisser,* loom.

métière, *f.* (*Salt*) an evaporating basin.

métis, *a.* (*Metal.*) red-short; hybrid, half-breed, mongrel.

métisser, *v.t.* cross-breed, cross.

métol, *m.* (*Photog.*) metol.

métoquinone, *f.* (*Photog.*) a combination of metol and hydroquinone, used as a developer.

métrage, *m.* measurement or length in meters.

mètre, *m.* meter, metre (39.37 inches).

— *carré,* square meter (1.196 sq. yd.).

— *courant,* running meter, linear meter.

— *cube,* cubic meter (1.308 cu. yd.).

— *anglais,* yard.

métré, *m.* measurement in meters.

mètre-kilogramme, *m.* kilogrammeter.

métrer, *v.t.* measure in meters.

métrique, *a.* metric; metrical.

métrologie, *f.* metrology.

métrologique, *a.* metrological.

métropole, *f.* mother country; metropolis.

métropolitain, *a.* metropolitan; of the mother country, home; archiepiscopal.

mets, *m.* food (prepared for the table), dish.

— *pr. 2 sing.* of mettre.

mettant, *p.a.* putting, etc. (see mettre).

mettre, *v.t.* put; place, set; put on; lay on; put down, write; bring, reduce; use, employ; spend; cause to consist (in). — *v.r.* put, place, or set oneself; go, get, stand, sit, etc.; begin; dress; be put, be set; be worn; break out.

— *chauffer de l'eau,* put water to heat, heat water (and so for similar phrases).

— *en liberté,* set free, liberate.

— *hors,* put out; expend; put (a blast furnace) out of blast.

(For other phrases commencing with *mettre* see the next important word in the phrase.)

méture, *f.* corn bread (from maize flour).

meuble, *m.* piece of furniture; set of furniture; (*pl.*) furniture; article; (*Law*) chattel.

— *a.* movable; (of property) personal; (of soil) easily worked.

meubler, *v.t.* furnish (as a house), stock, store.

meule, *f.* stone or wheel for grinding (as a millstone, grindstone, emery wheel); stack, rick (as of hay); heap, pile, (for bricks) clamp, (for charcoal) meiler; a round, flat cheese (esp. Gruyère); (*Gunpowder*) runner, roller.

— *d'émeri,* emery wheel.

meulier, *a.* millstone, mill-.

meulière, *f.* millstone, buhrstone, burrstone; millstone quarry.

meunerie, *f.* milling, milling industry.

meunier, *m.* miller.

meurent, *pr. 3 pl. indic. & subj.* of mourir.

meurt, *pr. 3 sing.* of mourir.

meurtrier, *a.* murderous, deadly. — *m.* murderer.

meurtrir, *v.t.* bruise.

meurtrissure, *f.* bruise, (*Med.*) contusion.

meut, *pr. 3 sing.* of mouvoir.

meuvent, *pr. 3 pl. indic. & subj.* of *mouvoir.*
mexicain, *a.* Mexican. — *m.* (*cap.*) Mexican.
Mexique, *m.* Mexico.
mézéréon, *m.* (*Bot.*) mezereon (*Daphne mezereum*); (*Pharm.*) mezereum.
mi-, half, semi-, mid-, middle; in halves.
mica, *m.* (*Min.*) mica.
micacé, *a.* micaceous.
micaschiste, *m.* (*Petrog.*) mica schist.
micaschisteux, *a.* of the nature of a mica schist.
micellaire, *a.* micellar.
micelle, *f.* micella.
mi-chaud, *a.* (*Soap*) made by a process intermediate between the cold process and the usual boiling process (the boiling lasting a short time only).
mi-chemin. In the phrase *à mi-chemin,* halfway.
microbe, *m.* microbe, microörganism.
microbie, *f.* microbiology.
microbien, *a.* microbic, microbial.
microbiologie, *f.* microbiology.
microbiologique, *a.* microbiological.
microbique, *a.* microbic.
microchimie, *f.* microchemistry.
microchimique, *a.* microchemical.
micrococcus, microcoque, *m.* (*Bact.*) micrococcus.
microcosmique, *a.* microcosmic.
microcristallin, *a.* microcrystalline.
micrographe, *m.* micrographer.
micrographie, *f.* micrography.
micrographique, *a.* micrographic.
microlithe, *m.* (*Petrog.*) microlite.
microlithique, *a.* (*Petrog.*) microlitic.
micromètre, *m.* micrometer.
micrométrie, *f.* micrometry.
micrométrique, *a.* micrometric(al).
micrométriquement, *adv.* micrometrically.
micromillimètre, *m.* micromillimeter.
micron, *m.* micron (one thousandth millimeter).
micro-organisme, *m.* microörganism.
microphotographie, *f.* microphotography; microphotograph, photomicrograph.
microrganisme, *m.* microörganism.
microscope, *m.* microscope.
— *composé,* compound microscope.
microscopie, *f.* microscopy.
microscopique, *a.* microscopic, microscopical.
microscopiste, *m.* microscopist.
microspectroscope, *m.* microspectroscope.
microtome, *m.* microtome.
midi, *m.* noon; south (specif., southern France).
mi-doux, *a.* half-soft.
mi-dur, *a.* half-hard, semi-hard.
mie, *f.* crumb (the soft part of bread).

miel, *m.* honey.
— *boraté,* (*Pharm.*) borax honey.
— *despumé,* clarified honey.
— *rosat,* (*Pharm.*) honey of rose.
mielat, miellat, *m.* honeydew.
miellé, *a.* honey, like honey (specif., honey-colored); honeyed.
miellée, *f.* honeydew.
mielleux, *a.* honey, like honey, honeyed; luscious.
miellure, *f.* honeydew.
mien, mienne, *pron. & a.* my, mine, of mine.
miette, *f.* crumb; morsel; small piece.
mieux, *adv., a. & m.* better; (with *le*) best.
au —, as well as possible.
mi-fin, *a.* designating a grade of Para rubber inferior to that called "fine."
mi-fixe, *a.* semiportable.
mignon, *a.* tiny, small, pretty, delicate.
mignonette, *f.* coarsely powdered pepper; any of several plants, as the garden pink or the mignonette; fine pebbles; minion type; mignonette lace.
migraine, *f.* sick headache, megrim, migraine.
migrateur, *a.* migrating, migratory.
migration, *f.* migration.
migratoire, *a.* migratory, migrational.
mijoter, *v.i.* simmer, boil gently. — *v.t.* cause to simmer.
mil, *a.* one thousand. — *m.* millet.
mi-laine, *m.* half-wool fabric, half-wool.
milanais, *a.* Milanese.
mildiou, mildew, *m.* mildew.
mildiousé, mildewsé, *a.* mildewed.
milieu, *m.* medium; specif., atmosphere; middle; midst; middle course; (fig.) heart, center.
— *réducteur,* reducing medium, reducing atmosphere.
militaire, *a.* military. — *m.* soldier; military.
militairement, *adv.* in a military way.
militer, *v.i.* militate.
mille, *a.* thousand, a thousand, one thousand.
— *m.* thousand, a thousand, one thousand; mile.
mille-feuille, *f.* (*Bot.*) milfoil, yarrow.
mille-pertuis, *m.* St.-John's-wort (*Hypericum*).
millet, *m.* millet.
milliard, *m.* billion, thousand millions (specif., a billion francs).
millième, *m. & a.* thousandth.
millier, *m.* thousand; 500 kilograms, half a metric ton (or, a thousand pounds).
milligramme, *m.* milligram, milligramme.
millilitre, *m.* milliliter, millilitre.
millimètre, *m.* millimeter, millimetre.
million, *m.* million.
millionième, *a. & m.* millionth.

millistère, *m.* millistere, liter, cubic decimeter.

mince, *a.* thin; slender; slight.

mincer, *v.t.* mince, cut fine.

minceur, *f.* thinness; slenderness; slightness.

mine, *f.* mine; ore; minium; graphite; look, appearance, aspect, mien.

— *anglaise,* red lead, minium.

— *brute,* raw ore.

— *d'acier,* siderite, spathic iron.

— *d'aimant,* magnetite.

— *de cuivre,* copper mine; copper ore.

— *de fer,* iron mine; iron ore.

— *de fer blanche,* siderite, spathic iron.

— *de fer spéculaire,* specular iron ore.

— *de marais,* bog iron ore.

— *de platine,* — *du platine,* crude platinum.

— *de plomb,* lead mine; lead ore; graphite, black lead; hence, pencil.

— *d'étain,* tin mine; tin ore.

— *douce,* sphalerite, zinc blend.

— *grasse,* ore containing little or no gang.

— *orange,* orange lead, orange minium (a pigment).

mines grosses, ore in pieces, lump ore.

mine-orange, *f.* orange lead, orange minium.

miner, *v.t.* mine; undermine; (fig.) destroy.

mineral, *m.* ore.

— *brut* raw ore.

— *d'alun,* alum stone, alunite.

— *d'argent,* silver ore.

— *de cuivre,* copper ore.

— *de fer,* iron ore.

— *d'étain,* tin ore.

— *de vitriol,* pyrite, iron pyrites.

— *en roche,* ore in compact masses.

— *pauvre,* poor ore, low-grade ore.

minerais menus, small ore, smalls.

minéral, *a.* mineral; specif., inorganic. — *m.* mineral.

— *d'Hermès,* mercury.

minéralisable, *a.* mineralizable.

minéralisateur, *a.* mineralizing — *m.* mineralizer.

minéralisation, *f.* mineralization.

minéraliser, *v.t.* mineralize.

minéralité, *f.* mineral or inorganic state.

minéralogie, *f.* mineralogy.

minéralogique, *a.* mineralogical.

minéralogiquement, *adv.* mineralogically.

minéralogiste, *m.* mineralogist.

minéralurgie, *f.* the application of mineralogical knowledge to industrial use.

minéralurgique, *a.* relating to *minéralurgie.*

minéraux, *pl. masc.* of minéral.

minerie, *f.* rock-salt mine.

minette, *f.* (*Petrog.*) minette (either mica trap or oölitic iron ore).

mineur, *m.* miner; minor. — *a.* minor; mining.

minier, *a.* mining, relating to mines.

minière, *f.* ore-bearing material; surface mine, diggings.

minimant, *a.* minimum, at a minimum.

minime, *a.* very small; trifling, unimportant.

minimer, minimiser, *v.t.* minimize.

minimum, *m. & a.* minimum.

au —, at the minimum; in the lowest (or lower) state of valence.

ministère, *m.* ministry; minister's office.

ministre, *m.* minister.

minium, *m.* minium.

— *de fer,* red ocher.

minofor, *m.* an alloy of tin 68.6, antimony 17, zinc 10 and copper 4.4.

minoratif, *a. & m.* (*Med.*) laxative.

minorité, *f.* minority.

Minorque, *f.* Minorca.

minoterie, *f.* flour mill; milling business.

minuit, *m.* midnight.

minuscule, *f.* small letter, lower-case letter. — *a.* small, very small.

minute, *f.* minute; first draft; original.

minutieusement, *adv.* minutely, in detail.

minutieux, *a.* minute, particular.

miocène, *a.* (*Geol.*) Miocene.

mi-parti, *a.* equally divided, half and half, half . . . and half.

mi-partir, mipartir, *v.t.* divide into halves.

mi-partition, *f.* division into halves.

miracle, *m.* miracle.

à —, admirably.

miraculeusement, *adv.* miraculously.

miraculeux, *a.* miraculous; astonishing.

mi-raffiné, *a.* half-refined, partially refined.

mirbane, *f.* mirbane.

essence de —, essence of mirbane (nitrobenzene).

mire, *f.* sight (specif., front sight, fore sight); leveling rod; directing mark.

mirent, *p.def. 3 pl.* of mettre.

mire-œuf, mire-œufs, *m.* candling apparatus.

mirer, *v.t.* look at; specif., examine by transmitted light; candle (eggs).

miroir, *m.* mirror, (looking) glass, speculum.

— *ardent,* burning mirror.

— *plan,* plane mirror.

— *réflecteur,* reflector.

miroitant, *p.a.* glittering, shining, specular.

miroitement, *m.* glitter, glittering.

miroiter, *v.i.* glitter, shine. — *v.t.* polish.

miroiterie, *f.* mirror manufacture; mirror factory; mirror trade.

miroitier, *m.* one who makes or sells mirrors.

mis, *p.p. & p.def. 1 & 2 sing.* of mettre.

miscellannées, *m.pl.* miscellanies.

miscibilité, *f.* miscibility.

miscible, *a.* miscible.

mise, *f.* putting, placing, etc. (see mettre); (*Soap*) frame; (*Metal.*) layer; share (of capital); stake; bid; dress, dressing; (of money) circulation.

de —, admissible; suitable; fashionable; presentable.

— *au mille,* (*Metal.*) the gross weight needed to make 1000 kg. of useful metal.

— *de feu,* setting on fire, lighting, ignition, firing.

— *de feu chimique,* chemical ignition.

— *de fonds,* expenditure, outlay.

— *en feu,* = mise de feu.

— *en liberté,* setting free, liberation.

— *hors,* outlay; putting out; putting (a blast furnace) out of blast.

For the meaning of other phrases see the corresponding phrases of *mettre*, e.g. mettre au point (under point), mettre en suif (under suif), etc.

misérable, *a.* miserable, wretched.

misérablement, *adv.* miserably, wretchedly.

misère, *f.* misery; annoyance; trifle.

miséricorde, *f.* mercy.

mi-soie, *m.* half-silk fabric, half-silk.

mi-solide, *a.* semi-solid.

mispickel, *m.* (*Min.*) mispickel, arsenopyrite.

missent, *imp. subj. 3 pl.* of mettre.

mission, *f.* mission.

mistelle, *f.* (*Wine*) a must to which alcohol has been added to check fermentation.

mit, *p.def. 3 sing.* of mettre.

mitaine, *f.* mitten; mitt.

mithridate, *m.* (*Old Pharm.*) mithridate.

mitiger, *v.t.* mitigate. — *v.r.* be mitigated.

mi-toile, *m.* half-linen fabric, half-linen.

mitonner, *v.i.* simmer.

mitoyen, *a.* middle, intermediate, partition.

mitraille, *f.* scrap metal, scrap; specif., scrap iron; (*Mil.*) any material (as shrapnel) for scattering effect.

— *cuivre,* scrap copper, old copper.

— *d'étain,* scrap tin, old tin.

— *jaune,* scrap brass, old brass.

— *rouge,* scrap copper, old copper.

mitrailleuse, *f.* machine gun, mitrailleuse.

mitre, *f.* cowl, hood (for a chimney); thick paving block; cone valve; miter.

mi-vitesse, *f.* half speed.

mixte, *a.* mixed.

mixtion, *f.,* **mixtionnage,** *m.* mixture; mixtion (a mordant for gold leaf); (*Ceram.*) a copal varnish used in transferring designs; (*Engraving*) a mixture of tallow and oil for covering the etched parts.

mixture, *f.* mixture (esp. in *Pharm.*); = mixtion.

— *de gomme ammoniaque,* (*Pharm.*) ammoniacum mixture.

mixture *de réglisse,* (*Pharm.*) compound mixture of glycyrrhiza, brown mixture.

— *de résine de gaïac,* (*Pharm.*) guaiacum mixture.

— *ferrugineuse de Griffith,* (*Pharm.*) compound iron mixture, Griffith's mixture.

mobile, *a.* mobile; movable, moving; changeable. — *m.* moving body; motive power; mover; motive.

mobilier, *a.* (of property) personal, movable. — *m.* furniture, household goods.

mobiliser, *v.t.* mobilize.

mobilité, *f.* mobility.

modalité, *f.* circumstance, details; modality.

mode, *m.* mode; method. — *f.* mode; manner; fashion; (*pl.*) millinery.

à la —, in the fashion, fashionable.

— *d'emploi,* method of use, directions for use.

— *opératoire,* method of operation.

modelage, *m.* modeling.

modèle, *m.* model; pattern. — *a.* model.

— *de fonte,* foundry pattern.

— *de poche,* pocket model.

modeler, *v.t. & i.* model.

modèle-type, *m.* (*Ceram.*) a plaster cast from which the working molds are made.

modeleur, *m.* modeler.

modérateur, *m.* governor; regulator; moderator; damper. — *a.* governing, regulating, moderating.

modération, *f.* moderation; diminution, reduction.

modéré, *a.* moderate; moderated; reduced.

modérément, *adv.* moderately.

modérer, *v.t.* moderate; slacken; reduce.

moderne, *a.* modern.

moderniser, *v.t.* modernize.

modeste, *a.* modest.

modestement, *adv.* modestly.

modicité, *f.* smallness.

modifiable, *a.* modifiable.

modifiant, *p.a.* modifying.

modificateur, *a.* modifying. — *m.* modifier; (*Mach.*) engaging and disengaging gear.

modification, *f.* modification.

modifier, *v.t.* modify. — *v.r.* be modified.

modique, *a.* small, unimportant.

modiquement, *adv.* in a small way, unimportantly.

module, *m.* modulus; module.

moelle, *f.* marrow; pith; (*Anat.*) medulla; (of a hair) medulla, pith.

— *allongée,* (*Anat.*) medulla oblongata.

— *de sureau,* elder pith.

— *d'os,* — *des os,* bone marrow, marrow.

— *épinière,* spinal cord.

moelleux, *a.* mellow; soft, sweet or gentle; marrowy; pithy. — *m.* mellowness; softness; grace.

moellon, *m.* (*Leather*) degras, dégras, moellon; sandstone for grinding mirrors; ashlar, rough or partly dressed stone.

— *aggloméré*, artificial stone, cement block.

moeurs, *f.pl.* manners, customs; morals; habits.

mofétisé, *a.* mephitic, noxious.

mofette, *f.* noxious gas, specif. choke damp.

moi, *pron.* me, to me, I. — *m.* self; egoism.

moignon, *m.* stump.

moi-même, *pron.* myself.

moindre, *a.* less, lesser, smaller, inferior; least, slightest, smallest.

moindres carrés, (*Math.*) least squares.

moindrement, *adv.* least, the least.

moine, *m.* monk; (*Metal.*) blister.

moineau, *m.* sparrow.

moins, *adv.* less, not so, not as; least. — *prep.* less, minus; except. — *m.* least; minus sign, dash.

à — que, (with *ne* and the subj.) unless, if not; (with *de* and infinitive) unless, without.

au —, at least, at the least, above all, however.

de —, less; missing.

de — en —, less and less.

du —, at least, at any rate, still, however, nevertheless.

en —, minus, less, as a deduction.

en — de, in less than.

moins-value, *f.* loss in value, diminution in value; deduction, dockage.

moirage, *m.* watering (of fabrics), moiréeing.

moire, *f.* water, watering, moiré (on fabrics, etc.); watered fabric, moire; honeysuckle.

moiré, *p.a.* watered, moiré, chatoyant. — *m.* watering, water, moiré; watered or moiré material, moiré.

— *métallique*, a watered or moiré appearance on tin; also, the tin or tinplate itself.

moirer, *v.t.* water, moiré (fabrics, etc.).

moirure, *f.* watering, water, moiré.

mois, *m.* month; month's pay or allowance; (*Physiol.*, *pl.*) menses.

moise, *f.* tie, brace, couple, connecting piece; an implement for handling plate glass.

moisi, *p.a.* moldy, molded. — *m.* moldiness, mold.

moisir, *v.t.* make moldy, mold. — *v.i.* become moldy, mold.

moisissure, *f.* mold; moldiness, mustiness.

moisson, *f.* harvest.

moissonner, *v.t.* harvest, reap, mow; gather.

moite, *a.* moist, damp, humid.

moiteur, *f.* moistness, moisture, dampness, humidity.

moitié, *f. & adv.* half.

à —, half, at half.

moitir, *v.t.* moisten, dampen, make damp.

moka, *m.* Mocha coffee, Mocha.

mol, *a.* Used instead of *mou* before a vowel.

molaire, *a. & f.* molar.

molarite, *f.* a kind of millstone or buhrstone.

molasse, *f.* (*Geol.*) molasse.

moléculaire, *a.* molecular.

moléculairement, *adv.* molecularly.

molécularisation, *f.* reduction or conversion into molecules.

molécularisé, *a.* constituted of molecules.

moléculariser, *v.t.* arrange in molecules.

molécule, *f.* molecule.

molène, *f.* (*Bot.*) mullein, mullen (*Verbascum*).

— *médicinale*, great mullein, common mullein (*V. thapsus*).

molester, *v.t.* molest.

molet, *m.* pincers, nippers; (*pl.*) boggy ground.

moletage, *m.* milling; polishing.

moleter, *v.t.* mill; polish.

moletoir, *m.* a glass-polishing instrument.

molettage, *m.* milling; polishing.

molette, *f.* any of various small wheels for milling, tracing, cutting, etc.; muller (for pigments); rubber, mealer (for powder); pulley; rowel (of a spur); (*Bot.*) shepherd's purse.

moletter, *v.t.* mill; polish.

moliant, *a.* pliable, soft.

mollasse, *a.* flabby, soft, lacking consistence. — *f.* (*Geol.*) molasse.

molle, *a. fem.* of mou.

mollement, *adv.* softly, soft; gently; feebly; lazily; effeminately; gracefully.

mollesse, *f.* softness; (of fruit) mellowness; (of climate) mildness; feebleness, sluggishness; effeminacy; grace, ease.

mollet, *a.* soft; light; tender. — *m.* calf (of the leg); pincers, nippers.

molleter, *v.t.* mill; polish.

molleterie, *f.* in the phrase *cuir de molleterie*, soft leather (esp. chamois leather).

mollir, *v.i.* soften, become soft; (of fruit) become mellow; slacken; weaken; yield.

mollusque, *m.* (*Zoöl.*) mollusk, mollusc.

Moluques, *f.pl.* Moluccas.

molybdate, *m.* molybdate.

molybdaté, *a.* molybdate of.

molybdène, *m.* molybdenum.

molybdénite, *f.* (*Min.*) molybdenite.

molybdénocre, *m.* molybdic ocher, molybdite.

molybdeux, *a.* molybdous.

molybdine, *f.* (*Min.*) molybdite.

molybdique, *a.* molybdic.

molybdoïde, *f.* a kind of graphite.

molybdoménite, *f.* (*Min.*) molybdomenite.

molybdurane, *m.* (*Min.*) uranium molybdate.

moment, *m.* moment; momentum.

dans ce —, just now, at present.

moment, *dans le* —, in a moment, instantly, immediately.

dans le — *où* (or *que*), at the moment when, just as, as soon as.

du — *que,* as soon as, when once, since.

en ce —, at present, just now.

par moments, at intervals.

momentané, *a.* momentary.

momentanément, *adv.* momentarily, a moment.

momie, *f.* mummy.

momificateur, *a.* mummifying.

momification, *f.* mummification.

momifier, *v.t.* mummify.

mon, *a.* my.

monade, *f.* monad.

monamide, *f.* monamide.

monamine, *f.* monamine.

monarchie, *f.* monarchy.

monarde, *f.* (*Bot.*) Monarda.

monarque, *m.* monarch.

monazite, *f.* (*Min.*) monazite.

monazoïque, *a.* monoazo. — *m.* monoazo compound.

monceau, *m.* heap, mound, stack, pile.

mondation, *f.* cleaning, cleansing, peeling.

monde, *m.* world; people, persons; company; men, hands, crew; servants.

monder, *v.t.* clean, cleanse, peel, hull.

mondial, *a.* world, of the world.

mondificatif, *a.* (*Med.*) mundificant, cleansing.

monétaire, *a.* monetary, money.

monétiser, *v.t.* monetize.

mongolien, mongolique, *a.* Mongolian.

moniteur, *m.* monitor.

monnaie, *f.* money; change; mint.

monnayage, *m.* coinage, coining, mintage, minting.

monnayer, *v.t.* coin, mint.

monoalcoolique, *a.* monoalcoholic, containing one alcohol group.

monoaminé, *a.* monoamino.

monoargentique, *a.* monosilver.

monoatomique, *a.* monatomic, monoatomic.

monobasique, *a.* monobasic.

monobromé, *a.* monobromo, monobromo-.

monobromure, *m.* monobromide.

— *de naphtaline,* monobromonaphthalene, bromonaphthalene.

monocalcique, *a.* monocalcium.

monocarbonique, *a.* (*Org. Chem.*) monocarboxylic.

monochloré, *a.* monochloro, monochloro-.

monochlorhydrique, *a.* In the phrase *éther monochlorhydrique,* ethyl chloride, chloroethane.

monochromatique, *a.* monochromatic.

monochrome, *a.* monochrome, monochromic. — *m.* monochrome.

monoclinique, *a.* (*Cryst.*) monoclinic.

monocotylédone, *a.* (*Bot.*) monocotyledonous. — *f.* monocotyledon.

monoferreux, *a.* monoferrous.

monoformine, *f.* monoformin.

monographe, *m.* monographist, monograph writer.

monographie, *f.* monograph.

monohalogéné, *a.* monohalogenated, monohalo.

monohydrate, *m.* monohydrate.

monohydraté, *a.* monohydrated.

monoléine, *f.* monoölein, monolein.

monomagnésique, *a.* monomagnesium.

monométallique, *a.* monometallic.

monométhylique, *a.* monomethyl.

mononitré, *a.* mononitrated, mononitro, mononitro-. — *m.* mononitro compound.

mononitrine, *f.* mononitrin (glycerol mononitrate).

monophasé, *a.* (*Elec.*) single-phase, monophase.

monophénolique, *a.* monophenolic, containing one phenol group.

monopole, *m.* monopoly.

monopoler, *v.i.* have a monopoly.

monopolisation, *f.* monopolizing.

monopoliser, *v.t.* monopolize.

monopotassé, *a.* monopotassium.

monosaccharide, *n.* monosaccharide.

monoséléniure, *m.* monoselenide.

monosodé, *a.* monosodium.

monostéarine, *f.* monostearin.

monosubstitué, *a.* monosubstituted.

monosulfure, *m.* monosulfide.

monotellurure, *m.* monotelluride.

monotérébénique, *a.* monoterpene.

monotone, *a.* monotonous.

monotonement, *adv.* monotonously.

monotonie, *f.* monotony.

monovalent, *a.* univalent, monovalent.

monsieur, *m.* sir, master, Mr., Esq., gentleman.

monstrueux, *a.* monstrous.

mont, *m.* mount, mountain.

montacide, *m.* = monte-acide.

montage, *m.* rising, rise, ascent, increase, lifting, hoisting, mounting, erecting, etc. (see monter); (*Elec.*) connection, wiring.

montagne, *f.* mountain.

montagnes Rocheuses, Rocky Mountains.

montagneux, *a.* mountainous.

montant, *p.a.* rising, ascending, etc. (see monter); steep; high. — *m.* upright, standard, post; stem (of plants); amount, total; pungent taste; strong odor; rise, rising.

mont-dore, *m.* a kind of Auvergne cheese.

monté, *p.a.* lifted, raised, mounted, erected, set, etc. (see monter); deep-colored, strongly colored; excited.

monte-acide, *m.* acid pump, monte-acide (specif. an acid egg).

monte-charge, *m.* freight elevator, lift, hoist, hoisting machine.

montée, *f.* ascent; rise, rising; gradient; height; stairs; step, stair.

monte-jus, *m.* monte-jus, an apparatus for pumping liquids, esp. in the sugar industry (where it is also called juice pump).

monter, *v.i.* rise; ascend, mount, go up, come up; increase; embark. — *v.t.* ascend; lift, raise, hoist, take up; mount; set; set up, erect; put together, assemble; fit up; (*Elec.*) connect up; establish; get up; raise, increase; deepen, strengthen (colors); wind (timepieces). — *v.r.* rise, mount; amount; supply oneself, stock up; ride; be ascended, be raised, etc.; become excited.

monteur, *m.* mounter, erecter, etc. (see monter); fitter (of pipes, engines, etc.).

monticule, *m.* mound, hillock.

montmartrite, *m.* (*Min.*) a kind of gypsum containing some calcium carbonate.

montrable, *a.* capable of being shown.

montre, *f.* watch; sample (of goods); display, show; show case, show window, show place; (*Ceram.*) trial piece or other temperature indicator (esp. a fusible cone).

— *à repos,* stop watch.

— *de Séger,* (*Ceram.*) Seger cone.

— *d'or,* gold watch; (*Ceram.*) a gilded piece which by its color indicates the temperature of the kiln.

— *marine,* chronometer.

montrer, *v.t.* show, exhibit, display; teach. — *v.r.* show oneself, appear, be seen.

monture, *f.* mounting, setting, fitting, support, frame, socket, etc.

monument, *m.* monument; building, structure.

morailles, *f.pl.* pincers, nippers, forceps.

moraillon, *m.* hasp, clasp.

moraine, *f.* skin wool; (*Geol.*) moraine. — *a.* (of wool) skin.

moral, *a.* moral. — *m.* morale; moral faculties.

morale, *f.* morals, ethics; moral; reprimand.

moralement, *adv.* morally.

morceau, *m.* piece; bit, morsel, fragment, part.

morceler, *v.t.* divide, subdivide, parcel out.

morcellement, *m.* division, subdivision.

mordache, *f.* clamp; jaw; tongs, pincers.

mordacité, *f.* corrosiveness; (fig.) causticity.

mordançage, *m.* mordanting.

mordancer, *v.t.* mordant.

mordanceur, *m.* mordanter (also called: (*Dyeing*) mordant maker, mordant-liquor man; (*Calico*) padding machine operator).

mordant, *a.* mordant; corrosive; keen, caustic. — *m.* mordant (as for dyeing or for gilding); cutting power (of a file); pincers; holder, clamp; keenness, causticity.

mordicant, *a.* acrid, biting, corrosive.

mordoré, *a. & m.* reddish brown with golden reflection, mordoré.

mordorure, *f.* = mordoré, *m.*

mordre, *v.t.* bite; corrode, eat away; seize, catch, nip; cut, incise; penetrate; effect; attack. — *v.i.* bite; corrode; have effect, take effect, act, work; take hold; engage; fit; (of a flame) play; encroach; carp (at), reflect (on).

mords, *m.* jaw (as of a clamp); (*Salt*) compartment of an evaporating basin.

mordu, *p.p.* of mordre.

moré, *m.* mulberry wine.

morelle, *f.* (*Bot.*) nightshade (*Solanum*).

— *grimpante,* bittersweet (*Solanum dulcamara*).

— *noire,* black nightshade, common nightshade (*Solanum nigrum*).

moresque, *a.* Moorish.

morfil, *m.* wire edge (on tools).

morin, *m.* morin.

morindine, *f.* morindin.

morindone, *f.* morindone.

morintannique, *a.* morintannic.

acide —, morintannic acid (maclurin).

morne, *a.* sad, dull.

morphine, *f.* morphine.

morphiné, *a.* (*Pharm.*) morphinated, morphiated.

morphique, *a.* morphine, of morphine.

morpholine, *f.* morpholine.

morphologie, *f.* morphology.

morphologique, *a.* morphologic(al).

morruol, morrhuol, *m.* (*Pharm.*) morrhuol.

mors, *m.* jaw (of a vise, pincers, etc.); (*Glass*) end of the blowpipe to which the glass is attached; bit (for a horse).

morsure, *f.* biting, etc. (see mordre); bite.

mort, *a.* dead; dull, faded, lifeless, still; (of water) stagnant; (of capital) unemployed. — *p.p.* of mourir. — *m.* dead person, dead body. — *f.* death.

à —, to death; mortally; exceedingly.

— *au chien,* (*Pharm.*) colchicum.

— *aux mouches,* fly poison, specif. arsenic.

— *aux poules,* henbane (*Hyoscyamus niger*).

— *aux rats,* ratsbane.

mortaise, *f.* mortise, hole, recess.

mortaiser, *v.t.* mortise.

mortalité, *f.* mortality.

mort-chien, *m.* (*Pharm.*) colchicum.

morte, *fem.* of mort.

mortel, *a.* mortal; fatal.

mortellement, *adv.* mortally.

morte-saison, *f.* full season, dead season.

mort-gage, *m.* mortgage.

mortier, *m.* mortar (in various senses).

— *à chaux et à sable,* sand-lime mortar, ordinary mortar.

— *à prise lente,* slow-setting mortar.

— *clair,* thin mortar.

— *d'Abich,* Abich mortar (a kind of diamond mortar).

— *de chaux,* lime mortar.

— *de ciment,* cement mortar.

— *de tranchée,* (*Mil.*) trench mortar.

— *dur,* stiff mortar.

— *en acier,* steel mortar.

— *en fonte,* cast iron mortar.

— *en verre,* glass mortar.

— *liquide,* grouting, grout, fluid mortar.

mortifier, *v.t.* mortify; make (meat) tender.

mortine, *f.* leaves of the myrtle and other plants, used in tanning.

mort-plain, *m.* (*Leather*) old lime liquor.

morts-murs, *m.pl.* walls of a smelting furnace.

mort-terrain, *m.* ground containing no ore.

morue, *f.* cod, codfish.

— *salée,* — *verte,* salt cod.

morve, *f.* glanders; nasal mucus; rot (of certain plants).

mosaïque, *f.* mosaic.

Moscou, *f.* Moscow.

moscouade, moscovade, *f.* muscovado, raw sugar.

mot, *m.* word; saying; motto; countersign.

— *à* —, word for word, verbatim, literally.

moteur, *m.* motor; engine; mover; author. — *a.* motor, motive, moving, driving, actuating.

— *à air carburé,* internal-combustion engine.

— *à air chaud,* hot-air engine, hot-air motor.

— *à deux temps,* two-cycle engine.

— *à essence,* gasoline engine, petrol engine.

— *à explosion,* explosion engine, explosion motor.

— *à gaz,* gas engine.

— *à pétrole,* (usually) kerosene engine, oil engine. (See pétrole.)

— *à vapeur,* steam engine.

— *électrique,* electric motor.

— *thermique,* heat engine.

motif, *m.* motive, reason. — *a.* determining.

motiver, *v.t.* be the motive of; justify; state the motive of or reason for.

motrice, *a. fem.* of moteur.

motricité, *f.* motility, motivity.

motte, *f.* lump, clod, mass (as of earth, sod, peat or the like); pat (of butter); ball (around roots); hillock.

— *à brûler,* a small flat or round mass of tan, turf or other combustible.

motton, *m.* lump (as of flour in water).

mou, *a.* soft; (of fruit) mellow; (of weather) sultry, close; slack; feeble; lax, sluggish, lazy, spiritless; effeminate. — *m.* lights (of animals); slack (of a rope).

mouche, *f.* fly; spot, dark spot; bull's-eye; (*Mech.*) point (of a bit).

moucher, *v.t.* square or trim the end of; snuff (a candle).

moucheron, *m.* gnat, small fly.

moucheter, *v.t.* spot; speckle; dot.

mouchette, *f.* residue from sifted plaster.

moucheture, *f.* spot, speckle, dot; slight scarification.

mouchoir, *m.* handkerchief.

mouchure, *f.* nasal mucus; snuff (of a candle).

moud, *pr. 3 sing.* of moudre.

moudre, *v.t.* grind, mill.

moufette, *f.* = mofette.

moufle, *m.* muffle. — *f.* mitten; tackle, system of pulleys; anchor, tie iron.

moufler, *v.t.* put in a muffle.

mouillade, *f.* wetting, moistening.

mouillage, *m.* wetting, etc. (see mouiller); anchorage; depth.

mouillé, *p.a.* wetted, etc. (see mouiller); wet; watery.

mouillement, *m.* wetting, etc. (see mouiller).

mouiller, *v.t.* wet; steep; soak; water; moisten; cast (anchor). — *v.i.* anchor, moor.

mouilleur, *m.* wetter, etc. (see mouiller).

mouilleux, *a.* wet (as land).

mouilloir, *m.* moistening cup or vessel.

mouillure, *f.* wetting, etc. (see mouiller); moisture.

moulage, *m.* grinding, milling; molding; casting; modeling; mold; (*Mach.*) millwork.

— *à la presse,* (*Ceram.*) pressing.

— *à sec,* dry shaping, dry molding, dry pressing.

— *au tour,* (*Ceram.*) jiggering.

— *creux,* hollow casting.

— *en argile,* loam molding.

— *en châssis,* flask molding.

— *en sable,* sand casting.

moulait, *imp. 3 sing.* of moudre and mouler.

moulant, *p.a.* grinding, milling; molding, etc. (see mouler).

moule, *m.* mold; cast, form; model. — *f.* (*Zoöl.*) mussel.

— *à compression,* (*Ceram.*), pressing mold.

— *à écrasement,* (*Ceram.*) mold for dry pressing, die.

— *à fonte,* casting mold.

— *de fusée,* (*Pyro.*) rocket mold.

— *de gueuse,* channel, pig.

— *de travail,* (*Ceram.*) working mold.

— *en sable gras,* dry-sand mold.

— *en sable maigre,* green-sand mold.

moulé, *p.a.* molded, etc. (see mouler). — *m.* printed characters, print.

mouleau, *m.* mold (rectangular, for fatty acids); small cake of wax (sufficient to make a candle).

moulée, *f.* swarf, grindings.

moule-mère, *n.* (*Ceram.*) block mold.

moulent, *pr. 3 pl.* of moudre and of mouler.

mouler, *v.t.* mold; cast; model, form; print. — *v.r.* be molded, cast or modeled; (of metals) cast; fit exactly; model oneself.

moulerie, *f.* founding, casting; foundry.

mouleur, *m.* molder; caster, founder.

moulin, *m.* mill.

— *à beurre,* churn.

— *à bis,* a mill where brown flour only is made.

— *à blanc,* a mill where white flour only is made.

— *à canne,* cane mill, sugar mill.

— *à cylindre,* (*Paper*) hollander.

— *à eau,* water mill.

— *à farine,* flour mill.

— *à meules,* (*Gunpowder*) incorporating mill.

— *à minerais,* ore mill.

— *à papier,* paper mill.

— *à pilon,* stamping mill, stamp.

— *à poudre,* powder mill.

— *à sucre,* sugar mill.

— *à vapeur,* steam mill.

— *à vent,* windmill.

— *concasseur,* crushing mill, crusher.

mouliner, *v.t.* throw, twist (silk); grind, polish (as marble); eat away (wood).

moulinet, *m.* twirl, turn, whirl; windlass, capstan, winch, reel, drum; chuck; turnstile.

mouloir, *m.* a kind of covered casserole.

moulu, *p.a.* ground, milled; finely divided (as gold).

moulure, *f.* molding (the decoration).

mour, *m.* (*Metal.*) muzzle, nose (of a twyer).

mourant, *p.a.* dying, etc. (see mourir); fading; (of colors) pale; languid.

mourir, *v.i.* die; die away; die out; be spent. — *v.r.* be dying; die away; die out.

faire —, cause to die; kill; distress, worry.

mouron, *m.* any of several primulaceous plants.

— *rouge,* scarlet pimpernel (*Anagallis arvensis*).

mourra, *fut. 3 sing.* of mourir.

mourut, *p.def. 3 sing.* of mourir.

moussache, *f.* cassava starch, cassava, tapioca meal.

moussant, *p.a.* foaming, foamy, frothing, frothy.

mousse, *f.* foam, froth; lather; whipped cream; any light porous mass, sponge; (*Bot.*) moss. — *a.* blunt, dull.

mousse *de platine,* platinum sponge (specif., unpurified material containing iridium as distinguished from *éponge de platine*).

— *marine perlée,* carrageen.

mousseline, *f.* muslin. — *a.* muslin, mousseline.

mousseliner, *v.t.* decorate (glass) in imitation of muslin.

mousser, *v.i.* foam, froth; lather; (of wine) sparkle, fizz.

mousseron, *m.* mushroom.

mousseux, *a.* foamy, frothy; (of soap) lathering; mossy, moss.

moussu, *a.* mossy, moss-grown, moss.

moustique, *m.* mosquito.

moût, *m.* must (of grapes, etc.); (*Brewing*) wort.

— *sucré,* (*Brewing*) sweet wort.

moutarde, *f.* mustard.

— *blanche,* white mustard.

— *des moines,* horseradish.

— *noire,* black mustard.

moût-levain, *m.* a nutrient liquid for the cultivation of yeast.

mouton, *m.* sheep; mutton; sheep leather, sheep; ram, rammer, monkey, driver, pile driver; maul, beetle; punch.

mouture, *f.* grinding; fee for grinding; a mixture of wheat, rye and barley.

mouvant, *p.a.* moving; motive; unstable, (of sand) quick.

mouve-chaux, *m.* (*Sugar*) stirring pole, stirrer.

mouvement, *m.* motion, movement, moving; impulse; commotion, disturbance; variation, fluctuation, change; action, activity, animation.

mettre en —, set in motion, start.

— *brownien,* Brownian movement.

— *croissant,* accelerated motion.

— *de roulis,* rocking motion.

— *de va-et-vient,* reciprocal motion, back-and-forth motion.

— *en arrière,* retrograde motion.

— *en avant,* forward motion.

— *moyen,* mean motion.

— *rotatif,* rotary motion, rotation.

se mettre en —, start, stir, move.

mouveron, *m.* a wooden stirrer.

mouvoir, *v.t. & i.* move. — *v.r.* be moved, move.

faire —, cause to move, move, set in motion.

moyau, *m.* beam (of a press).

moyen, *a.* middle; mean, intermediate; (*Math.*) mean, average; medium; mediocre. — *m.* means; middle quality or grade, middlings; (*Math.*) mean; abilities; reasons, grounds.

au — *de,* by means of.

le Moyen âge, the Middle Ages.

moyen de contrôle, means of control, check.
— goût, (Alcohol) the next to the last fraction in rectification, preceding the "mauvais goûts de queue."
— yens goûts, (Alcohol) the second fraction in rectification, following the "mauvais goûts de tête."
— teint, dye or color of medium fastness; moderately fast.
moyennant, prep. by means of, in exchange for, for.
— que, on condition that.
moyenne, f. (Math.) mean (specif., average); (pl.) middle quality or grade, middlings.
moyennement, adv. moderately, middlingly.
moyeu, m. hub, nave; (egg) yolk; (fruit) stone; core (as in molding).
mû, p.p. of mouvoir (to move).
mucate, m. mucate.
mucédine, f. mucedin.
mucilage, m. mucilage.
— adragant, — de gomme adragante, (Pharm.) mucilage of tragacanth.
— arabique, — de gomme, (Pharm.) mucilage of acacia (or gum arabic).
mucilagineux, a. mucilaginous.
mucine, f. mucin.
mucinoïde, n. mucoid.
mucique, a. mucic.
mucoïde, n. & a. mucoid.
muconique, a. muconic.
mucosité, f. mucus; mucosity.
mucus, m. (Physiol.) mucus.
mue, fem. of mû, p.p. of mouvoir. — f. cage, coop; molting; cast skin.
muet, a. mute, silent, dumb.
muguet, m. lily of the valley; (Med.) thrush.
muid, m. hogshead.
muire, m. salt water, brine (in the salines).
mulet, m. mule.
mulot, m. field mouse.
multangulaire, multangulé, a. multangular.
multicolore, a. multicolor, multicolour.
multiforme, a. multiform.
multiple, a. & m. multiple.
multipliant, p.a. multiplying.
multiplicateur, m. multiplier. — a. multiplying, multiplicative.
multiplicatif, a. multiplicative.
multiplication, f. multiplication.
multiplicité, f. multiplicity.
multiplier, v.t. & i. multiply. — v.r. be multiplied; multiply, increase.
multipolaire, a. multipolar.
multirotation, f. multirotation.
multirotatoire, a. multirotatory.
multitérébénique, a. polyterpene.
multitude, f. multitude.
mundick, m. (Mining) mundic.

muni, p.p. of munir.
municipal, a. municipal. — m. municipal guard.
munir, v.t. provide, supply, furnish.
munition, f. providing, provision, etc. (see munir); munition, munitions; (pl.) ammunition; (pl.) provisions, stores, supplies.
munjeestine, f. munjistin.
munjeet, m. munjeet, Indian madder.
munjistine, f. munjistin.
munster, m. Münster cheese.
muqueuse, f. mucous membrane.
muqueux, a. mucous.
mur, m. wall.
— de refend, partition wall.
mûr, a. ripe, mature; (of clothes) shabby.
murage, m. walling, walling up, masonry.
muraille, f. wall.
couleur de —, couleur —, a gray color like that of stone walls.
muraillement, m. walling, walls.
mûral, a. (Med.) mulberry.
mûre, f. mulberry (the fruit).
— sauvage, blackberry.
mureau, m. (Metal.) twyer wall.
mûrement, adv. ripely, maturely.
murent, p.def. 3 pl. of mouvoir; pr. 3 pl. of murer.
murer, v.t. wall, wall in, wall up, stop up, block up, block.
muret, m. small wall.
murexane, m. murexan (uramil).
murexide, f. murexide.
murexoïne, f. murexoïn.
mûri, p.a. ripened, ripe, matured, mature.
muriate, m. muriate (chloride, or hydrochloride).
— d'ammoniaque, ammonium chloride.
— oxygéné, (Old Chem.) hypochlorite.
— suroxygéné, (Old Chem.) chlorate.
muriaté, a. muriated (combined with hydrochloric acid or a chloride). Archaic.
muriatique, a. muriatic (hydrochloric).
muride, m. (Old Chem.) muride (bromine).
mûrier, m. mulberry tree, mulberry.
— à papier, paper mulberry (Broussonetia papyrifera).
mûrir, v.t., r. & i. ripen, mature; (Med.) maturate.
mûrissant, p.a. ripening, maturing.
murmure, m. murmur.
murmurer, v.i. & t. murmur.
mûron, m. blackberry.
murrayine, f. murrayin.
mus, masc. pl. of mû, moved.
musc, m. musk; musk deer.
muscade, f. nutmeg.
muscadet, m. muscatel (the wine).
muscadier, m. nutmeg tree, nutmeg.
muscarine, f. muscarine.

muscat, *m. & a.* muscat, muscatel (wine and grape); muscatel (the pear).

muscle, *m.* (*Anat.*) muscle.

muscovite, *f.* (*Min.*) muscovite.

musculaire, *a.* muscular.

musculine, *f.* musculin.

musée, *m.* museum.

musical, *a.* musical.

musif, *a.* mosaic.

musique, *f.* music; band; sediment, settlings.

musqué, *a.* musk, musky, musked, scented with or like musk; (of language) affected.

musquer, *v.t.* musk, perfume with musk.

mussif, *a.* mosaic.
or —, mosaic gold.

mut, *p.def. 3 sing.* of mouvoir.

mutage, *m.* mutage (checking of fermentation).

mutation, *f.* change; transfer; mutation.

muté, *a.* transferred, of changed ownership.
— *p.p.* of muter.

muter, *v.t.* mute, check the fermentation of.

muteuse, *f.* muting apparatus.

mutilation, *f.* mutilation, mutilating.

mutiler, *v.t.* mutilate.

mutuel, *a.* mutual, reciprocal.

mutuellement, *adv.* mutually, reciprocally.

mycoderme, (*Bact.*) *m.* mycoderma.

mycodermique, *a.* (*Bact.*) mycodermic.

mycologie, *f.* mycology.

mycoprotéine, *f.* mycoprotein.

mycose, *f.* mycose (trehalose); (*Med.*) mycosis.

mydriase, *f.* (*Med.*) mydriasis.

mydriatique, *a. & m.* (*Med.*) mydriatic.

myéline, *f.* myelin.

mylabre, *m.* (*Zoöl.*) member of the genus *Mylabris.*

myocaillot, *m.* muscle coagulum.

myoplasma, *m.* muscle plasma.

myose, *f.* (*Med.*) myosis.

myosérum, *m.* muscle serum.

myosine, *f.* myosin.

myosinogène, *m.* myosinogen.

myotique, *a. & m.* (*Med.*) myotic.

myrcène, *m.* myrcene.

myriade, *f.* myriad.

myriagramme, *m.* myriagram (10 kilograms).

myrica, *m.* (*Bot.*) wax myrtle (*Myrica*).
— *cérifère,* wax myrtle (esp. *M. cerifera*).
— *galé,* sweet gale (*M. gale*).

myricacées, *f.pl.* (*Bot.*) Myricaceæ.

myricine, *f.* myricin.

myricique, *a.* myricyl.

myrico-mélissique, *a.* In the phrase *éther myrico-mélissique* (myricyl melissate?).

myristicacées, myristicées, *f.pl.* Myristicaceæ.

myristicine, *f.* myristicin.

myristicinique, *a.* myristicic, myristicinic.

myristicol, *m.* myristicol.

myristine, *f.* myristin.

myristique, *a.* myristic.

myristone, *f.* myristone.

myrobalan, myrobolan, *m.* (*Bot.*) myrobalan.

myrobalanier, *m.* myrobalan tree (*Terminalia*).

myronique, *a.* myronic.

myrosine, *f.* myrosin.

myroxyle, *m.* (*Bot.*) any member of *Myroxylon.*

myrrhe, *f.* myrrh (the gum resin).

myrrhé, *a.* perfumed with myrrh, myrrhy.

myrrhine, *f.* myrrhin.

myrrhique, *a.* myrrhic.

myrrhol, *m.* myrrhol.

myrsite, *m.* (*Pharm.*) wine flavored with myrtle leaves.

myrtacées, *f.pl.* (*Bot.*) Myrtaceæ.

myrte, *m.* (*Bot.*) myrtle (*Myrtus*).

myrtil, *m.*, **myrtille,** *f.* (*Bot.*) bilberry.

mystère, *m.* mystery.

mystérieusement, *adv.* mysteriously.

mystérieux, *a.* mysterious.

mystifier, *v.t.* mystify.

myxome, *m.* (*Med.*) myxoma.

N

n'. Contraction of *ne*.

nacarat, *a. & m.* nacarat, light red with an orange tinge.

nacelle, *f.* boat; weighing scoop; car (of a balloon).
— *à fond plat,* flat-bottomed boat.
— *en platine,* platinum boat.
— *en porcelaine,* porcelain boat.
— *en silice,* silica boat.
— *en verre pour pesées,* glass weighing scoop.

nacre, *f.* nacre, mother-of-pearl.

nacré, *a.* nacreous, pearly.

nacrer, *v.t.* give a nacreous luster to.

nacrure, *f.* pearly whiteness.

nage, *f.* swimming, swim; rowing; rowlock.

nageoire, *f.* fin (of fishes); float.

nager, *v.i.* swim; row.

naguère, naguères, *adv.* only a short time ago, lately, but now.

naïf, *a.* artless, natural, naive.

nain, *m., f. & a.* dwarf.

naissance, *f.* source, origin, beginning, rise; birth.
donner — *à,* give rise to, produce.
prendre —, be formed, originate, arise, appear.

naissant, *a.* nascent; incipient, beginning, rising; (of colors) faint, pale; newborn.

naître, *v.i.* be born; spring up, come up, grow; spring, proceed, arise, originate.
faire —, give birth to; give rise to; cause to grow, raise, produce.

nantir, *v.t.* secure (a creditor); provide.

nantissement, *m.* security (for debt).

napel, *m.* monkshood, aconite (*Aconitum napellus*).

napelline, *f.* napelline.

napht-. naphth-.

naphtadil, naphtagil, *m.* a kind of ozocerite.

naphtalène, *m.* naphthalene.

naphtalénique, *a.* of or pertaining to naphthalene, naphthalene, naphthalenic.

naphtaline, *f.* naphthalene.
— *bichlorée,* dichloronaphthalene.

naphtaline-disulfoné, naphtalinodisulfoné, *a.* naphthalenedisulfonic.

naphtalino-sulfoné, *a.* naphthalenesulfonic.

naphtalino-monosulfoné, *a.* naphthalenemonosulfonic.

naphtazine, *f.* naphthazine.

naphte, *m.* naphtha (in its various senses).
— *acétique,* ethyl acetate (old name).
— *de charbon,* crude naphtha, light oil (from coal tar).
— *de pétrole,* petroleum naphtha.
— *d'os,* bone oil, Dippel's oil.

naphtène, *m.* naphthene.

naphtionique, *a.* naphthionic.

naphtoate, *m.* naphthoate.

naphtofurfurane, naphtofurane, *m.* naphthofuran.

naphtoïque, *a.* naphthoic.

naphtol, *m.* naphthol.

naphtolage, *m.* naphtholization.

naphtoler, *v.t.* naphtholize, treat or impregnate with naphthol.

naphtolsulfoné, *a.* naphtholsulfonic.

naphtophénazine, *f.* naphthophenazine (for the four-ring compound benzophenazine is a better name).

naphtoquinone, *f.* naphthoquinone.

naphtylamine, *f.* naphthylamine.

naphtyle, *m.* naphthyl.

naphtylène, *m.* naphthylene.

naphtylènediamine, *f.* naphthylenediamine.

naphtylénique, *a.* of or pertaining to naphthalene (an improper use).

naphtylique, *a.* naphthylic, naphthyl.

naphtylol, *m.* naphthol.

naphtylurée, *f.* naphthylurea.

napoléon, *m.* napoleon (20-franc piece).

napolitain, *a.* Neapolitan, of Naples.

nappe, *f.* sheet (of cloth, of water, etc.); surface, level (of water, etc.); sheet, bed, stratum, deposit (as of petroleum); (*Math.*) nappe, sheet; table cloth.

naquit, *p.def. 3 sing.* of naître.

narcéine, *f.* narceine.

narcisse, *m.* (*Bot.*) narcissus.
— *des prés,* daffodil.

narcose, *f.* (*Med.*) narcosis.

narcotico-âcre, *a. & m.* (*Med.*) narcotico-acrid.

narcotine, *f.* narcotine.

narcotinique, *a.* narcotinic.

narcotique, *a. & m.* narcotic.

narcotiser, *v.t.* narcotize.

narcotisme, *m.* narcotism.

nard, *m.* nard (the ointment); (*Bot.*) any of several plants, specif. (1) nard, matgrass (*Nardus stricta*); (2) nard, spikenard (*Nardostachys jatamansi*).
— *indien,* nard (the ointment); spikenard.

narine, *f.* nostril.

narrateur, *m.* narrator.

narré, *m.* narrative, account, story.

narrer, *v.t.* narrate.

nasal, *a. & m.* nasal.

nasitor, nasitort, *m.* (*Bot.*) garden cress.

nasturce, *m.* (*Bot.*) Nasturtium (Rorippa).
— *officinal,* water cress (*Rorippa nasturtium*).

natatoire, *a.* natatory.

natif, *a. & m.* native.

nationalement, *adv.* nationally.

nationaliser, *v.t.* nationalize.

nationalité, *f.* nationality.

natron, natrum, *m.* (*Min.*) natron, native soda.

natte, *f.* mat, matting; plait, plaiting, braid.

natter, *v.t.* plait, braid; cover with mats.

naturaliser, *v.t.* naturalize.

nature, *f.* nature.

naturel, *a.* natural; native. — *m.* nature; naturalness; native.
au —, naturally, according to nature or life; cooked and served simply.

naturellement, *adv.* naturally; plainly.

naufrage, *m.* shipwreck; wreck, ruin.

nauséabond, *a.* nauseous, sickening.

nausée, *f.* nausea.

nauséeux, *a.* nauseous, causing nausea.

nautique, *a.* nautical.

naval, *a.* naval.

navet, *m.* turnip (specif., *Brassica rapa*).

navette, *f.* rape, colza (see *colza,* note); rape oil, rapeseed oil, colza oil; pig (of lead); shuttle; shift (of workers).

naviguer, *v.i.* navigate, sail.

navire, *m.* ship, vessel.
— *à vapeur,* steamship, steamer.
— *à voiles,* sailing vessel.
— *de charge,* freight vessel, freighter.

ne, *adv.* (ordinarily accompanied by *pas, point, personne,* or a similar word) not, no. (*Ne* is omitted in translating certain clauses beginning with *que,* esp. in expressions of fearing, doubting, etc., e.g.: *je crains qu'il ne vienne,* I fear he will come; but, *je crains qu'il ne vienne pas,* I fear he will *not* come; also, *mieux que vous ne pensez,* better than you think.)
ne . . . guère, scarcely, hardly.
ne . . . jamais, never.
ne . . . pas, not, no.
ne . . . pas encore, not yet.
ne . . . personne, no one, nobody.
ne . . . plus, no more, no longer.
ne . . . que, only, but.

ne . . . *rien,* nothing.

né, *p.a.* born; well born.

néanmoins, *adv.* nevertheless, however.

néant, *m.* nothing, nothingness.
mettre à —, annul.

nébulaire, *a.* (*Astron.*) nebular.

nébuleuse, *f.* (*Astron.*) nebula.

nébuleux, *a.* nebulous.

nébulosité, *f.* nebulosity.

nécessaire, *a.* necessary. — *m.* necessary; case, box, kit, outfit (of necessary or useful objects).

nécessairement, *adv.* necessarily.

nécessité, *f.* necessity.
de —, of necessity, necessarily.
par —, thru necessity.

nécessiter, *v.t.* necessitate; compel.

nécrologie, *f.* necrology, obituary.

nécromancie, *f.* necromancy.

nécromancien, nécromant, *m.* necromancer.

nécrose, *f.* (*Med. & Bot.*) necrosis.

nécroser, *v.t.* (*Med.*) necrose, mortify.

nectaire, *m.* (*Bot.*) nectary.

néerlandais, *a.* Dutch. — *m. cap.* Dutchman, (*pl.*) Dutch.

nèfle, *f.* medlar (the fruit).

néflier, *m.* medlar, medlar tree (*Mespilus germanica*).

neftgil, *m.* (*Min.*) a kind of ozocerite.

négatif, *a. & m.* negative.

négation, *f.* negation, denial; negative (word).

négative, *f.* negative (proposition).

négativement, *adv.* negatively.

négativité, *f.* negativeness, negativity.

négligé, *p.a.* neglected; careless, negligent.

négligeable, *a.* negligible.

négligemment, *adv.* carelessly, negligently.

négligence, *f.* negligence; neglect.

négligent, *a.* negligent, careless.

négliger, *v.t.* neglect. — *v.r.* be careless, be negligent.

négoce, *m.* trade, commerce, traffic.

négociable, *a.* negotiable.

négociant, *m.* merchant.

négociateur, *m.* negotiator.

négociation, *f.* negotiation.

négocier, *v.t.* negotiate. — *v.i.* trade.

nègre, *a. & m.* negro.

neige, *f.* snow.

neiger, *v.i.* snow.

neigeux, *a.* snowy.

néodyme, *m.* neodymium.

néon, *m.* neon.

néoytterbium, *m.* neoytterbium, ytterbium.

Népaul, Népâl, *m.* Nepal.

népérien, *a.* (*Math.*) Napierian, Naperian.

néphélémètre, *m.* nephelometer.

néphélémétrique, *a.* nephelometric.

néphéline, *f.* (*Min.*) nephelite, nepheline.

néphélinite, *f.* (*Petrog.*) nephelinite.

néphrectomie, *f.* nephrectomy.
néphrétique, *a.* nephritic. — *f.* nephrite, jade.
    — *m.* kidney remedy; nephritic person.
néphrite, *f.* (*Min.*) nephrite; (*Med.*) nephritis.
néradol, *n.* neradol.
nerf, *m.* nerve; fiber; sinew, tendon.
    — *de fer,* fibrous texture of well forged
    iron.
néroli, *m.* neroli (orange-flower oil).
    essence de —, oil of neroli, neroli.
    — *bigarade,* oil of bitter-orange flowers.
    — *pétale,* oil of sweet-orange flowers.
nerprun, *m.* (*Bot.*) buckthorn (*Rhamnus*).
    — *purgatif,* purging buckthorn (*R. cathar-
    tica*).
nerveux, *a.* nervous; fibrous; sinewy; strong.
nervure, *f.* vein, nerve (as of a leaf); rib; web,
    fin, feather, flange, fillet; groove.
nervuré, *a.* veined; ribbed; flanged; grooved.
nesslérisation, *f.* nesslerization.
nesslériser, *v.t.* nesslerize.
net, *a.* clean; pure; clear; distinct, sharp;
    neat, unmixed; (of prices, etc.) net; frank,
    open. — *adv.* all at once; clean; quite;
    clearly, flatly.
    faire —, make clean, cleanse, clean.
    mettre au —, make a fair copy of.
nettement, *adv.* cleanly, neatly; clearly, dis-
    tinctly, sharply; frankly, flatly.
netteté, *f.* cleanness, etc. (see net).
nettoiement, nettoyage, *m.* cleaning, cleans-
    ing, etc. (see nettoyer).
nettoyer, *v.t.* clean, cleanse; scour, sweep,
    wipe, etc.; clear, clear out.
nettoyeur, *m.*, **nettoyeuse,** *f.* cleaner, etc.
    (see nettoyer); specif., cleaning machine.
    nettoyeur de tan, (*Leather*) bark cleaner.
neuf, *a. & m.* new; nine; ninth.
    à —, anew, afresh, like new.
neurine, *f.* neurine.
neutralement, *adv.* neutrally.
neutraline, *f.* neutraline (a fixed oil).
neutralisation, *f.* neutralization.
neutraliser, *v.t.* neutralize. — *v.r.* be neutral-
    ized; become neutral; neutralize each
    other.
neutralité, *f.* neutrality.
neutre, *a.* neutral; (*Grammar*) neuter.
neutrement, *adv.* neutrally.
neuve, *a. fem.* of neuf (nine, ninth).
neuvième, *a. & m.* ninth.
neveu, *m.* nephew; (*pl.*) descendants.
névr-. neur-.
névralgie, *f.* (*Med.*) neuralgia.
névrine, *f.* neurine.
névrinique, *a.* of or pertaining to neurine.
névrite, *f.* (*Med.*) neuritis.
newtonien, *a.* Newtonian.
nez, *m.* nose; beak, projection, end; promon-
    tory.

ni, *conj.* nor, or.
ni . . . ni, neither . . . nor, (following a
    negative) either . . . or.
niaouli, *f.* (*Bot.*) cajuput (*Melaleuca*).
    essence de —, cajuput oil.
niccolique, *a.* nickel, of nickel.
nickel, *m.* nickel.
    — *arsenical blanc,* (*Min.*) chloanthite.
    — *gris,* (*Min.*) gersdorffite, nickel glance.
nickelage, *m.* nickeling, nickel-plating, nickel-
    age.
nickeler, *v.t.* nickel, nickel-plate, nickelize.
nickéleux, *a.* nickelous.
nickelglanz, *m.* (*Min.*) nickel glance, gers-
    dorffite.
nickélifère, nickelifère, *a.* nickeliferous.
nickéline, *f.* (*Min.*) niccolite, kupfernickel.
nickélique, *a.* nickelic.
nickélisation, *f.*, **nickélisage,** *m.* = nickelage.
nickéliser, *v.t.* = nickeler.
nickelure, *f.* nickeling, nickle-plating.
nicol, *m.* (*Optics*) nicol, Nicol prism.
nicotéine, *f.* nicoteine.
nicotianine, *f.* oil of tobacco; nicotianin.
nicotianique, *a.* nicotinic, nicotic.
nicotine, *f.* nicotine.
nicotineux, *a.* containing nicotine.
nicotinique, *a.* of or pertaining to nicotine;
    (of the acid) nicotinic, nicotic.
nicotinisme, *m.* (*Med.*) nicotinism.
nicotique, *a.* of or pertaining to tobacco or
    nicotine; (of the acid) nicotinic, nicotic.
nicotiser, *v.t.* nicotinize, nicotize.
nid, *m.* nest; berth, place.
    — *d'abeilles,* honeycomb.
nie, *pr. 3 sing.* of nier.
nielle, *f.* niello; smut, blight, rust; (*Bot.*) a
    plant of the genus *Agrostemma.*
    — *des blés,* corn cockle (*A. githago*).
nier, *v.t.* deny. — *v.r.* be denied.
nigrine, *f.* (*Min.*) nigrine (a variety of rutile).
nigrite, *f.* nigrite (an insulator).
nigrosine, *f.* (*Dyes*) nigrosine.
Nil, *m.* Nile.
niobate, *m.* niobate (columbate).
niobé, *n.* Niobe oil (methyl benzoate).
niobifère, *a.* columbiferous.
niobique, *a.* niobic (columbic).
niobium, *m.* niobium (columbium).
niton, *m.* niton.
nitragine, *f.* (*Agric.*) nitragin.
nitramidine, *f.* (*Expl.*) nitramidin(e).
nitranilate, *m.* nitranilate.
nitranilique, *a.* nitranilic.
nitrant, *p.a.* nitrating.
nitratation, *f.* nitration.
nitrate, *m.* nitrate.
    — *d'argent,* silver nitrate.
    — *de baryte,* nitrate of baryta (barium ni-
    trate).

nitrate *de chaux*, nitrate of lime (calcium nitrate).

— *de Chili*, Chile saltpeter, sodium nitrate.

— *de cuivre*, copper nitrate.

— *de fer*, iron nitrate (specif., ferric nitrate).

— *de plomb*, lead nitrate.

— *de potasse*, nitrate of potash (potassium nitrate).

— *de soude*, nitrate of soda (sodium nitrate).

nitrater, *v.t.* nitrate.

nitration, *f.* nitration.

nitre, *m.* niter, nitre, saltpeter (specif., potassium nitrate).

— *cubique*, cubic niter, sodium nitrate.

— *de houssage*, wall saltpeter (calcium nitrate).

— *du Chili*, Chile niter (sodium nitrate).

— *du commerce*, commercial niter, commercial saltpeter.

— *lunaire*, silver nitrate.

— *prismatique*, potassium nitrate.

nitré, *p.a.* nitrated; nitro.

nitrer, *v.t.* nitrate.

nitrésine, *f.* (*Expl.*) nitrated resin.

nitréthane, *m.* nitroethane.

nitreux, *a.* nitrous.

nitrière, *f.* niter bed, niter works.

nitrifiant, *p.a.* nitrifying.

nitrificateur, *m.* nitrifier. — *a.* nitrifying.

nitrification, *f.* nitrification.

nitrifier, *v.t.* nitrify. — *v.r.* be converted into or be covered with niter.

nitrile, *m.* nitrile.

— *acide*, acid nitrile, nitrile.

nitrique, *a.* nitric.

nitrite, *m.* nitrite.

— *de soude*, sodium nitrite.

nitrobenzène, *m.*, **nitrobenzine**, *f.* nitrobenzene.

nitrocalcite, *f.* (*Min.*) nitrocalcite.

nitrocellulose, *f.* nitrocellulose, cellulose nitrate.

nitrocolle, *f.* (*Expl.*) nitrated glue.

nitro-coton, *m.* (*Expl.*) guncotton.

nitrocumène, **nitrocumol**, *m.* nitrocumene.

nitroferrocyanure, **nitro-ferro-prussiate**, *m.* nitroferrocyanide (nitroprusside).

nitroforme, *m.* nitroform (trinitromethane).

nitro-gélatine, *f.* nitrogelatin.

— *ammoniacale*, nitrogelatin with an ammonium nitrate base.

nitrogène, *m.* nitrogen.

nitrogéné, *a.* nitrogenous.

nitrogéner, *v.t.* nitrogenize, azotize.

nitroglucose, *f.* (*Expl.*) nitroglucose.

nitroglycérine, *f.* nitroglycerin.

nitroglycol, *m.* nitroglycol.

nitro-houille, *f.* (*Expl.*) nitro-coal.

nitro-hydrochlorique, *a.* nitrohydrochloric.

nitroleum, *m.* nitroleum (nitroglycerin).

nitroline, *f.* (*Expl.*) nitrolin, nitroline.

nitrolique, *a.* nitrolic.

nitrolite, *m.* (*Expl.*) nitrolite.

nitromannite, *f.* (*Expl.*) nitromannite.

nitromélasse, *f.* (*Expl.*) nitrated molasses.

nitrométhane, *m.* nitromethane.

nitromètre, *m.* nitrometer.

nitromuriatique, *a.* nitromuriatic (nitro-hydrochloric).

nitronaphtaline, *f.* nitronaphthalene.

nitro-oxhydrilé, *a.* hydroxynitro.

nitrophénol, *m.* nitrophenol.

nitroprussiate, *m.* nitroprussiate, nitroprusside.

nitrosaccharose, *f.* (*Expl.*) nitrosaccharose.

nitrosamine, *f.* nitrosamine.

nitrosation, *f.* introduction of the nitroso group, conversion into a nitroso compound; conversion into a nitrite.

nitrosé, *p.a.* nitroso; converted into a nitroso compound; converted into a nitrite.

nitroser, *v.t.* introduce the nitroso group into, convert into a nitroso compound; convert into a nitrite.

nitrosité, *f.* nitrous quality or state.

nitrosobenzène, *m.* nitrosobenzene.

nitrosochlorure, *m.* nitrosochloride.

nitrosophénol, *m.* nitrosophenol.

nitrosubstitué, *a.* nitrosubstituted.

nitrosulfate, *m.* nitrosulfate.

nitrosyle, *m.* nitrosyl.

nitrosylsulfurique, *a.* nitrosylsulfuric.

nitrotartrique, *a.* nitrotartaric.

nitrotoluène, **nitrotoluol**, *m.* nitrotoluene.

nitrure, *m.* nitride.

nitryle, *m.* nitryl (NO_2).

niveau, *m.* level; gage.

au —, *de* —, on a level, level, horizontal.

mettre au —, *mettre de* —, level, bring to a level.

— *à bulle d'air*, spirit level.

— *d'eau*, water level.

— *de la mer*, sea level.

— *de l'eau*, water level, water line.

— *flotteur*, float gage.

niveler, *v.t.* level.

niveleur, *a.* leveling. — *m.* leveler.

nivellement, *m.* leveling, levelling.

N.O., *abbrev.* (nord-ouest) northwest.

nobélite, *f.* (*Expl.*) dynamite.

noble, *a. & m.* noble.

noblement, *adv.* nobly.

noblesse, *f.* nobility; nobleness.

noce, *f.* (also *pl.*) wedding, marriage; wedding feast; wedding party.

nochère, *f.* gutter, channel; skylight.

nocif, *a.* noxious, harmful, injurious.

nocivité, *f.* noxiousness, harmfulness.

nocturne, *a.* nocturnal. — *m.* nocturne.

nocturnement, *adv.* nocturnally, by night.

nocuité, *f.* nocuousness, harmfulness.

nodule, *m.* nodule.

noduleux, *a.* nodular; noduled.

Noël, *m.* Christmas.

nœud, *m.* knot; node (as of a curve); joint; knuckle; difficulty, knotty point; bond, tie.

noie, *pr. 3 sing.* of noyer.

noir, *a.* black; dark; (of rope) tarred; dirty. — *m.* black; (*Sugar*) char, charcoal; bull's-eye; bruise; rust (on plants). — *adv.* black.

  mettre au —, black (as leather), blacken.

  — *animal*, animal black, animal charcoal.

  — *au soufre*, sulfur black.

  — *colorant*, black dye or pigment.

  — *d'acétylène*, acetylene black.

  — *d'Allemagne*, German black.

  — *d'aniline*, aniline black.

  — *de bougie*, candle black.

  — *de charbon*, coal black.

  — *de Chine*, India ink.

  — *de diamine*, diamine black.

  — *de fer*, iron black (precipitated antimony).

  — *de Francfort*, Frankfort black.

  — *de fumée*, lampblack.

  — *de lampe*, lampblack, esp. the best quality.

  — *de liège*, cork black.

  — *de naphtol*, naphthol black.

  — *de pêche*, peach black.

  — *de platine*, platinum black.

  — *de Prusse*, Prussian black (obtained by calcining Prussian blue).

  — *de raffinerie*, sugarhouse black, animal charcoal used in sugar refining; specif., the spent char (sometimes used as a fertilizer).

  — *d'Espagne*, Spanish black, cork black.

  — *de vigne*, vine black.

  — *diamant*, diamond black.

  — *d'imprimerie*, printer's ink.

  — *d'ivoire*, ivory black.

  — *d'os*, bone black.

  — *foncé*, deep black.

  — *jais*, jet black.

  — *minéral*, mineral black.

  — *pour fonderie*, founder's black, foundry black (powdered charcoal or coal).

  — *végétal*, vegetable black.

noirâtre, *a.* blackish, dark.

noirceur, *f.* blackness; black spot.

noircir, *v.t.*, *r. & i.* blacken, turn black; dye black; darken; (*Founding*) blackwash.

noircissement, *m.* blackening, turning black; dyeing in black; (*Founding*) blackwashing.

noircisseur, *m.* blackener, blacker; (*Dyeing*) dyer in black.

noircissure, *f.* black spot.

noisetier, *m.* (*Bot.*) hazel, hazel tree (*Corylus*).

noisette, *f.* hazelnut; hazel, hazel color. — *a.* hazel.

noix, *f.* nut; walnut; sleeve, clamp (as that for fastening a rod to an iron stand); plug, key (of a cock); stud; chain pulley; drum (of a capstan or winch); cone (of a mill); semicircular groove.

  — *d'Alep*, Aleppo gall.

  — *d'Amérique*, Brazil nut.

  — *d'arec*, betel nut, areca nut.

  — *d'eau*, water chestnut.

  — *de banda*, nutmeg.

  — *de Bengale*, myrobalan.

  — *de coco*, coconut, cocoanut.

  — *de cola*, kola nut.

  — *de galle*, — *de Galle*, gallnut, nutgall.

  — *de gourou*, kola nut, guru nut.

  — *de palme*, — *de palmier*, palm nut.

  — *de sassafras*, pichurim bean, sassafras nut.

  — *des Barbades*, physic nut, Barbados nut.

  — *des jésuites*, water chestnut, Jesuit nut.

  — *des Moluques*, nux vomica.

  — *de vomique*, nux vomica.

  — *d'Inde*, coconut, cocoanut.

  — *du Congo*, shea nut.

  — *du Soudan*, kola nut.

  — *muscade*, nutmeg.

  (For the meaning of other phrases, as *noix d'acajou*, see the next important word in the phrase).

nom, *m.* name; noun.

  au — *de*, in the name of, for the sake of.

  de —, by name; in name, nominally.

  — *de famille*, family name, surname.

  — *de guerre*, assumed name; nickname.

  — *social*, firm name.

  petit —, given name, Christian name, first name.

nombrable, *a.* numerable, that can be counted.

nombre, *m.* number.

  au — *de*, among.

  dans le —, in the number, among the number.

  du — *de*, in the number of, in the category of.

  en —, in numbers, numerously.

  faire —, count.

  — *atomique*, atomic number.

  — *entier*, whole number, integer.

  — *impair*, odd number.

  — *pair*, even number.

  — *rond*, round number.

  — *suffisant*, sufficient number; quorum.

  sans —, without number, countless, innumerable.

nombrer, *v.t.* number, count.

nombreusement, *adv.* numerously.

nombreux, *a.* numerous.

nombril, *m.* navel; bull's-eye (in glass); eye (of fruit).

  — *de Vénus*, navelwort (*Cotyledon umbilicus*).

nomenclateur, *m.* nomenclator, nomenclaturist; vocabulary.

nomenclature, *f.* nomenclature; vocabulary; catalog, list.
— *chimique,* chemical nomenclature.
nomenclaturer, *v.t.* name methodically.
nominal, *a.* nominal.
nominalement, *adv.* nominally.
nommément, *adv.* by name; particularly.
nommer, *v.t.* name; call, denominate; appoint. — *v.r.* be named; tell one's name.
non, *adv.* no; (sometimes with *pas*) not, non-, un-, in-.
— *plus,* neither, either.
— *plus que,* not more than, no more than.
— *que,* — *pas que,* not that.
non-activité, *f.* nonactivity.
nonane, *m.* nonane.
nonantième, *a. & m.* ninetieth.
non-conducteur, *m.* nonconductor; nonconducting.
non-disponibilité, *f.* nonavailability.
non-disponible, *a.* nonavailable.
non-dosé, *a.* undetermined. — *m.* undetermined matter or quantity.
non-isolé, *a.* (*Elec.,* etc.) uninsulated.
nononique, *a.* nononic.
nonose, *f.* nonose.
non-ouvré, *a.* unworked, unwrought; (of fabrics) uncut.
non-réussite, *f.,* **non-succès,** *m.* failure.
non-tanin, *m.* nontannin, non-tan.
non-toxicité, *f.* nontoxicity, nonpoisonousness.
nonuple-effet, *m.* nonuple effect (apparatus).
non-usage, *m.* nonusage.
non-valeur, *f.* waste, loss, deficiency; deduction; bad debt; unproductive property.
nonyle, *m.* nonyl.
nonylène, *m.* nonylene.
nonylique, *a.* nonyl; (of acids) nonylic, nonoic.
noper, *v.t.* burl (cloth).
nord, *m.* north. — *a.* north, northern.
nord-est, *m.* northeast; northeast wind. — *a.* northeast, northeastern.
nord-ouest, *m.* northwest; northwest wind. — *a.* northwest, northwestern.
normal, *a.* normal; (of gages, etc.) standard.
normale, *f.* (*Geom.*) normal.
normalement, *adv.* normally.
normand, *a.* Norman; crafty, evasive.
Normandie, *f.* Normandy.
normaux, *masc. pl.* of normal.
norme, *f.* norm.
noropianique, *a.* noropianic.
Norvège, *f.* Norway.
norvégien, *a.* Norwegian. — *m.* (*cap.*) Norwegian.
nos, *pl.* of notre (our).
nota, *m.* note.
notabilité, *f.* notability.
notable, *a.* notable; considerable.

notablement, *adv.* notably; considerably, much.
notaire, *m.* notary.
notamment, *adv.* notably, especially, particularly, for example.
notation, *f.* notation.
— *chimique,* chemical notation.
note, *f.* note; mark, grade, rating; (*Com.*) bill, statement.
noter, *v.t.* note; mark. — *v.r.* be noted, be marked.
notice, *f.* notice.
notification, *f.* notification.
notifier, *v.t.* notify.
notion, *f.* notion; information (esp. elementary).
notoire, *a.* well known, notorious.
notoirement, *adv.* notoriously.
notoriété, *f.* notoriety, notoriousness.
notre, *a.* our.
nôtre, *pron.* ours. — *m.* ours, our own; (*pl.*) ours, our people, our side.
nouage, *m.* knotting, etc. (see nouer); (*Med.*) rickets, rachitis.
nouer, *v.t.* knot; tie; tie up; knit; form, make up; engage in.
nouet, *m.* little bag or cloth containing a substance to be steeped.
noueux, *a.* knotty, knotted, gnarled.
nougat, *m.* nougat.
nouilles, *f.pl.* noodles.
— *aux œufs,* egg noodles.
nouillettes, *f.pl.* fine noodles.
nourri, *p.a.* nourished, etc. (see nourir); full, thick, strong, steady, rich; accustomed, hardened.
nourrice, *f.* nurse.
nourricier, *a.* nourishing, nutrient, nutritive.
nourrir, *v.t.* nourish; feed; support; rear; raise; produce; improve; cherish; nurse. — *v.r.* feed, live; thrive; improve oneself; board oneself.
nourrissage, *m.* nourishing, etc. (see nourrir); specif., cattle feeding.
nourrissant, *p.a.* nourishing, etc. (see nourrir); nutritive, nutritious.
nourrisseur, *m.* feeder; dairyman. — *a.* feeding, feed.
nourrisson, *m.* nursing infant.
nourriture, *f.* nourishment, nutriment, food, diet, (for animals) feed; feeding, fattening (of animals); nursing, suckling; (*Leather*) tawing paste.
nous, *pron.* we, us, to us, ourselves, each other, one another.
nous-mêmes, *pron.* ourselves.
nouveau, *a.* new; novel. — *m.* new; novelty; newcomer. — *adv.* newly, new.
à —, anew, again.
la nouvelle Galles du Sud, New South Wales.

nouveau-né, *a.* newborn. — *m.* newborn child.

nouveauté, *f.* novelty; newness; new fashion; new book; early fruit; (*pl.*) fancy goods.

nouvel, *a. masc.* Replaces *nouveau* before a vowel.

nouvelle, *a. fem.* of nouveau. — *f.* news; novel.

Nouvelle-Écosse, *f.* Nova Scotia.

Nouvelle-Galles du Sud, *f.* New South Wales.

nouvellement, *adv.* newly, recently, lately.

Nouvelle-Orléans, *f.* New Orleans.

Nouvelle-Zélande, *f.* New Zealand.

Nouvelle-Zemble, *f.* Nova Zembla.

novaculite, *f.* (*Petrog.*) novaculite.

novale, *f.* new land. — *a.* new, newly cleared.

novembre, *m.* November.

novice, *a.* inexperienced. — *m.* novice.

noyau, *m.* nucleus; core; stone (of a fruit); stem, shank (as of a bolt); plug, key (of a cock); hub (of a wheel); (fig.) heart.
— *atomique,* atomic nucleus.
— *benzénique,* benzene nucleus.
— *naphtalénique,* naphthalene nucleus.
— *pentagonal,* five-membered nucleus.
— *pyridique,* pyridine nucleus.

noyer, *v.t.* drown; flood; sink; immerse; bed (as in cement); bury (in something); countersink; blend, confuse (colors); overslake (lime). — *m.* walnut (wood and tree).
— *commun,* English walnut, common European walnut (*Juglans regia*).
— *gris,* white walnut, butternut (*Juglans cinerea*).
— *noir,* black walnut (*Juglans nigra*).

nu, *a.* bare; exposed; naked; nude. à —, bare, exposed.

nuage, *m.* cloud.

nuageux, *a.* cloudy; clouded.

nuance, *f.* shade, tint, gradation, nuance.

nuancement, *m.* shading, gradation; variation.

nuancer, *v.t.* shade, gradate; vary, variegate.
— *v.r.* be shaded; be blended; be variegated.

nucine, *f.* nucin (juglone).

nucléal, nucléaire, *a.* nuclear.

nucléase, *f.* nuclease.

nuclée, *a.* nucleate, nucleated.

nucléine, *f.* nuclein.

nucléique, *a.* nucleic.

nucléo-albumine, *f.* nucleoalbumin.

nucléohistone, *f.* nucleohistone.

nucléole, *m.* (*Biol.*) nucleolus.

nucléoprotéide, *m.* nucleoprotein.

nucléus, *m.* nucleus.

nudité, *f.* bareness; exposed state; nakedness.

nue, *f.* cloud. — *a. fem.* of nu.

nuée, *f.* (large) cloud; swarm, host.

nui, *p.p.* of nuire.

nuire, *v.i.* do harm, be harmful, be injurious.
— à, harm, injure, interfere with.

nuisibilité, *f.* harmfulness, injuriousness.

nuisible, *a.* harmful, injurious, hurtful, noxious.

nuisiblement, *adv.* harmfully, injuriously.

nuisit, *p.def. 3 sing.* of nuire.

nuit, *f.* night. — *pr. 3 sing.* of nuire.

nul, *a.* no, not a, not any, (after a negative) any; null, void; of no value; (of letters) silent. — *pron.* no one, nobody, none.

nullement, *adv.* not at all, by no means.

nullité, *f.* nullity; incapacity; nonentity.

nûment, *adv.* plainly, frankly; barely; nakedly.

numéraire, *m.* specie, coin. — *a.* legal (value).

numéral, *a.* numeral.

numérateur, *m.* (*Math.*) numerator.

numération, *f.* numeration, numbering, notation.

numérique, *a.* numerical.

numériquement, *adv.* numerically.

numéro, *m.* number.
— *de fabrique,* factory number.
— *d'ordre,* serial number.

numérotage, *m.* numbering.

numéroter, *v.t.* number.

nuque, *f.* nape (of the neck).

nutrescibilité, *f.* nutritive value.

nutrescible, *a.* nutrient, nutritive.

nutricier, *a.* nourishing, nutritious, nutrient.

nutriment, *m.* nutriment.

nutrimentaire, *a.* nutrimental, nutritive.

nutrimentif, *a.* pertaining to nutriment.

nutritif, *a.* nutritive, nutritious, nutrient.

nutrition, *f.* nutrition.

nutritivité, *f.* nutritiveness, nutritiousness.

O

U., *abbrev.* (ouest) west.

ô, *interj.* O, oh, ah.

obéir, *v.i.* be obedient, obey; yield (to a force). — *à,* obey.

obéissance, *f.* obedience.

obéissant, *a.* obedient.

objecter, *v.t.* object, object to, oppose.

objectif, *a. & m.* objective.

objection, *f.* objection.

objectionnable, *a.* objectionable.

objectivement, *adv.* objectively.

objet, *m.* object; article.
— *d'art,* an article valued for its artistic merit, objet d'art.
— *de luxe,* article of luxury.
— *de parure,* ornament.
— *de rechange,* spare part.
objets de consommation, articles for consumption, goods, stores.

obligataire, *m.* bondholder.

obligation, *f.* obligation; specif., bond.

obligatoire, *a.* obligatory.

obligé, *p.a.* obliged; necessary. — *m.* one under obligation.

obligeamment, *adv.* obligingly.

obligeance, *f.* kindness, obligingness.

obligeant, *p.a.* obliging, courteous, kind.

obliger, *v.t.* oblige; obligate.

obliquangle, *a.* oblique-angled.

oblique, *a. & m.* oblique.
en —, obliquely.

obliquement, *adv.* obliquely.

obliquité, *f.* obliqueness, obliquity.

oblitération, *f.* obliteration; cancellation; (*Med.*) stopping, stoppage.

oblitérer, *v.t.* obliterate; cancel; (*Med.*) stop.

oblong, *a.* oblong.

oboval, obové, *a.* (*Bot.*) obovate.

obscur, *a.* dark; obscure.

obscurateur, *m.* darkener, obscurer. — *a.* darkening, obscuring.

obscurcir, *v.t.* obscure; darken, dim. — *v.r.* become obscure, dark or dim; be obscured.

obscurcissement, *m.* obscuring, obscurement, darkening.

obscurément, *adv.* obscurely, darkly, dimly.

obscurité, *f.* darkness; obscurity, obscureness.

obséder, *v.t.* beset, besiege; obsess.

observable, *a.* observable.

observateur, *m.* observer. — *a.* observing.

observation, *f.* observation.

observatoire, *m.* observatory.

observer, *v.t.* observe. — *v.r.* be observed; be circumspect; observe each other.

obsidiane, obsidienne, *f.* obsidian.

obsonine, *f.* opsonin.

obsonique, *a.* opsonic.

obstacle, *m.* obstacle.

obstination, *f.* obstinacy.

obstiné, *a.* obstinate; persistent.

obstinément, *adv.* obstinately, stubbornly.

obstructeur, *m.* obstructer, obstructor.

obstruction, *f.* obstruction.

obstruer, *v.t.* obstruct.

obtenir, *v.t.* obtain. — *v.r.* be obtained.

obtention, *f.* obtaining, obtainment.

obtenu, *p.a.* obtained.

obtient, *pr. 3 sing.* of obtenir.

obtint, *p.def. 3 sing.* of obtenir.

obturateur, *m.* obturator, check, cut-off, valve, stopper, plug, cap, shutter; specif.: (1) a valve in the stopper of a bottle to release excess pressure; (2) a gas check for a gun, (3) a camera shutter. — *a.* obturator, obturating, stopping, closing.

obturation, *f.* stopping, closing, obturation.

obturatrice, *a. fem.* of obturateur.

obturer, *v.t.* stop, close, obturate.

obtus, *a.* obtuse.

obtusangle, *a.* obtuse-angled.

obtusement, *adv.* obtusely.

obtusion, *f.* obtuseness; making obtuse.

obus, *m.* (*Mil.*) shell; (of a calorimeter) bomb.
— *à balles,* shrapnel.
— *à étoile,* star shell.
— *à explosif,* high-explosive shell, H.E. shell.
— *à fusée à temps,* time shell.
— *à fusée percutante,* percussion shell.
— *à gaz,* gas shell.
— *armé,* fused shell, live shell.
— *chargé,* loaded shell, filled shell.
— *de rupture,* armor-piercing shell, A.P. shell.
— *de semi-rupture,* semi-armor-piercing shell.
— *éclairant,* light shell, star shell.
— *en acier,* steel shell.
— *explosif,* high-explosive shell, H.E. shell.
— *fusant,* time shell.
— *incendiaire,* incendiary shell.
— *lacrymogène,* tear shell, lachrymator shell.
— *torpille,* torpedo shell.

245

obuser, *v.t.* (*Mil.*) shell.
obusier, *m.* (*Mil.*) howitzer.
obvers, *m.* obverse.
obvier, *v.i.* (with *à*) obviate, prevent.
occasion, *f.* occasion; bargain.
occasionnel, *a.* occasional.
occasionellement, *adv.* occasionally.
occasionner, *v.t.* occasion, cause.
occidental, *a.* western, west, occidental.
occlure, *v.t.* occlude.
occlus, *p.a.* occluded.
occlusif, *a.* occlusive.
occlusion, *f.* occlusion; (*Steam*) cut-off.
occulte, *a.* occult.
occultement, *adv.* occultly.
occupant, *p.a.* occupying. — *m.* occupant.
occupation, *f.* occupation.
occuper, *v.t.* occupy.
occurrence, *f.* occurrence.
  dans l'—, as it happens.
occurrent, *a.* occurring, happening.
océan, *m.* ocean.
  grand —, Pacific Ocean.
  — Glacial du Nord, Arctic Ocean.
  — Glacial du Sud, Antarctic Ocean.
Océanie, *f.* Oceania.
océanique, *a.* oceanic, ocean.
ochracé, *a.* ocherous, ochreous.
ocre, *f.* ocher, ochre.
  — brune, brown ocher.
  — de nickel, (*Min.*) nickel ocher, annabergite.
  — de ru, *— de ruisseau*, a yellow ocher col-
    lected from water that has been used in
    washing iron ore.
  — jaune, yellow ocher.
  — rouge, red ocher.
  — verte, green ocher.
ocreux, *a.* ocherous, ochreous.
octaèdre, *m.* octahedron. — *a.* octahedral.
octaédrique, *a.* octahedral.
octane, *m.* octane.
octave, *f.* octave.
octobre, *m.* October.
octocarbure, *m.* octacarbide, octocarbide.
octogonal, *a.* octagonal.
octogone, *m.* octagon. — *a.* octagonal.
octohydrure, *m.* octahydride, octohydride.
octonique, *a.* octonic.
octonitrique, *a.* octanitro, octanitro-.
octovalent, *a.* octavalent, octovalent.
octroi, *m.* octroi; specif., (1) tax paid on goods
    brought into a town or city, (2) grant.
octroiement, *m.* granting, concession.
octroyer, *v.t.* grant, concede.
octuple, *a.* octuple, eight-fold, eight times.
octyle, *m.* octyl.
octylène, *m.* octylene.
octylique, *a.* octyl, octylic.
oculaire, *a.* ocular, eye. — *m.* ocular, eye-
    piece.

oculaire *à projection*, projection ocular.
  — chercheur, a low-power eyepiece used for
    locating what is to be observed.
  — compensateur, compensating eyepiece,
    compensating ocular.
  — de travail, an eyepiece for detailed obser-
    vation, as contrasted with *oculaire cher-
    cheur*.
  — d'Huyghens, Huygenian eyepiece, Huy-
    genian ocular.
odeur, *f.* odor.
  — animalisée, animal odor.
  — de rance, rancid odor.
odieux, *a.* odious. — *m.* odium.
odorabilité, *f.* capability of being smelled.
odorable, *a.* capable of being smelled.
odorant, *a.* having an odor, odorous; specif.,
    having a good odor, fragrant, odorous.
odorat, *m.* smell.
odoratif, *a.* pertaining to smell.
odoration, *f.* smelling.
odorer, *v.t.* smell; smell of.
odoriférant, *a.* odoriferous, (usually) fra-
    grant.
odorifique, *a.* odoriferous, odorific, odorous.
œdémateux, *a.* edematous, œdematous.
œdème, *m.* (*Med. & Bot.*) edema, œdema.
œil, *m.* eye (in various senses); hole, opening,
    aperture, loop, etc.; luster (as of silk or
    pearls); shade, tint; drop of fat or oil on
    the surface of a liquid; face (of type).
  à l'—, by the eye, with the eye.
  à l'— nu, with the naked eye.
  — de chat, cat's-eye.
  — de coulée, (*Metal.*) tap hole.
  — d'écrevisse, (*Pharm.*) crab's-eye, crab's-
    stone.
  — de perdrix, bird's-eye.
œil-de-bœuf, *m.* bull's-eye.
œil-de-chat, *m.* cat's-eye.
œillade, *f.* look, glance.
œillet, *m.* eyelet, eye, eyehole, little hole; (*Bot.*)
    pink; (formerly) poppy.
œilleton, *m.* eyehole, peephole; (*Bot.*) sucker.
œillette, *f.* poppy, opium poppy (*Papaver som-
    niferum*); poppy oil, poppy-seed oil.
œnanthal, *m.* enanthal, œnanthal (enanthalde-
    hyde).
œnanthe, *f.* (*Bot.*) water dropwort (*Œnanthe*).
œnanthine, *f.* enanthin, œnanthin.
œnanthique, *a.* enanthic, œnanthic.
œnanthol, *m.* enanthole, œnanthole (enanthal-
    dehyde).
œnanthylique, *a.* enanthylic, œnanthylic
    (same as *enanthic* in the latter's modern
    sense).
œnobaromètre, *m.* enobarometer, œnoba-
    rometer.
œnobarométrique, *a.* enobarometric, œnoba-
    rometric.

œnolé, m., **œnolature**, f. (Pharm.) medicated wine, wine (of so-and-so).
(For the meaning of phrases see the corresponding phrases under vin.)

œnolique, a. (of certain acids) œnolic, enolic; (incorrectly, of unsaturated hydroxy compounds) enol, enolic; (Pharm.) pertaining to wine, having wine as excipient. — m. a medicine having wine as excipient.

œnologie, f. enology, œnology.

œnologique, a. enological, œnological.

œnologiste, œnologue, m. enologist, œnologist.

œnolotif, a. (Pharm.) containing wine. — m. medicated wine (esp. for external use).

œnomètre, m. enometer, œnometer.

œnométrie, f. enometry, œnometry.

œnométrique, a. enometric(al), œnometric(al).

œnoscope, m. enoscope, œnoscope.

œnoxydase, f. enoxidase, œnoxydase.

œsophage, m. (Anat.) esophagus, œsophagus.

œuf, m. egg.
— de Pâques, Easter egg.
— de poule, hen's egg.

œuvre, f. work; action; setting (of a stone). — m. work, production.
grand —, (Alchemy) the transmutation of metals.
hors d'—, outside the main work, accessory; (of a stone) not set.
mettre à l'—, set to work.
mettre en —, put into practice, employ, use, work up, (of stones) set.

œuvrer, v.i. & t. work.

offenser, v.t. offend.

offensif, a. & f. offensive.

offensivement, adv. offensively.

offert, p.a. offered.

office, m. office; duty; charge; service. — f. pantry.
faire l'— de, discharge the duty of, perform the function of, serve as.

officiel, a. official.

officiellement, adv. officially.

officier, m. officer; official.
— de santé, a medical man of lower grade than an M.D.

officier-chimiste, m. (Mil.) chemical expert who inspects ammunition.

officieux, a. obliging, kind; semi-official.

officinal, a. officinal.

officine, f. pharmacy, apothecary's shop (including its laboratory); laboratory.

offrant, m. bidder. — p.pr. of offrir.

offre, f. offer; tender; supply. — pr. 1 & 3 sing. indic. & subj. of offrir.

offrir, v.t. offer. — v.r. offer oneself, offer.

offusquer, v.t. obscure; dazzle; offend.

ognon, m. = oignon.

oie, f. goose.

oignant, p.pr. of oindre.

oignon, m. onion; bulb, bulbous root; bunion.

oindre, v.t. anoint.

oing, m. grease (for an ointment or for lubrication); specif., lard.

oint, p.p. & pr. 3 sing. of oindre.

oiseau, m. bird; hod (of a mason).

oiseux, oisif, a. idle.

oisiveté, f. idleness.

-ol. (1) in names of alcohols and phenols (i.e., when the ending signifies hydroxyl), -ol. (2) in the case of other compounds, preferably -ole; as, scatol, skatole.

oléacées, f.pl. (Bot.) Oleaceæ.

oléagineux, a. oleaginous, oily. — m. oleaginous substance.

oléandre, m. (Bot.) oleander.

oléate, m. oleate.
— de mercure, oleate of mercury, specif. (as in Pharm.) mercuric oleate.

oléfine, f. olefin, (less desirably) olefine.

oléicole, a. pertaining to olive culture.

oléiculture, f. culture of the olive.

oléifère, a. oil-producing.

oléifiant, a. olefiant.

oléiforme, a. of the consistency of oil, oily.

oléine, f. olein.

oléinées, f.pl. (Bot.) Oleaceæ.

oléique, a. oleic.

oléographie, f. oleography; oleograph.

oléomargarine, f. oleomargarine.

oléomètre, m. oleometer, eleometer, elæometer.

oléoréfractomètre, m. oleorefractometer.

oléorésine, f. oleoresin, oleo-resin.
— de capsique, (Pharm.) oleoresin of capsicum.
— de copahu, copaiba, balsam of copaiba.
— de cubèbe, (Pharm.) oleoresin of cubeb.
— de gingembre, (Pharm.) oleoresin of ginger.
— de lupuline, (Pharm.) oleoresin of lupulin.
— de poivre noir, (Pharm.) oleoresin of pepper, fluid extract of black pepper.

oléorésineux, a. oleoresinous.

oléoricinique, a. ricinoleic.

oléosaccharure, oléosaccharat, oléosaccharolé, oléosaccharum, oléosucre, m. (Pharm.) oleosaccharum.

oléracé, a. oleraceous, esculent.

oléum, m. oleum; specif., fuming sulfuric acid.

olfactif, a. olfactory, olfactive, of smell.

olfaction, f. olfaction, smell, smelling.

oliban, m. olibanum, frankincense.

olibène, m. olibene.

olide, m. lactone.

oligiste, a. (Min.) oligist, oligistic. — m. oligist, oligist iron (hematite).

oligocène, a. & m. (Geol.) Oligocene.

oligoclase, f. (Min.) oligoclase.

olivacé, a. olivaceous, olive-green.

olivaire, *a.* (*Anat.*) olivary.

olivâtre, *a.* inclined to olive in color.

olive, *f.* olive; something resembling an olive, as a button or knob; specif., the bulbous end of a pipe to which a rubber tube is attached. — *a.* olive, olive-green.

petite huile d'—, poppy oil.

olivenite, *f.* (*Min.*) olivenite.

oliverie, *f.* olive-oil mill; olive-oil works.

olivète, *f.* = œillette.

olivier, *m.* olive tree, olive (*Olea*); olive, olive wood.

olivine, *f.* (*Min.*) olivine.

ollaire, *a.* In the phrase *pierre ollaire*, potstone (a kind of steatite).

ombelle, *f.* (*Bot.*) umbel.

ombellifère, *a.* umbelliferous. — *f.* umbellifer.

ombellifères, *f.pl.* (*Bot.*) Umbelliferæ (Apiaceæ).

ombelliférone, *f.* umbelliferone.

ombellique, *a.* umbellic.

ombilic, *m.* umbilicus.

ombrage, *m.* shade; umbrage, offense.

ombrager, *v.t.* shade; screen.

ombrant, *p.a.* shading.

ombre, *f.* shadow; shade; umber.

à l'— *de*, under the shade of; under the protection of.

faire —, cast a shadow; make shade.

— *brûlée*, burnt umber.

— *de Cassel*, Cassel earth.

— *de Cologne*, Cologne earth.

— *naturel*, raw umber.

sous l'—, *sous* —, under pretense.

ombrer, *v.t.* shade.

oméga, *m.* omega (Greek letter, ω).

omet, *pr. 3 sing.* of omettre.

omettre, *v.t.* omit.

— *de*, omit to.

omis, *p.a.* omitted.

omission, *f.* omission.

on, *pron.* one, people, they, we, someone, somebody; (sometimes) I, you. (*On* may be preceded by *l'* for the sake of euphony.)

onagraire, **onagre**, *m.* (*Bot.*) evening primrose.

once, *f.* ounce.

oncle, *m.* uncle.

onction, *f.* unction, anointing.

onctueux, *a.* unctuous, greasy, fatty, oily.

onctuosité, *f.* unctuousness, unctuosity.

onde, *f.* wave; (*Glass*) streak; corrugation.

ondé, *a.* wavy, waved, wavelike, (of fabrics, etc.) watered; (of glass) streaked.

ondoiement, *m.* undulation, waving.

ondoyant, *p.a.* undulating, waving, wavy.

ondoyer, *v.i.* undulate, wave.

ondulant, *p.a.* undulating, waving; fluttering.

ondulation, *f.* undulation; waving.

ondulatoire, *a.* undulatory.

ondulé, *a.* undulated, waved, wavy; corrugated.

onduler, *v.i.* undulate; wave; flutter.

onduleux, *a.* undulating, wavy, undulous.

onéreux, *a.* burdensome, onerous.

ongle, *m.* nail; claw; hoof.

— *du pouce*, thumb nail.

onglet, *m.* miter, miter joint; notch, nail cut, hollow; flat graver; ungula.

onguent, *m.* ointment, unguent.

— *basilicum*, rosin cerate, basilicon ointment.

— *blanc de Rhazés*, lead carbonate ointment.

— *citrin*, ointment of mercuric nitrate, citrine ointment.

— *de staphisaigre*, stavesacre ointment.

— *diachylon*, diachylon ointment.

— *mercuriel double*, — *napolitain*, mercurial ointment, blue ointment, Neapolitan ointment.

onique, *a.* designating acids of the sugar group whose names end in -onic.

ont, *pr. 3 pl.* of avoir (to have).

onze, *a. & m.* eleven; eleventh.

onzième, *a. & m.* eleventh.

oolithe, *m.* oölite.

oolithique, *a.* oölitic.

opacifiant, *p.a.* opaquing, rendering opaque. — *m.* opaquing agent, something that causes opaqueness.

opacifier, *v.t.* opaque, render opaque. — *v.r.* become opaque.

opacimétrique, *a.* pertaining to the measurement of opacity.

opacité, *f.* opacity, opaqueness.

opalage, *m.* (*Sugar*) stirring under of the crust called *opale* (which see).

opale, *f.* (*Min.*) opal (the mineral and the color); (*Sugar*) an opaline crust of sugar crystals which forms on the surface of the liquor. — *a.* opal.

opaler, *v.t.* stir under the crust called *opale*.

opalescence, *f.* opalescence.

opalescent, *a.* opalescent.

opalin, *a.* opaline.

opaline, *f.* opaline, specif. opal glass.

opalisant, *p.a.* opalizing; opalescent.

opaliser, *v.t.* opalize.

opaque, *a.* opaque.

opérateur, *m.* operator; (*Mach.*) motor.

opératif, *a.* operative.

opération, *f.* operation.

opératoire, *a.* operative, of operating.

opercule, *m.* (*Biol.*) operculum; (*Mach.*) diaphragm.

opérer, *v.t.* operate, work, effect, do; operate upon. — *v.i.* operate, act, work. — *v.r.* be operated, be effected.

ophite, *m.* (*Petrog.*) ophite.

opiacé, *a.* containing opium.

opiacer, *v.t.* add opium to, opiate.

opianate, *m.* opianate.
opianique, *a.* opianic.
opianyle, *m.* opianyl.
opiat, *m.* (*Pharm.*) electuary; tooth paste; (formerly) opiate.
opiatique, *a.* opiatic.
opiner, *v.i.* give one's opinion, opine.
opiniâtre, *a.* obstinate; opinionated.
opinion, *f.* opinion.
opium, *m.* opium.
opobalsamum, *m.* opobalsam, opobalsamum.
opodeldoch, *m.* (*Pharm.*) opodeldoc.
opolé, opolite, *m.* (*Pharm.*) juice, sap.
opopanax, *m.* (*Bot. & Pharm.*) opopanax.
opothérapie, *f.* organotherapy, opotherapy.
opothérapique, *a.* organotherapeutic.
opportun, *a.* opportune.
opportunément, *adv.* opportunely.
opportunité, *f.* opportunity; opportuneness.
opposant, *p.a.* opposing. — *m.* opponent.
opposé, *p.a.* opposed; opposite. — *m.* opposite.
opposer, *v.t.* oppose; compare; juxtapose; urge (against), object (to). — *v.r.* be opposed; object; (with *à*) oppose.
opposite, *m.* opposite, contrary.
opposition, *f.* opposition; (for patents) caveat.
oppresser, *v.t.* oppress.
oppressif, *a.* oppressive.
oppression, *f.* oppression.
oppressivement, *adv.* oppressively.
opprimer, *v.t.* oppress.
opsonine, *f.* opsonin.
opsonique, *a.* opsonic.
opter, *v.t. & i.* choose.
optima, *a. fem.* optimum, best. — *m. pl.* optima.
optimum, *m.* optimum.
option, *f.* option.
optique, *a.* optic, optical. — *f.* optics.
optiquement, *adv.* optically.
or, *m.* gold. — *conj.* now.
 — *affiné,* refined gold.
 — *blanc,* white gold (gold alloyed with silver; also, formerly, platinum).
 — *brillant,* (*Ceram.*) a kind of gold luster.
 — *de coupelle,* refined gold.
 — *de Manheim,* Mannheim gold (a brass).
 — *en coquille,* powdered gold formerly used for gilding, ormolu.
 — *en feuille,* gold leaf.
 — *faux,* imitation gold, false gold.
 — *fulminant,* fulminating gold, gold fulminate.
 — *graphique,* (*Min.*) sylvanite.
 — *musif,* — *mussif,* mosaic gold.
 — *potable,* potable gold, aurum potabile.
 — *telluré,* (*Min.*) sylvanite.
 — *telluré plombifère,* (*Min.*) nagyagite.

or *vierge,* virgin gold.
orage, *m.* storm.
orageux, *a.* stormy.
oral, *a.* oral.
oralement, *adv.* orally.
orange, *f.* orange (the fruit). — *m.* orange (the color). — *a.* orange, orange-colored.
 — *amère,* bitter orange.
 — *douce,* sweet orange.
orangé, *a. & m.* orange.
 — *d'alizarine,* alizarin orange.
 — *de chrome,* chrome orange.
 — *Mikado,* mikado orange.
oranger, *m.* orange tree, orange; orange man.
 — *a.* orange, pertaining to oranges.
orateur, *m.* orator.
oratoire, *a.* oratorical.
orbe, *m.* orb.
orbiculaire, *a.* orbicular.
orbite, *f.* orbit.
orcanette, orcanète, *f.* orcanet, alkanet (dyer's bugloss, *Anchusa tinctoria,* or its root used as a dyestuff).
orcéine, *f.* orcein.
orcine, *f.* orcinol, orcin.
ordinaire, *a. & m.* ordinary.
 à l'—, as usual.
 d'—, ordinarily, usually.
ordinairement, *adv.* ordinarily, commonly.
ordonnance, *f.* ordering, disposition, etc. (see ordonner); order; (*Med.*) prescription.
ordonnancer, *v.t.* authorize, pass (for payment).
ordonnée, *f.* (*Geom.*) ordinate.
ordonner, *v.t.* order, dispose, arrange; order, direct, command; (*Med.*) prescribe; ordain. — *v.i.* dispose (of).
ordre, *m.* order (in various senses).
ordure, *f.* excrement, dung, ordure; dirt, dust, rubbish; (fig.) filth.
ore, *f.* (*Metal.*) twyer plate. — *adv.* now.
oreille, *f.* ear; lug, handle, projection, corner, lip, foot, etc.; (mountain) peak.
oreiller, *m.* pillow.
oreillon, *m.* cutting, paring (as of leather); (*Med., pl.*) mumps.
orfèvre, *m.* goldsmith (worker in precious metals).
orfévré, orfévri, *a.* wrought by the goldsmith.
orfèvrerie, *f.* goldsmith's art; goldsmithery; goldsmith's work, articles of gold, silver, etc.
organe, *m.* organ; (of a machine or apparatus) part, organ.
 — *actif,* (*Mach.*) working part.
organicien, *m.* organic chemist; (*Med.*) organicist. — *a.* (of chemists) organic.
organique, *a.* organic, (rarely) organical.
organiquement, *adv.* organically.
organisable, *a.* organizable.

organisant, *p.a.* organizing.
organisateur, *a.* organizing.
organisation, *f.* organization.
organiser, *v.t.* organize.
organisme, *m.* organism.
organoleptique, *a.* (*Physiol.*) organoleptic.
organo-magnésien, *a.* organic magnesium.
organo-métallique, *a.* organometallic.
organo-métalloïdique, *a.* organometal-
loïd(al).
organsin, *m.* organzine (a silk thread).
orge, *f. & m.* barley.
— *perlé,* pearl barley.
orgue, *m.* organ.
orgueil, *m.* pride; fulcrum.
orgueilleusement, *adv.* proudly.
orgueilleux, *a.* proud.
oribus, *m.* resin torch.
orichalque, *m.* orichalch (an ancient alloy).
orient, *m.* east, orient; Orient, East; (of
pearls) luster; beginning, rise.
oriental, *a. & m.* Oriental.
orientation, *f.* orientation, orienting.
orienter, *v.t.* orient; direct, guide.
orifice, *m.* orifice; mouth, aperture, etc.
origan, *m.* (*Bot.*) origanum.
originaire, *a.* native; proceeding; original.
originairement, *adv.* originally.
original, *a. & m.* original.
d'—, from the original source.
en —, in the original; in reality; in person.
originalement, *adv.* originally.
originalité, *f.* originality.
origine, *f.* origin.
dans l'—, originally, at first.
dès l'—, from the beginning, from the first.
originel, *a.* original.
originellement, *adv.* originally, from the first.
orillon, *m.* ear, lug, handle, projection.
oripeau, *m.* tinsel.
orme, *m.* elm (tree and wood).
— *à trois feuilles,* hop tree (*Ptelea trifoliata*).
— *fauve,* slippery elm (*Ulmus fulva* or its
inner bark).
orné, *p.a.* ornamented, etc. (see orner); ornate.
ornement, *m.* ornament.
ornemental, *a.* ornamental.
ornementation, *f.* ornamentation, ornament-
ing.
ornementer, *v.t.* ornament.
orner, *v.t.* ornament, adorn, decorate, embellish, deck.
ornière, *f.* rut.
ornithine, *f.* ornithine.
ornithurique, *a.* ornithuric.
orobanche, *f.* a plant of the genus *Orobanche*
or certain allied genera.
— *de Virginie,* beechdrops (*Leptamnium vir-
ginianum*).
orogénie, *f.* (*Geol.*) orogeny.

orogénique, *a.* orogenic.
oronge, *f.* (*Bot.*) amanita.
orpailleur, *m.* gold washer.
orphelin, *m. & a.* orphan.
orpiment, *m.* orpiment.
orpimenter, *v.t.* mix, treat or color with orpiment.
orpin, *m.* orpiment; (*Bot.*) orpine, orpin.
orseille, *f.* archil, orchil (plant and dye).
— *de terre,* cudbear.
orsellinique, *a.* orsellinic.
orsellique, *a.* orsellic, lecanoric.
ort, *a. & adv.* (*Com.*) gross.
orteil, *m.* toe, specif. great toe.
ortho-. ortho-, ortho.
ortho-antimonique, *a.* orthoantimonic.
orthochromatique, *a.* orthochromatic.
orthoclase, *m.* (*Min.*) orthoclase.
orthodérivé, *m.* ortho derivative.
orthographe, *f.* orthography, spelling.
orthophosphorique, *a.* orthophosphoric.
orthophtalate, *m.* orthophthalate, *o*-phthalate.
orthophtalique, *a.* orthophthalic, *o*-phthalic.
orthoptère, *a.* orthopterous. — *m.pl.* Orthop-
tera.
orthoquinone, *f.* orthoquinone, *o*-quinone.
orthose, *m.* (*Min.*) orthoclase, orthose.
orthosérie, *f.* ortho series.
orthosubstitué, *a.* ortho-substituted.
ortho-titanique, *a.* (of an acid) orthotitanic
(normal titanic).
orthovanadique, *a.* orthovanadic.
orthoxy-. orthohydroxy-, *o*-hydroxy-.
orthoxybenzoïque, *a.* *o*-hydroxybenzoic.
ortie, *f.* (*Bot.*) nettle (esp. the genus *Urtica*).
— *brûlante,* nettle (*Urtica*).
— *de Chine,* china grass, ramie.
ortol, *m.* (*Photog.*) ortol.
orvale, *f.* (*Bot.*) clary (*Salvia sclarea* and related
genera).
os, *m.* bone.
— *de seiche,* — *de sèche,* cuttlefish bone.
— *dissous,* (*Fertilizers*) dissolved bone.
osazone, *f.* osazone.
oscillant, *p.a.* oscillating.
oscillation, *f.* oscillation.
oscillatoire, *a.* oscillatory.
osciller, *v.i.* oscillate; fluctuate.
oseille, *f.* (*Bot.*) sorrel; specif., common sorrel
(*Rumex acetosa*).
oser, *v.t.* dare, venture.
osier, *m.* osier; wicker.
osmazome, *m.* (*Old. Chem.*) osmazome.
osmiamique, *a.* osmiamic.
osmiate, *m.* osmate, osmiate.
osmié, *a.* combined with or containing
osmium.
osmieux, *a.* osmious, osmous.
osmique, *a.* osmic.
osmium, *m.* osmium.

osmiure, *m.* a binary compound of osmium — usually in the phrase *osmiure d'iridium*, iridosmium, osmiridium.

osmogène, *n.* osmogene, osmotic apparatus.

osmomètre, *m.* osmometer.

osmométrie, *f.* osmometry.

osmométrique, *a.* osmometric.

osmose, *f.* osmosis, osmose.

osmoser, *v.t.* subject to or obtain by osmosis.

osmotique, *a.* osmotic.

osone, *f.* osone.

osotétrazine, *f.* osotetrazine.

osotriazol, *m.* osotriazole (1, 2, 5-triazole).

osséine, *f.* ossein.

osseux, *a.* osseous, bony, bone.

ossifier, *v.t. & r.* ossify.

ostéine, *f.* ostein (ossein).

ostensible, *a.* ostensible.

ostensiblement, *adv.* ostensibly.

ostéocolle, *f.* osteocolla.

ostracés, *m.pl.* (*Zoöl.*) Ostracea, ostraceans.

ôté, *p.a.* removed, etc. (see ôter). — *prep.* except, with the exception of.

ôter, *v.t.* remove; take away, take off, take out, take, displace, except. — *v.r.* move away, get away, retire; rid oneself (of); deprive oneself (of).

ou, *conj.* or.

ou . . . ou, either . . . or.

ou bien, or else, or otherwise.

ou bien . . . ou bien, either . . . or else.

où, *adv.* where; to what; when; at, in, into, on, of, from or with which (or whom).

d'où, whence, from where.

où que, wherever, wheresoever.

par où, by what way, by which, thru which, how, where.

vers où, to which, whereto.

ouabaïne, *f.* ouabain.

ouate, *f.* wadding; (carded) cotton, cotton wool.

ouater, *v.t.* wad, stuff with wadding, pack.

oubli, *m.* forgetfulness; forgetting.

oublie, *f.* wafer.

oublier, *v.t.* forget; omit.

ouest, *m.* west; (*cap.*) West.

oui, *adv. & m.* yes.

ouï, *p.a.* heard.

ouï-dire, *m.* hearsay.

ouïe, *f.* hearing; gill (of a fish).

ouillage, *m.* filling up (of a cask with wine).

ouiller, *v.t.* fill up (a cask by adding wine).

ouïr, *v.t.* hear.

Oural, *m.* Ural, Ural River. — *a.* Ural.

ourda, *f.* a Rumanian cheese.

ourdir, *v.t.* plait, twist; warp; hatch, concoct.

ourlet, *m.* hem; (of metals) seam; edge, border.

ours, *m.* bear.

ourse, *f.* (she) bear.

outil, *m.* tool, implement, instrument.

— *tranchant,* cutting tool, edged tool.

outillage, *m.* equipment, outfit, set of tools, tools, implements, plant.

outiller, *v.t.* equip, fit out, supply with tools or implements.

outil-machine, *n.* machine tool.

outrance, *f.* In the phrase *à outrance,* to excess, to the uttermost, to the limit.

outrancier, *a.* excessive, extravagant.

outre, *prep.* beyond; besides, beside. — *adv.* beyond, further, on. — *f.* leather bottle, skin.

d'— en —, clear thru, thru and thru.

en —, further, besides, in addition.

— que, only not . . . but.

outré, *p.a.* overdone; excessive; exasperated.

outre-Manche, *adv.* beyond the Channel (i.e., in the British Isles, in England).

outremer, *m.* ultramarine. — *a.* ultramarine (specif., beyond the sea, oversea, foreign).

— *lapis,* ultramarine from lapis lazuli, genuine ultramarine.

outre-mer, *adv.* overseas.

outre-monts. In the phrase *d'outre-monts,* ultramontane (referring esp. to Italy, or to Italy and Spain).

outrepasser, *v.t.* go beyond, exceed, overstep.

outrer, *v.t.* overdo; overstrain; provoke.

outre-Rhin, *adv.* beyond the Rhine.

ouvert, *p.a.* open; quick, ready; (of soap paste) separate from the liquid; opened, etc. (see ouvrir).

ouvertement, *adv.* openly.

ouverture, *f.* opening; aperture, orifice, hole, mouth, gap, etc.; width, span; overture; opportunity; aptness, readiness.

— *à la scorie,* (*Metal.*) slag hole.

— *de coulée,* (*Metal.*) tap hole.

— *de la chauffe,* stokehole.

— *de la vapeur,* steam port.

ouvrable, *a.* workable; (of time) working.

ouvrage, *m.* work; production, product; workmanship; hearth (of a furnace); construction.

— *blanc,* tool making or other fine smithing.

— *de fabrique,* factory product, manufacture.

— *noir,* blacksmithing, blacksmith work.

ouvragé, *p.a.* worked, wrought; figured.

ouvrager, *v.t.* work; figure.

— *la mine,* (*Metal.*) scum the ore.

ouvrant, *p.pr.* of ouvrir and of ouvrer.

ouvre, *pr. 3 sing.* of ouvrir and of ouvrer.

ouvré, *p.a.* worked, wrought; figured.

ouvreau, *m.* (*Glass,* etc.) working hole; air vent.

ouvrer, *v.t.* work; make. — *v.i.* work.

ouvreur, *m.* opener; (*Paper*) vatman, dipper.

ouvrier, *m.* workman, worker, artisan, laborer.

— *a.* working, laboring, operative.

ouvrier *fondeur,* (a workman who is a) melter, founder, smelter.

— *marcheur,* treader.

— *tiseur,* stoker.

— *verrier,* glass blower.

ouvrir, *v.t.* open; cut open; unfold; begin work on; propose. — *v.i.* open, be open. — *v.r.* open; unbosom oneself.

ouvroir, *m.* workshop; work room.

ovaire, *m.* (*Biol.*) ovary.

ovalbumine, *f.* ovalbumin.

ovale, *a.* & *m.* oval.

ové, *a.* egg-shaped, ovoid.

ovin, *a.* ovine, pertaining to sheep.

ovoïde, *a.* ovoid, egg-shaped. — *m.* ovoid.

ovomucoïde, *m.* ovomucoid.

ovovitelline, *f.* ovovitellin.

ovule, *m.* (*Biol.*) ovule.

oxacétique, *a.* hydroxyacetic, glycolic.

oxacétylurée, *f.* hydantoic acid.

oxacide, *m.* oxacid.

oxalacétique, *a.* oxalacetic.

oxalantine, *f.* oxalantine, leucoturic acid.

oxalate, *m.* oxalate.

— *de fer,* oxalate of iron, specif. (as in *Pharm.*) ferrous oxalate.

oxalaté, *a.* combined with or converted into an oxalate.

oxaldéhyde, *m.* hydroxyaldehyde; glyoxal, oxaldehyde.

oxalide, *f.* (*Bot.*) oxalis, wood sorrel.

oxalidées, *f.pl.* (*Bot.*) Oxalidaceæ.

oxalimide, *f.* oxalimide.

oxalique, *a.* oxalic.

oxalurate, *m.* oxalurate.

oxalurie, *f.* (*Med.*) oxaluria.

oxalurique, *a.* oxaluric.

oxalyle, *m.* oxalyl.

oxalylurée, *f.* oxalyl urea (parabanic acid).

oxamide, *f.* oxamide.

oxamique, *a.* oxamic.

oxanilide, *f.* oxanilide.

oxanilique, *a.* oxanilic.

oxazine, *f.* oxazine.

oxazinique, *a.* oxazine.

oxazol, *m.* oxazole.

oxazoline, *f.* oxazoline.

oxéolat, oxéolé, *m.* (*Pharm.*) acetolatum.

— *simple,* distilled vinegar.

oxhydrile, *m.* hydroxyl.

oxhydrilé, *a.* hydroxylated, hydroxyl.

oxhydrilique, *a.* hydroxylic.

oxhydrique, *a.* oxyhydrogen.

oxhydryle, *m.* hydroxyl.

oxide, *m.* oxide. (For phrases see oxyde.)

oximation, *f.* conversion into an oxime.

oxime, *m.* oxime.

oxindol, *m.* oxindole.

oxy-. oxy-, hydroxy-. (In organic names this indicates, in very many cases, the hydroxyl

group and should then be translated hydroxy- to conform to the best usage in English; e.g., *oxybutyrique,* hydroxybutyric. If the group is known to be ketonic, the translation keto- is preferable.)

oxyacétique, *a.* hydroxyacetic, glycolic.

oxyacétylénique, *a.* oxyacetylene.

oxyacide, *m.* (*Org. Chem.*) hydroxy acid.

oxyammoniaque, *m.* hydroxylamine, oxyammonia.

oxyanthraquinone, *f.* hydroxyanthraquinone.

oxyazo-. hydroxyazo-; azoxy-.

oxyazoté, *a.* combined with oxygen and nitrogen or with an oxide of nitrogen.

oxybenzoate, *m.* hydroxybenzoate.

oxybenzoïque, *a.* hydroxybenzoic.

oxybromure, *m.* oxybromide.

oxybutyrique, *a.* hydroxybutyric.

oxycarboné, *a.* combined with oxygen and carbon or with an oxide of carbon (specif. CO).

oxycétone, *f.* hydroxyketone.

oxychlorique, *a.* oxychloric (old syn. of perchloric).

oxychlorure, *m.* oxychloride.

— *ammoniacal de mercure,* ammoniated mercury, white precipitate.

— *de carbone,* carbon oxychloride, phosgene.

oxycinchonique, *a.* hydroxycinchonic; hydroxycinchoninic (less correct use).

oxycoumarine, *f.* hydroxycoumarin.

oxycyanure, *m.* oxycyanide.

oxydabilité, *f.* oxidizability, oxidability.

oxydable, *a.* oxidizable, oxidable.

oxydant, *p.a.* oxidizing. — *m.* oxidizing agent, oxidizer.

oxydase, *f.* oxidase.

oxydation, *f.* oxidation, oxidizing.

oxyde, *m.* oxide.

— *alcoolique,* alkyl oxide, ether.

— *d'antimoine,* antimony oxide, specif. (as in *Pharm.*) antimony trioxide.

— *d'argent,* silver oxide.

— *de carbone,* oxide of carbon, specif. carbon monoxide.

— *de cuivre,* copper oxide, specif. cupric oxide.

— *de fer,* iron oxide, oxide of iron.

— *de fer bihydraté,* ferric hydroxide.

— *de fer magnétique* (or *noir*), magnetic oxide of iron (Fe_3O_4).

— *de mercure jaune,* — *de mercure précipité,* yellow mercuric oxide.

— *de méthyle,* methyl oxide (methyl ether).

— *de phényle,* phenyl oxide (phenyl ether).

— *de plomb fondu,* litharge.

— *d'éthyle,* ethyl oxide (ethyl ether).

— *d'éthylène,* ethylene oxide.

— *magnétique,* magnetic oxide of iron, ferroso-ferric oxide.

— *mercurique jaune,* yellow mercuric oxide.

oxyde *mercurique rouge*, red mercuric oxide.
— *nitreux*, nitrous oxide, nitrogen monoxide.
— *noir de cuivre*, black copper oxide, cupric oxide.
— *noir de manganèse*, black oxide of manganese (manganese dioxide).
— *puce de plomb*, puce-colored oxide of lead, lead peroxide.
— *rouge de plomb*, red lead, minium.
— *salin*, an oxide which is regarded as a salt, compound oxide.
— *salin de chrome*, the oxide Cr_3O_4.
— *salin de fer*, ferroso-ferric oxide.
— *salin de plomb*, minium, red lead.
oxyder, *v.t. & r.* oxidize.
oxydeur, *m.* oxidizer.
oxydule, *m.* lower or -ous oxide, (formerly) protoxide.
— *de cuivre*, cuprous oxide.
oxydulé, *a.* in the form of a lower oxide, oxidulated. (*Archaic.*)
oxygénable, *a.* oxidizable, oxygenizable.
oxygénant, *p.a.* oxygenating, oxidizing.
oxygénase, *f.* oxygenase.
oxygénation, *f.* oxygenation, oxidation.
oxygène, *m.* oxygen.
oxygéner, *v.t.* oxygenate, oxidize.
oxygénifère, *a.* carrying oxygen.
oxygras, *a.* hydroxyfatty.
oxyhémoglobine, *f.* oxyhemoglobin, oxyhæmoglobin.
oxyhydrile, *m.* hydroxyl.

oxyhydrogène, *m.* oxyhydrogen.
oxypyridine, *f.* hydroxypyridine.
oxyiodure, *m.* oxyiodide.
oxyioduré, *a.* combined with oxygen and iodine, oxyiodide of.
oxylithe, *m.* oxylith, oxylithe.
oxyme, *m.* oxime.
oxymel, oxymellite, *m.* (*Pharm.*) oxymel.
  oxymellite simple, oxymel.
oxyquinoléine, *f.* hydroxyquinoline.
oxysaccharum, *m.* (*Pharm.*) oxysaccharum.
oxysel, *m.* oxysalt, salt of an oxacid.
oxysulfure, *m.* oxysulfide.
oxytoluique, *a.* hydroxytoluic.
oxyvalérique, *a.* hydroxyvaleric.
ozocérite, ozokérite, *f.* ozocerite, ozokerite.
ozonateur, *m.* ozonizer. — *a.* ozonizing.
ozonatrice, *a. fem.* of ozonateur.
ozone, *m.* ozone.
ozoner, *v.t.* ozonize.
ozoneur, *m.* ozonizer.
ozonique, *a.* ozonic.
ozonisation, *f.* ozonization.
ozoniser, *v.t.* ozonize.
ozoniseur, *m.* ozonizer.
ozonomètre, *m.* ozonometer.
ozonométrie, *f.* ozonometry.
ozonométrique, *a.* ozonometric.
ozonoscope, *m.* ozonoscope.
ozonoscopique, *a.* ozonoscopic.
ozotypie, *f.* (*Photog.*) ozotype.

P

p. 100, *abbrev.* (pour cent) per cent.

P.A., *abbrev.* (poids atomique) atomic weight.

pacifier, *v.t.* pacify.

pacifique, *a.* pacific.

pacifiquement, *adv.* pacifically, peacefully.

packfond, packfong, *m.* paktong, pakfong (an alloy).

padelin, *m.* (*Glass*) crucible.

P. AE., *abbrev.* (partes aequales) equal parts.

page, *f. & m.* page.

paginer, *v.t.* page, paginate.

pagodite, *f.* (*Min.*) pagodite, agalmatolite.

paiement, *m.* payment.

païen, *a. & m.* pagan.

paillasse, *f.* flask stand; masonry support, bed, foundation (as for a furnace); straw mattress, husk mattress.

paillasson, *m.* mat.

paille, *f.* straw; scale (of iron); flaw (in metals, glass, etc.). — *a.* straw-colored, straw-yellow.

menue —, chaff.

— *de bois,* wood fiber.

— *de fer,* hammer scale, forge scale, scale.

— *de maïs,* cornstalks, maize stalks.

paillé, *a.* straw-colored; (of metals) flawy.

pailler, *m.* farmyard; straw pile, straw stack; straw loft. — *v.t.* cover or wrap with straw.

paillet, *a.* pale red wine, verging on straw color; mat, straw mat. — *a.* pale red, verging on straw color.

pailleté, *a.* lamellar, laminar; spangled.

paillette, *f.* small plate or scale, lamella, platelet, leaflet, lamina; shining particle (as of gold dust); spangle; scale (of iron), forge scale; flaw (in gems).

pailleux, *a.* straw, strawy; (of metals) flawy.

paillon, *m.* scale, plate; leaf, foil (of metals); tin plate; large spangle; link; bundle of straw used as a filter; straw envelope for a bottle.

— *de soudure,* leaf solder.

pailloner, *v.t.* cover with sheet metal, esp. tin.

pail-mel, *n.* feed made of straw and molasses.

pain, *m.* bread; loaf; wafer; cake (as of wax or grease); lump; pig (of metal); block (of soap as it comes from the frame).

en —, in the form of a cake or lump or the like, in the cake, in the lump.

fonte en — *de sucre,* see under *fonte.*

pain *à blanchir,* cake of whiting.

— *à cacheter,* wafer.

— *à café,* coffee bread.

— *à crétons,* greaves cake.

— *azyme,* unleavened bread.

— *bis,* brown bread.

— *biscuité,* bread baked longer than usual.

— *blanc,* white bread.

— *de cire,* cake of wax; wax wafer.

— *de commission,* ration bread.

— *de coucou,* (*Bot.*) wood sorrel.

— *de froment,* wheat bread.

— *de grenouilles,* (*Bot.*) water plantain.

— *de liquation,* (*Copper*) liquation cake.

— *de mer,* ship biscuit.

— *de munition,* ration bread, ammunition bread (esp. a grayish bread made from second and third grade flours).

— *de pourceau,* (*Bot.*) sow bread.

— *de seigle,* rye bread.

— *de sucre,* sugar loaf.

— *levé,* raised bread.

— *noir,* black bread.

— *rassis,* stale bread.

— *tendre,* new bread.

— *viennois,* Vienna bread.

— *visqueux,* ropy bread.

pair, *a.* (of numbers) even. — *m.* par; peer.

au —, at par.

de —, on a par, on an equality.

hors de —, *sans* —, without a peer, peerless.

paire, *f.* pair.

pairement, *adv.* (of numbers) evenly, even.

paisible, *a.* peaceful, peaceable.

paisiblement, *adv.* quietly, peacefully.

paix, *f.* peace.

palais, *m.* palate; palace; court (of justice); bar, law, legal profession.

palan, *m.* tackle.

pale, *f.* pile; pale, stake; bung, stopper; vane, blade, board; sluice gate.

pâle, *a.* pale.

paléozoïque, *a.* (*Geol.*) Paleozoic, Palæozoic.

paletot, *m.* overcoat, greatcoat.

palette, *f.* paddle; spatula; (*Glass*) a tool for rounding off the mass of metal gathered on the pipe; (*Ceram., Gilding,* etc.) pallet; (*Mach.*) palette, drill plate; (painter's) palette.

254

palétuvier, *m.* (*Bot.*) mangrove (also, any of several other tropical trees).

pâleur, *f.* paleness, pallor.

palier, *m.* (*Mach.*) bearing; landing (as of a stairway); floor; level stretch.
— *à billes*, ball bearing.
— *graisseur*, self-oiling bearing.

palier-graisseur, *m.* self-oiling bearing.

pâlir, *v.i.* pale, grow pale, turn pale. — *v.t.* make pale, blanch.

palisson, *m.* (*Leather*) (1) stake, (2) stretching machine.

palissoner, *v.t.* stretch (leather).

palladate, *m.* palladate.

palladeux, *a.* palladious.

palladique, *a.* palladic.

palladium, *m.* palladium.

palladure, *f.* palladide.

palliatif, *a. & m.* palliative.

palma-christi, *m.* (*Bot.*) castor-oil plant, palma Christi (*Ricinus communis*).

palmarosa, palma-rosa, *f.* (*Bot.*) palmarosa, ginger grass (*Andropogon schœnanthus*).

palme, *f.* palm.

palmer, *m.* micrometer gage.

palmier, *m.* (*Bot.*) palm tree, palm.
— *à chanvre*, hemp palm.
— *à cire*, wax palm.
— *à huile*, oil palm (*Elœis guineënsis*).
— *nain*, dwarf fan palm (*Chamœrops humilis*).

palmiste, *m.* (*Bot.*) cabbage palm.

palmitate, *m.* palmitate.

palmitine, *f.* palmitin.

palmitique, *a.* palmitic.

palon, *m.* wooden spatula (for wax, tallow, etc.).

pâlot, *a.* palish, rather pale.

palper, *v.t.* feel.

palpiter, *v.i.* palpitate.

paludéen, *a.* marshy; (*Med.*) malarial.

paludier, *m.* worker in a salt garden.

palus, *m.* marsh.

palustre, *a.* marsh.

pamplemousse, *m.* shaddock, grapefruit, pomelo (*Citrus decumana*).

pan, *m.* face (of a crystal or other solid); facet; side, flat surface; pane; panel; partition.

panache, *m.* plume, panache; finial.

panaché, *a.* variegated; mixed; plumed. *cuivre* —, (*Min.*) bornite.

panacher, *v.t.* variegate; plume.

panachure, *f.* variegation, spot or stripe of a different color.

panaire, *a.* panary, of bread or bread making.

panais, *m.* parsnip.

panama, *m.* soapbark, quillaja bark, Panama bark (from *Quillaja saponaria*); Panama hat.

pancarte, *f.* placard, bill; (written or printed) sheet; cover (for papers), folder.

panclastite, *f.* (*Expl.*) panclastite.

pancréas, *m.* pancreas.

pancréatine, *f.* pancreatin.

pancréatique, *a.* pancreatic.

panerée, *f.* basketful.

panier, *m.* basket; pannier; beehive.

panifiable, *a.* capable of being made into bread; bread-making.

panification, *f.* bread making, conversion into bread, panification.

panifier, *v.t.* convert into bread.

panne, *f.* fat (esp. of swine); plush; (of a hammer) peen; pantile; (of roofs) purlin, rib; (of boats) heaving to; stop, halt.

panneau, *m.* panel; pane (as of glass); face (of a stone); template; snare, trap.

panneresse, *f.* (*Masonry*) stretcher.

panse, *f.* belly, paunch.

pansement, *m.* dressing (of wounds); grooming.

panser, *v.t.* dress (wounds); groom (animals).

pansu, *a.* big-bellied.

pantage, *m.* bundling (of hanks).

pantalon, *m.* trousers.

pante, *n.* bundle (of hanks).

panteler, *v.i.* pant, gasp.

pantenne, pantène, *f.* wicker tray.

panter, *v.t.* bundle (hanks).

pantoufle, *f.* slipper.

paon, *m.* peacock, peafowl (*Pavo*).

papaïne, *f.* papain.

papavéracées, *f.pl.* (*Bot.*) Papaveraceæ.

papavéraldine, *f.* papaveraldine.

papavérine, *f.* papaverine.

papavérique, *a.* papaveric.

papaye, *f.* papaya, papaw (the fruit).

papayer, *m.* papaya tree, papaya (*Carica papaya*).

pape, *m.* pope.

paperasse, *f.* old paper, used or waste paper.

papeterie, *f.* paper making; paper mill, paper factory; paper business or trade; stationery; writing case.

papetier, *m.* paper maker; paper seller, specif. stationer; stationer's shop. — *a.* paper, papermaking; stationery.

papier, *m.* paper.
— *à calquer*, — *à calque*, tracing paper.
— *à cartouches*, cartridge paper.
— *à copier*, carbon paper, copying paper.
— *à crayons*, crayon paper, crayon board.
— *à décalquer*, transfer paper.
— *à dessin*, — *à dessiner*, drawing paper.
— *à dialyse*, dialyzing paper, parchment paper.
— *à écrire*, writing paper.
— *à envelopper*, wrapping paper.
— *à étoupilles*, primer paper.
— *à filigrane*, watermarked paper.
— *à filtrer*, filter paper.

papier à *gargousses*, cartridge paper.
— à *journaux*, newspaper stock, news.
— à *la cuve*, vat paper, handmade paper.
— à *la main*, handmade paper.
— à *la phtaléine du phénol*, phenolphthalein paper.
— *albuminé*, albumenized (or albuminized) paper.
— *albumineux*, albumen paper.
— à *l'émeri*, emery paper.
— à *lettre*, letter paper.
— à *patrons*, cross-section paper.
— à *poncer*, tracing paper, pounce paper.
— *apprêté*, paper subjected to calendering, coating, oiling, marbling, goffering or any other finishing process.
— à *réactif*, test paper.
— *argenté*, silver paper, specif. a fine white tissue paper.
— à *sucre*, the dark blue paper in which sugar loaves are wrapped.
— *au cyano-fer*, blueprint paper.
— *au ferro-prussiate*, blueprint paper.
— *autocopiste*, copying paper.
— *blanc*, white paper; blank paper.
— *Bristol*, Bristol board, Bristol paper.
— *brouillard*, blotting paper.
— *bulle*, scratch paper, scribbling paper.
— *buvard*, bibulous paper, absorbent paper, blotting paper.
— *calque*, tracing paper.
— *carbone*, carbon paper.
— *chimique*, chemically impregnated paper, as for tests; carbon paper.
— *ciré*, waxed paper.
— *Colas*, black-print paper (black lines on a white ground).
— *collé*, sized paper.
— *coquille*, letter paper.
— *couché*, coated paper.
— *court*, (Com.) short-term paper, short-term note.
— *cuir*, leather paper.
— *curcuma*, turmeric paper.
— *d'amiante*, asbestos paper.
— *d'argent*, tin foil.
— *de bois*, wood paper, paper made from wood pulp.
— *de Chine*, Chinese paper.
— *de coton*, paper made from cotton.
— *de fantaisie*, fancy paper.
— *de Fernambouc*, test paper impregnated with extract of Pernambuco wood.
— *dégraissé*, fat-extracted paper.
— *de Hollande*, Dutch paper.
— *d'emballage*, wrapping paper.
— *d'émeri*, emery paper.
— *de montagne*, (Min.) a papery form of asbestos, mountain paper.
— *de paille*, straw paper.

papier *de poste*, letter paper.
— *d'épreuve*, test paper.
— *de rebut*, waste paper.
— *de riz*, rice paper.
— *de serpente*, fine tissue paper.
— *de soie*, a fine white tissue paper.
— *de sûreté*, safety paper.
— *d'étain*, tin foil.
— *de tenture*, wall paper, paper hangings.
— *de tournesol*, litmus paper.
— *de verre*, glass paper.
— *d'oiseau*, very thin paper for messages by pigeon.
— *doré*, gilt paper.
— *du Japon*, Japanese paper.
— *durci*, hardened paper, hard paper.
— *écolier*, white paper of medium quality used for school work.
— *en bloc*, paper in the form of a block or pad.
— *entoilé*, cloth-mounted paper.
— *filigrane*, watermarked paper.
— *filtre*, filter paper.
— *fossile*, asbestos paper.
— *gauffré*, goffered paper.
— *Gilot*, a kind of drawing paper.
— *glacé*, glazed paper.
— *gommé*, gummed paper.
— *goudronné*, tar paper, tarred paper.
— *huilé*, oiled paper.
— *hygiénique*, toilet paper.
— *indien*, India paper.
— *iodoamidonné*, — *ioduré amidonné*, starch-iodide paper.
— *ioduré*, iodide paper, iodized paper.
— *joseph*, filter paper; a bibulous white or light gray tissue paper, called in full *papier joseph de soie*.
— *lacmoïde*, lacmoid paper.
— *lavé*, washed paper, extracted paper.
— *libre*, unstamped paper.
— *lissé*, glazed paper.
— *long*, (Com.) long-term paper, long-term note.
— *mâché*, papier-mâché (paper pulp with size, etc.).
— *marbré*, marbled paper.
— *Marion*, blueprint paper (white lines on a blue ground).
— *maroquiné*, morocco paper.
— *marqué*, stamped paper.
— *mécanique*, machine-made paper.
— *mi-collé*, half-sized paper.
— *moiré*, watered paper, moiré paper.
— *mort*, unstamped paper.
— *moutarde*, (Pharm.) mustard paper.
— *ozonométrique*, ozonometric paper, esp. starch-iodide paper.
— *parchemine*, — *parcheminé*, parchment paper.
— *peigne*, grained paper.

papier *peint*, wall paper, paper hangings.
— *Pellet*, blueprint paper (white lines on a blue ground).
— *pelure*, onionskin, foreign post paper.
— *photographique*, photographic paper.
— *plissé*, crêpe paper.
— *plombaginé*, paper coated with graphite.
— *plombique*, lead paper (impregnated with a lead salt for the detection of hydrogen sulfide).
— *poudre*, gunpaper.
— *quadrillé*, cross-section paper.
— *réactif*, test paper.
— *réglé*, ruled paper.
— *Rigollot*, (*Pharm.*) mustard paper.
— *sablé*, sandpaper.
— *satiné*, satined paper, satin paper, plated paper.
— *sensible*, sensitive paper, sensitized paper.
— *sinapisé*, (*Pharm.*) mustard paper.
— *sparadrapique*, (*Pharm.*) paper, medicated paper.
— *suédois*, Swedish paper, specif. Swedish filter paper.
— *témoin*, (*Expl.*) standard-tint paper.
— *timbré*, stamped paper.
— *toile*, tracing cloth.
— *tournesol*, litmus paper.
— *tue-mouches*, fly paper.
— *végétal*, tracing paper.
— *vélin*, a superior wove paper, vellum paper.
— *vergé*, laid paper.
— *volant*, loose sheet of paper.
papier-carton, *m.* paper board.
papier-goudron, *m.* tar paper.
papier-monnaie, *m.* paper money.
papier-parchemin, *m.* parchment paper.
papier-tenture, *n.* wall paper, paper hangings.
papilionacé, *a.* papilionaceous.
papilionacées, *f.pl.* (*Bot.*) Papilionaceæ.
papille, *f.* papilla.
papillon, *m.* butterfly; insert, extra sheet; butterfly valve; damper. — *a.* butterfly.
papilloter, *v.i.* (of colors) dazzle, be dazzling.
paprika, *n.* paprika, paprica.
papyrus, *m.* papyrus.
pâque, *f.* passover. — *m.* (*cap.*) Easter.
paquebot, *m.* packet, mail boat.
paquerette, *f.* self-heal (*Prunella vulgaris*); daisy (*Bellis*).
Pâques, *m.* Easter.
paquet, *m.* packet, package; parcel, bundle; (*Metal.*) pile, fagot; mail; packet, mail boat.
paquetage, *m.* packing; pack; (*Metal.*) piling.
paqueter, *v.t.* pack; (*Metal.*) pile, fagot.
paqueteur, *m.*, **paqueteuse**, *f.* packer.
par, *prep.* thru; by; about, over; on, upon; per; (of division) by, into, in; (of time or weather) in.

par (continued).
de —, by the order of, in the name of.
— *entre*, between.
(For other phrases beginning with *par*, see the next important word in the phrase.)
para-. para-, (in some names) *p*-.
parabanique, *a.* parabanic.
parabole, *f.* (*Geom.*) parabola; parable.
parabolicité, *f.* parabolicness.
parabolique, *a.* parabolic, parabolical.
paraboliquement, *adv.* parabolically.
paraboliser, *v.t.* parabolize, make parabolic.
paraboloïde, *m.* (*Geom.*) paraboloid.
paracaséine, *f.* paracasein.
paracentrique, *a. & f.* paracentric.
parachèvement, *m.* finishing, completion.
parachever, *v.t.* finish, complete.
parachloré, *a.* parachloro, *p*-chloro.
paracrésol, *m.* paracresol, *p*-cresol.
paracyanure, *m.* paracyanide.
paracyanogène, *m.* paracyanogen.
parade, *f.* parade; show.
paradérivé, *m.* para derivative.
paradiamido-. paradiamino-, *p*-diamino-.
paradis, *m.* paradise.
paradoxal, *a.* paradoxical.
paradoxalement, *adv.* paradoxically.
paradoxe, *m.* paradox.
paraffinage, *m.* paraffining.
paraffine, *f.* paraffin, (*Com.*) paraffine.
— *brute*, crude paraffin.
— *fossile*, fossil paraffin, ozocerite.
— *liquide*, (*Pharm.*) liquid petrolatum, liquid paraffin.
paraffiner, *v.t.* paraffin, paraffine.
parage, *m.* dressing, etc. (see *parer*); lineage, degree; locality, quarter, waters, parts.
paragraphe, *m.* paragraph; (*Printing*) section.
paraison, *f.* (*Glass*) a preliminary blowing and shaping done by the blower's assistant, also the mass of glass so shaped.
paraisonner, *v.t.* blow (glass) in a preliminary manner.
paraisonnier, *m.* = grand gamin, under *grand*.
paraissant, *p.pr.* of paraître.
paraitre, *v.i.* appear; shine, be conspicuous.
— *m.* appearance.
faire —, show, exhibit, display; cause to appear; publish (books, etc.).
laisser —, allow to be seen, show.
parajour, *m.* shade (as for a lamp).
paralactique, *a.* paralactic.
paraldéhyde, *m.* paraldehyde.
parallactique, *a.* parallactic, parallactical.
parallaxe, *f.* parallax.
parallèle, *a. & f.* parallel. — *m.* parallel; parallel ruler.
parallèlement, *adv.* in a parallel manner.
parallélépipède, **parallélipipède**, *m.* (*Geol.*) parallelepiped, parallelepipedon.

parallélépipédique, parallélipipédique, *a.* parallelepipedal, parallelepipedonal.

paralléli que, *a.* parallel.

paralléliser, *v.t.* parallelize, parallel.

parallélisme, *m.* parallelism.

parallélogrammatique, *a.* parallelogram-matic(al), parallelogrammic.

parallélogramme, *m.* (*Geom.*) parallelogram. — *de Wheatstone,* (*Elec.*) Wheatstone bridge.

paralysant, paralysateur, *a.* paralyzing.

paralyser, *v.t.* paralyze.

paralysie, *f.* paralysis.

paralytique, *a.* & *m.* paralytic.

paramagnétique, *a.* paramagnetic.

paramagnétisme, *m.* paramagnetism.

paramètre, *m.* parameter.

paramino-. para-amino-, *p*-amino-.

parangon, *m.* paragon; comparison; flawless gem. — *a.* flawless.

paranitré, *a.* paranitro, *p*-nitro.

parant, *p.a.* ornamental; dressing, etc. (see parer).

paranucléine, *f.* paranuclein.

paranucléique, *a.* paranucleic.

paranucléoprotéide, *f.* paranucleoprotein.

paraoxybenzoïque, *a.* *p*-hydroxybenzoic.

paraoxybenzylique, *a.* *p*-hydroxybenzyl.

paraphtalique, *a.* paraphthalic, *p*-phthalic (terephthalic).

parapluie, *f.* umbrella; hood (as for a chimney).

pararosaniline, *f.* pararosaniline.

parasemidique, *a.* parasemidine, *p*-semidine.

parasérie, *f.* para series.

parasitaire, *a.* parasitic(al).

parasite, *m.* parasite. — *a.* parasitic(al).

parasiticide, *a.* & *m.* parasiticide.

parasitique, *a.* parasitic(al).

parasoleil, *m.* lens hood.

parasorbique, *a.* parasorbic.

parastannique, *a.* parastannic.

parasulfoné, *a.* parasulfonated, parasulfo, *p*-sulfo.

paratartrique, *a.* paratartaric (racemic).

paratonnerre, *m.* lightning rod; lightning arrester.

paravent, *m.* screen, folding screen.

paraxanthine, *f.* paraxanthine.

parc, *m.* park; inclosure; pen; fold; depot.

parcelle, *f.* small part, particle; (of land) parcel, plot.

parceller, *v.t.* parcel, parcel out, subdivide.

parce que, *conj.* because. — *m.* cause, reason.

parchemin, *m.* parchment. — *végétal,* vegetable parchment, parchment paper.

parcheminé, *p.a.* parchmentized; parchment-like; parchment.

parcheminer, *v.t.* parchmentize.

parcheminerie, *f.* parchment making or trade.

parchemineux, *a.* of the nature or appearance of parchment.

parcheminier, *m.* parchment maker.

par-ci, *adv.* here, hither; now. —, *par-là,* here and there, now and then.

parcimonie, *f.* parsimony, sparingness.

parcimonieux, *a.* parsimonious, niggardly.

parcourir, *v.t.* go over, wander over, go thru, run thru, survey, examine.

parcours, *m.* course; line; trip, journey.

par-dessous, *adv.* & *prep.* underneath, beneath, under.

par-dessus, *adv.* & *prep.* above, over. — *m.* (usually spelled **pardessus**) overcoat.

par-devant, *adv.* before, in front. — *prep.* before.

pardonnable, *a.* pardonable.

pardonner, *v.t.* & *i.* pardon. — *à,* pardon; spare.

paré, *p.a.* dressed, etc. (see parer).

pare-éclats, *m.* shield or guard to protect from flying fragments, splinter-proof.

pare-étincelles, *m.* spark arrester.

parégorique, *a.* & *m.* (*Med.*) paregoric.

pareil, *a.* like, alike, similar, equal; same (esp. of time); such. — *m.* like, equal, match. *sans* —, without equal, unequaled, exceptional.

pareille, *f.* like, same, equal. *sans* —, without equal, unequalled, exceptional.

pareillement, *adv.* in like manner, similarly, equally, also, too.

parement, *m.* dressing, etc. (see parer); ornament, adornment; face, facing (as of brick); curb, curbstone.

parenchymateux, *a.* parenchymatous.

parenchyme, *m.* parenchyma.

parent, *m.* & *f.* relative, relation. — *m.* parent; ancestor.

parenté, *f.* relationship; kindred, relatives.

parenthèse, *f.* parenthesis.

parer, *v.t.* dress; trim; finish; prepare; adorn, embellish; ward, ward off; avert, avoid; parry; shelter, protect; clear (as an anchor). — *v.i.* (with *à*) guard against. — *v.r.* ripen, mature; dress oneself; be dressed; make a show; protect oneself.

paresseux, *a.* idle, lazy; sluggish, slow.

pare-torpilles, *m.* torpedo net.

pareur, *m.* dresser, etc. (see parer).

pareuse-encolleuse, *f.* dressing and sizing machine.

parfaire, *v.t.* complete, fill out; perfect; finish. — *v.r.* be completed, be finished.

parfaisant, *v.pr.* of parfaire.

parfait, *p.a.* perfect; completed, perfected, finished. — *m.* perfection; perfect (tense); ice cream (esp. with coffee flavor).

parfaitement, *adv.* perfectly.

parfera, *fut. 3 sing.* of parfaire.
parfit, *p.def. 3 sing.* of parfaire.
parfois, *adv.* sometimes, at times, occasionally.
parfondre, *v.t.* fuse thoroly, incorporate by fusion (as colors in an enamel).
parfont, *pr. 3 pl.* of parfaire.
parfum, *m.* perfume.
parfumer, *v.t.* perfume, scent; fumigate.
parfumerie, *f.* perfumery.
parfumeur, *m.* perfumer.
pari, *m.* bet, wager.
pariétaire, *f.* (*Bot.*) pellitory (*Parietaria*). — *officinale,* wall pellitory (*P. officinalis*).
parisien, *a.* Parisian, Paris. — *m.* (*cap.*) Parisian.
parité, *f.* parity, equivalence, correspondence; evenness (of numbers).
pari-valence, *f.* even valence.
parjure, *m.* perjury; perjurer. — *a.* perjured.
parlant, *p.a* speaking; expressive; lifelike.
parlement, *m.* parliament.
parlementaire, *a.* parliamentary.
parler, *v.i.* speak, talk. — *v.t.* speak of, talk of; speak, talk. — *v.r.* be spoken; be spoken of. — *m.* speaking, speech.
sans — de, to say nothing of.
parménie, *f.* (*Bot.*) stinking hellebore (*Helleborus fœtidus*).
parmesan, *a.* Parmesan, of Parma. — *m.* Parmesan cheese; (*cap.*) Parmesan.
parmi, *prep.* among, amid, in the midst of.
paroi, *f.* wall, side, surface (esp. interior); face, facing (of masonry); lining; shell (of a boiler); partition wall, partition. — *extérieure,* exterior surface, outside.
paroir, *m.* scraper; paring knife.
parole, *f.* word; speech; tone of voice; leave to speak; eloquence; credit, trust.
parou, *m.* dressing (of cloth).
parquer, *v.t.* pen, pen up; park.
parquet, *m.* floor (esp. in parquetry); platform; back of a mirror frame; parquet.
parrain, *m.* sponsor.
parsemer, *v.t.* strew, sprinkle, scatter, disseminate.
part, *f.* part; portion; share; information; source; allowance; side; place. — *m.* birth, childbirth. — *pr. 3 sing.* of partir.
à —, apart, aside, separately, alone; odd, singular; except, except for.
d'autre —, on the other side, on the other hand, as an offset.
de — en —, from one side to the other, right thru.
de — et d'autre, on both sides; on all sides, in all directions; reciprocally.
des deux parts, on both sides.
de toute —, de toutes parts, on all sides, in all quarters.
faire la — de, make allowance for.

part (continued).
faire — de, communicate, tell, announce (to someone).
partage, *m.* division, partition; portion, part, share, lot.
partagé, *p.a.* divided; supplied, endowed.
partageable, *a.* divisible.
partager, *v.t.* divide; share; supply, endow (with a share of something). — *v.r.* be divided; part. — *v.i.* partake, share.
partant, *p.a.* starting, etc. (see partir). — *adv.* therefore, consequently.
parténement, *m.* (*Salt*) = chauffoir.
partent, *pr. 3 pl.* of partir.
parthénogenèse, *f.* (*Biol.*) parthenogenesis.
parti, *m.* party; treatment; determination, decision; means, way, measure; profit, advantage. — *p.p.* of partir.
participant, *a.* participating. — *m.* participant.
participation, *f.* participation.
participe, *m.* participle.
participer, *v.i.* (with *à*) participate in, share in, take part in; (with *de*) partake of.
particulariser, *v.t. & r.* particularize.
particule, *f.* particle.
particulier, *a.* particular; separate; special; private; odd. — *m.* particular; privacy, private life; private person, individual.
en —, in private, privately; in particular, particularly.
particulièrement, *adv.* particularly, in particular, specially.
partie, *f.* part; party; project, plan; specialty, line; lot (of goods); entry (in bookkeeping); game.
faire — de, form part of, be a constituent of.
— du milieu, middle part, middle piece.
— d'about, end part, end piece.
partiel, *a.* partial.
partiellement, *adv.* partially, in part.
partir, *v.i.* start; set out, depart, leave; proceed; dart off, dart out, dart, shoot, (of guns, etc.) go off; come off. — *m.* departure.
à — de, starting from, beginning with or at, from, from . . . on.
faire —, make go, run, send, etc.; fire (as a gun); remove (as a spot).
partition, *f.* partition.
partout, *adv.* everywhere.
de —, everywhere, on all sides, from all sides.
— où, wherever.
paru, *p.a.* appeared; (of books) published.
parure, *f.* paring; adornment; ornament; dress; set (as of gems); match.
parurent, *p.def. 3 pl.* of paraître.
parut, *p.def. 3 sing.* of paraître.
parvenir, *v.i.* arrive, attain, come; succeed. — *à,* arrive at, reach, obtain; succeed in.

parvenu, *p.a.* arrived, come; successful. — *m.* parvenu, upstart.

parvient, *pr. 3 sing.* of parvenir.

parvint, *p.def. 3 sing.* of parvenir.

parvoline, *f.* parvoline.

pas, *m.* step; pace; passage, way, pass; strait; thread (of a screw); pitch (as of a screw); precedence. — *adv.* (with or without *ne*) not, no.

   de ce —, this instant.

   — à gaz, gas thread.

   — à —, step by step.

   — d'âne, coltsfoot (*Tussilago farfara*).

   — de, no.

   — de Calais, Straits of Dover.

   — de clerc, blunder.

   — du tout, not at all.

   — un, not one, none, not a, no.

passable, *a.* passable, tolerable.

passablement, *adv.* passably, tolerably, rather.

passage, *m.* passing, etc. (see passer); passage; road, way; crossing (as of a railway).

passager, *a.* transient, transitory, temporary, shortlived; passing. — *m.* traveler; passenger.

passagèrement, *adv.* transiently, temporarily.

passant, *p.a.* passing, etc. (see passer); public; crowded. — *m.* passer-by; traveler.

passe, *f.* pass; specif., (*Soap*) a single milling.

passé, *a.* past; passed, etc. (see passer); passé, out of fashion. — *m.* past. — *prep.* after; beyond.

   — maître, past master.

passementerie, *f.* passementerie, trimmings; also, passementerie factory, art or trade.

passe-méteil, *m.* maslin (⅔ wheat, ⅓ rye).

passe-partout, *m.* saw (of several kinds); master key; passport; plain frame; etc.

passe-perle, *m.* very fine iron wire.

passeport, *m.* passport.

passer, *v.i.* pass; pass away; get, go, come; (of colors) fade; serve, last. — *v.t.* pass; strain; sift, screen; pass over, omit; excuse; satisfy; go beyond, exceed, be over; surpass; outlast; put; put on; sharpen (tools); spend (time); make (a contract); sell. — *v.r.* pass; fade; happen; go, proceed; be content (with); dispense (with).

   en — par, submit to, put up with.

   faire —, cause to pass, make go, usher, bring over, pass, hand, transmit, remove, get rid of, cure.

   laisser —, let pass, let go.

   — à l'encre, ink (as a drawing).

   — au bleu, blue.

   — au noir, black, blacken, (*Founding*) blackwash.

   — au tamis, sift.

   — par, pass by, pass thru, go thru.

passerelle, *f.* footbridge.

passerie, *f.* (*Leather*) plumping liquor.

passe-temps, *m.* pastime.

passe-violet, *m.* (*Metal.*) violet color.

passible, *a.* sensible; (*Law*) liable.

passif, *a.* passive. — *m.* (*Com.*) liabilities.

passivement, *adv.* passively.

passiveté, passivité, *f.* passivity, passiveness.

passoire, *f.* colander, strainer.

pastel, *m.* pastel, woad (the dye or plant); pastel (the crayon or drawing). — *a.* pastel.

pastèque, *f.* (*Bot.*) watermelon.

pasteurien, *a.* of Pasteur, Pasteurian.

pasteurisateur, *m.* pasteurizer.

pasteurisation, *f.* pasteurization.

pasteuriser, *v.t.* pasteurize.

pastillage, *m.* (*Ceram.*) ornamentation in relief, made by pouring slip or by modeling and attaching.

pastillageur, *m.* (*Ceram.*) ornamenter in relief.

pastille, *f.* pastil, pastille, lozenge, troche, tablet; pellet.

   — à brûler, fumigating pastil.

   — du sérail, aromatic pastil.

   — fulminante, pellet primer.

   pastilles de Vichy, pastilles digestive, (*Pharm.*) troches of sodium bicarbonate.

pastilleur, *m.,* **pastilleuse,** *f.* a worker or machine that makes pastils or lozenges.

paston, *m.* (*Ceram.*) a piece of paste of convenient size for use in modeling, e.g. a cylinder about 15 × 4 cm. used in making glass pots.

patate, *f.* sweet potato (*Batatas batatas*).

patchouli, *m.* (*Bot.*) patchouli, patchouly.

pâte, *f.* paste; dough; (*Paper*) pulp, stuff; (*Ceram.*) paste, pâte; (*Powder*) mill cake; stuff, material; sort, kind; (*Printing*) pi.

   — à autocopier, copying paste.

   —alimentaire, = pâte d'Italie.

   — à papier, paper pulp.

   — à potage, = pâte d'Italie.

   — à savon, soap paste.

   — au bisulfite, (*Paper*) sulfite pulp.

   — chimique, (*Paper*) chemical pulp.

   — d'alfa, alfa pulp, esparto pulp.

   — de bois, (*Paper*) wood pulp.

   — de chiffons, rag pulp.

   — de fourneau, furnace lute.

   — demi-chimique, (*Paper*) semichemical pulp.

   — dentifrice, tooth paste.

   — de paille, straw pulp.

   — de papier, papier-mâché.

   — de riz, rice paste; milky crystal glass.

   — de sculpteur, a hard white porcelain paste for statuettes.

   — de sparte, esparto pulp.

   — d'Italie, Italian paste, edible paste (the paste from which macaroni, vermicelli, etc. are made, also the products themselves).

pâte *dure,* (*Ceram.*) hard paste.
— *ferme,* firm paste, stiff paste; stiff dough.
— *flamande,* a paste consisting of minium and polishing wax.
— *liquide,* (*Ceram.*) liquid paste, slip.
— *mécanique,* (*Paper*) mechanical pulp.
— *molle,* soft paste.
— *pour chaussures,* shoe polish.
— *sèche,* (*Ceram.*) clay dust from which articles are molded by pressure.
— *tendre,* (*Ceram.*) soft paste.

pâté, *m.* pasty, meat pie, patty; blot; mass (of buildings, of earth, etc.).

patent, *a.* patent.

patentable, *a.* liable to the license tax.

patente, *f.* license; (ship's) bill of health.
— *de santé,* bill of health.
— *nette,* clean bill (of health).

patenté, *p.a.* licensed; (in names of certain dyes) patent. — *m.* licensee.

patenter, *v.t.* license.

patère, *f.* support attached to a wall, as a coat hook or a block for a gas fixture.

pâteux, *a.* pasty; doughy; viscous; sticky; (of gems) milky, cloudy.

pathogénique, pathogène, *a.* pathogenic.

pathologie, *f.* pathology.

pathologique, *a.* pathological.

pathologiquement, *adv.* pathologically.

pathologiste, *m.* pathologist.

patiemment, *adv.* patiently.

patience, *f.* patience; specif., (*Bot.*) a coarse European dock (*Rumex patientia*).

patient, *a. & m.* patient.

patin, *m.* support, block, plate, foot, shoe, sole; slide block, guide block; skate.

patine, *f.* patina.

patiner, *v.t.* patinate. — *v.i.* skate; skid, slip.

patio, *m.* (*Metal.*) patio, paved floor.

pâtir, *v.i.* suffer.

pâtissage, *m.* pastry making.

pâtisser, *v.t.* work (flour) for pastry, make into pastry. — *v.i.* make pastry.

pâtisserie, *f.* pastry; pastry making; pastry shop; pastry business.

pâtissier, *m.* pastry cook or seller.

pâton, *m.* lump of dough, esp. the amount necessary for a loaf; (*Paper*) lump, ball.

patouille, *f.* ore separator.

patouillet, *m.* mixing or stirring apparatus, as an ore washer or separator, a pug mill for clay, or an apparatus for preparing milk of lime.

patouilleux, *a.* splashy, sloppy, muddy.

patrie, *f.* country, native land, native place.

patriote, *m.* patriot. — *a.* patriot, patriotic.

patriotiquement, *adv.* patriotically.

patron, *m.* pattern; model; sample; template; stencil; patron; master.

patronage, *m.* stenciling; patronage; association for benevolence, protection or the like.

patronner, *v.t.* stencil; trace by a pattern; stencil.

patrouiller, *v.i.* dabble, paddle; patrol.

patte, *f.* paw; foot; claw, claws; leg (of insects); (*Tech.*) foot; cramp, clamp; flange; flap, tongue, tab; catch, pawl; bottom (of a sugar loaf).
— *de coq,* (*Ceram.*) cockspur.
— *d'oie,* (*Bot.*) goosefoot; = patte-d'oie (below).

patte-d'oie, *f.* junction of three or more passages, cords, timbers or the like.

pattinsonage, *m.* (*Metal.*) pattinsonizing, Pattinson process.

pattinsoner, *v.t.* pattinsonize.

pâturage, *m.* pasturage, pasture.

pâturant, *p.a.* pasturing, grazing, herbivorous.

pâture, *f.* feed, food; forage; pasture.

pâturer, *v.i. & t.* pasture, graze.

paucité, *f.* paucity.

paume, *f.* palm (of the hand); tennis.

paupière, *f.* eyelid.

pause, *f.* pause; stop, rest, suspension.

pauser, *v.i.* pause.

pauvre, *a.* poor. — *m.* poor material, specif. poor ore; poor person, poor man.

pauvrement, *adv.* poorly.

pauvreté, *f.* poverty, poorness.

pavage, *m.* paving.
— *en asphalte,* — *en bitume,* asphalt paving.
— *en béton,* concrete paving.

pavé, *m.* paving block; paving, pavement; paved road, paved street, etc.; streets, public ways. — *p.a.* paved.
— *aggloméré,* concrete paving block.
— *de bois,* — *en bois,* wood paving block; wood paving.
— *en brique*(*s*), brick paving, brick pavement.
— *en pierre,* paving stone; stone paving.

pavement, *m.* paving; pavement (esp. ornamental).

paver, *v.t.* pave.

pavillon, *m.* flaring opening, as the mouth of a funnel, the mouthpiece of a telephone or the bell of a cornet; pavilion; flag.

pavot, *m.* (*Bot.*) poppy (ordinarily *Papaver*).
— *cornu,* horn poppy, horned poppy (*Glaucium glaucium*).
— *rouge,* corn poppy (*Papaver rhœas*).
— *somnifère,* opium poppy (*Papaver somniferum*).

payant, *p.a.* paying. — *m.* payer.

paye, *f.* pay; payment.

payement, *m.* payment.

payer, *v.t.* pay; pay for. — *v.i.* pay; give proof. — *v.r.* pay oneself; be paid; be satisfied.
— *de,* prove, show; expose (one's person).

pays, *m.* country; land; region; home.

paysage, *m.* landscape.

paysan, *m.* peasant, countryman. — *a.* peasant, country, rural, rustic.

Pays-Bas, *m.pl.* Netherlands, Low Countries.

p.c.c., *abbrev.* (pour copie conforme) true copy.

P. C. N., *abbrev.* (physique, chimie, naturelle) a certificate given on the completion of studies in physics, chemistry and natural history.

P.E., *abbrev.* (parties égales) equal parts; (point d'ébullition) boiling point, b. p.

péage, *m.* toll; tollhouse, tollgate.

peau, *f.* skin; hide.

— *à saucisses,* sausage skin.

— *brute,* — *crue,* raw hide, raw skin.

— *chamoisée,* chamoised skin, chamois.

— *d'agneau,* lambskin.

— *de bœuf,* oxhide.

— *de buffle,* buff, buff leather; buffalo hide.

— *de chamois,* chamois, chamois skin, chamois leather.

— *de cheval,* horsehide.

— *de chèvre,* goatskin.

— *de chien,* dogfish skin; dog skin.

— *de daim,* buckskin.

— *de lapin,* rabbit skin.

— *de mouton,* sheepskin.

— *des batteurs d'or,* goldbeater's skin.

— *de vache,* cowhide.

— *de veau,* calfskin.

— *de vélin,* vellum.

— *en merlus,* — *en merlut,* a skin limed and dried.

— *en poil,* skin with the hair or wool on, pelt, fur.

— *en poudre,* hide powder.

— *en tripe,* skin or hide ready for tanning.

— *en vert,* = peau en tripe.

— *mégis,* — *mégissée,* tawed skin, tawed leather.

— *passée,* — *passée en mégie,* tawed skin, tawed leather.

— *pleine,* unsplit skin or hide.

— *préparée,* dressed skin, dressed hide.

— *refendue,* split skin, split hide.

— *tannée,* tanned skin, tanned hide.

— *verte,* green (or raw) skin or hide, pelt.

peau-croûte, *f.* split skin, split.

peau-fleur, *f.* (*Leather*) hair side of a split.

peaussier, *m.* dealer in skins, hides or pelts.

peautre, *m.* tin; pewter. (*Obs.*).

peaux, *pl.* of peau.

pechblende, *f.* (*Min.*) pitchblende.

pêche, *f.* peach; fishing; fishery.

pêcher, *v.t.* draw out, fish out (as crystals from a liquid); fish, fish for; pick up, get. — *v.i.* fish. — *m.* peach tree, peach.

pêcherie, *f.* fishery.

pêcheur, *m.* fisher, fisherman. — *a.* fishing.

pêchurane, *f.* (*Min.*) pitchblende.

pectate, *m.* pectate.

pecteux, *a.* pectous, pectinous.

pectine, *f.* pectin.

pectineux, *a.* pectinous, pectous.

pectique, *a.* pectic.

pectoral, *a.* & *m.* pectoral.

pectose, *f.* pectose.

pécuniaire, *a.* pecuniary.

pédale, *f.* pedal; treadle.

pédalier, *m.* pedal (or treadle) mechanism.

pédoncule, *f.* peduncle.

pédonculé, *a.* pedunculate, pedunculated.

pegmatite, *f.* (*Petrog.*) pegmatite.

peignage, *m.* combing; carding; combing (or carding) works.

peignant, *p.pr.* of peigner and of peindre.

peigne, *m.* comb; card, chaser.

peignée, *f.* cardful.

peignent, *pr. 3 pl.* of peigner and of peindre.

peigner, *v.t.* comb; card.

peignerie, *f.* = peignage.

peigneuse, *f.* combing machine.

peille, *f.* (*Paper*) rag.

peinchebec, *m.* pinchbeck (a kind of brass).

peindre, *v.t.* & *i.* paint.

— *à la chaux,* whitewash.

— *à l'huile,* paint in oil.

peine, *f.* pain; pains; penalty; trouble, difficulty.

à —, scarcely, hardly.

avec —, with difficulty; with regret.

sans —, without difficulty, easily.

peint, *p.a.* painted; colored. — *pr. 3 sing.* of peindre.

peintre, *m.* painter. — *a.* painting.

— *en batiments,* house painter.

peintre-émailleur, *m.* enamel painter.

peintre-verrier, *m.* glass painter.

peinture, *f.* painting; coating; paint.

— *à la barbotine,* (*Ceram.*) slip painting.

— *à la chaux,* whitewashing.

— *à la cire,* wax painting, encaustic painting.

— *à la colle,* sizing; kalsomining.

— *à la gouache,* — *à l'aquarelle,* water-color painting.

— *à l'huile,* oil painting.

— *au vernis,* varnishing.

— *d'apprêt,* — *en apprêt,* priming.

— *en détrempe,* distemper painting.

peinturer, *v.t.* paint, coat.

pelage, *m.* unhairing; hair, coat (of animals).

pelain, *m.* (*Leather*) lime vat, lime pit, also lime liquor.

pelaineur, *m.* (*Leather*) limer.

pelan, *m.* bark.

pelanage, *m.* unhairing, depilation (esp. by lime).

— *à la chaux,* liming.

pelaner, *v.t.* unhair, depilate (esp. with lime).

pélargonique, *a.* pelargonic.

pelé, *p.a.* peeled, etc. (see peler); bald; bare.

pêle-mêle, *m. & adv.* pell-mell.

pêle-mêler, *v.t.* jumble, mix confusedly.

peler, *v.t.* peel; strip; skin; pare; unhair, depilate (hides or skins); clean, scrape (as a crucible). — *v.r.* peel, peel off, come off. — *v.i.* peel, come off; (of hair) come out.

pèlerinage, *m.* pilgrimage.

pélican, *m.* (*Zoöl. & Old Chem.*) pelican.

pelin, *m.* = pelain.

pellage, *m.* shoveling.

pellagre, *f.* (*Med.*) pellagra.

pelle, *f.* shovel.

pellée, *f.* shovelful.

peller, *v.t.* shovel.

pellerée, *f.* shovelful.

pelletage, *m.* shoveling.

pelletée, *f.* shovelful.

pelleter, *v.t.* shovel.

pelleterie, *f.* furs; fur making; fur trade; fur establishment.

pelletier, *m.* furrier.

pelletiérine, *f.* pelletierine.

pelliculage, *m.* (*Photog.*) stripping (of plates).

pelliculaire, *a.* pellicular, in the form of a pellicle or film.

pellicule, *f.* pellicle, thin skin or film; skin (of a grape); (*Photog.*) film.

pelliculé, *a.* covered with a pellicle or film, pelliculate.

pelliculer, *v.t.* (*Photog.*) strip (a plate).

pellucide, *a.* pellucid, limpid.

pellucidité, *f.* pellucidness, pellucidity.

pelotage, *m.* making into a ball (as thread by winding); (*Soap*) plotting.

pelote, *f.* ball; pellet; (*Glass*) a sole or plate on which the disk of crown glass is laid.

— *à feu,* fire ball, light ball.

— *de glaise,* (*Pyro.*) clay wad.

peloter, *v.t.* make into a ball; (*Soap*) plot.

peloteur, *m.* winder (of thread); (*Soap*) plotter.

peloton, *m.* ball; lump; cluster; group; (*Mil.*) platoon.

pelotonner, *v.t.* wind. — *v.r.* be wound or rolled up; roll up; gather, cluster.

peluche, *f.* plush.

pelure, *f.* peel, skin, rind; peeling, paring; pelt wool.

— *d'oignon,* onion peel, onion skin; (*Paper*) a thin writing paper, onionskin.

pemmican, *m.* pemmican.

pénalité, *f.* penalty, fine.

penchant, *p.a.* inclining, etc. (see pencher). — *m.* inclination; slope; verge; decline.

pencher, *v.t. & i.* incline, lean, bend. — *v.r.* bend, lean, stoop.

— *pour,* favor.

pendant, *p.a.* hanging, etc. (see pendre), suspended, pendent; pending; (of crops) (standing). — *m.* pendant; counterpart. — *prep.* pending, during. — *que,* while.

pendiller, *v.i.* dangle, swing.

pendoir, *m.* (*Leather*) a chamber in which hides are suspended to be steamed; hook or cord for suspending meat.

pendre, *v.t.* hang; hang up, suspend. — *v.r.* hang. — *v.i.* hang down, depend; be pending.

pendule, *m.* pendulum. — *f.* clock.

pêne, *m.* bolt. — *f.* pitch mop, tar brush.

pénétrabilité, *f.* penetrability.

pénétrable, *a.* penetrable.

pénétrant, *p.a.* penetrating.

pénétration, *f.* penetration.

— *du bois,* impregnation of wood.

pénétrer, *v.t.* penetrate; touch deeply. — *v.r.* penetrate each other, mix; saturate oneself. — *v.i.* penetrate.

pénible, *a.* painful, distressing, hard.

péniblement, *adv.* painfully, laboriously.

péninsule, *f.* peninsula.

la Péninsule, The Peninsula, Spain and Portugal.

penne, *f.* (long) feather; (*Weaving*) warp end.

pénombre, *f.* penumbra; twilight, dim light.

pensée, *f.* thought; mind; meaning; outline; (*Bot.*) pansy.

penser, *v.t.* think, think of. — *v.r.* think oneself; be thought. — *v.i.* think; be near, be in great danger.

penseur, *m.* thinker. — *a.* thoughtful, thinking.

pension, *f.* board, board and lodging; boarding house, pension; boarding school (or the tuition paid); pension, allowance.

pensionnaire, *m. & f.* boarder; pensioner.

pensionner, *v.t.* pension.

Pensylvanie, *f.* Pennsylvania.

pensylvanien, *a.* Pennsylvanian, Pennsylvania.

pentacarboné, *a.* pentacarbon, containing five carbon atoms.

pentachloré, *a.* pentachloro.

pentachlorure, *m.* pentachloride.

pentagonal, *a.* pentagonal; (of nuclei) five-membered.

pentagone, *m.* pentagon. — *a.* pentagonal.

pentaméthylène, *m.* pentamethylene (radical and compound; the latter is better called cyclopentane).

pentaméthylène-diamine, *f.* pentamethylenediamine (1, 5-pentanediamine).

pentaméthylénique, *a.* pentamethylene.

pentasulfure, *m.* pentasulfide.

pentathionique, *a.* pentathionic.

pentatomique, *a.* pentatomic.

pentavalent, *a.* pentavalent, quinquivalent.

pente, *f.* slope, inclination, incline, gradient, pitch, declivity, fall; bent, tendency.

aller en —, slope, incline.

— ascendante, ascent, acclivity.

— descendante, descent, declivity.

penthiophène, *m.* penthiophene (1,4-thio-pyran).

pentite, *f.* pentitol, pentite.

pentosane, *m.* pentosan.

pentose, *m.* pentose.

penture, *f.* hinge, hinge strap.

pénurie, *f.* scarcity, dearth, lack; penury.

péonine, *f.* (*Dyes*) peonin(e), pæonin(e).

péperin, peperino, *m.* (*Petrog.*) peperine, peperino (a volcanic tuff).

pépin, *m.* pip, small seed (of a fruit).

— de coing, quince seed, quince seeds.

pépite, *f.* nugget.

pepsine, *f.* pepsin.

— liquide, (*Pharm.*) solution of pepsin, liquid pepsin.

pepsinogène, *m.* pepsinogen.

peptide, *m.* peptide.

peptogène, *m.* peptogen.

pepto-glycériné, *a.* containing, or treated with, peptone and glycerol.

peptonate, *m.* peptonate.

peptone, *f.* peptone.

peptoné, *a.* peptonized, peptone.

peptonification, *f.* conversion into peptone.

peptonifier, *v.t.* convert into peptone.

peptonisation, *f.* peptonization.

peptoniser, *v.t.* peptonize.

peptonurie, *f.* peptonuria.

peptotoxine, *f.* peptotoxine.

per ascensum. per ascensum, by ascent.

pérat, *m.* (*Mining*) lump coal.

perçage, *m.* = percement.

perçant, *p.a.* piercing, etc. (see percer); sharp, keen, penetrating. — *m.* piercer, borer, puncher.

perce, *f.* piercing tool, piercer, borer, drill, punch, awl, etc.; opening, hole; size of opening (as in a sieve).

mettre en —, tap, broach (as a cask of wine).

percé, *p.a.* pierced, etc. (see percer). — *m.* = percée.

perce-bouchon, *m.* cork borer.

percée, *f.* opening; (*Metal.*) tapping bar.

perce-fournaise, *m.* (*Metal.*) tapping bar.

percement, *m.* piercing, etc. (see percer), also the opening or openings so made.

perce-muraille, *f.* (*Bot.*) wall pellitory.

percepteur, *m.* collector, gatherer (of taxes); perceiver.

perceptibilité, *f.* perceptibility.

perceptible, *a.* perceptible.

perception, *f.* perception; collection (of taxes); collectorship.

percer, *v.t.* pierce; bore, drill, punch, tap; perforate; penetrate; open (as a street); open up (as a country); drive (as a tunnel); wet thoroly. — *v.r.* be pierced, etc.; (of a fire grate) be uncovered by coals at some spot. — *v.i.* open, burst, break; pierce, penetrate; leak, leak out; come out, break thru, appear; rise, get on, succeed.

— à foret, drill, bore.

percerette, percette, *f.* borer, drill; specif., cork borer.

perceur, *m.* piercer, borer, etc. (see percer).

perceuse, *f.* boring machine, drilling machine.

percevable, *a.* perceivable; collectible.

percevant, *p.a.* perceiving; collecting.

percevoir, *v.t.* perceive; collect (as taxes).

perchage, *m.* (*Metal.*, etc.) poling.

perche, *f.* pole; rod; shaft (of wood); perch.

percher, *v.t.* (*Metal.*, etc.) pole.

perchlorate, *m.* perchlorate.

perchloré, *a.* perchloro, perchloro-.

perchlorique, *a.* perchloric.

perchlorure, *m.* perchloride.

— de carbone, carbon tetrachloride.

— de fer, perchloride of iron (ferric chloride).

perchromique, *a.* perchromic.

perclus, *p.a.* crippled.

perçoir, *m.,* **perçoire,** *f.* piercer, borer, drill, punch, awl, etc.; (*Metal.*) tapping bar.

perçoit, *pr. 3 sing.* of percevoir.

percolateur, *m.* percolator.

perçu, *p.a.* perceived; (of taxes) collected.

percussion, *f.* percussion.

percutant, *p.a.* percussive, percussion, striking, percussing.

percuter, *v.t.* strike, percuss.

percuteur, *m.* striker; hammer, plunger, firing pin.

perdant, *p.a.* losing, etc. (see perdre). — *m.* loser.

per descensum. per descensum, by descent.

perdre, *v.t.* lose; spoil; ruin. — *v.r.* lose oneself; be lost; disappear; (of colors) blend; be ruined; be damaged. — *v.i.* lose; fail; fall; ebb.

perdrix, *f.* (*Zoöl.*) partridge.

perdu, *p.a.* lost, etc. (see perdre); random; leisure; remote, out of the way; invisible; (of cases, etc.) not returnable; (of a bolt head) flush, countersunk.

perdurable, *a.* very durable.

perdurablement, *adv.* very durably.

perdurer, *v.i.* be very durable, last long.

père, *m.* father.

péreirine, *f.* pereirine.

péremptoire, *a.* peremptory.

péremptoirement, *adv.* peremptorily.

pérenne, *a.* perennial.

peréquation, *f.* equalization.

perfection, *f.* perfection.

à la —, en —, to perfection, perfectly.

perfectionnement, *m.* improvement; perfecting, perfection.

perfectionner, *v.t.* improve; improve upon; perfect. — *v.r.* improve; improve oneself.

perfectionneur, *m.* improver; perfecter.

perforage, *m.* perforation, perforating.

perforateur, *m.* drill; borer; perforator. — *a.* perforating.

perforatif, *a.* perforative.

perforation, *f.* perforation; perforating.

perforatrice, *f.* drill, borer. — *a.fem.* of perforateur.

perforer, *v.t.* perforate; pierce, bore, drill.

perhydrure, *m.* perhydride.

péricarpe, *m.* (*Bot.*) pericarp.

périclase, *f.* (*Min.*) periclase, periclasite.

péricliter, *v.i.* be endangered, be in danger.

péridérivé, *m.* peri derivative.

périderme, *m.* (*Bot. & Zoöl.*) periderm.

péridot, *m.* (*Min.*) peridot, chrysolite.

péridotite, *n.* (*Petrog.*) peridotite.

périer, *m.* (*Founding*) tapping bar.

péril, *m.* peril, danger.

mettre en —, imperil, endanger.

périlleusement, *adv.* perilously, dangerously.

périlleux, *a.* perilous, dangerous.

périmer, *v.i.* lapse.

périmètre, *m.* (*Geom.*) perimeter.

période, *f.* period. — *m.* period, stage; point, pitch, degree; height.

periodé, *a.* periodo, periodo-.

périodicité, *f.* periodicity.

periodique, *a.* (of the acid) periodic.

périodique, *a.* periodic(al). — *m.* periodical.

périodiquement, *adv.* periodically.

périoste, *m.* (*Anat.*) periosteum.

péripétie, *f.* vicissitude.

périphérie, *f.* periphery.

périphérique, *a.* peripheral.

périr, *v.i.* perish; (*Law*) lapse.

périsperme, *m.* (*Bot.*) perisperm.

périssable, *a.* perishable.

péritoine, *m.* (*Anat.*) peritoneum.

péritonéal, *a.* (*Anat.*) peritoneal.

péritonite, *f.* (*Med.*) peritonitis.

perlasse, *f.* pearlash.

perle, *f.* bead; pearl.

— de verre, — en verre, glass bead.

— fausse, artificial pearl, false pearl.

— métallique, metal bead.

perlé, *p.a.* pearled; pearly; pearl; beaded, beady; (of liquids) thick, rich; (of sugar) twice boiled; finished, perfect.

perler, *v.t.* pearl; bead; do exquisitely or perfectly. — *v.i.* bead, form beads.

perlier, *a.* pearl-bearing, pearl.

perlite, *f.* (*Micros. & Petrog.*) perlite.

perlitique, *a.* perlitic.

permanence, *f.* permanence, permanency.

en —, permanent, permanently.

permanent, *a.* permanent; (of an army) standing.

permanganate, *m.* permanganate.

— de potasse, potassium permanganate.

— de soude, sodium permanganate.

permanganique, *a.* permanganic.

perméabilité, *f.* permeability, permeableness.

perméable, *a.* permeable.

permettre, *v.t.* permit, allow, let.

permis, *p.a.* permitted; permissible. — *m.* permit.

permission, *f.* permission.

permissioner, *v.t.* authorize, license.

permit, *p.def. 3 sing.* of permettre.

permixtion, *f.* permixtion, intimate mixture.

permutabilité, *f.* exchangeability, interchangeability; permutability.

permutable, *a.* exchangeable, interchangeable; (*Math.*, etc.) permutable.

permutation, *f.* exchange, interchange; (*Math.*, etc.) permutation.

permuter, *v.t.* exchange, interchange; permute.

permutite, *f.* permutite, "permutit."

Pernambouc, *n.* Pernambuco, Recife (Brazil).

pernette, *f.* a prism-shaped support for pottery in the sagger; support for a sugar mold.

pernicieux, *a.* pernicious.

pernitrate, *m.* pernitrate.

— de fer, pernitrate of iron (ferric nitrate).

— de fer liquide, (*Pharm.*) solution of ferric nitrate.

— de mercure, pernitrate of mercury (mercuric nitrate).

— de mercure liquide, (*Pharm.*) solution of mercuric nitrate.

Pérou, *m.* Peru.

peroxydase, *f.* peroxidase.

peroxydation, *f.* peroxidation.

peroxyde, *m.* peroxide.

— d'azote, nitrogen peroxide.

— de fer, peroxide of iron (ferric oxide).

— de fer hydraté, hydrous peroxide of iron (ferric hydroxide).

— de mercure, peroxide of mercury (mercuric oxide, esp. the red form).

— d'hydrogène, hydrogen peroxide.

peroxydé, *p.a.* peroxidized.

peroxyder, *v.t.* peroxidize.

peroxydiastase, *f.* peroxydiastase.

perpendiculaire, *a. & f.* perpendicular.

perpendiculairement, *adv.* perpendicularly.

perpétuation, *f.* perpetuation; perpetuity.

perpétuel, *a.* perpetual.

perpétuellement, *adv.* perpetually.

perpétuer, *v.t.* perpetuate.

perpétuité, *f.* perpetuity.

  à —, endlessly, perpetually; in perpetuity, for ever; for life.

perplexe, *a.* perplexed; perplexing.

perquinone, *f.* perquinone (cyclohexane-hexone).

perrier, *m.* quarryman; (*Metal.*) tapping bar.

perrière, *f.* quarry (esp. of slate).

perron, *m.* steps and landing; platform.

perroquet, *m.* (*Zoöl.*) parrot.

perruthénique, *a.* perruthenic.

perrotine, *f.* (*Calico*) perrotine.

pers, *a.* blue-green, greenish blue.

persan, *a.* Persian.

perse, *f.* chintz; (*cap.*) Persia.

perséite, *f.* perseitol, perseite, (*d*-mannohepti-tol, *d*-mannoheptite).

persel, *m.* persalt.

persévérer, *v.i.* persevere.

persicaire, *f.* (*Bot.*) persicary.

persicot, *m.* persico, persicot (a cordial).

persil, *m.* (*Bot.*) parsley.

  — *des marais,* marsh parsley (*Peucedanum palustre*).

persillé, *a.* spotted with green (as cheese).

persistance, *f.* persistence; persistency.

persistant, *a.* persistent.

persister, *v.i.* persist.

personnalité, *f.* personality; egoism.

personne, *f.* person. — *m.* any one, anybody; (with a negative, expressed or understood) no one, nobody.

personnel, *a.* personal; selfish. — *m.* person-nel.

personnellement, *adv.* personally.

perspective, *f.* perspective.

perspiratoire, *a.* perspiratory.

persuader, *v.t.* persuade.

persuasif, *a.* persuasive.

persuasion, *f.* persuasion.

persulfate, *m.* persulfate.

  — *de fer,* persulfate of iron (ferric sulfate).

  — *de fer liquide,* (*Pharm.*) solution of ferric sulfate.

persulfure, *m.* persulfide.

persulfuré, *a.* persulfide of.

perte, *f.* loss.

  à —, at a loss, losing.

  à — *de vue,* out of sight, very far, very greatly; (of talking) at random.

  — *sèche,* sheer loss, absolute loss.

pertinnement, *adv.* pertinently, to the pur-pose.

pertuis, *m.* hole; opening; drain; strait, nar-rows; (*Metal.*) tap hole.

perturbateur, *a.* disturbing. — *m.* disturber.

perturbation, *f.* perturbation, disturbance.

perturber, *v.t.* disturb, perturb.

péruvien, *a.* Peruvian.

pervers, *a.* perverse.

perversion, *f.* perversion.

pervertir, *v.t.* pervert.

pervertissement, *m.* perversion.

pesable, *a.* weighable.

pesage, *m.* weighing.

pesamment, *adv.* heavily.

pesant, *a.* heavy; possessing weight; weighty; slow; dull. — *p.p.* of peser. — *m.* weight — *adv.* in weight.

pesanteur, *f.* gravity; heaviness; weight; slowness; dullness.

  — *spécifique,* specific gravity.

  — *universelle,* gravitation.

pèse, *m.* densimeter, hydrometer. — *1 & 3 sing. indic. & subj.* of peser.

pèse-acide, *m.* acid hydrometer, acidimeter.

pèse-alcali, *m.* alkali hydrometer, alkalimeter.

pèse-alcool, *m.* spirit hydrometer, alcoholo-meter.

pèse-bière, *m.* beer hydrometer.

pèse-cidre, *m.* cider hydrometer.

pèse-colle, *m.* glue hydrometer.

pesée, *f.* weighing; amount weighed out at one time; lift, effort, thrust.

pèse-esprit, *m.* = pèse-alcool.

pèse-essence, *m.* gasoline hydrometer.

pèse-éther, *m.* a hydrometer for liquids lighter than water, specif. one for ether.

pèse-filtre, *m.* weighing bottle (for filter papers with their precipitates).

pèse-flegme, *m.* low-wines hydrometer.

pèse-gouttes, *m.* dropper, drop counter.

pèse-lait, *m.* milk hydrometer, lactometer.

pèse-lessive, *m.* lye hydrometer.

pèse-lettre, *m.* postal scales, letter scales.

pèse-liqueur, *m.* hydrometer; = pèse-alcool.

pèse-mélasse, *m.* molasses hydrometer (37–45° Bé.).

pèse-moût, *m.* (*Wine*) must hydrometer.

pèse-nitre, *m.* a hydrometer for niter solutions.

pèse-papier, *m.* paper weight.

peser, *v.t.* weigh. — *v.r.* be weighed. — *v.i.* weigh; be heavy; have weight; bear down; lay stress; be a burden.

pèse-sel, *m.* hydrometer for salt solutions, sal-imeter, salinometer.

pèse-sirop, *m.* sirup hydrometer.

pèse-sucre, *m.* (*Wine*) must hydrometer.

pèse-tannin, *m.* tan-liquor hydrometer, bark-ometer.

pesette, *f.* a small balance of precision.

pèse-tube, *m.* weighing-tube support.

peseur, *m.* weigher.

pèse-urine, *m.* urine hydrometer, urinometer.

pèse-vin, *m.* wine hydrometer, enometer, œnometer.

pèse-vinaigre, *m.* vinegar hydrometer, ace-timeter, acetometer.

peson, *m.* (hand) balance, (hand) scales.

  — *à ressort,* spring balance, spring scales.

peste, *f.* pest, plague; specif., bubonic plague.

pétale, *m.* (*Bot.*) petal.

pétalite, *f.* (*Min.*) petalite.

pétard, *m.* (*Pyro.*) petard, cartridge, fire-cracker, cracker; (*Railways*) torpedo; (*Mining*) blast hole.

pétard-amorce, *m.* priming cartridge.

pétarder, *v.t.* blast, blow up, destroy by blasting or by explosives.

péter, peter, *v.i.* burst, explode, blow up; crack, make a loud report; crackle.

péterolle, *f.* (*Pyro.*) firecracker, cracker.

pétillant, *p.a.* crackling; (of wine, etc.) sparkling; (fig.) lively, animated.

pétillement, *m.* crackling; sparkling.

pétiller, *v.i.* crackle, (of salt) decrepitate; sparkle; (fig.) be transported (with); be eager.

petit, *a.* small, little; short; mean, low; petty; weak. — *adv.* little. — *m.* little; little one; young (of animals).

en —, on a small scale; in miniature.

— *à* —, little by little.

— *autel,* flue bridge (in a reverberatory furnace).

— *cardamome,* cardamom (*Elettaria cardamomum* or its seed).

— *chevai* (*Mach.*) donkey engine, donkey.

— *chêne,* wall germander (*Teucrium chamœdrys*).

petite bière, small beer, weak beer.

petite eau, petites eaux, weak solution, weak liquor.

petite vitesse, small velocity; low speed; (*Railways*) freight, dispatch by goods train, slow dispatch.

— *feu,* (*Ceram.*) slow heating (following the expulsion of moisture) up to about 600°.

— *gamin,* (*Glass,* etc.) boy, assistant, apprentice.

— *métal,* minor metal.

— *nard,* wild sarsaparilla (*Aralia nudicaulis*).

— *nom,* given name, Christian name, first name.

petits blés, oats and barley.

— *teint,* fugitive dye or color; fugitive.

un — *peu,* a very little, a little.

petit-cheval, *m.* (*Mach.*) donkey engine, donkey.

petit-deux, *m.* sugar loaf weighing a kilogram.

petitement, *adv.* little, but little; meanly.

petitesse, *f.* smallness, littleness, diminutiveness, minuteness; weakness; slenderness (of means); pettiness, meanness.

petit-grain, *m.* small immature orange.

essence de —, petitgrain oil (formerly made from the small immature fruits but now from the leaves and young shoots), orange-leaf oil.

petit-gris, *m.* a kind of squirrel.

pétition, *f.* petition.

— *de principe,* begging the question, petitio principii.

pétitionner, *v.i.* petition.

petit-lait, *m.* whey, serum of milk.

pétrifiant, *p.a.* petrifying, petrifactive.

pétrification, *f.* petrifaction.

pétrifier, *v.t. & r.* petrify.

pétrin, *m.* kneading trough, kneader.

— *mécanique,* kneading machine.

pétrir, *v.t.* knead, work; mold; form, make.

pétrissable, *a.* capable of being kneaded or worked; capable of being molded.

pétrissage, pétrissement, *m.* kneading, working; molding.

pétrisseur, *m.* kneader. — *a.* kneading.

pétrisseuse, *f.* kneading machine.

pétrofracteur, *m.* (*Expl.*) petrofracteur.

pétrographe, *m.* petrographer.

pétrographie, *f.* petrography.

pétrographique, *a.* petrographic(al).

pétrographiquement, *adv.* petrographically.

pétrolage, *m.* kindling or oiling with petroleum or kerosene.

pétrole, *m.* petroleum; any liquid petroleum product; specif., kerosene. (As a fuel for burners, etc. *pétrole* commonly means kerosene, *essence* being used to denote gasoline or naphtha.)

— *à brûler,* (usually) kerosene, burning oil.

— *brut,* crude petroleum, crude oil.

— *d'Écosse,* Scotch petroleum, shale oil.

— *lampant,* kerosene, illuminating oil.

— *ordinaire,* kerosene.

— *rectifié,* refined petroleum, specif. kerosene.

pétroléen, *a.* petroleum, of petroleum, petrolic.

pétroléine, *f.* (*Pharm.*) petrolatum.

pétrolène, *m.* petrolene.

pétroler, *v.t.* kindle or set fire to with petroleum or kerosene; oil (with petroleum or kerosene).

pétrolerie, *f.* petroleum works, oil refinery.

pétrolette, *f.* motorcycle.

pétrolien, *a.* petroleum, of petroleum, petrolic.

pétrolier, *a.* petroleum, of petroleum, oil. — *m.* tank steamer, oil steamer.

pétrolifère, *a.* petroliferous, petroleum.

pétrolique, *a.* petroleum, of petroleum, petrolic.

pétrologie, *f.* petrology.

pétrosilex, *m.* (*Petrog.*) petrosilex, felsite.

pétrosiliceux, *a.* petrosiliceous.

pétunsé, pétunzé, *m.* (*Ceram.*) petuntse.

peu, *adv.* little; not very, not, un-, in-; but a short time, not long. — *m.* little, a little; few, a few.

à — *près, à* — *de chose près,* almost, nearly, about.

dans —, soon, presently.

depuis —, recently, lately, a short time ago.

peu à —, little by little, by degrees, gradually.
— *après*, soon after, not long after.
— *de chose*, little thing, small matter, not much.
— *importe*, no matter.
pour — *que*, if only, if . . . only a little.
si — *que rien*, a very little, a mere nothing.

peucédan, *m.* (*Bot.*) Peucedanum.
— *officinal*, hog's fennel (*Peucedanum officinale*).

peuplade, *f.* tribe.

peuple, *m.* people. — *a.* of the (common) people; common, vulgar.

peupler, *v.t.* people; stock. — *v.i.* multiply.

peuplier, *m.* poplar (tree and wood).

peur, *f.* fear.
avoir — *de*, be afraid of.
de —, for fear.
de — *que*, for fear that, for fear lest.

peut, *pr. 3 sing.* of pouvoir.

peut-être, *adv.* perhaps, it may be.

peuvent, *pr. 3 pl.* of pouvoir.

p.ex., *abbrev.* (par exemple) for example, e.g.

Ph. Symbol for *phosphore* (phosphorus).

phagédénique, *a.* (*Med.*) phagedenic.

phagocyte, *m.* (*Physiol.*) phagocyte.

phagocytose, *f.* (*Physiol.*) phagocytosis.

phagolyse, phagocytolyse, *f.* (*Physiol.*) phagolysis, phagocytolysis.

phanérogame, *a.* (*Bot.*) phanerogamic, phanerogamous. — *m.* phanerogam.

phare, *m.* lighthouse; headlight; (fig.) beacon.

pharmaceutique, *a.* pharmaceutical, pharmaceutic. — *f.* pharmaceutics.

pharmacie, *f.* pharmacy; dispensary; medicine case, medicine chest.

pharmacien, *m.* pharmacist; druggist, apothecary, chemist and druggist. — *a.* pharmacy, of pharmacy.
— *chimiste*, pharmaceutical chemist.

pharmacochimie, *f.* pharmaceutical chemistry.

pharmacochimique, *a.* of or pertaining to pharmaceutical chemistry.

pharmacodynamique, *f.* pharmacodynamics. — *a.* of or pertaining to pharmacodynamics.

pharmacognosie, *f.* pharmacognosy, -cognosis.

pharmacologie, *f.* pharmacology.

pharmacologique, *a.* pharmacologic(al).

pharmacologiste, pharmacologue, *m.* pharmacologist.

pharmaconyme, *m.* pharmaceutical name.

pharmacopée, *f.* pharmacopeia, pharmacopœia.

pharmacosidérite, *f.* (*Min.*) pharmacosiderite.

pharmacothèque, *f.* medicine chest.

phase, *f.* phase.

phaséoline, *f.* phaseolin.

phaséolunatine, *f.* phaseolunatin.

phellandrène, *m.* phellandrene.

phellosine, *f.* an agglomerated cork.

phénacétine, *f.* (*Pharm.*) phenacetin(e).

phénacite, phénakite, *f.* (*Min.*) phenacite.

phénanthraquinone, *f.* phenanthraquinone (phenanthrenequinone).

phénanthrazine, *f.* phenanthrazine.

phénanthrène, *m.* phenanthrene.

phénanthrénique, *a.* phenanthrene, of or pertaining to phenanthrene.

phénate, *m.* phenolate, phenate.
— *de soude*, sodium phenolate.

phénazine, *f.* phenazine.

phène, *m.* phene (little used name for benzene).

phénéthol, *m.* phenetole.

phénétidine, *f.* phenetidine.

phénétol, *m.* phenetole.

phénique, *a.* of or pertaining to phenol, phenic, phenylic.
acide —, *alcool* —, carbolic acid, phenol.

phéniquer, *v.t.* treat or impregnate with phenol (carbolic acid), carbolate.

phénol, *m.* phenol.

phénolique, *a.* of, pertaining to or containing phenol, phenolic.

phénol-phtaléine, *f.* phenolphthalein.

phénol-sulfoné, *a.* phenolsulfonic.

phénoménal, *a.* phenomenal.

phénoménalement, *adv.* phenomenally.

phénomène, *m.* phenomenon.

phénone, *f.* a ketone of phenyl and some other radical (RCOC₆H₅).

phénosafranine, *f.* phenosafranine.

phénylacétamide, *f.* phenylacetamide (α-toluamide).

phénylacétique, *a.* phenylacetic (α-toluic).

phénylacétylène, *m.* phenylacetylene (ethinylbenzene).

phénylalanine, *f.* phenylalanine.

phénylallylène, *m.* phenylallylene (phenylpropine).

phénylamide, *f.* phenylamide, anilide.

phénylamine, *f.* phenylamine; specif., aniline.

phénylcarbylamine, *f.* phenyl carbylamine (isocyanobenzene, phenyl isocyanide).

phényle, *m.* phenyl.

phénylène, *m.* phenylene.

phénylène-diamine, *f.* phenylenediamine.

phényléthane, *m.* phenylethane (ethylbenzene).

phényléthylène, *m.* phenylethylene; specif., styrene.

phényléthylique, *a.* (of an alcohol) phenethyl.

phénylglycérine, *f.* phenylglycerol.

phénylglycocolle, *m.* phenylglycocoll (phenylglycine).

phénylglycol, *m.* phenylglycol.

phénylglycolique, *a.* phenylglycolic (mandelic).

phénylhydrazine, *f.* phenylhydrazine.
phénylhydrazone, *f.* phenylhydrazone.
phénylique, *a.* phenyl, (less often) phenylic.
phénylméthane, *m.* phenylmethane; specif., toluene.
phénylpyrazol, *m.* phenylpyrazole.
phénylpropiolique, *a.* phenylpropiolic.
phenylsulfate, *m* phenylsulfate.
phénylsulfite, *m.* benzenesulfonate.
phénylsulfureux, *a.* benzenesulfonic.
phénylsulfurique, *a.* phenylsulfuric; phenolsulfonic (improper use).
phénylurée, *f.* phenylurea.
philosophale, *a.* In the phrase *pierre philosophale,* philosophers' stone.
philosophaliste, *m.* a searcher for the philosophers' stone.
philosophe, *m.* philosopher. — *a.* philosophic(al).
philosopher, *v.i.* philosophize.
philosophie, *f.* philosophy.
philosophique, *a.* philosophic, philosophical.
philosophiquement, *adv.* philosophically.
phlegme, *m.* phlegm.
phlobaphène, *m.* phlobaphene.
phlogistication, *f.* (*Old Chem.*) phlogistication.
phlogisticien, *m.* phlogistian, phlogistonist.
phlogistique, *m.* phlogiston. — *a.* phlogistic.
phlogistiquer, *v.t.* (*Old Chem.*) phlogisticate.
phloorrhétine, *f.* phloretin.
phloorrhétique, *a.* phloretic.
phloridzine, phloorrhizine, *f.* phlorhizin, phloridzin.
phloroglucine, *f.* phloroglucinol, phloroglucin.
phloroglucite, *f.* phloroglucitol, phloroglucite, 1, 3, 5-cyclohexanetriol.
phloxine, *f.* (*Dyes*) phloxin, phloxine.
phlyctène, *f.* (*Med.*) phlyctena, phlyctæna.
phocénine, *f.* phocenin (trivalerin, glyceryl valerate).
phœnicine, *f.* (*Dyes*) phenicin(e), phœnicin(e).
phoque, *m.* (*Zoöl.*) seal.
phormium, phormion, *m.* (*Bot.*) Phormium, esp. *P. tenax* or New Zealand flax.
phosgène, *m.* phosgene, carbonyl chloride.
phosphame, *n.* phospham, phosphame.
phosphatage, *m.* (*Agric.*) application of phosphatic fertilizers; (*Wine*) addition of calcium phosphate to the grapes.
phosphate, *m.* phosphate.
— *ammoniaco-magnésien,* ammonium magnesium phosphate.
— *bicalcique,* dicalcium phosphate.
— *d'ammoniaque,* ammonium phosphate.
— *de chaux,* phosphate of lime (calcium phosphate).
— *de chaux hydraté,* (*Pharm.*) precipitated calcium phosphate.
— *de fer,* iron phosphate.

phosphate *de magnésie,* phosphate of magnesia (magnesium phosphate).
— *de potasse,* phosphate of potash (potassium phosphate).
— *de soude,* phosphate of soda (sodium phosphate).
— *mielleux,* a sirupy form of monocalcium phosphate.
— *monocalcique,* monocalcium phosphate.
— *tricalcique,* tricalcium phosphate.
phosphaté, *a.* containing phosphate or phosphoric acid, phosphatic, phosphated.
phosphaterie, *f.* phosphate mining or preparation or works.
phosphatide, *n.* phosphatide.
phosphatique, *a.* phosphatic.
phosphaturie, *f.* (*Med.*) phosphaturia.
phosphaturique, *a.* (*Med.*) phosphaturic.
phosphine, *f.* phosphine.
phosphite, *m.* phosphite.
phosphoglycérate, *m.* phosphoglycerate (glycerophosphate).
phosphoglycérique, *a.* phosphoglyceric (glycerophosphoric).
phosphomolybdique, *a.* phosphomolybdic.
phosphore, *m.* phosphorus, phosphor.
— *amorphe,* amorphous phosphorus.
— *blanc,* white phosphorus.
— *de Bologne,* Bologna phosphorus (a phosphorescent sulfide, esp. barium sulfide).
— *de Canton,* Canton's phosphorus (phosphorescent calcium sulfide).
— *rouge,* red phosphorus.
phosphoré, *a.* phosphorus, containing phosphorus; phosphide of, phosphuretted.
phosphore-bronze, *n.* phosphor bronze.
phosphoreferrosilicium, *m.* an alloy of phosphorus, iron and silicon.
phosphorer, *v.t.* treat, impregnate, coat or combine with phosphorus, phosphorize, phosphorate.
phosphorescence, *f.* phosphorescence.
phosphorescent, *a.* phosphorescent.
phosphoreux, *a.* phosphorous.
phosphoride, *m.* phosphoride.
phosphorique, *a.* phosphoric; phosphorus, of or containing phosphorus.
phosphorisation, *f.* phosphorization, phosphoration.
phosphoriser, *v.t.* phosphorize, phosphorate.
phosphorisme, *m.* (*Med.*) phosphorism.
phosphorite, *f.* (*Min. & Petrog.*) phosphorite.
phosphoritique, *a.* phosphoritic.
phosphorogénique, *a.* (*Physics*) phosphorogenic.
phosphoroscope, *m.* (*Physics*) phosphoroscope.
phosphotungstique, *a.* phosphotungstic.
phosphovanadique, *a.* phosphovanadic.
phosphure, *m.* phosphide.

phosphuré, *a.* containing phosphide.
photocéramique, *f.* photoceramics.
photochimie, *f.* photochemistry.
photochimique, *a.* photochemical.
photochromie, *f.* photochromy, color photography.
photochromogravure, *f.* photographic color printing, three-color process.
photochromotypographie, photochromotypie, *f.* photochromotypy, photographic color process.
photocollographie, *f.* photocollography (a collotype process).
photocopie, *f.* photographic reproduction (either the process or a print so made).
photo-électrique, *a.* photo-electric(al).
photogène, *a.* photogenic. — *m.* photogen, photogene.
photogénèse, photogénie, *f.* production of light.
photogénétique, *a.* photogenetic, photogenic.
photogénique, *a.* relating to the chemical effects of light, actinic; photogenic.
photogéniquement, *adv.* actinically; photogenically.
photoglyptie, *f.* photogravure; woodburytype.
photoglyptographie, *f.* photogravure.
photogramme, *m.* photographic print (esp. on paper).
photographe, *m.* photographer.
photographie, *f.* photography; photograph.
photographies animées, animated pictures, motion pictures.
photographier, *v.t.* photograph.
photographique, *a.* photographic, photographical.
photographiquement, *adv.* photographically.
photograveur, *m.* photoengraver, process engraver.
photogravure, *f.* photoengraving (whether relief or intaglio).
— *en creux,* photogravure, heliogravure.
— *en relief,* photoengraving (in relief), phototypography.
photolithographie, *f.* photolithography.
photomagnétique, *a.* photomagnetic.
photomécanique, *a.* photomechanical.
photomètre, *m.* photometer.
— *à éclats,* flicker photometer.
— *de Bunsen,* Bunsen photometer, grease-spot photometer.
photométrer, *v.t.* measure photometrically.
photométrie, *f.* photometry.
photométrique, *a.* photometric, photometrical.
photométriquement, *adv.* photometrically.
photomicrographie, *f.* photomicrography; photomicrograph, microphotograph.
photoniellure, *f.* a photographic niello process.

photopeinture, *f.* coloring of photographs.
photophore, *m.* photophore.
photoplastographie, *f.* photoplastography, woodburytype.
photopoudre, *f.* flash-light powder.
photo-sel, *m.* photosalt.
photosphère, *n.* (*Astron.*) photosphere.
photosphérique, *a.* (*Astron.*) photospheric.
photosynthèse, *f.* photosynthesis.
phototactique, *a.* (*Biol.*) phototactic.
phototaxie, *f.* (*Biol.*) phototaxis, phototaxy.
phototégie, *f.* (*Photog.*) a process for transforming a negative into a positive.
phototeinture, *f.* dyeing with the chemical aid of light.
photothérapie, *f.* phototherapy, phototherapeutics.
photothérapique, *a.* phototherapic, phototherapeutic.
photothermomètre, *m.* a thermometer with a photographic registering device.
phototype, *m.* an original photographic image (formed in a camera).
phototypie, *f.* phototypy, prototype.
phototypographie, *f.* phototypography, photoengraving (in relief).
photozincographie, *f.* photozincography, zinc photoengraving.
phrase, *f.* phrase.
phtal-. phthal-.
phtalate, *m.* phthalate.
phtaléine, *f.* phthalein.
— *du phénol,* phenolphthalein.
phtalidéine, *f.* phthalidein.
phtalidine, *f.* phthalidin, phthalidinol.
phtalimide, *f.* phthalimide.
phtaline, *f.* phthalin.
phtalique, *a.* phthalic.
phtalophénone, *f.* phthalophenone.
phtalyle, *m.* phthalyl.
phtanite, *f.* (*Petrog.*) phthanite, chert.
phtartique, *a.* deleterious, deadly.
phtisie, *f.* (*Med.*) phthisis.
phycochrome, *m.* (*Bot.*) phycochrome.
phylloporphyrine, *f.* phylloporphyrin.
phylloxanthine, *f.* phylloxanthin (xanthophyll).
phylloxéra, phylloxera, *m.* (*Zoöl.*) phylloxera.
phylloxéré, *a.* attacked by the phylloxera.
phylloxérien, phylloxérique, *a.* of or pertaining to the phylloxera.
physaline, *f.* physalin.
physalite, *f.* physalite (a variety of topaz).
physicien, *m.* physicist. — *a.* physics.
physico-chimie, *f.* physical chemistry.
physico-chimique, *a.* physicochemical.
physico-mathématique, *a.* physicomathematical.
physico-mécanique, *a.* physicomechanical.
physiographe, *m.* physiographer.

physiographie, *f.* physiography.
physiologie, *f.* physiology.
physiologique, *a.* physiological, physiologic.
physiologiquement, *adv.* physiologically.
physiologiste, physiologue, *m.* physiologist.
physique, *a.* physical. — *f.* physics. — *m.* physique.
physiquement, *adv.* physically.
physostigmine, *f.* physostigmine.
phytochimie, *f.* phytochemistry, plant chemistry.
phytochimique, *a.* phytochemical.
phytogène, *a.* phytogenic, phytogenous.
phytolaccine, *f.* phytolaccin.
phytolaque, *f.* (*Bot.*) Phytolacca; specif. (as in *Pharm.*) *P. decandra* or common poke.
phytosterine, *f.* phytosterol, phytosterin.
piano, *m.* piano.
pic, *m.* pick, pickax; poker; (*Glass*) an iron rod for detaching the cylinder from the blowpipe; (mountain) peak; woodpecker.
  à —, perpendicularly, perpendicular, precipitous.
  — *à feu,* poker.
picadil, *m.*, **picadils,** *m.pl.* (*Glass*) material which escapes from the pot and runs down thru the hearth.
picadon, *m.* (*Soap*) place where the soda is broken up.
picamare, *f.* picamar.
picéine, *f.* picein.
picène, *m.* picene.
pickelage, *m.* pickling.
pickeler, *v.t.* pickle.
picklage, *m.* pickling.
picnomètre, *m.* pycnometer.
picnométrique, *a.* pycnometric, pycnometrical.
picoline, *f.* picoline.
picolique, *a.* picolinic.
picotement, *m.* pricking, etc. (see picoter).
picoter, *v.t.* prick; cause to tingle, sting; peck; provoke.
picramique, *a.* picramic.
picrate, *m.* picrate.
picraté, *a.* of or containing picrate or picric acid, picrated.
picrique, *a.* picric.
picriqué, *a.* treated or mixed with picric acid, picrated.
picrol, *m.* (*Pharm.*) picrol.
picromel, *m.* picromel.
picrotine, *f.* picrotin.
picrotoxine, *f.* picrotoxin.
picrotoxique, *a.* picrotoxic.
pièce, *f.* piece; part; room; cask, barrel (as of wine); document, paper; apiece, each; head (as of cattle); tip, gratuity; (*Metal.*) bloom; (*Mach.*) work.
  mettre en pièces, break, tear, cut, etc., in pieces.

pièce *à* —, bit by bit, piecemeal.
  — *de coulée,* casting.
  — *de frottement,* friction piece.
  — *de rechange,* spare part.
  — *maîtresse,* in salt gardens, an elevated reservoir from which the brine is distributed to the *tables salantes.*
  — *moulée,* (*Metal.*) casting.
pied, *m.* foot; base, support; leg (as of a table); stem, stalk; slope, inclination; footing; footprint; foot rule.
  à —, on foot.
  au — *de la lettre,* in a literal sense, literally; to the letter, exactly.
  de — *ferme,* firmly, resolutely.
  — *à coulisse,* slide caliper.
  — *à* —, step by step.
  — *de bœuf,* neat's foot.
  — *de chat,* cat's paw, cat's foot; (*Bot.*) cudweed (*Gnaphalium*).
  — *de cuve,* sediment, dregs, foot; (*Alcohol*) fermenting liquid placed in the bottom of a vat which is then filled with unfermented liquid.
  sur —, on foot; up; well; in use, in fashion; (of crops) standing; (of soldiers) ready.
pied-d'alouette, *m.* (*Bot.*) larkspur, delphinium.
pied-de-chèvre, *m.* crowbar.
pied-de-corneille, *m.* (*Bot.*) spotted crane's-bill.
piédestal, *m.* pedestal.
piège, *m.* trap, snare.
pie-mère, *f.* (*Anat.*) pia mater.
pierraille, *f.* broken stone, pieces (or piece) of stone, rubble.
pierre, *f.* stone; hard mass (as of mercuric sulfide); (*Med.*) calculus.
  — *à aiguiser,* whetstone; grindstone.
  — *à bâtir,* building stone.
  — *à brunir,* burnishing stone, specif. hematite.
  — *à cautère,* (*Med.*) chemical caustic, potential cautery, specif. potassium hydroxide.
  — *à chaux,* limestone.
  — *à ciment,* cement stone, cement rock.
  — *à feu,* firestone.
  — *à filtrer,* filter stone, filtering stone.
  — *à laver,* sink.
  — *à l'huile,* oilstone.
  — *à meules,* millstone; grindstone.
  — *à paver,* paving stone.
  — *à plâtre,* gypsum, plaster stone.
  — *à rat,* (*Min.*) witherite.
  — *arénacée,* sandstone.
  — *à repassage,* — *à repasser,* hone, oilstone.
  — *artificielle,* artificial stone.
  — *blanche,* white stone; white stoneware.
  — *calaminaire,* (*Min.*) calamine.
  — *calcaire,* calcareous stone.

pierre d'aigle, (Min.) eaglestone, ætites.

— d'aimant, loadstone, lodestone.

— d'alum, (Min.) alum stone, alunite.

— d'Arménie, Armenian stone.

— de Bologne, (Min.) Bologna stone (a variety of barite).

— de ciment, cement block.

— d'écrevisse, (Pharm.) crab's-eye, crab's-stone.

— de fond, bottom stone; (Metal.) sole.

— de Labrador, (Min.) Labrador stone, labradorite.

— de l'air, meteoric stone, aërolite.

— de lard, soapstone, steatite.

— de lune, (Min.) moonstone.

— demi-doublée, a gem having the crown of different material from the rest of the stone.

— de mine, ore-bearing rock; ore.

— de savon, soapstone.

— de soleil, (Min.) sunstone, aventurine feldspar.

— de taille, freestone; ashlar, squared stone, hewn stone.

— d'étain, (Min.) tinstone, cassiterite.

— de touche, touchstone.

— de verre, a kind of vitreous porcelain; a kind of devitrified glass.

— d'évier, sink.

— de vin, tartar, wine stone.

— divine, (Pharm.) a collyrium composed of copper sulfate, potassium nitrate, alum and camphor (called also cuprum aluminatum).

— factice, — fausse, imitation stone, artificial gem.

— filtrante, = pierre à filtrer.

— fine, gem stone, gem.

— fondamentale, foundation stone.

— infernale, lapis infernalis (fused silver nitrate).

— infernale diluée, (Pharm.) mitigated silver nitrate.

— meulière, millstone, buhrstone.

— murale, calcium oxalate.

— ollaire, potstone (a kind of steatite).

— ophthalmique, crystallized copper sulfate.

— philosophale, philosophers' stone.

— plate, flat stone, stone slab.

— ponce, pumice stone, pumice.

— pourrie, rottenstone.

— précieuse, precious stone.

— réfractaire, refractory stone, firestone.

— simili, an imitation gem underlaid by a reflecting surface, esp. of silver.

— spéculaire, selenite; mica.

pierreries, f.pl. precious stones, gems.

pierrette, f. small stone.

pierreux, a. stony.

piétinement, m. tramping, etc. (see piétiner).

piétiner, v.t & i. tramp, trample, tread, stamp.

pieu, m. stake; pile.

piézochimique, a. piezochemical.

piézo-électricité, f. piezo-electricity.

piézomètre, m. piezometer.

pigeon, m. pigeon; thick plaster; trowelful, handful (of plaster); a piece of stone in lime.

pigeonnier, m. dovecote, pigeon house; damper (in a glass furnace).

pigment, m. pigment.

— biliaire, bile pigment.

— sanguin, blood pigment.

— urinaire, urinary pigment.

pigmentaire, a. pigmentary, pigmental.

pigmentation, f. pigmentation.

pigmenté, a. pigmented.

pigmenteux, a. pigmentary.

pigmentogène, a. pigment-forming.

pignon, m. (Mach.) pinion; gear; piñon, pine nut; inferior wool; gable.

— des Barbades, — d'Inde, physic nut, Barbados nut.

pilage, m. pounding, etc. (see piler).

pile, f. pile; (Elec.) cell, battery, pile; beating trough, stamping trough (as for crushing sugar); bed (of a powder mill); (Paper) hollander, pulping machine, beater; (Soap) oil cistern; pier, mole.

— à auges, trough battery.

— à charbon, charcoal pile; (Elec.) carbon cell.

— à déchiqueter, beater, pulper (as for gun-cotton).

— à densité, gravity cell, gravity battery.

— à deux liquides, two-fluid battery.

— à gravité, gravity cell, gravity battery.

— à immersion, plunge battery.

— à liquide, fluid cell.

— au bichromate, bichromate cell.

— à un liquide, single-fluid cell.

— défileuse, (Paper) breaker.

— de polarisation, storage battery.

— d'épreuve, — d'essai, testing cell, testing battery.

— de Volta, voltaic pile or cell.

— étalon, standard cell, normal cell.

— galvanique, galvanic pile (electric cell or battery).

— plongeante, plunge battery.

— primaire, primary cell, primary battery.

— sèche, dry cell, dry battery.

— secondaire, secondary battery, storage battery.

— thermo-électrique, thermoelectric pile, thermopile.

— voltaïque, voltaic pile or cell, electric cell.

piler, v.t. pound, crush, bray, bruise, grind (as in a mortar), stamp. — v.r. be pounded, etc.

pile-tan, m. (Leather) bark grinder.

pileur, m. pounder, etc. (see piler).

— de tan, (Leather) bark grinder.

pileux, a. pilose, pileous, hairy, hair.

pilier, *m.* pillar; pier; post.
pilocarpidine, *f.* pilocarpidine.
pilocarpine, *f.* pilocarpine.
piloir, *m.* a stick used in manipulating the hides in the vat in tawing.
pilon, *m.* pestle; stamp, crusher; rammer; tamper; hammer; (*Metal.*) shingling hammer.
— *atmosphérique,* compressed-air hammer.
— *à vapeur,* steam hammer.
— *mécanique,* power hammer.
pilonnage, pilonage, *m.* ramming, etc. (see pilonner).
pilonner, piloner, *v.t.* ram, tamp; beat, pound, stamp, hammer.
pilot, *m.* conical salt heap; rags for making paper; pile (the stake).
piloter, *v.t.* pilot; drive piles into.
pilotis, *m.* pile work, piling, piles.
pilulaire, *a.* pilular, pertaining to or like pills.
— *f.* (*Bot.*) pillwort.
pilule, *f.* pill.
— *de mercure,* mass of mercury, blue mass, mercury pill.
pilules aloétiques savoneuses, pills of aloes.
pilules altérantes composées, pilules antidartreuses, compound pill of mercurous chloride.
pilules au phosphore, pills of phosphorus.
pilules bleues, mass of mercury, blue mass.
pilules chalybées de Blaud, pills of ferrous carbonate, Blaud's pills.
pilules cochées mineures, compound pill of colocynth.
pilules d'acétate de plomb et d'opium, pill of lead with opium.
pilules d'aloès et de savon, pills of aloes.
pilules de Blancard, pills of ferrous iodide.
pilules de carbonate ferreux, mass of ferrous carbonate.
pilules de copahu, mass of copaiba.
pilules de Plummer, compound pill of mercurous chloride, Plummer's pills.
pilules de Rufus, pills of aloes and myrrh, Rufus's pills.
pilules ferrugineuses, mass of ferrous carbonate.
pilules ferrugineuses de Blaud, pills of ferrous carbonate, Blaud's pills.
pilules mercurielles simples, mass of mercury, mercury pill.
pilules phosphorées, pills of phosphorus.
pilulier, *m.* pill machine.
pimarique, *a.* pimaric.
pimélique, *a.* pimelic.
piment, *m.* red pepper, Cayenne pepper, (*Pharm.*) capsicum; (*Bot.*) any plant of the genus Capsicum, esp. *C. annuum,* or Guinea pepper; (in certain phrases) pepper.
— *annuel,* Guinea pepper (*C. annuum*).

piment *de Cayenne,* Cayenne pepper.
— *de la Jamaïque,* — *des Anglais,* — *poivre,* allspice, pimento (*Pimenta pimenta*).
— *des jardins,* = piment annuel.
— *rouge,* red pepper.
pimenta, *f.* (*Bot.*) Pimenta.
pimprenelle, *f.* (*Bot.*) burnet (*Sanguisorba*); (in certain phrases) pimpernel.
pin, *m.* pine (tree and wood).
— *à résine,* pitch pine, resin pine.
— *à trochets,* cluster pine (*Pinus pinaster*).
— *blanc,* white pine.
— *d'Alep,* Aleppo pine (*Pinus halepensis*).
— *de mâture,* Norway pine.
— *de poix,* pitch pine.
— *jaune,* yellow pine.
— *maritime,* maritime pine, cluster pine.
— *rouge,* red pine.
pinacle, *m.* pinnacle.
pinacoline, *f.* pinacolin.
pinacone, *f.* pinacol, (less desirably) pinacone.
pinastre, *m.* (*Bot.*) pinaster, cluster pine (*Pinus pinaster*).
pinçade, *f.* pinching.
pince, *f.* (pair of) pincers, pinchers, pliers, nippers, tongs, forceps, tweezers; pinchcock; clasp, clamp, clip, catch; crowbar, pinch bar, lever; rim, flange; grip, hold, purchase; pinch, pinching.
— *à bouts ivoire,* ivory-tipped forceps (for handling weights).
— *à bouts platine,* platinum-tipped forceps or tongs.
— *à bras,* tongs with ends curved like arms, as for holding a crucible.
— *à charbon,* charcoal tongs; coal tongs.
— *à coupelles,* cupel tongs.
— *à creusets,* crucible tongs.
— *à cuillère,* tongs with spoon-shaped ends, as for handling sodium.
— *à mâchoires,* pincers, pliers.
— *à ressort,* spring pinchcock; spring clamp; spring clip, spring catch; spring pliers.
— *à scorificatoires,* scorifier tongs.
— *à tourmalines,* tourmaline tongs.
— *à vis,* screw pinchcock; screw clamp; hand vise.
— *brucelle,* spring forceps, spring pincers.
— *coupante,* cutting pliers, nippers, cutter.
— *de chimiste,* chemical forceps, esp. steel spring forceps, usually platinum-tipped and often provided with adjusting screws.
— *de minéralogiste,* a spring holder (somewhat like a wire test-tube holder) for mineral specimens.
— *en bois,* wooden clamp.
— *plate,* flat-nosed pliers.
— *ronde,* round-nosed pliers.
— *universelle,* pincers combining several uses, as that of a wire cutter, pipe wrench, etc.

pinceau, *m.* brush (for painting or dusting), painter's pencil, hair pencil; (*Elec.*) brush; (*Anat.*) fasciculus.

pincée, *f.* pinch.

pincement, *m.* pinching, etc. (see pincer).

pincer, *v.t.* pinch; nip; grip; catch. — *v.i.* pinch.

pincettes, *f.pl.*, **pincette,** *f.* (small) pincers, pliers, nippers, tweezers, tongs.

pinchbeck, *m.* pinchbeck (a kind of brass).

pinçure, *f.* crease; pinching.

pinéal, *a.* (*Anat.*) pineal.

pineau, *m.* a variety of red grape, rich in sugar and much used in making wine.

pinène, *m.* pinene.

pinipicrine, *f.* pinipicrin.

pinique, *a.* pinic.

pinitannique, *a.* pinitannic.

pinite, *f.* pinitol, pinite; (*Min.*) pinite.

piochage, *m.* digging.

pioche, *f.* pickax, pick, mattock.

piochement, *m.* digging.

piocher, *v.t.* dig; work hard at. — *v.i.* dig.

pionnier, *m.* pioneer.

pipe, *f.* pipe (cask and smoking instrument).

pipéracées, *f.pl.* (*Bot.*) Piperaceæ.

pipérazine, *f.* piperazine.

pipéridine, *f.* piperidine.

pipéridone, *f.* piperidone.

pipérin, *m.*, **pipérine,** *f.* piperine.

pipérique, *a.* piperic.

pipéroïde, *m.* (*Pharm.*) piperoid.
 — *de gingembre,* piperoid of ginger.

pipéronylique, *a.* (of the acid) piperonylic; (of the alcohol) piperonyl.

pipéryle, *m.* piperyl (radical of piperic acid).

pipette, *f.* pipet, pipette.
 — *à ballon,* a pipet with a large flask-like reservoir, used esp. for bacterial cultures.
 — *à boule,* bulb pipet (pipet with spherical enlargement).
 — *à cylindre,* cylindrical pipet, pipet with cylindrical bulb.
 — *compte-globules,* blood-counting pipet.
 — *compte-gouttes,* dropping pipet, dropper.
 — *de précaution,* a volumetric pipet with graduations at the "full" and "empty" marks.
 — *divisée,* graduated pipet.
 — *effilée,* pipet with drawn-out point.
 — *étalon,* standard pipet, normal pipet.
 — *jaugée,* calibrated pipet.

pipette-mélangeur, *n.* mixing pipet.

piquant, *p.a.* pickling, etc. (see piquer); (of taste, etc.) piquant, sharp, tart, pungent, strong. — *m.* prickle, prick, spine; piquancy, sharpness, pungency, (of lye) causticity.

piqué, *p.a.* pricked, etc. (see piquer); (of wine, etc.) sour. — *m.* piqué (the fabric).

pique-feu, *m.* poker.

piquer, *v.t.* prick; sting; prickle; cause to smart, bite; sour (as wine); puncture, pierce; poke (as a fire); scale (a boiler); dig, quarry, mine; dot, spot; pit (as metal); mold, spot with mold; quilt; backstitch; lard; interest; pique. — *v.r.* sour, turn sour; be piquant, be strong, be tainted; mold; pride oneself (on); be piqued; prick oneself. — *v.i.* sour, turn sour.

piquet, *m.* stake, peg, pin; picket.

piqueter, *v.t.* spot, dot; stake out, mark out.

piquette, *f.* piquette (a weak beverage made by pouring water on the marc of grapes and fermenting).

piqueur, *m.* overseer, inspector; pricker, etc. (see piquer).

piqûre, *f.* prick, pricking, puncture; sting; pitting, pit (as in metals); perforation, hole; quilting.

pire, *a.* worse; worst. — *m.* worst.

piriforme, *a.* pear-shaped, pyriform.

pis, *adv.* worse; worst. — *m.* worst; dug, teat.
 au — *aller,* at the worst.
 de — *en* —, worse and worse.
 — *aller,* worst, worst possible; last resort; makeshift.

Pise, *f.* Pisa.

pisé, *m.* lining material, lining (fireclay, ganister, carbon or the like); pisé, rammed earth or clay.

pisolithe, *f.* (*Geol.*) pisolite.

pisolithique, *a.* pisolitic.

pissasphalte, *m.* pissasphalt, maltha, mineral tar.

pissée, *f.* (*Metal.*) slag duct, slag channel.

pissement, *m.* urination.

pissenlit, *m.* (*Bot.*) dandelion (*Taraxacum*).

pissette, *f.* washing bottle, wash bottle (for washing precipitates, etc.); (in general) any apparatus for delivering liquid in a fine stream.
 — *à eau,* washing bottle, wash bottle (for water).
 — *à eau chaude,* hot-water washing bottle.

pisseux, *a.* of or resembling urine.

pissote, *f.* escape pipe, waste pipe (of wood).

pistache, *f.* pistachio nut, pistachio.
 — *de terre,* peanut, earth nut.

pistachier, *m.* pistachio tree.

piste, *f.* track; trail.

pistolet, *m.* pistol.

piston, *m.* piston.
 — *plongeur,* plunger.

pitchpin, pitch-pin, *m.* pitch pine, esp. *Pinus palustris;* resinous pine, yellow pine (the wood).

pite, *f.* pita (fiber from *Agave americana* and other plants, also the plant). — *m.* pita, pita fiber, pita hemp.

piteux, *a.* pitiable.

pitié, *f.* pity.

piton, *m.* screw ring, screw eye; screw hook; eyebolt; peak, crag.

pitoyable, *a.* pitiful; pitiable.

pitoyablement, *adv.* pitiably, pitifully.

pittacal, *m.* pittacal (blue coloring matter).

pitte, *f.* = pite.

pittoresque, *a.* pictorial; picturesque.

pittoresquement, *adv.* pictorially; picturesquely.

pituitaire, *a.* (*Anat.*) pituitary.

pituite, *f.* (*Med.*) phlegm, mucus, pituite.

pivoine, *f.* (*Bot.*) peony (*Pæonia*).

pivot, *m.* pivot; pin, stud, axis; (*Bot.*) taproot.

pivotant, *p.a.* pivoting; (*Bot.*) tap-rooted.

pivoter, *v.i.* pivot; (*Bot.*) have a taproot.

placage, *m.* plating, etc. (see plaquer); (fig.) patchwork.

plaçage, *m.* placing; selling.

placard, *m.* placard; galley proof; cupboard.

place, *f.* place (in various senses); ground; (public) square; market; change, exchange; merchants, bankers; (*Glass*) chair, gang (of workmen).

faire —, make a place, make room, give way.

placement, *m.* placing; investment; selling.

placenta, *m.* placenta.

placentaire, *a.* & *m.* placental.

placer, *v.t.* place; put; invest; sell; plant. — *v.r.* be placed; station oneself; get a position; be invested; (of goods) sell. — *m.* (*Mining*) placer.

plafond, *m.* ceiling; top, crown; (*Mining*) roof; bottom (of a tank, a canal, etc.).

plage, *f.* region; shore, coast; beach.

plagièdre, *a.* (*Cryst.*) plagihedral.

plagioclase, *m.* (*Min.*) plagioclase.

plagionite, *f.* (*Min.*) plagionite.

plaider, *v.t.* plead; sue. — *v.i.* litigate; plead.

plaie, *f.* wound; sore; plague.

plaignant, *m.* plaintiff, complainant. — *p.pr.* of plaindre.

plain, *a.* plain, plane, level, even, flat; flush.

plaindre, *v.t.* pity; complain of; grudge. — *v.r.* & *i.* complain.

plaine, *f.* plain.

plainer, *v.t.* (*Leather*) unhair (esp. with lime).

plain-pied, *m.* flat, suite (on one floor).

de —, on the same level; as a matter of course, of course.

plaint, *p.p.* of plaindre.

plainte, *f.* complaint.

plaire, *v.i.* be pleasing, be pleasant, be agreeable. — *v.r.* be pleased (with), be fond (of), take pleasure (in). — *à,* please.

plaisamment, *adv.* pleasantly; ridiculously.

plaisant, *p.a.* pleasing, pleasant, agreeable; humorous; ridiculous. — *m.* humor, fun; joker.

plaisent, *pr. 3 pl. indic. & subj.* of plaire.

plaisir, *m.* pleasure.

plait, *pr. 3 sing.* of plaire.

plamage, *m.* (*Leather*) unhairing (esp. by lime).

plamer, *v.t.* (*Leather*) unhair (esp. by lime).

plan, *a.* plane. — *m.* plane, plane surface; plate; plan; map, plot, plat, chart, drawing.
— *coupe,* sectional plan.
— *de construction,* working drawing.
— *de joint,* parting plane.
— *de symétrie,* plane of symmetry.
— *du foyer,* focal plane.
— *incliné,* inclined plane.

planage, *m.* smoothing, etc. (see planer).

planche, *f.* board; plank; (of metal) plate; block (for printing).

planchéier, *v.t.* floor; board, plank.

plancher, *m.* floor, flooring; ceiling; platform, stage, stand.

planchette, *f.* (small) board; tray; plane table.

plan-concave, *a.* (*Optics*) plano-concave.

plan-convexe, *a.* (*Optics*) plano-convex.

plancton, *m.* (*Biol.*) plankton.

plane, *f.* any of several cutting or smoothing tools, as a paring knife, a drawing knife, a turning chisel. — *m.* plane, plane tree; Norway maple.

plané, *m.* gold foil. — *p.p.* of planer.

planer, *v.t.* smooth; planish (metals); plane, shave (as wood). — *v.i.* soar; look down.

planétaire, *a.* planetary.

planète, *f.* planet.

planotage, *m.* (*Sugar*) scraping smooth the base of a sugar loaf.

plansichter, *m.* plansifter.

plant, *m.* plant (for setting out); planting, bed.

plantage, *m.* planting, etc. (see planter).

plantain, *m.* (*Bot.*) plantain (*Plantago*).
— *d'eau,* water plantain (*Alisma,* esp. *A. plantago*).

plantanier, *m.* plantain (the fruit).

plantation, *f.* plantation; planting, etc. (see planter).

plante, *f.* (*Bot.*) plant; sole (of the foot).
— *fourragère,* forage plant.
— *herbacée,* herbaceous plant, herb.
— *industrielle,* a plant which has industrial uses.
— *ligneuse,* woody plant, ligneous plant (shrub or tree).
— *potagère,* vegetable.
— *vivace,* perennial plant.

planté, *p.a.* planted, etc. (see planter); erect, standing.

planter, *v.t.* plant; set, set up; put, place.

plantureux, *a.* abundant, copious; fertile.

planure, *f.* shavings; turnings.

plaque, *f.* plate; slab; (rigid) sheet; diaphragm; veneer; badge; decoration; (*Med.*) patch, spot.

plaque à *cavités*, spot-test plate.
— *alvéolée*, (*Elec.*) grid.
— *au gélatinobromure*, (*Photog.*) gelatino-bromide plate.
— *chauffante*, heating plate, hot plate.
— *de blindage*, armor plate.
— *de céramique*, pottery plate, earthenware plate.
— *de fond*, base plate, bottom plate.
— *de tuyère*, (*Metal.*) twyer plate.
— *filière*, drawplate.
— *filtrante*, filter plate.
— *pelliculaire*, (*Photog.*) film.
— *photographique*, photographic plate.
— *sensible*, (*Photog.*) sensitive plate.
— *souple*, (*Photog.*) film.

plaqué, *m.* plated ware, plated metal; veneer.
— *p.a.* plated, etc. (see plaquer).

plaqueminier, *m.* (*Bot.*) ebony tree (*Diospyros*).
— *de Virginie*, persimmon tree, persimmon (*D. virginiana*).

plaquer, *v.t.* plate; overlay; veneer; plaster; line, face; lay on (as plaster); lay down (as sods).

plaquette, *f.* small plate; thin book.

plasma, plasme, *m.* plasma, plasm.
— *musculaire*, muscle plasma.
— *sanguin*, blood plasma.

plasmolyse, *f.* (*Biol.*) plasmolysis.

plasticité, *f.* plasticity.

plastifiant, *p.a.* rendering plastic. — *m.* a substance which imparts plasticity.

plastique, *a.* plastic. — *f.* modeling.

plat, *a.* flat; (of vessels) silver. — *m.* flat, flat part; plate; (rigid) sheet; dish; (of a balance) pan, scale. — *adv.* flat, flatly.
à —, flatwise, flat.

platane, *m.* plane tree, plane (*Platanus*) or its wood.
— *d'Amérique*, American plane, buttonwood, sycamore (*P. occidentalis*).

plateau, *m.* pan, scale (of a balance); plate; disk; slab; board; (*Mach.*) bed, table; tray; plateau, table land; platform; shoal; (*Glass*) the brick bed on which the pot rests.
— *poucette*, a movable balance pan with thumb piece.

plate-forme, *f.* platform; low truck (esp. on rails); tram; roadbed (of a railway); beam.

platement, *adv.* flatly, flat.

platinage, *m.* platinizing, platinization; specif., platinum plating.

platine, *m.* platinum. — *f.* plate; platen; stage (of a microscope); flat, flat part; lock (of a gun).
— *à chariot*, (*Micros.*) mechanical stage.
— *en fil*, platinum wire.
— *en lame*, platinum foil.
— *en livret*, platinum leaf (in books).

platine *iridié*, platinum alloyed with iridium, iridioplatinum.
— *spongieux*, spongy platinum, platinum sponge.

platiner, *v.t.* platinize, platinate, plate or otherwise cover or treat with platinum.

platineux, *a.* platinous.

platinifère, *a.* platiniferous.

platinique, *a.* platinic.

platiniser, *v.t.* platinize.

platinocyanure, *m.* platinocyanide.

platinoïde, *m.* platinoid (the alloy).

platinotypie, *f.* (*Photog.*) platinotype.

platinotypique, *a.* (*Photog.*) platinotype.

platinure, *f.* platinizing, platinization.

plâtrage, *m.* plastering.

plâtras, *m.* old plaster.

plâtre, *m.* plaster (esp. plaster of Paris); plaster cast; plaster object.
— *aluné*, Keene's cement.
— *blanc*, white plaster, specif. a pure variety of plaster of Paris.
— *cru*, unburned gypsum.
— *de Paris*, plaster of Paris.

plâtreau, *m.* plaster stone in broken pieces.

plâtre-ciment, *m.* hydraulic lime.

plâtrer, *v.t.* plaster.

plâtreux, *a.* (of soil, etc.) gypseous; chalky.

plâtrière, *f.* plaster kiln; gypsum quarry.

plausible, *a.* plausible.

pléiade, *f.* pleiad.

plein, *a.* full; solid; whole, entire; (of wood) close-grained; (of air, fields, etc.) open; the middle of; pregnant. — *m.* full, fullness; full amount; plenum (as opposed to a vacuum); thickness; solid part; full stroke; middle.
de — *saut*, all at once, suddenly.
en —, full, in the middle.
en — *jour*, in open (or broad) daylight.
en — *vent*, in the open air.

pleinement, *adv.* fully, completely, thoroly.

plenum, *m.* plenum.

pléomorphisme, *m.* pleomorphism.

pléthore, *f.* plethora.

pleurer, *v.i.* weep; (of the eyes) water; bleed (as vines); become covered with moisture (as hides). — *v.t.* weep for, mourn.

pleurésie, *f.* (*Med.*) pleurisy.

pleurétique, *a.* (*Med.*) pleuritic.

pleuvoir, *v.i.* rain.

plèvre, *f.* (*Anat.*) pleura.

pli, *m.* fold; plait; crease; wrinkle; envelope; cover; letter; document; paper; bend (as of the arm); bent, direction; inequality (in the ground).
— *cacheté*, sealed communication, sealed document (e.g. one deposited with a scientific society to establish the priority for a discovery without revealing it at the time).

pliable, *a.* pliable.

pliage, *m.* folding, bending, etc. (see plier).

pliant, *a.* pliant, pliable. — *p.a.* folding, etc. (see plier). — *m.* folding chair.

plier, *v.t.* fold; bend; plait, double; fold up. — *v.r.* be folded; be bent; bend; yield. — *v.i.* bend; curve; sink; sag; yield. — *en quatre,* fold in quarters, fold twice.

plieur, *m.,* **plieuse,** *f.* folder.

Pline, *m.* Pliny.
— *l'Ancien,* Pliny the Elder.
— *le Jeune,* Pliny the Younger.

plinger, *v.t.* dip, plunge.

pliocène, *m. & a.* (*Geol.*) Pliocene.

plissage, *m.* plaiting; folding.

plissement, *m.* plaiting; corrugation; (*Geol.*) folding, fold.

plisser, *v.t.* plait; corrugate (as metals). — *v.r.* be plaited; be corrugated; crease, wrinkle. — *v.i.* form plaits; crease.

plissure, *f.* plaiting.

ploc, *m.* hair (from the cow, the dog, etc.); waste wool; sheathing felt (for ships).

plocage, *m.* carding (of wool).

ploie, *pr. 3 sing.* of ployer.

plomb, *m.* lead; shot; lump of lead; plummet, lead; lead seal; (*Printing*) forms; poisonous gas, specif. hydrogen sulfide.

à —, plumb, vertically, upright.

— *aigre,* hard lead (lead containing antimony).

— *antimonié,* lead containing antimony, hard lead.

— *blanc,* white lead.

— *blanchi,* lead alloyed with tin.

— *brulé,* lead ashes, litharge.

— *carbonaté,* carbonate of lead.

— *corné,* (*Min.*) phosgenite.

— *de chasse,* shot.

— *de garantie,* lead, lead seal.

— *de mer,* (*Min.*) molybdenite.

— *de sonde,* plummet.

— *de sûreté,* safety plug, fusible plug.

— *d'œuvre,* raw lead (usually containing silver); lead goods, lead ware.

— *doux,* soft lead, refined lead.

— *durci,* hardened lead.

— *en feuille,* sheet lead.

— *en fil,* lead wire.

— *en grains,* finely granulated lead; bird shot.

— *en lame,* lead in plates, plated lead.

— *en rouleaux,* sheet lead in rolls.

— *en saumon,* pig lead.

— *en tôle,* sheet lead.

— *filé,* lead wire.

— *frais,* refined lead.

— *fusible,* (*Elec.*) fuse.

— *gomme,* (*Min.*) plumboresinite.

— *granulé,* granulated lead; shot.

plomb *laminé,* sheet lead, rolled lead.

— *marchand,* merchant lead.

— *mou,* soft lead.

— *raffiné,* refined lead.

— *rouge,* red lead, minium.

— *rouge de Sibérie,* (*Min.*) red lead ore, crocoite, crocoisite (lead chromate).

— *spathique.* (*Min.*) cerussite.

— *sulfaté,* sulfate of lead.

plombage, *m.* leading; plumbing; sealing (with lead); (*Ceram.*) application of lead glaze.

plombagine, *f.* plumbago, graphite, black lead.

plombaginer, *v.t.* blacklead.

plombate, *m.* plumbate.

plombé, *p.a.* leaded; plumbed; leaden; livid.

plomber, *v.t.* lead (apply lead or lead glaze to, etc.); plumb, sound; seal (with lead). — *v.r. & i.* become livid or lead-colored.

plomberie, *f.* lead working, plumbing; lead works; plumber's shop.

plomb-éthyle, *m.* lead ethyl.

plombeur, *m.* sealer.

plombeux, *a.* plumbous; lead, of lead.

plombi-argentifère, *a.* containing lead and silver.

plombier, *a.* of or pertaining to lead or lead working. — *m.* plumber.

plombifère, *a.* containing lead, plumbiferous.

plombique, *a.* plumbic; lead, of lead.

plomb-mère, *m.* lead rich in silver, obtained in the Pattinson process.

plombo-argentifère, *a.* = plombi-argentifère.

plombo-cuivreux, *a.* of or containing lead and copper.

plombo-cuprifère, *a.* containing lead and copper.

plongeant, *p.a.* plunging, etc. (see plonger); downward, descending.

plongée, *f.* plunge, dive.

plongement, *m.* plunging; dip.

plonger, *v.t.* plunge; dip (as candles); (*Metal.*) quench. — *v.r.* plunge; dive. — *v.i.* plunge, dive; look down.

plongeur, *m.* plunger; dipper; diver; diving apparatus; dishwasher. — *a.* plunging; diving.

— *à barrette,* plunging bar, plunging rod, bar plunger.

ploque, *f.* = ploc.

plot, *m.* block; plug.

ployable, *a.* flexible.

ployer, *v.t.* bend; bow; fold. — *v.r.* be bent; bend; yield; be folded. — *v.i.* bend; bow; yield.

ployure, *f.* fold; bend.

plu, *p.p.* of plaire and of pleuvoir.

pluche, *f.* plush.

pluie, *f.* rain; shower.

plume, *f.* pen; quill; feather; feathers; plume.

plumeux, *a.* feathery, plumose.

plumotage, *m.* (*Sugar*) dressing the clay.

plupart, *f.* most part, greatest part, most, majority, most people.

  la — *du temps,* most of the time, generally.

  pour la —, for the most part, mostly.

pluralité, *f.* plurality; greatest number; multiplicity; (*Grammar*) plural.

plurent, *p. def. 3 pl.* of plaire.

pluriel, *a. & m.* (*Grammar*) plural.

plurivalent, *a.* multivalent, polyvalent.

plurivariant, *a.* multivariant.

plus, *adv.* more; most; longer; besides, in addition, also; (with an implied negative) no more. — *m.* most, maximum, utmost; more.

  au —, at most, at best, scarcely.

  de —, more, besides, in addition, moreover.

  de — *en* —, more and more.

  en —, besides.

  en — *de,* in addition to.

  — *haut,* higher; above.

  — *loin,* farther; farther on.

  — *ou moins,* more or less.

  qui —, *qui moins,* some more, some less.

  sans —, without more, without further; and no more.

plusieurs, *a. & m. pl.* several, many.

  à — *reprises,* several times, repeatedly.

plus-value, *f.* increase, increase in value; gain; excess; premium.

plut, *p.def. 3 sing.* of plaire and of pleuvoir.

plutonien, plutonique, *a.* (*Geol.*) Plutonic, Plutonian, (of rocks) igneous.

plutôt, *adv.* rather.

pluvial, *a.* of rain, rain; (*Geol.*) pluvial.

pluvieux, *a.* rainy.

P.M., *abbrev.* (poids moléculaire) molecular weight, mol. wt.

pneu, *m.* pneumatic tire.

pneumatique, *a.* pneumatic. — *m.* pneumatic tire. — *f.* pneumatics.

pneumatochimie, *f.* the chemistry of gases.

pneumatochimique, *a.* relating to the chemistry of gases, pneumatochemical.

pneumonie, *f.* (*Med.*) pneumonia.

poche, *f.* pocket; pouch; bag, sack; (*Anat.,* etc.), sac; ladle.

  — *à huile,* (*Mach.*) oil pan.

  — *de coulée,* casting ladle, foundry ladle.

pochette, *f.* small pocket; pocket case or kit.

pochon, *m.* ladle; (*Distilling*) a cup having a dish-like base to which a handle is attached.

podophylle, *m.* (*Bot.*) Podophyllum.

podophyllin, podophylline, *f.* podophyllin.

poêle, *m.* stove; pall. — *f.* shallow pan with handle (as for frying).

poêlon, *m.* small pan with handle.

poésie, *f.* poetry.

poids, *m.* weight; gravity; load, burden.

  — *atomique,* atomic weight.

  — *brut,* gross weight.

  — *cavalier,* rider.

  — *curseur,* sliding weight.

  — *excédant,* excess weight, overweight.

  — *mobile,* sliding weight.

  — *moléculaire,* molecular weight.

  — *moléculaire élevé,* high molecular weight.

  — *mort,* dead weight, dead load.

  — *net,* net weight.

  — *spécifique,* specific gravity.

poignée, *f.* handful; handle; grip, stock, hilt, ear, etc.; holder (for hot things); shake (of the hand).

poignent, *pr. 3 pl.* of poindre.

poignet, *m.* wrist; wristband.

poil, *m.* hair; nap, pile; coat (of animals).

  — *de chameau,* camel's hair.

  — *de chèvre,* goat's hair.

  — *de scorie,* slag wool.

poileux, poilu, *a.* hairy, shaggy.

poinçon, *m.* point, awl, bodkin, pricking punch; stamp, die, punch; puncheon; (*Mach.*) vertical arbor, post.

poinçonnage, poinçonnement, *m.* stamping, marking, punching.

poinçonner, *v.t.* stamp, mark, punch.

poinçonneuse, *f.* punching machine.

poindre, *v.i.* appear; dawn.

poing, *m.* fist; hand.

point, *m.* point (in various senses); dot; period; speck, spot; fine bubble (in glass); mark; position; point of view; stitch; (*Mach.*) center. — *adv.* (with or without *ne*) not at all, no indeed, no, none, not any.

  — *pr. 3 sing.* of poindre.

  à —, on the dot, just in time; to a nicety.

  de — *en* —, exactly, in every detail.

  mettre au —, bring to a focus; focus; adjust; make practicable, perfect, effect.

  — *critique,* critical point.

  — *d'appui,* supporting point; support; fulcrum (of a lever); center of motion (as of a balance).

  — *d'aspect,* point of view.

  — *d'ébullition,* boiling point.

  — *de concours,* (*Optics*) point of convergence.

  — *de congélation,* freezing point.

  — *de départ,* point of departure, starting point, initial point.

  — *de fuite,* vanishing point.

  — *de fusion,* melting point.

  — *de jonction,* meeting point.

  — *de mouvement,* center of motion.

  — *de repère,* reference point, datum point or mark; bench mark.

  — *de rosée,* dew point.

  — *d'eutexie,* eutectic point.

  — *de vaporisation,* vaporization point.

point de vue, point of view; focus.
— directeur, guide point.
— du tout, not at all.
— figuratif, a point which represents something in a diagram.
— fixe, fixed point; fulcrum.
— géométrique, geometrical point.
— mort, dead point, dead center.
— radieux, (Optics) radiant point.
points courants, dotted line.

pointage, m. check, check mark; pointing, aiming.

pointe, f. point (in various senses); top (as of a tree); peak; tip; nail; brad; tack; pin; (Mach.) bit; small quantity, dash, touch; break (of day).
en —, in a point, in the form of a point, pointed.
— de Paris, wire nail.

pointeau, m. center punch, prick punch; needle valve; needle (of a needle valve).

pointement, m. appearance; outcrop (as of ores); check, check mark.

pointer, v.t. prick, pierce, stick, stab; check, check off, mark; mark out; point (as a telescope). — v.i. spring up, shoot, appear; rise, soar; rear.

pointillage, pointillement, m. dotting; stippling, stipple.

pointillé, p.a. dotted; stippled. — m. dotting, dotted line, dots; stippling.

pointiller, v.t. dot; stipple. — v.i. stipple; cavil.

pointu, a. pointed.

poire, f. pear; pear-shaped object, specif. a rubber bulb.
en —, pear-shaped, pyriform.
— à feu, (Pyro.) smoke ball.
— d'aspiration, aspirating bulb.
— en caoutchouc, rubber bulb; rubber bottle.

poiré, m. perry, fermented pear juice.

poireau, m. leek; (Med.) wart.

poirier, m. pear (tree or wood).

pois, m. pea; pellet.
— à cautères, (Pharm.) issue pea.
— à gratter, = pois velus.
— cassés, split peas.
— d'iris, orris root pellet, issue pea.
— fulminant, priming, pellet primer.
— velus, cowhage (Stizolobium pruriens).

poison, m. poison.
— des flèches, arrow poison.
— du cœur, cardiac poison.
— sagittaire, arrow poison.

poissement, m. pitching, coating with pitch.

poisser, v.t. pitch, coat with pitch; soil.

poisseux, a. sticky; pitchy, pitchlike.

poisson, m. fish.

poitrine, f. chest; breast; lungs; breast (of a furnace).

poivre, m. pepper.
— de Cayenne, Cayenne pepper.
— de Guinée, Guinea pepper (Capsicum annuum).
— de la Jamaïque, Jamaica pepper, allspice.
— des murailles, wall pepper, common stonecrop (Sedum acre).
— d'Espagne, — d'Inde, Cayenne pepper, red pepper.
— noir, black pepper.

poivré, p.a. peppery, pungent; peppered.

poivrette, f. black caraway (Nigella sativa) or its carminative seeds.

poivrier, m. (Bot.) pepper (Piper); pepper box.

poix, f. pitch.
— blanche, white pitch, Burgundy pitch (esp. a kind made from galipot).
— de Bourgogne, Burgundy pitch.
— de cordonnier, shoemaker's pitch (or wax).
— de houille, coal-tar pitch.
— des Vosges, Burgundy pitch.
— grasse, liquid pitch (a mixture of wood tar and rosin).
— juive, Jew's pitch, bitumen of Judea.
— minérale, mineral pitch, asphalt.
— noire, black pitch, common pitch.
— sèche, hard pitch, stone pitch.

poix-résine, f. rosin.

polaire, a. polar.

polarimètre, m. polarimeter.

polarisant, p.a. polarizing.

polarisateur, a. polarizing. — m. polarizer.

polarisation, f. polarization, polarizing.
— rotatoire, rotatory polarization.

polariscope, m. polariscope.

polariscopique, a. polariscopic.

polarisé, p.a. polarized.

polariser, v.t. polarize. — v.r. be polarized.

polariseur, m. polarizer. — a. polarizing.

polarite, m. polarite (a water purifier).

polarité, f. polarity.

pôle, m. pole.
pôles de même nom, similar poles.
pôles de nom contraire, opposite poles.

polémique, a. polemic(al). m. polemic.

poli, p.a. polished; bright; polite. — m. polish; polishing.

polianite, f. (Min.) polianite.

police, f. policy; police; public order; rules of order; bill, list.
— d'assurance, insurance policy.

poliment, m. polishing; polish. — adv. politely.

polir, v.t. polish; burnish. — v.r. be polished.

polissable, a. polishable.

polissage, polissement, m. polishing; polish.

polisseur, m. polisher, one who polishes.

polissoir, m. polisher, burnisher, polishing tool or machine; (Powder) glazing barrel.

polissure, f. polishing; polish.

politique, *a.* political; politic. — *f.* policy; politics. — *m.* politician.
politiquement, *adv.* politically; politicly.
pollen, *m.* (*Bot.*) pollen.
polluer, *v.t.* pollute.
pollution, *f.* pollution.
Pologne, *f.* Poland.
polonais, *a.* Polish, of Poland or the Poles.
polonifère, *a.* containing polonium.
polonium, *m.* polonium.
polyacide, *a.* polyacid.
polyamine, *f.* polyamine.
polyarsénié, *a.* containing more than one arsenic atom.
polyatomicité, *f.* polyatomicity.
polyatomique, *a.* polyatomic.
polyazoté, *a.* containing more than one nitrogen atom.
polybasique, *a.* polybasic.
polybasite, *f.* (*Min.*) polybasite.
polychloré, *a.* polychloro.
polychroïque, *a.* pleochroic, polychroic.
polychroïsme, *m.* pleochroism, polychroism.
polychrome, *a.* polychrome.
polycopie, *f.* a gelatin copying contrivance similar to the hectograph, also the process of making copies with it.
polyédral, *a.* (*Geom.*) polyhedral.
polyèdre, *m.* (*Geom.*) polyhedron. — *a.* polyhedral.
polyédrique, *a.* (*Geom.*) polyhedral, polyhedric.
polygale, **polygala**, *m.* (*Bot.*) milkwort (*Polygala*).
— *de Virginie*, Senega root, snakeroot (*Polygala senega* or its root).
polygonal, *a.* polygonal.
polygone, *m.* polygon. — *a.* polygonal.
polyhalite, *f.* (*Min.*) polyhalite.
polymère, *m.* polymer. — *a.* polymeric.
polymérie, *f.* polymerism.
polymérique, *a.* polymeric.
polymérisation, *f.* polymerization.
polymériser, *v.t. & i.* polymerize.
polymérisme, *m.* polymerism.
polyméthylénique, *a.* polymethylene.
polymorphe, *a.* polymorphous.
polymorphisme, *m.*, **polymorphie**, *f.* polymorphism, polymorphy.
polynitré, *a.* polynitro, polynitro-.
polynitrile, *m.* polynitrile.
polynôme, *m.* polynomial.
polynucléaire, *a.* polynuclear.
polyol, *m.* polyhydric alcohol.
polyolal, *m.* aldose.
polypeptide, *n.* polypeptide.
polyphasé, *a.* (*Elec.*) polyphase.
polysaccharide, *m.* polysaccharide.
polysulfure, *m.* polysulfide.
polysulfuré, *a.* polysulfide of.
polytechnique, *a.* polytechnic.

polytérébénique, *a.* polyterpene.
pommade, *f.* (*Pharm.*, etc.) pomade, pomatum, ointment.
— *belladonée*, belladonna ointment.
— *camphrée*, camphor cerate.
— *citrine*, ointment of mercuric nitrate, citrine ointment.
— *créosotée*, creosote ointment.
— *d'acide phénique*, ointment of phenol.
— *de baleine*, spermaceti ointment.
— *de calomel*, mercurous chloride ointment, ointment of calomel.
— *de céruse*, lead carbonate ointment.
— *de chloramidure de mercure*, ointment of ammoniated mercury.
— *de deutoiodure de mercure*, mercuric iodide ointment.
— *de goudron*, tar ointment.
— *de grande ciguë*, conium ointment.
— *de Lyon*, ointment of red mercuric oxide.
— *de noix de galle*, nutgall ointment.
— *de piment des jardins*, capsicum ointment.
— *de potassium ioduré*, iodine ointment.
— *de poudre de Goa*, chrysarobin ointment.
— *de précipité rouge*, ointment of red mercuric oxide, red precipitate ointment.
— *de tannin*, ointment of tannic acid.
— *d'iode*, iodine ointment.
— *d'oléate d'aconitine*, aconitine ointment.
— *d'oléate de cocaïne*, cocaine ointment.
— *épispastique*, cantharides ointment; cantharides cerate.
— *épispastique au garou*, an ointment prepared from the bark of the spurge flax and equivalent to mezereum ointment.
— *mercurielle à parties égales*, mercurial ointment, blue ointment.
— *mercurielle composée*, compound mercurial ointment.
— *napolitaine*, mercurial ointment.
— *phéniquée*, ointment of phenol.
— *simple*, ointment, simple ointment.
— *soufrée*, sulfur ointment.
— *stibié*, antimonial ointment, tartar emetic ointment.
pomme, *f.* apple; (*Bot.*) pome; head (as of lettuce); knob, ball; rose (of a sprinkler); pommel.
— *d'arrosoir*, sprinkling rose, rosehead.
— *de pin*, pine cone.
— *de reinette*, pippin.
— *de terre*, potato.
— *de terre sucrée*, sweet potato.
— *épineuse*, thorn apple (*Datura*, esp. *D. stramonium*).
pommelle, *f.* grating (of a pipe); strainer.
pommier, *m.* apple tree.
pompe, *f.* pump; syringe; pomp.
— *à air*, air pump.
— *à bière*, beer pump.

pompe *à bras*, hand pump.
— *à huile*, oil pump.
— *à incendie*, fire engine.
— *à jet*, jet pump.
— *alimentaire*, feed pump.
— *à mercure*, mercury pump.
— *à moteur*, motor-driven pump.
— *à piston*, piston pump.
— *à pneumatique*, tire pump.
— *à pression*, pressure pump, compression pump.
— *aspirante*, suction pump.
— *à vapeur*, steam pump.
— *à vide*, vacuum pump.
— *à volant*, pump with flywheel (esp. a vacuum pump worked by hand).
— *centrifuge*, centrifugal pump.
— *d'alimentation*, feed pump.
— *de Gay-Lussac*, a simple hand pump for suction and pressure.
— *de robinet*, glass-blowing machine.
— *dickmaische*, (*Brewing*) thick-mash pump.
— *élévatoire*, lift pump.
— *foulante*, pressure pump, force pump, blast pump, compressor, blower.
— *rotative*, rotary pump.
pomper, *v.t.* pump; suck, suck up. — *v.i.* pump.
pompeux, *a.* stately; pompous.
pompier, *m.* fireman; pumper, pump maker or man.
ponçage, *m.* pumicing; pouncing.
ponce, *f.* pumice, pumice stone; pounce; ink made from lampblack and oil. — *a.* pumice.
— *phosphorique*, pumice moistened with phosphoric acid.
— *potassique*, pumice moistened with caustic potash solution.
— *sulfurique*, pumice moistened with sulfuric acid.
ponceau, *m.* (*Dyes*) ponceau; (*Bot.*) corn poppy; scarlet color like that of the corn poppy; small bridge, culvert. — *a.* corn-poppy red, poppy-colored.
poncer, *v.t.* pumice, rub with pumice; pounce.
ponceux, *a.* pumiceous.
ponchon, *m.* puncheon, large cask.
ponction, *f.* puncture, piercing, tapping.
ponctionner, *v.t.* tap, puncture, pierce.
ponctuage, *m.* (*Ceram.*) dotting, spottiness.
ponctuation, *f.* punctuation; punctuation mark; dotting, dot.
ponctuel, *a.* punctual.
ponctuer, *v.t.* punctuate; dot.
pondérabilité, *f.* ponderability.
pondérable, *a.* ponderable.
pondéral, *a.* of or pertaining to weight, by weight, ponderal; (of analytical methods) gravimetric.

pondéralement, *adv.* by weight; gravimetrically.
pondération, *f.* balancing, balance, equilibrium, poise.
pondéré, *p.a.* well balanced.
pondérer, *v.t.* balance, poise.
pondéreux, *a.* ponderous.
pondre, *v.t. & i.* lay (eggs).
pont, *m.* bridge (in various senses); deck.
— *d'échappement*, (*Metal.*) flue bridge.
— *de la chauffe*, (*Metal.*) fire bridge.
— *de Wheatstone*, (*Elec.*) Wheatstone bridge.
pontil, *m.* punty, pontee, pontil (an iron rod used in handling glass); small mass of glass used to fasten the main mass to the punty; small glass used in polishing mirrors; flare (as of a beaker).
populaire, *a.* popular. — *m.* populace.
populairement, *adv.* popularly.
populariser, *v.t.* popularize.
population, *f.* population.
populeux, *a.* populous.
populine, *f.* populin.
porc, *m.* hog, pig, swine; pork; (*Metal.*) slag containing ore.
porcelaine, *f.* porcelain, china; cowrie (shell).
— *à feu*, fireproof porcelain, fire-resisting porcelain.
— *d'amiante*, asbestos porcelain (a hard preparation of asbestos with a binding agent, used esp. for filters).
— *dégourdie*, biscuit porcelain, unglazed porcelain.
— *de Réaumur*, devitrified glass.
— *de Saxe*, Saxon porcelain.
— *de Sèvres*, Sèvres porcelain.
— *dure*, hard porcelain.
— *frittée*, frit porcelain, soft porcelain.
— *mate*, unglazed porcelain.
— *tendre*, soft porcelain.
porcelainier, *m.* porcelain maker. — *a.* of or pertaining to porcelain.
porcelané, **porcelanique**, *a.* porcelaneous, porcelanic.
pore, *m.* pore.
porée, *f.* molding earth.
poreux, *a.* porous.
porosité, *f.* porosity, porousness.
porphyre, *m.* (*Petrog.*) porphyry; (*Pharm.*) muller or slab of porphyry for triturating drugs.
porphyré, *a.* resembling porphyry.
porphyrique, *a.* porphyritic.
porphyrisation, *f.* grinding (with a muller), trituration.
porphyriser, *v.t.* grind (with a muller).
porreau, *m.* leek; (*Med.*) wart.
port, *m.* carrying; carriage; tonnage, burden (of boats); port, harbor.
portabilité, *f.* portability.

portable, *a.* portable.

portage, *m.* carriage; (*Mach.*) bearing.

portant, *m.* handle; support; prop; (of a magnet) armature. — *p.a.* bearing, etc. (see porter).

portatif, *a.* portable; hand; pocket.

porte, *f.* door; gate; entrance. — *a.* (*Anat.*) portal.

— *à coulisse,* sliding door.

— *arrière,* rear door, back door.

— *avant,* front door.

— *battante,* swinging door.

— *de charge,* — *de chargement,* charging door.

— *de chauffe,* fire door.

— *de travail,* working door, working hole.

— *du foyer,* fire door.

— *vitrée,* glazed door, glass door.

porté, *p.a.* borne, etc. (see porter).

porte-. A combining form used freely in forming compound words and meaning (1) in nouns, *holder, carrier, support, case, stand,* etc.; (2) in adjectives, *holding, supporting, carrying, bearing.*

porte-allumettes, *m.* match box, match case.

porte-amorce, *m.* primer holder.

porte-balai, *m.* broomstick; (*Elec.*) brush holder.

porte-bec, *m.* burner support, burner holder.

porte-caoutchouc, *a.* suitable for the attachment of a rubber tube.

porte-charbon, *m.* (*Elec.*) carbon holder.

porte-cigare, *m.* cigar holder.

porte-creusets, *m.* crucible support.

portée, *f.* range; reach, extent, distance, span; importance, significance; position; discharge, delivery (as of water); (*Mach.*) bearing, bearing surface; litter, brood; gestation; (musical) staff.

à — de, à la — de, within range (or reach) of.

porte-feu, *m.* flame passage; igniter. — *a.* fire-transmitting.

porte-filtre, *m.* filter support, specif. an annular funnel holder to be laid on a beaker. — *a.* carrying or supporting a filter.

portefeuille, *m.* portfolio; pocketbook.

en —, in manuscript, unpublished.

porte-graines, *a.* seed-bearing.

porte-lames, *m.* (*Micros.*) slide support.

porte-loupe, porte-loupes, *m.* lens holder, magnifying-glass stand.

porte-manchon, *m.* mantle holder (in a burner).

porte-mèche, *m.* wick holder; (*Mach.*) bit holder.

porte-miroir, *m.* mirror holder, mirror support.

porte-nitrate, *m.* a holder for silver nitrate used in cauterizing, caustic holder.

porte-objet, *m.* object holder; microscopic slide; stage (of a microscope). — *a.* object-holding.

porte-outil, *m.* tool holder.

porte-pièce, *m.* work holder. — *a.* work-holding.

porte-pierre, *m.* = porte-nitrate.

porte-plume, *m.* penholder.

porter, *v.t.* bear; support, hold; carry; convey, take, bring; wear; turn, direct; put; give, deliver; measure; cause; influence, incline; enter, inscribe; state, say; rate; elect; enact (a law); pass (judgment. — *v.r.* be borne, be supported, be carried, etc.; go, move; turn; tend; be (as to health); present oneself; act (as). — *v.i.* bear; rest; carry; reach; strike; hit; fall; happen, occur; take effect; win. — *m.* porter (the beverage); wear.

— *à faux,* be out of plumb.

— *à la tête,* (of wine, etc.) go to the head.

— *au rouge,* heat to redness.

— *coup,* take effect, tell; strike a blow.

porte-tubes, *m.* tube holder, tube support, esp. a basket to hold tubes to be sterilized.

porteur, *m.* carrier (e.g., of oxygen); bearer; holder; porter.

porte-vases, *m.* beaker holder, vessel holder, specif. a vessel for holding the beakers for an immersion refractometer.

porte-vent, *m.* blast pipe; twyer, tuyère.

porte-verre, *m.* glass holder; lens holder.

portillon, *m.* small door or gate; specif., the working door of a puddling furnace.

portion, *f.* portion.

portland, *m.* Portland cement.

— *naturel,* natural hydraulic cement

porto, *m.* port wine, port.

portugais, *a. & m.* Portuguese.

posage, *m.* placing, etc. (see poser).

pose, *f.* placing, laying, setting, setting up, putting up; stationing; (*Photog.*) exposure, specif. time exposure; (*Mach.*) seating; attitude, pose; shift (of workmen).

poser, *v.t.* place, lay, put, set; set up, put up; set down, put down; suppose, grant; establish. — *v.r.* pose. — *v.i.* bear, rest; pose.

positif, *a. & m.* positive.

position, *f.* position; status.

— *favorisée,* favored position.

positivement, *adv.* positively.

positivisme, *m.* positivism.

posologie, *f.* (*Med.,* etc.) posology.

posséder, *v.t.* possess; have; know well, be master of.

possession, *f.* possession.

possibilité, *f.* possibility.

possible, *a.* possible. — *m.* possible; best, utmost. — *adv.* possibly, perhaps.

au —, extremely.

possiblement, *adv.* possibly.

postal, *a.* postal.

poste   283   potion

poste, *f.* mail; post office; post. — *m.* post, station, situation; shift, turn (for workmen).

faire sa —, *(Glass)* finish gathering the metal and give it a preliminary blowing (said of a glassblower's assistant).

poster, *v.t.* post, station, place.

postérieur, *a.* posterior; hind, back; later.

postérité, *f.* posterity.

postface, *f.* notice at the end of a book.

post-glaciaire, *a.* *(Geol.)* postglacial.

posthume, *a.* posthumous.

postiche, *a.* artificial, false; temporary, provisional; superadded; superfluous.

postulant, *m.* applicant; candidate.

posture, *f.* posture, position.

pot, *m.* pot; pitcher, jug; jar, crock.

— *à boulet,* a clay or clay-lined receiver for the fused antimony sulfide in one process of separating it by liquation.

— *à brai,* pitch pot, pitch kettle.

— *à colle,* glue pot.

— *à eau,* water pitcher; *(Pyro.)* water cartridge.

— *à feu,* any of several old pyrotechnic devices, fire pot, stinkpot, etc.

— *à fumée,* *(Pyro.)* smoke compartment.

— *de fusée,* *(Pyro.)* rocket cylinder; rocket head.

— *de fusion,* melting pot, crucible.

— *Ruggieri,* *(Pyro.)* Bengal light.

potabilité, *f.* potability, drinkableness.

potable, *a.* potable, drinkable.

potage, *m.* soup.

potager, *a.* kitchen, culinary. — *m.* kitchen garden, vegetable garden.

potasse, *f.* potash; specif., caustic potash.

— *à la chaux,* caustic potash prepared by decomposing the carbonate with lime.

— *à l'alcool,* caustic potash purified by solution in alcohol, filtration and evaporation.

— *artificielle,* potassium carbonate made by the Leblanc process, as from carnallite.

— *carbonatée,* carbonate of potash, potassium carbonate.

— *caustique,* caustic potash.

— *caustique liquide,* *(Pharm.)* solution of potassium hydroxide, liquor potassæ.

— *de cendres,* potassium carbonate made from ashes.

— *de suint,* potash from suint.

— *de vinasses,* potash from beet (or beet-root) vinasse.

— *du commerce,* commercial potash, impure carbonate of potassium.

— *fondue,* fused potassium hydroxide, solid caustic potash.

— *en cylindres,* caustic potash in sticks.

— *factice,* soda.

— *nitratée,* nitrate of potash, potassium nitrate.

potasse *perlasse,* pearlash.

— *vitriolée,* potassium sulfate.

potassé, *a.* containing potash; combined with or containing potassium, potassium derivative of.

potasser, *v.t.* work at, study. *(Slang.)*

potasserie, *f.* potash factory.

potassico-. A combining form denoting the presence of potassium in a compound.

potasside, *m.* potassium or any of its compounds.

potassié, *a.* containing potassium.

potassimètre, *m.* an instrument (hydrometer?) for determining the purity of potash.

potassique, *a.* potassium, of or containing potassium, (sometimes) potassic; of or containing potash.

potassium, *m.* potassium.

pot-ban, *n.* a sort of wide-mouthed bottle.

pot-de-vin, *m.* bonus.

poteau, *m.* post, pole.

potée, *f.* potful; putty; *(Founding)* luting loam.

— *de fer,* — *de rouille,* iron putty, rust putty.

— *d'émeri,* emery dust, emery slime.

— *de montagne,* ground pumice.

— *d'étain,* putty powder (stannic oxide or a mixture of this with lead oxides); an alloy of tin and lead which is burned to form putty powder.

potence, *f.* (gallows-like) support, arm, crosspiece, bracket; *(Micros.)* tube carrier; gallows, gibbet; (T-shaped) crutch; crane.

potentiel, *a. & m.* potential.

potentiellement, *adv.* potentially.

potentille, *f.* *(Bot.)* Potentilla.

potentiomètre, *m.* *(Elec.)* potentiometer.

poterie, *f.* pottery; (in general) vessels, ware.

— *campanienne,* Campanian pottery, specif. luster ware.

— *de grès,* stoneware.

— *d'étain,* tinware.

— *de terre,* earthenware, pottery.

— *émaillée,* glazed pottery (with opaque glaze), enameled pottery.

— *en fonte,* cast-iron vessels, cast-iron ware.

— *lustrée,* lustered pottery, luster ware.

— *mate,* unglazed pottery.

— *sanitaire,* sanitary ware.

— *vernissée,* glazed pottery (with transparent glaze).

pothos fétide. *(Bot.)* skunk cabbage.

potiche, *f.* flask, bottle (for shipping mercury); Oriental vase; potiche; test cut in timber.

potier, *m.* potter; *(d'étain)* tinman.

potin, *m.* pot metal, potin (cast iron and any of several alloys, as pewter, brass, cock metal, pinchbeck).

potion, *f.* potion, draft.

— *à la rhubarbe alcaline,* *(Pharm.)* mixture of rhubarb and soda.

potion *effervescente,* — *gazeuse,* (*Pharm.*) mixture of citrate of potassium.

potiron, *m.* pumpkin.

pou, *m.* louse.

pouce, *m.* thumb; great toe; inch.

— *carré,* square inch.

— *cube,* cubic inch.

poucette, *a.* See plateau poucette, at *plateau.*

poud, *m.* pood (16.38 kg., or 36.113 lbs.).

poudingue, *m.* conglomerate, pudding stone.

poudre, *f.* powder; explosive; dust; pounce; (*Leather*) the operation of dusting (sprinkling with tanning material) followed by a period during which the hide lies undisturbed in the pit.

— *à bronzer,* bronze powder.

— *à canon,* gunpowder; cannon powder.

— *aciéreuse,* (*Metal.*) cementation powder.

— *à combustion lente,* slow-burning powder.

— *à combustion rapide,* quick-burning powder.

— *aërophore,* = poudre gazeuse.

— *à fumée,* ordinary gunpowder.

— *à lessiver,* washing powder.

— *à lever,* baking powder.

— *amide,* — *amidée,* (*Expl.*) amide powder.

— *à nettoyer,* cleaning powder, polishing powder.

— *antimildiou,* anti-mildew powder.

— *antimoniale de James,* (*Pharm.*) antimonial powder, James's powder.

— *à polir,* polishing powder.

— *à tremper,* (*Metal.*) tempering powder.

— *au bois,* (*Expl.*) nitrated sawdust or wood dust.

— *au charbon,* ordinary gunpowder (made with charcoal).

— *azotée,* any explosive containing nitrogen, specif. one containing an organic nitrogen compound.

— *B,* a nitrocellulose powder invented by Vieille in 1886.

— *blanche,* white powder; specif., (*Expl.*) Augendre's powder.

— *brisante,* disruptive powder, high explosive.

— *brune,* brown powder.

— *caillou,* pebble powder.

— *cémentante,* — *cémentatoire,* (*Metal.*) cementation powder.

— *chloratée,* (*Expl.*) chlorate powder.

— *chocolat,* (*Expl.*) cocoa powder.

— *comprimée,* compressed powder.

— *d'amidon,* starch powder.

— *d'artifice,* pyrotechnic powder, specif. colored fire.

— *de bois,* wood dust, wood powder; (*Expl.*) nitrated sawdust.

— *de carrière,* blasting powder.

— *de cashcuttie composée,* (*Pharm.*) compound powder of catechu.

— *de charbon,* coal dust, powdered coal.

poudre *de chasse,* sporting powder.

— *de coton,* guncotton.

— *de craie opiacée,* (*Pharm.*) aromatic powder of chalk with opium.

— *de Dover,* (*Pharm.*) powder of ipecac and opium, Dover's powder.

— *de fusil,* rifle powder, small-arms powder.

— *de fusion,* fluxing powder.

— *de Goa,* Goa powder, araroba.

— *de guerre,* military powder, service powder.

— *de Jean de Vigo,* (*Pharm.*) red mercuric oxide.

— *de kino opiacée,* (*Pharm.*) compound powder of kino.

— *de Knox,* chloride of lime.

— *de lait,* milk powder, powdered milk.

— *de liège,* cork dust, cork powder.

— *de mercure crayeux,* (*Pharm.*) mercury with chalk.

— *de mine,* blasting powder.

— *dentifrice,* tooth powder.

— *de plomb,* dust shot.

— *de réglisse composée,* (*Pharm.*) compound powder of glycyrrhiza.

— *de riz,* rice powder.

— *des aromates,* (*Pharm.*) aromatic powder.

— *de savon,* soap powder.

— *des Chartreux,* kermes mineral.

— *de Sedlitz,* (*Pharm.*) compound effervescing powder, Seidlitz powder.

— *de Seltz,* = poudre gazeuse.

— *des épices,* (*Pharm.*) aromatic powder.

— *de sûreté,* safety explosive, safety powder.

— *de Tennant,* chloride of lime.

— *de traite,* powder for colonial trade.

— *de Tully,* (*Pharm.*) compound powder of opium, Tully's powder.

— *de vente,* commercial powder.

— *de Vienne,* (*Pharm.*) Vienna caustic, potassa with lime.

— *de Vulcain,* Vulcan powder.

— *d'exercice,* blank-cartridge powder.

— *de zinc,* zinc dust.

— *d'os,* bone meal, bone dust.

— *en grains,* grained powder.

— *en roche,* caked powder.

— *en vrac,* loose powder.

— *faible,* low explosive.

— *Favier,* Favier explosive.

— *forte,* high explosive.

— *fulminante,* fulminating powder, detonating powder, priming powder.

— *gazeuse,* (*Pharm.*) effervescing powder, soda powder.

— *gazifère purgative,* — *gazogène laxative,* (*Pharm.*) compound effervescing powder, Seidlitz powder.

— *géante,* giant powder (a dynamite).

— *insecticide,* insect powder.

— *J,* an explosive consisting of guncotton 83, potassium dichromate 17.

poudre *lente,* slow-burning powder.
— *lissée,* glazed powder.
— *météorifuge,* (*Med.*) antiflatulent powder.
— *nitratée,* nitrated powder.
— *noire,* black powder.
— *normale,* standard powder.
— *pectorale,* (*Pharm.*) compound powder of glycyrrhiza.
— *picratée,* — *picrique,* picrate powder.
— *prismatique,* prismatic powder.
— *pyroxylée,* a sporting powder composed of cellulose nitrates and barium and potassium nitrates.
— *rapide,* quick-burning powder.
— *ronde,* round-grain powder.
— *salpêtrée,* any explosive containing saltpeter or potassium nitrate (applied esp. to ordinary gunpowders).
— *sans fumée,* smokeless powder.
poudres et salpêtres, a subdivision of the French War Department having to do with explosives and their manufacture.
— *type,* standard powder.
— *verte,* green powder, specif. an explosive not thoroly dried.
— *vive,* quick-burning powder.
poudre-coton, *m.* guncotton.
poudrement, *m.* powdering.
poudrer, *v.t.* powder; pounce.
poudrerie, *f.* powder mill, powder factory.
poudrette, *f.* fine dust; (*Fertilizers*) dried and powdered night soil.
poudreux, *a.* dusty; powdery.
poudrier, *m.* worker in a powder mill.
poudroyer, *v.i.* be dusty, get dusty. — *v.t.* cover with dust. — *v.r.* fall to dust.
poulain, *m.* colt, young horse.
poule, *f.* hen; fowl. — *a.* (*Metal.*) blistered.
poulet, *m.* chicken; fowl; letter.
poulevrin, *m.* priming powder, mealed powder.
poulie, *f.* pulley, block.
— *à gorge*(*s*), grooved pulley.
— *fixe,* fixed pulley, fast pulley.
— *folle,* loose pulley, idle pulley.
pouliot, *m.* (*Bot.*) pennyroyal (ordinarily *Mentha pulegium*).
— *américain,* American pennyroyal (*Hedeoma,* esp. *H. pulegioides*).
— *commun,* European pennyroyal (*Mentha pulegium*).
pouls, *m.* pulse (the beating).
poumon, *m.* (*Anat.*) lung.
poupée, *f.* doll; manikin; puppet; (*Mach.*) upright bar or rod; specif., poppet, headstock; (*Elec.*) binding post.
pour, *prep.* for; per, by; to, in order to; about to; as for; — *m.* pro.
pour 100, per cent.
— *ainsi dire,* so to speak.
— *autant que,* in so far as.

pour *lors,* then.
— *que,* in order that, that.
pour . . . que, however.
pour-cent, *m.* per cent.
pourcentage, *m.* percentage.
pourpre, *f., m., & a.* purple.
— *de Cassius,* purple of Cassius.
— *rétinien,* — *visuel,* retinal purple, visual purple.
pourpré, pourpreux, pourprin, *a.* purple.
pourquoi, *conj., adv. & m.* why, wherefore.
pourra, *fut. 3 sing.* of pouvoir.
pourrait, *cond. 3 sing.* of pouvoir.
pourri, *p.a.* rotten, rotted, spoiled. — *m.* rottenness; rotten part.
pourrir, *v.i* rot, decay, spoil, putrefy; age, ripen (as clay). — *v.t.* rot, cause to decay, putrefy; (fig.) corrupt.
pourrissable, *a.* liable to decay, putrefiable.
pourrissage, *m.* rotting, steeping (of rags for paper); aging (of ceramic clay).
pourrissant, *p.a.* rotting, putrefying.
pourrissoir, *m.* (*Paper*) steeping vat.
pourriture, *f.* rottenness; decay; putrefaction; rotten part; (*Bot.,* etc.) rot; (*Med.*) gangrene.
— *noble,* (*Wine*) a change produced in ripe grapes by the fungus *Botrytis cinerea,* which is in part responsible for the peculiar bouquet of sauterne.
— *sèche,* dry rot.
poursuit, *pr. 3 sing.* of poursuivre.
poursuite, *f.* pursuit; (*Law,* etc.) suit.
poursuivre, *v.t.* pursue; prosecute; sue; court. — *v.i.* pursue, continue; sue.
pourtant, *conj.,* however, nevertheless, yet.
pourtour, *m.* circumference; circuit, compass.
pourverra, *fut. 3 sing.* of pouvoir.
pourvit, *p.def. 3 sing.* of pouvoir.
pourvoi, *m.* (*Law*) appeal.
pourvoir, *v.i.* provide (for), attend (to).
— *v.r.* provide oneself (with); (*Law*) appeal. — *v.t.* provide, furnish, supply; provide for.
pourvoyant, *p.pr.* of pourvoir.
pourvu, *p.a.* provided, etc. (see pourvoir).
— *que,* provided (that), on condition that, if.
pouset, *m.* a red color obtained with cochineal.
pousse, *f.* growth; sprouting; shoot, sprout; a disease of wine characterized by the development of pressure from carbon dioxide and by loss of strength and flavor.
poussée, *f.* pressure; push; thrust.
pousser, *v.t.* push; drive; thrust, shove; produce (by growth), send forth; extend, carry; continue; pursue; show; utter; help on; stir up; actuate. — *v.r.* be pushed, etc.; make one's way; go on. — *v.i.* push; push on, go on; spring up; grow; project, bulge.

poussier, *m.* dust; specif., gunpowder dust.
— *de charbon*, charcoal dust; coal dust.
— *de houille*, coal dust.
— *de minerai*, pulverized ore, ore slime.
poussière, *f.* dust; (of a liquid) fine spray.
— *de charbon*, charcoal dust; coal dust.
— *de la poudre*, gunpowder dust.
poussiéreux, *a.* dusty.
poussoir, *m.* pusher, push button; driver, punch.
poussolane, *f.* pozzuolana, pozzolana.
poutre, *f.* beam, girder.
poutrelle, *f.* small beam, joist.
pouvant, *p.pr.* of pouvoir.
pouvoir, *v.i.* be able, can, may. — *v.t.* be able to do, can do. — *v.r.* be possible, can be done, may be. — *m.* power.
— *absorbant*, absorbent power, absorptivity.
— *calorifique*, calorific power, heating power.
— *conducteur*, conducting power.
— *éclairant*, illuminating power.
— *floculant*, flocculating power.
— *réducteur*, reducing power.
— *rotatoire*, rotatory power.
— *séparateur*, (*Optics*) defining power.
pouzzolane, *f.* pozzuolana, pozzolana.
— *en pierre*, trass.
pouzzolanique, *a.* pozzuolanic, pozzolanic.
pr., *abbrev.* (précédent) preceding.
prairie, *f.* meadow; prairie.
pralin, *m.* (*Agric.*) a mixture of fertilizer with earth and water, used for coating tree roots or seeds.
pralinage, *m.* (*Agric.*) treatment with the mixture called *pralin*; praline making.
praline, *f.* praline, almond roasted in sugar; (*Sugar*) a mass of impure calcium or strontium sucrate formed by treating molasses with lime or strontia.
praliner, *v.t.* make into or like pralines; (*Agric.*) treat with *pralin* (which see).
prase, *f.* (*Min.*) prase.
praséodyme, *m.* praseodymium.
praséolite, *f.* (*Min.*) praseolite.
prasin, *a.* light green.
praticabilité, *f.* practicability, -cableness.
praticable, *a.* practicable.
praticien, *m.* practitioner. — *a.* practicing; practical.
pratique, *f.* practice; experience; routine; intercourse; method of procedure; customer. — *a.* practical; experienced.
pratiquement, *adv.* practically, in practice.
pratiquer, *v.t.* practice; do, perform, conduct; make; associate with; obtain. — *v.r.* be practiced, be done, etc.; be customary. — *v.i.* practice.
pré, *m.* meadow.
préachat, *m.* prepayment.
préacheter, *v.t.* prepay.

préalable, *a.* previous, preliminary. — *m.* preliminary.
au —, previously, first.
préalablement, *adv.* previously, first.
préaviser, *v.t.* forewarn.
préc., *abbrev.* (précédent) preceding.
précaire, *a.* precarious.
précairement, *adv.* precariously.
précambrien, *a.* (*Geol.*) Pre-Cambrian.
précarité, *f.* precariousness, uncertainty.
précaution, *f.* precaution.
précautionner, *v.t.* warn, caution.
précédemment, *adv.* previously, already, above.
précédence, *f.* precedence, priority.
précédent, *a.* preceding, previous. — *m.* precedent.
précéder, *v.t.* precede; take precedence of. — *v.i.* have precedence or priority.
prêcher, *v.t. & i.* preach.
précieusement, *adv.* very carefully; affectedly.
précieux, *a.* precious; valuable; useful; affected; elaborate.
précipitable, *a.* precipitable.
précipitant, *m.* precipitant. — *p.pr.* of précipiter.
précipitation, *f.* precipitation; acceleration (of the pulse).
— *fractionnée*, fractional precipitation.
précipité, *m.* precipitate. — *p.a.* precipitated; precipitate, hasty; precipitous, steep.
— *galvanique*, electrodeposit.
— *rouge*, red precipitate, red mercuric oxide.
précipiter, *v.t.* precipitate. — *v.r.* be precipitated, precipitate; throw oneself; rush. — *v.i.* precipitate, be precipitated.
précipitine, *f.* precipitin.
précis, *a.* precise, exact. — *m.* abstract, summary, epitome, compendium.
précisément, *adv.* precisely, exactly.
préciser, *v.t.* state precisely, specify.
précision, *f.* precision, preciseness, exactness.
précité, *a.* previously cited, aforesaid, above-mentioned.
précoce, *a.* precocious; early.
préconçu, *p.a.* preconceived.
préconiser, *v.t.* recommend, commend, praise.
précurseur, *m.* precursor, forerunner. — *a.* precursory.
prédécesseur, *m.* predecessor.
prédiction, *f.* prediction.
prédire, *v.t.* predict, foretell.
prédit, *p.p. & pr. & p.def. 3 sing.* of prédire.
prédominant, *p.a.* predominating, predominant.
prédominer, *v.i.* predominate.
prééminent, *a.* preëminent.
préexistant, *a.* preëxisting, preëxistent.
préexister, *v.i.* preëxist.
préface, *f.* preface.

préférable, *a*. preferable.

préférablement, *adv*. preferably.

préférence, *f*. preference; mark of preference. de —, preferably, rather; from choice.

préférer, *v.t*. prefer.

préfet, *m*. prefect; specif., in France, civil administrator of a department.

prehnite, *f*. (*Min*.) prehnite.

prehnitène, *m*. prehnitene.

prehnitique, *a*. prehnitic.

préjudice, *f*. prejudice.

préjudiciable, *a*. prejudicial, injurious.

préjudiciel, *a*. preliminary; (*Law*) interlocutory.

préjudicier, *v.i*. be prejudicial or injurious.

préjugé, *m*. prejudice; presumption; precedent.

préjuger, *v.t*. prejudge; foresee.

prélart, **prélat**, *m*. tarpaulin.

prèle, *f*. (*Bot*.) horsetail (*Equisetum*).

prélèvement, *m*. retention, etc. (see prélever); retained portion; withdrawn portion, sample; previous deduction.

prélever, *v.t*. retain, save out (as a portion of a sample); remove, withdraw, draw out, take (as a sample); select, pick up; exclude; deduct previously.

préliminaire, *a*. & *m*. preliminary.

préliminairement, *adv*. preliminarily, first.

prématuré, *a*. premature.

prématurément, *adv*. prematurely.

prémices, *f.pl*. first fruits; beginnings.

premier, *a*. first; (of materials) raw; (*Math*.) prime. — *m*. first; former; leader. du — coup, du — jet, at the first attempt, the first time, at once; in one step or operation, directly.

— *choix*, first choice, best grade.

première poudre, (*Leather*) first dusting and lying (see poudre).

— *étage*, first floor, first story (above the ground floor).

— *jet*, first casting, first cast; rough sketch; (*Sugar*) first spinning (in the centrifugal); (*Steam*) priming.

— *titre*, (for gold) a standard of 0.920 fine (about 22 carats).

premièrement, *adv*. in the first place, first.

prémunir, *v.t*. caution, forewarn. — *v.r*. provide.

prenait, *imp. 3 sing*. of prendre.

prenant, *p.a*. taking, etc. (see prendre).

prendre, *v.t*. take; catch; seize; get; receive; take on; adopt; deduct. — *v.r*. be taken, be caught, etc.; solidify, congeal, freeze; curdle, coagulate; harden, set; set about, begin; catch (at); lay the blame (on). — *v.i*. solidify, congeal, freeze; curdle; set; take; take fire; take root; take hold; have effect; result, be.

prendre *feu*, take fire, catch fire.

— *garde*, take care, take heed.

— *l'habitude de*, form the habit of, become accustomed to.

— *l'heure*, set a watch (or clock).

— *naissance*, be formed, originate, arise, appear.

— *pied*, get a footing; base one's action (on).

— *place*, take place, occur.

— *sans vert*, take by surprise.

se — *en masse*, solidify; coagulate.

preneur, *m*. taker, catcher, etc. (see prendre); buyer; (*Law*) lessee; captor.

prennent, *pr. 3 pl. indic. & subj*. of prendre.

prénom, *m*. first name, Christian name.

prenons, *pr. 1 pl*. of prendre. — *imperative*. let us take, etc. (see prendre).

préoccuper, *v.t*. preoccupy. — *v.r*. be preoccupied; busy oneself, pay attention.

préparant, *p.a*. preparing, etc. (see préparer).

préparateur, *m*. preparer, preparator (as of specimens); mixer, worker, maker, manufacturer.

— *de bains de teinture*, (*Dyeing*) liquor man.

— *de laboratoire*, laboratory assistant; lecture assistant.

préparatifs, *m.pl*. preparations.

préparation, *f*. preparation; dressing.

— *par impression*, (*Micros*.) impression preparation.

préparatoire, *a*. preparatory.

préparer, *v.t*. prepare; prepare for; dress (as leather or ore). — *v.r*. be prepared; be preparing, be coming; prepare oneself, prepare.

prépareur, *m*. = préparateur.

prépondérant, *p.a*. preponderant; deciding.

préposé, *m*. officer; agent. — *p.p*. of préposer.

préposer, *v.t*. set, place (over); charge (with).

près, *adv*. near, close. — *prep*. near, near to, close to; to.

— *de*, near, close to; on the point of; in comparison with; nearly, almost, about.

de —, near, close.

à . . . *près*, with the exception of.

présage, *m*. omen, presage.

présager, *v.t*. presage; indicate, show.

prescription, *f*. prescription.

prescrire, *v.t*. & *i*. prescribe. — *v.r*. be prescribed.

prescrivant, *p.pr*. of prescrire.

préséance, *f*. precedence.

présence, *f*. presence.

présent, *a*. & *m*. present.

présentation, *f*. presentation.

présentement, *adv*. now, at present.

présenter, *v.t*. present.

préservateur, *a*. preservative, preserving.

préservatif, *a*. & *m*. preservative.

préservation, *f.* preservation.

préserver, *v.t.* preserve.

présidence, *f.* presidency.

président, *m.* president; presiding judge.

présider, *v.i.* preside. — *v.t.* preside over.

présomption, *f.* presumption.

présomptivement, *adv.* presumptively.

presque, *adv.* almost, nearly; (with negative) scarcely, hardly.

presqu'île, *f.* peninsula.

pressage, *m.* pressing.

pressant, *p.a.* pressing.

presse, *f.* press; clamp; crowd; eagerness; urgency; embarrassment.

— *à bras,* hand press.

— *à coins,* wedge press.

— *à fusée,* (*Pyro.*) rocket press.

— *à galeter,* gunpowder press.

— *à huile,* oil press.

— *à imprimer,* printing press.

— *à jus,* press for extracting juices.

— *à main,* clamp.

— *à serrer,* clamp.

— *à vis,* screw press.

sous —, in press.

pressé, *p.a.* pressed; crowded; in a hurry; very busy; eager; urgent; rapidly repeated; concise.

presse-artère, *m.* artery forceps.

presse-caillot, *m.* an apparatus for separating the solid part of a coagulated substance from the liquid.

pressée, *f.* pressing, pressure; amount pressed at one time; amount of juice expressed at one operation.

presse-étoupe, *m.* (*Mach.*) stuffing box.

pressentir, *v.t.* have a presentiment of; sound.

presse-papiers, *m.* paper weight.

presser, *v.t.* press; compress; squeeze. — *v.r.* press; hurry. — *v.i.* press, be urgent, be severe.

pressette, *f.* small press.

presseur, *a.* pressing, pressure. — *m.* presser; pressman.

pressin, *m.* beets (beetroots) rasped and ready to be pressed to extract the juice.

pression, *f.* pressure.

basse —, low pressure.

en —, under pressure.

haute —, high pressure.

moyenne —, mean pressure; medium pressure.

— *critique,* critical pressure.

— *d'éclatement,* bursting pressure.

— *de la vapeur,* steam pressure; vapor pressure.

— *de l'eau,* water pressure, hydrostatic pressure.

— *en marche,* working pressure.

— *osmotique,* osmotic pressure.

pressis, *m.* expressed juice (as of meat).

pressoir, *m.* press (for making wine, oil or the like); press room.

— *à cidre,* cider press.

— *à vin,* wine press.

pressurage, *m.* pressing; (fig.) pressure.

pressurer, *v.t.* press, express; squeeze.

preste, *a.* quick.

prestement, *adv.* quickly.

présumer, *v.t.* presume.

présupposer, *v.t.* presuppose.

présure, *f.* rennet.

présurer, *v.t.* curdle with rennet.

présurier, *m.* rennet maker.

prêt, *a.* ready. — *m.* loaning; loan; (*Mil.*) pay; (fig.) exchange.

prétendre, *v.t.* pretend, claim; maintain; want, require; mean, intend; hope. — *v.i.* aspire; lay claim (to).

prétendu, *p.a.* pretended; claimed; so-called; supposed; false.

prête-nom, *m.* person lending his name, dummy.

prétention, *f.* pretension.

prêter, *v.t.* lend; adapt; take (an oath). — *v.r.* be lent; lend oneself; be adapted; yield; consent. — *v.i.* lend; stretch (as cloth); furnish material; give rise (to). — *m.* lending, loan.

préterminal, *a.* next to the end, last but one.

prêteur, *m.* lender. — *a.* lending.

prêtre, *m.* priest.

preuve, *f.* proof.

faire — *de,* prove.

prévalant, *p.pr.* of prévaloir.

prévaloir, *v.i.* prevail; gain currency, come into use. — *v.r.* take advantage (of), avail oneself (of); be proud (of).

prévalu, *p.p.* of prévaloir.

prévaut, *pr. 3 sing.* of prévaloir.

prévenant, *p.pr.* of prévenir.

prévenir, *v.t.* inform; prevent; anticipate; forestall; precede; prepossess.

préventif, *a.* preventive.

prévenu, *p.a.* informed, etc. (see prévenir); (*Law*) accused.

préviendra, *fut. 3 sing.* of prévenir.

prévient, *pr. 3 sing.* of prévenir.

prévint, *p.def. 3 sing.* of prévenir.

prévision, *f.* anticipation, conjecture.

prévit, *p.def. 3 sing.* of prévoir.

prévoir, *v.t.* foresee; provide for.

prévoyance, *f.* foresight.

prévoyant, *p.pr.* of prévoir.

prévu, *p.a.* foreseen; provided for.

primaire, *a.* primary.

primauté, *f.* primacy, preëminence.

prime, *f.* premium; bounty, subsidy; prism; semi-transparent stone corresponding to a precious stone; prime wool. — *a.* prime; first (*Obs.* except in certain phrases).

prime (continued).

  de — *abord, de* — *face,* at first sight, at first.

  de — *saut,* at the first attempt, at once, all at once.

primer, *v.i.* lead, take the lead, excel; precede.

  — *v.t.* take precedence of; excel; prime; give a premium to.

primerose, *f.* (*Dyes*) primrose; (*Bot.*) hollyhock.

primeur, *f.* early season, newness; early product; (fig.) freshness, bloom.

primevère, *f.* (*Bot.*) primrose (*Primula*).

primitif, *a.* original; primitive; primary (as colors).

primitivement, *adv.* originally, at first; primitively.

primo-secondaire, *a.* both primary and secondary (as a glycol).

primuline, *f.* (*Dyes*) primuline, primulin.

principal, *a.* principal. — *m.* principal; principal thing.

principalement, *adv.* principally.

principe, *m.* principle.

  — *actif,* active principle.

  — *amer,* bitter principle.

  — *doux,* principium dulce, sweet principle (Scheele's name for glycerol).

  — *gommeux,* gummy principle, gum (as a chemical individual).

  — *gras,* fatty principle, fat (in the sense of a chemical individual).

  — *immédiat,* immediate principle.

printanier, *a.* spring, vernal.

printemps, *m.* spring (the season).

priorité, *f.* priority.

prirent, *p. def. 3 pl.* of prendre.

pris, *p.a.* taken, caught, solidified, etc. (see prendre); shaped, proportioned.

prisable, *a.* estimable, valuable.

prise, *f.* solidification, congealing; setting (as of cement); coagulation; taking, etc. (see prendre); hold, grasp, purchase; intake (as of air); supply (as of water); valve, cock; amount taken at one time, specif. dose; capture, prize; quarrel, fighting.

  en —, (*Mach.*) in gear.

  faire —, (of cement, etc.) set.

  — *de courant,* current supply, place of attachment to an electric circuit.

  — *d'essai,* sample taking, sampling; sample.

priser, *v.t.* appraise, value; snuff.

prismatique, *a.* prismatic, prismatical.

prisme, *m.* prism.

  — *droit,* right prism.

prismoïde, *a.* prismoidal. — *m.* prismoid.

prit, *p.def. 3 sing.* of prendre.

privé, *a.* private; tame. — *m.* private; privy.

  — *p.p.* of priver.

priver, *v.t.* deprive (of); tame, domesticate.

privilège, *m.* privilege.

privilégié, *a.* privileged.

prix, *m.* price; prize; reward.

  au — *de,* at the price of, at the cost of; in comparison with.

  — *à forfait,* job price.

  — *courant,* market price, current price.

  — *d'achat,* purchase price, first cost.

  — *de fabrique,* factory price (cost of material plus cost of manufacture).

  — *de facture,* invoice price, purchase price.

  — *de marché,* market price.

  — *de revient,* cost price, cost.

  — *de vente,* selling price.

  — *fixe,* fixed price, set price.

prix-courant, *m.* price list.

probabilité, *f.* probability.

probable, *a.* probable.

probablement, *adv.* probably.

probe, *a.* upright, honest.

problématique, *a.* problematical, problematic.

problème, *m.* problem.

procédé, *m.* process; procedure, proceeding; behavior, conduct.

  — *à la cloche,* bell process.

  — *à la petite chaudière,* the process of soap making from fats and caustic soda in a single operation (without separation of the curd from the lye).

  — *au charbon,* (*Photog.*) carbon process.

  — *d'application,* process of application (as distinguished from process of manufacture).

  — *de la grande chaudière,* (*Soap*) Marseilles process.

  — *des chambres de plomb,* lead-chamber process.

  — *d'Orléans,* Orleans process (for vinegar).

  — *expéditif,* quick process, rapid process.

  — *Grignard,* Grignard process, Grignard method.

  — *hollandais,* Dutch process.

  — *marseillais,* Marseilles process, Marseilles method.

  — *par contact,* contact process.

procéder, *v.i.* proceed.

procédure, *f.* procedure; proceedings.

procès, *m.* process; (*Law*) suit, process, trial.

processus, *m.* process.

procès-verbal, *m.* report, proceedings, minutes.

prochain, *a.* next, early, near, coming. — *m.* neighbor.

prochainement, *adv.* presently, soon.

proche, *a., prep. & adv.* near. — *m.* relative.

  de — *en* —, gradually, by degrees; from place to place.

procuration, *f.* proxy, power of attorney.

procurer, *v.t.* procure. — *v.r.* be procured.

prodiastase, *f.* proenzyme, zymogen.

prodigalement, *adv.* prodigally.

prodigalité, *f.* prodigality.

prodigieusement, *adv.* prodigiously.

prodigieux, *a.* prodigious.
prodigue, *a.* prodigal, lavish.
prodiguer, *v.t.* lavish, be prodigal of.
prodrome, *m.* introduction; (*Med.*) prodrome.
producteur, *a.* producing, **productive**; generating. — *m.* producer.
productible, *a.* producible.
productif, *a.* productive.
production, *f.* production.
productivité, *f.* productivity.
produire, *v.t.* produce; introduce.
produisant, *p.a.* producing; introducing.
produit, *m.* product; produce; proceeds. — *p.a.* produced; introduced. — *pr. 3 sing.* of produire.
— *accessoire*, accessory product, by-product.
— *alimentaire*, food product.
— *à meuler*, abrasive.
— *brut*, raw product.
— *chimique*, chemical product, chemical.
— *colorant*, coloring matter (pigment or dye).
— *courant*, staple product, ordinary product.
— *de déchet*, waste product.
— *intermédiaire*, intermediate product, intermediate.
— *principal*, principal product, chief product.
proéminence, *f.* prominence.
proéminent, *a.* prominent.
proenzyme, *f.*, **proferment**, *m.* proenzyme, zymogen.
professer, *v.t.* profess; practice; teach.
professeur, *m.* professor.
profession, *f.* profession; occupation.
professionnel, *a.* professional.
profil, *m.* profile; section.
profit, *m.* profit.
profitable, *a.* profitable.
profiter, *v.i.* profit, make a profit, be profitable; take advantage (of); grow, thrive.
profond, *a.* deep; profound. — *m.* depth.
profondément, *adv.* deeply, profoundly.
profondeur, *f.* depth.
de —, in depth, deep.
— *de foyer*, (*Optics*) depth of focus.
profus, *a.* profuse.
profusion, *f.* profusion.
progrès, *m.* progress.
progresser, *v.i.* progress, make progress.
progressif, *a.* progressive.
progression, *f.* progression.
progressivement, *adv.* progressively.
progressivité, *f.* progressiveness.
prohiber, *v.t.* prohibit.
prohibitif, *a.* prohibitive.
proie, *f.* prey.
projecteur, *m.* projector; searchlight.
— *de flammes*, flame projector.
projectif, *a.* projective.
projectile, *m. & a.* projectile.
— *asphyxiant*, asphyxiating projectile.

projectile *creux*, hollow projectile, specif. **shell.**
— *lacrymogène*, lachrymatory projectile.
— *suffocant*, asphyxiating projectile.
— *toxique*, poisonous projectile.
projection, *f.* projection; spattering, scattering.
projet, *m.* project, design; first draft.
projeter, *v.t.* project; throw out, throw. — *v.r.* project.
proliférer, *v.t. & i.* (*Biol.*) proliferate.
prolifique, *a.* prolific.
proline, *f.* proline.
prolongation, *f.* **prolongement**, *m.* prolongation, extension.
prolonger, *v.t.* prolong; (*Geom.*) produce. — *v.r.* be prolonged; extend.
promener, *v.t.* lead, direct, drive, move, guide. — *v.r.* wander; go walking, riding, etc.
promesse, *f.* promise.
promettre, *v.t. & i.* promise.
promeut, *pr. 3 sing.* of promouvoir.
promis, *p.a.* promised.
promit, *p.def. 3 sing.* of promettre.
promoteur, *m.* promoter. — *a.* promoting.
promouvoir, *v.t.* promote.
prompt, *a.* prompt; rapid; quick; (of cement) quick-setting.
promptement, *adv.* promptly; quickly.
promptitude, *f.* promptness; quickness; readiness.
promu, *p.p.* of promouvoir.
prononcer, *v.t. & i.* pronounce.
prononciation, *f.* pronunciation; pronouncement.
pronostic, *m.* prognostic, sign; conjecture.
pronostiquer, *v.t.* prognosticate, predict.
propagateur, *a.* propagating.
propagation, *f.* propagation, propagating.
propager, *v.t.* propagate; extend. — *v.r.* be propagated, spread.
propane, *m.* propane.
propanoïque, *a.* propanoic (propionic).
propargylique, *a.* (of the alcohol) propargyl, propargylic; (of the acid) propargylic (propiolic).
propényle, *m.* propenyl.
propénylé, *a.* propenyl.
propeptone, *f.* propeptone.
propeptonique, *a.* propeptone.
prophétiser, *v.t. & i.* prophesy.
prophylactique, *a.* (*Med.*) prophylactic(al).
prophylaxie, *f.* (*Med.*) prophylaxis.
propice, *a.* propitious, favorable.
propiolique, *a.* propiolic.
propione, *m.* propione (3-pentanone).
propionique, *a.* propionic.
propionitrile, *m.* propionitrile.
propolis, *f.* propolis, bee glue.
proportion, *f.* proportion.
à —, *en* —, in proportion, **proportionately.**

proportion (continued).

à — de, en — de, in proportion to.

à — que, in proportion as.

proportionnalité, *f.* proportionality.

proportionné, *p.a.* proportioned; proportionate.

proportionnel, *a.* proportional.

proportionnellement, *adv.* proportionally.

proportionnément, *adv.* proportionately.

proportionner, *v.t.* proportion. — *v.r.* be proportioned, be in proportion.

propos, *m.* discourse, talk; gossip; remark; purpose.

à —, apropos, opportune, opportunely; opportuneness.

à — de, apropos of, in connection with.

hors de —, inopportune(ly).

proposer, *v.t.* propose. — *v.r.* offer oneself; propose; be proposed.

proposition, *f.* proposition.

propre, *a.* proper; own; peculiar; clean; same, very; apt; fitted; pretty, nice, neat. — *m.* property; nature, characteristic.

proprement, *adv.* properly; cleanly; nicely, neatly.

— dit, properly so called, properly speaking.

propreté, *f.* cleanness, cleanliness; neatness.

propriétaire, *m.* proprietor. — *a.* proprietary.

propriété, *f.* property; ownership; estate.

propulseur, *a.* propelling, propulsive. — *m.* propeller.

propulsif, *a.* propulsive, propelling.

propulsion, *f.* propulsion.

propylamine, *f.* propylamine.

propylbenzine, *f.* propylbenzene.

propyle, *m.* propyl.

propylène, *m.* propylene.

propylénique, *a.* propylene, of propylene.

propylglycol, *m.* propylglycol (by some used improperly for propylene glycol or propane-diol).

propylique, *a.* propyl, (sometimes) propylic.

prorata, *m.* proportional part, share.

proscrire, *v.t.* proscribe, prohibit, reject.

proscrivant, *p.p.* of proscrire.

prosécrétine, *f.* prosecretin.

prospecter, *v.t.* prospect.

prospère, *a.* prosperous.

prospèrement, *adv.* prosperously.

prospérer, *v.i.* prosper.

prospérité, *f.* prosperity.

prostate, *f.* (*Anat.*) prostate, prostate gland.

prosthétique, *a.* prosthetic.

protagon, *m.* protagon.

protamine, *f.* protamine.

protaminique, *a.* protamine.

protane, *m.* methane.

protargol, *m.* (*Pharm.*) protargol.

protéase, *f.* protease.

protecteur, *m.* protector. — *a.* protecting, protective.

protection, *f.* protection.

protectrice, *fem.* of protecteur.

protéger, *v.t.* protect.

protéide, *f.* protein; specif., conjugated protein.

protéine, *f.* protein.

— conjuguée, conjugated protein.

protéique, *a.* protein, of or pertaining to protein; protean. — *m.* protein substance.

protéoïde, *n.* albuminoid, scleroprotein.

protéolyse, *f.* proteolysis.

protéolytique, *a.* proteolytic.

protéose, *f.* proteose.

protester, *v.t. & i.* protest.

protêt, *m.* protest.

prothrombine, *f.* prothrombin.

protiodure, *m.* protiodide (*Archaic*).

— de mercure, protiodide of mercury (mercurous iodide).

protobromure, *m.* protobromide (*Archaic*).

protocarbure, *m.* (*Old Chem.*) protocarburet.

protocarburé, *a.* (*Old Chem.*) protocarburetted.

protocatéchique, *a.* protocatechuic.

protochlorure, *m.* protochloride (*Archaic*).

— de mercure, mercurous chloride.

protochloruré, *a.* protochloride of (*Archaic*).

protocole, *m.* record, minutes, protocol.

protocyanure, *m.* protocyanide (*Archaic*).

— jaune de fer et de potassium, potassium ferrocyanide.

protofluorure, *m.* protofluoride (*Archaic*).

protoiodure, *m.* protiodide (*Archaic*).

protone, *f.* protone.

protophosphure, *m.* protophosphide, protophosphuret.

protoplasma, protoplasme, *m.* (*Biol.*) protoplasm.

protoplasmique, *a.* protoplasmic.

protoprotéose, *f.* protoproteose.

protosel, *m.* protosalt (*Archaic*).

protoséléniure, *m.* protoselenide (*Archaic*).

protosulfate, *m.* protosulfate (*Archaic*).

— de fer, ferrous sulfate.

protosulfure, *m.* protosulfide (*Archaic*).

protoxyde, *m.* protoxide (*Archaic*).

— d'azote, nitrous oxide.

— de plomb, lead monoxide.

— d'étain, stannous oxide.

protoxyder, *v.t.* protoxidize (*Archaic*).

protozoaire, *m.* (*Zoöl.*) protozoan; (*pl.*) Protozoa.

prouesse, *f.* prowess; exploit, feat.

proustite, *f.* (*Min.*) proustite.

prouvable, *a.* provable.

prouver, *v.t.* prove. — *v.r.* be proved.

provenance, *f.* origin; production.

provenant, *p.a.* proceeding, coming, arising.

— de, from, proceeding from, coming from.

provençal, *a.* Provençal, of Provence.

provende, *f.* provisions; provender, specif. a mixture of chopped forage and crushed grain.

provenir, *v.i.* proceed, come, arise.

proviendra, *fut. 3 sing.* of provenir.

provient, *pr. 3 sing.* of provenir.

province, *f.* province; all France except Paris.

provint, *p.def. 3 sing.* of provenir.

provision, *f.* provision.

    par —, provisionally.

provisoire, provisionnel, *a.* provisional.

provisoirement, *adv.* provisionally.

provoquer, *v.t.* provoke.

proximité, *f.* proximity, nearness.

prudemment, *adv.* prudently, cautiously.

prudence, *f.* prudence, caution.

prudent, *a.* prudent, cautious.

prune, *f.* plum. — *m.* (*Dyes*) prune.

pruneau, *m.* prune, dried plum.

prunelle, *f.* sloe, wild plum; a fermented liquor made from sloes; pupil (of the eye); prunella (the fabric); (*Bot.*) Prunella.

prunier, *m.* plum tree, plum.

Prusse, *f.* Prussia.

prussiate, *m.* prussiate (cyanide).

    — *jaune,* — *jaune de potasse,* potassium ferrocyanide.

    — *rouge,* — *rouge de potasse,* potassium ferricyanide.

prussien, *a.* Prussian. — *m.* (*cap.*) Prussian.

prussique, *a.* prussic (hydrocyanic).

psammite, *m.* (*Petrog.*) psammite, sandstone.

pseudoaconitine, *f.* pseudaconitine.

pseudocubique, *a.* (*Cryst.*) pseudocubic.

pseudocyanure, *m.* pseudocyanide (a name applied by some to organic cyanides or nitriles).

pseudoglobuline, *f.* pseudoglobulin.

pseudoionone, *f.* pseudoionone.

pseudomorphe, *a.* (*Min.*) pseudomorphous.

pseudomorphique, *a.* (*Min.*) pseudomorphic.

pseudomorphisme, *m.* (*Min.*) pseudomorphism.

pseudomucine, *f.* pseudomucin.

pseudonitrol, *m.* pseudonitrole.

pseudopelletiérine, *f.* pseudopelletierine.

pseudoxanthine, *f.* pseudoxanthine.

psilomélane, *m.* (*Min.*) psilomelane.

p. suiv., *abbrev.* (page suivante) following page.

psychotrine, *f.* psychotrine.

psychromètre, *m.* psychrometer.

psychrométrie, *f.* psychrometry, hygrometry.

psychrométrique, *a.* psychrometric(al).

ptomaïne, *f.* ptomaïne.

ptyaline, *f.* ptyalin.

pu, *p.p.* of pouvoir.

puant, *p.a.* stinking, offensive.

puanteur, *f.* stink, stench, offensive smell.

public, *a. & m.* public.

publicateur, *m.* publisher.

publication, *f.* publication.

publicité, *f.* publicity; public nature.

publier, *v.t.* publish. — *v.r.* be published.

publique, *a. fem.* of public.

publiquement, *adv.* publicly.

puce, *a.* puce, dark brown. — *f.* flea.

puddlage, *m.* (*Metal.*) puddling.

    — *au gaz,* gas puddling.

    — *chaud,* — *gras,* — *humide,* wet puddling, pig boiling.

    — *maigre,* dry puddling.

    — *mécanique,* mechanical puddling.

    — *sec,* dry puddling.

puddler, *v.t.* (*Metal.*) puddle.

puddleur, *m.* (*Metal.*) puddler.

pudrolithe, *f.* (*Expl.*) pudrolithe.

puer, *v.i.* stink, smell. — *v.t.* smell of.

puis, *adv.* then, next, afterward. — *pr. 1 sing.* of pouvoir.

    et —, and then; and besides.

puisage, *m.* drawing, etc. (see puiser).

puisard, *m.* cesspool; trap, sink; draining well; sump.

puiselle, *f.* ladle.

puisement, *m.* drawing, etc. (see puiser).

puiser, *v.t.* draw, draw off; draw out; take; obtain, drive.

    — *dans,* draw from, take from.

puisoir, *m.* ladle.

puisque, *conj.* since, as.

puissamment, *adv.* powerfully; extremely.

puissance, *f.* power; specif., horse power; effectiveness; amount of explosive in a projectile; authority.

    — *bactéricide,* bactericidal power.

    — *calorifique,* calorific power, heating power.

    — *de cheval,* horse power.

    — *de la vapeur,* steam power.

    — *dilacératrice,* rending power, rending strength.

    — *effective,* effective (or actual) power, effective horse power.

    — *en bougies,* candle power.

    — *en chevaux,* horse power.

    — *indiquée,* indicated horse power.

    — *motrice,* motive power.

puissant, *a.* powerful; great; large; rich.

puisse, *pr. subj. 1 & 3 sing.* of pouvoir.

puits, *m.* well; shaft, pit; water tank.

    — *à combustion,* flue.

    — *d'éclatement,* explosion chamber (of guns).

    — *de sondage,* bored well.

pulégone, *f.* pulegone.

pulluler, *v.i.* multiply rapidly, pullulate; swarm.

pulmonaire, *a.* pulmonary. — *f.* (*Bot.*) lungwort.

pulpation, *f.* pulping.

pulpe, *f.* pulp.

pulper, *v.t.* pulp, reduce to pulp.

pulpeux, *a.* pulpy.

pulque, *m.* pulque (fermented drink)

pulsateur, *m.* pulsator. — *a.* pulsating, pulsatory.

pulsation, *f.* pulsation.

pulsatoire, *a.* pulsatory.

pulvérin, *m.* mealed powder, priming powder; fine spray.

pulvérisable, *a.* pulverizable.

pulvérisateur, *m.* atomizer, sprayer; pulverizer. — *a.* atomizing, spraying; pulverizing.

pulvérisation, *f.* pulverization, powdering; atomizing, spraying.

pulvériseur, *m.* pulverizer, grinder.

pulvériser, *v.t.* pulverize, powder, grind (to powder); atomize, spray; meal (gunpowder).

pumicin, *m.* palm oil.

pumiqueux, *a.* pumiceous, of or like pumice.

punaise, *f.* bug, specif. bedbug; thumb tack.

punir, *v.t.* punish.

punition, *f.* punishment.

pupe, *f.* (*Zoöl.*) pupa.

pupillaire, *a.* pupillary.

pupille, *m. & f.* pupil.

pupitre, *n.* rack (as for champagne bottles); small desk.

pur, *a.* pure; free (from); clear.

en pure perte, uselessly, to no purpose.

purée, *f.* puree, thick soup; thick absinthe.

purement, *adv.* purely.

purent, *p.def. 3 pl.* of pouvoir.

pureté, *f.* purity, pureness; freedom (from); clearness, clarity.

purette, *f.* a black ferruginous sand.

purgatif, *a. & m.* purgative.

purgation, *f.* purging, purgation, etc. (see purger); purgative.

purge, *f.* purging, cleansing; (in tawing) elimination of alum with warm water; (*Med.*) purge, purgative.

— *de chaux,* (*Leather*) deliming.

purgeoir, *m.* purifying or filtering tank.

purger, *v.t.* purge; cleanse, clean; clear; free; refine (metals).

purgeur, *m.* purger, cleanser; purifier; specif., a vessel thru which vapors are passed to rid them of solid or liquid particles; (*Steam*) blow-off gear. — *a.* purging, cleansing.

purifiant, *p.a.* purifying.

purificateur, *m.* purifier. — *a.* purifying.

purification, *f.* purification, purifying.

purifier, *v.t.* purify. — *v.r.* be purified, become pure.

purin, *m.* liquid manure.

purine, *f.* purine.

purique, *a.* purine, of or pertaining to purine.

puron, *m.* sweet whey.

purpurase, *f.* purpurase.

purpurate, *m.* purpurate.

purpurin, *a.* purplish.

purpurine, *f.* purpurin.

purpurique, *a.* purpuric.

purpurogalline, *f.* purpurogallin.

purpuroxanthine, *f.* purpuroxanthin.

pus, *m.* (*Med.*) pus.

pussent, *imp.subj. 3 pl.* of pouvoir.

pustule, *f.* (*Med.*) pustule.

pustuleux, *a.* (*Med.*) pustulous.

put, *p.def. 3 sing.* of pouvoir.

putréfactif, *a.* putrefactive.

putréfaction, *f.* decay, decomposition, rotting, putrefaction.

putréfait, *a.* decayed, rotten, putrefied.

putréfier, *v.t. & r.* rot, decompose, putrefy.

putrescent, *a.* putrescent.

putrescible, *a.* putrescible.

putride, *a.* putrid.

putridité, *f.* putridity.

putzen, *m.* (*Metal.*) unfused ore on the sides of the furnace.

pyémie, *f.* (*Med.*) pyemia, pyæmia.

pyine, *f.* pyin.

pyocyanine, *f.* pyocyanin.

pyohémie, *f.* (*Med.*) pyemia, pyæmia.

pyoxanthose, *f.* pyoxanthose.

pyramidal, *a.* pyramidal.

pyramide, *f.* pyramid.

pyramidon, *m.* pyramidone.

pyrane, *m.* pyran.

pyrargyrite, *f.* (*Min.*) pyrargyrite.

pyrazine, *f.* pyrazine.

pyrazol, *m.* pyrazole.

pyrazolcarbonique, *a.* pyrazolecarboxylic.

pyrazolidine, *f.* pyrazolidine.

pyrazolidone, *f.* pyrazolidone.

pyrazoline, *f.* pyrazoline.

pyrazolique, *a.* pyrazole, of pyrazole.

pyrazolone, *f.* pyrazolone.

pyrène, *m.* pyrene.

pyrèthre, *m.* (*Bot.*) Pyrethrum (Chrysanthemum); (*Pharm.*) pyrethrum, pellitory.

— *officinal,* — *vrai,* pellitory (*Anacyclus pyrethrum*).

pyridazine, *f.* pyridazine.

pyridine, *f.* pyridine.

pyridinecarbonique, *a.* pyridinecarboxylic.

pyridique, *a.* pyridine, of pyridine.

pyridone, *f.* pyridone.

pyrimidine, *f.* pyrimidine.

pyrimidinique, pyrimidique, *a.* pyrimidine, of pyrimidine.

pyrique, *a.* pyrotechnic(al).

pyrite, *f.* (*Min.*) pyrites, specif. iron pyrites or pyrite.

— *blanche, pyrites blanches,* white iron pyrites, marcasite.

— *cuivreuse,* — *de cuivre,* copper pyrites, chalcopyrite.

— *de fer,* iron pyrites, pyrite.

— *jaune,* pyrite; chalcopyrite.

— *martiale,* iron pyrites, pyrite.

pyriteux, *a.* (*Min.*) pyritic.

pyritifère, *a.* (*Min.*) pyritiferous.

pyritiser, *v.t.* (*Min.*) pyritize.
pyroacétique, *a.* pyroacetic.
pyro-antimonique, *a.* pyroantimonic.
pyrocatéchine, *f.* pyrocatechol, pyrocatechin.
pyrochimie, *f.* pyrochemistry (*Obs. or Rare*).
pyrochimique, *a.* pyrochemical.
pyrochlore, *m.* (*Min.*) pyrochlore.
pyrocolle, *m.* pyrocoll.
pyrocoton, *m.* (*Expl.*) pyrocotton.
pyroélectricité, *f.* pyroelectricity.
pyroélectrique, *a.* pyroelectric, pyroelectrical.
pyrogallate, *m.* pyrogallate.
pyrogallique, *a.* pyrogallic.
  acide —, pyrogallic acid (pyrogallol).
pyrogallol, *m.* pyrogallol.
pyrogallolphtaléine, *f.* pyrogallolphthalein (gallein).
pyrogénation, *f.* subjecting to the action of fire, heating.
pyrogène, pyrogéné, *a.* pyrogenous, igneous, involving or produced by fire or heat; (*Med.*) pyrogenic.
pyrogénèse, *f.* pyrogenesis.
pyrogénésique, pyrogénétique, *a.* pyrogenetic.
pyrognostique, *a.* pyrognostic.
pyrogranit, *m.* a kind of stoneware.
pyroïde, *a.* phosphorescent.
pyroligneux, *a.* pyroligneous.
pyrolignite, *m.* pyrolignite.
pyrolite, *f.* (*Expl.*) pyrolite.
pyrolusite, *f.* (*Min.*) pyrolusite.
pyromellique, *a.* pyromellitic.
pyromètre, *m.* pyrometer.
pyrométrie, *f.* pyrometry.
pyrométrique, *a.* pyrometric, pyrometrical.
pyromorphite, *f.* (*Min.*) pyromorphite.

pyromucique, *a.* pyromucic.
pyrone, *f.* pyrone.
pyronine, *f.* (*Dyes*) pyronine.
pyronique, *a.* pyrone, of pyrone.
pyronome, *n.* (*Expl.*) pyronome.
pyrope, *m.* (*Min.*) pyrope.
pyrophore, *m.* pyrophorus.
pyrophorique, *a.* pyrophoric.
pyrophosphate, *m.* pyrophosphate.
  — *de fer et de soude,* (*Pharm.*) soluble ferric pyrophosphate, pyrophosphate of iron with sodium citrate.
  — *de soude,* sodium pyrophosphate.
pyrophosphorique, *a.* pyrophosphoric.
pyrotartrique, *a.* pyrotartaric.
pyrotechnicien, *m.* pyrotechnist.
pyrotechnie, *f.* pyrotechnics, pyrotechny.
pyrotechnique, *a.* pyrotechnic, pyrotechnical.
pyrothèque, *m.* (*Expl. & Elec.*) exploder.
pyrotritarique, *a.* pyrotritaric.
pyrovanadique, *a.* pyrovanadic.
pyroxanthine, *f.* pyroxanthin.
pyroxène, *m.* (*Min.*) pyroxene.
pyroxyle, pyroxile, *m.* pyroxylin.
pyroxylé, *a.* pyroxylin, nitrocellulose.
pyroxyline, *f.* pyroxylin.
pyroxylique, *a.* pyroxylic (pyroligneous).
pyrrodiazol, *m.* pyrrodiazole (triazole).
pyrrol, *m.* pyrrole, (less desirably) pyrrol.
pyrrolcarbonique, *a.* pyrrolecarboxylic.
pyrrolidine, *f.* pyrrolidine.
pyrrolidinique, *a.* pyrrolidine, of pyrrolidine.
pyrrolidone, *f.* pyrrolidone.
pyrroline, *f.* pyrroline.
pyrrolique, *a.* pyrrole, of pyrrole.
pyrrotriazol, *m.* pyrrotriazole (tetrazole).
pyruvique, *a.* pyruvic.
pyurie, *f.* (*Med.*) pyuria.

Q

q., *abbrev.* of quintal (which see).

q.m., *abbrev.* of quintal métrique (which see).

q.s., *abbrev.* (quantum sufficit) a sufficient quantity.

quadrangulaire, *a.* quadrangular.

quadratique, *a.* (*Cryst.*) quadratic (tetragonal); (*Math.*) quadratic.

quadrichloré, *a.* tetrachloro, tetrachloro-.

quadriennal, *a.* quadrennial.

quadrilatéral, *a.* quadrilateral.

quadrilatère, *a. & m.* quadrilateral.

quadrillage, *m.* cross ruling, etc. (see quadriller); grid.

quadrillé, *p.a.* cross-ruled; checkered. — *m.* cross ruling; checkering, checkerwork.

quadriller, *v.t.* cross-rule (as paper); mark in squares or rectangles; checker.

quadrivalence, *f.* quadrivalence.

quadrivalent, *a.* quadrivalent.

quadroxalate, *m.* quadroxalate (tetroxalate).

quadroxyde, *m.* quadroxide (tetroxide). *Obs.*

quadruple, *a. & m.* quadruple.

quadruple-effect, *m.* (*Sugar*, etc.) quadruple effect.

quadrupler, *v.t. & i.* quadruple.

quai, *m.* quay, wharf; (railroad) platform.

qualificatif, *a.* (of analysis) qualitative; qualifying.

qualification, *f.* title; qualification.

qualifier, *v.t.* qualify; call, style.

qualitatif, *a.* qualitative.

qualitativement, *adv.* qualitatively.

qualité, *f.* quality; title.

  en — de, in the character (or capacity) of.

quand, *adv.* when. — *conj.* when; whenever; when instead; altho, even tho, even if.

  — *même,* in any case, in spite of all; even tho, even if.

quant à. as for, as to, with respect to.

quanta, *m.pl.* quanta.

quantitatif, *a.* quantitative.

quantitativement, *adv.* quantitatively.

quantité, *f.* quantity.

  en —, in quantity; wholesale; in bulk; (*Elec.*) in parallel, in quantity.

quantum, *m.* quantum.

quarantaine, *f.* forty (as a round number); quarantine; Lent; age of forty.

quarante, *a.* forty; fortieth. — *m.* forty.

quarantième, *a. & m.* fortieth.

quarré, quarrément, etc. See carré, etc.

quart, *m.* quarter; quarter of a pound; small cask; cup holding about a quarter of a liter; (nautical) watch.

  — *de cercle,* quadrant.

  — *de rond,* quarter round.

quartation, *f.* (*Assaying*) quartation.

quartaut, *m.* a cask holding 57 to 137 liters.

quartelette, *f.* quarter of a ton (as of soap); a kind of slate.

quarteron, *m.* twenty-five; book of 25 leaves of gold or silver foil.

quartier, *m.* quarter; large block; lodging.

quartz, *m.* (*Min.*) quartz.

  — *en fil,* quartz thread.

quartzeux, *a.* quartzose, quartzous.

quartzifère, *a.* quartziferous.

quartzique, *a.* quartz, of quartz.

quartzite, *f.* (*Petrog.*) quartzite.

quasi, *adv.* almost. — *combining form.* quasi.

quassia, quassier, *m.* (*Bot.*) Quassia.

quassie, *f.* (*Pharm.*) quassia.

  — *amère,* bitter quassia, esp. Surinam quassia (wood of *Quassia amara*).

  — *de la Jamaïque,* Jamaica quassia.

quassine, *f.* quassin.

quaternaire, *a.* quaternary.

quatorze, *a. & m.* fourteen; fourteenth.

quatorzième, *a. & m.* fourteenth.

quatre, *a. & m.* four; fourth.

quatre-vingt-, eighty-.

quatre-vingt-dix, *a.* ninety.

quatre-vingtième, *a. & m.* eightieth.

quatre-vingt-onze, *a.* ninety-one.

quatre-vingts, *a.* eighty; eightieth.

quatre-vingt-un, *a.* eighty-one.

quatrième, *a.* fourth. — *m.* fourth; fourth floor. — *f.* fourth class.

  — *titre,* (for gold) a standard of 0.583 fine (14 carats).

quayage, *m.* quayage, wharfage.

que, *pron.* whom, which, that; what; when, that. — *adv.* how; how much, how many; why. — *conj.* that; O that; let, may; if; whether; (with negative) lest; yet; altho; when; since; so that; as far as; (with negative) unless; until, till; (in comparisons) as, than; but, only; as; why; because.

quebracho, *m.* (*Bot.*) quebracho.

295

quel, *a.* what.

　quel (or *quelle*) *que soit, quels* (or *quelles*) *que soient,* whatever may be, whoever may be.

quelconque, *a.* whatever, whatsoever, any.

quelle, *a. fem.* of quel.

quelque, *a.* some, any, a few. — *adv.* about, some.

　en — *sorte,* so to speak.

　— *chose,* something.

　— *peu,* a little, somewhat.

　quelque . . . que, whatever, however.

quelquefois, *adv.* sometimes.

quelques-uns, *pron. pl.* some.

quelqu'un, *pron.* someone, somebody.

quercitannique, *a.* quercitannic.

quercite, *f.* quercitol, quercite.

quercitrine, *f.* quercitrin.

quercitron, *m.* (*Bot.*) quercitron, quercitron oak (*Quercus velutina*).

querelle, *f.* quarrel.

quereller, *v.t.* quarrel with. — *v.i. & r.* quarrel.

querir, *v.t.* Only in the phrases *aller querir* (go and get, go for), *envoyer querir* (send for) and *venir querir* (come for).

question, *f.* question.

questionner, *v.t. & i.* question.

quetche, *f.* = quetsche.

quête, *f.* quest, search; (church) collection.

quetsche, *f.* a large plum used for making prunes and brandy.

quetsche-wasser, *m.* a kind of plum brandy.

queue, *f.* tail; tailings (as of ore or of spirit); end; handle; queue; cue.

queue-de-cheval, *f.* (*Bot.*) horsetail (*Equisetum*).

queue-de-morue, *f.* flat brush; (*Glass*) a kind of coarse window glass no longer made.

queue-de-rat, *m.* rat-tail file; small taper.

queursage, queurçage, *m.* (*Leather*) slating.

queurse, queurce, *f.* (*Leather*) slater.

queurser, queurcer, *v.t.* (*Leather*) slate.

queux, *f.* hone; whetstone.

qui, *pron.* who, whom, which, that; he who, he that, whoever, what; any one, any.

　— *que,* whoever, whatever.

　— *que ce soit,* whoever it may be, anyone.

　qui . . . qui, some . . . some, one . . . the other.

quiconque, *pron.* whoever, whichever.

quiètement, *adv.* quietly.

quillaja, quillaya, *m.* (*Bot.*) Quillaja, esp. *Q. saponaria* (soapbark tree or quillai).

quille, *f.* post, support; cap, string piece; keel; skittle, ninepin; tall skittle-shaped bottle.

quina, *m.* cinchona (tree and bark).

quinaldinique, *a.* quinaldic, quinaldinic.

quinamine, *f.* quinamine.

quinate, *m.* quinate.

quincaille, *f.* hardware, piece of hardware.

quincaillerie, *f.* hardware; hardware trade.

quincaillier, *m.* hardware merchant, ironmonger.

quinhydrone, *f.* quinhydrone.

quinicine, *f.* quinicine.

quinidine, *f.* quinidine.

quinimétrie, *f.* measurement of the amount of quinine in cinchona bark.

quinine, *f.* quinine.

quininique, *a.* quininic.

quinique, *a.* quinic.

quinite, *n.* quinitol, quinite.

quinizarine, *f.* quinizarin.

quinizarol, *m.* quinizarol.

quinoïdine, *f.* (*Pharm.*) quinoidine.

quinoléine, *f.* quinoline.

quinoléine-carbonique, *a.* quinolinecarboxylic.

quinoléique, *a.* quinoline, of quinoline; (of the acid) quinolinic.

quinone, *f.* quinone.

quinonimide, *f.* quinonimine, (less desirably) quinonimide.

quinonique, *a.* quinone, of quinone; quinoid, quinonoid.

quinoniser, *v.t.* convert into a quinone, quinonize.

quinotannique, *a.* quinotannic.

quinovine, *f.* quinovin.

quinoxaline, *f.* quinoxaline.

quinoyle, *m.* quinoyl.

quinquennal, *a.* quinquennial.

quinquina, *m.* cinchona (tree and bark).

　— *africain,* doundaké, doundaké bark (from the country fig (*Sarcocephalus esculentus*).

　— *jaune,* cinchona, yellow cinchona bark.

　— *rouge,* red cinchona, red cinchona bark.

quintal, *m.* quintal, hundredweight, 100 pounds.

　— *métrique,* metric quintal, 100 kilograms (not to be confused with the "metric hundredweight" which equals 50 kg. or 110.23 lbs.).

quintaux, *pl.* of quintal.

quintessence, *f.* quintessence.

quintessenciation, *f.* reduction to a quintessence.

quintessenciel, *a.* quintessential.

quintessencier, *v.t.* extract the quintessence from; (fig.) refine, subtilize.

quintuplation, *f.* quintupling, quintuplication.

quintuple, *a.* quintuple.

quintupler, *v.t.* quintuple.

quinzaine, *f.* fifteen (or about that); fortnight.

quinze, *a. & m.* fifteen; fifteenth.

quinzième, *a. & m.* fifteenth.

quittance, *f.* receipt.

quittancer, *v.t.* receipt.

quitte, *a.* free (as from an obligation), out of debt; quit, rid.

quitte *à*, — *pour*, save for, it being only necessary to, the sole inconvenience being.

quitter, *v.t.* quit; leave; give up; release, free; leave off; take off. — *v.r.* be left. — *v.i.* leave, go away.

quoi, *pron.* what; which, that. — *interj.* what! how!

— *que, quoi . . . que,* whatever.

quoi *qu'il en soit,* however it may be, in any case, no matter what.

quoique, *conj.* tho, altho.

quote-part, *f.* share, quota.

quotidien, *a.* daily.

quotidiennement, *adv.* daily, every day.

quotient, *m.* quotient.

— *respiratoire,* respiratory quotient.

quotité, *f.* quota, share.

R

rabais, *m.* reduction; discount, rebate.

rabaissement, *m.* lowering, reduction.

rabaisser, *v.t.* lower, reduce; depreciate.

rabat, *m.* discount, reduction. — *pr. 3 sing.* of rabattre.

rabattage, rabattement, *m.* lowering, etc. (see rabattre).

rabattre, *v.t.* lower; diminish, lessen; put down, beat down, press down, turn down, etc.; flatten; beat out, rake; (*Metal.*) smooth; polish; soften, sadden (a color); suppress; humble. — *v.r.* be lowered, etc.; come down, fall; turn, turn off. — *v.i.* turn off; come down, diminish.

rabette, *f.* rape, colza (see *colza,* note); kohl-rabi.

rabique, *a.* (*Med.*) rabic, rabietic.

râble, *m.* stirrer; skimmer (as for molten glass or metals); fire hook, rake; (*Metal.*) rabble.

râbler, *v.t.* stir; (*Metal.*) rabble.

rabonnir, *v.t. & i.* improve.

rabot, *m.* plane; shaver (as for soap); scraper; cutter (as for glass); stirrer; spreader (for gunpowder); smoother; polisher.

raboter, *v.t.* plane; dress; smooth; polish; chip (wood for paper pulp).

raboteuse, *f.* planing machine, planer.

raboteux, *a.* rough, uneven; (of wood) knotty.

raccommoder, *v.t.* mend, repair; right; correct.

raccord, raccordement, *m.* joining, junction, joint; coupling, connection, union; adjustment; (*Opt.*) coincidence (of images); reconciliation.

raccorder, *v.t.* join, unite; level, smooth; adjust; harmonize, reconcile.

raccourci, *p.a.* shortened, etc. (see raccourcir). — *m.* abridgment, summary; short cut.

raccourcir, *v.t.* shorten; abridge; contract. — *v.r.* grow shorter; contract, shrink.

raccourcissement, *m.* shortening; contraction.

raccours, *m.* shrinking, shrinkage.

race, *f.* race.

racème, *m.* (*Bot.*) raceme.

racémique, *a.* racemic.

rache, *f.* dregs of tar, tar of inferior quality.

racheter, *v.t.* redeem; buy, buy up.

râcheux, racheux, *a.* (of wood) knotty.

rachevage, rachèvement, *m.* finishing.

rachever, *v.t.* finish.

rachitique, *a.* (*Med.*) rachitic.

racinage, *m.* edible roots; (*Dyeing*) decoction of root, bark, etc. of the walnut.

racinal, *m.* sill, sleeper.

racine, *f.* root.

— *alimentaire,* edible root.

— *brésilienne,* ipecac.

— *carrée,* (*Math.*) square root.

— *cubique,* (*Math.*) cube root.

— *d'actée à grappes,* (*Pharm.*) cimicifuga, black cohosh.

— *d'arnique,* (*Pharm.*) arnica rhizome.

— *de leptandra,* (*Pharm.*) leptandra, Culver's root.

— *de phytolaque,* (*Pharm.*) phytolacca, poke root.

— *de Saint Christophe,* common baneberry, herb Christopher (*Actæa spicata*).

— *de sumbul,* (*Pharm.*) sumbul, musk root.

— *de trioste,* (*Pharm.*) triosteum, fever root.

— *douce,* licorice root.

raciner, *v.i.* take root, root. — *v.t.* dye yellow.

rack, *m.* arrack, rack (the liquor).

raclage, *m.* scraping.

racler, *v.t.* scrape, scrape off; (*Leather*) skive.

raclette, *f.* (small) scraper.

racloir, *m.* scraper.

raclure, *f.,* **raclon,** *m.* scrapings.

raclure de brique, brick dust.

raconter, *v.t.* relate, recount, tell, narrate.

racornir, *v.t. & r.* harden (like horn).

racornissement, *m.* hardening, hardness.

radeau, *m.* raft.

radiable, *a.* radiable. — *m.* a soapmaker's tool.

radiaire, *a.* radiate, radiated.

radial, *a.* radial.

radialement, *adv.* radially.

radiance, *f.* radiance, radiation.

radiant, *a.* radiant.

radiateur, *m.* radiator. — *a.* radiating.

radiation, *f.* radiation; erasure, striking out.

radical, *m. & a.* radical.

— *composé,* compound radical.

— *hydrocarboné,* hydrocarbon radical.

radicalement, *adv.* radically.

radicaux, *pl.* of radical.

radicelle, *f.* rootlet.

radicule, *f.* (*Bot.*) radicle.

radié, *a.* radiated, radiate, radial.

radier, *v.t.* radiate; erase, strike off. — *v.i.* radiate. — *m.* frame, apron, floor, bed; strengthening (of masonry or concrete).

radieux, *a.* radiant.

radifère, *a.* radiferous, containing radium.

radio-actif, *a.* radioactive.

radioactiver, *v.t.* render radioactive.

radio-activité, *f.* radioactivity.

radiographie, *f.* radiography; radiograph.

radiographier, *v.t.* radiograph.

radiographique, *a.* radiographic(al).

radiolaire, *m.* (*Zoöl.*) radiolarian.

radiologie, *f.* radiology.

radiomètre, *m.* radiometer.

radiométrie, *f.* radiometry.

radiométrique, *a.* radiometric.

radioscopie, *f.* radioscopy.

radioscopique, *a.* radioscopic(al).

radiothérapie, *f.* radiotherapy.

radiothérapique, *a.* radiotherapeutic.

radiothorium, *m.* radiothorium.

radis, *m.* radish.
— *de cheval,* horseradish.

radium, *m.* radium.

radouber, *v.t.* (*Expl.*) work over, remanufacture; repair, refit (vessels).

radoucir, *v.t.* soften, render mild; (*Metal.*) anneal. — *v.r.* soften, grow mild.

raffermir, *v.t.* harden, make harder or firmer; strengthen, improve.

raffermissement, *m.* hardening, making firmer; strengthening, improvement.

raffinade, *f.* refined sugar.

raffinage, *m.* refining.

raffinase, *f.* raffinase.

raffiné, *p.a.* refined; (of cheese) ripened. — *m.* (*Paper*) pulp, stuff (ready for the paper-making machine).

raffinerie, *f.* refinery; refining.

raffineur, *m.* refiner; (*Paper*) = raffineuse.
— *a.* refining.

raffineuse, *f.* (*Paper*) refining engine, beating engine.

raffinose, *f.* raffinose.

rafle, *f.* stalk (of grapes); cob (of maize); carrying off.

rafler, *v.t.* carry off, carry away.

rafleux, *a.* rough, uneven.

rafraîchir, *v.t.* cool; refresh; renew, renovate, restore, freshen; trim; thin (mortar); sharpen (as a saw); refine (metals); alloy (copper with excess of lead). — *v.r.* cool, become cool; be refreshed. — *v.i.* cool, grow cool.

rafraîchissant, *p.a.* cooling, etc. (see rafraîchir). — *m.* (*Med.*) refrigerant.

rafraîchissement, *m.* cooling, etc. (see rafraîchir); refreshment; (*Med.*) refrigerant.

rafraîchissoir, rafraîchisseur, *m.* cooler; (*Sugar*) crystallizer.

rage, *f.* rage, madness; (*Med.*) rabies.

raide, *a.* stiff; rigid; tight, taut; steep; inflexible. — *adv.* quickly, promptly.

raideur, *f.* stiffness, etc. (see raide).

raidir, *v.t.* stiffen; tighten, make taut, stretch.
— *v.r.* stiffen; bear up, offer resistance.
— *v.i.* stiffen.

raidissement, *m.* stiffening, etc. (see raidir).

raie, *f.* line (as of a spectrum); scratch; stroke; stripe; streak; groove; ray (the fish).

raifort, *m.* horseradish (*Rorippa armoracia*).
— *cultivé,* a kind of winter radish.
— *sauvage,* horseradish.

rail, *m.* rail.

rainer, *v.t.* groove; score; flute.

rainure, *f.* groove; channel, furrow, keyway.

raisin, *m.* grape, grapes.
grain de —, grape.
grappe de —, bunch of grapes.
pepin de —, grape seed, grapestone.
— *de cuve,* wine grape.
— *d'ours,* bearberry (*Arctostaphylos uvaursi*).
— *sec,* dried grape, raisin.

raison, *f.* reason; ratio; (firm) name; satisfaction, reparation; (*Law*) claim.
à — de, at the rate of, at the price of.
avoir —, be right.
avoir — de, get the better of.
en — de, in consideration of, by reason of; in the ratio of, in ratio to.
— *d'être,* reason for existence, raison d'être.
— *directe,* direct ratio.
— *inverse,* inverse ratio.
— *sociale,* firm name.

raisonnable, *a.* reasonable.

raisonnablement, *adv.* reasonably.

raisonné, *p.a.* reasoned, etc. (see raisonner); rational; logical; methodical, systematic.

raisonnement, *m.* reasoning; arguing.

raisonner, *v.i.* reason; argue. — *v.t.* reason; consider, discuss; plan.
— *sur,* discuss, consider.

rajeunir, *v.t.* rejuvenate, revive, make young.

rajouter, *v.t.* add again.

rajustement, *m.* readjustment.

rajuster, *v.* . readjust; restore; reconcile.

ralentir, *v.t.* slacken, ease, slow up, retard.
— *v.r. & i.* slacken, slow up, abate.

ralentissement, *m.* slackening, slowing up, retardation, abatement.

ralentisseur, *m.* slackener, retarder; (*Sugar*) a space into which the vapors from the evaporator are allowed to pass, the resulting slackening of current causing liquid particles to deposit.

rallier, *v.t. & r.* rally; join; approach.

rallonge, *f.* extension piece, eking piece.

rallongement, *m.* lengthening, extension, eking.

rallonger, *v.t.* lengthen, extend, eke out.
— *v.r.* lengthen, be extended.

rallumer, *v.t.* relight, light again, rekindle.

rallumeur, *m.* relighter, specif. a pilot burner.

ramas, *m.* collection, assemblage, heap.

ramassé, *p.a.* collected, etc. (see ramasser); thickset, compact.

ramasser, *v.t.* gather, collect; pick up; sum up.
— *v.r.* collect, gather; be collected, be gathered; gather oneself up; double up.

ramasseur, ramassoir, *m.* collector, gatherer.

rame, *f.* ream (of paper); paddle; oar.

rameau, *m.* branch; small branch, shoot, twig.

ramener, *v.t.* bring back, put back, draw back; restore; reduce, convert; win over.

rameux, *a.* branching, branched, ramose.

ramie, *f.* (*Bot.*) ramie.

ramification, *f.* ramification.

ramifier, *v.r.* branch, ramify.

ramille, *f.* twig, shoot.

ramoitir, *v.t.* remoisten, redampen.

ramollir, *v.t. & r.* soften.

ramollissable, *a.* capable of being softened, softening.

ramollissement, *m.* softening.

rampant, *p.a.* sloping; creeping; crawling.
— *m.* slope, sloping part, specif. (of a furnace) flue (a sloping passage between furnace and chimney).

rampe, *f.* row of burners (or, by extension, of cocks or other fixtures); slope, incline, inclined plane, ramp; flight (of stairs); (stair) railing; footlights.
— *à éclipse,* a row of burners with a device for extinguishing a number at once.
— *à gaz,* row of gas burners.
— *à vide,* row of vacuum cocks.
— *de becs à alcool* (*essence,* *pétrole*), row of alcohol (gasoline, kerosene) burners.
— *distillatoire,* row or battery of stills.

ramper, *v.i.* slope, incline; wind; creep; crawl.

ramure, *f.* branching, branches.

rance, *a.* rancid. — *m.* rancidness, rancidity.

rancidité, *f.* rancidity, rancidness.

rancir, *v.i.* become rancid.

rancissement, *m.* becoming rancid; rancidness.

rancissure, *f.* rancidness, rancidity.

rang, *m.* rank; row; tier; range; course (as of bricks).
au — de, in the number of, among.

rangé, *p.a.* ranged, etc. (see ranger); steady.

rangée, *f.* row, range.

ranger, *v.t.* range; arrange; rank; set in order; back (as cube sugar); bring back, put back, keep back; subdue. — *v.r.* range oneself; side (with); make way; submit.

ranimer, *v.t.* revive, restore; brisk up (a fire).

râpage, *m.* rasping; grating; grinding (of wood).

râpe, *f.* rasp; grater; stalk (of grapes).
— *à tambour,* a rasping machine with casing or hopper.

râpé, *p.a.* rasped; grated; threadbare. — *m.* rape wine (made from marc or fresh grapes by adding water and fermenting); grapes added to wine to improve it, also the wine so treated; shavings or chips added to wine to clarify it; (in cafés) a mixture of wine leavings.

râper, *v.t.* rasp; grate; grind (wood for pulp).

râperie, *f.* (*Sugar*) rasping station, beet-juice mill; (*Paper*) mechanical pulp mill.

rapetisser, *v.t.* make smaller, lessen, shorten, diminish; minimize. — *v.r. & i.* diminish, shorten, shrink.

rapide, *a.* rapid; swift; steep. — *m.* rapid (in a river); express (train).

rapidement, *adv.* rapidly, swiftly.

rapidité, *f.* rapidity.

rapiécer, *v.t.* patch, piece.

rapointir, *v.t.* repoint, resharpen.

rappeler, *v.t.* recall; call back; call again; repeal, revoke. — *v.r.* remember.

rapport, *m.* relation; ratio; report; return; production; harmony, agreement.
— *fonctionnel,* functional relation.
par — à, with relation to, with regard to, on account of, in comparison with, in proportion to.
sous le — de, with respect to, in respect of.

rapporter, *v.t.* refer; relate; report; report on; quote, cite; return; bring back, take back; produce (as a crop or revenue); revoke, repeal; add, join, insert; transport; turn, direct; protract, plot; ascribe, attribute; compare. — *v.r.* agree; relate; refer; rely (on); be referred, etc. — *v.i.* produce, bear; report; (of the tide) rise.

rapporteur, *m.* reporter; recorder; secretary; protractor (the instrument).

rapprêter, *v.t.* dress again.

rapproché, *p.a.* brought nearer, etc. (see rapprocher); close together, near; closely related.

rapprochement, *m.* bringing nearer, etc. (see rapprocher); rapprochement; comparison.

rapprocher, *v.t.* bring nearer; lessen (distance); bring together; compare; reconcile; unite; (*Old Chem.*) condense. — *v.r.* be brought nearer, etc.; draw nearer, approach; be allied (to); bear a resemblance.

râpure, *f.* rasping, raspings.

rapuroir, *m.* copper boiler (for saltpeter).

rare, *a.* rare; thin.

raréfactibilité, *f.* capability of being rarefied.

raréfactif, *a.* rarefactive.

raréfaction, *f.* rarefaction; depletion.

raréfiable, *a.* rarefiable.

raréfiant, *a.* rarefying, rarefactive.

raréfier, *v.t.* rarefy. — *v.r.* be rarefied.
rarement, *adv.* rarely.
rarescence, *f.* becoming rarefied.
rarescent, *a.* becoming rarefied.
rarescibilité, *f.* capability of being rarefied.
rarescible, *a.* rarefiable.
rareté, *f.* rarity, rareness.
ras, *a.* flat, level; smooth; short-napped; short-haired; close-cropped; shaven. — *m.* level.
  à —, *au* —, even.
  à — de, au — de, to a level with, even with, close to, up to, down to.
rasage, *m.* shaving.
rasance, *f.* flatness (as of a trajectory).
rase, *f.* turpentine; oil of turpentine; rosin; rosin oil; a resinous mixture applied to ship bottoms. — *a. fem.* of ras.
raser, *v.t.* shave; raze; graze.
rasoir, *m.* razor; (microtome) knife.
rassasier, *v.t.* satisfy; satiate; surfeit.
rassemblement, *m.* collecting, collection, etc. (see rassembler); assemblage, crowd.
rassembler, *v.t. & r.* collect, gather, assemble; reassemble, collect again.
rasseoir, *v.t.* replace; reseat; calm.
rassied, *pr. 3 sing.* of rasseoir.
rassis, *p.a.* replaced, etc. (see rasseoir); (of bread) stale; (of earth) settled; calm, staid. — *m.* (of spirit) settled state.
  prendre son —, settle (said of spirits).
rassortiment, *m.* reassortment.
rassortir, *v.t.* reassort.
rassurer, *v.t.* strengthen, secure; reassure.
rat, *m.* rat.
  — d'Amérique, guinea pig.
  — de cave, a long, thin wax taper.
  — musquée, muskrat.
ratafia, *m.* ratafia (the liqueur).
ratanhia, *m.* (*Bot.*) rhatany, ratanhia (*Krameria triandra*), or any of several allied species.
  — officinal, Krameria triandra.
ratatiner, *v.t. & r.* shrivel, shrivel up, dry up.
rate, *f.* (*Anat.*) spleen; female rat.
raté, *p.a.* missed; failed. — *m.* misfire.
râteau, *m.* rake.
râtelage, *m.* raking.
râteler, *v.t.* rake.
râtelier, *m.* rack; set of teeth.
rater, *v.i.* miss fire; miscarry, fail. — *v.t.* miss.
ratifier, *v.t.* ratify.
ration, *f.* ration, allowance.
rationnel, *a.* rational.
rationnellement, *adv.* rationally.
ratissage, *m.* scraping; raking.
ratisser, *v.t.* scrape; rake.
ratissoire, *f.* scraper; rake.
ratissure, *f.* scrapings; rakings.
rattachage, rattachement, *m.* connection, etc. (see rattacher).

rattacher, *v.t.* connect (with); attach (to); tie, fasten; tie again. — *v.r.* be connected; be attached; be tied, be fastened.
rattrapage, *m.* recovering, recovery, overtaking, etc. (see rattraper).
rattraper, *v.t.* recover; make up for; overtake; retake; take up. — *v.r.* be recovered, etc.; seize hold (of).
rature, *f.* scrapings; erasure.
raturer, *v.t.* scrape; erase.
ravaler, *v.t.* rough-cast, rough-coat; lay on (as gold foil); clean, dress, finish (a surface); level (soil); swallow; disparage. — *v.i.* (of metals) develop cavities, pipe.
rave, *f.* turnip (*Brassica rapa*); rape; radish.
ravi, *p.a.* ravished; delighted.
ravir, *v.t.* ravish; delight.
  à —, admirably.
ravison, *m.* ravison, Black Sea rape (wild *Brassica campestris*).
ravitailler, *v.t.* revictual.
ravivage, *m.* reviving, etc. (see raviver).
raviver, *v.t.* revive; brighten, brighten up (as metal surfaces); freshen.
ravoir, *v.t.* have again, get again, get back.
rayage, *m.* scratching, etc. (see rayer).
rayer, *v.t.* scratch; rifle; erase, strike off; rule; stripe, streak.
rayon, *m.* ray; radius; region; shelf; spoke; furrow.
  — calorifique, heat ray.
  — chimique, chemical ray, actinic ray.
  — de lumière, light ray.
  — de miel, honeycomb.
  — lumineux, luminous ray, light ray.
  — photogénique, actinic ray.
  rayons cathodiques, cathode rays.
  rayons durs, hard rays.
  rayons Röntgen, Röntgen rays.
  rayons ultra-violets, ultraviolet rays.
  rayons X, X-rays, Röntgen rays.
  rayons α, α-rays.
  — visuel, visual ray; line of sight.
rayonnage, *m.* shelving, shelves; furrowing.
rayonnant, *p.a.* radiating; radiant.
rayonné, *p.a.* radiated, radiate.
rayonnement, *m.* radiation; radiance.
rayonner, *v.i.* radiate; beam, shine.
rayure, *f.* scratch; rifling; erasure; ruling; stripe, streak; groove; furrow.
raze, *f.* = rase.
réabonnement, *m.* renewal (of a subscription).
réabsorber, *v.t.* reabsorb.
réabsorption, *f.* reabsorption, reabsorbing.
réacquérir, *v.t.* recover.
réactibilité, *f.* reactivity.
réactif, *m.* reagent. — *a.* reactive.
  — cupro-ammoniacal, c u p r a m m o n i u m, Schweitzer's reagent.
  — cupropotassique, Fehling's solution.

réactif *de*, (with name of a substance) reagent for.

— *de Fritzsche*, a dinitroanthraquinone used as a reagent for polycyclic hydrocarbons.

— *de Millon*, Millon's reagent.

— *de Nessler*, Nessler reagent, Nessler solution.

— *de Schweitzer*, Schweitzer's reagent.

réaction, *f.* reaction.

— *de soudure*, coupling reaction.

réactionnel, *a.* of or pertaining to reaction, reactional, reactive.

réadmettre, *v.t.* readmit.

réaffirmer, *v.t.* reaffirm.

réagglomérer, *v.t.* reagglomerate, reform.

réagglutiner, *v.t.* reagglutinate.

réagir, *v.i.* react.

réagissant, *p.a.* reacting.

réaimanter, *v.t.* remagnetize.

réajuster, *v.t.* readjust.

réalgar, *m.* (*Min.*) realgar.

— *jaune*, orpiment.

réalisable, *a.* realizable.

réalisation, *f.* realization.

réaliser, *v.t.* realize. — *v.r.* be realized.

réalité, *f.* reality.

réapparaître, *v.i.* reappear.

réapparition, *f.* reappearance.

réapprêter, *v.t.* dress again.

réargenter, *v.t.* resilver.

réassortir, *v.t.* reassort.

réassurer, *v.t.* reinsure.

rebaisser, *v.t.* lower again.

rebâtir, *v.t.* rebuild.

rebattre, *v.t.* beat again, beat, hammer; say over and over.

rebattu, *p.a.* beaten again, etc. (see rebattre); hackneyed, trite; stunned; tired.

rebelle, *a.* refractory; (*Med.*) obstinate; rebellious, rebel.

rebeller, *v.r.* rebel.

reblanchir, *v.t.* rebleach; re-whiten. — *v.i.* become white again.

rebond, *m.* rebound.

rebondir, *v.i.* rebound.

rebord, *m.* rim; flange; edge, border; ledge; hem. *à* —, rimmed, flanged, with raised edge.

reboucher, *v.t.* stopper again, etc. (see boucher).

rebouillir, *v.i.* boil again.

rebours, *m.* wrong way (of the grain, of the nap); contrary, reverse. *à* —, *au* —, against the nap (or grain), the wrong way, backward.

rebout, *pr. 3 sing.* of rebouillir.

rebrouiller, *v.t.* mix again, stir up again.

rebrousser, *v.t.* turn back; retrace (one's way). — *v.i.* turn back, go back.

rebroyer, *v.t.* regrind, etc. (see broyer).

rebrûler, *v.t.* reburn, etc. (see brûler); soften (glass).

rebrunir, *v.t.* brown again; reburnish. — *v.i.* become brown again.

rebut, *m.* refuse, rejected object or material, waste, rubbish, trash; repulsion; repulse. *de* —, refuse, waste.

mettre au —, reject, discard.

rebuter, *v.t.* reject; refuse; discourage.

recalciner, *v.t.* recalcine.

recalculer, *v.t.* recalculate.

recaler, *v.t.* relevel, reset, etc. (see caler); smooth, level, polish.

recalescence, *f.* recalescence.

récapituler, *v.t.* recapitulate.

recarbonisation, *f.* recarbonization.

recarboniser, *v.t.* recarbonize, recarburize.

recarburation, *f.* recarburization.

recarburer, *v.t.* recarburize, recarbonize.

recarburisation, *f.* recarburization.

recasser, *v.t.* break again; soften (hides).

receler, *v.t.* conceal.

récemment, *adv.* recently.

recense, *f.* re-pressing of the marc (as of olives); restamping (of gold and silver).

recensement, *m.* census; inventory.

récent, *a.* recent; (of memory) distinct, fresh.

récépissé, *m.* receipt.

réceptacle, *m.* receptacle.

récepteur, *m.* receiver. — *a.* receiving.

réception, *f.* receiving, receipt; reception.

réceptivité, *f.* receptivity.

recette, *f.* receipt; recipe, formula, prescription; acceptance.

recevable, *a.* receivable.

recevant, *p.pr.* of recevoir.

receveur, *m.* collector, receiver (of money).

recevoir, *v.t.* receive; accept; get; take.

rechange, *m.* spare part, duplicate, change. *de* —, extra, spare, duplicate.

réchapper, *v.i.* escape; recover. — *v.t.* deliver.

rechargement, *m.* recharging, reloading.

recharger, *v.t.* recharge, reload; replenish; renew, repair.

réchaud, *m.* small portable stove (such as that of a chafing dish), hot plate; heater; hot-bed (for plants).

réchauffage, *m.* reheating.

réchauffement, *m.* reheating; (*Agric.*) new manure.

réchauffer, *v.t.* reheat; warm over.

réchauffeur, *m.* heater; warmer; forewarmer, preheater; reheater; (*Chlorine*, etc.) superheater; (*Steam*) feed-water heater.

réchauffoir, *m.* (*Sugar*) heater; plate warmer.

rêche, *a.* harsh.

recherche, *f.* research; investigation; inquiry; search; prospecting (for ore); affectation, elegance.

recherché, *p.a.* sought after, etc. (see rechercher); in demand, in request; far-fetched; elaborate, finished; studied, affected.

rechercher, *v.t.* seek after, seek for, search for; test for; inquire into; (*Tech.*) finish off.

rechercheur, *m.* researcher, research worker; seeker; inquirer.

rechinser, *v.t.* wash (wool).

récidive, *f.* repetition, recurrence.

récipé, *m.* recipe.

récipient, *m.* receiver; container; vessel, dish, cistern, cylinder (for gases), etc.
— *florentin,* Florentine receiver.

réciproque, *a.* reciprocal; inverse (ratio); reciprocating; reversible. — *f.* like, same thing; (*Logic*) converse.

réciproquement, *adv.* reciprocally; conversely.

recirer, *v.t.* wax again, rewax.

récit, *m.* account, statement, narration.

réciter, *v.t.* relate, recite.

réclamation, *f.* claim, demand; complaint.

réclamer, *v.t.* demand; beg. — *v.i.* protest.

reçoit, *pr. 3 sing.* of recevoir.

recoller, *v.t.* reglue, etc. (see coller).

recolorer, *v.t.* recolor. — *v.r.* become colored again, have one's color restored.

récolte, *f.* collection, gathering; harvest; crop.

récolter, *v.t.* gather, collect; harvest.

récolteur, *m.* collector; harvester. — *a.* collecting, gathering; harvesting.

recombiner, *v.i. & t.* recombine.

recommandable, *a.* to be recommended; estimable.

recommandation, *f.* recommendation; (postal) registration; esteem; counsel.

recommander, *v.t.* recommend; register (as a letter); instruct; request; exhort.

recommencer, *v.t. & i.* recommence, begin again.

réconciliable. *a.* reconcilable.

réconcilier, *v.t.* reconcile.

réconfortant, *p.a.* strengthening, tonic; comforting, cheering. — *m.* tonic.

réconforter, *v.t.* strengthen; comfort, cheer.

reconnaissable, *a.* recognizable.

reconnaissance, *f.* recognition; examination; verification; exploration; acknowledgment; gratitude.

reconnaissant, *p.pr.* of reconnaître.

reconnaître, *v.t.* recognize; examine; verify; discover; survey; explore; acknowledge; return, repay. — *v.r.* be recognized; find oneself; reflect; acknowledge one's error.

reconnu, *p.p.* of reconnaître.

reconstruire, *v.t.* reconstruct.

reconvertir, *v.t.* reconvert.

recoquillement, *m.* curling up, turning up.

recoquiller, *v.t. & r.* curl up, turn up.

recouchage, *m.* laying again, etc. (see coucher).

recouler, *v.t.* recast, etc. (see couler); scrape (hides) to remove lime, etc., scud. — *v.i.* flow again, etc. (see couler).

recoupage, *m.* recutting, cutting again, etc. (see couper and recouper).

recoupe, *f.* clippings, cuttings, scraps; chips (of stone); blended spirit; an inferior grade of flour; second cutting (of hay).

recouper, *v.t.* recut, cut again, etc. (see couper); mix (wines) a second time.

recoupette, *f.* a very coarse flour used for industrial purposes.

recourant, *p.pr.* of recourir.

recourber, *v.t.* bend back, bend; bend again.
— *v.r.* bend, be bent.

recourbure, *f.* bend; bent state.

recourir, *v.i.* have recourse, resort (to), make use (of); run again.

recourra, *fut. 3 sing.* of recourir.

recours, *m.* recourse; resort; resource.

recourt, *pr. 3 sing.* of recourir.

recouru, *p.p.* of recourir.

recouvert, *p.a.* covered; covered again.

recouvrant, *p.pr.* of recouvrir.

recouvrement, *m.* covering; cover; hood, cap, etc.; overlapping, lap; coat (as of plaster); recovery.
à —, by overlapping.

recouvrer, *v.t.* recover; collect (as taxes).

recouvrir, *v.t.* cover, cover over; mask; coat; cover again, re-cover. — *v.r.* become covered or coated.

récrément, *m.* recrement, specif. dross.

recrépir, *v.t.* regrain, etc. (see crépir).

récrire, *v.t.* rewrite. — *v.i.* reply (in writing).

recrobiller, *v.r.* = recroqueviller.

recroiser, *v.t.* recross, cross again.

recroissant, *p.pr.* of recroître.

recroître, *v.i.* grow again; get longer.

recroqueviller, *v.r.* curl up, shrivel up.

récrouir, *v.t.* reheat; anneal.

recru, *p.p.* of recroître.

recrutement, *m.* recruiting.

recruter, *v.t.* recruit.

rectangle, *a.* rectangular. — *m.* rectangle.

rectangulaire, *a.* rectangular.

recteur, *a.* directing. — *m.* rector.

rectifiable, *a.* rectifiable.

rectificateur, *m.* rectifier. — *a.* rectifying.

rectificatif, *a.* rectifying.

rectification, *f.* rectification, rectifying.

rectifier, *v.t.* rectify. — *v.r.* be rectified.

rectiligne, *a.* rectilinear.

recto, *m.* right side, right-hand page.

reçu, *p.a.* received, etc. (see recevoir). — *m.* receipt.

recueil, *m.* collection, assemblage, compilation.

recueillement, *m.* collection, collecting; gathering; collectedness.

recueillir, *v.t.* collect; gather; harvest; receive.
— *v.r.* be collected, etc.; collect oneself.

recuire, *v.t.* anneal (as glass); reheat; reburn, etc. (see cuire).

reculsant, *p.pr.* of recuire.
reculseur, *m.* (*Glass,* etc.) annealer.
reculsson, *m.* annealing, etc. (see recuire).
recuit, *p.a.* annealed, etc. (see recuire). — *m.* annealing; reheating; reburning, reboiling, etc. (see cuire). — *pr. 3 sing.* of recuire.
— *passé,* (*Soap*) a used lye of about 23° Baumé.
recuite, *f.* = recuit, *m.*
recuiteur, *m.* (*Glass,* etc.) annealer.
recul, *m.* recoil.
reculade, *f.* retreat, backward movement.
reculé, *a.* remote, distant.
reculer, *v.t.* move back, put back, withdraw; extend (as boundaries); delay, postpone. — *v.i.* go back, recede, retreat; shrink (from), recoil; give way; delay.
reculons. In the phrase *à reculons,* backward.
récupérable, *a.* recoverable.
récupérateur, *m.* recuperator.
récupération, *f.* recuperation, recovery.
— *avec inversion,* alternating system of recuperation.
— *simple,* continuous system of recuperation.
récupérer, *v.t. & r.* recover, recuperate.
récurer, *v.t.* scour, clean.
reçurent, *p.def. 3 pl.* of recevoir.
récuser, *v.t.* challenge, object to.
reçut, *p.def. 3 sing.* of recevoir.
rédacteur, *m.* editor; one who draws up or puts into written form.
rédaction, *f.* editing; editorial staff; editorial office; drawing up (of a writing); wording.
rédactionnel, *a.* editorial.
redan, *m.* projection; check, recess; notch.
redécouvrir, *v.t.* rediscover.
redent, *m.* = redan.
redescendre, *v.i.* descend again; redescend.
redevable, *a.* indebted; debtor. — *m.* debtor.
redevenir, *v.i.* become again.
rédiger, *v.t.* edit; draw up, write out, word.
redire, *v.t.* repeat; blame, censure.
redissolution, *f.* redissolving, re-solution.
redissolvant, *p.a.* redissolving.
redissoudre, *v.t. & r.* redissolve.
redistillation, *f.* redistillation.
redistiller, *v.t.* redistill.
redistribuer, *v.t.* redistribute.
rédo, *m.* redo (a hydrosulfite preparation).
redonner, *v.t.* give again, give back, return; restore. — *v.i.* begin again; give oneself up.
redorer, *v.t.* regild.
redoublé, *p.a.* redoubled; doubled, double.
redouter, *v.t.* dread, fear.
redressage, redressement, *m.* straightening, resetting, etc. (see redresser).
redresser, *v.t.* straighten; reset, reërect; right, rectify; correct, adjust, redress; (*Elec.*) commutate, rectify; dress, dress again; (*Leather*) stake a second time.

redresseur, *a.* straightening, etc. (see redresser). — *m.* straightener, etc.; (*Elec.*) commutator.
redû, *m.* balance due. — *a.* still owing.
réducteur, *m.* reducer, reducing agent. — *a.* reducing.
réductibilité, *f.* reducibility, reducibleness.
réductible, *a.* reducible.
réductif, *a.* reducing.
réduction, *f.* reduction.
réductrice, *a. fem.* of réducteur.
réduire, *v.t.* reduce. — *v.r.* be reduced; come, amount (to).
réduisant, *p.a.* reducing.
réduit, *p.a.* reduced. — *m.* retreat; redoubt. — *pr. 3 sing.* of réduire.
réel, *a.* real.
réellement, *adv.* really, in reality.
refaire, *v.t.* make again, remake, do again, etc. (see faire); repair, mend; begin again; restore; revive. — *v.i.* begin again. — *v.r.* be made again, etc.; recover; reëstablish oneself. — *passer,* cause to pass again, repass, etc. (see faire passer, under *passer*).
refaisage, *m.* (*Leather*) tanning anew, steeping anew.
refaisant, *p.pr.* of refaire.
refait, *p.p. & pr. 3 sing.* of refaire.
refasse, *pr. 1 & 3 sing. subj.* of refaire.
réfection, *f.* repair, reconstruction, making over, overhauling; recovery.
refendre, *v.t.* split, divide lengthwise.
refente, *f.* splitting, lengthwise division.
refera, *fut. 3 sing.* of refaire.
référence, *f.* reference.
référer, *v.t.* refer; attribute. — *v.r.* refer; rely (on), refer the matter (to). — *v.i.* report.
refermer, *v.t. & r.* close again, close.
refiltrer, *v.t.* refilter, filter again.
refit, *p.def. 3 sing.* of refaire.
réfléchir, *v.t. & r.* bend again.
réfléchir, *v.t. & i.* reflect. — *v.r.* be reflected.
réfléchissement, *m.* reflection, reflecting.
réflecteur, *a.* reflecting. — *m.* reflector.
reflet, *m.* reflection.
refléter, *v.t. & i.* reflect. — *v.r.* be reflected.
réflexe, *a. & m.* reflex.
réflexible, *a.* capable of being reflected.
réflexion, *f.* reflection.
refluer, *v.i.* flow back, reflow, (of tide) ebb.
reflux, *m.* reflux, flowing back.
à —, (of condensers) so that the condensed liquid flows back.
refondre, *v.t.* remelt, fuse again, recast, etc. (see fondre); correct, remodel, reform.
refont, *pr. 3 pl.* of refaire.
refonte, *f.* remelting, etc. (see fondre); remodeling; reform.
réforme, *f.* reform, reformation.
reformer, *v.t. & r.* form again, re-form.

réformer, *v.t.* reform, correct; (*Mil.*) retire.

refouillement, *m.* hollowing out, hollow.

refouiller, *v.t.* hollow out; carve; dig again.

refoulement,*m.* compressing,pressing back,etc. (seerefouler,*v.t.*); delivery(ofapipe);flowing back, backflow, backing up; turning back.

refouler, *v.t.* compress; press back, drive back, force back; ram, drive, force; upset, jump (pieces of metal); deliver (as water); stem (as a current); (fig.) check, restrain; press again, remill, etc. (see fouler). — *v.i.* flow back; turn back.

refouleur, *m.* compressor, driver, etc. (see refouler); specif., force pump.

refournir, *v.t.* refurnish, supply again.

réfractaire, *a.* refractory; resistant.

réfracter, *v.t.* refract.

réfracteur, *a.* refracting. — *m.* refractor.

réfractif, *a.* refractive.

réfraction, *f.* refraction.

réfractoire, *a.* refractional, of refraction.

réfractomètre, *m.* refractometer.

— *à beurre,* butyrorefractometer.

réfranger, *v.t.* refract.

réfrangibilité, *f.* refrangibility.

réfrangible, *a.* refrangible.

refrapper, *v.t.* strike again, etc. (see frapper).

refrayer, *v.t.* (*Ceram.*) planish.

refréner, *v.t.* restrain, check.

réfrigérant, *m.* condenser; cooler, refrigerator, refrigerating apparatus; (*Med.*) refrigerant. — *p.a.* cooling, refrigerating, (of mixtures) freezing; (*Med.*) refrigerant.

— *à air,* air condenser.

— *à boules,* bulb condenser.

— *à colonne,* column condenser.

— *à reflux,* — *ascendant,* reflux condenser, return condenser.

— *descendant,* ordinary condenser (as distinguished from a reflux condenser).

— *en cuivre,* copper condenser.

— *en verre,* glass condenser.

réfrigérateur, *m.* refrigerator.

réfrigératif, *a. & m.* (*Med.*) refrigerant.

réfrigération, *f.* refrigeration, cooling.

réfrigérer, *v.t.* refrigerate, cool.

réfringence, *f.* refractivity, refringency.

réfringent, *a.* refractive, refracting, refringent.

refroidir, *v.t.* cool; chill. — *v.r. & i.* cool, grow cool or cold.

refroidissement, *m.* cooling; chilling; coolness; cold, chill.

refroidisseur, *m.* cooler. — *a.* cooling.

refroidissoir, *m.* cooler.

refuge, *m.* refuge.

réfugier, *v.r.* take refuge, find shelter.

refus, *m.* refusal; refuse.

à —, (of passing gases, etc.) to excess, till no more is retained; (of driving piles, etc.) as far as possible, till progress ceases.

refuser, *v.t.* refuse. — *v.r.* be refused; (with *à*) refuse to, refuse, refuse to do, not permit. — *v.i.* refuse; (of tools, etc.) fail to work.

réfutation, *f.* refutation.

réfuter, *v.t.* refute. — *v.r.* be refuted.

regagnage, *m.* recovery (as of acids in making explosives); regaining.

regagner, *v.t.* regain; recover.

régale, *a.* In the phrase *eau régale,* aqua regia.

régaler, *v.t.* level, spread evenly (as earth); regale, entertain.

régalin, *a.* pertaining to aqua regia.

régalisation, *f.* (*Old Chem.*) action of aqua regia.

regard, *m.* look, eye, glance; regard; attention; opening, hole (as in a sewer); peephole, sighthole.

au — de, with regard to, regarding; in comparison with.

en —, opposite, face to face, on the opposite page.

regarder, *v.t.* look at; regard; concern; face, be directed toward. — *v.i.* look; be directed; (with *à*) regard, pay attention to, look to.

regarnir, *v.t.* refurnish, etc. (see garnir).

regel, *m.,* **regélation,** *f.* freezing again.

regeler, *v.t. & i.* freeze again.

régénérateur, *m.* regenerator. — *a.* regenerating, regenerative.

régénération, *f.* regeneration.

régénérer, *v.t.* regenerate.

régie, *f.* excise (or internal revenue) administration; excise office; administration (of property). — *p.a.* from régir.

régime, *m.* normal or usual operation, regular running (as of a furnace or motor); regulation, management, government; régime; regimen; system; rules, regulations.

— *carné,* meat diet.

— *végétarien,* vegetarian diet.

région, *f.* region.

régir, *v.t.* govern, rule, direct.

registre, *m.* register; record; damper, regulator.

registrer, *v.t.* register.

réglable, *a.* regulable, adjustable, etc. (see régler).

réglage, *m.* regulation; adjustment; ruling.

règle, *f.* rule; (*Mach.*) guide bar; rod; strip, batten; model, example; order, discipline; (*pl.*) menses.

— *à calcul,* computing rule, slide rule.

— *des phases,* phase rule.

— *de trois,* rule of three, proportion.

— *divisée,* scale.

— *glissante,* slide rule.

— *réduite,* reduced scale.

réglé, *p.a.* regulated, etc. (see régler); regular, steady.

règlement, *m.* regulation, adjustment, etc. (see régler); rule, regulation; settlement.

réglementaire, *a.* regular, prescribed, regulation; pertaining to regulations.

réglementer, *v.t.* regulate, make rules for.

régler, *v.t.* regulate; adjust; control, order, manage; fix, determine; settle (as an account); rule (as paper). — *v.r.* be regulated, etc.

réglette, *f.* small rule or scale; reglet.
— *à curseur,* sliding rule or scale.

régleur, *m.* regulator, etc. (see régler).

réglisse, *f.* licorice, liquorice.

reglisser, *v.t. & i.* slide again, slip again.

réglisserie, *f.* licorice manufacture.

réglure, *f.* ruling.

regne, *m.* reign; kingdom.
— *végétal,* vegetable kingdom.
— *vivant,* organic kingdom (either vegetable or animal kingdom).

régner, *v.i.* reign; extend, run.

regommer, *v.t.* regum, gum again.

regonflement, *m.* refilling; swelling.

regonfler, *v.t. & r.* refill; swell.

regorgement, *m.* overflowing, overflow.

regorger, *v.i.* overflow, run over; abound.
— *v.t.* vomit; disgorge; restore.

regoûter, *v.t. & i.* taste again.

regratter, *v.t.* scrape again, etc. (see gratter).

regrêler, *v.t.* repass thru the *grêloir* (see this).

régressif, *a.* regressive.

régression, *f.* regression, retrogression.
être en —, be going back, be receding.

regret, *m.* regret.

regrettablement, *adv.* regrettably.

regretter, *v.t.* regret.

regriller, *v.t.* roast again, etc. (see griller).

régularisation, *f.* regulation; regularization.

régulariser, *v.t.* regulate; set right; make regular, regularize.

régularité, *f.* regularity.

régulateur, *m.* regulator; specif., governor.
— *a.* regulating, regulative.
— *de chaleur,* heat (or temperature) regulator, thermoregulator.

régulation, *f.* regulation.

régule, *m.* regulus.

régulier, *a.* regular.

régulièrement, *adv.* regularly.

régulin, *a.* reguline.

régurgiter, *v.t.* regurgitate.

réhabiliter, *v.t.* rehabilitate, reëstablish.

rehaussement, *m.* raising, etc. (see rehausser).

rehausser, *v.t.* raise, make higher; heighten, intensify, enhance, set off; touch up; extol.

rehaut, *m.* retouch (in a design).

rehumecter, *v.t.* remoisten, dampen again.

réhydrater, *v.t. & i.* hydrate again.

réhydrogéniser, *v.t.* hydrogenize again.

réimporter, *v.t.* reimport.

réimpression, *f.* reprinting; reprint.

réimprimer, *v.t.* reprint. — *v.r.* be reprinted.

rein, *m.* kidney; (*pl.*) loins, back.

reine, *f.* queen.
— *des prés,* (*Bot.*) meadowsweet.

reine-Claude, *f.* greengage (plum).

reinette, *f.* pippin (apple).

réinsérer, *v.t.* reinsert.

réintégration, *f.* reinstatement, restoration.

réintégrer, *v.t.* reinstate, restore; reintegrate.

réitérer, *v.t.* reiterate, repeat.

rejaillir, *v.i.* rebound; reflect, be reflected; (of liquids) gush, gush out, spurt out, spring up; spring, arise.

rejaillissement, *m.* rebounding, etc. (see rejaillir).

rejauger, *v.t.* gage or calibrate again.

rejaunir, *v.t. & i.* yellow again.

réjection, *f.* rejection.

rejet, *m.* rejection; outthrow; thrown-out material, as earth; shoot (of a tree); escape pipe; (*Dyeing*) second dipping; transfer (of accounts).

rejetable, *a.* rejectable.

rejeter, *v.t.* reject; throw back, throw again, throw up, throw out, throw away; throw, cast (as blame); transfer. — *v.r.* be rejected, etc.; fall back (on).

rejeton, *m.* shoot (of a plant); descendant.

rejettement, *m.* rejection, etc. (see rejeter).

rejoignant, *p.pr.* of rejoindre.

rejoindre, *v.t. & r.* rejoin, reunite.

réjouir, *v.t.* rejoice; delight; amuse.

relâchant, *p.a.* relaxing, etc. (see relâcher); (*Med.*) laxative. — *m.* (*Med.*) laxative.

relâche, *m.* relaxation, rest, respite.

relâché, *p.a.* relaxed, etc. (see relâcher); loose, slack, lax.

relâchement, *m.* relaxing, etc. (see relâcher); looseness, laxity, laxness.

relâcher, *v.t.* relax, loosen, slacken; make limp (as paper by moisture); abate; divert; release; give up, yield. — *v.r.* be relaxed, relax, loosen, get loose; abate; (of weather) moderate. — *v.i.* relax, yield; put into port.

relai, relais, *m.* relay; (*Salt*) second washing (of saliferous sand) or the water used for it.

relaisser, *v.t.* leave again, etc. (see laisser).

relaminer, *v.t.* roll again, etc. (see laminer).

relancer, *v.t.* throw again, etc. (see lancer).

relargage, *m.* (*Soap*) salting out, graining.

rélargir, *v.t.* widen.

rélargissement, *m.* widening.

relarguer, *v.t.* (*Soap*) salt out, grain.

relater, *v.t.* relate, state.

relatif, *a.* relative.

relation, *f.* relation; connection.

relativement, *adv.* relatively.

relativité, *f.* relativity.

relaver, *v.t.* rewash, wash again.

relaxer, *v.t.* relax; release.

relayer, *v.t.* relieve, change.

reléguer, *v.t.* relegate.

relent, *m.* musty or moldy taste or odor.

rêler, *v.r.* be fissured, cracked or split (as a cake of tallow or a sugar loaf).

relevage, *m.* raising, etc. (see relever).

relevant, *a.* dependent.

relevé, *p.a.* raised, etc. (see relever); piquant, pungent; exalted, noble, lofty. — *m.* raising, etc. (see relever); summary, account, statement.

relèvement, *m.* raising, etc. (see relever); summary, statement, account; bearing, position.

relever, *v.t.* raise; raise up, take up, pick up, draw up, turn up; raise again; restore; rebuild; enhance, heighten; relieve; replace; release; remark on, point out; note, record; take the bearings of; exalt; extol. — *v.r.* rise; recover; be raised. — *v.i.* depend (on); be under (the orders of); recover (from).

relief, *m.* relief; scrap, leaving.

relien, *m.* (*Pyro.*) coarse gunpowder.

relier, *v.t.* join, unite; bind (books); hoop (casks); bind again.

religieusement, *adv.* religiously.

relimer, *v.t.* file again; (fig.) polish up.

reliquat, *m.* remainder; balance (of an account).

reliquataire, *a.* remaining, left. — *m.* debtor.

reliquéfier, *v.t. & r.* reliquefy, liquefy again.

relire, *v.t.* reread, read again.

relisant, *p.pr.* of relire.

reliure, *f.* binding, bookbinding.

relu, *p.a.* reread, read again.

reluire, *v.i.* shine, glitter, gleam.

reluisant, *p.a.* shining, glittering, glossy.

relut, *p.def. 3 sing.* of relire.

remaillage, *m.* removal of the last traces of epidermis from chamois leather, shaving.

remailler, *v.t.* subject to *remaillage*, shave.

rémailler, *v.t.* reënamel, enamel again.

remanent, rémanent, *a.* remaining, remanent, residual.

remaniement, remaniment, *m.* alteration, modification, change; repairing; rehandling.

remanier, *v.t.* alter, modify, change, revise; repair; rehandle, handle again.

remarquable, *a.* remarkable.

remarquablement, *adv.* remarkably.

remarque, *f.* remark.

remarquer, *v.t.* remark; mark again, re-mark.

remballer, *v.t.* repack, pack again.

remblai, *m.* filling; fill; embanking; embankment.

remblayer, *v.t.* fill, fill up; embank.

remboîter, *v.t.* fit together again; rebind (a book).

rembouger, *v.t.* fill up.

rembourrer, *v.t.* stuff, pad.

remboursement, *m.* reimbursement, repayment.

contre —, C.O.D.

rembourser, *v.t.* reimburse, repay, refund.

remède, *m.* remedy.

remédier, *v.t.* remedy.

remémorer, *v.t. & r.* recall (to mind).

remener, *v.t.* lead back, etc. (see mener).

remerciement, *m.* thanks, acknowledgment.

remercier, *v.t.* thank; decline; dismiss.

remercîment, *m.* = remerciement.

remesurage, *m.* remeasurement.

remesurer, *v.t.* remeasure, measure again.

remet, *pr. 3 sing.* of remettre.

remettre, *v.t.* put back, put again, replace, set back, etc. (see mettre); take back; recall (to mind); return; remit (as money); deposit; deliver; give up, resign; commit, entrust; refer; delay; restore; calm, compose; reconcile. — *v.r.* go back, return; recall; set oneself again; begin again, start again; be postponed; commit oneself (to); leave the matter (to); rely (on); recover; be reconciled.

remis, *p.p. & p.def. 1 & 2 sing.* of remettre.

remise, *f.* putting back, etc. (see remettre); delivery; remittance; allowance, deduction, discount, rebate; commission; delay.

remit, *p.def. 3 sing.* of remettre.

remmailler, *v.t.* mend.

remmancher, *v.t.* put a new handle on.

remodeler, *v.t.* remodel.

rémois, *a.* of Rheims.

remonder, *v.t.* reclean, etc. (see monder).

remontage, *m.* ascending, etc. (see remonter, *v.i.*).

remonter, *v.i.* go up, rise, ascend, mount; go up again, rise again; go back; (of ink) deepen in color. — *v.t.* ascend, go up; reascend; raise; take (or carry) up again; set up again, put together again, reassemble; refurnish, restock; strengthen (as brandy); wind up; revive. — *v.r.* provide oneself with a fresh supply; be reassembled; be wound up; revive, rouse.

remontrer, *v.t.* show, point out; show again. — *v.i.* remonstrate; teach.

remorque, *f.* towing.

se mettre à la —, follow in the wake.

remotis. In the phrase *à remotis,* aside.

remoudre, *v.t.* regrind, grind again.

rémoudre, *v.t.* resharpen, sharpen again.

remouillage, *m.* wetting again, remoistening.

remouiller, *v.t.* wet again, remoisten.

remouillure, *f.* (*Baking*) renewal of yeast.

remoulage, *m.* regrinding, remilling; remolding; reassembling of a mold; bran.

remoulant, *p.pr.* of remoudre and of remouler.

rémoulant, *p.pr.* of rémoudre.

remouler, *v.t.* remold, recast; remodel.

remoulu, *p.p.* of remoudre.

rémoulu, *p.p.* of rémoudre.

rempart, *m.* rampart.

remplaçable, *a.* replaceable.

remplacement, *m.* replacement, substitution; reinvestment (of funds).

remplacer, *v.t.* replace; take the place of; succeed; reinvest (funds). — *v.r.* be replaced; replace each other.

remplage, *m.* filling up, filling.

rempli, *p.a.* filled, etc. (see remplir); full.

remplir, *v.t.* fill, fill up; fill out; fill again, refill; complete, make up; take up, occupy; accomplish, fulfill; stock, supply; repay. — *v.r.* fill, be filled; repay.

remplissage, *m.* filling; filling up; filling in.

remplisseur, *m.* filler. — *a.* filling.

remporter, *v.t.* bear off, carry off, take away; carry back, take back.

remuable, *a.* movable; capable of being stirred.

remuage, *m.* moving; stirring; digging.

remuant, *p.a.* moving; stirring; restless.

remuement, *m.* moving; removing, removal; stirring; disturbance.

remuer, *v.t.* move; stir, stir up; dig; (fig.) agitate. — *v.r. & i.* stir, move.

remûment, *m.* = remuement.

rémunérateur, *a.* remunerative, profitable.

rémunération, *f.* remuneration.

rémunérer, *v.t.* remunerate, reward, pay.

remunir, *v.t.* refurnish, supply again.

renaissance, *f.* revival, renewal, renaissance.

renaissant, *a.* reviving, renascent.

renaître, *v.i.* revive, return, be restored.

rénal, *a.* (*Anat.*) renal.

renaquit, *p.def. 3 sing.* of renaître.

renard, *m.* fox; leak, opening, gap; (*Metal*) bloom.

rencaisser, *v.t.* put into a case or box again.

renchérir, *v.t. & i.* increase in price.

renchérissement, *m.* rise in price.

rencontre, *f.* meeting; encounter; chance; occasion; coincidence.

rencontrer, *v.t.* find, meet, meet with, encounter, hit on, happen on. — *v.r.* meet; be found, be met with, be; agree.

rendement, *m.* yield; efficiency (of a machine, etc.); produce; product; return, profit; (*Soap*) specif., the yield of soap from 100 kg. of fatty acids.

rendre, *v.t.* return; render; yield, produce; give back; restore; repay; pay; carry, transport; deliver; give out, emit; throw out, eject; give up, surrender; express. — *v.r.* be returned, etc.; return; go; lead; run, flow; give up, submit; be exhausted; become. — *v.i.* yield; give, stretch; discharge; go, lead. — *m.* repayment.
— *compte de,* give an account of, account for, answer for.
se — *compte de,* ascertain; understand.

rendu, *p.a.* returned, etc. (see rendre); arrived. — *m.* return; returned article.

renduire, *v.t.* coat again, cover again.

rendurcir, *v.t.* harden, make harder.

rendurcissement, *m.* hardening.

rené, *p.p.* of renaître.

renettoyer, *v.t.* clean again, reclean.

renfermé, *m.* close or musty smell.

renfermer, *v.t.* contain; include, comprise; confine; shut up; shut up again. — *v.r.* be contained; be included; be confined; confine oneself.

renflammer, *v.t.* rekindle. — *v.r.* take fire again.

renflé, *p.a.* swelled, swollen, swelling, inflated.

renflement, *m.* swelling, enlargement; specif., bulb; swell; shoulder; boss.

renfler, *v.t., r. & i.* swell, swell out, inflate.

renfoncement, *m.* hollow, depression; recess.

renforçage, *m.* strengthening, etc. (see renforcer).

renforçateur, *m.* (*Photog.*) intensifier.

renforcé, *p.a.* strengthened, etc. (see renforcer); strong, substantial; downright.

renforcement, *m.* strengthening, etc. (see renforcer).

renforcer, *v.t.* strengthen; reënforce; increase; (*Photog.*) intensify. — *v.r.* be strengthened or reënforced, grow stronger.

renfort, *m.* reënforce, strengthening piece; strengthening, reënforcement; supply.

rengréner, *v.t.* recoin; (*Mach.*) reëngage.

reniflard, *m.* air valve, snifting valve.

renifler, *v.t. & n.* snuff, sniff.

reniveler, *v.t.* level again.

renne, *m.* reindeer.

renom, *m.* renown, fame, reputation.

renommé, *a.* renowned, famous, celebrated.

renommée, *f.* renown, fame, reputation; report.

renommer, *v.t.* name again; reëlect; celebrate.

renoncer, *v.i.* (with *à*) abandon; renounce; forbear (to), desist (from).

renonculacées, *f.pl.* (*Bot.*) Ranunculaceæ.

renoncule, *f.* (*Bot.*) crowfoot, buttercup (*Ranunculus*).

renouée, *f.* (*Bot.*) polygonum.

renouer, *v.t.* renew; resume; join; tie again.

renouveau, *m.* renascence, revival.

renouveler, *v.t.* renew; revive; recall (to mind); recommence. — *v.r.* be renewed, etc. — *v.i.* increase; revive.

renouvellement, *m.* renewal, etc. (see renouveler).

renseignement, *m.* (piece or body of) information, intelligence; indication; hint.

renseigner, *v.t.* inform, give information to; teach again. — *v.r.* obtain information.

rentamer, *v.t.* cut again; resume.

rente, *f.* revenue, income; interest; rent; annuity; pension.

rentonner, *v.t.* tun again, barrel again.

rentraîner, *v.t.* carry away again, carry over again, etc. (see entraîner).

rentrant, *p.a.* reëntering, reëntrant.

rentrée, *f.* reëntrance, reëntering; return; re-opening; receipt (as of money).

rentrer, *v.i.* reënter, enter again; return; resume; reopen; (of money) come in. — *v.t.* take in, get in, drive in; take back.

renverra, *fut. 3 sing.* of renvoyer.

renversable, *a.* invertible; overturnable.

renversement, *m.* inversion, etc. (see renverser).

renverser, *v.t.* invert; turn upside down; overturn, upset; spill; overthrow; astound; reverse (as an engine or a current). — *v.i.* upset, fall over.

renvoi, *m.* return; dismissal, discharge; reference; explanation; adjournment; (*Mach.*) gear, gearing; cross reference.

renvoie, *pr. 3 sing. indic. & subj.* of renvoyer.

renvoyer, *v.t.* return; reflect (as light); dismiss, discharge; refer; adjourn. — *v.r.* be returned, etc.

réorganiser, *v.t.* reorganize.

réoxydable, *a.* capable of being reoxidized.

réoxydation, *f.* reoxidation.

réoxyder, *v.t.* reoxidize.

repaître, *v.t. & r.* feed.

répandre, *v.t.* spread, diffuse, send forth, emit, give out, exhale; shed, pour out, effuse, spill; utter. — *v.r.* be spread, etc.; spread, extend.

répandu, *p.a.* spread, etc. (see répandre); (of minerals, etc.) widely distributed; common.

réparage, *m.* repairing, etc. (see réparer).

reparaissant, *p.pr.* of reparaître.

reparaître, *v.i.* reappear.

réparateur, *a.* repairing; restoring; invigorating. — *m.* repairer.

réparation, *f.* repairing, repair, repairs; reparation.

réparer, *v.t.* repair, mend; work over; restore; make up for. — *v.r.* be repaired, etc.

repartir, *v.i.* start again, go off again; reply.

répartir, *v.t.* distribute; portion out, divide. — *v.r.* be distributed; be divided.

répartiteur, *m.* distributer, divider.

répartition, *f.* distribution, division.

reparu, *p.p.* of reparaître.

repassage, *m.* sharpening, etc. (see repasser).

repasser, *v.t.* sharpen, whet; polish; treat again, dye, dress, temper, etc. again; iron (cloth); go over again; rehearse; pass again. — *v.i.* repass, pass again, return.

repayer, *v.t.* repay.

repeignant, *p.pr.* of repeindre.

repeindre, *v.t.* repaint, paint again.

repérage, *m.* marking, etc. (see repérer).

repercer, *v.t.* pierce, perforate; pierce again.

répercussion, *f.* repercussion; reaction, rebound.

répercuter, *v.t.* reflect; (*Med.*) check.

repère, *m.* mark, reference mark, adjusting mark; reference scale; basis of reference.

repérer, *v.t.* mark; fix by reference marks, register; adjust.

répertoire, *m.* index, register, table, catalog; repertory.

repeser, *v.t.* reweigh, weigh again.

répéter, *v.t.* repeat; reflect (optical images).

répétition, *f.* repetition.

repétrir, *v.t.* knead again, etc. (see pétrir).

rephosphorer, *v.t.* rephosphorize.

repiler, *v.t.* pound again, etc. (see piler).

repiquer, *v.t.* transplant; repair (a road).

répit, *m.* respite.

replacer, *v.t.* replace; reinvest.

repli, *m.* fold, plait; turn, winding.

repliement, *m.* folding, etc. (see replier).

replier, *v.t.* fold, fold up, fold again; bend, bend up, turn up, coil up; bend back. — *v.r.* fold up, coil up, turn up, bend back, wind; fall back, retire.

répliquer, *v.i.* reply. — *v.t.* reply to, answer.

replonger, *v.t.* plunge (dip, immerse) again.

repolir, *v.t.* repolish, reburnish.

repolissage, *m.* repolishing, reburnishing.

répondre, *v.i.* answer, respond, reply; correspond; take an examination; lead (to). — *v.t.* answer, reply to. — *v.r.* correspond, agree.

répons, *m.* answer, response, reply.

reporter, *v.t.* carry back; carry, transport, place, put. — *v.r.* be carried back; refer, turn.

repos, *m.* rest, repose.

reposer, *v.i.* rest; stand; (of liquids) settle. — *v.t.* rest; lay, put, place; replace, put again, set again; calm, relieve. — *v.r.* rest; (of liquids) settle.

reposoir, *m.* (*Dyeing*) vat, tub (for indigo); (*Paper*) washing trough.

repoudrer, *v.t.* repowder, powder again.

repous, *m.* a kind of mortar or concrete.

repoussant, *p.a.* repelling, etc. (see repousser); repulsive.

repousser, *v.t.* repel, push back, thrust back, drive back; reject; be repulsive to; emboss, chase. — *v.i.* be repulsive; sprout again, grow again; (of a color) strengthen, deepen; (of a gun) recoil; (of a spring) resist.

repoussoir, *m.* drift, driver, starter, rammer.

repoustage, *m.* shaking, agitation (see repouster).

repouster, *v.t.* shake, agitate (gunpowder) to break up lumps.

reprécipitable, *a.* capable of being reprecipitated.

reprécipitation, *f.* reprecipitation.

reprécipiter, *v.t.* reprecipitate.

reprenant, *p.pr.* of reprendre.

reprendre, *v.t.* take up, dissolve, absorb; take again, take back, retake; get again, get back; resume; recover; repair; reprove; censure. — *v.r.* be taken up, etc.; congeal again; (of a wound) heal; recover oneself. — *v.i.* congeal again; begin again; grow again; revive; recover; (of a wound) heal; reply.

représentant, *a. & m.* representative.

représentation, *f.* representation.

représenter, *v.t.* represent; present; present again; remind of. — *v.r.* reappear; appear; occur; be represented.

réprimer, *v.t.* repress.

repris, *p.p.* of reprendre.

reprise, *f.* taking up, etc. (see reprendre); repetition, return.

à *plusieurs reprises,* several times, repeatedly, again and again.

reprit, *p.def. 3 sing.* of reprendre.

reproche, *m.* reproach.

reproducteur, *a.* reproductive, reproducing.

reproductibilité, *f.* reproducibility.

reproductible, *a.* reproducible.

reproductif, *a.* reproductive.

reproduire, *v.t.* reproduce; breed (animals). — *v.r.* be reproduced; reappear, recur.

reproduisant, *p.pr.* of reproduire.

republier, *v.t.* republish.

république, *f.* republic.

répudier, *v.t.* repudiate.

répugner, *v.i.* be repugnant; be reluctant.

répulsif, *a.* repulsive, repelling.

répulsion, *f.* repulsion.

repurger, *v.t.* purge again, etc. (see purger).

réputation, *f.* reputation, repute.

réputé, *p.a.* reputed; in high repute.

réputer, *v.t.* repute, consider, deem, think.

requérir, *v t.* require; request; demand; summon; claim; look for again.

requerra, *fut. 3 sing.* of requérir.

requete, *f.* request; petition.

requiert, *pr. 3 sing.* of requérir.

requis, *p.a.* required, etc. (see requérir); requisite. — *m.* requisite.

résacétophénone, *f.* resacetophenone (2, 4-dihydroxyacetophenone).

resceller, *v.t.* seal again.

réseau, *m.* network, net, netting; system (as of lines, wires or ore veins); (*Optics*) grating; screen (as of wires); (*Anat.*) plexus.

resécher, *v.t.* dry again, re-dry.

réséda, *m.* (*Bot.*) any plant of the genus *Reseda,* mignonette.

— *gaude,* yellowweed, dyer's weed (*R. luteola*).

— *odorant,* garden mignonette (*R. odorata*).

réservage, *m.* (*Calico*) reserving.

réserve, *f.* reservation; reserve; claim (as for damaged freight); (*Calico*) reserve, resist.

à *la* — *de,* with the exception of.

à *la* — *que,* excepting that.

en —, in reserve, aside, spare.

réserver, *v.t.* reserve. — *v.r.* save oneself; wait.

réservoir, *m.* reservoir; tank, well, cistern, chamber, holder, container, vessel, etc.

— à *gaz,* gas holder, gasometer.

— *de chasse,* flushing tank.

résider, *v.i.* reside, live, dwell.

résidu, *m.* residue, residuum; remainder.

résiduaire, *a.* residual, residuary.

résiduel, *a.* residual.

résigner, *v.t.* resign.

résilier, *v.t.* cancel, annul.

résinage, *m.* extraction of turpentine or rosin.

résine, *f.* resin; specif., rosin.

— *blanche,* white resin, specif. white rosin.

— *commune,* rosin.

— *copal,* copal resin, copal.

— *de gaïac,* guaiacum resin, guaiacum.

— *de jalap,* resin of jalap.

— *de scammonnée,* resin of scammony.

— *de térébenthine,* rosin.

— *de terre,* mineral resin.

— *élastique,* rubber, caoutchouc.

— *élémi,* elemi, gum elemi.

— *jaune,* yellow resin, specif. yellow rosin.

résiner, *v.t.* extract rosin or turpentine from; rosin, treat with rosin.

résineusement, *adv.* resinously.

résineux, *a.* resinous, resiny.

résinier, *m.* resin (or rosin) collector or maker.

résinifère, *a.* resiniferous.

résinifiable, *a.* resinifiable.

résinification, *f.* resinification, resinifying.

résinifier, *v.t. & r.* resinify.

résiniforme, *a.* resiniform.

résini-gomme, *f.* gum resin.

résino-amer, *m.* aloes.

résino-gommeux, *a.* gum-resinous.

résinoïde, *a.* resinoid.

résinylie, *m.* rosin oil.

résistance, *f.* resistance; strength (of materials).

— à *la tension* (or *traction*), tensile strength.

— à *l'écrasement,* crushing strength.

— *de frottement,* frictional resistance.

— *de mise en marche,* (*Elec.*) starting resistance, starting rheostat.

— *d'isolement,* (*Elec.*) insulation resistance.

résistant, *a.* resisting, resistant; strong.

résister, *v.i.* resist, offer resistance.

— à, resist, withstand; support, endure.

résolu, *p.a.* solved, etc. (see résoudre); resolute, fixed.

résoluble, *a.* capable of being solved, etc. (see résoudre).

résolument, *adv.* resolutely.

résolut, *p.def. 3 sing.* of résoudre.

résolutif, *m. & a.* (*Med.*) resolvent, discutient.

résolution, *f.* solution; resolution; reduction, conversion; cancelling.

résolvant, *p.a.* solving, etc. (see résoudre); (*Med.*) resolvent. — *m.* (*Med.*) resolvent.

résolvent, *pr. 3 pl.* of résoudre.

résonner, *v.i.* resound; be resonant, sonorous.

résorber, *v.t.* resorb, reabsorb.

résorcine, *f.* resorcinol, resorcin.

résorption, *f.* resorption, reabsorption.

résoudre, *v.t.* solve; resolve; reduce, convert, transform; dissolve; cancel; persuade, induce; decide upon. — *v.r.* be solved, etc.; separate (into); change (into); resolve. — *v.i.* resolve, decide.

résous, *p.a.* resolved; converted; dissolved.

résout, *pr. 3 sing.* of résoudre.

respect, *m.* respect.

respecter, *v.t.* respect.

respectif, *a.* respective.

respectivement, *adv.* respectively.

respirabilité, *f.* respirability.

respirable, *a.* respirable.

respirateur, *m.* respirator. — *a.* respiratory.

respiration, *f.* respiration, breathing, breath.

respiratoire, *a.* respiratory.

respirer, *v.i.* breathe, respire; long (after). — *v.t.* breathe; long for.

resplendir, *v.i.* shine brightly, be resplendent.

responsabilité, *f.* responsibility.

responsable, *a.* responsible.

ressasser, *v.t.* sift again; examine again, examine, scrutinize; repeat over and over.

ressaut, *m.* projection, shoulder, lug; jog.

ressayer, *v.t.* try again.

ressemblance, *f.* resemblance.

ressemblant, *a.* similar, alike, like. — *p.pr.* of ressembler.

ressembler, *v.i.* be like, be alike; (with *à*) resemble, be like. — *v.r.* be alike.

ressense, *f.* = recense.

ressentir, *v.t.* feel, experience; resent. — *v.r.* be felt; feel the effects (of), suffer (from).

resserrement, *m.* contraction, etc. (see resserrer); tightness (as of money); oppression.

resserrer, *v.t.* contract; compress, condense; tighten, draw tighter, draw closer; shorten; narrow; confine; restrain; put away, put back (in a close place); tie again; (*Med.*) constipate. — *v.r.* be contracted, etc.; contract, shrink; get tight or tighter; (of weather) get colder. — *v.i.* (*Med.*) be constipating.

resservir, *v.i. & t.* serve again.

ressort, *m.* spring; elasticity, springiness; (fig.) force, energy; province, jurisdiction. — *pr. 3 sing.* of ressortir.

à —, spring, with spring.

— *à boudin,* — *à hélice,* helical spring.

— *d'attache,* spring catch.

ressortir, *v.i.* stand out; be evident; spring, result; go out again; be under the jurisdiction (of).

ressoudre, *v.t.* resolder; reweld; reunite.

ressoudure, *f.* resoldering; rewelding.

ressource, *f.* resource; resources, means.

ressouvenir, *v.r.* (with *de*) recollect, remember again. — *m.* recollection, reminiscence.

ressuage, *m.* sweating; (*Metal.*) liquation, eliquation.

ressuer, *v.i.* sweat; (of a metal) be liquated, run out in liquation; sweat again. — *v.t.* sweat; sweat again; (*Metal.*) liquate.

ressui, *m.* (*Ceram.*) sweating.

ressuiement, *m.* drying.

ressuivre, *v.t.* follow again, follow back.

ressusciter, *v.t.* resuscitate, revive, restore.

ressuyage, *m.* drying.

ressuyé, *p.a.* dried, etc. (see ressuyer); (*Ceram.*) affected by weathering.

ressuyer, *v.t.* dry; wipe again; dry again.

restant, *p.a.* remaining, left, residual. — *m.* remainder, residue.

restaurant, *a.* restorative. — *m.* restorative; restaurant.

restauration, *f.* restoration.

restaurer, *v.t.* restore.

reste, *m.* residue, remainder, rest, remain(s), remnant; leavings; refuse.

au —, *du* —, besides, moreover, also.

de —, left, remaining; to spare, more than enough.

— *d'acide,* acid residue.

rester, *v.i.* remain; stay, continue; stop.

en — *à,* leave off, discontinue.

restituer, *v.t.* restore.

restitution, *f.* restoration; restitution.

restreignant, *p.pr.* of restreindre.

restreindre, *v.t.* restrict, restrain, limit.

restreint, *p.a.* restricted, limited. — *pr. 3 sing.* of restreindre.

restriction, *f.* restriction; reservation.

restringent, *a. & m.* astringent, restringent.

résultant, *p.a.* resulting, resultant.

résultat, *m.* result.

résulter, *v.i.* result.

résumé, *m.* résumé, summary.

au —, *en* —, to sum up, on the whole, after all.

résumer, *v.t & r.* sum up, recapitulate.

rétablir, *v.t.* reëstablish; restore.

rétablissement, *m.* reëstablishment; restoration.

retaille, *f.* cutting, clipping, shred, scrap, paring, bit.

retaillement, *m.* cutting again, etc. (see verb).

retailler, *v.t.* cut again, recut, etc. (see tailler); dig up (marc before re-pressing).

rétaler, *v.t.* spread again, etc. (see étaler).

rétamage, *m.* retinning; replating.

rétamer, *v.t.* retin; replate.

retanner, *v.t.* tan again, retan.

retard, *m.* delay; lag; lateness; slowness (of timepieces).

en —, late, behind time, slow.

retardateur, *a.* retarding, retardative.

retardation, *f.* retardation.

retarder, *v.t.* retard; delay; set back (a timepiece). — *v.i.* delay; be slow, be behind.

retassement, *m.* (*Metal.*) (1) piping, (2) pipe.

retasser, *v.i.* (*Metal.*) develop cavities, pipe.

retassure, *f.* (*Metal.*) pipe (the cavity).

retâter, *v.t.* feel again, etc. (see tâter).

reteignant, *p.pr.* of reteindre.

reteindre, *v.t.* redye, etc. (see teindre).

réteindre, *v.t.* reëxtinguish, etc. (see éteindre).

reteint, *p.p. & pr. 3 sing.* of reteindre.

retenant, *p.pr.* of retenir.

retendre, *v.t.* restretch, etc. (see tendre, *v.t.*).

rétendre, *v.t.* dilute again, etc. (see étendre).

retène, *m.* retene.

retenir, *v.t.* retain; hold; keep; hold back; restrain; hold up, sustain; get back, get again. — *v.r.* hold, cling; hold back, stop, refrain; be retained, etc.

retenter, *v.t.* try again.

rétenteur, *a.* retaining; restraining.

rétention, *f.* retention; reservation.

retentir, *v.i.* resound, echo, ring.

retentissement, *m.* renown; publicity; sound, echo, noise; resounding, echoing.

retenu, *p.a.* retained, etc. (see retenir); reserved, discreet.

retenue, *f.* deduction, docking, stoppage (of pay); reservoir; guy, guy rope; reserve, discretion.

réticulaire, *a.* reticular.

réticule, *m.* (*Optics*) reticle, reticule, cross wires, cross hairs.

réticulé, *a.* reticulated; reticulate.

réticulum, *m.* (*Biol.*, etc.) reticulum.

retient, *pr. 3 sing.* of retenir.

rétinalite, *f.* (*Min.*) retinalite.

rétinasphalte, *m.* (*Min.*) retinasphaltum.

rétine, *f.* (*Anat.*) retina.

rétinien, *a.* (*Anat.*) retinal.

rétinolé, *m.* (*Pharm.*) a solid preparation having a resin or resins as a base.

retint, *p.def. 3 sing.* of retenir.

retirement, *m.* withdrawing, withdrawal, etc. (see retirer); (*Ceram.*) crawling (of the glaze).

retirer, *v.t.* withdraw, draw back, take back; get, obtain, derive; take in, receive; draw in; contract; draw out, take out; draw

again. — *v.r.* retire; contract, shrink; subside; be withdrawn, etc.

retirure, *f.* (*Metal.*) hollow due to shrinkage.

retomber, *v.i.* fall again, fall back, fall; sink again; relapse; attribute. — *v.t.* (*Soap*) transfer (the paste) to another boiler.

retordre, *v.t.* twist again; twist; wring again.

retors, *a.* twisted; wrung; bent; far-fetched; crafty. — *m.* twist; second twist.

retorte, *f.* retort. (*Obs.*)

retoucher, *v.t.* dress, finish; retouch.

retour, *m.* return; turn, winding; angle, corner; backing up (of a flowing liquid); change; repetition; wane, decline; caprice.

en —, in return; at an angle; (of titration) back, in the opposite direction.

en — *d'équerre,* at right angles.

sans —, forever.

retourner, *v.t.* turn; turn over; examine carefully; turn again; return; affect, move. — *v.r.* turn, turn round; manage; be turned; (with *en*) return. — *v.i.* return.

rétracter, *v.t. & r.* contract; retract.

retrait, *m.* contraction, shrinking, shrinkage; withdrawal; retreat (the place); repurchase. — *a.* contracted; shrunken; (of wood) warped.

retraite, *f.* contraction, shrinkage; retirement; retiring pension; retreat; (*Com.*) redraft.

retraiter, *v.t.* treat again, retreat; remove (as hides from the pit); retire.

retranchement, *m.* subtraction, etc. (see retrancher).

retrancher, *v.t.* subtract, deduct; cut off; take away, take out; leave out; stop; curtail; retrench; intrench.

retravailler, *v.t.* work over again, finish.

retraverser, *v.t.* retraverse.

rétrécir, *v.t. & r.* narrow; contract; shrink; shorten.

rétrécissement, *m.* narrowing; shrinking, shrinkage; contraction; narrowness; (*Med.*) stricture.

retreindre, *v.t.* hammer out, shape (by hammering).

retremper, *v.t.* soak again, temper again, etc. (see tremper); reinvigorate.

rétribuer, *v.t.* remunerate, pay.

rétribution, *f.* remuneration, pay, fee.

rétroactivement, *adv.* retroactively.

rétrogradation, *f.* reversion (to an insoluble state); retrogradation, retrogression.

rétrograde, *a.* retrograde, backward.

rétrograder, *v.i.* revert (to an insoluble state); retrograde.

retrousser, *v.t.* turn up, tuck up, hold up.

retrouver, *v.t.* find, find again; meet with; meet again. — *v.r.* be found, etc.; find one's way; meet again.

rets, *m.* net.

rétudier, *v.t.* study again.

réunion, *f.* union: junction; connection; meeting; reunion.

réunir, *v.t.* reunite; unite, join; collect; call together. — *v.r.* reunite; unite; meet, assemble.

réusiner, *v.t.* machine again.

réussir, *v.i.* succeed, be successful; prosper; (of plants) thrive; result, turn out, end. — *v.t.* execute well, perform successfully.

réussite, *f.* success; issue, result.

revanche, *f.* return, requital; revenge.
en —, in return, in compensation, on the other hand.

revaporisation, *f.* revaporization.

revaporiser, *v.t.* revaporize.

rêve, *m.* dream.

revêche, *a.* harsh, rough (to taste or feel); cross-grained; (of iron) becoming hard on annealing; (of diamond) not polishing uniformly.

revécu, *p.p.* of revivre.

réveil, *m.* waking, awakening; alarm; reveille.

réveille-matin, *m.* alarm clock.

réveiller, *v.t.* wake, awaken; rouse; revive. — *v.r.* wake, awake; revive.

révélateur, *a.* serving for the detection (of); indicative (of); revealing, disclosing. — *m.* detector, indicator; (*Photog.*) developer; revealer, discloser.

révéler, *v.t.* reveal, disclose; (*Photog.*) develop. — *v.r.* manifest itself; behave; be disclosed.

revenant, *p.a.* returning, etc. (see revenir). — *m.* ghost, apparition.

revendication, *f.* claim.

revendiquer, *v.t.* claim.

revenir, *v.i.* return, come back; recur; recover; amount (to); result, arise; give in, yield; be suited (to); be pleasing (to).
faire —, restore (steel) to its former condition, as by tempering; restore; recover; bring back, call back; cook slightly.
— *à,* please, suit, agree with, (of colors) harmonize with; match; cost; return to; succeed in.
— *sur,* return to; reconsider; retract; retrace; fall back on.

revenu, *m.* revenue; income; (*Metal.*) refined state. — *pp.* of revenir.

revenue, *f.* return; new growth (of wood).

rêver, *v.i.* dream. — *v.t.* dream, dream of.

réverbération, *f.* reverberation, reflection.

réverbératoire, *a.* reverberatory.

réverbère, *m.* reflector; dome of a reverberatory furnace; street lamp.
à —, reverberatory.

réverbérer, *v.t. & i.* reverberate, reflect.

reverdir, *v.t.* make green again; clean and soften (hides). — *v.i.* become green again; revive.

reverdissage, *m.* cleaning and softening (hides).

reverdissement, *m.* growing green again; new life.

reverra, *fut. 3 sing.* of revoir.

revers, *m.* reverse; back, wrong side, other side.

reverser, *v.t.* pour again; pour back; transfer.

réversibilité, *f.* reversibility.

réversible, *a.* reversible.

revêtement, *m.* facing, casing, covering, coating, lining, sheathing, lagging, revetment.

revêtir, *v.t.* face, case, cover, coat, line, sheathe; put on, assume; fit, equip; clothe; dress; draw up (a paper). — *v.r.* (with *de*) put on; assume.

revêtu, *p.a.* faced, etc. (see revêtir).

revidage, *m.* re-emptying, etc. (see revider and vider); enlarging (of a hole).

revider, *v.t.* re-empty, etc. (see vider); enlarge (a hole).

revient, *pr. 3 sing.* of revenir. — *m.* cost.

revint, *p.def. 3 sing.* of revenir.

reviser, réviser, *v.t.* revise; examine, inspect.

revision, révision, *f.* revision; examination, inspection.

revisser, *v.t.* screw again, screw back.

revit, *p.def. 3 sing.* of revivre and revoir.

revivant, *p.pr.* of revivre.

revivificateur, *a.* revivifying, reviving.

revivification, *f.* revivification, reviving.

revivifier, *v.t.* revivify, revive.

revivre, *v.i.* revive.
faire —, revive.

revoir, *v.t.* revise; see again; examine again; inspect; oversee.

révolution, *f.* revolution.

revolver, *m.* revolver (the weapon, or the revolving changer of a microscope).

revoyant, *p.pr.* of revoir.

revu, *p.a.* revised, etc. (see revoir).

revue, *f.* review; inspection; seeing again.

révulsif, *a. & m.* (*Med.*) revulsive.

rez, *prep.* even with, level with.
à — *de,* on a level with.

rez-de-chaussée, *m.* ground level; ground floor.

rhabillage, *m.* repair, mending; dressing again.

rhabiller, *v.t.* repair, mend; dress again.

rhamnétine, *f.* rhamnetin.

rhamnose, *f.* rhamnose.

rhénan, *a.* Rhenish, Rhine, of the Rhine.

rhéostat, *m.* (*Elec.*) rheostat.
— *à clef,* plug rheostat.

rhigolène, *m.* rhigolene.

Rhin, *m.* Rhine.

rhizome, *m.* (*Bot.*) rhizome, rhizoma.

rhodamine, *f.* rhodamine.

rhodanien, *a.* Rhone, of the Rhone.

rhodéose, *m.* rhodeose.

rhodique, *a.* rhodic.

rhodium, *m.* rhodium.

rhodizonique, *a.* rhodizonic.

rhombe, *m.* (*Cryst.*) rhomb, rhombohedron; (*Geom.*) rhomb, rhombus. — *a.* rhombic.

rhombique, *a.* rhombic.

rhomboèdre, *m.* rhombohedron.

rhomboédrique, *a.* rhombohedral.

rhomboïdal, *a.* rhomboidal.

rhomboïde, *m. & a.* rhomboid.

rhubarbe, *f.* rhubarb (*Rheum*). — *de Chine,* Chinese rhubarb.

rhum, *m.* rum.

rhumatisant, *m.* rheumatic patient.

rhumatismal, *a.* rheumatic.

rhumatisme, *m.* rheumatism.

rhume, *m.* cold (the indisposition).

ri, *p.p.* of rire (to laugh).

riblon, *m.* scrap (of iron or steel).

ribonique, *a.* ribonic.

ribose, *n.* ribose.

riche, *a.* rich. — *m.* rich material, esp. ore; rich person.

richement, *adv.* richly.

richesse, *f.* richness; specif., concentration; riches, wealth.

ricin, *m.* castor-oil plant, Ricinus; (sometimes) ricin.

ricine, *f.* ricin.

riciné, *a.* impregnated with castor oil.

ricinine, *f.* ricinine.

ricinoléique, ricinolique, *a.* ricinoleic.

ricocher, *v.i.* ricochet, rebound.

ricochet, *m.* ricochet; chain of events. *par —,* indirectly.

ride, *f.* wrinkle; fold; ripple; lanyard.

ridé, *p.a.* wrinkled, etc. (see rider).

rideau, *m.* curtain; screen.

rider, *v.t.* wrinkle; shrivel; corrugate, rib, flute; ripple; tighten, haul taut.

ridicule, *a.* ridiculous. — *m.* ridiculous, ridiculousness; ridicule.

ridiculement, *m.* ridiculously.

rien, *m.* (with or without *ne*) nothing; anything, something. — *m.* nothing. *en —* (with *ne*) not at all, in no degree; somewhat, in some degree. *— que,* nothing but, only.

riflard, *m.* coarse file (for metals); toothed chisel; jack plane.

rifler, *v.t.* scratch; file; rasp; plane.

rifloir, *m.* curved file, riffler.

rigaud, *m.* stone, hard lump (in lime).

rigide, *a.* rigid.

rigidement, *adv.* rigidly.

rigidité, *f.* rigidity, rigidness.

rigole, *f.* channel, groove, furrow, trench, trough, gutter, drain, ditch; rill, rivulet.

rigoureusement, *adv.* rigorously, strictly.

rigoureux, *a.* rigorous; severe.

rigueur, *f.* rigor, rigour. *à la —,* at the worst, if necessary; rigorously; strictly; strictly speaking.

rinçage, rincement, *m.* rinsing.

rince-bouteilles, *m.* bottle rinser.

rincer, *v.t.* rinse.

ringage, *m.* cinders, scoria.

ringard, *m.* an iron tool for stirring, poking, cleaning, etc.; specif., poker.

ringarder, *v.t.* stir, poke (as a fire).

riper, *v.i. & t.* slip, slide, shift; scrape.

riquette, *f.* scrap iron.

rire, *v.i. & r.* laugh; jest. — *m.* laughing.

ris, *m.* sweetbread; laughter; reef (in a sail).

risque, *m.* risk.

risquer, *v.t.* risk.

rit, *pr. & p.def. 3 sing.* of rire.

rivage, *m.* riveting; clinching; shore, bank.

rival, *m. & a.* rival.

rivaliser, *v.i.* compete, vie.

rivalité, *f.* rivalry.

rive, *f.* border, edge; shore, bank.

rivement, *m.* riveting; clinching.

river, *v.t.* rivet; clinch (nails).

rivet, *m.* rivet; clinch.

rivetage, *m.* riveting.

riveter, *v.t.* rivet.

riveur, *m.,* **riveuse,** *f.* riveter.

rivière, *f.* river; (*cap.*) Riviera.

rivure, *f.* riveting; rivet joint; rivet head; pin (of a hinge); pin joint.

riz, *m.* rice.

rizerie, *f.* rice mill.

rob, *m.* (*Pharm.*) rob.

robe, *f.* robe, gown, dress; coat (as of horses); fleece; skin, husk; wrapper, envelope.

rober, *v.t.* bark (madder); cover, envelop.

robinet, *m.* stopcock; cock, faucet, tap; valve; key, plug (of a cock). *— à cadran,* cock with pointer and scale, for fine adjustment of flow. *— à clef creuse,* hollow-plug cock. *— à deux voies,* two-way stopcock, two-way cock. *— à eau,* water cock. *— à flotteur,* a constant-level device consisting of a cock to which a float is connected. *— à gaz,* gas cock. *— à huile,* oil cock. *— alimentaire,* feed cock. *— à pointeau,* a cock with a key ending in a point which fits in a conoidal seat when the cock is closed. *— à quatre voies,* four-way cock. *— à raccord,* cock to which a pipe may be joined, esp. by soldering. *— à tétine,* cock with corrugated or nipple-shaped tip for the attachment of rubber tubes. *— à trois voies,* three-way stop-cock, three-way cock. *— à vapeur,* steam cock. *— à vis,* screw cock, screw faucet.

robinet d'arrêt, stopcock.
— d'aspiration, suction cock.
— d'eau, water cock, water tap; gage cock.
— de communication, a cock in the middle of a line (as contrasted with one for delivery), connecting cock, communicating cock.
— de décharge, discharge cock.
— d'épreuve, try cock.
— de purge, purging cock, clearing cock, waste cock.
— de sorbonne, cock for use in a hood, specif. a gas cock to be operated from the outside.
— d'essai, try cock.
— de sûreté, safety cock.
— de vidange, drain cock, outlet cock, discharge cock.
— graisseur, lubricating cock.
— mélangeur, mixing cock, specif. one for mixing and delivering hot and cold water.
— souffleur, blast cock.
robinetterie, f. cocks, cocks and valves, also their manufacture.
roburite, f. (Expl.) roburite.
robuste, a. strong, robust, stout.
robustesse, f. strength, stoutness, robustness.
roc, m. rock.
roccella, roccelle, f. (Bot.) Roccella.
roccelline, f. roccellin, roccelline (a dye).
roccellique, a. roccellic.
rochage, m. frothing, etc. (see rocher).
roche, f. rock; very hard stone; overburned brick; crude borax (Obs.); (Cement, pl.) clinker; (Metal.) furnace sow; (Mining) gang, gangue.
en —, caked.
— à feu, (Pyro.) Valenciennes.
— à sel, rock salt.
rocher, v.i. froth, foam; (of silver, etc.) expel oxygen on solidifying, spit, sprout; — v.t. powder with borax (in soldering). — m. rock.
rochet, m. ratchet.
rochette, f. (military) rocket. (Obs.)
rocheux, a. rocky.
rocou, m. annatto.
rocouer, v.t. color or dye with annatto.
rocouerie, f. annatto factory.
rocouyer, m. (Bot.) annatto tree, achiote (Bixa orellana).
rodage, m. grinding (as with emery).
roder, v.t. grind (as glass with emery); rub; polish.
rôder, v.i. roam, rove; (Mach.) work freely.
rodoir, m. grinder; polisher; tanning vat.
rognage, m. clipping, etc. (see rogner).
rogne, f. outer, fissured part of bark; scab (of plants); mange, itch.
rognement, m. clipping, etc. (see rogner).
rogner, v.t. clip, trim, pare; cut off; curtail.
rognoir, m. clipper, trimmer, parer, cutter.

rognon, m. reniform mass, nodule; specif., kidney stone; kidney (of animals).
rognure, f. clipping, cutting, trimming, paring, scrap, shred, remnant.
roi, m. king.
roide, roideur, etc. = raide, raideur, etc.
rôle, m. roll; list; (Mach.) roll, roller; rôle.
romain, a. Roman; (of type) roman. — m. roman.
romaine, f. steelyard; testing machine; roman.
roman, a. Romance. — m. romance, novel.
romanée, f. Romanée (a red Burgundy wine).
romarin, m. rosemary (Rosmarinus officinalis).
— des marais, marsh rosemary, inkroot (Limonium carolinianum).
— sauvage, marsh tea, wild rosemary (Ledum palustre).
romite, f. (Expl.) romite.
rompre, v.t. break; break up; break off; shatter; rupture; tear; refract (as light); blend (colors); interrupt; give up; tire out. — v.r. be broken, etc.; break, break up, break off. — v.i. break, break off, break down.
rompu, p.a. broken, etc. (see rompre); used, accustomed, experienced.
ronce, f. bramble, briar (any species of Rubus)
— noire, blackberry.
ronces artificielles, barbed wire.
rond, a. round. — m. round, circle, disk, (round) plate; ring; wheel; bar of round cross section; washer.
en —, round, roundly, in a ring, in a circle, circularly.
— de serviette, napkin ring.
ronde, f. round, rounds.
à la —, round, around, all around.
rondelle, f. (small) round, disk, ring, circlet; ring (of a water bath); washer; collar; dial; (Ordnance) obturator.
— à cuvette, cup washer.
— de caoutchouc, rubber washer, ring or disk.
— de carton, pasteboard disk or ring.
— de poudre, powder pellet; powder ring.
— en cuir, leather washer, leather; leather disk or ring.
— fusible, fusible plug.
rondement, adv. roundly, briskly, frankly.
rondeur, f. roundness.
rondin, m. (cylindrical) stick; billet; small log; roll (as of copper gauze).
ronfler, v.i. roar, rumble, boom, hum; snore.
rongeage, m. (Dyeing) discharging.
rongeant, p.a. corroding, etc. (see ronger).
— m. (Dyeing) discharge.
rongement, m. corroding, etc. (see ronger).
ronger, v.t. corrode, eat away; etch, pit; (Dyeing) discharge; gnaw, nibble, bite.
rongeur, a. corroding, etc. (see ronger). — m. (Zoöl.) rodent.

roquefort, *m.* Roquefort cheese, Roquefort.

roquet, *m.* (*Mil.*) rocket.

rorqual, *m.* (*Zoöl.*) rorqual, finback.
— *rostré,* dœgling, beaked whale.

rosace, *f.* rosette.

rosacées, *f.pl.* (*Bot.*) Rosaceæ.

rosaginine, *f.* rosaginin.

rosaniline, *f.* rosaniline.

rosat, *a.* rose, of roses.

rose, *m. & f.* rose. — *a.* rose, rosy.
— *bengale,* (*Dyes*) rose bengale.

rosé, *a.* rose, rose-colored, rosy, pink.

roseau, *m.* (*Bot.*) reed.

rosée, *f.* dew.
— *du soleil.* (*Bot.*) sundew.

roséochromique, *a.* roseochromic.

roser, *v.t.* rose, make pink.

rosette, *f.* rosette, rose; disk; washer; (*Metal.*) disk of purified copper, rosette; red ink; red chalk.

rosier, *m.* (*Bot.*) rose bush, rose.

rosinduline, *f.* rosinduline.

rosolique, *a.* rosolic.

rossolis, *m.* rosolio, rossolis (a liquor).

rotang, *m.* (*Bot.*) rattan.

rotateur, *a.* rotating, rotatory.

rotatif, *a.* rotary, rotative.

rotation, *f.* rotation.

rotationnel, *a.* rotational.

rotatoire, *a* rotatory.

rotin, *m.* (*Bot.*) rattan.

rôtir, *v.t.,* *r. & i.* roast.

rôtissage, *m.* roasting.

rotondité, *f.* rotundity.

rotule, *m.* swivel joint; ball-and-socket joint; small roll, roller or wheel; kneecap, patella.

rouable, *m.* hook, rake (as for stirring salt).

rouage, *m.* wheelwork, wheels, machinery, mechanism; wheel, esp. a toothed wheel.

roucou, *m.* annatto.

roue, *f.* wheel; coil (of rope).
— *à cames,* cam wheel.
— *à dents,* = roue dentée.
— *à feu,* (*Pyro.*) (1) Catherine wheel, (2) pin-wheel.
— *à gorge,* wheel with grooved rim, double-flanged wheel.
— *à rochet,* ratchet wheel.
— *d'angle,* bevel wheel, miter wheel.
— *dentée,* toothed wheel, cogwheel, gear wheel.
— *folle,* idle wheel.
— *motrice,* driving wheel, drivewheel.
— *volante,* flywheel.

rouelle, *f.* round slice, round cut.

rouennais, *a.* Rouen, of Rouen.

rouet, *m.* sheave, pulley wheel, pulley.

rouge, *a.* red; red hot. — *m.* red; red heat; rouge.
au —, at red heat.

rouge (continued).
au — blanc, at white heat.
— *anglais,* English red (an iron oxide).
— *à peine naissant,* faintest red heat.
— *blanc,* white heat.
— *brique,* brick-red.
— *carmin,* carmine red.
— *cerise,* cherry red.
— *cinchonique,* cinchona red.
— *Congo,* Congo red.
— *cramoisi,* crimson-red.
— *d'Angleterre,* English red.
— *de Bordeaux,* Bordeaux red, Bordeaux.
— *de chair,* flesh-colored.
— *de chrome,* chrome red.
— *de cuivre,* copper-red, coppery red.
— *de flammés,* (*Ceram.*) a red flambé glaze containing copper.
— *de grand feu,* (*Ceram.*) a red glaze produced at high heat with copper or a copper compound.
— *de Japon,* a red lake made from eosin and lead acetate.
— *de Saint-Denis,* an azo dye of the formula $C_{24}H_{24}N_6Na_2O_9S_2$.
— *de sang,* blood red.
— *de Saturne,* Saturn red, minium.
— *de stilbène,* stilbene red.
— *de toluylène,* toluylene red.
— *de Venise,* Venetian red.
— *d'indigo,* indigo red.
— *groseille,* currant red.
— *indien,* Indian red.
— *naissant,* nascent red, incipient red heat.
— *neutre,* neutral red.
— *para,* para red, paranitraniline red.
— *ponceau,* corn-poppy red, poppy red.
— *pourpre,* purple.
— *rose,* rose-red.
— *sang,* blood red.
— *sombre,* dark red; dull red heat.
— *suant,* white heat.
— *turc,* (*Dyeing*) Turkey red; (*Pigments*) an orange-red basic chromate of lead.
— *vif,* bright red; bright red heat.
— *vineux,* wine red.

rougeâtre, *a.* reddish.

rouge-chair, *m.* (literally, flesh red) a kind of red enamel.

rougeole, *f.* (*Med.*) measles.

rougeur, *f.* redness; red spot; flush, blush.

rougir, *v.t.* redden, turn red; heat red-hot; tan. — *v.i.* redden, turn red.
faire —, = rougir, *v.t.*
— *à blanc,* — *en blanc,* heat white-hot, raise to white heat.

roui, *p.a.* retted. — *m.* retting; staleness, musty odor.

rouillage, *m.* rusting.

rouille, *f.* rust.

rouille *de cuivre*, a green coating on copper or copper alloys, verdigris.

rouillé, *p.a.* rusted; rusty.

rouiller, *v.t. & r.* rust.

rouilleux, *a.* rusty.

rouillure, *f.* rustiness.

rouir, *v.t.* ret. — *v.i.* be retted, steep.

rouissage, *m.* retting.

rouissoir, *m.* retting place.

roulage, *m.* rolling, etc. (see rouler); carriage, transportation (in wheeled vehicles).

roulaison, *f.* the work of sugar manufacture.

roulant, *p.a.* rolling, etc. (see rouler); easy (for riding); running (fire).

rouleau, *m.* roller; roll; cylinder.

— *alimentaire,* feed roller.

— *compresseur,* — *de pression,* (*Mach.*) press roll.

roulement, *m.* rolling, etc. (see rouler); roll; rolling mechanism; working season, campaign (as of a furnace).

— *à billes,* ball bearing.

rouler, *v.t.* roll; roll up; (fig.) revolve; pass (one's life). — *v.i.* roll; run (on wheels); ride (in a wheeled vehicle); revolve; rotate; turn; (of money) circulate; be plentiful; wander. — *v.r.* be rolled, etc.; roll.

roulette, *f.* small wheel (of various kinds); roller; roll; runner, caster; tapeline.

rouleur, *m.* roller; workman who rolls or wheels something. — *a.* rolling.

roulier, *m.* carter, wagoner, trucker. — *a.* of or pertaining to carriage in wagons, trucks, etc.

roulis, *m.* rolling, rolling motion.

rouloir, *m.* roller, roll, cylinder.

roulon, *m.* round, rung.

roulure, *f.* rolling, etc. (see rouler); roll; (of wood) shake.

roumain, *a.* Roumanian. — *m.* (*cap.*) Roumanian.

Roumanie, *f.* Roumania.

roure, *m.* = rouvre.

roussâtre, *a.* russet, russety.

rousse, *a. fem.* of roux.

rousseur, *f.* russet color or quality.

roussi, *p.a.* turned russet, browned, reddened; scorched. — *m.* scorching, scorched odor; Russia leather.

roussiller, *v.i.* scorch.

roussir, *v.t. & r.* turn russet, brown, redden; scorch; singe.

roussissage, *m.* dyeing russet.

roussissement, *m.* turning russet, etc. (see roussir).

route, *f.* road; way; path, track, route, course.

routier, *a.* road, of roads.

routine, *f.* routine.

routinier, *a.* routine.

rouverin, rouverain, *a.* (of iron) hot-short, red-short.

rouvert, *p.a.* reopened.

rouvre, *m.* British oak (*Quercus robur*). — *pr.* 1 & 3 *sing. indic. & subj.* of rouvrir.

rouvrir, *v.t.* reopen.

roux, *a. & m.* russet, reddish (or yellowish) brown; (of hair) red, auburn.

royal, *a.* royal.

royalement, *adv.* royally.

royaume, *m.* kingdom, realm.

Royaume-Uni, *m.* United Kingdom.

royauté, *f.* royalty.

ru, *m.* brook, steamlet, rill.

rubace, rubacelle, *f.* rubicelle (an orange-red or yellow variety of ruby spinel); quartz colored red.

ruban, *m.* ribbon; tape; band, strip, strap.

rubané, *a.* ribboned; striped, streaked; (of cast iron) mottled.

rubaner, *v.t.* ribbon.

rubasse, *f.* = rubace.

rubéfaction, *f.* (*Med.*, etc.) rubefaction.

rubéfiant, *a. & m.* (*Med.*) rubefacient.

rubéfier, *v.t.* redden.

rubéole, *f.* (*Med.*) rubella, German measles.

rubéosine, *f.* (*Dyes*) rubeosin.

rubiacées, *f.pl.* (*Bot.*) Rubiaceæ.

rubicelle, *f.* = rubace.

rubidium, *m.* rubidium.

rubigineux, *a.* rusty; rust-colored, rubiginous.

rubine, *f.* (*Old Chem.*) ruby: as, *rubine d'arsenic,* ruby of arsenic (realgar).

rubis, *m.* (*Min.*) ruby.

— *balais,* balas ruby.

— *de Bohême,* Bohemian ruby (red quartz).

— *oriental,* Oriental ruby, true ruby.

— *spinelle,* spinel ruby.

rubrique, *f.* red chalk, red earth, ruddle; rubric, title; method, practice; ruse, trick.

ruche, *f.* hive, beehive.

rude, *a.* rough; harsh; hard, arduous, severe; rude; rigid, strict; formidable.

rudement, *adv.* roughly, harshly, etc. (see rude).

rudesse, *f.* roughness, harshness, etc. (see rude).

rudiment, *m.* rudiment.

rudimentaire, *a.* rudimentary.

rudimentairement, *adv.* in a rudimentary way.

rue, *f.* street; (*Bot.*) rue.

— *de chèvre,* goat's-rue (*Galega officinalis*).

— *odorante,* common rue (*Ruta graveolens*).

rufigallique, *a.* rufigallic (designating an acid known also as rufigallol).

rufigallol, ruffigallol, *m.* rufigallol.

rugir, *v.i.* roar.

rugissement, *m.* roar, roaring.

rugosité, *f.* rugosity, roughness, wrinkle.

rugueux, *a.* wrinkled, rough, rugose.

ruine, *f.* ruin.

ruiner, *v.t.* ruin; destroy. — *v.r.* be ruined, go to ruin.

ruineusement, *adv.* ruinously.

ruineux, *a.* ruinous.

ruisseau, *m.* stream, small stream, rivulet, brook; gutter.

ruisseler, *v.i.* stream, stream down, run down, run, flow, trickle, drip.

ruissellement, *m.* streaming, etc. (see ruisseler); (*Geol.*) stream action.

rumeur, *f.* murmur; hum; din, clamor; rumor.

rumex, *m.* (*Bot.*) Rumex (sorrels and docks).

ruolz, *m.* electroplated metal or ware.

rupture, *f.* rupture, breaking, break.

rusque, *f.* (*Leather*) bark of the kermes oak.

russe, *a.* Russian. — *m.* Russian.

Russie, *f.* Russia.

rustine, *f.* back (of a blast furnace).

ruthénique, *a.* ruthenic.

ruthénium, *m.* ruthenium.

rutilance, *f.* redness.

rutilant, *a.* deep or glowing red, rutilant.

rutile, *m.* (*Min.*) rutile.

rutiler, *v.i.* shine.

rutique, *a.* rutic (capric).

S

s., *abbrev.* (son, sa) his, her; (sud) south; (sauf) save, except; (stère) stere.

s', Contraction of *se* (which see).

sa, *a. fem.* his, her, its.

sabadilline, *f.* sabadilline.

sabine, *f.* (*Bot.*) savin, savine, sabine (*Juniperus sabina*, also *J. virginiana*).

sable, *m.* sand; (sometimes) gravel; (*Med.*) gravel.
— *à fabriquer le verre*, glass sand.
— *à noyaux*, core sand.
— *à souder*, welding sand.
— *aurifère*, auriferous sand, gold-bearing sand.
— *de fer*, (*Pyro.*) iron filings.
— *de mer*, sea sand.
— *de mine*, pit sand.
— *de moulage*, molding sand.
— *de rivière*, river gravel (or sand).
— *fin*, fine sand.
— *gravier*, gravelly sand, coarse sand.
— *gros*, coarse sand.
— *phosphaté*, phosphate sand.
— *pour verreries*, glass sand.
— *sec*, dry sand.
— *vert*, green sand.

sablé, *p.a.* sanded, etc. (see sabler); speckled.

sabler, *v.t.* sand; gravel; grind, cut or clean with the sand blast; cast in a sand mold.

sablerie, *f.* place where sand molds are made.

sableur, *m.* sander; sand dresser, sand screener; maker of sand molds.

sableuse, *f.* sand blast (the apparatus). — *a. fem.* of sableux.

sableux, *a.* sandy, arenaceous.

sablier, *m.* sandglass; sand box; (*Bot.*) sandbox tree (*Hura crepitans*).

sablière, *f.* sand pit; gravel pit; (*Building*) plate, ground plate, wall plate.

sablon, *m.* very fine sand.

sablonner, *v.t.* scour with sand; sand.

sablonnette, *f.* (*Glass*) sand case.

sablonneux, *a.* sandy, gritty.

sablonnière, *f.* sand pit; sand box.

sabord, *m.* porthole, port.

sabot, *m.* shoe; socket; (*Glass*) a device for seizing the bottom of a bottle; hoof (of a horse); (whipping) top; wooden shoe, sabot.

sac, *m.* sack, bag; pouch; sac; (*Mining*) pocket.
— *à gaz*, gas bag.

sac *à poudre*, powder bag.
— *à sable*, sand bag.
— *de nuit*, — *de voyage*, traveling bag.
— *de soldat*, knapsack.

saccade, *f.* jerk.

saccadé, *a.* jerky, irregular.

saccager, *v.t.* sack, pillage.

saccharate, *m.* saccharate.
— *de fer*, iron saccharate.

secchareux, *a.* saccharine.

saccharide, *m.* saccharide.

saccharidé, *a.* resembling sugar.

saccharifère, *a.* containing or yielding sugar.

saccharifiable, *a.* saccharifiable.

saccharifiant, *p.a.* saccharifying.

saccharificateur, *m.* saccharifier.

saccharification, *f.* saccharification.

saccharifier, *v.t.* saccharify, convert into, or impregnate with, sugar. — *v.r.* be saccharified.

saccharigène, *a.* producing sugar.

saccharimètre, *m.* saccharimeter.

saccharimétrie, *f.* saccharimetry.

saccharimétrique, *a.* saccharimetric(al).

saccharin, *a.* saccharine.

saccharine, *f.* saccharin.

saccharique, *a.* saccharic.

saccharoïde, *a.* saccharoid, saccharoidal.

saccharolé, *m.* (*Pharm.*) a preparation containing sugar.
— *mou*, confection.

saccharolique, *a.* (*Pharm.*) containing sugar.

saccharomètre, *m.* saccharimeter, -rometer.

saccharomyces, saccharomyce, saccharomycète, *m.* (*Bot.*) Saccharomyces, yeast fungus, yeast.

saccharone, *f.* saccharone.

saccharose, *m. & f.* saccharose.

saccharure, *m.* (*Pharm.*) a preparation made with sugar.
— *de carbonate ferreux*, saccharated ferrous carbonate.
— *d'oxyde de fer soluble*, ferric saccharate.

sachant, *p.pr.* of savoir.

sache, *pr. subj. 1 & 3 sing.* of savoir.

sachet, *m.* small sack or bag; sachet.

sacré, *a.* sacred; (*Anat.*) sacral.

sacrifier, *v.t. & i.* sacrifice.

saffre, *m.* zaffer.

safran, *m.* saffron.

319

safran *artificiel*, artificial saffron (Victoria yellow).
— *bâtard*, safflower, false saffron (*Carthamus tinctorius*).
— *de Vénus*, cupric oxide.
safrané, *a.* saffron-colored, saffron; of or pertaining to saffron.
safraner, *v.t.* color with saffron.
safranière, *f.* saffron plantation.
safranine, *f.* (*Dyes*) safranine, safranin.
safranone, *f.* safranone.
safranum, *m.* safflower.
safre, *m.* zaffer (impure oxide of cobalt).
safrol, *m.* safrole.
sagapénum, *m.* sagapenum.
sage, *a.* wise; discreet, modest, good, gentle. — *m.* wise man, sage.
sagement, *adv.* wisely, prudently, sensibly.
sagesse, *f.* wisdom; sobriety, goodness.
sagou, *m.* sago.
sagoutier, sagouier, *m.* sago palm, sago tree.
saignement, *m.* bleeding.
saigner, *v.t.* bleed; drain. — *v.i.* bleed.
saigneux, *a.* bloody.
saillant, *p.a.* projecting; outstanding, salient; striking, remarkable.
saillie, *f.* projection; ledge, cheek, lug, flange, etc.; start, fit; sally; (*Painting*) relief.
saillir, *v.i.* project, protrude, stand out.
sain, *a.* sound; sane; healthful.
sainbois, *m.* (*Bot.*) spurge flax (*Daphne gnidium*), (*Pharm.*) mezereum.
saindoux, *m.* lard.
sainegrain, *m.* (*Bot.*) fenugreek.
sainement, *adv.* soundly; sanely; wholesomely.
sainfoin, *m.* (*Bot.*) sainfoin (*Onobrychis sativa*).
saint, *a.* holy. — *m.* saint.
sais, *pr. 1 & 2 sing.* of savoir.
saisie, *f.* seizure.
saisir, *v.t.* seize; take hold of, catch, grasp; understand, catch, perceive; put in possession (of). — *v.r.* (with *de*) lay hold of, seize, take; be seized; be nonplussed.
saisissable, *a.* capable of being seized.
saisissant, *p.a.* seizing, etc. (see saisir); piercing (cold); startling.
saisissement, *m.* seizing, seizure, etc. (see saisir); chill; shock.
saison, *f.* season.
sait, *pr. 3 sing.* of savoir.
saké, saki, *m.* sake, rice beer.
salade, *f.* salad.
salage, *m.* salting.
salaire, *m.* pay, wages, hire, salary; reward.
salaison, *m.* salting; salt meat or other salted article of food, (*pl.*) salt provisions.
salange, *m.* (*Salt*) season, campaign.
salanque, *f.* salt garden (for making sea salt).
salant, *a.* salt. — *m.* saline ground.

salarier, *v.t.* pay, pay wages or salary to.
sale, *a.* dirty; (of colors) dull, grayish.
salé, *p.a.* salted; salt; piquant, keen; (of prices) dear, excessive. — *m.* salt pork.
salep, *m.* salep.
saler, *v.t.* salt.
saleté, *f.* dirtiness, dirt; filth.
saleur, *m.* salter.
salicaire, *f.* (*Bot.*) loosestrife (*Lythrum*).
— *commune*, purple loosestrife (*L. salicaria*).
salicine, *f.* salicin.
salicole, *a.* salt-producing, salt.
salicor, salicorne, *f.* glasswort (*Salicornia*).
saliculture, *f.* the operation of salt gardens.
salicylate, *m.* salicylate.
— *d'ammoniaque*, ammonium salicylate.
— *de lithine*, lithium salicylate.
— *de phényle*,— *de phénol*, phenyl salicylate, salol.
— *de potasse*, potassium salicylate.
— *de soude*, sodium salicylate.
salicyle, *m.* salicyl (used preferably of the radical *o*-HOC$_6$H$_4$; for the radical HOC$_6$H$_4$CO the name *salicylyl* is preferred).
salicyler, *v.t.* treat with salicylic acid or a salicylate, esp. for preservation.
salicyleux, *a.* salicylous (salicylal was formerly called "salicylous acid ").
salicylique, *a.* salicylic.
salicylol, *m.* salicylal, salicylic aldehyde.
salifère, *a.* saliferous.
salifiable, *a.* salifiable.
salification, *f.* salification; esterification (incorrect use).
salifier, *v.t.* salify.
saligénine, *f.* saligenol, saligenin.
salignon, *m.* a lump of salt, saltcat.
salin, *a.* saline. — *m.* salin, saline (crude potash obtained by evaporating the lye from ashes); salt works; salt garden.
— *de betteraves*, potash from sugar-beet vinasse.
salinage, *m.* (*Salt*) concentration of the brine after removing sulfates; salt works; salt, saline (crude potash).
saline, *f.* salt works; salt pit; salt garden; salt meat; salt fish.
salinier, *m.* salt maker, salt boiler; salt dealer. — *a.* relating to salt production.
salinité, *f.* salinity, salineness, saltness.
salinomètre, *m.* salinometer, salimeter.
salipyrine, *f.* (*Pharm.*) salipyrin, salipyrine.
salir, *v.t.* soil, make dirty, foul, befoul; stain, tarnish; defile, pollute. — *v.r.* become soiled or dirty, soil.
salitre, *m.* Chile saltpeter, sodium nitrate; (according to one authority) magnesium sulfate.
salivaire, *a.* salivary. — *m.* (*Pharm.*) pyrethrum.

salivation, f. (*Med.*) salivation.
salive, f. (*Physiol.*) saliva.
salle, f. hall; room; ward (of a hospital).
 — *aux compositions,* (*Gunpowder*) mixing room, mixing shop, mixing house.
 — *d'artifice*(*s*), pyrotechnic laboratory.
 — *d'attente,* waiting room.
 — *de lecture,* reading room.
 — *de préparation,* a room where lecture experiments are prepared.
 — *des machines,* engine room; machine room.
 — *de travail,* workroom, laboratory.
 — *d'expériences,* a room in which experiments are made, laboratory.
 — *d'humidité,* (*Gunpowder*) damping room.
salmiac, m. sal ammoniac, ammonium chloride.
salmine, f. salmine.
salol, m. (*Pharm.*) salol.
salophen, salophène, m. (*Pharm.*) salophen, salophene.
salpêtrage, m. formation of saltpeter.
salpêtre, m. saltpeter, niter.
 — *brut,* crude saltpeter.
 — *cubique,* — *de Chili,* cubic saltpeter, Chile saltpeter (sodium nitrate).
 — *de conversion,* conversion saltpeter (potassium nitrate made from sodium nitrate).
 — *de deux eaux.* — *de deux cuites,* saltpeter refined by two boilings.
 — *de houssage,* wall saltpeter (calcium nitrate obtained as an efflorescence).
 — *de potasse,* potash saltpeter (potassium nitrate).
 — *de roche,* rock saltpeter.
salpêtrer, v.t. cover or treat with saltpeter.
salpêtrerie, f. saltpeter (or niter) factory.
salpêtreux, a. saltpetrous.
salpêtrier, m. saltpeter maker.
salpêtrière, f. saltpeter works, niter works.
salpêtrisation, f. covering or treating with saltpeter; conversion into saltpeter.
salplicat, m. gold lacquer.
salse, f. (*Geol.*) salse, mud volcano.
salsepareille, f. (*Bot.*) sarsaparilla.
salsifis, m. (*Bot.*) salsify.
salubre, a. healthful, salubrious.
salubrité, f. health; healthfulness, salubrity.
saluer, v.t. salute; greet.
salure, f. saltness, saltiness.
salut, m. safety; salutation, greeting; salute.
salutaire, a. salutary; wholesome; beneficial.
samarium, m. samarium.
samarskite, f. (*Min.*) samarskite.
sambunigrine, f. sambunigrin.
samedi, m. Saturday.
samienne, a. *fem.* Samian.
sanction, f. sanction; penalty.
sanctionner, v.t. sanction, approve; confirm.
sandal, m. sandalwood. (See santal.)
sandaraque, f. sandarac (the resin).

sang, m. blood.
 — *artériel,* arterial blood.
 — *desséché,* dried blood.
 — *veineux,* venous blood.
sang-dragon, sang-de-dragon, m. dragon's blood.
sanglant, a. bloody; blood-red; cruel, cutting.
sangle, f. strap, band; girth.
sangler, v.t. bind; strap; girth.
sangsue, f. (*Zoöl.*) leech.
sanguin, a. blood, of blood; sanguine.
sanguinaire, f. (*Bot.*) bloodroot (*Sanguinaria canadensis*). — a. sanguinary.
sanguine, f. (*Min.*) bloodstone; red chalk.
sanitaire, a. sanitary.
sans, prep. without.
 — *que,* without; as, *sans qu'il le sût,* without his knowing it.
 (For phrases in which *sans* governs a noun, see the respective nouns).
santal, m. sandalwood.
 — *blanc,* white sandalwood, sandalwood proper (*Santalum album* or its wood).
 — *citrin,* yellow sandalwood.
 — *rouge,* red sandalwood (specif. *Lingoum santalinum* or its wood).
santaline, santaléine, f. santalin.
santalol, m. santalol.
santé, f. health.
santonine, f. santonin; (*Bot.*) santonica.
santonique, a. santonic.
sapa, m. (*Pharm.*) inspissated grape juice.
sapan, m. sapanwood.
saphène, f. (*Anat.*) saphenous vein, saphena.
saphir, m. (*Min.*) sapphire.
saphirin, a. sapphirine.
sapide, a. savory, sapid.
sapidité, a. savoriness, savor, sapidity.
sapin, m. fir (tree and wood), specif. any plant of the genus *Abies* or its wood; spruce (see the phrases).
 — *argenté,* — *commun,* — *des Vosges,* — *en peigne,* — *pectiné,* silver fir (*Abies picea* or *pectinata*).
 — *baumier,* balsam fir (*Abies balsamea*).
 — *blanc,* = sapin argenté; white spruce (*Picea canadensis* or *alba*).
 — *du Canada,* hemlock spruce (*Tsuga canadensis*).
 — *épicéa,* Norway spruce (*Picea abies*).
 — *noir,* black spruce (*Picea mariana* or *nigra*).
sapindus, m. (*Bot.*) soapberry (*Sapindus*).
sapinette, f. spruce beer; (*Bot.*) spruce, specif. hemlock spruce (*Tsuga canadensis*).
 — *blanche,* white spruce (*Picea canadensis* or *alba*).
 — *noire,* black spruce (*Picea mariana* or *nigra*).
saponacé, a. saponaceous, soapy.
saponaire, f. (*Bot.*) Saponaria.
 — *officinale,* soapwort (*S. officinalis*).

saponé, *m.* (*Pharm.*) a medicated soap.
saponifiable, *a.* saponifiable.
saponification, *f.* saponification.
saponifier, *v.t.* saponify.
saponiforme, *a.* resembling soap.
saponine, *f.* saponin.
saponite, *f.* (*Min.*) saponite.
saponule, *f.* a soapy deposit obtained by cooling a solution of tallow soap; (*Old Chem.*) saponul, saponule.
saponulé, *m.* an alcoholic soap solution, solidifying to a transparent mass.
saponure, *m.* (*Pharm.*) a combination of powdered soap with extractive matter.
sapotacées, *f.pl.* (*Bot.*) Sapotaceæ.
sapotille, sapote, *f.* sapodilla (the fruit).
sapotillier, sapotier, *m.* sapodilla (the tree).
saprophyte, *a.* (*Bot.*) saprophytic.
sarcine, *f.* sarcine; (*Bact.*) sarcina.
sarcler, *v.t.* weed, weed out.
sarcocolle, *f.* sarcocolla (the exudate).
sarcocollier, *m.* (*Bot.*) Sarcocolla.
sarcolactique, *a.* sarcolactic, *d*-lactic.
sarcomateux, *a.* (*Med.*) sarcomatous.
sarcome, *m.* (*Med.*) sarcoma.
sarcosine, *f.* sarcosine.
sarde, *f.* (*Min.*) sard. — *a.* Sardinian.
sardoine, *f.* (*Min.*) sardonyx.
sariette, *f.* = sarriette.
sarment, *m.* shoot, twig (esp. of the grapevine); (*Bot.*) sarmentum, runner.
  sarments de vignes, vine shoots, vine twigs.
sarracénie, *f.* (*Bot.*) pitcher plant (*Sarracenia*).
sarrasin, *m.* buckwheat (*Fagopyrum*); (*Metal.*) dross, refuse. — *a.* Saracenic; in the phrase *blé sarrasin,* buckwheat.
sarrau, *m.* frock blouse, specif. a laboratory or surgeon's blouse.
sarriette, *f.* (*Bot.*) savory (*Satureia,* esp. *S. hortensis*).
  — *commune,* — *des jardins,* summer savory. (*S. hortensis*).
  — *de montagne,* winter savory (*S. montana*).
sas, *m.* sieve; lock chamber.
sassafras, *m.* (*Bot.*) sassafras.
sassement, *m.* sifting, bolting.
sassenage, *m.* Sassenage cheese.
sasser, *v.t.* sift, bolt.
sasset, *m.* little sieve.
sasseur, *m.* sifter, bolter.
sassoline, *f.* (*Min.*) sassolite.
satin, *m.* satin.
satinage, *m.* satining.
satiné, *a.* satin, satiny; satined. — *m.* satiny appearance, gloss.
satiner, *v.t.* satin, give a satiny or glossy surface to. — *v.i.* look like satin.
satisfaction, *f.* satisfaction.
satisfaire, *v.t.* satisfy. — *v.r.* be satisfied. — *v.i.* satisfy, render satisfaction; (with *à*) satisfy.

satisfaisant, *p.a.* satisfying, satisfactory.
satisfait, *p.a.* satisfied; pleased, content.
saturabilité, *f.* saturability.
saturable, *a.* saturable.
saturant, *p.a.* saturating, saturant.
saturateur, *m.* saturator.
saturation, *f.* saturation.
saturer, *v.t.* saturate. — *v.r.* be or become saturated.
satureur, *m.* saturator, one who saturates.
saturne, Saturne, *m.* (*Old Chem.*) Saturn, lead. *extrait de saturne,* (*Pharm.*) Goulard's extract.
saturnin, *a.* (*Old Chem.*) saturnine, of lead.
saturnisme, *m.* (*Med.*) saturnism, plumbism.
sauce, *f.* sauce.
saucer, *v.t.* soak, steep, sop, souse.
saucisse, *f.* (small) sausage; (*Pyro.*) powder hose, powder bag, or powder.
saucisson, *m.* (large) sausage; powder hose, saucisson; roll of gamboge.
sauf, *prep. & conj.* save, saving, except, excepting, but, reserving. — *a.* safe, unharmed.
sauge, *f.* (*Bot.*) sage (*Salvia*).
  — *des jardins,* — *officinale,* garden sage (*S. officinalis*).
  — *des prés,* meadow sage (*S. pratensis*).
  — *sclarée,* clary (*S. sclarea*).
saugé, *a.* containing sage.
saugie, *f.* sage tea.
saule, *m.* willow (tree and wood).
saumâtre, *a.* somewhat salty, brackish.
saumon, *m.* salmon (fish and color); (*Metal.*) pig; (*Founding*) head, sullage piece; charge of coal in a coke furnace; melting trough for wax. — *a.* salmon, salmon-colored.
saumurage, *m.* pickling (in brine), brining.
saumure, *f.* brine; specif., brine in which meat or fish has been pickled.
  — *ammoniacale,* (*Soda*) ammoniacal brine.
  — *vierge,* (*Soda*) fresh brine (not yet ammoniated).
saumuré, *p.a.* pickled (in brine), brined.
saunage, *m.,* **saunaison,** *f.* salt making, esp. in salt gardens; salt trade.
sauner, *v.i.* make salt; deposit salt.
saunerie, *f.* salt works, saltern.
saunier, *m.* salt maker; salt merchant.
saupoudrage, *m.,* **saupoudration,** *f.* powdering, dusting, sprinkling; salting.
saupoudrer, *v.t.* powder, dust, sprinkle; specif., sprinkle with salt, salt.
saur, saure, *a.* (of herrings) smoked.
saura, *fut. 3 sing.* of savoir.
saurait, *cond. 3 sing.* of savoir.
saurer, *v.t.* smoke (herrings).
saurien, *m. & a.* (*Zoöl.*) saurian.
saurin, *m.* smoked herring, red herring.
sauris, *m.* herring brine.
saussurite, *f.* (*Min. & Petrog.*) saussurite.

saut, *m.* jump; leap, vault, spring, bound; fall.

sautage, *m.* exploding, explosion, blowing up.

sauter, *v.i.* jump, leap, spring, bound, vault; explode, blow up; burst; fly off, fly out; rush (at). — *v.t.* skip, omit; jump, jump over, spring over.

faire — , blow up, explode; make jump; get rid of.

sauterelle, *f.* grasshopper, locust.

sauternes, *m.* sauterne, Sauternes wine.

sautiller, *v.i.* hop, skip.

sauvage, *a.* wild; (of oil) bitter; savage; shy. — *m.* savage; unsociable person.

sauvagerie, *f.* wildness; savagery; shyness.

sauvagin, *a.* fishy, marshy. — *m.* fishy or marshy odor or taste.

sauve, *a. fem.* of sauf.

sauvegarde, *f.* safeguard.

sauvegarder, *v.t.* safeguard.

sauvement, *m.* salvage.

sauver, *v.t.* save; justify; spare; conceal. — *v.r.* be saved; escape; take refuge; indemnify oneself, make up (for).

sauvetage, *m.* saving, esp. lifesaving; salvage.

savait, *imp. indic. 3 sing.* of savoir.

savamment, *adv.* learnedly; knowingly.

savant, *a.* learned; well informed; skillful, clever. — *m.* learned person, savant.

savent, *pr. 3 pl.* of savoir.

saveur, *f.* savor, flavor, relish.

savinier, *m.* = sabine.

Savoie, *f.* Savoy.

savoir, *v.t.* know; know how; be able, can; (infinitive used adverbially) to wit, namely. — *v.r.* be known. — *v.i.* be learned; be well informed. — *m.* knowledge, learning.

à — , to wit, that is to say, that is, namely.

faire — , inform, notify.

— *gré,* be thankful, be pleased.

savoir-faire, *m.* skill; judgment; wits.

savoisien, *a.* of Savoy. — *m.* (*cap.*) Savoyard

savon, *m.* soap; cake of soap.

— *à barbe,* shaving soap.

— *à base de potasse,* potash soap.

— *à base de soude,* soda soap.

— *à détacher,* soap for removing spots or stains, specif. oxgall soap.

— *à foulon,* fulling soap.

— *à froid,* cold-process soap.

— *ammoniacal,* (*Pharm.*) ammonia liniment.

— *amygdalin,* almond oil soap, amygdaline soap.

— *animal,* soap made from animal fat; specif., one made from fine or purified fat for medical purposes, as the beef's-marrow soap of French pharmacists.

— *arsenical,* arsenical soap.

— *au miel,* a kind of glycerin soap containing beeswax.

savon *au procédé de la grande* (or *petite*) *chaudière,* see under *procédé.*

— *au sable,* soap containing sand, sand soap.

— *au soufre,* soap containing sulfur, sulfur soap.

— *blanc,* white soap.

— *calcaire,* lime soap, calcium soap; (*Pharm.*) lime liniment, carron oil.

— *chaud,* boiled soap.

— *de cire,* soap made from or containing wax, wax soap.

— *de cuivre,* copper soap; = savon métallique.

— *de glycérine,* glycerin soap.

— *de goudron,* tar soap.

— *de Marseille,* Marseilles soap; (more broadly) Castile soap (either white or mottled).

— *de Menotti,* a waterproofing soap containing alumina and gelatin.

— *d'empâtage,* soap made in a single operation without separation of the curd from the lye.

— *dentifrice,* soap for cleansing the teeth.

— *de plomb,* lead soap.

— *de potasse,* potash soap.

— *de résine,* rosin soap; (in general) resin soap.

— *de soude,* soda soap.

— *d'Espagne,* Castile soap.

— *des verriers,* glass soap, glassmaker's soap (manganese dioxide or any substance used for decolorizing glass).

— *de toilette,* toilet soap.

— *de Venise,* Venetian soap (essentially the same as Castile soap).

— *dijonnais,* Dijon soap (a kind of silicate soap).

— *dur,* hard soap.

— *encaustique,* soap containing wax, wax soap.

— *en feuilles,* soap in the form of thin sheets or leaves.

— *en poudre,* powdered soap, soap powder.

— *épuré,* purified soap, white soap.

— *grenu,* soft soap of granular texture made from certain mixtures of fats.

— *hydrofuge,* a kind of soap used in waterproofing fabrics.

— *jaune,* yellow soap.

— *léger,* light soap, floating soap.

— *levé sur lessive en une seule opération,* soap boiled on a lye.

— *marbré,* marbled soap, mottled soap.

— *médicinal,* (*Pharm.*) white Castile soap; medicated soap.

— *métallique,* (*Ordnance*) soap containing copper sulfate, used as a lubricant.

— *mi-chaud,* see mi-chaud.

— *minéral,* soap containing mineral matter, specif, one containing powdered pumice.

savon *mixte,* mixed soap; specif., a red or blue mottled soap made by boiling two soaps together, with addition of coloring matter.

— *mou,* soft soap.

— *mousseux,* free-lathering soap.

— *noir,* black soap, specif. a dark-colored soft soap made from impure linseed or colza oil and oleic acid.

— *oriental,* a kind of toilet soap made from coconut oil and containing salt.

— *parfumé,* scented soap.

— *par refonte,* remelted soap.

— *ponce,* soap containing powdered pumice.

— *pour bain,* bath soap, specif. a floating soap.

— *pour la barbe,* shaving soap.

— *relargué,* grained soap, settled soap.

— *résineux,* rosin soap; resin soap, resinous soap.

— *roux,* brown soft soap made with wood ashes.

— *silicaté,* silicate soap, silicated soap.

— *unicolore,* unmottled soap, specif., white Marseilles soap.

— *vert,* green soap.

savonnage, *m.* soaping, etc. (see savonner).

savonner, *v.t.* soap; wash (with soap); lather; (*Glass*) rub with fine emery to remove imperfections.

savonnerie, *f.* soap making; soap factory.

savonnette, *f.* cake or ball of toilet soap.

savonneux, *a.* soapy.

savonnier, *m.* soap maker, soap boiler; soapberry tree, soapberry (*Sapindus*). — *a.* soap, of soap, relating to the manufacture of or trade in soap.

savonnière, *f.* soapwort (*Saponaria officinalis*).

savons, *pr. 1 pl.* of savoir. — *pl.* of savon.

savonule, *f.* a substance composed of an essential oil and an alkali (according to Larousse).

savourer, *v.t.* taste; relish, savor.

savoureux, *a.* savory, savoury.

saxe, *m.* Saxon porcelain, Saxony porcelain. — *f.* (*cap.*) Saxony.

saxifrage, *f.* (*Bot.*) saxifrage.

saxon, *a.* Saxon. — *m.* Saxon.

scabieuse, *f.* (*Bot.*) scabious (*Scabiosa*).

scabreux, *a.* rough; difficult, perilous.

scalpel, *m.* scalpel.

scammonée, *f.* scammony (plant and gum resin).

— *d'Alep,* Aleppo scammony.

scandinave, *a. & m.* (*cap.*) Scandinavian.

scandium, *m.* scandium.

scarificateur, *m.* scarifier.

scarifier, *v.t.* scarify.

scarlatine, *f.* (*Med.*) scarlatina, scarlet fever.

scatol, *m.* skatole, scatole.

scatolcarbonique, *a.* skatolecarboxylic.

scatoxylsulfate, *m.* skatoxylsulfate.

scatoxylsulfurique, *a.* skatoxylsulfuric.

sceau, *m.* seal.

— *de Salomon,* (*Bot.*) Solomon's seal.

— *d'or,* (*Bot.*) goldenseal (*Hydrastis canadensis*).

scellage, *m.* sealing.

scellé, *p.a.* sealed. — *m.* seal.

scellement, *m.* sealing.

à —, sealed in, sealed.

sceller, *v.t.* seal.

scelleur, *m.* sealer.

scène, *f.* scene; stage, theater.

sceptique, *a.* skeptical, skeptic. — *m.* skeptic.

sceptiquement, *adv.* skeptically.

schaff, *m.* (*Glass*) one of the tiers on which window-glass cylinders are placed.

scheelite, *f.* (*Min.*) scheelite.

scheidage, *m.* hand sorting (of ore, e.g. tin ore).

schelling, *m.* shilling.

schelotage, *m.* = schlotage.

schéma, schème, *m.* graphic or structural formula; scheme, schema, diagram.

schématique, *a.* diagrammatic, schematic.

schématiquement, *adv.* diagrammatically, schematically.

schématiser, *v.t.* represent graphically or diagrammatically, diagram.

schilling, *m.* shilling.

schiste, *m.* (*Petrog.*) schist.

— *aluné,* alum schist, alum shale.

— *ardoisier,* slate.

— *argileux,* argillaceous schist.

— *bitumineux,* bituminous schist.

— *calcaire,* calcareous schist.

— *pyriteux,* pyritic schist.

schisteux, *a.* schistose, schistous.

schlamm, *m.* (*Ore Dressing*) slime.

schlammpeter, *m.* slimes (e.g. of zinc ores).

schlich, *m.* (*Ore Dressing*) slimes, schlich.

schlot, *m.* = curain.

schlotage, *m.* (*Salt*) removal of the first deposit (sulfates) from the brine.

schloter, *v.t.* (*Salt*) remove the first deposit from (the brine).

sciage, *m.* sawing; cutting.

scie, *f.* saw; sawfish.

— *à main,* handsaw.

— *à métaux,* saw for cutting metals, as a hack saw.

— *anglaise,* scroll saw, jig saw.

— *sans fin,* band saw.

sciemment, *adv.* knowingly, consciously.

science, *f.* science.

scientifique, *a.* scientific.

scientifiquement, *adv.* scientifically.

scier, *v.t.* saw; cut (as a block of rubber, or as standing grain). — *v.r.* be sawn; be cut. — *v.i.* saw.

scierie, *f.* sawmill; (power) saw.

scieur, *m.* sawer, sawyer; cutter; reaper.

scille, f. (Bot.) squill.
— maritime, officinal squill, sea onion (Urginea maritima).
scinder, v.t. & r. split, decompose, (sometimes) dissociate; divide.
scission, f. scission, fission, splitting, division, (sometimes) dissociation; separation (of a union); schism.
scissiparité, f. (Biol.) scissiparity (schizogenesis).
sciure, f. sawdust, saw cuttings.
— de bois, sawdust (from wood).
scléroprotéine, f. scleroprotein.
sclérose, f. (Med.) sclerosis.
sclérotique, a. & r. sclerotic.
scolaire, a. school, scholastic, academic.
scombre, m. (Zoöl.) mackerel (Scomber).
scombrine, f. scombrine.
scopolamine, f. scopolamine.
scorbut, m. (Med.) scurvy.
scorbutique, a. (Med.) scorbutic.
scoriacé, a. scoriaceous.
scorie, f. slag, scoria, dross. (See note at laitier.)
— crue, raw slag, ore slag.
— de déphosphoration, basic slag, Thomas slag.
— de fer, iron slag, iron cinder.
— de forge, forge scale, hammer slag.
— de verre, glass gall, sandiver.
— douce, refining slag, rich slag.
— pauvre, poor slag.
— phosphatée, phosphate slag.
— riche, rich slag.
— (or scories) Thomas, Thomas slag.
scorification, f. scorification.
scorificatoire, m. scorifier.
scorifier, v.t. scorify. — v.r. be scorified.
scourtin, m. a press bag of coarse material. Cf. étreindelle.
scrofulaire, f. (Bot.) figwort (Scrophularia).
scrofule, f. (Med.) scrofula.
scrofuleux, a. scrofulous.
scrupule, m. scruple; scrupulousness.
scrupuleusement, adv. scrupulously.
scrupuleux, a. scrupulous.
scruter, v.t. scrutinize.
scrutin, m. ballot, balloting.
scubac, m. (Irish or Scotch) whisky, usquebaugh.
sculptage, m. carving, sculpturing.
sculpter, v.t. carve, sculpture.
sculpteur, m. sculptor.
se, pron. oneself, himself, herself, itself, themselves; to oneself, to himself, etc.; each other, one another; to each other, to one another. (A verb with se as its object may often be translated by use of the passive voice or by an intransitive verb; as, un précipité se fait, a precipitate is formed, or forms. See the various definitions of words marked v.r.)

séance, f. sitting, session, meeting, séance.
séant, p.a. sitting; becoming, fitting.
seau, m. pail, bucket; pailful, bucketful.
sébacé, a. sebaceous.
sébacique, a. sebacic.
sébate, m. sebacate, sebate.
sébifère, a. sebiferous.
sébilation, séboulation, f. grinding or mixing in a sort of bowl (sébile) in which rolls a ball.
sébile, f. wooden bowl, wooden vessel.
séborrhée, f. (Med.) seborrhea, seborrhœa.
sébum, m. (Physiol.) sebum.
sec, a. dry; dried (as fruit); (of coal) dry, lean; (of metals) dry, brittle, short; (of pitch) hard; net; short, sharp, hard, harsh; lean, spare. — m. dryness; dry material, esp. dry food. — adv. dryly; hard, sharply, harshly, etc.
à —, to dryness; dryly, dry.
mettre à —, drain, dry.
sécante, f. (Math.) secant.
sécateur, m. pruning shears, cutter.
séchage, m. drying; seasoning (as of wood).
— à l'air, air drying.
— à la vapeur, steam drying.
— au four, kiln drying.
sèche, a. fem. of sec. — f. = seiche.
séchée, f. drying; duration of drying.
sèchement, adv. dryly; sharply, harshly, coldly.
sécher, v.t. dry; dry up; season (wood); drain; heat up, warm (as a furnace). — v.r. & i. dry, become dry; dry up; wither; wither away.
sécheresse, f. dryness; harshness, coldness.
sécherie, f. drying place, drying room, drying shed; drying.
sécheur, m., **sécheuse,** f. drier, dryer.
séchoir, m. drying place, drying room; drying apparatus, drier.
second, a. second. — m. second; second floor.
en —, in second place, second.
seconde poudre, (Leather) second dusting and lying (see poudre).
secondaire, a. secondary.
seconde, f. second; second class; second proof.
secondement, adv. secondly.
seconder, v.t. second, assist, support.
secouage, secouement, m. shaking, shake.
secouer, v.t. shake; shake off; reprimand.
— v.r. be shaken; rouse oneself.
secoueur, m. shaker; (Founding) form breaker.
secoûment, m. shaking, shake.
secourir, v.t. help, aid, assist, succor.
secours, m. help, aid, assistance, relief.
secousse, f. shake, shaking, shock, jolt, jerk, blow, concussion.
secret, a. secret; discreet. — m. secret; secrecy, secretness; (Furriery) carroting agent (as a mixture of mercury and nitric acid).
en —, in secret, secretly, privately.

secrétaire, *m.* secretary; clerk.

secrètement, *adv.* secretly, privately.

sécréter, *v.t.* secrete.

sécréteur, *a.* secreting, secretory.

sécrétine, *f.* secretin.

sécrétion, *f.* secretion.

sécrétoire, *a.* secretory.

sécrétrice, *a. fem.* of sécréteur.

secteur, *m.* sector; specif., one of the units into which a city is divided for lighting, etc.; hence, a company or circuit which supplies such a unit; (*Mach.*) sector, quadrant.

— *d'alimentation,* (*Elec.*) feed circuit, city circuit.

— *électrique,* electric company; electric supply.

section, *f.* section.

sectionnel, *a.* sectional.

sectionnement, *m.* division, sectioning.

sectionner, *v.t.* cut or divide into sections, section.

secundo, *adv.* secondly.

sécurité, *f.* security, secureness; safety.

sédatif, *a. & m.* (*Med.*) sedative.

sédiment, *m.* sediment.

sédimentaire, *a.* sedimentary.

Sedlitz, *m.* Seidlitz.

séducteur, *a.* seductive. — *m.* seducer.

séduire, *v.t.* seduce; attract.

séduisant, *a.* seductive; attractive.

segment, *m.* segment.

segmentaire, *a.* segmental, segmentary.

segrégatif, *a.* segregative.

ségrégation, *f.* segregation.

seiche, *f.* (*Zoöl.*) cuttlefish.

seigle, *m.* rye (plant and grain).

— *ergoté,* — *noir,* spurred rye, rye affected with ergot.

seigneur, *m.* lord.

seille, *f.* pail, bucket, vessel.

seillon, *m.* small tub for wine drippings; wooden vessel for milk.

sein, *m.* midst; bosom; breast.

au — de, in; in the midst of.

au — de l'eau, in water, in aqueous solution.

du — de, from; from the midst of.

seing, *m.* signature.

seizaine, *f.* sixteen (or about that).

seize, *a. & m.* sixteen; sixteenth.

seizième, *a., m. & f.* sixteenth.

séjour, *m.* stay, sojourn, continuance; abode.

séjourner, *v.i.* stay, remain, continue, stand; sojourn; dwell, reside.

sel, *m.* salt.

en —, (of a solution) on the point of depositing salt.

— *admirable,* sal mirabile, Glauber's salt (sodium sulfate).

— *alembroth,* (*Old Chem.*) sal alembroth, alembroth (ammonium mercuric chloride).

sel *amer,* bitter salt, Epsom salt.

— *ammoniac,* sal ammoniac, ammonium chloride.

— *ammoniacal,* ammonium salt, ammonia salt.

— *ammoniac martial,* (*Pharm.*) ammoniated iron (iron and ammonium chlorides).

— *anglais,* Epsom salt.

— *basique,* basic salt.

— *biliaire,* bile salt.

— *blanc,* white salt.

— *calcaire,* lime salt, calcium salt.

— *chalybé,* (*Old Chem.*) iron salt.

— *chromique,* chromic salt; crystallized chromic acid.

— *commun,* common salt (sodium chloride).

— *d'absinthe,* (*Old Chem.*) salt of wormwood (an impure potassium carbonate).

— *d'argent,* silver salt.

— *de benjoin,* (*Old Chem.*) benzoic acid.

— *de canal,* (*Old Chem.*) Epsom salt.

— *de chaux,* lime salt (calcium salt).

— *de cuisine,* common salt, esp. a grade for culinary use.

— *de cuivre,* copper salt.

— *de duobus,* (*Old Chem.*) sal de duobus (potassium sulfate).

— *de fer,* iron salt.

— *de Glauber,* Glauber's salt (sodium sulfate).

— *d'Egra,* Epsom salt.

— *de magnésie,* magnesium salt, specif. Epsom salt.

— *de mercure,* mercury salt.

— *de Mohr,* Mohr's salt (ferrous ammonium sulfate).

— *de nitre,* (*Old Chem.*) saltpeter.

— *de Perse,* borax.

— *de phosphore,* microcosmic salt.

— *de plomb,* lead salt.

— *de potasse,* potassium salt, potash salt.

— *de prunelle,* sal prunelle, fused saltpeter.

— *d'Epsom,* Epsom salt (magnesium sulfate).

— *d'Epsom de Lorraine,* sodium sulfate crystallized in needles resembling those of Epsom salt.

— *de roche,* rock salt.

— *de sagesse,* (*Old Chem.*) salt of wisdom (ammonium mercuric chloride).

— *de saline,* common salt produced by evaporation as distinguished from rock salt; specif., salt prepared from deposits as distinguished from sea salt.

— *de Saturne,* (*Old Chem.*) salt of Saturn (lead acetate).

— *de Sedlitz,* Epsom salt.

— *de Seignette,* Seignette salt, Rochelle salt (sodium potasium tartrate).

— *de soude,* sodium salt, soda salt; sal soda.

— *de soude carbonaté,* sodium carbonate.

— *de soude raffiné caustique,* sodium carbonate that has been partially causticized by lime.

sel *des tombeaux,* Rochelle salt.
— *de strontiane,* strontium salt.
— *d'étain,* tin salt (the term is used specif. of crystallized stannous chloride).
— *de tartre,* salt of tartar (purified potassium carbonate).
— — *de varech,* kelp salt, kelp soda.
— *de Vichy,* sodium bicarbonate.
— *d'hiver,* Glauber's salt.
— *digestif,* digestive salt (potassium chloride).
— *d'indigo, (Dyeing)* indigo salt (a salt from which indigo is produced on the fiber by the action of caustic soda).
— *diurétique, (Old Pharm.)* diuretic salt (potassium acetate).
— *d'oseille,* salt of sorrel (acid potassium oxalate).
— *excitateur, (Elec.)* exciting salt.
— *fin,* fine salt.
— *fin-fin,* very fine salt, powdered salt.
— *fossile,* rock salt.
— *gemme,* rock salt.
— *grimpant,* a salt that " climbs " by evaporation of its solution.
— *gris,* crude common salt, specif. bay salt.
— *ignigène,* salt (sodium chloride) obtained by evaporation with artificial heat.
— *infernal, (Old Chem.)* saltpeter.
— *interne,* inner salt.
— *marin,* common salt, esp. that obtained from sea water.
— *microcosmique,* microcosmic salt ($HNaNH_4PO_4 . 4H_2O$).
— *minéral,* mineral salt, inorganic salt.
— *neutre,* neutral salt.
— *polychreste, (Old Pharm.)* polychrest salt (potassium sulfate, also Rochelle salt).
sels *de déblai,* abraum salts.
— *secret de Glauber,* ammonium sulfate.
— *sédatif, (Old Pharm.)* sedative salt (boric acid).
— *végétal, (Old Chem.)* potassium tartrate.
— *volatil d'Angleterre,* sal volatile (ammonium carbonate).
sélectif, *a.* selective.
sélection, *f.* selection.
sélénhydrique, *a.* hydroselenic.
séléniate, *m.* selenate, seleniate.
sélénié, *a.* of or combined with selenium, seleniuretted.
sélénieux, *a.* selenious, selenous.
sélénifère, *a.* seleniferous.
séléniocyanate, *m.* selenocyanate.
séléniocyanique, *a.* selenocyanic.
sélénique, *a.* selenic.
sélénite, *f. (Chem. & Min.)* selenite.
séléniteux, *a.* selenitic, (of water) containing calcium sulfate.
sélénium, *m.* selenium.
séléniure, *m.* selenide.

selle, *f.* bench; *(Med.)* stool; *(Metal.)* slacken; saddle.
sellette, *f.* rack (for rubber tubing); seat; saddle; bed, bedplate.
selon, *prep.* according to.
  c'est —, that depends.
  — *moi,* in my opinion.
  — *que,* according as.
Seltz, Selters, *n.* Selters.
  eau de —, Seltzer water.
seltzogène, *m.* seltzogene, gazogene.
semage, *m.* sowing, etc. (see semer).
semaille, *f.* sowing; seed; seed time.
semaine, *f.* week; week's work; week's pay.
semblable, *a.* like, similar; such. — *m. & f.* like, equal; fellow creature, fellow.
semblablement, *adv.* similarly; likewise, also.
semblant, *m.* appearance, semblance. — *p.pr.* of sembler.
sembler, *v.i.* appear, seem.
semé, *p.a.* sown, etc. (see semer).
semelle, *f.* sole; sill, sleeper, bed, bedplate, bottom, foot, shoe; step, pace.
semence, *f.* seed; seed pearl; small nail, tack.
  — *de canarie,* canary seed.
  — *de lin,* linseed, flaxseed.
  — *du médicinier,* physic nut.
semenceau, *m.* a sugar beet raised for the seed.
semen-contra, semencine, *f. (Pharm.)* santonica, Levant wormseed, semencontra, semencine.
semer, *v.t.* sow; scatter, strew, sprinkle, spread.
semestre, *m.* half year, six months; six months' revenue, pay, service or leave; semester.
  — *a.* of or for six months.
semestriel, *a.* semiannual; for six months.
semi-annuel, *a.* semiannual.
semicarbazide, *n.* semicarbazide.
semi-circulaire, *a.* semicircular, half round.
semidique, *a.* semidine.
semi-fluide, *a.* semifluid.
semi-hebdomadaire, *a.* semiweekly.
semi-liquide, *a.* semiliquid.
semi-mensuel, *a.* semimonthly.
séminose, *m.* seminose (*d*-mannose).
semi-organisé, *a.* semi-organized.
semis, *m.* seedling; sowing; seed bed.
semi-sphérique, *a.* hemispheric, hemispherical.
semoule, *f.* semolina; a food paste made from potatoes.
séné, *m. (Bot. & Pharm.)* senna.
  — *américain,* wild senna (*Cassia marilandica*).
  — *de la palthe,* Alexandria senna.
  — *de la pique,* India senna, Mocha senna.
  — *indigène,* bladder senna (*Colutea arborescens*).
seneçon, *m. (Bot.)* groundsel (*Senecio*).
sénegré, *m. (Bot.)* fenugreek.
sénevé, *m.* mustard (*Sinapis*), esp. black mustard (*S. nigra*).

sénevol, *m.* mustard oil, isothiocyanate.

sénilisation, *f.* a process of seasoning wood in which the sap is expelled by injection of water.

sens, *m.* sense; way, direction, side; opinion, judgment.

  de — contraire, in the opposite direction.

  en — divers, in various directions.

  en tout —, in every direction.

  — dessus dessous, upside down.

  — devant derrière, hind part before, back part first.

sensation, *f.* sensation.

sensationnel, *a.* sensational.

sensé, *a.* sensible, judicious.

sensément, *adv.* sensibly, judiciously.

sensibilisateur, *m.* sensitizer. — *a.* sensitizing.

sensibilisation, *f.* sensitization.

sensibiliser, *v.t.* sensitize.

sensibilité, *f.* sensitiveness, sensibility.

sensible, *a.* sensitive; sensible.

sensiblement, *adv.* sensitively; sensibly.

sensitif, *a.* sensitive.

sent, *pr. 3 sing.* of sentir.

sentence, *f.* sentence, decree; maxim.

senteur, *m.* odor, perfume, fragrance.

senti, *p.p.* of sentir.

sentier, *m.* path.

sentiment, *m.* feeling; sensation; sense; sentiment; interest, concern.

  au —, by estimate, by guess.

sentinelle, *f.* sentinel, sentry.

sentir, *v.t.* feel, perceive; smell; smell of; taste of; smack of, suggest, indicate. — *v.r.* be felt; feel; be conscious; feel the effects (of). — *v.i.* feel; smell.

seoir, *v.i.* sit, sit down; be becoming, become, fit. — *v.r.* sit, sit down.

séparabilité, *f.* separability.

séparable, *a.* separable.

séparage, *m.* separation; sorting.

séparant, *p.a.* separating.

séparateur, *m.* separating funnel; separator. — *a.* separating, separatory, separative.

séparatif, *a.* separative.

séparation, *f.* separation; partition.

séparatoire, *m.* separating funnel. *Obs.*

séparé, *p.a.* separated; separate; distinct.

séparément, *adv.* separately.

séparer, *v.t.* separate; sort; divide; distinguish. — *v.r.* be separated, separate.

sépia, *f.* sepia.

sept, *a. & m.* seven; seventh.

septembre, *m.* September.

septentrional, *a.* north, northern.

septicémie, *f.* (*Med.*) septicemia, septicæmia.

septicité, *f.* septicity, septic quality.

septième, *a. & m.* seventh.

septique, *a.* septic.

septum, *m.* septum.

séquelle, *f.* series.

sera, *fut. 3 sing.* of être (to be).

serait, *cond. 3 sing.* of être (to be).

sérancer, *v.t.* hackle, hatchel (as flax).

serbe, *a.* Serbian. — *m. & f.* (*cap.*) Serbian, Serb.

Serbie, *f.* Serbia, Servia.

serein, *a.* serene. — *m.* serein, night mist.

sereinage, *m.* dew retting.

séreux, *a.* serous.

sergent, *m.* cramp, clamp, screw frame; sergeant.

sériaire, *a.* serial.

séricicole, *a.* sericultural, silkworm.

sériciculture, *f.* sericulture, silkworm raising.

séricine, *f.* sericin.

série, *f.* series.

  en —, (*Elec.*) in series.

  — aliphatique, aliphatic series.

  — aromatique, aromatic series.

  — benzénique, benzene series (usually equivalent to aromatic series).

  — grasse, fatty series (aliphatic series or, specif., saturated aliphatic series).

  — purique, purine series.

  — quinoléique, quinoline series.

sériel, *a.* serial.

sérieusement, *adv.* seriously.

sérieux, *a.* serious. — *m.* seriousness.

serin, *m.,* **serine,** *f.* canary.

sérine, *f.* serine.

seringue, *f.* syringe.

seringuer, *v.t.* syringe; spray (with a pump).

sérique, *a.* of serum, serous.

serment, *m.* oath.

séro-albumine, *f.* serum albumin, seralbumin.

séro-diagnostic, *m.* serodiagnosis.

séro-globuline, *f.* serum globulin, seroglobulin.

seront, *fut. 3 pl.* of être (to be).

sérosité, *f.* serosity.

sérothérapie, *f.* serotherapy, serum therapy.

serpent, *m.* serpent, snake.

serpentaire, *f.* (*Bot.*) any of various plants reputed to cure snake bites or having a snakelike part.

  — de Virginie, Virginia snakeroot (*Aristolochia serpentaria*); (*Pharm.*) serpentaria.

serpente, *f.* silver paper, fine tissue paper.

serpenté, *a.* winding, serpentine.

serpenteau, *m.* (*Pyro.*) serpent.

serpenter, *v.i.* wind, meander.

  faire — (*Metal.*) pass back and forth thru a set of rolls.

serpentin, *m.* worm, coil. — *a.* serpentine.

  — à vapeur, steam coil.

  — de chauffage, heating coil.

  — fusant, time train.

  — rafraîchisseur, worm condenser; refrigerating coil.

  — réchauffeur, reheating coil.

serpentineux, *a.* (*Min.*) serpentinous.

serpolet, *m.* wild thyme (*Thymus serpyllum*).

serrage, *m.* tightening, etc. (see serrer); tightness; tension, strain; pressure.

de — (*Mach.*) tightening, adjusting.

serre, *f.* pressure, pressing; greenhouse.

serré, *p.a.* tightened, etc. (see serrer); tight; snug; close; compact, dense; narrow; concise. *— adv.* hard; cautiously.

serre-écrous, *m.* nut setter.

serre-étoupe, *m.* stuffing box.

serre-feu, *m.* (*Metal.*) fire screen.

serre-fil, serre-fils, *m.* (*Elec.*) binding screw, binding post.

serre-joints, *m.* clamp, cramp.

serrement, *m.* tightening, etc. (see serrer).

serrer, *v.t.* tighten, fasten, secure; clamp; screw down; close, shut; squeeze, press, compress, wedge in, lock, grip; pinch; crowd; force; contract; put away, put up, lock up; keep close to, hug; oppress; (*Metal.*) shingle. *— v.r.* be tightened, etc.; contract, shrink; crowd, press close.

serrière, *f.* (*Founding*) plug, stopple.

serrure, *f.* lock.

serrurerie, *f.* locksmithing; ironwork.

serrurier, *m.* locksmith; blacksmith.

sert, *pr. 3 sing.* of servir.

sertir, *v.t.* set, seat.

sérum, *m.* serum.

— musculaire, muscle serum.

— sanguin, blood serum.

sérumalbumine, *f.* serum albumin, seralbumin.

sérumglobuline, *f.* serum globulin.

sérumthérapie, *f.* serum therapy, serotherapy.

servante, *f.* any of several tools, as a screen for protecting a glassworker from the fire, a scraper in a powder mill, a prop or support, a bench vise, etc.; servant.

service, *m.* service; duty; branch, department; set; course; (*Soap*) addition of lye (at any one time).

serviette, *f.* towel; filter cloth; napkin.

servir, *v.t.* serve; work, operate; pay interest or rent on. *— v.r.* (with *de*) make use of, use, employ, profit by; be served, etc.; help oneself or one another. *— v.i.* serve; be in use; (with *de*) serve as, serve for.

serviteur, *m.* servant.

servomoteur, *m.* auxiliary engine or motor.

ses, *a. pl.* his, her, its.

sésame, *m.* sesame (plant or grain).

sesquifluorure, *m.* sesquifluoride (M₂F₆).

sesquiiodure, *m.* sesquiiodide (M₂I₆).

sesquioxyde, *m.* sesquioxide.

— de fer, sesquioxide of iron, ferric oxide.

— de fer hydraté, ferric hydroxide.

sesquisiliciure, *m.* sesquisilicide (apparently applied to compounds of the formula M₂Si₃).

sesquisulfure, *m.* sesquisulfide.

sesquitérébénique, *a.* sesquiterpene.

session, *f.* session, sitting.

seuil, *m.* threshold, sill, ground sill, sole.

seul, *a.* alone, only, sole, single; solitary; lonely.

— m. one, one only, only one.

seulement, *adv.* only; even; yet, but.

sève, *f.* sap; (fig.) vigor, life; (*Wine*) strength, body (or, according to some, fine quality).

sévère, *a.* severe.

sévèrement, *adv.* severely.

sévérité, *f.* severity.

séveux, *a.* of or pertaining to sap.

sévir, *v.i.* be severe, rage, prevail.

sevrage, *m.* weaning.

sevrer, *v.t.* wean; separate; deprive.

Sèvres, *m.* Sèvres porcelain or ware, Sèvres.

sexangulaire, sexangulé, *a.* hexangular, hexagonal.

sexe, *m.* sex.

sexuel, *a.* sexual.

seyait, *imp. 3 sing.* of seoir.

seyant, *p.a.* becoming, fitting.

seybertite, *m.* (*Min.*) seybertite.

S.F., *abbrev.* (sans frais) without charges.

S.G.D.G., *abbrev.* (sans garantie du gouvernement) not guaranteed by the government.

shellac, *m.* shellac.

shérardisation, *f.* sherardizing, -ization.

shérardiser, *v.t.* sherardize.

si, *conj. & m.* if. *— adv.* so; however; yes.

— bien que, — que, so that.

— fait, yes, yes indeed.

si . . . que, so . . . that; however.

sialagogue, *m.* (*Med.*) sialagogue. *— a.* sialagogic.

siamois, *a.* Siamese. *— m.* (*cap.*) Siamese.

Sibérie, *f.* Siberia.

sibérien, *a.* Siberian. *— m.* (*cap.*) Siberian.

siccatif, *a.* drying, siccative. *— m.* drying agent, drier, dryer, siccative.

— flamand, Flemish drier.

siccativité, *f.* drying power or property.

siccité, *f.* dryness.

Sicile, *f.* Sicily.

sicilien, *a.* Sicilian. *— m.* (*cap.*) Sicilian.

sidérin, *a.* of or pertaining to iron.

sidérique, *a.* pertaining to iron; sidereal.

sidérite, *f.* (*Min.*) siderite.

sidérochrome, *m.* (*Min.*) chromite.

sidérolithe, *f.* (*Min.*) siderolite.

sidérose, *f.* (*Min.*) siderite, ferrous carbonate; (*Med.*) siderosis.

sidérotechnie, *f.* siderotechny.

sidérotechnique, *a.* siderotechnic(al).

sidéroxyle, sidéroxylon, *m.* (*Bot.*) ironwood (*Sideroxylon*).

— cendré, Sideroxylon cinereum.

sidérurgie, *f.* siderurgy.

sidérurgique, *a.* siderurgical.

siècle, *m.* century; age; world.

sied, *pr. 3 sing.* of seoir.

siéent, *pr. 3 pl.* of seoir.

siège, *m.* seat; chair, stool, bench; siege; see.
— *de soupape,* valve seat.
— *social,* headquarters (of a company).

siéger, *v.i.* sit; lie, be.

sien, sienne, *pron.* one's own, his, his own, hers, her own, its, its own.
le —, one's own, his own, her own, one's property.
les siens, one's (his, her) own kindred, friends or followers.

Sienne, *n.* Siena (Italian province).

sifflement, *m.* whistling; hissing.

siffler, *v.i. & t.* whistle; hiss.

sifflet, *m.* whistle; hiss, hissing; (*Metal.*) spot, defect; bevel.

sigillé, *a.* In the phrase *terre sigillée,* terra sigillata, Lemnian earth.

signal, *m.* signal.

signalé, *p.a.* signal, remarkable; pointed out, etc. (see signaler).

signalement, *m.* description.

signaler, *v.t.* call attention to, point out, announce; describe; signalize; signal.

signalisation, *f.* signaling.

signataire, *a.* signatory, signing. — *m.* signatory, signer.

signature, *f.* signing; signature.

signaux, *m.pl.* of signal.

signe, *m.* sign; mark.
— *inverse,* opposite sign.

signer, *v.t.* sign; stamp, mark.

significatif, *a.* significant, significative.

signification, *f.* signification; legal notice.

significativement, *adv.* significantly.

signifier, *v.t. & i.* signify.

sil, *m.* ocher.

silence, *m.* silence.

silencieusement, *adv.* silently.

silencieux, *a.* silent.

Silésie, *f.* Silesia.

silésien, *a.* Silesian. — *m.* (*cap.*) Silesian.

silex, *m.* flint.
— *corné,* hornstone.
— *meulière,* burrstone, buhrstone.
— *noir,* black flint.
— *pyromaque,* flint for striking fire.
— *xyloïde,* petrified wood.

silicate, *m.* silicate.
— *de chaux,* calcium silicate, lime silicate.
— *de plomb,* lead silicate.
— *de potasse,* potassium silicate.
— *de soude,* sodium silicate.

silicaté, *a.* silicated (as soap); siliceous.

silicatisation, *f.* silicating; silicification.

silicatiser, *v.t.* silicate; silicify.

silice, *f.* silica.

silicé, *a.* containing or combined with silica.

siliceux, *a.* siliceous, silicious.

silicié, *a.* containing or combined with silicon, (of compounds and derivatives) silicon.

silicifère, *a.* siliciferous.

silicification, *f.* silicification.

silicifier, *v.t.* silicify.

silicique, *a.* silicic.

silicium, *m.* silicon.

silicium-éthyle, *m.* silicon ethyl (tetraethylsilicane, Si(C₂H₅)₄).

siliciuration, *f.* conversion into a silicide.

siliciure, *m.* silicide.

silicophényle, *m.* silicophenyl (tetraphenylsilicane, Si(C₆H₅)₄).

silico-tungstique, *a.* silicotungstic.

sillon, *m.* furrow; groove; streak; track, trail, train.

sillonner, *v.t.* furrow, plow; groove; streak.

silo, *m.* silo (usually more or less below the ground; in sugarbeet silos the material is covered with straw and earth and provided with air circulation).

silurien, *a.* (*Geol.*) Silurian.

silvine, *f.* (*Min.*) sylvite, sylvine.

simaruba, simarouba, *m.* (*Bot.*) Simarouba.

similaire, *a.* similar.

similamètre, *m.* an instrument for determining the proportion of wheat flour in a mixture.

similargent, *m.* an imitation silver.

similarité, *f.* similarity.

simili, *a. & m.* imitation. — *f.* half-tone.
en — imitation.

simili-. imitation, artificial.

similibronze, *m.* a brass imitating bronze.

similifer, *m.* a ferriferous zinc.

similigravure, *f.* process engraving, half-tone.

similimarbre, *m.* an imitation marble.

similipierre, *m.* a kind of artificial stone.

similisage, *m.* the process of imparting a silk finish, as by mercerizing or rolls.

similiser, *v.t.* give a silk finish to; specif., mercerize.

similitude, *f.* similarity, likeness, similitude.

similor, *m.* a brass imitating gold.

simple, *a.* simple; elementary; single; mere; private (soldier). — *m.* (*Pharm.*) simple.
à — *effet,* single-effect, single-acting.

simplement, *adv.* simply; plainly.

simplicité, *f.* simplicity, simpleness.

simplificateur, *m.* simplifier. — *a.* simplifying.

simplification, *f.* simplification.

simplifier, *v.t.* simplify. — *v.r.* be or become simplified.

simulacre, *m.* imitation; semblance; shadow.

simuler, *v.t.* simulate, feign, pretend.

simultané, *a.* simultaneous.

simultanéité, *f.* simultaneousness, -taneity.

simultanément, *adv.* simultaneously.

sinalbine, *f.* sinalbin.

sinapine, *f.* sinapine.

sinapique, *a.* sinapic.

sinapisé, *a.* (*Pharm.*) containing mustard.

sinapiser, *v.t.* mix with mustard, sinapize.

sinapisme, *m.* (*Pharm.*) sinapism.
— *en feuille,* mustard paper.

sinapoline, *f.* sinapoline.

sincère, *a.* sincere.

sincèrement, *adv.* sincerely.

sincérité, *f.* sincerity, sincereness.

singe, *m.* monkey; hoist, crab; windlass.

singularité, *f.* singularity.

singulier, *a.* singular.

singulièrement, *adv.* singularly.

sinistre, *a.* sinister. — *m.* disaster, loss.

sinistrorsum, *a. & adv.* counterclockwise, from right to left.

sinon, *conj.* if not, or else, otherwise, unless, except.

sinueux, *a.* sinuous, winding.

sinuosité, *f.* sinuosity; bend, turn.

sinus, *m.* sinus; (*Math.*) sine.

siphoïde, *a.* in the form of a siphon.

siphon, *m.* siphon.

siphonal, *a.* siphonal.

siphoner, *v.t.* siphon.

siphonnement, *m.* siphoning.

siphonner, *v.t.* siphon.

sippage, *m.* leather dressing (Danish process).

sirop, *m.* sirup, syrup.
— *antiscorbutique,* (*Pharm.*) compound sirup of horseradish.
— *aromatique,* (*Pharm.*) aromatic sirup.
— *balsamique,* (*Pharm.*) sirup of Tolu.
— *composé,* (*Pharm.*) compound sirup.
— *de baume de Tolu,* (*Pharm.*) sirup of Tolu.
— *de betterave,* beet sirup.
— *de capillaire,* (*Pharm.*) a sirup prepared from maidenhair fern.
— *de chaux,* (*Pharm.*) sirup of lime, sirup of calcium hydroxide.
— *de coquelicot,* (*Pharm.*) sirup of red poppy.
— *d'écorce de cerisier,* (*Pharm.*) sirup of wild cherry.
— *d'écorce de ronce,* (*Pharm.*) sirup of rubus, sirup of blackberry bark.
— *d'écorce d'orange amère,* (*Pharm.*) sirup of orange.
— *de cuisinier,* (*Pharm.*) a compound sirup of sarsaparilla.
— *de fécule,* starch sirup, glucose.
— *de fleurs d'oranger,* (*Pharm.*) sirup of orange flowers.
— *de glycose,* (*Pharm.*) sirup of glucose.
— *de gomme,* (*Pharm.*) sirup of acacia.
— *d'égout,* — *d'égouttage,* (*Sugar*) drips.
— *de hypophosphite de chaux composé,* (*Pharm.*) sirup of hypophosphites.

sirop *de phosphate de fer,* (*Pharm.*) sirup of ferrous phosphate.
— *de polygala,* (*Pharm.*) sirup of senega.
— *de ratanhia,* (*Pharm.*) sirup of krameria, sirup of rhatany.
— *de rose rouge,* (*Pharm.*) sirup of rose, sirup of red rose.
— *des hypophosphites composé,* (*Pharm.*) compound sirup of hypophosphites.
— *de suc de citron* (or *limon*), (*Pharm.*) sirup of lemon.
— *de sucre,* sugar sirup, (*Pharm.*) simple sirup, sirup.
— *d'iodure de fer,* (*Pharm.*) sirup of ferrous iodide.
— *d'ipécacuanha,* (*Pharm.*) sirup of ipecac.
— *d'orgeat,* (*Pharm.*) sirup of orgeat (sirup of almond).
— *Gibert,* (*Pharm.*) Gibert's sirup (contains mercuric and potassium iodides).
— *simple,* (*Pharm.*) simple sirup, sirup.
— *sudorifique,* (*Pharm.*) compound sirup of sarsaparilla.
— *tonique d'Easton,* (*Pharm.*) sirup of the phosphates of iron, quinine and strychnine, Easton's sirup.
— *vert,* (*Sugar*) green sirup, greens.

sirupeux, *a.* sirupy, syrupy.

sis, *p.a.* seated; situated.

sitomètre, *m.* an instrument for determining the density of cereals.

sitôt, *adv.* so soon, so quickly; as soon.
— *que,* as soon as.

situation, *f.* situation; state; report, account.

situé, *p.a.* situated; placed.

situer, *v.t.* place, locate.

six, *a. & m.* six; sixth.

sixième, *a. & m.* sixth.

sizain, sixain, *m.* half dozen, package of six.

slave, *a.* Slavic. — *m.* (*cap.*) Slav.

smalt, *m.* smalt.

smaltine, *f.* (*Min.*) smaltite, smaltine.

smaragdite, *f.* (*Min.*) smaragdite.

smectique, *a.* smegmatic, detersive.

smectite, *f.* fuller's earth.

smegma, *m.* (*Physiol.*) smegma.

smithsonite, *f.* (*Min.*) smithsonite.

S.-O., *abbrev.* (Sud-Ouest) southwest.

sobre, *a.* sober, temperate, moderate, sparing.

social, *a.* social; (*Com.*) company, firm.

sociétaire, *m.* member (of a society or association), associate, partner.

société, *f.* society; association; (*Com.*) company, firm, partnership, house.
— *anonyme,* stock company (so called because usually conducted under a name indicating its object rather than its members).
— *civile,* an association not engaged in business.

société *commerciale*, commercial company or partnership.

— *de capitaux*, stock company.

— *de personnes*, partnership.

— *en commandite*, see commandite.

— *en nom collectif*, a firm doing business under the name of its members (as, *Jean, Louis et C*ᵗᵉ), who are personally responsible.

— *en participation*, coöperative society or association.

— *par actions*, any company or partnership the capital of which is divided into transferable shares.

socle, *m.* base, bottom, stand, foot, footing, foundation, pedestal.

socotrin, *a.* (of aloes) Sokotrine.

soda, *m.* soda water, soda (the drink).

sodalite, *f.* (*Min.*) sodalite.

sodé, *a.* combined with sodium, sodium, sodio-.

sodico-. sodium, sodio-.

sodico-ammonique, *a.* sodium ammonium.

sodique, *a.* sodium, of sodium; containing soda.

sodium, *m.* sodium.

sœur, *f.* sister.

soi, *pron.* self, oneself, himself, herself, itself, themselves.

chez —, at home, home, in one's own house or country.

soi-disant, *a.* so-called, pretended, self-styled.

— *adv.* supposedly, pretendedly.

soie, *f.* silk; bristle; (*Mach.*) journal pin.

— *à bluter*, bolting cloth.

— *artificielle*, artificial silk.

— *cuite*, boiled silk.

— *de porc*, hog bristle.

— *écrue*, — *grège*, raw silk.

— *légis*, Persian silk, Persian.

— *marine*, byssus silk.

— *moulinée*, — *ouvrée*, thrown silk.

soient, *pr. 3 pl. subj.* of être (to be).

soierie, *f.* silk fabric, silk (*pl.*) silk goods; silk mill; silk manufacture; silk trade.

soif, *f.* thirst.

soigné, *p.a.* cared for; attended to; well made, finished; (*Med.*) treated.

soigner, *v.t.* care for; attend to; do or execute well or carefully; (*Med.*) treat.

soigneur, *m.* caretaker, attendant.

soigneusement, *adv.* carefully, with care.

soigneux, *a.* careful.

soi-même, *pron.* oneself, self, itself, himself, herself.

soin, *m.* care; attention; (*Med., pl.*) care, attendance.

soir, *m.* afternoon; evening; period from noon to midnight.

soirée, *f.* evening; evening party.

soit, *pr. 3 sing. subj. & imper.* of être: be, may be; be it so; let . . . be, suppose that . . . is. — *conj.* either, or; say, let us say.

soit . . . *soit*, either . . . or, whether . . . or.

soixantaine, *f.* sixty (or about that), threescore.

soixante, *a.* sixty; sixtieth. — *m.* sixty.

— *et onze*, seventy-one.

soixante-dix, *a.* seventy.

soixante-dixième, *a. & m.* seventieth.

soixante-douze, *a.* seventy-two.

soixante-quatorze, *a.* seventy-four.

soixante-quinze, *a.* seventy-five.

soixante-seize, *a.* seventy-six.

soixante-treize, *a.* seventy-three.

soixantième, *a. & m.* sixtieth.

soja, *n.* soy bean, soja bean, soja.

sol, *m.* soil; ground; bottom; (*Physical Chem.*) sol.

solaire, *a.* solar.

solamire, *f.* sieve cloth.

solanées, **solanacées**, *f.pl.* (*Bot.*) Solanaceæ, nightshade family.

solanine, *f.* solanine.

solanomètre, *m.* an instrument for determining the starch content of potatoes.

solanum, *m.* (*Bot.*) solanum, nightshade.

solariser, *v.t.* (*Photog.*) solarize.

soldat, *m.* soldier. — *a.* soldierly.

solde, *m.* remainder; balance; job lot. — *f.* pay.

solder, *v.t.* pay; settle; sell at a reduction.

sole, *f.* sole; bottom, bed, hearth (of a furnace), ground plate, sleeper, sill; plot, field.

soleil, *m.* sun; (*Bot.*) sunflower.

soleilleux, *a.* sunny.

soléine, *f.* a resin spirit used as an illuminant.

solennel, *a.* solemn.

solénoïde, *m.* (*Elec.*) solenoid.

solfatare, *f.* (*Geol.*) solfatara; sulfur mine.

solfatarien, *a.* (*Geol.*) solfataric.

solidaire, *a.* integral, in one piece (with); (*Law*) jointly and severally responsible.

solidarité, *f.* solidarity, community; (*Law*) joint and several responsibility.

solide, *a.* solid; (of colors) lasting, fast; steadfast, firm. — *m.* solid.

solidement, *adv.* solidly.

solidificateur, *m.* solidifier. — *a.* solidifying.

solidification, *f.* solidification, -fying.

solidifier, *v.t. & r.* solidify.

solidité, *f.* solidity.

solitaire, *a.* solitary.

solive, *f.* joist.

solliciter, *v.t.* solicit; (of forces, etc.) influence, act on; excite, urge.

solubilisation, *f.* rendering soluble.

solubiliser, *v.t.* render soluble.

solubilité, *f.* solubility.

soluble, *a.* soluble.

soluté, *m.* (*Pharm.*) solution.

— *de Burnett*, solution of zinc chloride.

soluté *de chaux*, lime water.

— *de chlore*, compound solution of chlorine, chlorine water.

— *de citrate de bismuth ammoniacal*, solution of bismuth and ammonium citrate.

— *de pancréatine*, pancreatic solution.

— *de perchlorure de fer*, solution of ferric chloride.

— *d'hypochlorite de chaux*, solution of chlorinated lime.

— *d'iodo-arsénite de mercure*, solution of arsenous and mercuric iodides.

— *ioduré de Lugol*, compound solution of iodine, Lugol's solution.

solutif, *a.* capable of dissolving; laxative.

solution, *f.* solution.

mettre en —, dissolve.

— *empirique*, a standard solution of arbitrary strength (as contrasted with a normal solution).

— *de continuité*, breach of continuity.

— *normale*, normal solution.

— *type*, standard solution.

solutol, *m.* solutol (an antiseptic).

solvabilité, *f.* solvency.

solvant, *m.* solvent.

solvent-naphta, *m.* solvent naphtha, benzine (from coal tar).

sombre, *a.* dark; dull; somber; melancholy. — *m.* darkness; sadness.

sommaire, *a.* summary, brief. — *m.* summary, abstract.

sommairement, *adv.* summarily, briefly.

sommation, *f.* summons; (*Math.*) summation.

somme, *f.* sum; amount; epitome; burden, load. — *m.* sleep; nap.

en —, — *toute*, on the whole, in short, after all.

sommeil, *m.* sleep.

sommer, *v.t.* summon; (*Math.*) sum up, sum, add.

sommes, *pr. 1 pl.* of être (to be).

sommet, *m.* summit; vertex; apex; top; crown.

sommier, *m.* support (of various kinds); bed (of a machine); beam (of a balance); girder; mattress; record book; pack animal.

sommité, *f.* top, summit; extremity (of a plant); eminent person.

somnifère, *a. & m.* (*Med.*) soporific.

somptueux, *a.* sumptuous.

son, *m.* bran; sound. — *pron.* his, her, its, one's.

sondage, *m.* boring, etc. (see sonder).

sonde, *f.* borer, boring tool, drill; sampler (as for soil or flour); taster (as for cheese); proof stick; sound, probe; sounding line, lead; sounding.

sonder, *v.t.* bore; examine by boring; sample (as soil); taste (as cheese); sound; probe; examine, investigate, test, try.

songe, *m.* dream.

songer, *v.i.* dream; (with *à*) think of. — *v.t.* dream.

sonnant, *p.a.* sounding, ringing, sonorous; (of clocks) striking.

sonner, *v.i. & t.* sound, ring, strike. *faire* —, sound, ring.

sonnerie, *f.* bell; set of bells; ringing; call.

sonnette, *f.* (small) bell; pile driver: monkey, ram; shaker (as in making explosives).

sonore, *a.* sonorous.

sonorité, *f.* sonorousness.

sont, *pr. 3 pl.* of être (to be).

sophistication, *f.* sophistication, adulteration.

sophistiquer, *v.t.* sophisticate, adulterate.

soporatif, soporifère, soporifique, *a. & m.* (*Med.*) soporific.

sorbe, *f.* sorb, sorb apple (berry of any species of *Sorbus*, see sorbier).

sorbet, *m.* sherbet.

sorbier, *m.* (*Bot.*) any plant of the genus *Sorbus*, sorb.

— *domestique*, service tree (*S. domestica*).

— *des oiseaux*, — *des oiseleurs*, mountain ash, rowan tree (*S. aucuparia*).

sorbiérite, *f.* sorbieritol, sorbierite (*d*-iditol).

sorbine, *f.* sorbin (sorbose).

sorbique, *a.* sorbic.

sorbite, *f.* sorbitol, sorbite.

sorbonne, *f.* (laboratory) hood; (*cap.*) Sorbonne.

sorbose, *m.* sorbose.

sorcellerie, *f.* sorcery.

sorcier, *m.* sorcerer.

sorgho, sorgo, *m.* sorghum.

— *sucré*, the variety of sorghum from which sorghum sirup is made.

— *commun*, broom corn.

sorne, *f.* (*Metal.*) dross, scoria.

sort, *m.* fate; lot; chance; charm. — *pr. 3 sing.* of sortir.

sorte, *f.* sort, kind; way, manner.

de — *que, en* — *que*, so that, so as.

sortent, *pr. 3 pl.* of sortir.

sorteur, *m.* sorter.

sortie, *f.* going out, coming out; outlet; mouth; issue; exit, egress; leaving, departure; retirement; export, exportation; sortie, sally; outburst.

sortir, *v.i.* go out, come out; get out; be out; issue; escape; depart; leave; retire; proceed; come; project, stand out; to have just gone or left; — *v.t.* take out, bring out, get out. — *m.* leaving.

au — *de*, on leaving, at the end of.

faire —, bring out, take out, get out, draw out, call out, turn out, put out.

sot, *a.* stupid, foolish, silly. — *m.* fool, dolt.

sottise, *f.* foolishness, folly, stupidity.

Souabe, *f.* Swabia.

soubassement, *m.* basement.

soubresaut, *m.* bump (as in boiling), start, shock, jolt, jerk.

soubresauter, *v.i.* bump (as in boiling), start, jolt, jerk.

souche, *f.* stump; stock; stub; stem.

souchon, *m.* small stump; short thick bar of iron.

souci, *m.* care, anxiety; (*Bot.*) marigold.

soucier, *v.r.* care, concern oneself.

soucoupe, *f.* saucer.

soudabilité, *f.* weldability; capability of being soldered.

soudable, *a.* weldable; capable of being soldered.

soudage, *m.* welding; soldering; joining, union.

soudain, *a.* sudden. — *adv.* at once, forthwith.

soudainement, *adv.* suddenly, all of a sudden.

soudaineté, *f.* suddenness.

soude, *f.* soda; (*Bot.*) any plant of the genus Salsola.

— *à la chaux,* caustic soda obtained by treating the carbonate with lime.

— *à l'alcool,* caustic soda purified by solution in alcohol, soda from alcohol.

— *à l'ammoniaque,* soda made by the ammonia process, ammonia soda, Solvay soda.

— *artificielle,* artificial soda (not occurring as a mineral or made from plant ashes).

— *boratée,* borax.

— *brute,* crude soda, black ash.

— *carbonatée,* carbonate of soda, sodium carbonate.

— *caustique,* caustic soda, sodium hydroxide.

— *caustique liquide,* (*Pharm.*) solution of sodium hydroxide.

— *d'Alicante,* Alicante soda, barilla.

— *nitratée,* nitrate of soda, sodium nitrate.

— *sulfatée,* sulfate of soda, sodium sulfate.

— *tartarisée,* potassium sodium tartrate, Rochelle salt.

— *végétale,* soda obtained from the ashes of plants, vegetable soda.

souder, *v.t.* weld; solder; unite, join (as atoms). — *v.r.* be welded, etc.; weld; unite; join; coalesce; consolidate.

— *à bout,* butt-weld, jump-weld.

— *à chaud,* weld.

— *à l'autogène,* burn on, solder autogenically.

— *à recouvrement,* lap-weld.

— *au cuivre,* braze, hard-solder.

soudeur, *m.* welder, etc. (see souder).

soudier, *a.* soda, of soda. — *m.* soda maker.

soudière, *f.* soda factory, soda works.

soudoir, *m.* soldering iron.

soudure, *f.* solder; welding; soldering; union; joining; weld; soldered place; seam, suture, joint; stiff plaster.

sans —, without weld or seam.

soudure *à la résine,* rosin soldering.

— *au plomb,* lead solder.

— *autogène,* autogenic (or autogenous) soldering or welding.

— *d'argent,* silver solder.

— *de cuivre,* hard solder.

— *de laiton,* brass solder, hard solder.

— *des plombiers,* plumbers' solder (2–3 parts of lead to 1 of tin).

— *d'étain,* tin solder, soft solder.

— *électrique,* electric welding.

— *fondante,* soft solder.

— *forte,* hard solder; hard soldering.

— *grasse,* a solder relatively rich in tin.

— *maigre,* a solder relatively poor in tin.

— *non-autogène,* soldering proper (i.e., with the use of other material as a solder).

— *tendre,* soft solder; soft soldering.

souffert, *p.p.* of souffrir.

soufflage, *m.* blowing, etc. (see souffler); specif., glass blowing; blast.

— *en plat,* (*Glass*) crown-glass blowing.

soufflant, *p.a.* blowing, etc. (see souffler).

soufflante, *f.* blowing engine, blower.

soufflard, *m.* volcanic steam jet.

souffle, *m.* breath; breathing; exhalation; blowing; blast; blasting; puff.

soufflement, *m.* blowing, etc. (see souffler).

souffler, *v.i.* blow; breathe; pant. — *v.t.* blow; blow out; blow away, blow off; inflate, blow up; breathe; prompt; sheathe (a ship). — *v.r.* be blown, etc.

— *sur,* blow upon; blow out; destroy.

soufflerie, *f.* blowing engine, blowing machine, blower; bellows; blast; alchemical work.

— *hydraulique,* hydraulic blower, water blast.

soufflet, *m.* bellows; blowing machine, blower; fanner, fan; slap.

— *à bras,* hand bellows, hand blower.

— *à pédale,* foot bellows, foot blower.

soufflette, *f.* (*Ceram.*) air hole, air bubble.

souffleur, *m.* blower, etc. (see souffler); specif., glassblower; alchemist (*Obs.*).

— *de bouteilles,* bottle blower.

— *de boyaux,* catgut blower.

— *de grande place,* (*Glass*) head blower.

— *de paraison,* = grand gamin, under *grand.*

— *de verre,* glass blower.

soufflure, *f.* air hole, blowhole, blister, bleb.

souffrance, *f.* suffering; suspense.

souffrir, *v.t.* suffer; endure, bear; experience. — *v.i.* suffer. — *v.r.* be suffered, etc.

soufrage, *m.* sulfuring, sulphuring.

soufre, *m.* sulfur, sulphur.

— *brut,* crude sulfur.

— *de mine,* native sulfur.

— *doré d'antimoine,* (*Pharm.*) sulfurated antimony.

— *en bâtons,* stick sulfur, roll sulfur.

— *en canons,* roll sulfur, stick sulfur.

soufre *en fleur(s)*, flowers of sulfur.
— *précipité*, precipitated sulfur, (*Pharm.*)
milk of sulfur.
— *rouge*, realgar (old name).
— *sublimé*, sublimed sulfur, flowers of sulfur.
— *sublimé lavé*, (*Pharm.*) washed sulfur.
— *végétal*, vegetable sulfur (lycopodium
powder).
— *vierge*, — *vif*, virgin sulfur, native sulfur.
soufrer, *v.t.* sulfur, sulphur.
soufrière, *f.* sulfur mine, sulfur pit; (*Matches*)
place where the sulfur is kept.
soufroir, *m.* sulfuring stove, sulfur stove.
souhait, *m.* wish, desire.
à —, according to one's wishes, as one would
wish.
souhaitable, *a.* desirable.
souhaiter, *v.t.* wish, wish for, desire.
souillarde, *f.* (*Soap*) lye tub, lye vat.
souiller, *v.t.* contaminate; soil; (fig.) stain.
souillure, *f.* contamination; stain, spot.
soûl, *a.* surfeited, satiated, full.
soulagement, *m.* alleviation, relief, aid.
soulager, *v.t.* alleviate, relieve, aid; lighten.
soulèvement, *m.* rising; breaking up; upris-
ing.
soulever, *v.t.* raise; lift, lift up; break up (as
the contents of a furnace); rouse. — *v.r.*
rise, rise up; be raised, etc.
soulier, *m.* shoe.
soulignement, *m.* underlining; emphasis.
souligner, *v.t.* underline; emphasize.
soumettre, *v.t.* subject; submit; refer; force;
subdue, overcome. — *v.r.* be subjected,
etc.; submit, yield; consent.
soumis, *p.a.* subjected, etc. (see soumettre);
subject; submissive.
soumission, *f.* submission; engagement; bond;
tender; bid; submissiveness.
soumissionaire, *m.* bidder, tendering party.
soupape, *f.* valve.
à —, valved, with valve(s).
— *d'arrêt*, check valve, cut-off valve.
— *d'arrivée*, inlet valve.
— *de sûreté*, safety valve.
soupçon, *m.* suspicion; conjecture; touch, dash.
soupçonnable, *a.* suspicious.
soupçonner, *v.t.* suspect.
soupçonneusement, *adv.* suspiciously.
soupçonneux, *a.* suspicious.
soupe, *f.* soup; meal, dinner.
soupeser, *v.t.* try the weight of.
soupière, *f.* soup tureen.
soupirail, *m.* air hole, vent hole, vent.
soupirer, *v.i.* sigh.
souple, *a.* pliable, pliant, flexible, limber, supple, tough.
souplesse, *f.* flexibility, etc. (see souple).
source, *f.* source; spring (as of water or oil).
sourcil, *m.* eyebrow.

sourd, *a.* deaf; dull; (of gems) cloudy; secret,
dark. — *m.* deaf person.
sourdement, *adv.* low, with a rumbling or hollow sound; secretly.
sourdre, *v.i.* spring, spring up, spring forth.
sourire, *v.i. & m.* smile.
souris, *m.* mouse; smile.
sous, *prep.* under, beneath, below.
sous-. sub-, under-, assistant.
sous-acétate, *m.* subacetate.
— *de cuivre*, copper subacetate.
— *de plomb*, lead subacetate.
— *de plomb liquide*, (*Pharm.*) solution of lead
subacetate.
sous-azotate, *m.* subnitrate.
— *de bismuth*, bismuth subnitrate (bismuth
oxynitrate).
sous-carbonate, *m.* subcarbonate, basic carbonate.
sous-chlorure, *m.* subchloride.
sous-classe, *f.* subclass.
sous-comité, *m.* subcommittee.
sous-couche, *f.* substratum, underlying layer.
souscripteur, *m.* subscriber.
souscription, *f.* subscription.
souscrire, *v.t. & i.* subscribe.
sous-cutané, *a.* subcutaneous.
sous-diviser, *v.t.* subdivide.
sous-domaine, *m.* subdomain.
sous-égaliser, *v.t.* pass thru a fine sieve.
sous-élément, *m.* secondary element.
sous-épidermique, *a.* subepidermal.
sous-espèce, *f.* subspecies.
sous-fluorure, *m.* subfluoride.
sousgallate, *m.* subgallate.
sous-genre, *m.* subgenus.
sous-granitique, *a.* subgranitic.
sous-groupe, **sous-groupement**, *m.* sub-
group.
sous-jacent, *a.* subjacent, underlying.
sous-marin, *a. & m.* submarine.
sous-multiple, *a. & m.* submultiple.
sous-muriate, *m.* subchloride.
— *de mercure*, mercurous chloride.
sous-nitrate, *m.* subnitrate, basic nitrate.
— *de bismuth*, bismuth subnitrate (bismuth
oxynitrate).
sous-oxyde, *m.* suboxide.
sous-préfect, *m.* subprefect.
sous-pression, *f.* pressure from below.
sous-produit, *m.* by-product.
sous-région, *f.* subregion.
sous-salicylate, *m.* subsalicylate.
sous-sel, *m.* subsalt, (usually) basic salt.
sous-siliciure, *m.* subsilicide.
sous-sol, *m.* subsoil; basement.
sous-sulfure, *m.* subsulfide.
sous-tangente, *f.* (*Math.*) subtangent.
sous-tendre, *v.t.* (*Math.*) subtend.
sous-titre, *m.* subtitle, subhead.

soustraction, *f.* removal, etc. (see soustraire).

soustraire, *v.t.* remove, abstract, withdraw; save, protect; (*Arith.*) subtract. — *v.r.* (with *à*) escape, avoid.

sous-traitant, *m.* subcontractor.

soustrate, *m.* substrate.

soute, *f.* storeroom, room, bunker, locker, compartment, tank, magazine.

soutenir, *v.t.* sustain, support, maintain, bear; resist. — *v.r.* be sustained, etc.; be firm, be stiff; last; (of colors) hold; succeed; continue; keep up; resist; stand up; sit up; support oneself.

soutenu, *p.a.* sustained, etc. (see soutenir).

souterrain, *a.* underground, subterranean. — *m.* underground room or passage.

soutien, *m.* support.

soutient, *pr. 3 sing.* of soutenir.

soutirage, *m.* drawing off, racking off, racking.

soutirer, *v.t.* draw off, rack off, rack.

souvenir, *v.r.* (often with *de*) remember, recollect. — *v.i.* come to mind. — *m.* remembrance, recollection; memory; memorial; souvenir; memorandum book.

il me souvient, I remember, that reminds me.

souvent, *adv.* often, frequently.

souverain, *a. & m.* sovereign.

souveraineté, *f.* sovereignty.

souvient, *pr. 3 sing.* of souvenir.

soya, *m.* soy bean, soja, soya.

soyeux, *a.* silky; silken, silk.

sozoiodol, *m.* sozoiodol.

spacieux, *a.* spacious, large, ample, roomy.

spadelle, spadèle, *f.* stirring tool, stirrer.

spagirie, *f.* alchemy. (*Obs.*)

spagirique, *a.* spagyric(al), alchemical.

spagirisme, *m.* alchemy, chemistry, specif. iatrochemistry. (*Obs.*)

spagiriste, *m.* spagyrist. (*Obs.*)

spalt, *m.* asphalt; (*Metal.*) fluor spar.

sparadrap, *m.* (*Med.*) sparadrap, adhesive plaster.
— *de capsique,* capsicum plaster.

sparadrapier, *m.* an instrument for making sparadraps or adhesive plasters.

spart, sparte, *m.* esparto, esparto grass, (more generally) matweed.

spartéine, *f.* sparteine.

spasmodique, *a.* spasmodic.

spath, *m.* (*Min.*) spar, specif. calc-spar.
— *brunissant,* ankerite.
— *calcaire,* calc-spar (calcite).
— *d'Islande,* Iceland spar (transparent calcite).
— *fluor,* fluor spar.
— *pesant,* — *lourd,* heavy spar (barite).

spathification, *f.* conversion into spar or stony material (petrifaction?).

spathifier, *v.t.* convert into spar (petrify?).

spathique, *a.* (*Min.*) spathic.

spatial, *a.* spatial.

spatialiser, *v.t.* spatialize.

spatialité, *f.* spatiality.

spatulage, *m.* stirring, etc. (see spatuler).

spatule, *f.* spatula, spattle; paddle; stirrer.
— *en acier,* steel spatula or stirrer.
— *en argent,* silver spatula.
— *en bois,* wooden spatula, paddle or stirrer.
— *en corne,* horn spatula.
— *en os,* bone spatula.
— *en platine,* platinum spatula.
— *en porcelaine,* porcelain spatula.
— *en verre,* glass spatula.

spatulé, *p.a.* stirred, worked; (of flax, etc.) hatcheled; spatulate, spoon-shaped.

spatuler, *v.t.* stir or work with a spatula or paddle; hatchel (as flax).

spécial, *a.* special; professional.

spécialement, *adv.* specially, especially.

spécialiser, *v.t.* specify. — *v.r.* specialize.

spécialiste, *m.* specialist. — *a.* specializing.

spécialité, *f.* specialty; specialist; speciality, particularity; special expense.

spécieux, *a.* specious.

spécification, *f.* specification.

spécificité, *f.* specificity, specificness.

spécifier, *v.t.* specify.

spécifique, *a. & m.* specific.

spécifiquement, *adv.* specifically.

spécimen, *m.* specimen, sample. — *a.* sample.

spectral, *a.* spectral.

spectre, *m.* spectrum; specter, spectre.
— *continu,* continuous spectrum.
— *d'absorption,* absorption spectrum.
— *de bandes,* band spectrum.
— *de haute fréquence,* high-frequency spectrum.
— *d'émission,* emission spectrum.
— *discontinu,* discontinuous spectrum.
— *solaire,* solar spectrum.
— *stellaire,* stellar spectrum.

spectrographe, *m.* spectrograph.

spectromètre, *m.* spectrometer.

spectrométrie, *f.* spectrometry.

spectrométrique, *a.* spectrometric(al).

spectrophotomètre, *m.* spectrophotometer.

spectrophotométrie, *f.* spectrophotometry.

spectrophotométrique, *a.* spectrophotometric(al).

spectroscope, *m.* spectroscope.
— *à main,* hand spectroscope.

spectroscopie, *f.* spectroscopy.

spectroscopique, *a.* spectroscopic(al).

spectroscopiquement, *adv.* spectroscopically.

spectroscopiste, *m.* spectroscopist.

spéculaire, *a.* specular.

spéculatif, *a.* speculative.

spéculation, *f.* speculation.

spéculer, *v.i.* speculate.

speiss, *m.* (*Metal.*) speiss.

speisscobalt, *m.* (*Min.*) speisscobalt (smaltite).

speltre, *m.* spelter.

spermaceti, spermacéti, sperma coeti, *m.* spermaceti.

spermatine, *f.* spermatin.

spermatique, *a.* (*Physiol.*) spermatic.

spermatozoaire, *m.* (*Zoöl.*) spermatozoön.

sperme, *m.* (*Physiol.*) sperm.

spermine, *f.* spermine.

sphène, *m.* (*Min.*) sphene (titanite).

sphère, *f.* sphere.

sphéricité, *f.* sphericity.

sphérique, *a.* spherical, spheric.

sphériquement, *adv.* spherically.

sphéroïdal, *a.* spheroidal.

sphéroïde, *m.* spheroid.

sphérosidérite, *m.* (*Min.*) spherosiderite, sphærosiderite.

spic, *m.* (*Bot.*) spike lavender, spike (*Lavandula spica*).

spiculaire, *a.* spicular, dartlike.

spiegel, *m.* (*Metal.*) spiegeleisen, spiegel.

spigélie, *f.* (*Bot.*) Spigelia.
— *du Maryland,* pinkroot, Carolina pink (*S. marilandica*).

spinelle, *m.* & *a.* (*Min.*) spinel.

spiral, *a.* spiral. — *m.* spiral spring.

spirale, *f.* (*Math.*) spiral.

spire, *f.* spire, turn, winding, coil.

spirée, *f.* (*Bot.*) spirea, spiræa.

spiritualiser, *v.t.* (*Old Chem.*) spiritualize.

spiritueux, *a.* spirituous. — *m.* spirituous liquor.

spirituosité, *f.* spirituousness, spirituosity.

spiroïdal, spiroïde, *a.* spiroid.

spirol, *m.* phenol. (*Obs.*)

spitzkasten, *m.* (*Mining*) funnel box, spitzkasten.

splendeur, *m.* splendor.

splendide, *a.* splendid.

spode, *f.* spodium (old name for various ashes and sublimates; still in use for boneblack).

spodite, *f.* volcanic ash.

spongieux, *a.* spongy.

spongine, *f.* spongin.

spongiosité, *f.* sponginess.

spontané, *a.* spontaneous.

spontanéité, *f.* spontaneity, spontaneousness.

spontanément, *adv.* spontaneously.

sporadique, *a.* sporadic.

sporadiquement, *adv.* sporadically.

sporange, *m.* (*Bot.*) sporangium.

spore, *f.* (*Bot.*) spore.

sporuler, *v.i.* (*Biol.*) sporulate.

spumeux, *a.* spumous, spumy, frothy, foamy.

squame, *f.* (small) scale.

squameux, squammeux, *a.* squamous, scaly.

squelette, *f.* skeleton.

squine, *f.* chinaroot (rootstock of *Smilax china*).

stabilisateur, *m.* stabilizer. — *a.* stabilizing.

stabilisation, *f.* stabilization.

stabiliser, *v.t.* stabilize. — *v.r.* be stabilized.

stabilité, *f.* stability.

stable, *a.* stable.

stablement, *adv.* stably, firmly, fixedly.

stade, *m.* stage; stadium.

stage, *m.* term (as of probation); course.
faire un —, take a course; pass thru a term.

stagnant, *a.* stagnant, standing.

stagner, *v.i.* stagnate, stand.

stalactite, *f.* stalactite.

stalactitique, *a.* stalactitic.

stalagmite, *f.* stalagmite.

stalagmomètre, *m.* stalagmometer.

stalagmométrique, *a.* stalagmometric(al).

stalle, *f.* (*Mining*) stall.

stampe, *f.* (*Mining*) interval between two veins.

standiéthyle, *m.* tin diethyl, $Sn(C_2H_5)_2$.

stanéthyle, *m.* = stannéthyle.

stanméthyle, *m.* any methyl compound of tin, esp. tetramethylstannane, $Sn(CH_3)_4$.

stannage, *m.* (*Dyeing*) tin mordanting.

stannate, *m.* stannate.

stannéthyle, *m.* any ethyl compound of tin, esp. tetraethylstannane, $Sn(C_2H_5)_4$.

stanneux, *a.* stannous.

stannifère, *a.* stanniferous.

stannine, *f.* (*Min.*) stannite.

stannique, *a.* stannic.

stantétraéthyle, *m.* tin tetraethyl (tetraethylstannane, $Sn(C_2H_5)_4$.

stantriéthyle, *m.* tin triethyl.

staphisaigre, *f.* (*Bot.*) stavesacre (*Delphinium staphisagria*).

staphylocoque, *m.* (*Bact.*) staphylococcus.

staticité, *f.* static condition or quality.

statif, *m.* stand (of a microscope). — *a.* of or pertaining to a station.

station, *f.* station; stop; short stay; standing.

stationnaire, *a.* stationary.

stationnement, *m.* stopping; standing.

stationner, *v.i.* stop; stand.

statique, *a.* static, statical. — *f.* statics.

statistique, *f.* statistics, statistic. — *a.* statistical.

statuaire, *a.*, *f.* & *m.* statuary.

staurolithe, staurotide, *f.* (*Min.*) staurolite.

stéapsine, *f.* steapsin.

stéarane, *m.* normal octadecane, $C_{18}H_{38}$.

stéarate, *m.* stearate.

stéarine, *f.* stearin.

stéariner, *v.t.* coat or impregnate with stearin.

stéarinerie, *f.* stearin manufacture or factory.

stéarinier, *m.* stearin manufacturer.

stéarique, *a.* stearic (of candles, etc.) stearin.

stéarolé, *m.* (*Pharm.*) a preparation having a solid fat as base.

stéarone, *f.* stearone ($C_{17}H_{35}$)$_2CO$.

stéaroptène, *f.* stearoptene.

stéatite, *f.* (*Min.*) steatite.
stéatiteux, *a.* steatitic.
stellaire, *a.* stellar.
sténographe, *m.* stenographer.
stercoraire, *a.* stercoraceous, relating to dung.
stère, *m.* stere, cubic meter (1.308 cu. yds.).
stéréochimie, *f.* stereochemistry.
stéréochimique, *a.* stereochemical.
stéréoisomère, *a.* stereoisomeric. — *m.* stereo-isomer.
stéréoisomérie, *f.* stereoisomerism.
stéréoscopique, *a.* stereoscopic, stereoscopical.
stéréotypage, *m.* stereotyping.
stéréotype, *a.* stereotype, stereotyped.
stéréotyper, *v.t.* stereotype.
stéréotypie, *f.* stereotypy, stereotyping.
stérile, *a.* sterile. — *m.* sterile material, esp. in ore dressing.
stérilement, *adv.* sterilely.
stérilisateur, *m.* sterilizer.
stérilisation, *f.* sterilization.
stériliser, *v.t.* sterilize.
stérilité, *f.* sterility.
stérilobaril, *m.* a barrel-shaped glass receptacle for sterilized water.
sternutatif, sternutatoire, *a.* & *m.* (*Med.*) sternutative, sternutatory.
sthénosage, *m.* strengthening (as of artificial silk by formaldehyde).
stibial, *a.* stibial, antimonial.
stibié, *a.* stibiated, combined with antimony.
stibieux, *a.* stibious (antimonious).
stibine, *f.* stibine; (*Min.*) stibnite.
stibiure, *m.* antimonide.
stigmate, *m.* stigma; mark; scar.
stigmatiser, *v.t.* stigmatize.
stilbène, *m.* stilbene.
stilbénique, *a.* stilbene, of stilbene.
stilbite, *f.* (*Min.*) stilbite.
stil-de-grain, *m.*, **stil de grain.** stil de grain (a yellow lake).
stillant, *a.* dropping, dripping.
stillation, *f.* dropping, dripping.
stimulant, *p.a.* stimulating, stimulant. — *m.* stimulant.
stimulation, *f.* stimulation, stimulating.
stimuler, *v.t.* stimulate.
stipuler, *v.t.* stipulate.
stock, *m.* (*Com. & Finance*) stock.
stockfish, *m.* stockfish.
stœchiogénie, *f.* stoichiogeny, origin of the elements.
stœchiogénique, *a.* stoichiogenic.
stœchiologie, *a.* stoichiology.
stœchiologique, *a.* stoichiological.
stœchiométrie, *f.* stoichiometry.
stœchiométrique, *a.* stoichiometric(al).
stomacal, *a.* (*Anat.*) of the stomach, stomachic.
stomachique, *a.* & *m.* (*Med.*) stomachic.
stomate, *m.* (*Biol.*) stoma.

stoppage, *m.* stopping.
stopper, *v.t.* stop.
stoquer, *v.t.* stoke, tend (a fire or furnace).
stoqueur, *m.* poker; fire rake; stoker.
storax, *m.* storax.
— *liquide,* liquid storax.
stovaïne, *f.* (*Pharm.*) stovaine, stovain.
stramoine, *f.* (*Bot.*) stramonium, thorn apple.
stramonium, *m.* (*Bot.*) white thorn apple (*Datura stramonium*).
strass, stras, *m.* strass.
strasse, *f.* refuse silk; coarse packing paper.
strate, *f.* stratum, layer.
stratégie, *f.* strategy.
stratégique, *a.* strategic.
stratification, *f.* stratification.
stratifier, *v.t.* stratify.
streptocoque, (*Bact.*) streptococcus.
striation, *f.* striation.
strict, *a.* strict.
strictement, *adv.* strictly.
strie, *f.* stria; streak.
strié, *p.a.* striated.
strier, *v.t.* striate.
striure, *f.* striation, stria.
strobile, *m.* (*Bot.*) strobile; (*Zoöl.*) strobila.
stroma, *m.* (*Biol.*) stroma.
strontiane, *f.* strontia.
— *sulfatée,* sulfate of strontia (strontium sulfate).
strontianique, *a.* of or containing strontia.
strontianite, *f.* (*Min.*) strontianite.
strontique, *a.* strontium, of strontium.
strontium, *m.* strontium.
strophante, strophantus, *m.* (*Bot.*) Strophanthus.
strophantine, *f.* strophanthin.
structural, *a.* structural.
structure, *f.* structure.
strychnée, *f.* a plant of the genus *Strychnos.*
strychnine, *f.* strychnine.
strychnique, *a.* strychnic.
strychniser, *v.t.* strychninize (treat with strychnine or affect with strychninism).
strychnisme, *m.* strychninism, strychnism.
strychnos, *m.* (*Bot.*) Strychnos.
— *vomiquier,* nux vomica (*S. nux-vomica*).
stuc, *m.* stucco.
stucage, *m.* stuccoing.
stucatine, *f.* a coating of lime and phosphoric and salicylic acids used to give the appearance of cut stone.
studieusement, *adv.* studiously.
studieux, *a.* studious.
stupéfactif, *a.* & *m.* (*Med.*) stupefactive.
stupéfait, *p.a.* stupefied.
stupéfiant, *p.a.* stupefying, stupefacient, stupefactive. — *m.* (*Med.*) stupefacient, -factive.
stupéfier, *v.t.* stupefy.
stupeur, *f.* stupor.

stupide, *a.* stupid.
stupidement, *adv.* stupidly.
stupidité, *f.* stupidity.
stuquer, *v.t.* stucco.
sturine, *f.* sturine.
stycérine, *f.* stycerol, stycerin.
style, *m.* style.
stylet, *m.* stylet; small style.
styphinique, *a.* styphnic.
stypticine, *f.* (*Pharm.*) stypticin.
stypticité, *f.* stypticity, astringency.
styptique, *a. & m.* styptic.
styracées, *f.pl.* (*Bot.*) Styracaceæ.
styracine, *f.* styracin.
styrax, *m.* (*Bot.*) Styrax; storax.
— *officinale, Styrax officinalis.*
Styrie, *f.* Styria.
styrolène, *m.* styrene, styrolene.
styrolénique, *a.* styrene, of styrene (styrolene).
styrone, *f.* styrone.
su, *p.p.* of savoir. — *m.* knowledge.
suage, *m.* sweating (as of wood); paying, coating (with tallow); paying stuff, tallow; swage.
suager, *v.t.* pay, coat (with tallow or other grease); swage.
suant, *p.a.* sweating, sweaty; (of heat) welding.
suave, *a.* sweet, agreeable; (of colors) soft.
suavement, *adv.* sweetly, agreeably, softly.
suavité, *f.* sweetness, agreeableness, softness.
subcutine, *f.* (*Pharm.*) subcutin.
subdiviser, *v.t.* subdivide.
subdivision, *f.* subdivision.
subdivisionnaire, *a.* subdivisional.
subérate, *m.* suberate.
subéreux, *a.* (*Bot.*) suberose, suberous.
subérification, *f.* suberization, suberification.
subérine, *f.* suberin.
subérique, *a.* (of the acid, etc.) suberic; (of the alcohol) suberyl.
subérisation, *f.* suberization, suberification.
subérone, *f.* suberone (cycloheptanone).
subéryle, *m.* suberyl.
subérylique, *a.* suberyl, of suberyl.
subir, *v.t.* undergo. — *v.r.* be undergone.
subit, *a.* sudden.
subitement, *adv.* suddenly.
subjuguer, *v.t.* subjugate.
sublimable, *a.* sublimable.
sublimation, *f.* sublimation.
sublimatoire, *a. & m.* sublimatory.
sublime, *a. & m.* sublime.
sublimé, *m.* sublimate. — *p.a.* sublimed.
— *corrosif,* corrosive sublimate (mercuric chloride).
sublimer, *v.t. & i.* sublime. — *v.r.* be sublimed, sublime.
submerger, *v.t.* submerge.
submersion, *f.* submersion.
subordonné, *p.a.* subordinate; subordinated.
— *m.* subordinate.

subordonner, *v.t.* subordinate.
subséquemment, *adv.* subsequently.
subséquent, *a.* subsequent.
subside, *m.* subsidy.
subsidiaire, *a.* subsidiary, auxiliary.
subsidiairement, *adv.* subsidiarily, in addition.
subsidier, *v.t.* subsidize.
subsistance, *f.* subsistence; (*pl.*) provisions.
subsister, *v.i.* subsist, exist, continue.
substance, *f.* substance.
— *isolante,* (*Elec.*) insulating substance, insulator.
— *rebelle,* refractory substance.
substantiel, *a.* substantial.
substantiellement, *adv.* substantially.
substituant, *p.a.* substituting, substituent.
— *m.* substitute.
substituer, *v.t.* substitute.
substitutif, *a.* substitutive.
substitution, *f.* substitution.
substrat, substratum, *m.* substratum, substrate.
subterrané, *a.* subterranean.
subtil, *a.* fine (as a powder); subtle, subtile.
subtilement, *adv.* subtly.
subtilisation, *f.* volatilization; subtilization.
subtiliser, *v.t.* volatilize; subtilize.
subtilité, *f.* subtlety, subtleness, subtility.
subvenir, *v.i.* provide (for), suffice (for).
subventionner, *v.t.* subventionize, subsidize.
suc, *m.* juice; (fig.) substance, pith.
— *de citron,* lemon juice.
— *de genêt à balais,* (*Pharm.*) juice of broom.
— *de grande ciguë,* (*Pharm.*) juice of conium.
— *de jusquiame,* (*Pharm.*) juice of hyoscyamus.
— *de levure,* yeast juice.
— *de pissenlit,* (*Pharm.*) juice of taraxacum.
— *de réglisse,* licorice extract, (*Pharm.*) extract of glycyrrhiza.
— *gastrique,* gastric juice.
— *pancréatique,* pancreatic juice.
— *végétal,* vegetable juice, plant juice.
succédané, *m.* substitute, succedaneum. — *a.* substituting, substitute, succedaneous.
succéder, *v.i.* succeed (to); (with *à*) succeed, follow. — *v.r.* succeed one another.
succès, *m.* success; issue, result.
successeur, *m.* successor.
successif, *a.* successive; (*Law*) successional.
succession, *f.* succession.
successivement, *adv.* successively.
succin, *m.* amber (the resin).
succinate, *m.* succinate.
— *d'ammoniaque,* ammonium succinate.
— *de potasse,* potassium succinate.
— *de soude,* sodium succinate.
succinct, *a.* succinct, brief.
succinctement, *adv.* succinctly, briefly.
succiné, *a.* amber, resembling amber.

succinimide, *f.* succinimide.
succinique, *a.* succinic.
succinonitrile, *m.* succinonitrile.
succion, *f.* suction, sucking.
succomber, *v.i.* succumb.
succotrin, *m.* Sokotrine (or Socotrine) aloes.
succulence, *f.* succulence, succulency.
succulent, *a.* succulent, juicy.
succursale, *f.* branch, branch establishment.
sucement, *m.* sucking, suction.
sucer, *v.t.* suck, suck in, suck up.
sucette, *f.* an apparatus for sucking off the sirup from sugar loaves.
suceur, *a.* sucking. — *m.* sucker.
sucotrin, *a.* (of aloes) Sokotrine.
sucrage, *m.* sugaring.
sucrase, *f.* sucrase (invertase).
sucratage, *m.* extraction of sugar, desugaring (of molasses).
sucrate, *m.* sucrate, saccharate.
sucraterie, *f.* place or establishment where saccharates are made, as in desugaring molasses; extraction of sugar, desugaring.
sucratier, *m.* saccharate maker, sugar extractor.
sucre, *m.* sugar.
— *blanc,* white sugar.
— *brut,* raw sugar, brown sugar, unrefined sugar.
— *candi,* sugar candy, rock candy.
— *d'amidon,* starch sugar, glucose.
— *de bambou,* bamboo sugar, tabasheer.
— *de betterave,* beet sugar, beetroot sugar.
— *de canne,* cane sugar.
— *de deuxième jet,* (*Sugar*) a second crop of crystals obtained by boiling down the sirup from the first crop and centrifuging.
— *de fécule,* starch sugar, glucose.
— *de foie,* diabetic sugar.
— *de fruit,* fruit sugar, d-fructose.
— *de gélatine,* (old name for) glycine, glycocoll.
— *de houille,* saccharin.
— *de lait,* milk sugar, lactose.
— *de plomb,* sugar of lead, lead acetate.
— *de pomme,* sugar candy flavored with cider.
— *de premier jet,* (*Sugar*) the first crop of crystals removed by the centrifugals.
— *d'érable,* maple sugar.
— *de raisin,* grape sugar, d-glucose.
— *de réglisse,* licorice sugar (glycyrrhyzic acid).
— *de Saturne,* sugar of lead, lead acetate.
— *des colonies,* colonial sugar (formerly a synonym of cane sugar).
— *de troisième jet,* (*Sugar*) the third crop of crystals (see sucre de deuxième jet).
— *d'orge,* barley sugar.
— *en bâtons,* stick candy.
— *en pain(s),* loaf sugar.

sucre *interverti,* — *inverti,* invert sugar.
— *liquide,* fruit sugar, d-fructose (old name).
— *neutre,* neutral sugar (i.e., not showing rotation).
— *noir,* licorice extract; licorice.
— *non réducteur,* non-reducing sugar.
— *pilée,* crushed sugar.
— *pur,* pure sugar, refined sugar.
— *raffiné,* refined sugar.
— *réducteur,* reducing sugar.
sucré, *p.a.* sugared, treated or sweetened with sugar; of or containing sugar; sugary, saccharine, sweet.
— *couleur,* caramel brown.
sucrerie, *f.* sugar factory or refinery, sugar works, sugarhouse; sweet, confection.
sucreur, *m.* sugarer, sweetener (as of wine).
sucrier, *a.* sugar, of sugar, relating to the manufacture of sugar. — *m.* sugar maker or manufacturer; sugar bowl.
— *de montagne,* = bois cochon.
sucro-carbonate, *m.* (*Sugar*) sucro-carbonate.
— *de chaux,* "sucro-carbonate of lime" (a gelatinous mixture of calcium sucrate, calcium carbonate, lime, etc., obtained during carbonation).
sud, *a. & m.* south.
sud-américain, *a.* South American.
sud-est, *a. & m.* southeast.
sudoral, *a.* sudoral, of sweat.
sudorifique, *a. & m.* sudorific.
sudoripare, *a.* sudoriferous, sudoriparous.
sud-ouest, *a. & m.* southwest.
Suède, *f.* Sweden.
suédois, *a.* Swedish. — *m.* (*cap.*) Swede.
suée, *f.* sweat, sweating.
suer, *v.i. & t.* sweat.
suerie, *f.* sweating; sweating house (as for tobacco).
sueur, *f.* sweat, perspiration.
suffioni, *m.pl.* volcanic steam jets, suffioni.
suffire, *v.i.* suffice, be sufficient. — *v.r.* be self-supporting.
suffisamment, *adv.* sufficiently, enough.
suffisance, *f.* sufficiency; self-conceit.
à —, en —, sufficiently, enough.
suffisant, *p.a.* sufficient; sufficing; conceited.
suffocant, *p.a.* suffocating.
suffocation, *f.* suffocation.
suffoquer, *v.t. & i.* suffocate.
suggérer, *v.t.* suggest.
suggestif, *a.* suggestive.
suggestion, *f.* suggestion.
suie, *f.* soot.
— *d'arsenic,* arsenical soot.
— *de bois,* wood soot.
— *d'encens,* a kind of lampblack made from frankincense.
suif, *m.* tallow; suet; tallow composition for ship bottoms.

suif (continued).

mettre en —, (*Leather*) dress with tallow, stuff, tallow.

— *de place*, = dégraisse.

— *d'os*, bone fat, bone grease.

— *en branche(s)*, suet.

— *minéral*, (*Min.*) mineral tallow, hatchettite.

— *noir*, an impure tallow extracted from cracklings.

suiffer, suifer, *v.t.* tallow.

suiffeux, *a.* tallowy.

suiffier, *m.* tallow maker.

suint, *m.* suint, grease (of wool).

— *de laine*, wool grease, wool fat, suint.

suintement, *m.* oozing, etc. (see suinter).

suinter, *v.i. & t.* ooze, ooze out, exude; sweat; leak. — *m.* an impure calcium soap obtained by precipitating soapy waste waters with lime.

suintine, *f.* an impure fatty mixture obtained by treating with sulfuric acid and steam the water in which wool has been scoured.

suis, *pr. 1 sing.* of être (to be). — *pr. 1 & 2 sing.* of suivre.

suisse, *a.* Swiss. — *m.* a small cream cheese; porter; (*cap.*) Swiss. — *f.* (*cap.*) Switzerland.

suit, *pr. 3 sing.* of suivre.

suite, *f.* succession; series; sequence; sequel, continuation, "continued"; result; consequence; connection, order; set, collection; perseverance; suite.

de —, in succession; in line.

par la —, later.

par —, consequently.

par — *de*, in consequence of, as a result of, on account of.

suivant, *prep.* according to; in proportion to; following, along. — *p.a.* following, etc. (see suivre); next. — *m.* follower; attendant.

— *que*, according as, in proportion as.

suivi, *p.a.* followed, etc. (see suivre); continued, constant; connected, coherent; popular.

suivre, *v.t.* follow; pursue; keep up with; frequent; attend; watch. — *v.i.* follow; go on; attend (to). — *v.r.* follow one another; be connected; be continuous.

faire —, cause to follow; (on mail) please forward; (on proof) run on.

sujet, *m.* subject; reason, cause; person, fellow. — *a.* subject; liable.

sujétion, *f.* obligation; constraint; subjection.

sulfacide, *m.* sulfacid, thio acid.

sulfanilique, *a.* sulfanilic, sulphanilic.

sulfarsénique, *a.* thioarsenic, sulfarsenic.

sulfatage, *m.*, **sulfatation**, *f.* sulfating (see sulfater).

sulfate, *m.* sulfate, sulphate; specif., (*Soda*) sodium sulfate.

— *d'alumine*, aluminium sulfate.

sulfate *d'ammoniaque*, ammonium sulfate.

— *de baryte*, barium sulfate.

— *de chaux*, calcium sulfate.

— *de cuivre*, copper sulfate.

— *de fer*, iron sulfate.

— *de magnésie*, magnesium sulfate.

— *de potasse*, potassium sulfate.

— *de soude*, sodium sulfate.

— *de strontiane*, strontium sulfate.

— *ferreux*, ferrous sulfate.

— *ferrique*, ferric sulfate.

— *manganeux*, manganous sulfate.

— *mercureux*, mercurous sulfate.

— *mercurique*, mercuric sulfate.

sulfaté, *p.a.* sulfated; sulfate of; sulfatic.

sulfater, *v.t.* sulfate, treat or combine with sulfuric acid or a sulfate, convert into sulfate; specif., treat with copper sulfate (as vines).

sulfatique, *a.* sulfatic, sulphatic.

sulfatisation, *f.* sulfatizing, sulfatization, etc. (see sulfatiser).

sulfatiser, *v.t.* sulfatize, convert into sulfate; specif., convert (a sulfide) into a sulfate by oxidation; treat with copper sulfate.

sulfhydrate, *m.* sulfhydrate, sulphydrate (hydrosulfide).

sulfhydrique, *a.* sulfhydric, sulphydric.

acide —, sulfhydric acid (hydrogen sulfide).

sulfhydrométrie, *f.* determination of hydrogen sulfide (as in mineral waters).

sulfhydrométrique, *a.* pertaining to the determination of hydrogen sulfide.

sulfimétrie, *f.* determination of sulfurous acid or sulfites (as in wine).

sulfimétrique, *a.* pertaining to *sulfimétrie*.

sulfine, *f.* sulfine, sulphine.

sulfiné, *a.* sulfinated, sulfinic.

sulfinique, *a.* (of acids) sulfinic; (of SH_3 derivatives) sulfine, sulfonium.

sulfitage, *m.* treatment with a sulfite or sulfurous acid.

sulfite, *m.* sulfite, sulphite.

— *de potasse*, potassium sulfite.

— *de soude*, sodium sulfite.

— *sulfuré de soude*, sodium thiosulfate.

sulfiter, *v.t.* treat with a sulfite or sulfurous acid.

sulfitique, *a.* sulfite, or of pertaining to sulfites or the sulfite process.

sulfitomètre, *m.* an apparatus for determining sulfur dioxide or sulfites.

sulfo, *a.* sulfo, sulpho.

sulfoantimoniate, *m.* thioantimonate, sulfantimonate.

sulfoarsénite, *m.* thioarsenite, sulfarsenite.

sulfo-arséniure, *m.* (*Min.*) sulfarsenide.

sulfobase, *f.* (*Old Chem.*) sulfur base.

sulfobenzide, *m.* sulfobenzide (phenyl sulfone).

sulfocarbimide, *f.* sulfocarbimide (isothiocyanic acid or an organic isothiocyanate).

sulfocarbolique, *a.* sulfocarbolic.

sulfocarbonate, *m.* thiocarbonate, sulfocarbonate.

sulfocarboné, *a.* containing or combined with thiocarbonic acid or thiocarbonates.

sulfocarbonique, *a.* thiocarbonic, sulfocarbonic.

sulfochlorure, *m.* sulfochloride (better translated sulfonyl chloride or sulfonic chloride when used of organic compounds of the formula RSO_2Cl).

sulfoconjugué, *a.* (*Dyes*, etc.) sulfonated, sulfo; (*Biol. Chem.*, etc.) designating an acid (acid ester) of the formula RSO_4H (e.g. phenylsulfuric acid, $C_6H_5SO_4H$). — *m.* sulfo derivative.

sulfocyanate, *m.* thiocyanate, sulfocyanate.

sulfocyanhydrique, *a.* thiocyanic, sulfocyanic.

sulfocyanique, *a.* thiocyanic, sulfocyanic.

sulfocyanogène, *m.* thiocyanogen, sulfocyanogen.

sulfocyanure, *m.* sulfocyanide (thiocyanate, sulfocyanate).

sulfoglycérique, *a.* (*Expl.*) sulfoglyceric.

sulfo-indigotique, *a.* sulfindigotic.

sulfoléique, *a.* sulfoleic.

sulfonal, *m.* (*Pharm.*) sulfonal, sulphonal.

sulfoné, *p.a.* sulfonated, sulfonic, sulfo.

sulfonique, *a.* sulfonic; of or pertaining to a sulfone.

sulfonitrique, *a.* sulfonitric (designating a mixture of sulfuric and nitric acids, or a mixed anhydride of them).

sulfophénate, *m.* phenolsulfonate.

sulforganique, *a.* designating organic compounds which contain sulfur.

sulforiciné, *a.* (*Pharm.*) containing sulforicinoleic acid or its salts.

sulfosel, *m.* sulfo salt (esp. a salt of a thio acid).

sulfostéatite, *m.* sulfosteatite (a fungicide).

sulfotellurure, *m.* sulfotelluride.

sulfo-urée, *f.* thiourea, sulfourea.

sulfovinique, *a.* sulfovinic (ethylsulfuric).

sulfurabilité, *f.* capability of being sulfurized.

sulfurable, *a.* capable of being sulfurized.

sulfurage, *m.* (*Wine*) treatment of the ground with carbon disulfide to destroy phylloxera; (*Med.*) external application of sulfur.

sulfuration, *f.* sulfurization, sulfuration.

sulfure, *m.* sulfide, sulphide.

— *d'antimoine,* antimony sulfide.

— *de carbone,* carbon disulfide.

— *de plomb,* lead sulfide.

— *salin,* a sulfide which is regarded as a salt, compound sulfide.

sulfuré, *p.a.* combined with sulfur, sulfurized, sulfurated, sulfuretted, sulfide of, sulfur.

alcool —, mercaptan.

sulfurer, *v.t.* combine or treat with sulfur (or, sometimes, with a sulfur compound), sulfurize, sulfurate, sulfur.

sulfureur, *a.* sulfuring, sulfurizing.

sulfureux, *a.* sulfurous, sulphurous.

sulfurifère, *a.* containing sulfur.

sulfurique, *a.* sulfuric, sulphuric.

sulfurisation, *f.* sulfurization, sulfuration.

sulfuriser, *v.t.* sulfurize; specif., treat with sulfuric acid (as paper).

sulfuroïde, *a.* resembling or analogous to sulfur.

sulfuryle, *m.* sulfuryl, sulphuryl.

sumac, *m.* sumac, sumach.

— *à perruque,* — *des teinturiers,* fustet (*Cotinus cotinus*).

— *des corroyeurs,* tanner's sumac (*Rhus coriaria*).

— *du Japon,* Japanese varnish tree (*Rhus vernicifera*).

sûmes, *p.def. 1 pl.* of savoir.

super, *v.i.* be stopped up, be plugged up. — *v.t.* suck up. — *m.* Short for *superphosphate.*

superbe, *a.* superb; lofty; proud. — *f.* pride.

superbement, *adv.* superbly; proudly.

superficiaire, *a.* superficial.

superficie, *f.* surface; superficies; area.

superficiel, *a.* superficial.

superficiellement, *adv.* superficially, on the surface.

superfin, *a.* superfine. — *m.* superfine article or quality.

superflu, *a.* superfluous. — *m.* superfluous; superfluity, excess.

superfluité, *f.* superfluity.

supérieur, *a.* higher; upper; superior. — *m.* superior.

supériorité, *f.* superiority.

superoxydation, *f.* superoxidation. (*Obs.*)

superpalite, *f.* superpalite, diphosgene, trichloromethyl chloroformate.

superphosphate, *m.* superphosphate.

— *de chaux,* superphosphate of lime, acid calcium phosphate.

— *d'os,* bone superphosphate.

superposable, *a.* superposable.

superposer, *v.t.* superpose.

superposition, *r.* superposition, superposing.

supersaturer, *v.t.* supersaturate.

supersécrétion, *f.* hypersecretion.

supplanter, *v.t.* supplant.

suppléant, *a. & m.* substitute; assistant.

suppléer, *v.t.* supply, make good; replace, take the place of; substitute. — *v.i.* (with *à*) make up for, take the place of.

supplément, *m.* supplement.

supplémentaire, *a.* supplementary, supplemental.

supplice, *m.* punishment; torment.

support, *m.* support; stand, rack, rest, stay, holder, bearing, bracket, socket, etc.

— *à agitateurs,* stirring-rod rack.

— *à anneau,* ring stand.

upport à *bascule*, rocking support, rocker; specif., a tilting stand for carboys.
— à *entonnoir(s)*, funnel stand.
— à *pince*, clamp stand.
— à *râtelier*, rack.
— à *trois pieds*, — à *trépied*, tripod.
— *égouttoir*, draining rack, draining stand.
— *en bois*, wooden support (or rack or stand).
— *en fil*, wire support (or rack or stand).
— *porte-tubes*, tube rack, tube stand.
— *pour burettes*, burette stand.
— *pour pèse-filtres*, weighing-tube support (or stand).
— *pour pipettes*, pipet stand (or rack).
— *pour tubes à essais*, test-tube rack (or stand).
supporter, *v.t.* support, stand, endure.
supposé, *p.a.* supposed; assumed; fictitious.
supposer, *v.t.* suppose. — *v.r.* suppose; be supposed.
supposition, *f.* supposition; forgery.
suppositoire, *m.* (*Pharm.*) suppository.
— *morphiné*, morphine suppository.
suppression, *f.* suppression; omission; abolition.
supprimer, *v.t.* suppress; remove, relieve (as pressure); do away with, abolish.
suppurant, *p.a.* suppurating.
suppuratif, *a. & m.* suppurative.
suppuration, *f.* suppuration.
suppurer, *v.i.* suppurate.
supputation, *f.* calculation, computation.
supputer, *v.t.* calculate, compute, reckon.
suprarénine, *f.* suprarenine (adrenaline).
suprématie, *f.* supremacy.
suprême, *a.* supreme; last.
au — degré, in the highest degree.
sur, *prep.* on, upon; over; above; out of, among; after, according to; about, toward; (of dimensions) by. — *a.* sour.
sur-. super-, supra-, sur-, over-, extra-.
sûr, *a.* sure, certain; secure, safe.
pour —, certainly, surely.
surabondamment, *adv.* superabundantly.
surabondance, *f.* superabundance.
surabondant, *a.* superabundant.
surabonder, *v.i.* be very abundant, superabound.
suraddition, *f.* superaddition.
suraffinage, *m.* overrefining.
suraffiner, *v.t.* overrefine, etc. (see affiner).
surajoutement, *m.* superaddition.
surajouter, *v.t.* superadd.
suralcooliser, *v.t.* add excess of alcohol to.
suranné, *a.* antiquated, old-fashioned, obsolete; expired, lapsed; superannuated.
surbaissé, *a.* low (in proportion to width), flat, depressed, surbased.
surcalciner, *v.t.* overcalcine, overburn.
surcapitalisation, *f.* overcapitalization.

sur-cémenter, *v.t.* face-harden (as armor).
surcharge, *f.* overload; overcharge; surcharge.
surcharger, *v.t.* overload; overcharge; surcharge.
surchauffage, *m.*, **surchauffe**, *f.* superheating; overheating.
surchauffer, *v.t.* superheat (as steam); overheat.
surchauffeur, *m.* superheater.
surchauffure, *f.* overheating, burning; burnt spot (as in steel).
surchoix, *m.* first choice, best quality.
surcroît, *m.* addition, increase.
surcuire, *v.t.* overburn; overboil; overcook; burn, boil or cook some more.
surcuisson, *m.* overburning, etc. (see surcuire).
surcuit, *p.a.* overburned, etc. (see surcuire).
— *m.* overburned material (as lime or cement).
surdité, *f.* deafness.
surdorer, *v.t.* double-gild.
sureau, *m.* (*Bot.*) elder (*Sambucus*).
surégalisage, *m.* (*Gunpowder*) coarse screening.
surégaliser, *v.t.* screen (gunpowder) coarsely.
surégalisoir, *m.* (*Gunpowder*) coarse sieve.
surélévation, *f.* raising; rise; increase.
surélèvement, *m.* raising, raise, increase.
surélever, *v.t.* raise; elevate; increase.
surelle, *f.* (*Bot.*) sorrel (esp. wood sorrel).
sûrement, *adv.* surely, certainly; securely.
surenchérissement, *m.* further increase (in price).
surent, *p.def. 3 pl.* of savoir.
surépaisseur, *m.* increased thickness.
surestimation, *f.* overvaluation, overestimate.
surestimer, *v.t.* overvalue, overestimate.
suret, *a.* sourish, somewhat sour.
sûreté, *f.* safety; security; sureness; certainty; reliability; safe keeping.
de —, (of valves, tubes, etc.) safety.
surette, *f.* sorrel, esp. wood sorrel; a coarse jute cloth. — *a. fem.* of suret.
surexciter, *v.t.* overexcite.
surface, *f.* surface.
— *courbe*, curved surface.
— *d'appui*, (*Mach.*) working surface.
— *de chauffe*, heating surface.
— *inégale*, uneven surface.
— *réfrigérante*, — *refroidissante*, cooling surface.
— *unie*, even surface, smooth surface.
surfaire, *v.t.* overvalue, overrate.
surfin, *a.* superfine.
surfondre, *v.t. & i.* superfuse.
surfondu, *p.a.* superfused.
surforce, *f.* excess of alcohol over the ordinary amount or over proof.
surfusibilité, *f.* superfusibility; extreme fusibility.
surfusible, *a.* superfusible; extremely fusible.

surfusion, *f.* superfusion.
surge, *f.* raw wool. — *a.* (of wool) raw; (of paper pulp) not retaining water.
surgir, *v.i.* rise, arise, appear.
surhaussement, *m.* raising, elevation.
surhausser, *v.t.* raise, elevate.
surimposer, *v.t.* increase the tax on; surtax.
surintendance, *f.* superintendence; superintendent's office or duties.
surintendant, *m.* superintendent, overseer
surir, sûrir, *v.t. & i.* sour, turn sour, turn.
surjacent, *a.* overlying.
sur-le-champ, *adv.* immediately, at once.
surlendemain, *m.* second day after, day after tomorrow.
surmener, *v.t. & r.* overwork.
surmontable, *a.* surmountable.
surmonter, *v.t.* surmount; top; rise above; be above or over; surpass.
surmoulage, *m.* molding from a casting.
surmoule, *f.* mold made from a casting.
surmouler, *v.t.* mold from or upon; mold from a casting.
surmoût, *m.* must, unfermented grape juice.
surnageant, *p.a.* supernatant, floating on the surface; surviving.
surnager, *v.i.* float on the surface; survive.
surnaturel, *a.* supernatural.
surnaturellement, *adv.* supernaturally.
surnom, *m.* surname.
surnuméraire, *a. & m.* supernumerary.
suroxydation, *f.* peroxidation, superoxidation.
suroxydé, *p.a.* peroxidized; peroxide of.
suroxyder, *v.t. & r.* peroxidize, superoxidize.
suroxygénation, *f.* peroxidation, superoxidation.
suroxygéner, *v.t.* peroxidize, superoxidize.
surpalite, *f.* (*Mil.*) superpalite ($ClCO_2CCl_3$).
surpasser, *v.t.* surpass; exceed; be higher than.
surplomber, *v.i.* be out of plumb. — *v.t.* overhang.
surplus, *m.* surplus; rest, remainder.
au —, besides.
surpoids, *m.* overweight, excess weight.
surprenant, *p.a.* surprising, etc. (see surprendre).
surprendre, *v.t.* surprise; astonish; catch; detect; deceive, abuse.
surpression, *f.* overpressure, excess pressure.
surpris, *p.a.* surprised, etc. (see surprendre).
surproduction, *f.* overproduction.
surrénal, *a.* (*Anat.*) suprarenal.
sursaturation, *f.* supersaturation.
sursaturer, *v.t.* supersaturate; saturate to excess (as a solution with ammonia).
sursel, *m.* supersalt (acid salt).
surseoir, *v.t.* suspend, delay, defer.
sursis, *p.a.* suspended, etc. (see surseoir). — *m.* suspension, delay; reprieve.
sursoufflage, *m.* (*Metal.*) afterblow.

surtaxe, *f.* surtax; excessive tax.
surtaxer, *v.t.* surtax; overtax.
surtension, *f.* (*Elec.*) overvoltage.
surtonte, *f.* clipping (as of wool).
surtout, *adv.* especially, principally, above all.
survécu, *p.p.* of survivre.
surveillance, *f.* supervision, inspection, superintendence, looking after, attention.
surveillant, *p.a.* supervising, etc. (see surveiller); vigilant. — *m.* supervisor, superintendent, overseer, inspector, watchman, keeper.
surveiller, *v.t.* supervise, superintend, oversee, watch, look after, see to.
survenir, *v.i.* happen; come unexpectedly.
survider, *v.t.* reduce the contents of (something that is too full).
survie, *f.* survival.
survit, *pr. 3 sing.* of survivre.
survivant, *p.a.* surviving. — *m.* survivor.
survivre, *v.t. & i.* (often with à) survive.
survolter, *v.t.* (*Elec.*) raise the voltage of.
sus, *adv.* above. — *interj.* come, now then. — *p.def. 1 & 2 sing.* of savoir.
— *à,* on, upon.
en —, besides, more.
en — *de,* over and above, in addition to.
sus-. above-, afore-, super-, supra-; sus-.
susceptibilité, *f.* susceptibility.
susceptible, *a.* susceptible; liable; capable (of).
susciter, *v.t.* raise up, stir up, cause.
suscription, *f.* address (on mail, etc.).
sus-dénommé, susdit, sus-mentionné, sus-nommé, *a.* above-mentioned, aforesaid.
suspect, *a.* suspicious; suspected; doubtful.
suspecter, *v.t.* suspect.
suspendre, *v.t.* suspend. — *v.r.* be suspended.
suspens, *m.* suspense, uncertainty, doubt.
suspension, *f.* suspension; suspense; hanging support.
— *bifilaire,* bifilar suspension.
— *par couteaux,* knife-edge suspension.
sus-relaté, *a.* above-related, above-mentioned.
sussent, *imp. subj. 3 pl.* of savoir.
sustentation, *f.* sustaining; support; sustenance.
sustenter, *v.t.* sustain; support.
sut, *p.def. 3 sing.* of savoir.
sût, *imp. subj. 3 sing.* of savoir.
sûtes, *p.def. 2 pl.* of savoir.
svelte, *a.* slender.
sveltesse, *f.* slenderness.
sycomore, *m.* sycamore (either *Ficus sycomorus* or *Acer pseudo-platanus,* not the American sense).
syénite, *f.* (*Petrog.*) syenite.
syénitique, *a.* (*Petrog.*) syenitic.
syllabe, *f.* syllable.
sylvane, *m.* sylvan.
sylvestre, *a.* forest, wood, woody, sylvan.

sylvestrène, *m.* sylvestrene.
sylvine, *f.* (*Min.*) sylvite, sylvine.
symbiose, *f.* (*Biol.*) symbiosis.
symbole, *m.* symbol.
symbolique, *a.* symbolic, symbolical.
symbolisation, *f.* symbolization, symbolizing.
symboliser, *v.t.* symbolize.
symbolisme, *m.* symbolism.
symétrie, *f.* symmetry.
symétrique, *a.* symmetrical, symmetric.
symétriquement, *adv.* symmetrically.
sympathie, *f.* sympathy.
sympathique, *a.* sympathetic.
sympathiquement, *adv.* sympathetically.
symptôme, *m.* symptom.
synchrone, synchronique, *a.* synchronous, synchronic(al).
synchroniser, *v.t.* synchronize.
syndic, *m.* syndic, magistrate, representative; assignee; receiver.
syndical, *a.* syndical. — *m.* member of a syndicate.
syndicat, *m.* syndicate; company, society.
syndiquer, *v.t.* syndicate.
synonyme, *a.* synonymous. — *m.* synonym.
synopsis, *f.* synopsis.

synovial, *a.* (*Anat.*) synovial.
synovie, *f.* (*Anat.*) synovia.
synthèse, *f.* synthesis.
synthétique, synthésique, *a.* synthetic(al).
synthétiquement, *adv.* synthetically.
synthétiser, *v.t.* synthesize, (rarely) synthetize.
syntonine, *f.* syntonin.
syphilitique, *a.* (*Med.*) syphilitic.
syphon, *m.* siphon.
Syrie, *f.* Syria.
syrien, *a.* Syrian. — *m.* (*cap.*) Syrian.
syringa, *m.* (*Bot.*) lilac (*Syringa*).
systématique, *a.* systematic, systematical.
— *m.* systematist. — *f.* systematics.
systématiquement, *adv.* systematically.
systématisation, *f.* systematizing. -tization.
systématiser, *v.t.* systematize.
système, *m.* system.
— *à vide,* vacuum system.
— *chimique,* chemical system.
— *cubique,* (*Cryst.*) cubic system (isometric system).
— *périodique,* periodic system.
— *rhomboédrique,* (*Cryst.*) rhombohedral system.
szekso, *m.* a Hungarian earth containing soda.

T

t., *abbrev.* (tonne) ton; (tome) volume; (transcrivez) copy, transcribe.

t'. Contraction of *te, toi* (thee, you).

-t-. A connective used for the sake of euphony between a verb ending in a vowel and *il, elle* or *on*.

ta, *a.fem.* thy, your.

tabac, *m.* tobacco; tobacco color. — *a.* tobacco-colored.
  — *à chiquer,* chewing tobacco.
  — *à fumer,* smoking tobacco.
  — *à mâcher,* chewing tobacco.
  — *à priser,* — *en poudre,* snuff.

tabacal, *a.* tobacco, of tobacco.

tabaschir, tabashir, *m.* tabasheer, tabashir.

tabellaire, *a.* in the form of tables or tablets.

tabiser, *v.t.* tabby, water.

tablant, *p.pr.* of tabler.

table, *f.* table; plate, slab; sheet (as of lead); face (as of an anvil); bed (as of a planer); index; (*Salt*) basin, bed.
  — *à écrire,* writing table.
  — *à égruger,* mealing tray (for powder); mixing table.
  — *alphabétique,* index; table arranged alphabetically.
  — *à ouvrage,* work table, work bench.
  — *à sécher,* drying table; drying tray.
  — *brisée,* folding table.
  — *d'appareil,* apparatus table, work table.
  — *de chimiste,* chemist's table, specif. one provided with a blast apparatus.
  — *de construction,* working plan, working drawing; table of specifications.
  — *de division,* (*Soap*) an apparatus for cutting soap into bars or cakes.
  — *de foyer,* hearth plate.
  — *de laboratoire,* laboratory table.
  — *d'émailleur,* enameler's table, blowpipe table.
  — *de manipulation,* (*Elec.*) switchboard.
  — *des matières,* table of contents.
  — *de tiroir,* valve face, slide face.
  — *en béton,* concrete bed.
  — *en fonte,* cast-iron plate.
  — *en lave,* lava-topped table.
  — *rase,* tabula rasa, " clean slate."
  — *salante,* (*Salt*) a shallow clay-lined evaporating basin in a salt garden.

tableau, *m.* table; list, catalog; diagram; painting; picture; scenery, scene; board; blackboard; bulletin board, billboard.
  — *de distribution,* (*Elec.*) switchboard.
  — *noir,* blackboard.

tabletier, *m.* a maker or seller of *tabletterie.*

tablette, *f.* shelf; tablet; troche, lozenge, pastil; cake; (*pl.*) notebook; (window) sill.
  — *de cachou,* (*Pharm.*) troche of gambir.
  — *de fer,* (*Pharm.*) reduced iron lozenge.
  — *de l'acide phénique,* (*Pharm.*) phenol lozenge.
  — *de ratanhia,* (*Pharm.*) troche of krameria, rhatany lozenge.
  — *de soufre,* (*Pharm.*) sulfur lozenge.
  — *de tannin,* (*Pharm.*) troche of tannic acid, tannin lozenge.

tabletterie, *f.* fancy goods, toys, games, etc., also the making or selling of them.

tablier, *m.* apron; floor, platform.
  — *de laboratoire,* laboratory apron.

tabouret, *m.* stool.

tabulaire, *a.* tabular.

tacamaque, *m.* tacamahac, tacamahaca.
  — *terreuse,* Brazilian elemi.

tache, *f.* spot; stain.
  — *d'étain,* (*Metal.*) tin spot.

tâche, *f.* task, job (specif., work to be done in a day or other fixed time).

tacher, *v.t.* spot; stain; (fig.) tarnish.

tâcher, *v.t.* try, endeavor, strive.

tacheter, *v.t.* spot, speckle.

tacheture, *f.* speck, speckle.

tachymètre, tachomètre, *m.* tachometer, tachymeter, speed gage.

tact, *m.* touch, sense of touch; tact.

taffetas, *m.* taffeta, taffety.

tafia, *m.* tafia, taffia (a kind of rum).

taillade, *f.* gash, cut.

taillanderie, *f.* edge tools; edge-tool business.

taillant, *p.a.* cutting, etc. (see tailler). — *m.* cutting edge.

taille, *f.* cutting; cut; height; figure, shape; tally; coppice, copse.

taille-douce, *f.* (*Engraving*) copperplate.

tailler, *v.t.* cut; trim, prune; form, shape (with cutting); carve; engrave.

taillerie, *f.* gem cutting, lapidary work (or shop).

tailleur, *m.* cutter; tailor.

tailleuse, *f.* cutter, cutting machine; tailoress.

taillis, *m.* coppice, copse.

tain, *m.* silvering, tin amalgam (for mirrors); tin bath (for tinning iron); tinfoil.

taire, *v.t.* be silent on, not tell; silence. — *v.r.* be silent, be quiet; (with *de*) be silent on, pass over; be passed over, not be mentioned.

talc, *m.* (*Min.*) talc.

— *de Venise,* Venetian talc.

talcaire, *a.* talcose, talcous.

talcique, *a.* talc, of talc.

taloche, *f.* (plasterer's) float.

talon, *m.* heel; stub; (*Tech.*) heel, nose, etc.

talose, *f.* talose.

talqueux, *a.* talcose, talcous.

talquer, *v.t.* talc, coat or impregnate with talc.

talus, *m.* slope, declivity; sloping embankment; talus.

tamarin, *m.* tamarind (tree and fruit); tamarisk.

tamarinier, *m.* (*Bot.*) tamarind tree, tamarind (*Tamarindus indica*).

tamaris, tamarisc, tamarix, *m.* (*Bot.*) tamarisk.

tambour, *m.* drum (in various senses); barrel, cylinder, roll, roller, etc.; tambour.

tamis, *m.* sieve; screen; sifter; bolter.

— *à tambour,* drum sieve.

— *de crin,* hair sieve.

— *de soie,* silk sieve.

— *de toile métallique,* wire-gauze sieve or screen.

— *en laiton,* — *en toile de laiton,* brass-gauze sieve or screen.

tamisage, *m.* sifting, etc. (see tamiser).

tamise, *f.* tammy (a fabric); (*cap.*) Thames.

tamiser, *v.t.* sift, sieve, screen, bolt. — *v.i.* pass thru a sieve or screen.

tamiseur, *m.,* **tamiseuse,** *f.* sifter, bolter.

tampon, *m.* plug; stopper; wad; pad; buffer; cushion; bung; tampon; tampion; tompion.

— *de lumière,* vent plug.

— *de sûreté,* safety plug.

tamponnement, *m.* plugging, etc. (see tamponner).

tamponner, *v.t.* plug, plug up, stop, stop up; tamp, pack; rub with a pad; knock against.

tan, *m.* tanbark, tan.

— *épuisé,* spent tanbark, spent tan.

tanaisie, *f.* (*Bot.*) tansy (*Tanacetum*).

tandis que. while; as long as; whereas.

tang., *abbrev.* (tangente) tangent.

tangence, *f.* tangency.

tangent, *a. & f.* tangent.

tangentiel, *a.* tangential.

tangentiellement, *adv.* tangentially.

Tanger, *n.* Tangiers.

tanghin, tanghen, *m.* tanghin (an arrow poison).

tangibilité, *f.* tangibility.

tangible, *a.* tangible.

tangrum, *m.* a kind of fish guano.

tanification, *f.* conversion into tannin.

tanin, *m.* tannin.

tanique, *a.* tannic.

tanisage *m.,* **tanisation,** *f.* addition of, or treatment with, tannin.

taniser, *v.t.* add tannin to, treat with tannin.

tannage, *m.* tanning, tannage.

— *à la flotte,* liquor tanning.

— *à la fosse,* — *à la jusée,* bark tanning, pit tanning.

— *à l'électricité,* electric tanning.

— *à l'huile,* oil tanning.

— *au chrome,* chrome tanning.

— *au tanin,* tanning with tannin, vegetable tanning.

— *aux écorces,* bark tanning.

— *aux extraits,* liquor tanning.

— *en fosse,* bark tanning, pit tanning.

— *mixte,* compound tanning.

— *rapide,* rapid tanning.

tannant, *p.a.* tanning.

tannate, *m.* tannate.

tanne, *f.* spot (on leather); (*Med.*) blackhead.

tanné, *p.a.* tanned; tan. — *m.* tan (the color).

tannée, *f.* spent tanbark, spent tan.

tanner, *v.t.* tan.

tannerie, *f.* tannery, tanyard; tanning.

tanneur, *m.* tanner.

tannin, *m.* tannin.

tannique, *a.* tannic.

tannisage, tanniser, etc. See tanisage, etc.

tannogélatine, *f.* tannogelatin.

tannoïde, *a. & m.* tannoid.

tant, *adv.* so much, so; so long, so far.

à — *de titres,* in so many respects, for so many reasons.

en — *que,* in so far as; considered as, as being, as; if, provided.

faire — *que de,* decide, resolve (to).

si — *est que,* if perchance, supposing.

— *de,* so much, so many, such.

— *mieux,* so much the better.

— *pis,* so much the worse.

— *plus,* so much the more, the more.

— *plus que moins,* more or less.

— *que,* as long as, while; as far as; so much as; until; so that.

tant . . . *que,* however, tho; both . . . **and;** as well . . . as.

— *s'en faut que,* far from, so far from.

— *soit peu,* ever so little, a very little.

— *y a que,* however it may be, at any rate.

tantalate, *m.* tantalate.

tantale, *m.* tantalum.

tantaleux, *a.* tantalous.

tantalifère, *a.* containing tantalum, tantaliferous.

tantalique, *a.* tantalic.

tantaliser, *v.t.* tantalize.

tantalite, *f.* (*Min.*) tantalite.

tantième, *a.* given, such a. — *m.* given part or per cent.

tantôt, *adv.* soon; a little later; just now, a short time ago.

   sur le —, in the afternoon, toward evening.

   tantôt . . . tantôt, now . . . now, sometimes . . . sometimes, at one time . . . at another time.

tapage, *m.* tapping; noise, racket.

tape, *f.* plug, stopper; bung; tompion; tap; rap.

tapé, *p.a.* stopped, etc. (see taper); (of fruit) dried; (of varnish) spread evenly.

taper, *v.t.* stop, stop up, plug, plug up; tap, rap, knock, stamp. — *v.i.* (*Metal.*) develop cracks internally. — *m.* (*Elec.*) short-circuit key.

tapette, *f.* plug, stopper; tap, light blow.

tapioca, tapioka, *m.* tapioca.

tapirer, *v.t.* redden, color red.

tapis, *m.* carpet; rug; (cloth) cover, cloth; tapestry; tapis.

tapisser, *v.t.* cover, coat, line; carpet; paper; hang; deck, adorn.

tapisserie, *f.* tapestry; hangings; upholstery.

tapissier, *m.* upholsterer.

tapure, *f.* (*Metal.*) interior crack or fissure.

taque, *f.* (*Metal.*) plate.

taqueret, *m.* fore plate of a forge hearth.

taquerie, *f.* charging hole of a reverberatory furnace.

taquet, *m.* cleat; step; catch, stop, detent.

   — d'arrêt, stop, detent, pawl.

tarage, *m.* taring; allowance for tare; calibration; damaging.

taraison, *m.* tile disk for closing the working hole of a glass furnace.

tarare, *m.* winnower, fanning machine.

taraud, *m.* screw tap, tap.

taraudage, *m.* tapping (with a screw tap).

tarauder, *v.t.* tap; cut (screws).

taraxacine, *f.* taraxacin.

tard, *a. & adv.* late. — *m.* late hour.

   au plus —, at the latest.

   sur le —, late (in the evening).

tarder, *v.i.* delay.

   il me tard de, I long to.

   ne — pas de (or *à*), not delay to, soon, quickly, immediately.

tardif, *a.* slow; tardy.

tardivement, *adv.* slowly; tardily.

tare, *f.* tare; loss, deficiency; defect.

tarer, *v.t.* tare; damage, deteriorate; calibrate (for weight).

targette, *f.* slide bolt, fastening.

tari, *p.a.* dried up, exhausted. — *m.* palm wine.

tarière, *f.* auger, borer.

tarif, *m.* price list; rate (of prices); tariff.

tarifer, *v.t.* fix or list (prices); tariff.

tarir, *v.t.* dry up, drain, exhaust. — *v.i.* dry up, go dry, be exhausted, be drained.

tarissable, *a.* exhaustible.

tarissant, *p.a.* drying up, near exhaustion.

tarissement, *m.* drying up, exhaustion.

tarmacadam, *m.* tar macadam, tar-bound macadam.

tarmacadamiser, *v.t.* pave with tar macadam.

tartareux, *a.* tartareous, tartarous.

Tartarie, *f.* Tartary, Tatary.

tartarique, *a.* tartaric.

tartariser, *v.t.* tartarize.

tartrage, *m.* addition of, or treatment with, tartaric acid or a tartrate (as in wine making).

tartrate, *m.* tartrate.

   — antimonio-potassique, antimony potassium tartrate, tartar emetic.

   — borico-potassique, soluble cream of tartar (cream of tartar mixed with boric acid or borax).

   — d'antimoine et de potasse, antimony potassium tartrate, tartar emetic.

   — de chaux, calcium tartrate.

   — de potasse, potassium tartrate.

   — de potasse et de soude, potassium sodium tartrate, Rochelle salt.

   — ferrico-ammonique, — ferrique ammoniacal, (*Pharm.*) iron and ammonium tartrate, ammonio-ferric tartrate.

   — ferrico-potassique, ferric potassium tartrate, (*Pharm.*) tartarated iron.

tartre, *m.* tartar; (in boilers, etc.) scale.

   — brut, crude tartar, argol.

   — blanc, white tartar (from white wines).

   — chalybé, (*Pharm.*) iron and potassium tartrate, tartarated iron.

   — dentaire, dental tartar.

   — des chaudières, boiler scale.

   — émétique, tartar emetic.

   — martial, = tartre chalybé.

   — rouge, red tartar (from red wines).

   — soluble, soluble tartar (potassium tartrate).

   — stibié, tartar emetic.

   — vitriolé, vitriolated tartar (old name for potassium sulfate).

tartreux, *a.* tartarous, tartareous.

tartrier, *m.* tartar maker.

tartrifuge, *m.* boiler compound.

tartrimètre, *m.* an instrument for determining the purity of cream of tartar.

tartrique, *a.* tartaric.

tartronique, *a.* tartronic.

tartrovinique, *a.* tartrovinic (ethyltartaric).

tas, *m.* heap, pile, mass; hand anvil, block (of metal); (*Construction*) part already built.

Tasmanie, *f.* Tasmania.

tasse, _f._ cup.
— _à dégustation,_ taster's cup, tasting cup.
— _à thé,_ tea cup.
tasseau, _m._ bracket, block, cleat, strip; clamp; hand anvil; catch, tappet.
tassement, _m._ packing, etc. (see tasser).
tasser, _v.t._ pack, tamp, ram, compress; heap up, pile up. — _v.r._ sink, settle, pack; heap up, pile up. — _v.i._ grow thick.
tâtage, _m._ feeling, etc. (see tâter).
tâte, _f._ sample.
tâter, _v.t._ feel; try, test; taste; sound. — _v.i._ (with _de_ or _à_) taste, try; feel.
— _le terrain,_ feel one's way.
tâte-vin, _m._ winetaster, wine sampler.
tâton, _n._ In the phrase _à tâtons,_ gropingly, tentatively.
tâtonnement, _m._ groping; trial, experiment.
tâtonner, _v.i._ grope, feel about.
tatouer, _v.t._ tattoo.
taupe, _f._ (Zoöl.) mole; moleskin.
taure, _f._ heifer.
taureau, _m._ bull.
taurine, _f._ taurine.
taurocholique, _a._ taurocholic.
taurocolle, _f._ taurocol, taurocolla (a glue).
tautomère, _m._ tautomer.
tautomérie, _f._ tautomerism, tautomery.
tautomérique, _a._ tautomeric.
tautomérisation, _f._ tautomerization.
tautomériser, _v.t. & r._ tautomerize.
taux, _m._ rate; proportion, amount, strength, dilution, etc.; price.
— _pour 100 parties,_ parts per hundred, per cent.
taveler, _v.t._ speckle, spot.
tavelure, _f._ speckling, spotting, spots.
taxation, _f._ taxation, etc. (see taxer).
taxe, _f._ tax; (officially fixed) price, rate.
taxer, _v.t._ tax; fix the price of; fix the pay of.
taxidermie, _f._ taxidermy.
tchèque, _a._ Czechic, Czechish. — _m._ Czech (language); (_cap._) Czech (person).
te, _pron._ thee, to thee, you, to you.
té, _m._ T, T-shaped object or arrangement (as a T square, T tube).
technicien, _m._ technician, technical man.
technicité, _f._ technicalness.
technique, _a._ technical. — _f._ technic, technique. — _m._ technical character.
techniquement, _adv._ technically.
technologie, _f._ technology.
— _chimique,_ chemical technology.
technologique, _a._ technological, -logic.
technologue, technologiste, _m._ technologist. — _a._ technological.
teck, _m._ teak (tree or wood).
tégumentaire, _a._ tegumentary.
teignant, _p.a._ dyeing, etc. (see teindre); suitable for dyeing.

teignent, _pr. 3 pl. indic. & subj._ of teindre.
teignit, _p.def. 3 sing._ of teindre.
teillage, _m._ (Glass) whitish spot due to excess of soda; breaking (as of hemp).
teindre, _v.t._ dye; color, tinge; stain; imbue, tincture. — _v.r._ be dyed; be colored; be tinctured. — _v.i._ dye.
teint, _m._ dye, (dyed) color; (manner of) dyeing; complexion. — _p.a._ dyed, etc. (see teindre). — _pr. 3 sing._ of teindre.
teinte, _f._ tint; tinge, hue, color; (fig.) tincture, shade, touch.
— _de passage,_ transition tint, intermediate tint.
— _dure,_ ground tint, ground color.
— _plate,_ flat tint.
— _vierge,_ unmixed tint, simple tint.
teinter, _v.t._ tint, tinge, color slightly.
teinture, _f._ dyeing; dye bath; dye; color, hue; (Pharm., etc.) tincture.
— _ammoniacale de quinine,_ ammoniated tincture of quinine.
— _ammoniacale de seigle ergoté,_ ammoniated tincture of ergot.
— _aromatique sulfurique,_ aromatic sulfuric acid.
— _avec réserve,_ dyeing with resists.
— _balsamique,_ compound tincture of benzoin.
— _composée,_ (Pharm.) compound tincture.
— _d'actée à grappes,_ tincture of cimicifuga.
— _de baume de Tolu,_ tincture of tolu.
— _de cachou,_ compound tincture of gambir, tincture of catechu.
— _de camphre concentrée,_ spirit of camphor.
— _de ciguë,_ tincture of conium.
— _de cochenille,_ (Dyeing) cochineal bath; (Pharm.) tincture of cochineal.
— _d'écorce de cerisier,_ tincture of Virginian prune (tincture of wild cherry bark).
— _de fève du Calabar,_ tincture of physostigma.
— _de fleur d'arnica,_ tincture of arnica.
— _de fleur de tous-les-mois,_ tincture of calendula.
— _de gaïac ammoniacale,_ ammoniated tincture of guaiac.
— _de lobélie enflée,_ tincture of lobelia.
— _de noix de galle,_ tincture of nutgall.
— _de perchlorure de fer,_ tincture of ferric chloride.
— _de polygala de Virginie,_ tincture of senega.
— _de quinquina jaune,_ tincture of cinchona.
— _de racine d'aconit,_ tincture of aconite.
— _de ratanhia,_ tincture of krameria, tincture of rhatany.
— _de résine de gaïac,_ tincture of guaiac.
— _de résine de podophyllum,_ tincture of podophyllum.
— _de sanguinaire,_ tincture of sanguinaria, tincture of bloodroot.

teinture *de savon vert*, liniment (or tincture) of green soap.

— *de séné aromatique* (or *composée*), compound tincture of senna.

— *de souci*, tincture of calendula, tincture of marigold.

— *d'essence*, (*Pharm.*) alcoholic solution of an essential oil, spirit (of so-and-so).

— *d'essence d'anis vert*, spirit of anise.

— *d'essence de cajeput*, spirit of cajuput.

— *d'essence de cannelle*, spirit of cinnamon.

— *d'essence de gaulthérie*, spirit of gaultheria, spirit of wintergreen.

— *d'essence de genièvre*, spirit of juniper.

— *d'essence de lavande*, spirit of lavender.

— *d'essence de menthe*, spirit of peppermint.

— *d'essence de muscade*, spirit of nutmeg, essence of nutmeg.

— *d'essence de romarin*, spirit of rosemary.

— *d'essence d'orange*, compound spirit of orange.

— *de strophanthus kombé*, tincture of strophanthus.

— *de tournesol*, tincture of litmus.

— *de valériane ammoniacale*, ammoniated tincture of valerian.

— *d'extrait d'opium*, tincture of opium, laudanum.

— *d'iode*, tincture of iodine.

— *d'opium ammoniacale*, ammoniated tincture of opium.

— *d'opium camphrée*, camphorated tincture of opium, paregoric.

— *d'opium sans odeur*, tincture of deodorized opium.

— *éthérée*, ethereal tincture.

— *grand teint*, fast dyeing.

— *simple*, (*Pharm.*) simple tincture.

— *thébaïque*, tincture of opium, laudanum.

— *vulnéraire*, a tincture of various herbs used as an astringent and stimulating lotion.

(For the meaning of other phrases see the next important word in the phrase.)

teinturerie, *f.* dyeing; dyehouse, dye works.

teinturier, *m.* dyer; a kind of dark red grape.
— *a.* dyeing, dye, relating to dyeing.

tek, *m.* teak (tree or wood).

tel, *a.* such, like, so, so great, so much. — *m.* such a one, so and so, many a one.

à — fin que de raison, at all events, in any case.

comme —, as such.

de telle sorte que, en telle sorte que, so that, in such a way that.

— *que*, just as; such that, so great that.

— *quel, telle quelle*, such as it is; just as it is, as it is, without modification; such as it was, just as it was; such as he (or she) is or was.

tel . . . tel, like . . . like, as . . . so.

telas, *m.* a kind of coarse sackcloth.

télégraphier, *v.t. & i.* telegraph.

télescope, *m.* telescope.

télescopique, *a.* telescopic.

telle, *a. fem.* of tel.

tellement, *adv.* so, in such a manner, of such a kind, to such a degree.

tellurate, *m.* tellurate.

tellure, *m.* tellurium.

telluré, *a.* combined with or containing tellurium, tellurized, telluretted; tellurium, of tellurium.

tellureux, *a.* tellurous.

tellurhydrique, *a.* tellurhydric, hydrotelluric.

tellurien, *a.* earth, of or from the earth.

tellurifère, *a.* telluriferous.

tellurique, *a.* telluric.

eau —, water in the earth or that has come out of the earth.

tellurite, *m.* tellurite.

tellurure, *m.* telluride.

témoignage, *m.* testifying, etc. (see témoigner); testimony, evidence; token, proof.

témoigner, *v.t.* testify, give evidence of, evidence, show. — *v.i.* testify, witness.

témoin, *m.* witness; mark; proof; evidence; (*Ceram.*) trial piece; (*Metal.*) trial rod; bead. — *a.* indicating, indicator; test, comparison; witnessing.

tempérament, *m.* temperament; temper; moderation.

tempérant, *p.a.* tempering, etc. (see tempérer); temperate; (*Med.*) sedative.

température, *f.* temperature.

— *absolue*, absolute temperature.

— *basse*, low temperature.

— *critique*, critical temperature.

— *de régime*, working temperature, right temperature for regular operation (as of a furnace).

— *du rouge*, red heat.

— *haute*, high temperature.

tempéré, *p. a.* tempered, etc. (see tempérer); temperate, moderate. — *m.* medium temperature.

tempérer, *v.t.* temper; moderate; mitigate.

tempêtueux, *a.* tempestuous.

templine, *n.* See essence de templine.

temporaire, *a.* temporary.

temporairement, *adv.* temporarily.

temps, *m.* time; weather; delay, pause; tense.

à —, in time, in season; for a time.

au — jadis, in former times, formerly.

dans le — que (or *où*), at the moment when, while.

de — à autre, de — en —, from time to time, now and then.

en — et lieu, at the proper time and place.

entre —, in the interval, meantime.

— *d'arrêt*, stoppage, delay.

— *de pose*, (*Photog.*) time of exposure.

temps *moyen*, mean time.
— *mort*, — *perdu*, lost time; (*Mach.*) lost motion, backlash.
selon le —, *suivant le* —, according to circumstances.
sur le —, at once, off-hand.
tenable, *a.* tenable.
tenace, *a.* tenacious; cohesive, (of metals, etc.) tough; adhesive, sticky; (of colors) persistent.
tenacement, *adv.* tenaciously; persistently.
tenacité, *f.* tenaciousness, tenacity, etc. (see tenace).
tenaille, *f.* (usually in *pl.*) pincers, tongs, pliers, nippers; vise.
tenaille(s) à creuset, crucible tongs.
tenaille(s) à feu, fire tongs.
tenaille(s) à loupes, (*Metal.*) bloom tongs.
— *à vis*, hand vise.
tenait, *imp. 3 sing.* of tenir.
tenant, *p.a.* holding, etc. (see tenir); sitting.
— *m.* champion, defender; part adjacent; relation, related thing.
tendance, *f.* tendency.
tendant, *p.a.* stretching, etc. (see tendre, *v.t.* and *v.i.*); directed (toward).
tendeur, *m.* stretcher, etc. (see tendre, *v.*); coupling screw; extension piece. — *a.* stretching, etc. (see tendre, *v.*).
tendineux, *a.* tendinous.
tendon, *m.* (*Anat.*) tendon.
tendre, *a.* soft; tender; brittle; mild; delicate; (of bread) fresh, new; open, impressionable. — *m.* soft part, tender part.
— *v.t.* stretch; stretch out, draw out; tighten; strain; spread, spread out; hold out; throw out. — *v.r.* be stretched, etc. — *v.i.* tend, be directed, contribute.
tendresse, *f.* softness, etc. (see tendre).
tendreté, *f.* tenderness (of meat, etc.).
tendu, *p.a.* stretched, etc. (see tendre, *v.*); tight, taut, tense; intent; (of trajectories) flat.
ténèbres, *f.pl.* darkness.
ténébreux, *a.* dark, gloomy, obscure, secret.
ténesme, *m.* (*Med.*) tenesmus.
tenette, *f.* (also pl.) small pincers, forceps.
teneur, *f.* content; tenor, purport, course.
— *m.* holder, etc. (see tenir).
— *pondérale*, content by weight.
tenir, *v.t.* hold; keep; have; contain; take; get; take, be on (a road); consider; oblige, bind; owe; affect. — *v.r.* be held, etc.; hold (by or to), hold on; be adjacent, adjoin; hold together; stick, adhere (to), stand (by); stand; stay, remain, keep; refrain, keep; consider oneself. — *v.i.* hold; hold fast, stick, adhere; hold out; keep up; hold good; hold together, stick together; be attached (to); care (for); be anxious, wish;

be adjacent (to), be next (to); arise, result; depend (on), be the fault (of); relate (to); be related; be akin (to), partake (of), smack (of); (with *de*) resemble; (of assemblies) sit; be held; stand; stay, remain; be contained, go (into).
— *bon*, hold fast, hold on, stand fast.
— *compte de*, take into account (or consideration), take advantage of.
tenon, *m.* tenon, stud, lug, projection.
tension, *f.* tension; pressure; tenseness; tightness; (of a trajectory) flatness.
en —, (*Elec.*) in series.
— *chimique*, chemical tension.
— *de la vapeur*, steam pressure.
— *de vapeur*, vapor tension, vapor pressure.
— *de régime*, normal tension.
— *superficielle*, surface tension.
tentant, *p.a.* attempting; tempting.
tentatif, *a.* tentative.
tentation, *f.* temptation.
tentative, *f.* attempt, effort, endeavor.
tente, *f.* tent; awning.
tenter, *v.t.* try; attempt; risk; tempt; tent.
tenture, *f.* hangings, hanging.
tenu, *p.a.* held, etc. (see tenir).
ténu, *a.* tenuous, slender, thin.
tenue, *f.* holding, etc. (see tenir); session, meeting; order; steadiness; behavior; bearing; dress, attire.
de —, steady, settled, lasting.
ténuité, *f.* tenuity.
tephrosie, *f.* (*Bot.*) Tephrosia (Cracca).
terbine, *f.* terbia.
terbium, *m.* terbium.
térébate, *m.* terebate.
térébène, *n.* terebene (a mixture of terpenes).
térébénique, *a.* terpene, of (or of the nature of) a terpene or terpenes.
térébenthène, *m.* terebenthene (*l*-pinene).
térébenthine, *f.* turpentine.
essence de —, oil of turpentine.
— *commune*, common European turpentine.
— *d'Alsace*, Strassburg turpentine.
— *de Bordeaux*, Bordeaux turpentine, common European turpentine.
— *de Boston*, white turpentine, American turpentine.
— *de Chio*, Chian turpentine.
— *de mélèze*, — *de Venise*, Venice turpentine, larch turpentine.
— *des Vosges*, Strassburg turpentine.
— *du Canada*, Canada turpentine, Canada balsam, balsam of fir.
térébenthiné, *a.* (*Pharm.*) containing turpentine.
térébinthacées, *f.pl.* (*Bot.*) Anacardiaceæ, Terebinthaceæ.
térébinthe, *m.* (*Bot.*) terebinth (*Pistacia terebinthus*).

térébinthiné, *a.* terebinthine, terebinthinate.
térébique, *a.* terebic.
térébrer, *v.t.* terebrate, bore.
téréphtalique, *a.* terephthalic.
terme, *m.* term; limit, end; end point; member (of a series); time; quarter, three months.
— *de métier*, technical term.
— *ultime*, end product; end member; last term.
terminaison, *f.* termination.
terminé, *p.a.* terminated, ended; finished, ready.
terminer, *v.t.* terminate; end; finish, complete. — *v.r.* terminate, end.
terminologie, *f.* terminology.
terminus, *m.* terminus. — *a.* terminal.
ternaire, *a.* ternary.
terne, *a.* dull, lusterless.
ternir, *v.t. & r.* tarnish; dull; soil; fade.
ternissement, *m.* tarnishing, etc. (see ternir).
ternissure, *f.* tarnishing, etc. (see ternir); dullness; tarnish.
terpène, *m.* terpene.
terpénique, *a.* terpene, of a terpene or terpenes.
terpine, *f.* terpinol, terpin.
terpinène, *m.* terpinene.
terpinéol, terpinol, *m.* terpineol, terpinol.
terpinolène, *m.* terpinolene.
terrage, *m.* scouring, etc. (see terrer).
terraille, *f.* a fine kind of earthenware.
terrain, *m.* ground; piece of ground, tract; (*Geol.*) terrane; formation; (*Mil.*) terrain.
— *d'alluvion*, alluvial ground; alluvial terrane.
terramare, *f.* terramara (earth as a fertilizer).
terras, *m.* resin mixed with earth collected at the foot of trees.
terrasse, *f.* earthwork; terrace; foreground.
terrasser, *v.t.* bank up, fill in; terrace; floor, confound.
terre, *f.* earth; clay; ground; soil; land.
mettre à la —, (*Elec.*) ground.
— *à briques*, brick earth, brick clay.
— *à ciment*, cement earth, cement clay.
— *à couleur*, any earth used as a pigment.
— *à faïence*, crockery earth, earthenware clay.
— *à foulon*, fuller's earth.
— *alcaline*, alkaline earth.
— *à mouler*, molding loam.
— *à pipe*, pipe clay.
— *à porcelaine*, porcelain clay, kaolin.
— *à pot, à poterie, — à potier*, potter's earth, potter's clay.
— *au feu*, fireclay.
— *bolaire*, bole.
— *calcaire*, calcareous earth.
— *colorée*, colored earth, specif. umber.
— *crue*, raw earth, raw clay.
— *cuite*, terra cotta, baked clay (esp. the porous variety, including tiles and bricks).

terre *d'alun*, aluminous earth.
— *de fer*, (*Min.*) iron clay, clay ironstone.
Terre de Feu, Tierra del Fuego.
— *de la Chine*, china clay, kaolin.
— *de la Nouvelle-Orléans*, annatto.
— *de Lemnos*, Lemnian earth.
— *de Paris*, a refractory clay used for making crucibles.
— *de pipe*, pipe clay; (*Ceram.*) a kind of glazed earthenware.
— *de Sienne*, sienna.
— *de Sienne brûlée*, burnt sienna.
— *de Vérone*, Verona earth, Veronese earth, Verona green.
— *d'infusoires*, infusorial earth (diatomaceous earth, kieselguhr).
— *d'Italie*, Italian earth, sienna.
— *d'ombre*, umber.
— *du Japon*, terra japonica (gambier, or, more widely, any catechu).
— *franche*, vegetable mold.
— *glaise*, clay (esp. one used for burning).
— *grasse*, clayey soil.
— *mérite*, turmeric.
— *pesante*, heavy earth, baryta.
— *pourrie*, rottenstone.
— *rare*, rare earth.
— *réfractaire*, fire clay.
— *tourbeuse*, peat soil, peat land.
— *végétale*, vegetable mold.
terreau, *m.* mold, humus; compost.
Terre-Neuve, *f.* Newfoundland.
terrer, *v.t.* scour, full (with fuller's earth); clay (sugar); earth, cover or treat with earth.
terrestre, *a.* terrestrial.
terreux, *a.* earthy; of an earth or earth metal, of the earths or earth metals; (of colors) dull; dirty.
terriblement, *adv.* terribly.
terrine, *f.* pan, dish (typically one of earthenware or stoneware with sloping sides).
— *pour évaporations*, evaporating dish.
territoire, *m.* territory.
terroir, *m.* soil, ground.
tertiaire, *a.* tertiary.
tertiobutylbenzène, *m. tert*-butylbenzene.
tertiobutyle, *m.* tertiary butyl, *tert*-butyl.
tes, *a.pl.* thy, your.
tesson, *m.* fragment (of glass or pottery).
test, *m.* = têt; (*Biol.*) test, testa.
testicule, *m.* (*Anat.*) testicle.
testif, *m.* camel's hair.
test-objet, *m.* (*Micros.*) test object.
têt, *m.* test (the vessel, esp. a small cup or dish of fire clay); fragment (of glass or pottery).
— *à combustion*, a small fireclay cup for the combustion of phosphorus.
— *à gaz*, a fireclay dish having a hole in the bottom and a notch in the side, and used inverted as a cover.

têt à *rôtir*, roasting dish (a vessel similar to a scorifier).

— *de coupellation*, cupel.

tétanine, *f.* tetanine.

tétanique, *a.* (*Physiol. & Med.*) tetanic.

tétanos, *m.* (*Med.*) tetanus.

tétanotoxine, *f.* tetanus toxin, tetanotoxin.

têtard, *m.* (*Zoöl.*) tadpole.

tétartoèdre, *m.* (*Cryst.*) tetartohedral.

tétartoédrie, *f.* (*Cryst.*) tetartohedrism.

tétartoédrique, *a.* (*Cryst.*) tetartohedral.

tête, *f.* head (in various senses); top, front, first part, etc.; (in distilling) first fraction; (of a sugar loaf) point, head.

de —, of the head; at the head, chief; from memory; by imagination.

en —, at the head.

— *de nègre*, (*Rubber*) negrohead, sernamby.

— *morte*, caput mortuum.

— *perdue*, countersunk head; flush head.

tétine, tetine, *f.* nipple.

à *tétine*, teat-shaped, corrugated (said of the end of a tube to which hose may be attached).

teton, téton, *m.* projection, elevation, stud, end, point; breast.

tétrabasique, *a.* tetrabasic.

tétrabromé, *a.* tetrabromo, tetrabromo-.

tétracarbonique, *a.* tetracarboxylic.

tétracétylénique, *a.* designating acids containing four triple bonds.

tétrachloré, *a.* tetrachloro, tetrachloro-.

tétrachlorure, *m.* tetrachloride.

— *de carbone*, carbon tetrachloride.

tétracuprique, *a.* tetracupric.

tétradymite, *f.* (*Min.*) tetradymite.

tétraédral, *a.* tetrahedral.

tétraèdre, *m.* tetrahedron. — *a.* tetrahedral

tétraédrique, *a.* tetrahedral.

tétraéthylé, *a.* tetraethyl, tetraethyl-.

tétraéthylique, *a.* tetraethyl.

tétrafluorure, *m.* tetrafluoride.

tétragonal, *a.* tetragonal.

tétragone, *a.* tetragonal. — *m.* tetragon.

tétrahalogéné, *a.* tetrahalogenated, tetrahalogen, tetrahalo, tetrahalo-.

tétrahydrobenzène, *m.* tetrahydrobenzene (cyclohexene).

tétrahydrogéné, *a.* tetrahydrogenated, tetrahydro, tetrahydro-.

tétrahydroglyoxaline, *f.* tetrahydroglyoxaline (tetrahydroimidazole, imidazolidine).

tétrahydroquinone, *f.* tetrahydroquinone (1, 4-cyclohexanedione).

tétrahydrure, *m.* tetrahydride.

tétraiodé, *a.* tetraiodo, tetraiodo-.

tétraméthylène, *m.* tetramethylene (either cyclopropane or the radical -$CH_2CH_2CH_2CH_2$-).

tétraméthylène-diamine, *f.* tetramethylenediamine (1, 4-butanediamine, putrescine).

tétraméthylène-imine, *f.* tetramethylenimine (pyrrolidine).

tétraméthylénique, *a.* tetramethylene.

tétrane, *m.* tetrane (butane).

tétranitré, *a.* tetranitrated, tetranitro-, tetranitro.

tétranol, *m.* tetranol (butanol).

tétraoxy-, tetrahydroxy-; tetraoxy-. (See note at *oxy-*).

tétraoxyquinone, *f.* tetrahydroxyquinone.

tétraséléniure, *m.* tetraselenide.

tétrasubstitué, *a.* tetrasubstituted.

tétratomicité, *f.* tetratomicity.

tétratomique, *a.* tetratomic.

tétravalence, *f.* tetravalence (quadrivalence).

tétravalent, *a.* tetravalent (quadrivalent).

tétrazine, *f.* tetrazine.

tétrazoïque, *a.* tetrazo.

tétrazol, *m.* tetrazole.

tétréthylé, *a.* tetraethyl, tetraethyl-.

tétronal, *m.* (*Pharm.*) tetronal.

tétrose, *f.* tetrose.

tétroxyde, *m.* tetroxide.

tétryl, *m.* (*Expl.*) tetryl (tetranitromethylaniline).

texte, *m.* text.

textile, *a. & m.* textile.

textuel, *a.* textual; literal.

textuellement, *adv.* textually; literally, verbatim.

textulaire, *a.* textural.

texture, *f.* texture; structure.

thalleux, thallieux, *a.* thallous, thallious.

thalline, *f.* thalline.

thallique, *a.* thallic.

thallium, *m.* thallium.

thapsie, *f.* (*Bot.*) Thapsia.

thé, *m.* tea.

— *de Terre- Neuve*, — *du Canada*, wintergreen.

théâtral, *a.* theatrical.

thébaïne, *f.* thebaine.

thébaïque, *a.* of or pertaining to opium.

théine, *f.* theine (same as caffeine).

thème, *m.* theme.

thénardite, *f.* (*Min.*) thenardite.

théobrome, *m.* (*Bot.*) Theobroma.

théobromine, *f.* theobromine.

théophylline, *f.* theophylline.

théorème, *m.* theorem.

théorétique, *a.* theoretical, theoretic.

théoricien, *m.* theorist.

théorie, *f.* theory.

— *atomique*, atomic theory.

— *chimique*, chemical theory.

— *cinétique*, kinetic theory.

théorique, *a.* theoretical, theoretic.

théoriquement, *adv.* theoretically.

théoriser, *v.i.* theorize.

thérapeutique, *a.* therapeutic, therapeutical.

— *f.* therapeutics.

thérapeutiquement, *adv.* therapeutically.
thériaque, *f.* (*Old Pharm.*) theriaca, theriac.
thermal, *a.* thermal.
thermique, *a.* thermic, thermal, of heat.
thermite, *f.* thermite, "thermit."
thermochimie, *f.* thermochemistry.
thermochimique, *a.* thermochemical.
thermochimiste, *m.* thermochemist.
thermochroïque, *a.* thermochroic.
thermodynamicien, *m.* thermodynamicist.
thermodynamique, *a.* thermodynamic, thermodynamical. — *f.* thermodynamics.
thermo-électrique, *a.* thermoelectric(al).
thermogène, *a.* thermogenic, thermogenous. — *m.* heat generator; (*Old Chem.*) thermogen.
thermogénèse, *f.* thermogenesis.
thermolabile, *a.* thermolabile.
thermomètre, *m.* thermometer.
— *à alcool,* alcohol thermometer, spirit thermometer.
— *à bulle mouillée* (*sèche*), wet-bulb (drybulb) thermometer.
— *à gaz,* gas thermometer.
— *à mercure,* mercury (or mercurial) thermometer.
— *contrôlé,* calibrated thermometer.
— *enregistreur,* recording thermometer.
— *étalon,* standard thermometer.
thermométrie, *f.* thermometry.
thermométrique, *a.* thermometric, thermometrical.
thermométriquement, *adv.* thermometrically.
thermophile, *a.* (*Bact.*) thermophilic, -philous.
thermorégulateur, *m.* thermoregulator, thermostat.
thermostabile, thermostable, *a.* thermostable.
thèse, *f.* thesis.
thialdéhyde, *m.* thioaldehyde.
thiazinique, *a.* thiazine, of thiazine.
thiazol, *m.* thiazole.
thiénone, *f.* thiénone.
thioacide, *m.* thio acid.
thioaldéhyde, *m.* thioaldehyde.
thiobenzénylique, *a.* thiobenzenyl (applied to dyes better called thiazole dyes).
thiocyanate, *m.* thiocyanate.
thiocyanique, *a.* thiocyanic.
thioflavine, *f.* (*Dyes*) thioflavin, thioflavine.
thiohydantoïne, *f.* thiohydantoïn.
thionaphtène, *m.* thionaphthene, benzothiophene.
thionine, *f.* thionine.
thionique, *a.* thionic.
thionyl, *m.* thionyl.
thionylique (*thiénylique?*), *a.* designating the alcohol $C_4H_3SCH_2OH$ (2-thiophenecarbinol).

thiophène, *m.* thiophene.
thiophénique, *a.* thiophene, of thiophene, (esp. of the acid and aldehyde) thiophenic.
thiophosgène, *m.* thiophosgene.
thiophtène, *m.* thiophthene.
thiosinamine, thiosinnamine, *f.* thiosinamine.
thiosulfate, *m.* thiosulfate.
thiosulfocarbamique, *a.* dithiocarbamic.
thio-urée, *f.* thiourea.
thioxène, *m.* thioxene.
thon, *m.* tunny (a fish).
thorine, *f.* thoria.
thorique, *a.* thorium, of thorium, thoric.
thorite, *f.* (*Min.*) thorite.
thorium, *m.* thorium.
thran, *m.* train oil.
thrombine, *f.* thrombin.
thrombogène, *m.* thrombogen (prothrombin).
thrombose, *f.* (*Med.*) thrombosis.
thulium, *m.* thulium.
thuya, thuia, *m.* (*Bot.*) thuja.
thym, *m.* (*Bot.*) thyme.
— *bâtard,* — *serpolet,* wild thyme (*Thymus serpyllum*).
— *commun,* — *vulgaire,* garden thyme (*Thymus vulgaris*).
thymélée, *f.* (*Bot.*) Thymelæa.
thymine, *f.* thymine.
thymique, *a.* thymic (pertaining either to the thymus gland or to thyme; thymic acid in the latter sense is better called *thymol*).
thymol, *m.* thymol.
thymonucléique, *a.* thymonucleic, thymus nucleic.
thymus, *m.* (*Anat.*) thymus gland, thymus.
thyréoglobuline, *f.* thyreoglobulin.
thyroïde, thyréoïde, *a. & m.* (*Anat.*) thyroid.
thyroïdien, thyréoïdien, *a.* thyroidal, thyroid.
thyroïdine, thyroïodine, *f.* thyroidin.
thyroxine, *f.* thyroxin.
tiède, *a.* moderately warm, tepid, lukewarm.
tiédeur, *f.* lukewarmness, tepidity, tepidness.
tiédir, *v.i.* become tepid or lukewarm (esp. by cooling). — *v.t.* make tepid or lukewarm.
tien, *pron.* thine, yours.
tiendra, *fut. 3 sing.* of tenir.
tienne, *pr. 1 & 3 sing. subj.* of tenir. — *fem.* of tien.
tiennent, *pr. 3 pl. indic. & subj.* of tenir.
tient, *pr. 3 sing.* of tenir.
tierce, *a.fem.* of tiers.
tiers, *a.* third. — *m.* third.
tiers-argent, *m.* tiers-argent (silver 1, aluminium or German silver 2).
tiers-point, *m.* three-cornered file; saw file.
tige, *f.* rod; stem, stalk; shaft, shank, spindle, bar, pin, etc.; stock, origin.

tige à cavaliers, rider rod (for a balance).
— conductrice, — de guide, — directrice, (Mach.) guide rod.
— de communication, (Mach.) connecting rod.
— en verre, glass rod.

tigelle, f. (Bot.) tigella, tigelle, tigellus.

tigre, m. tiger.

till, m. sesame.

tillage, m. = teillage.

tille, f. linden bast; hemp harl; a kind of hammer with an edged peen.

tilleul, m. (Bot.) linden, basswood (Tilia).

timbrage, m. stamping.

timbre, m. stamp; stamp office; stamp duty; bell; (of sounds) timbre.
— en caoutchouc, rubber stamp.

timbre-poste, m. postage stamp.

timbrer, v.t. stamp.

timide, a. timid.

timidement, adv. timidly.

timoré, a. timorous.

tincal, m. tincal (crude borax).

tinctorial, a. tinctorial, dyeing.

tine, f. a kind of cask; tub; = jalot; = tinne.

tinette, f. a keg or small cask; small tub.

tinkal, m. tincal (crude borax).

tinne, f. pug mill (for clay); = jalot; = tine.

tint, p.def. 3 sing. of tenir.

tinter, v.i. tinkle, ring, tingle; toll. — v.t. toll; prop, block.

tintin, m. (Silver) a large stone mortar; clink, tinkle.

tion, m. (Metal.) a tool for cleaning crucibles.

tip, n. a kind of vegetable butter.

tir, m. firing, fire, shooting; shooting practice, target practice; shooting grounds.
— à obus, shell fire; shell practice.
— rapide, rapid fire.

tirage, m. drawing, etc. (see tirer); draft; traction; impression (of a book); difficulty.
à bon —, à fort —, with good draft, with strong draft.
— à la poudre, (Mining) blasting.
— forcé, forced draft.

tiraille, f. (Mach.) connecting rod.

tiraillement, m. pulling about, pulling; twitching; trouble, discord.

tirailler, v.t. pull about, pull; importune.

tirant, p.a. drawing, etc. (see tirer); — m. something which draws or is drawn, as a V-shaped cutter for sugar beets, a tie beam, tie rod or stay rod, a shoe lace, etc.; draft (of a ship); tendon (in meat).

tire, f. drawing, draw, pull, etc. (see tirer).
d'une —, at one stroke, at a stretch.

tire-. drawer, extractor.

tiré, p.a. drawn, pulled, etc. (see tirer).

tire-bouchon, m. corkscrew.

tire-pièce, m. (Sugar) skimmer.

tire-point, m. a sharp-pointed piercing tool.

tirer, v.t. draw; pull; draw or pull out, in, up, down, off, on, etc.; stretch, tighten; extract; mine, quarry; derive, deduce; get, obtain; attract; absorb; shoot, fire, fire at; pass (a spark); print; strain (by pulling); wind (as silk); extricate; milk. — v.r. be drawn, be pulled, etc.; get out, away or thru. — v.i. draw; pull; shrink, contract; verge (on), tend (toward); shoot, fire; go off; fence.
— à clair, — au clair, draw off, bottle off (liquors); clear up.
— à point, draw to a point.
— parti de, take advantage of, profit by.

tireur, m. drawer, puller, etc. (see tirer); specif., wiredrawer; printer's or photographer's assistant; marksman.

tiroir, m. drawer (as of a table); (Mach.) slide valve, slide.

tisage, m. (Glass) stoking.

tisane, f. (Pharm.) (1) infusion, tea, (2) decoction.
— d'aloès composée, compound decoction of aloes.
— d'angosture, infusion of cusparia, infusion of angustura.
— de bois de Campêche, decoction of logwood.
— de buchu, infusion of buchu.
— de cascarille, infusion of cascarilla.
— de champagne, a very light champagne used for the sick.
— de chirette, infusion of chiretta.
— de columbo, infusion of calumba.
— d'écorce de cerisier sauvage, infusion of wild cherry.
— d'écorce d'orange, infusion of orange peel.
— de digitale, infusion of digitalis.
— de genêt à balais, infusion of broom.
— de gentiane composée, compound infusion of gentian.
— de girofle, infusion of cloves.
— de houblon, infusion of hops.
— de polygale de Virginie, infusion of senega.
— de quassie, infusion of quassia.
— de quinquina jaune, acid infusion of cinchona.
— de ratanhia, infusion of krameria, infusion of rhatany.
— de rhubarbe, infusion of rhubarb.
— de rose composée, acid infusion of roses, compound infusion of rose.
— de seigle ergoté, infusion of ergot.
— de séné, infusion of senna.
— de serpentaire, infusion of serpentary.
— d'uva-ursi, infusion of bearberry.
— par décoction, decoction.

tisard, m. fire door, fire hole (of a furnace).

tiser, v.t. (of fire) stir, stir up, poke, stoke.

tiseur, m. stoker, fireman.

tisoir, m. poker, fire iron.

tisonner, *v.t.* stir, stir up, poke (a fire).

tisonnier, *m.* poker, fire iron.

tissage, *m.* weaving; cloth mill.

tisser, *v.t.* weave.

tisserand, tisseur, *m.* weaver.

tissu, *m.* tissue; cloth, fabric. — *p.a.* woven.
— *adipeux,* adipose tissue.
— *collagène,* collagenous tissue.
— *conjonctif,* (*Anat.*) connective tissue.
— *en caoutchouc,* rubber cloth, rubberized cloth.
— *filtrant,* filter cloth.
— *imperméable,* waterproof cloth.
— *métallique,* wire gauze.

tissulaire, *a.* pertaining to organic tissues.

tissure, *f.* texture.

titanate, *m.* titanate.

titane, *m.* titanium.

titané, *a.* combined with or containing titanium.

titaneux, *a.* titanous.

titanferrosilicium, *m.* an alloy of titanium, iron and silicon.

titanifère, *a.* titaniferous.

titanique, *a.* titanic; Titanic, gigantic.

titanite, *f.* (*Min. & Expl.*) titanite.

titanium, *m.* titanium.

titrage, *m.* titration; testing.
— *en retour,* back titration.

titrant, *p.a.* titrating, etc. (see titrer).

titre, *m.* titer; standard; title; title page; head, heading; document; stock certificate.
— *pondéral,* gravimetric titer, titer as determined by weighing; standard by weight, specif. the percentage of sodium oxide in commercial caustic soda.
à — de, in the quality or capacity of, as; by virtue of, by right of.
— *de patent,* patent (the document).
à — principal, as the main product.

titré, *p.a.* titrated, standard; tested; titled.

titrer, *v.t.* titrate; standardize; test (compare with some standard); title. *v.i.* titrate; test.

titulaire, *a.* titular; titulary. — *m.* titular, incumbent, holder (of an office, etc.).

toc, *m.* stop, catch, tappet; imitation gold.

tocage, *m.* adding of fuel.

tocane, *f.* a wine made from unpressed must.

toddy, *m.* toddy.

toi, *pron.* thou, thee, to thee, you, to you.

toile, *f.* cloth; specif., linen; (of wire) gauze; canvas; web; curtain.
— *à calquer,* tracing cloth.
— *amiantine,* asbestos cloth.
— *caoutchoutée,* rubber cloth, rubberized cloth.
— *cirée,* oilcloth.
— *d'amiante,* asbestos cloth.
— *d'araignée,* spider web, cobweb.
— *de coton,* cotton cloth, calico.

toile *de crin,* hair cloth.
— *d'emballage,* packing cloth, sackcloth.
— *d'émeri,* emery cloth.
— *de platine,* platinum gauze.
— *émerisée,* emery cloth.
— *grasse,* waterproof canvas.
— *imperméable,* waterproof cloth.
— *métallique,* wire gauze.
— *métallique amiantée,* asbestos wire gauze.
— *peinte,* printed cloth, print.
— *vernie,* varnished cloth; oilskin.

toilette, *f.* fine cloth, cambric; toilet; toilet table; wrapper; light packing case.

toilier, *a.* linen. — *m.* linen-cloth maker.

toiser, *v.t.* measure; scrutinize.

toison, *f.* fleece.

toit, *m.* roof.

toiture, *f.* roofing, roof.

tokai, tokay, *m.* Tokay wine, Tokay.

tolane, *m.* tolan, diphenylacetylene.

tôle, *f.* sheet iron; iron plate; sheet (of metal); plate (of metal).
— *à chaudière,* boiler plate.
— *d'acier,* sheet steel; steel plate.
— *de cuivre,* sheet copper; copper plate.
— *de fer,* — *en fer,* sheet iron; iron plate.
— *émaillée,* enameled sheet iron, enameled ware.
— *étamée,* sheet tin, tinned iron, tin plate.
— *forte,* heavy plate, slab.
— *galvanisée,* galvanized sheet iron.
— *mince,* sheet metal, sheet (as distinguished from plate).
— *noir,* black sheet iron; black plate.
— *ondulée,* corrugated iron.
— *plombée,* sheet iron (or iron plate) coated with lead.
— *ridée,* corrugated iron.
— *vernie,* japanned sheet iron.
— *zinguée,* galvanized sheet iron.

tolérablement, *adv.* tolerably.

tolérance, *f.* tolerance, allowance; toleration.

tolérer, *v.t.* tolerate. — *v.r.* be tolerated.

tôlerie, *f.* sheet-iron (or iron-plate) making, factory, trade or ware.

tôlier, *m.* sheet-iron (or iron-plate) worker.

tolite, *f.* (*Expl.*) tolite.

tolu, *m.* tolu, balsam of Tolu.

toluène, *m.* toluene.

toluène-sulfoné, *a.* toluenesulfonic.

toluidine, *f.* toluidine.

toluifère, *a.* yielding balsam of Tolu.

toluique, *a.* toluic.

toluisation, *f.* coating with balsam of Tolu.

toluiser, *v.t.* coat (pills) with balsam of Tolu.

tolunitrile, *m.* tolunitrile.

toluol, *m.* toluene, toluol, toluole (should be translated toluene when the pure compound is referred to; cf. benzol).

toluquinoléine, *f.* toluquinoline.

toluylène, _m._ toluylene (in the sense either of stilbene or of tolylene).

tolylénique, _a._ tolylene, of tolylene.

tolysal, _m._ (_Pharm._) tolysal.

tomate, _f._ tomato (plant and fruit).

tombac, _m._ tombac, tombak (an alloy).

tombant, _p.a._ falling, etc. (see tomber).

tombe, _f._ tombstone; tomb, grave.

tombeau, _m._ tomb, grave.

tombée, _f._ fall; small amount of material added to tip the scales.

tomber, _v.i._ fall; fail; decline; lose vogue.

tome, _m._ main division, volume, part (refers to a division according to subject matter and may or may not correspond to a division for binding, which is called _volume_).

ton, _m._ tone. — _a._ thy, your.

tonalité, _f._ tonality.

tonca, _m._ tonka bean, tonka.

tondage, _m._ shearing; clipping.

tondaison, _f._ = tonte.

tondelle, _f._ shearings of cloth, flocks.

tondre, _v.t._ shear; clip, trim. — _f._ tinder.

tondure, _f._ shearing (of cloth, etc.); shearings.

tonicité, _f._ tonicity.

tonifier, _v.t._ tone, give tone to.

tonique, _a._ & _m._ tonic.

tonite, _f._ (_Expl._) tonite.

tonka, _m._ tonka bean, tonka.

tonnant, _p.a._ detonating; thundering.

tonne, _f._ ton (usually the metric ton of 1000 kg., equal to 2204.6 lbs.); tun, large cask; (_Gunpowder_) barrel, drum.
— _à poudre_, powder barrel.
— _de lissage_, — _lissoir_, (_Gunpowder_) glazing drum.
— _mélangeoir_, (_Gunpowder_) pulverizing barrel.
— _ternaire_, a barrel in which the three ingredients of gunpowder are pulverized together.

tonneau, _m._ cask, tun; barrel; (_Mech._) drum; (nautical) ton.
— _à fouler_, a large drum for agitating hides with water.
— _à mortier_, (barrel-shaped) mortar mixer.
— _mélangeur_, mixing barrel or drum.
— _rotatif_, rotating cask, barrel or drum, specif. (_Leather_) retanning mill.

tonnelet, _m._ small cask, barrel; (of metal) drum.

tonnelier, _m._ cooper.

tonnelle, _f._ (_Glass_) opening for introducing the pots into the furnace.

tonnellerie, _f._ cooperage.

tonner, _v.i._ detonate; thunder.

tonnerre, _m._ thunder; explosion chamber.

tonomètre, _m._ tonometer (specif., an apparatus for measuring vapor tension).

tonométrie, _f._ tonometry.

tonométrique, _a._ tonometric, tonometrical.

tonoscopique, _a._ tonoscopic.

tonsille, _f._ (_Anat._) tonsil.

tonte, _f._ shearing; clip (the wool sheared); shearing time; clipping; pruning.

tontisse, _a._ of or from shearing.

tonture, _f._ shearings (from cloth), flocks.

topaze, _f._ (_Min._) topaz.

topette, _f._ slender bottle or flask, phial.

topinambour, _m._ (_Bot._) Jerusalem artichoke.

topique, _m._ topic; (_Med._) local remedy or application, topic. — _a._ topical, local.

toquerie, _f._ fireplace, hearth (as of a forge).

toqueux, _m._ (_Sugar_) poker.

torche, _f._ torch; mat, pad; coil (as of wire).

torcher, _v.t._ wipe; build of _torchis_.

torchis, _m._ clay mixed with straw.

torchon, _m._ cloth (for wiping or dusting), rag, towel.

torchonner, _v.t._ wipe, clean, dust.

tordage, _m._ twisting; twist.

tordoir, _m._ oil mill, oil press; ore crusher; twisting machine.

tordre, _v.t._ twist; wring; wrest; contort; distort. — _v.r._ twist; be twisted.

tormentille, _f._ tormentil (_Potentilla tormentilla_).

torpeur, _f._ torpor.

torpide, _a._ torpid.

torpille, _f._ torpedo.

torpilleur, _m._ torpedo boat; torpedo man.

torque, _f._ coil; twist (as of tobacco); torque.

torquer, _v.t._ twist (tobacco).

torquette, _f._ twist (of tobacco).

torréfacteur, _m._ torrefier, roaster. — _a._ torrefying, roasting, parching.

torréfaction, _f._ torrefaction, roasting.

torréfier, _v.t._ torrefy, roast, parch.

torrent, _m._ torrent.

tors, _a._ twisted; contorted; crooked. — _m._ twisting; twist; torsion.

torsade, _f._ twist; coil, helix; twisted joint.
_en_ —, twisted.

torse, _f._ (_Leather_) wringing.

torsinage, _m._ twisting (of plastic glass).

torsiner, _v.t._ twist (in glassblowing).

torsion, _f._ torsion; twisting; twisted state.

torsoir, _m._ (_Leather_) wringing pole.

tort, _m._ wrong; injury; harm, damage.
_à_ —, wrongly, wrong, wrongfully.
_à_ — _et à travers_, at random, without discernment.
avoir —, be wrong, be in the wrong.
faire — _à_, injure, harm, damage.

torte, _a.fem._ of tors.

tortelle, _f._ hedge mustard (_Sisymbrium officinale_).

tortiller, _v.t._ twist.

tortu, _a._ tortuous, crooked, twisted.

tortue, _f._ tortoise.

tortueux, *a.* tortuous, winding, crooked.

toscan, *a.* Tuscan.

Toscane, *f.* Tuscany.

tôt, *adv.* soon; early; promptly.

— *ou tard,* sooner or later.

total, *a.* total. — *m.* total; whole.

au —, on the whole, after all.

totalement, *adv.* totally, entirely.

totalisateur, *a.* totaling. — *m.* totaler, adder.

totaliser, *v.t.* total, add up, sum up.

totalité, *f.* totality, whole, entirety.

totaux, *m.pl.* of total.

touage, *m.* towing.

touchant, *p.a.* touching, etc. (see toucher); tangent. — *prep.* touching, concerning, about.

touchau, touchaud, *m.* touch needle; (*Elec.*) contact, contact piece.

touche, *f.* touch; key (of instruments); (*Mach.*) contact piece, tracer, roller, etc.; (*Elec.*) contact; drove.

à la —, (*Analysis*) by the touch method, by the drop test; touch, drop.

toucher, *v.t.* touch; adjoin; affect; touch on; reach, hit; test (with or as with a touchstone); receive (money); be related to; drive; (*Printing*) ink. — *v.r.* touch each other, touch; adjoin; be related; be touched. — *v.i.* touch; (with *à*) touch (or touch on.) — *m.* touch.

touer, *v.t.* tow.

touffe, *f.* tuft; bunch; cluster; clump.

touffu, *a.* thick, bushy; wooded; full of detail.

touillage, *m.* mixing, stirring, beating.

touille, *f.* mixing shovel.

touiller, *v.t.* mix, stir, beat.

touilloir, *m.* (*Gunpowder*) mixing tool, spatula.

toujours, *adv.* always, ever; still; nevertheless.

touloucouna, *n.* (*Bot.*) tulucuna, kundah.

touloupe, *f.* lambskin.

toupie, *f.* top, spinning top; shaping machine.

toupillage, *m.* spinning, whirling.

toupiller, *v.i.* spin, whirl. — *v.t.* shape.

tour, *m.* turn; revolution; trip; round; circuit; tour; contour, outline; reel; feat; trick; (*Mach.*) lathe, turning lathe; (potter's) wheel; winch. — *f.* tower; turret.

— *à main,* hand lathe.

— *à potier,* potter's wheel.

— *à* —, by turns.

— *de Gay-Lussac,* Gay-Lussac tower.

— *de Glover,* Glover tower.

— *de main,* trick, dodge, wrinkle; sleight of hand; instant, trice.

— *de potier,* potter's wheel.

— *du chat,* small space or interval.

— *mécanique,* power lathe.

touraillage, *m.* kiln drying (of malt).

touraille, *f.* malt kiln.

tourailler, *v.t.* kiln-dry (malt).

touraillon, *m.,* **touraillons,** *m.pl.* malt dust.

tourangeau, *a.* of Touraine or of Tours.

tourbage, *m.* peat winning.

tourbe, *f.* peat (sometimes also called turf).

tourber, *v.t.* extract peat from. — *v.i.* dig peat.

tourbeux, *a.* peaty.

tourbier, *m.* peat worker. — *a.* peaty.

tourbière, *f.* peat bog, turf pit.

tourbillon, *m.* vortex; eddy; whirlwind; whirlpool; (*Pyro.*) tourbillon, tourbillion.

tourbillonnement, *m.* whirling, eddying.

tourbillonner, *v.i.* whirl, eddy.

tourelle, *f.* turret.

touret, *m.* (small) wheel; reel; drill.

tourie, *f.* carboy.

— *en grès,* stoneware carboy.

— *en verre,* glass carboy.

tourillon, *m.* axle, spindle, journal, bearing, pivot, pin, trunnion, gudgeon.

tourmaline, *f.* (*Min.*) tourmaline.

tourmenter, *v.t.* torment.

tournage, *m.* turning; (*Ceram.*) throwing.

tournant, *p.a.* turning; revolving. — *m.* turn; turning; turning point; bend, elbow, corner; wheel (of a mill); shift; (*Leather*) a salt solution applied to tawed leather after dyeing.

tournassage, tournasage, *m.* (*Ceram.*) finishing on the wheel, turning.

tournasser, tournaser, *v.t.* (*Ceram.*) finish on the wheel, turn.

tournassin, tournasin, *m.* (*Ceram.*) finishing tool.

tournassure, tournasure, *f.* (*Ceram.*) shavings, turnings.

tourne, *f.* a disease of wine which causes it to become turbid and insipid and partly lose its color.

tourné, *p.a.* turned, etc. (see tourner). — *m.* (*Leather*) = tournant.

tourne-à-gauche, *m.* handle, lever; wrench.

tournée, *f.* round, circuit, tour.

tourner, *v.t.* turn; examine; get around. — *v.r.* be turned; turn. — *v.i.* turn; revolve; (of milk, etc.) turn, become sour or rancid; (of wine) become turbid and flat (cf. tourne).

— *à la graisse,* (*Wine*) become ropy.

tournesol, *m.* litmus; (*Bot.*) turnsole.

— *des teinturiers, Chrozophora tinctoria.*

tournesolie, *f.* (*Bot.*) *Chrozophora tinctoria.*

tournette, *f.* reel; an instrument for cutting out circular pieces of glass.

tourneur, *m.* turner. — *a.* turning.

tournevis, *m.* screw driver.

tourniquet, *m.* any of several revolving objects, as a turnstile, a capstan, a pulley, a swivel, or (*Sugar*) an apparatus for measuring and delivering lime into molasses.

tournoiement, *m.* turning, whirling.

tournoyer, *v.i.* turn round and round, whirl.
tournure, *f.* turning, turnings; turn; shape.
— *de cuivre,* copper turning(s).
— *de fer,* iron turning(s).
touroir, *m.* malt kiln.
tourque, *f.* (*Metal.*) tub (as a measure for ore).
tourte, *f.* press cake (from fruits or seeds); cake of powdered silver ore; crucible stand (earthenware disk); (*Cooking*) pie.
tourteau, *m.* press cake, cake; specif., oil cake; disk.
— *de colza,* colza cake, rape cake.
— *de graines de coton,* cottonseed cake.
— *de lin,* linseed cake.
tous, *pl. masc.* of tout.
tous-les-mois, *m.* tous-les-mois (starch from *Canna edulis*); (*Bot.*) pot marigold (*Calendula officinalis*).
tousser, *v.i.* cough.
tout, *a.* all; whole; every; any. — *m.* all; whole; (*pl.,* tous) all, everyone. — *adv.* all, wholly, quite, completely, fully; very.
après —, after all, on the whole.
à toute force, absolutely; after all.
à toute heure, hourly, frequently, continually.
à — événement, at all events.
à — prendre, all things considered, on the whole.
à — propos, at every instant.
de toutes pièces, completely.
de — point, in all points, entirely.
de — temps, at all times, always.
du —, at all (usually with negative; e.g., *pas du tout,* not at all).
en —, in all.
en — sens, in every direction.
en — temps, at all times, always.
sur le —, above all.
tous les deux, tous deux, both.
tous les jours, every day.
tous les trois, all three.
— *à coup,* suddenly, all of a sudden.
— *à fait,* quite, altogether, wholly, entirely.
— *à l'heure,* presently, soon, in a moment; just now, a moment ago.
— *à plat,* flatwise, flat.
— *au moins,* at least, at the least.
— *au plus,* at most, at the most, at best.
— *court,* simply, only; suddenly, short.
— *courant,* cursorily; currently; readily; running.
— *d'abord,* from the first, from the very first.
— *de bon,* in earnest, seriously.
— *de go,* readily, immediately.
— *de même,* in the same way; all the same.
— *de suite,* at once, immediately, instantly.
— *du long,* all along; at full length; continuously; wholly, entirely.
— *du long de,* all along; all thru.
— *d'un coup,* all at once, at one time.

tout *d'une pièce,* all of one piece; all at once; stiff, unbending; stiffly; without interruption.
— *d'un temps,* at the same time, at once, without delay.
— *d'un tenant,* all in one piece, continuous.
— *en,* while, while at the same time.
— *en étant,* all in one piece, altho.
— *en gros,* in all, all together.
— *entier,* entire, whole, complete.
— *en un tenant,* all in one piece, continuous.
— *épice,* allspice, pimento (*Pimenta pimenta*).
toute proportion gardée, due allowance being made.
— *fait,* ready made.
— *le long de,* all along; all thru.
— *plat,* flat, flatly.
— *le monde,* everybody; all the world.
— *venant,* run of mine, mine run; unsorted material.
toute, *fem.* of tout.
toute-épice, *f.* allspice, pimento.
toutefois, *adv.* nevertheless, yet, still, however.
toutenague, *f.* tutenag, tutenague (the alloy).
toutes, *fem. pl.* of tout.
tout-venant, *m.* run of mine, mine run; unsorted material of any kind.
toux, *f.* cough, coughing.
toxalbumine, *f.* toxalbumin.
toxémie, *f.* (*Med.*) toxemia, toxæmia.
toxicité, *f.* toxicity, poisonousness.
toxicogène, *a.* toxicogenic.
toxicologie, *f.* toxicology.
toxicologique, *a.* toxicological.
toxicologue, *m.* toxicologist.
toxicose, *f.* toxicosis.
toxine, *f.* toxin.
toxique, *a.* toxic, poisonous. — *m.* poison; toxic substance; virus.
toxoïde, *m.* toxoid.
toxophore, *a.* toxophore, toxophoric.
trace, *f.* trace; track; mark.
tracé, *p.a.* traced, drawn. — *m.* outline, sketch, drawing, plan; tracing, copy; line, route.
tracement, *m.* tracing, drawing, marking.
tracer, *v.t.* trace; draw; sketch; mark out; plot; lay out.
trachée, *f.* (*Anat.*) trachea.
trachyte, *m.* (*Petrog.*) trachyte.
trachytique, *a.* (*Petrog.*) trachytic.
traçoir, *m.* tracing point, marking tool.
traction, *f.* traction.
traditionnel, *a.* traditional.
traditionnellement, *adv.* traditionally.
traducteur, *m.* translator.
traduction, *f.* translation.
traduire, *v.t.* translate; show; express; (*Law*) arraign. — *v.r.* be translated; be shown; be expressed, be manifested.
traduisible, *a.* translatable; expressible.

trafic, *m.* traffic.

trafiquer, *v.i.* traffic, trade, deal.

tragacanthe, *f.* (*Bot.*) tragacanth (the plant).

trahir, *v.t.* betray.

trahison, *f.* betrayal; treachery; treason.

train, *m.* train; motion, movement, action; pace, rate, speed; way, course; mood; series, succession; noise; (*Mach.*) train of rolls, rolls.

mettre en —, (*Mach.*) throw into gear.

— *de grande vitesse,* fast train.

— *de marchandises,* freight train, goods train.

— *de voyageurs,* passenger train.

— *ébaucheur,* roughing rolls.

— *finisseur,* finishing rolls.

traînant, *p.a.* dragging, etc. (see traîner); dull; slow; flagging.

traine, *f.* dragging, etc. (see traîner).

traineau, *m.* sledge; sled; drag; sleigh.

trainée, *f.* trail; track; train (as of powder).

trainement, *m.* dragging, etc. (see traîner).

trainer, *v.t.* drag, trail; haul, draw, pull; tow; carry along; drag out, delay, put off. — *v.r.* crawl, creep; be long drawn out. — *v.i.* drag, trail; be scattered about; lag; languish.

traire, *v.t.* milk.

trait, *m.* stroke; mark, dash, line, streak; cut; shaft, bolt, dart, arrow; ray, beam, flash; touch; piquancy, keenness; trait; feature; connection, relation; draft. — *a.* drawn, wire-drawn; milked.

— *de force,* heavy stroke, heavy line.

— *de jauge,* calibration mark.

— *de repère,* reference mark.

— *d'union,* hyphen.

— *plein,* solid line, unbroken line.

traite, *f.* trade, traffic; stage, stretch; (*Com.*) draft; milking.

traité, *m.* treatise; treaty; contract, agreement.

traitement, *m.* treatment; management; salary.

traiter, *v.t.* treat; negotiate; execute; (with *de*) call. — *v.r.* be treated; live. — *v.i.* treat; negotiate.

traître, traitreux, *a.* treacherous; traitorous.

trajectoire, *f.* trajectory. — *a.* of the nature of or pertaining to a trajectory.

trajet, *m.* passage; distance; path; course; journey, trip.

trame, *f.* woof, weft; tram (silk thread); plot.

tramer, *v.t.* weave; plot, contrive.

tramway, *m.* street railway, tramway; street car, tramcar.

tranchant, *p.a.* cutting, etc. (see trancher); sharp; decisive, peremptory; (of colors) glaring. — *m.* edge; cutter; tanner's knife; chisel.

tranche, *f.* slice; section; slab, plate; edge (of a book, coin, etc.); surface, face; chisel (for metal).

tranché, *p.a.* cut, etc. (see trancher); distinct.

tranchée, *f.* trench; ditch; drain; cutting, excavation; (*Med.*) griping pain.

trancher, *v.t.* cut, cut off, cut up; cut short, curtail, abridge; settle (a question). — *v.r.* be cut, be cut off, etc. — *v.i.* contrast; clash; decide; (with *de*) play, affect the part of.

tranchet, *m.* cutter, knife; chisel.

tranquille, *a.* quiet, tranquil, calm.

tranquillement, *adv.* quietly, tranquilly.

tranquilliser, *v.t.* quiet, calm.

tranquillité, *f.* quiet, quietness, calmness, calm, tranquility.

transaction, *f.* compromise; transaction.

transborder, *v.t.* transship, tranship.

transcendant, *a.* transcendent; transcendental.

transcrire, *v.t* transcribe, copy.

transférable, *a.* transferable.

transfèrement, *m.* transfer.

transférer, *v.t.* transfer.

transfert, *m.* transfer.

transformable, *a.* transformable.

transformateur, *m.* (*Elec.*) transformer.

transformation, *f.* transformation; (*Org. Chem.*) rearrangement; conversion.

— *de Beckmann,* Beckmann rearrangement.

transformer, *v.t.* transform. — *v.r.* be transformed.

transfuser, *v.t.* transfuse.

transgresser, *v.t.* transgress.

transiger, *v.i.* compromise; come to terms.

transir, *v.t.* chill, numb; paralyze. — *v.i.* be chilled; tremble.

transit, *m.* transit.

transiter, *v.t.* convey in transit. — *v.i.* pass in transit.

transition, *f.* transition.

transitionnel, *a.* transitional.

transitoire, *a.* transitory, transient.

transitoirement, *adv.* transitorily, transiently.

translation, *f.* transfer; (*Mech.*) translation.

translucide, *a.* translucent, translucid.

translucidité, *f.* translucency, translucidity.

transmetteur, *a.* transmitting, (of chemical agents) carrying. — *m.* transmitter; carrier.

transmettre, *v.t.* transmit. — *v.r.* be transmitted.

transmis, *p.a.* transmitted.

transmission, *f.* transmission.

transmuable, *a.* transmutable.

transmuer, *v.t.* transmute.

transmutabilité, *f.* transmutability.

transmutable, *a.* transmutable.

transmutateur, *m.* transmuter.

transmutation, *f.* transmutation.

transmutatoire, *a.* transmutative.

transmuter, *v.t.* transmute.

trou de remplissage, filling hole.
— de vidange, emptying hole.
— d'homme, manhole.
trouble, a. turbid, cloudy, troubled, muddy; dull, dim. — m. turbidity, cloudiness; trouble; disturbance, disorder; misunderstanding. — adv. confusedly, indistinctly. — pr. 3 sing. of troubler.
troubler, v.t. cloud, render turbid, trouble; disturb, trouble. — v.r. cloud, become turbid; become dull or dim; become troubled.
troué, a. having a hole or holes, holed, bored, pierced, perforated, full of holes.
trouée, f. gap, opening.
trouer, v.t. make a hole or holes in, hole, bore, pierce, perforate. — v.r. get full of holes.
troupe, f. troop; company, throng, band, body, party, number; herd, drove, flock, etc.
troupeau, m, flock; herd, drove.
trousse, f. case (as of instruments); bundle; packet.
— alcoométrique, case of alcoholometers.
— aréométrique, case of hydrometers.
trousseau, m. small bundle or packet; (Anat.) fasciculus; bunch (of keys); outfit; trousseau.
trousser, v.t. turn up; tuck up; pack up; tie up.
trouvable, a. that can be found, discoverable.
trouvaille, f. find.
trouvé, p.a. found, etc. (see trouver); new, original; happy, felicitous.
trouver, v.t. find; discover; detect; meet with; think. — v.r. be found; be present; be; prove to be; meet, meet with; find oneself; think oneself; feel; happen; happen to be.
— à, find a way (to), contrive.
— à dire à, find fault with.
truc, m. truck; (Railways) flat car; cleverness, skill; trick.
truck, m. truck.
truellage, m. troweling.
truelle, f. trowel.
truellée, f. trowelful.
trueller, v.t. trowel, work with a trowel.
truellette, f. small trowel.
truffe, f. truffle.
truie, f. sow.
truite, f. trout.
truité, a. spotted, speckled; mottled (as iron); (Ceram.) finely crackled so as to suggest fish scales, truité.
trust, m. (Com.) trust.
trustee, m. member of a trust.
trusteur, m. organizer of a trust.
truxilline, f. truxilline.
truxillique, a. truxillic.
trypanose, trypanosomiase, f. trypanosomiasis.
trypanosome, m. (Zoöl.) trypanosome.
trypsine, f. trypsin.

trypsinogène, f. trypsinogen.
tryptique, a. tryptic.
tryptophane, m. tryptophan.
tu, pron. thou, you. — p.a. kept secret; silent.
Tu. Symbol of tungstène (tungsten).
tub, m. bathtub; bath.
tubage, m. tubing, tubage, lining (as a bored well) with a tube.
tube, m. tube; pipe; (pl.) tubes, tubing.
— à acide carbonique, absorption tube for carbonic acid, potash bulbs.
— à air, air tube.
— à azote, nitrogen tube (specif. an absorption tube used in nitrogen determinations).
— abducteur, delivery tube.
— à boule, bulbed tube (e.g. a Peligot tube).
— à brome, bromine funnel (a small dropping funnel).
— absorbant, absorption tube.
— acidimétrique, acidimetric tube (esp. one for determining tartaric acid in wine).
— à combustion, combustion tube.
— à condensation, condensation tube (e.g. for liquefying gases).
— à couleur, color tube (thin tube for holding pigments).
— à dégagement, delivery tube (for gases).
— à dessécher, drying tube.
— à distillation fractionnée, fractional-distillation tube, distillation tube, distilling tube.
— à distiller, distilling tube, distillation tube.
— à eau, water tube, specif. a tube in which water is collected and weighed.
— à entonnoir, funnel tube.
— à essai, test tube.
— à extraction, extraction tube.
— à fumée, (Steam) fire tube.
— à gaz, gas tube, gas tubing.
— alimentaire, (Mach.) feed pipe.
— allonge, lengthening tube.
— à ponce, an absorption tube containing pumice soaked with some reagent, esp. sulfuric acid.
— à potasse, potash bulbs.
— à poudre, powder tube.
— à reaction, reaction tube.
— à rectifier, rectifying tube, distilling tube.
— à réduction, reduction tube.
— à robinet, tube with stopcock.
— atmosphérique, air tube; vacuum tube or pipe.
— à vaccin, vaccine tube.
— à vide, vacuum tube, vacuum tubing.
— à vis, a long narrow cylindrical bottle with a screw top.
— barboteur, bubbling tube (as a gas-washing bottle or a vessel for the bateriological analysis of air).
— capillaire, capillary tube.
— collecteur, collecting tube.

tube d'affluence, influx tube, inflow tube, inlet tube.

— d'alimentation, (Mach.) feed pipe.

— d'amorçage, — d'amorce, priming tube.

— d'arrivée, inlet tube, influx tube.

— d'aspersion, spraying tube.

— de charge, — de chargement, charging tube, loading tube.

— de communication, connecting tube, connection tube, connection.

— de cuivre, copper tube, copper pipe.

— de dégagement, delivery tube (for gases).

— de Frédéricq, a kind of ureameter.

— de Geissler, (Elec.) Geissler tube.

— de laiton, brass tube, brass pipe.

— de Liebig, Liebig bulbs.

— de Mohr, Geissler bulbs.

— de Nessler, Nessler tube.

— de Péligot, Peligot tube.

— de plomb, lead tube.

— d'épreuve, — d'essai, test tube.

— de sûreté, safety tube (specif. a safety funnel).

— d'étain, tin tube (of solid or block tin).

— distributeur, distributing tube.

— divisé, graduated tube.

— doseur, measuring tube (as for filling ampoules).

— d'Yvon, a kind of ureameter.

— en acier, steel tube, steel pipe.

— en biscuit, porcelain tube (usually unglazed or glazed on the interior only).

— en caoutchouc, rubber tube, rubber tubing.

— en carton, pasteboard tube.

— en fer, iron tube, iron pipe.

— en fer bouché à vis, iron tube with screw cap.

— en grès, stoneware tube.

— en Iéna, tube of Jena glass.

— en papier, paper tube.

— en porcelaine, porcelain tube.

— en silice, silica tube.

— en terre, earthenware tube, fireclay tube.

— en U, U tube, U-tube.

— en verre, glass tube.

— en verre de Bohême, tube of Bohemian glass.

— en verre d'Iéna, tube of Jena glass.

— étiré, drawn tube.

— fermé, closed tube.

— fuyant, leaky tube.

— indicateur, indicator tube.

— laboratoire, reaction tube.

— latéral, side tube, lateral tube; side neck.

— laveur, absorption tube (thru which gases are bubbled); gas-washing tube.

— mesureur, measuring tube.

— métallique, metal tube.

— pour cultures, culture tube.

— sans soudure, seamless tube.

— séparateur, separating tube (specif. a tube for collecting separately distillation fractions).

tube soudé, welded tube.

— sulfimétrique, a graduated tube for determining sulfurous acid in wines.

— témoin, indicator tube, indicating tube; comparison tube, tube used as a standard of comparison.

— tuteur, protecting tube.

— uréométrique, ureametric tube, ureameter. (See also the phrases under tuyau.)

tuber, v.t. tube (line with a tube, furnish with tubes, inclose in tubes, etc.).

tubercule, m. tuber; tubercle; nodule.

tuberculeux, a. tubercular, tuberculous.

tuberculose, f. (Med.) tuberculosis.

tubéreux, a. tuberous.

tubulaire, a. tubular.

tubule, m. small tube, tubule, tubulus.

tubulé, a. tubulated.

tubuleux, a. tubulous, tubulose.

tubulure, f. tubulure, tubulation; neck; tubulus, tubule, small tube.

— plongeante, a tubulure which is prolonged into the interior of the vessel.

tue-chien, m. (Bot.) meadow saffron (Colchicum autumnale).

tue-mouche, m. poison fly paper; fly agaric.

tuer, v.t. & i. kill.

tuf, m. (Geol.) tufa, tuff; (fig.) bottom.

tufacé, a. tufaceous, tuffaceous.

tuffeau, tufeau, m. calcareous tufa.

tufier, a. tufaceous. — m. tufa quarry.

tuile, f. tile; misfortune.

— creuse, gutter tile; hollow tile.

— écaille, a flat tile with one edge curved so as to resemble a fish scale when laid.

— faîtière, ridge tile.

— flamande, Flemish tile, pantile.

— panne, pantile.

— plate, flat tile.

tuileau, m. fragment of a tile.

tuilerie, f. tile works, tilery; tile making; (pl., cap.) Tuileries.

tuilette, f. small tile.

tuilier, a. of or pertaining to tile making. — m. tile maker.

tulipe, f. tulip.

tulipier, m. tulip tree (Liriodendron tulipifera).

tulle, m. tulle, fine silk net.

tuméfier, v.t. & r. tumefy, swell.

tumeur, f. (Med.) tumor, tumour.

tumulte, m. tumult, disturbance, agitation.

tumultueusement, adv. tumultuously, violently.

tumultueux, a. tumultuous, (of reactions) violent.

tungstate, m. tungstate.

— d'ammoniaque, ammonium tungstate.

— de potasse, potassium tungstate.

— de soude, sodium tungstate.

tungstène, m. tungsten.

tungstique, *a.* tungstic.

tuniciers, *m.pl.* (*Zoöl.*) Tunicata.

tunicine, *f.* tunicin.

tunique, *f.* tunic; coat, layer, envelope.

Tunisie, *f.* Tunis (the country).

tunisien, *a.* Tunisian.

turanose, *f.* turanose.

turbide, *a.* turbid.

turbinage, *m.* centrifugalizing, centrifugalization, centrifuging (called also, in particular cases, spinning, machining, turbinage, wringing).

turbine, *f.* centrifugal, centrifuge; turbine. (*Turbine* in the sense of "centrifugal" occurs in English but has not been generally recognized.)
— *à vapeur*, steam turbine.
— *essoreuse*, centrifugal, centrifuge, hydroextractor.

turbineur, *m.* centrifugal operator.

turbith, *m.* turpeth (also turbeth, turbith).
— *minéral*, turpeth mineral (basic mercuric sulfate).
— *nitreux*, a basic nitrate of mercury, $Hg_2(OH)NO_3$.
— *végétal*, vegetable turpeth, Indian jalap.

turbulence, *f.* turbulence, turbulency.

turbulent, *a.* turbulent. — *m.* a cubical rotating box in which skins are agitated with water to clean and soften them.

turc, *a.* Turkish. — *m.* Turkish; (*cap.*) Turk.

turgide, *a.* turgid, swollen.

turnep, **turneps**, *m.* turnip.

turque, *a.fem.* Turkish.

turquet, *m.* maize, (Indian) corn.

Turquie, *f.* Turkey.

turquin, *a.* (of blue) dark; designating a kind of blue marble veined with white.

turquoise, *f.* (*Min.*) turquoise.

tussah, tussau, *a.* (*Silk*) tussah, tusseh.

tussigène, *a.* (*Med.*) cough-provoking.

tussilage, *m.* (*Bot.*) coltsfoot (*Tussilago*).

tussore, tussor, *m.* tussah silk; a light silk fabric.

tute, *f.* assay crucible (small cup with foot).

tutelle, *f.* tutelage; guardianship.

tutenay, *n.* tutenag (crude zinc).

tuteur, *a.* protecting. — *m.* protector, specif. a support or guard for plants; guardian.

tutie, tuthie, *f.* tutty (crude zinc oxide).

tuyau, *m.* pipe, tube, hose, conduit; flue; (hollow) stem, stalk.
— *à eau*, water pipe.
— *à gaz*, gas pipe, gas tube.

tuyau *alimentaire*, feed pipe.
— *à vapeur*, steam pipe.
— *collecteur*, collecting pipe or tube.
— *coudé*, tube, elbow pipe or tube.
— *courbé*, bent pipe or tube.
— *d'arrivée*, supply pipe, intake pipe.
— *d'aspiration*, suction pipe, suction tube.
— *de cheminée*, chimney flue, flue, smoke pipe, stack.
— *de communication*, connecting pipe or tube.
— *de conduite*, conduit.
— *de décharge*, discharge pipe.
— *de dégagement*, delivery pipe; escape pipe.
— *de départ*, delivery pipe; waste pipe.
— *d'entrée*, inlet pipe, supply pipe.
— *de fonte*, cast-iron pipe.
— *de plomb*, lead pipe.
— *de refoulement*, delivery pipe; exhaust pipe; compression pipe.
— *de sortie*, delivery pipe.
— *de trop-plein*, overflow pipe.
— *de vidange*, escape pipe, discharge pipe, exhaust pipe, blow-off pipe, waste pipe.
— *de vide*, vacuum pipe or tube.
— *d'introduction*, admission pipe.
— *élastique*, hose, hose pipe.
— *en caoutchouc*, rubber hose, rubber tube.
— *flexible*, hose.
— *principal*, main pipe, main.
— *souffleur*, blast pipe.
(See also the phrases under *tube*.)

tuyautage, *m.* pipes, piping, tubes, tubing.

tuyauterie, *f.* pipes, piping, tubes, tubing; pipe works, pipe factory; pipe trade.

tuyère, *f.* (*Metal.*) tuyère, twyer; blast pipe.

tympe, *f.* (*Metal.*) tymp.

tyndalliser, *v.t.* tyndallize, Tyndallize.

type, *m.* type; standard; character. — *a.* standard.

typer, *v.t.* stamp. — *v.i.* reach a standard, be of a certain standard.

typhique, *a.* (*Med.*) of typhus, typhus, typhous.

typhoïde, *a.* (*Med.*) typhoid.

typique, *a.* typical; (of persons) original.

typographie, *f.* typography.

typographique, *a.* typographic, typographical.

typographiquement, *adv.* typographically.

tyrannie, *f.* tyranny.

tyranniser, *v.t.* tyrannize over.

tyrien, *a.* Tyrian.

tyrolien, *a.* Tyrolese.

tyrosinase, *f.* tyrosinase.

tyrosine, *f.* tyrosine.

U

ulcère, *m.* (*Med.*) ulcer.
ulcérer, *v.t. & r.* ulcerate.
ulcéreux, *a.* (*Med.*) ulcerous.
ulexine, *f.* ulexine (cytisine).
ulmaire, *f.* meadowsweet (*Spiræa salicifolia* or *ulmaria*).
ulmine, *f.* ulmin.
ulmique, *a.* ulmic.
ultérieur, *a.* ulterior; further; subsequent.
ultérieurement, *adv.* later, subsequently.
ultime, ultième, *a.* last.
ultimo, *adv.* lastly.
ultra-chimique, *a.* ultrachemical.
ultrafiltration, *f.* ultrafiltration.
ultrafiltre, *m.* ultrafilter.
ultra-marine, *f.* ultramarine.
ultramicroscope, *m.* ultramicroscope.
ultramicroscopie, *f.* ultramicroscopy.
ultramicroscopique, *f.* ultramicroscopic(al).
ultra-terrestre, *a.* ultraterrestrial.
ultra-violet, *a.* ultraviolet.
un, *a. & pron.* one. — *article.* a, an. — *m.* one.
l' — l'autre, each other, reciprocally.
— à —, one by one.
unanime, *a.* unanimous.
unanimement, *adv.* unanimously.
unanimité, *f* unanimity.
une, *fem.* of un.
uneicosane, *m.* henicosane, heneicosane.
uni, *p.a.* united; even, level, smooth; uniform, constant; calm, quiet; regular, usual; plain; simple. — *adv.* evenly. — *m.* fabric of solid (single) color.
unicellulaire, *a.* unicellular.
unicolore, *a.* of one color, unicolor, unicolour.
unième, *a.* (in compound numbers) first.
unification, *f.* unification.
unifier, *v.t.* unify.
uniforme, *a. & m.* uniform.
uniformément, *adv.* uniformly.
uniformiser, *v.t.* make uniform.
uniformité, *f.* uniformity.
uniment, *adv.* evenly, smoothly, uniformly; simply, plainly.
unimoléculaire, *a.* unimolecular, monomolecular.
union, *f.* union; unity.
unipolaire, *a.* unipolar.
unique, *a.* only, single, sole; unique; singular.

uniquement, *adv.* only, solely, alone; above all; uniquely.
unir, *v.t.* unite; level, smooth. — *v.r.* unite.
unissant, *p.a.* uniting.
unisson, *m.* unison; simultaneous action.
unité, *f.* unit; unity.
— *attractive*, unit of attraction.
— *de chaleur*, heat unit, unit of heat, thermal unit.
— *de longueur*, unit of length.
— *de masse*, unit of mass.
univers, *m.* universe.
universaliser, *v.t.* universalize.
universalité, *f.* universality; entirety.
universaux, *pl.* of universel.
universel, *a.* universal; (*Law*) residuary. — *m.* universal.
universellement, *adv.* universally.
universitaire, *a.* university, of a (or of the) university. — *m.* university professor.
université, *f.* university; specif., the University of France.
upas, *m.* upas (tree or juice).
Ur. Symbol of *urane* (uranium).
uracile, uracil, *m.* uracil.
ural, *m.* (*Pharm.*) ural.
uraminé, *a.* uramido, carbamido, combined with the radical $H_2NCONH-$.
uranate, *m.* uranate.
urane, *m.* uranium.
uraneux, *a.* uranous.
uranifère, *a.* uraniferous.
uranique, *a.* uranic.
uranite, *f.* (*Min.*) uranite.
uranium, *m.* uranium.
uranophane, *n.* (*Min.*) uranophane.
uranyle, *m.* uranyl.
urao, *m.* (*Min.*) urao.
urate, *m.* urate.
— *de lithine*, lithium urate.
urazol, *m.* urazole.
urbain, *a.* urban, city.
uréase, *f.* urease.
urée, *f.* urea.
— *composée*, substituted urea.
uréide, *m.* ureide.
uréique, *a.* urea, of urea, pertaining to urea.
urémie, *f.* (*Med.*) uremia, uræmia.
urémique, *a.* (*Med.*) uremic, uræmic.
uréomètre, *m.* ureameter, ureometer.

uréométrie, f. ureametry, ureometry.
uréométrique, a. ureametric, ureometric.
uretère, m. (Anat.) ureter.
uréthane, m. urethan.
urètre, m. (Anat.) urethra.
urgemment, adv. urgently.
urgence, f. urgency.
urgent, a. urgent.
urinaire, a. urinary.
urine, f. urine.
urineux, a. urinous, urinose.
urinomètre, m. urinometer.
urique, a. uric.
urobiline, f. urobilin.
urobilinogène, m. urobilinogen.
urochrome, m. urochrome.
urolithe, m. urolith, urinary calculus.
uromètre, m. urometer, urinometer.
uroscopie, f. (Med.) uroscopy.
uroscopique, a. uroscopic.
urotoxique, a. urotoxic.
urotropine, f. urotropine (hexamethylenetetramine).
urticaire, f. (Med.) urticaria.
usage, m. use; usage, custom, habit.
  d' —, customary.
  hors d' —, out of use, no longer in use.
  — courant, current use, common use.
usagé, p.a. used, old.
user, v.t. use up, use, consume; wear out, wear away, wear down; grind, polish, rub; impair. — v.r. be used up, etc.; wear out, wear away; rub; deteriorate. — v.i. (with de) use. — m. wear, wearing.
useur, m. grinder, polisher. — a. wearing; grinding.
  — de grain, (Ceram.) polisher, grinder, finisher.
usinage, m. machining, machine work.
usine, f. factory, manufactory, works, mill, plant, shop.

usine à gaz, gas works.
  — centrale, power house.
  — de finissage, finishing factory or mill, specif. (Dyes) a factory that starts with intermediates.
  — d'engrais, fertilizer factory.
  — d'impression, print works.
  — électrique, electric plant.
  — élévatoire pour les eaux, waterworks, pumping station.
  — frigorifique, refrigerating plant.
  — génératrice, power plant, power station.
usiner, v.t. machine, work with a machine tool.
usinier, m. manufacturer. — a. manufacturing.
usité, p.a. used, in use; customary, usual.
usquebac, m. whisky (esp. Scotch or Irish).
ustensile, m. utensil; implement, instrument.
  — chimique, chemical utensil.
  — de cuisine, kitchen utensil.
  — de laboratoire, laboratory utensil.
  — de ménage, household utensil.
usuel, a. usual; common; customary.
usufruit, m. (Law) usufruct.
usure, f. wear, wearing, wearing out; wear and tear; usury.
usurpateur, m. usurper. — a. usurping.
usurper, v.t. usurp, encroach on. — v.i. encroach.
utile, a. useful; beneficial, advantageous; convenient.
utilement, adv. usefully, etc. (see utile).
utilisable, a. utilizable, usable.
utilisation, f. utilization.
utiliser, v.t. utilize.
utilitaire, a. utilitarian.
utilité, f. utility; use, benefit, profit.
utopique, a. Utopian.
uva ursi. (Bot.) bearberry, uva-ursi.

V

v., *abbrev.* (voyez, voir) see; (votre) your; (volume) volume.

va, *pr. 3 sing. indic.* and *2 sing. imper.* of aller (to go). — *interj.* believe me!; done!

vacance, *f.* vacancy; vacation.

vacant, *a.* vacant.

vaccin, *m. & a.* vaccine.

vaccinateur, *m.* vaccinator. — *a.* vaccinating.

vaccination, *f.* vaccination.

vaccine, *f.* (*Med.*) vaccinia, cowpox.

vacciner, *v.t.* vaccinate.

vaccinique, *a.* vaccine.

vache, *f.* cow; cowhide; leather (from cowhide).

vachelin, vacherin, *m.* a kind of Gruyère cheese.

vachette, *f.* kip, kipskin (from cattle).

vacillant, *p.a.* wavering, etc. (see vaciller); unsteady, inconstant.

vacillation, *f.* wavering, etc. (see vacillation); unsteadiness.

vaciller, *v.i.* waver, flicker, tremble, shake, stagger, fluctuate; vacillate; falter.

vacillité, *f.* unsteadiness, waveringness.

vacuité, *f.* vacuity, emptiness.

vacuole, *f.* vacuole.

vacuum, *m.* vacuum.

va-et-vient, *m.* back-and-forth (or up-and-down) motion, reciprocating motion; (*Mach.*) reciprocating gear. — *a.* back-and-forth, reciprocating.

vagabonder, *v.i.* ramble, wander, rove.

vagin, *m.* (*Anat.*) vagina.

vagon, *m.* wagon.

vague, *f.* wave; (*Brewing*) an iron stirring tool, rake. — *m.* vagueness; waste land; empty space. — *a.* vague; empty; (of land) waste, untilled.

vaguement, *adv.* vaguely.

vaguer, *v.i.* wander. — *v.t.* (*Brewing*) stir or mix (mash) with the rake or rakes.

vaillant, *m.* possessions, property, fortune. — *adv.* worth, in value. — *a.* valiant, valorous.

vaille, *pr. 3 sing. subj.* of valoir.

vain, *a.* vain; (of land) waste, unoccupied. *en* —, in vain, vainly.

vaincre, *v.t.* overcome; surmount, vanquish, conquer, subdue; surpass, outdo.

vaincu, *p.p.* of vaincre.

vainement, *adv.* vainly, in vain.

vainquent, *pr. 3 pl. indic. & subj.* of vaincre.

vainqueur, *m.* victor, conqueror. — *a.* victorious.

vainquit, *p.def. 3 sing.* of vaincre.

vais, *pr. 1 sing.* of aller (to go).

vaisseau, *m.* vessel; structure, edifice.
— *de terre,* earthen vessel.

vaisselle, *f.* dishes, tableware.
— *d'argent,* silver plate.
— *de porcelaine,* table china.
— *d'étain,* tin tableware.
— *de terre,* earthenware for the table.
— *plate,* plate.

vaissellerie, *f.* tableware or its manufacture.

vake, *f.* (*Geol.*) wacke.

val, *m.* valley, vale.

valable, *a.* valid, good.

valablement, *adv.* validly, with legal force.

Valachie, *f.* Wallachia.

valait, *imp. 3 sing.* of valoir.

valant, *p.pr.* of valoir.

valaque, *a.* Wallachian.

valence, *f.* valence, valency; (*cap.*) Valencia (Spain); (*cap.*) Valence (France).
— *impaire,* odd valence.

valendré, *a.* (*Ceram.*) warped, out of shape.

valent, *pr. 3 pl.* of valoir.

valérianate, valérate, *m.* valerate, valerianate.
— *d'ammoniaque,* ammonium valerate.
— *de soude,* sodium valerate.

valériane, *f.* (*Bot.*) valerian (*Valeriana*).
— *américaine,* American valerian (any of several American species of *Cypripedium*).
— *officinale,* common valerian (*Valeriana officinalis*).

valérianique, valérique, *a.* valeric, valerianic.

valéroamylique, *a.* In the phrase *éther valéroamylique,* amyl valerate.

valérylène, *m.* valerylene, pentine.

valet, *m.* support, rest; specif., a ring of straw or other material on which to set balloon flasks, etc.; clamp, holdfast, claw, dog; door weight; servant, valet, groom.
— *élévateur,* a hydraulic lifting jack for heavy loads.
— *en bois,* wooden ring.
— *en jonc,* reed ring, rush ring.
— *en paille,* straw ring.

valeur, *f.* value, worth; bill, paper (of money value); security; valor.

valeur (continued).

   la — de, the equivalent of, what amounts to.

   mettre en —, enhance in value, improve.

   — approchée, approximate value.

   — limite, limiting value.

   — marchande, commercial or market value.

valide, *a.* healthy, vigorous; valid.

valider, *v.t.* validate, make valid.

validité, *f.* validity, validness.

valine, *f.* valine.

vallée, *f.* valley.

vallon, *m.* small valley, dell.

vallonée, *f.* valonia.

vallonier, *m.* valonia oak (*Quercus œgilops*).

valoir, *v.i.* be worth, be of value, have the value of, be as good as. *— v.t.* gain, win, procure.

   — vaudrait, it would be equally well.

   à —, on account.

   faire —, render valuable, improve, turn to account, make the most of; urge, emphasize, call attention to; commend, praise.

   vaille que vaille, come what may, at all events, for better or for worse.

   — mieux, be better, be preferable; be worth more.

valoné, *m.*, **valonée**, *f.* valonia.

valu, *p.p.* of valoir.

value, *f.* value.

valut, *p.def. 3 sing.* of valoir.

valve, *f.* (*Mach.*) valve.

valvé, *a.* valved, valvate.

valvoline, *f.* cylinder oil.

valvule, *f.* valvule, small valve; (*Anat.*) valve.

vanadate, *m.* vanadate.

   — d'ammoniaque, ammonium vanadate.

   — de soude, sodium vanadate.

vanadeux, *a.* vanadious, vanadous.

vanadinite, *f.* (*Min.*) vanadinite.

vanadique, *a.* vanadic.

vanadite, *m.* vanadite.

vanadium, *m.* vanadium.

vanadyle, *m.* vanadyl.

vanille, *f.* vanilla.

vanillé, *a.* flavored with vanilla.

vanillier, *m.* vanilla plant, esp. *V. planifolia.*

vanilline, *f.* vanillin.

vanillique, *a.* vanillic.

vanillisme, *m.* (*Med.*) vanillism.

vanillon, *m.* an inferior variety of vanilla or vanilla bean, vanillon.

vanité, *f.* vanity.

vannage, *m.* winnowing, fanning.

vanne, *f.* valve, gate valve; sluice gate, water gate. *— a.* See eaux vannes, under *eau.*

   — d'air chaud, (*Metal.*) hot-blast valve.

   — d'air froid, (*Metal.*) cold-blast valve.

vannée, *f.* material removed by winnowing.

vannelle, *f.* small valve; small sluice gate.

vanner, *v.t.* winnow, fan.

vanneur, *m.*, **vanneuse**, *f.* winnower.

vannure, *f.* material removed by winnowing.

vanter, *v.t.* praise, boast of, vaunt. *— v.r.* boast, brag.

vapeur, *f.* vapor, vapour; steam; fume, exhalation, damp; steamer, steamboat.

   à —, (usually) steam.

   en —, in the form of vapor; with steam on, with steam up.

   — à citerne, tank steamer.

   — d'eau, water vapor; steam.

   — d'eau surchauffée, superheated steam, superheated water vapor.

   — de décharge, exhaust steam.

   — humide, *— mouillée*, moist vapor; moist steam, wet steam.

   — saturée, saturated vapor; saturated steam.

   — sèche, dry vapor; dry steam.

   — surchauffée, superheated vapor; superheated steam.

vaporeux, *a.* vaporous, vapory.

vaporifère, *m.* steam generator.

vaporisage, *m.* steaming.

vaporisateur, *m.* vaporizer; atomizer.

vaporisation, *f.* vaporization, vaporizing.

vaporiser, *v.t.* vaporize; atomize; spray (with an atomizer). *— v.r.* vaporize, be vaporized.

vaporiseur, *m.* vaporizer; atomizer.

vaquer, *v.i.* be vacant; not meet, not sit; (with *à*) attend to, give attention to.

varaigne, *f.* (*Salt*) tide gate.

varech, *m.* (*Bot.*) varec, kelp, wrack.

   — vésiculeux, bladder wrack (*Fucus vesiculosus*).

vare-crue, *f.* underburned brick.

variabilité, *f.* variability.

variable, *a.* variable.

variablement, *adv.* variably.

variation, *f.* variation.

varié, *p.a.* varied.

varier, *v.t.*, *r. & i.* vary.

variété, *f.* variety.

variole, *f.* (*Med.*) variola, smallpox.

variolite, *f.* (*Petrog.*) variolite.

varlope, *f.* large plane, jointer.

varloper, *v.t.* plane, dress (with a plane).

varlopeuse, *f.* planer.

varme, *m.* twyer plate (of certain furnaces).

Varsovie, *f.* Warsaw.

vasais, *m.* reservoir (of a salt garden).

vasculaire, **vasculeux**, *a.* vascular, vasculose.

vase, *m.* vessel; vase. *— f.* mud, slime, ooze.

   — à anse, vessel with handle.

   — à chlorure de calcium, calcium chloride vessel.

   — à filtration chaude, beaker.

   — à lévigations, levigating jar, decanting jar.

   — à précipiter, precipitating vessel, precipitating jar (a sort of beaker without a flaring rim).

vase à *saturation*, saturating vessel, saturating jar (specif. a tall lipped vessel shaped like the frustum of a cone).
— *clos*, closed vessel.
— *de Bohême*, vessel of Bohemian glass.
— *de condensation*, condenser.
— *de Mariotte*, Mariotte bottle, Mariotte flask.
— *de sûreté*, safety vessel (as one for catching liquid that may come over in distillation).
— *en Bohême*, vessel of Bohemian glass.
— *en grès*, stoneware vessel.
— *en Iéna*, vessel of Jena glass.
— *en Krasna*, vessel of Krasna glass.
— *en tôle émaillée*, vessel of enameled iron.
— *gradué*, graduated vessel.
— *Griffin*, Griffin beaker (low and wide).
— *jaugé*, calibrated vessel.
— *poreux*, porous vessel, specif. (*Elec.*) porous cell.
vasé, *a.* muddy, covered with mud.
vaseline, *f.* (*Pharm.*) vaseline
vaseux, *a.* muddy, slimy.
vaso-dilatateur, *m.* (*Physiol.*) vasodilator.
vason, *m.* piece of clay from which a tile or brick is made.
vaste, *a.* vast.
vastement, *adv.* vastly.
vaucour, *m.* potter's bench.
vaudois, *a.* of Vaud (a Swiss canton).
vaudra, *fut. 3 sing.* of valoir.
vaudrait, *cond. 3 sing.* of valoir.
vaut, *pr. 3 sing.* of valoir.
vautrer, *v.t. & r.* roll, wallow.
vaux, *pl.* of val (valley).
veau, *m.* calf (animal and leather); veal.
— *ciré*, waxed calf.
— *marin*, seal.
veau-laq, *m.* a very flexible leather.
vecteur, *m. & a.* (*Math.*) vector.
vectoriel, *a.* vectorial.
vécu, *p.a.* lived; real, that has happened.
vécut, *p.def. 3 sing.* of vivre (to live).
végétal, *a. & m.* vegetable, plant.
— *supérieur*, higher plant.
végétaline, *f.* vegetaline (a coconut oil butter).
végétalisme, *m.* vegetarianism (strictest sense).
végétant, *p.a.* vegetating, vegetative.
végétarien, *a. & m.* vegetarian.
végétarisme, *m.* vegetarianism (excluding meats and meat products but permitting milk, etc.).
végétatif, *a.* vegetative.
végétation, *f.* vegetation.
végétaux, *pl. masc.* of végétal.
végéter, *v.i.* vegetate.
véhicule, *m.* vehicle; medium.
veillant, *p.a.* watching, etc. (see veiller); watchful; wakeful.

veille, *f.* watch, watching; wakefulness; night labor; eve; day before.
veillée, *f.* evening; evening work or party; night attendance (on the sick).
veiller, *v.i.* watch; see (to), attend (to), look (after); wake, be awake, keep awake; stay up, sit up; (of a buoy, etc.) be visible. — *v.t.* watch, watch over, look after; watch by, attend.
— à *ce que*, watch that, see to it that, see that.
veilleur, *m.* watcher; watchman.
veilleuse, *f.* low burner, low-burning attachment; night light; (*Bot.*) meadow saffron.
veillotte, *f.* (*Bot.*) meadow saffron.
veinage, *m.* veining.
veine, *f.* vein; jet, stream; (*pl.*) interior, bosom; luck, fortune.
— *porte*, (*Anat.*) portal vein.
veiné, *p.a.* veined, veiny.
veiner, *v.t.* vein.
veineux, *a.* venous; veined, veiny.
vélanède, *f.* valonia.
vélani, *m.* (*Bot.*) valonia oak (*Quercus ægilops*).
vélar, vélaret, *m.* hedge mustard (*Sisymbrium*).
— *officinale*, Sisymbrium officinale.
vélin, *m. & a.* vellum.
véloce, *a.* swift, rapid.
vélocité, *f.* velocity; swiftness, speed.
velours, *m.* velvet; velvety surface.
velouté, *a.* velvet; velvety, (of liquors) smooth to the taste; (of gems) rich. — *m.* velveting; velvety surface, bloom; (of gems) deep color; (*Cooking*) velouté.
velouter, *v.t.* give the appearance of velvet to.
veloutine, *f.* a rice powder cosmetic containing bismuth.
veltage, *m.* gaging (of casks).
velte, *f.* gaging stick (for gaging casks).
velter, *v.t.* gage (casks).
velu, *a.* hairy, shaggy, villous; (of stone, etc.) rough. — *m.* hairiness; roughness.
venaison, *f.* venison.
venait, *imp. 3 sing.* of venir (to come).
venant, *p.a.* coming, etc. (see venir); thriving; forthcoming. — *m.* comer.
vend, *pr. 3 sing.* of vendre.
vendable, *a.* salable, marketable.
vendage, *f.* vintage; grapes.
vendanger, *v.t.* gather the grapes from; ravage, devastate. — *v.i.* gather the grapes; (fig.) reap a harvest.
vendeur, *m.* seller, vender, dealer.
vendre, *v.t.* sell. — *v.r.* sell, be sold.
à —, for sale, to sell, to be sold.
vendredi, *m.* Friday.
vendu, *p.p.* of vendre.
vené, *p.a.* (of meat) slightly tainted, high.
vénéneux, *a,* poisonous.
vénénosité, *f.* poisonousness.

vénérer, *v.t.* venerate.
vénérien, *a.* venereal.
Vénétie, *f.* Venetia.
vénézuélien, *a.* Venezuelan.
venger, *v.t.* avenge.
venimeux, *a.* venomous.
venin, *m.* venom.
venir, *v.i.* come; be suited (to), harmonize (with), agree; consent (to); happen; (with *de* and an infinitive) to have just; grow, thrive.
  en —, arrive, get, come.
  en —, *à,* come to, go so far as, be reduced to; get off with.
  faire —, send for, summon; bring; have brought; suggest; raise, grow.
  s'en —, come, come away, come along.
  — à bout, succeed.
  — à bout de, overcome.
  — à point, (*Metal.*) come to nature.
  — de fonte, be cast on.
  — en prise, (*Mach.*) mesh, engage.
Venise, *f.* Venice.
vénitien, *a.* Venetian.
vent, *m.* wind; blast; air; (*Mach.*) clearance.
  mettre au —, expose to the air, air; spread out, spread (as leather); draw out, draw (as a sword).
  — chaud, hot blast; hot air; hot wind.
  — coulis, draft.
  — échauffé, hot blast; hot air, heated air.
  — froid, cold blast; cold air; cold wind.
vente, *f.* sale; cutting (of timber).
  de bonne —, *de* —, salable.
  en —, on sale.
venter, *v.i.* blow (said of the wind).
venteux, *a.* windy; (of metal) blistered.
ventilateur, *m.* ventilating machine or fan, ventilator.
  — électrique, electric fan.
ventilation, *f.* ventilation; (*Law*) valuation.
ventiler, *v.t.* ventilate; (*Law*) value.
ventillon, *m.* valve (of a bellows).
ventilocalorique, *a.* designating an apparatus for drying by means of an air current and heat.
ventouse, *f.* ventilator, air hole, vent; cylindrical tile; cupping glass; cupping; sucker.
ventre, *m.* belly (in various senses, e.g. swell, bulge, body).
  faire le —, *faire* —, bulge (said e.g. of an elliptical glass pot which tends to become round owing to the pressure of its contents).
ventricule, *m.* (*Anat.*) ventricle.
venu, *p.a.* come; arrived; grown; done; happened. *— m.* comer.
venue, *f.* coming, arrival; growth; rush (as of water); (*Leather*) lot of hides or skins stuffed at one time.

ver, *m.* worm.
  — à soie, silkworm.
  — luisant, glowworm.
véracité, *f.* veracity.
vératrate, *m.* veratrate.
vératre, *m.* white or false hellebore (*Veratrum*).
  — blanc, European white hellebore (*V. album*).
  — vert, American hellebore, Indian poke, green hellebore (*V. viride*).
vératrine, *f.* veratrine.
vératrique, *a.* veratric.
verbal, *a.* verbal. *— m.* official report.
verbalement, *adv.* verbally.
verbaliser, *v.i.* draw up a report; make a statement.
verbe, *m.* word; verb; voice, tone.
verdâtre, *a.* greenish.
verdaud, *a.* somewhat green, not fully ripe.
verdelet, *a.* (of wine) tart, slightly acid. *— m.* (*Leather*) a minute hole made by an insect.
verdet, *m.* verdigris.
  — crystallisé, crystallized verdigris (neutral copper acetate in crystals).
  — de Montpellier, true verdigris (basic copper acetate).
  — distillé, distilled verdigris (neutral copper acetate).
  — naturel, natural verdigris (basic copper carbonate).
verdeur, *f.* greenness; (of wine) tartness, sourness, acidity (sometimes, strength); (*fig.*) sharpness.
verdillon, *m.* crowbar, pinchbar.
verdir, *v.i.* turn green, become green, be turned green; (of copper) be coated with verdigris. *— v.t.* turn green, color green, green.
verdissage, *m.* coloring green.
verdissant, *p.a.* turning green, etc. (see verdir); verdant, fresh, green.
verdissement, *m.* turning green, etc. (see verdir).
verdoiement, verdoiment, *m.* growing green.
verdoyant, *p.a.* growing green; greenish; verdant.
verdoyer, *v.i.* become green, be green.
verdure, *f.* verdure, greenness; pot herbs.
véreux, *a.* wormy; (*fig.*) bad, untrustworthy.
verge, *f.* rod; stick, bar, staff, spindle; shaft, shank, handle, stock; (scale) beam.
  — d'or, (*Bot.*) goldenrod.
vergé, *a.* (of paper) laid; (of cloth) streaky.
vergeoise, *f.* sugarloaf mold; (*pl.*) sugar recovered from refinery waste.
verger, *m.* orchard.
vergeure, *f.* wire (on which laid paper is made); wire marks (on paper); streakiness (of cloth).
verglas, *m.* glaze of ice.
véricle, *m.* imitation gem, paste.

vérifiable, *a.* verifiable.

vérificateur, *m.* verifier; examiner, inspector; tester; gage.

vérification, *f.* verification, etc. (see vérifier).

vérifier, *v.t.* verify; test, check (as weights and measures); examine; inspect; adjust; audit.
— *v.r.* be verified, etc.
— *si,* see if, see that, make sure that.

vérin, *m.* jack, lifting jack.
— *a vis,* screw jack.

vérissime, *a.* very true.

véritable, *a.* true, genuine, actual, veritable.

véritablement, *adv.* truly, really, actually, in reality, in fact, indeed.

vérité, *f.* truth, verity.
à la —, in truth, it is true.
en —, truly, really, in reality, in truth, indeed.

verjus, *m.* verjuice, juice of green fruit or of crab apples or sour grapes; sour grape or grapes; very sour wine.

verjuté, *a.* sour, tart; made with verjuice.

vermeil, *m.* vermeil, gilded silver (sometimes, gilded copper or bronze). — *a.* vermilion, red.

vermicelier, *m.* vermicelli maker.

vermicelle, vermicel, *m.* vermicelli; vermicelli soup.

vermicellerie, *f.* vermicelli making or factory.

vermicide, *a.* vermicidal. — *m.* vermicide.

vermifuge, *a.* vermifugal. — *m.* vermifuge.

vermillon, *m.* vermilion (pigment and color).
— *d'antimoine,* antimony vermilion.

vermillonner, *v.t.* vermilion.

vermine, *f.* vermin.

vermoulu, *a.* worm-eaten.

vermout, vermouth, *m.* vermuth, vermouth.

verni, *p.a.* varnished; glazed; japanned; (of leather) enameled, patent.

vernier, *m.* vernier.

vernir, *v.t.* varnish; glaze (as porcelain); japan (as iron); enamel (leather).

vernis, *m.* varnish; glaze, glazing (as on pottery, esp. a thin and transparent glaze; cf. émail and glaçure); japan; (*Bot.*) varnish tree; polish, gloss.
— *à fer,* iron varnish.
— *à la copale,* copal varnish.
— *à la gomme-laque,* gum-lac varnish, shellac varnish, shellac.
— *à l'alcool,* spirit varnish.
— *à l'asphalte,* asphaltum varnish.
— *à l'esprit de vin,* spirit varnish.
— *à l'essence,* turpentine varnish.
— *à l'huile,* oil varnish.
— *au bitume de Judée,* asphaltum varnish.
— *au succin,* amber varnish.
— *à voitures,* carriage varnish.
— *d'ambre,* amber varnish.
— *d'ébéniste,* cabinet varnish.

vernis *de copal,* copal varnish.
— *de plomb,* lead glaze.
— *de silice,* silicate varnish, silicate paint.
— *du Japon,* (*Bot.*) Japanese varnish tree (*Rhus vernicifera*).
— *gras,* oil varnish.
— *hydrofuge,* waterproof varnish.
— *japonais,* Japan varnish.
— *siccatif,* quick-drying varnish.
— *spiritueux,* spirit varnish.

vernissage, *m.* varnishing, etc. (see vernir).

vernisser, *v.t.* glaze (ceramic ware); varnish.

vernisseur, *m.* varnisher; varnish maker.

vernissure, *f.* varnishing; glazing.

vérole, *f.* (*Med.*) pox (specif., syphilis).
petite —, smallpox.

vérolette, *f.* (*Med.*) varicella, chicken pox.

Vérone, *f.* Verona.

véronique, *f.* (*Bot.*) speedwell (*Veronica*).
— *de Virginie,* Culver's root (*Leptandra virginica*).
— *mâle,* — *officinale,* common speedwell (*V. officinalis*).

verpunte, *m.* a sugar loaf of inferior quality.

verra, *fut. 3 sing.* of voir (to see).

verraille, *f.* small glassware.

verrain, *m.* = vérin.

verrait, *cond. 3 sing.* of voir (to see).

verre, *m.* glass.
— *à boire,* drinking glass.
— *à boudine,* crown glass.
— *à bouteilles,* bottle glass.
— *à dégustation,* tasting glass.
— *à expérience,* a conical glass vessel with lip and foot.
— *à glaces,* plate glass.
— *à pied,* glass with a foot (esp. a wine glass).
— *à précipiter,* precipitating glass (cf. vase à précipiter, under *vase*).
— *ardent,* burning glass.
— *à réaction,* reaction glass, reagent glass (e.g. a conical glass vessel with lip and foot).
— *à reliefs,* figured rolled plate.
— *armé,* wire glass, armored plate.
— *à vitre(s),* window glass.
— *à vitre poli,* patent plate glass (blown sheet glass ground and polished).
— *blanc,* white glass (esp. ordinary colorless glass with a lime base as distinguished from crystal glass).
— *bombé,* convex glass.
— *cassé,* broken glass, cullet.
— *chevé,* a watch glass made by hollowing out a disk of crystal glass.
— *cloisonné,* a glass vessel with a partition in it, for the comparison of two samples of beer or other liquid.
— *coulé,* cast glass.
— *coulé et laminé,* rolled plate.

verre *de Bohême*, Bohemian glass; vessel of Bohemian glass; (formerly) crown glass.
— *de champ*, field lens.
— *de cobalt*, cobalt glass; specif., smalt.
— *de couleur*, colored glass, stained glass.
— *de couronne*, crown glass.
— *de lampe*, lamp chimney.
— *de lunette*, spectacle lens.
— *de montre*, watch glass.
— *de Moscovie*, Muscovy glass (mica).
— *de pendule*, clock glass.
— *de plomb*, lead glass.
— *dépoli*, ground glass.
— *de quartz*, quartz glass.
— *dévitrifié*, devitrified glass.
— *d'Iéna*, Jena glass.
— *d'œil*, (*Optics*) eyepiece, eyeglass.
— *d'optique*, optical glass.
— *doublé*, flashed glass.
— *d'urane*, uranium glass.
— *en canons*, cylinder glass.
— *en cylindres*, cylinder glass.
— *enfumé*, smoked glass.
— *en manchons*, cylinder glass.
— *en tables*, plate glass.
— *filigrané*, filigree glass.
— *grossissant*, magnifying glass.
— *imprimé*, figured glass.
— *moulé*, — *moulé et comprimé*, pressed glass.
— *mousseline*, muslin glass, mousseline glass, mousseline.
— *objectif*, (*Optics*) object glass, objective.
— *oculaire*, (*Optics*) eyepiece, eyeglass.
— *peu fusible*, difficultly fusible glass, hard glass.
— *pilé*, crushed glass.
— *pulvérisé*, powdered glass, glass powder.
— *sablé*, sanded glass.
— *soluble*, soluble glass, water glass.
— *soufflé*, blown glass.
— *vert*, green glass, specif. hard glass.
— *volcanique*, volcanic glass.
verré, *a.* (of paper) coated with glass particles.
verrée, *f.* glassful, glass.
verrerie, *f.* glassmaking; glass factory, glass works; glassware; glass trade.
— *artistique*, — *d'art*, artistic glass, decorative glass.
— *de Bohême*, Bohemian glassware.
— *de Krasna*, Krasna glassware.
— *de laboratoire*, laboratory glassware.
— *de pharmacie*, glassware for pharmacists or pharmaceutical chemists.
— *d'Iéna*, Jena glassware.
— *étalon*, normal glassware.
— *graduée*, graduated glassware.
— *ordinaire*, ordinary glassware, common glassware.
verrez, *fut. 2 pl.* of voir (to see).

verrier, *m.* glassmaker; tray for glasses. — *a.* glass, pertaining to glass or the glass industry.
verrière, *f.* glass (for pictures); stained glass window.
verrine, *f.* glass (for pictures); barometer tube; glass bell.
verrons, *fut. 1 pl.* of voir (to see).
verront, *fut. 3 pl.* of voir (to see).
verroterie, *f.* small glassware, glass trinkets.
verrou, *m.* bolt.
verrouiller, *v.t.* bolt.
verrucosité, *f.* verrucosity (wartiness, wart).
verrue, *f.* wart.
verruqueux, *a.* warty, verrucose.
vers, *prep.* toward, towards, to. — *m.* verse.
versage, *m.* emptying (as of material into a furnace); first plowing (of fallow land).
versant, *p.a.* pouring, etc. (see verser). — *m.* slope, sloping side, sloping ground.
verse, *f.* pouring, etc. (see verser); charcoal basket (holding 25–30 pounds); beating down (of grain). — *a.* (*Math.*) versed.
versé, *p.a.* poured, etc. (see verser); versed.
versement, *m.* pouring, etc. (see verser); specif., payment, deposit.
verser, *v.t.* pour; pour out; pour in; spill; empty; overturn, upset; beat down (grain); pay in, deposit, invest; plow; give, confer; lavish; issue. — *v.r.* be poured, etc.; discharge. — *v.i.* overturn, upset; (of grain) fall down.
verseur, *m.* pourer, teemer; (*Glass*) shearer.
versicolore, *a.* versicolor, versicolour.
version, *f.* version.
verso, *m.* left-hand page, back, reverse.
vert, *a.* green; fresh; unwrought; (of wine) tart, somewhat acid; (fig.) sharp, severe, firm. — *m.* green; green grass, fresh vegetation; (of wine) tartness, acidity.
prendre sans —, take by surprise.
— *acide*, (*Dyes*) acid green.
— *à l'aldéhyde*, aldehyde green.
— *à l'essence*, a name of malachite green.
— *anglais*, mixed chrome green (chrome yellow and Prussian blue).
— *antique*, verd antique (green mottled stone).
— *bouteille*, bottle green.
— *brillant*, brilliant green (a dye).
— *cantharide*, iridescent green.
— *céladon*, pale green, pale sea green.
— *clair*, light green.
— *d'eau*, sea green.
— *de Brême*, Bremen green.
— *de Brunswick*, Brunswick green.
— *de Chine*, Chinese green, lokao.
— *de chrome*, chrome green.
— *de cuivre*, copper green; verdigris; (*Min.*) chrysocolla.

vert *de gris*, verdigris (see verdet).
— *de Guignet*, Guignet's green.
— *d'Égypte*, verd antique.
— *de Hongrie*, = vert de montagne.
— *de mer*, sea green.
— *d'émeraude*, emerald green.
— *de Mitis*, Mitis green.
— *de montagne*, mountain green (esp. green verditer, basic copper carbonate).
— *de Montpellier*, true verdigris (basic copper acetate).
— *de naphtol*, naphthol green.
— *de Scheele*, Scheele's green.
— *de Schweinfurth*, Schweinfurt green, Paris green.
— *de sève*, sap green.
— *des feuilles*, leaf green (chlorophyll).
— *de vessie*, sap green.
— *d'herbe*, grass green.
— *d'olive*, olive green.
— *émeraude*, emerald green.
— *foncé*, deep green, dark green.
— *franc*, pure green; specif., a kind of mixed chrome green made with light chrome yellow.
— *Guignet*, Guignet's green.
— *iodé*, iodine green (a dye).
— *lumière*, (*Dyes*) light green.
— *malachite*, malachite green, benzaldehyde green.
— *mélangé*, mixed green, mixed chrome green.
— *métis*, Mitis green.
— *Milori*, = vert anglais.
— *minéral*, mineral green (specif., an arsenate of copper).
— *naissant*, faint green, pale green.
— *pomme*, apple green.
— *pré*, grass green.
— *rompu*, blended green, broken green; specif., a kind of mixed chrome green made with dark chrome yellow.
— *Véronèse*, Veronese green.
vert-de-gris, *m.* verdigris (see verdet).
vert-de-grisé, *a.* coated with verdigris.
vertèbre, *f.* (*Anat.*) vertebra.
vertébré, *a. & m.* vertebrate.
vert-émeraude, *a. & m.* emerald green.
— *fixe*, Guignet's green.
vertelle, *f.* sluice gate (in a salt garden).
vertement, *adv.* vigorously, energetically.
vert-gazon, *a.* grass-green.
vertical, *a.* vertical.
verticalement, *adv.* vertically.
verticalité, *f.* verticalness, verticality.
vertige, *m.* vertigo, dizziness.
vert-méthyle, *m.* methyl green.
vert-pomme, *a. & m.* apple green.
vert-soie, *m.* (literally, silk green) a mixed chrome green containing indigo carmine.
vertu, *f.* virtue.

vertu (continued).
en — *de*, by (or in) virtue of, in consequence of.
verveine, *f.* (*Bot.*) vervain (*Verbena*).
— *de l'Inde*, lemon grass (*Andropogon* sp.).
— *officinale*, common European vervain (*V. officinalis*).
vesce, *f.* (*Bot.*) vetch (*Vicia*).
— *commune*, common vetch (*V. sativa*).
vésicant, **vésicatoire**, *a. & m.* (*Med.*) vesicatory, vesicant.
vésiculaire, *a.* vesicular.
vésicule, *f.* vesicle.
vesou, *m.* cane juice, sugar-cane juice.
vespétro, *m.* vespetro (a liqueur).
vessie, *f.* (*Anat.*) bladder; blister, vesicle.
— *natatoire*, natatory vessel, air bladder, sound.
veste, *f.* jacket; vest.
vestige, *m.* vestige, trace.
vêtement, *m.* garment, clothing, clothes, dress.
vétérinaire, *a.* veterinary. — *m.* veterinarian.
vétille, *f.* trifle; (*Pyro.*) a small serpent.
vétilleux, *a.* particular; requiring great care.
vêtir, *v.t.* clothe, dress; put on. — *v.r.* dress.
vétiver, *m.* vetiver (*Andropogon squarrosus*).
vétuste, *a.* old, worn out, antiquated.
vétusté, *f.* age, decrepitude, decay.
vétyver, *m.* = vétiver.
veuf, *a.* widowed; bereft, deprived.
veuille, *pr. 3 sing. subj.* of vouloir.
veule, *a.* soft, weak; (of soil) too light.
veulent, *pr. 3 pl.* of vouloir.
veut, *pr. 3 sing.* of vouloir.
veux, *pr. 1 & 2 sing.* of vouloir.
vexatoire, *a.* vexatious.
vexer, *v.t.* vex.
viable, *a.* capable of living, (*Med.*) viable.
viaduc, *m.* viaduct.
viager, *a.* life, for life. — *m.* life interest.
viande, *f.* meat.
— *à la gelée*, meat jelly.
— *congelée*, frozen meat.
— *conservée*, = *de conserve*, preserved meat.
— *fraîche*, fresh meat.
— *frigorifiée*, refrigerated meat, (usually) frozen meat.
— *fumée*, smoked meat.
— *marinée*, pickled meat.
— *réfrigérée*, refrigerated meat, cold-storage meat (not frozen).
— *salée*, salt meat.
viandes de conserves, preserved meats.
vibrant, *p.a.* vibrating.
vibrateur, *m.* vibrator.
vibratile, *a.* vibratile.
vibration, *f.* vibration.
vibratoire, *a.* vibratory.
vibrer, *v.i.* vibrate.
vibreur, *m.* vibrator; (*Elec.*) make-and-break.

vibrion, *m.* (*Bact.*) vibrio.

vice, *m.* defect, imperfection, flaw; vice.

viciation, *f.* vitiation.

vicier, *v.t.* vitiate. — *v.r.* be vitiated.

vicieux, *a.* vicious.

victoire, *f.* victory.

victorieusement, *adv.* victoriously.

victorieux, *a.* victorious.

vidage, *m.* emptying, etc. (see vider).

vidange, *f.* emptying, etc. (see vider); discharge, vent; removing (as of earth); ditch; partial emptiness, state of not being full; state of not filling a vessel; amount which a vessel lacks of being full, ullage; removal of night soil; (*pl.*) night soil; sediment (as in a boiler).

   en —, (of a vessel) not full, partly empty.

vidanger, *v.t.* empty; discharge.

vide, *a.* empty; vacant; void; devoid (of). — *m.* vacuum; emptiness; empty space, void, gap, opening, interstice, hole, hollow; blank; chasm.

   à —, empty; in vacuo.

   faire le —, *faire un* —, create a vacuum.

vide-bouteille(s), *m.* siphon for bottles.

videment, *m.* emptying, etc. (see vider).

vider, *v.t.* empty; drain; clear; clean; hollow out, bore (as a cannon); blow off (a boiler); purge; stone (fruit); draw (fowls); vacate; leave, quit; settle (as accounts). — *v.r.* empty; become empty; be emptied, etc.

vide-tourie(s), *m.* carboy inclinator, carboy stand.

vidoir, *m.* sink.

vidure, *f.* material emptied; openwork.

vie, *f.* life; living; path, way (in a salt garden).

   à —, for life, life.

   en —, living, alive.

   la — *chère,* the high cost of living, H.C.L.

   pour la —, for life; for a lifetime.

vieil, *a.* Used for *vieux* before a vowel sound.

vieille, *a. fem.* of vieux.

vieillesse, *f.* age, old age; old people.

vieilli, *p.a.* grown old, aged; outworn.

vieillir, *v.i.* grow old, become old, age; become obsolete, go out of use. — *v.t.* age, make old.

vieillissement, *m.* aging; senescence, growing old; obsolescence, growing obsolete.

vieillot, *a.* somewhat old or antiquated.

viendra, *fut. 3 sing.* of venir (to come).

viendrait, *cond. 3 sing.* of venir (to come).

vienne, *pr. 1 & 3 sing. subj.* of venir (to come).

Vienne, *f.* Vienna (Austria); Vienne (France).

viennent, *pr. 3 pl. indic. & subj.* of venir.

viennois, *a.* Viennese, Vienna. — *m.* (*cap.*) Viennese.

viens, *pr. 1 & 2 sing.* of venir (to come).

vient, *pr. 3 sing.* of venir (to come).

vierge, *a. & f.* virgin.

   vigne —, (*Bot.*) Virginia creeper.

vieux, *a.* old; (of words, etc.) obsolete. — *m.* old thing; old man.

vif, *a.* live, alive, living; lively, brisk; bright, vivid, brilliant; sharp, keen; intense; strong; great; quick; hasty; spirited; ardent, zealous, earnest; severe, harsh. — *m.* living flesh, quick; heart, core, solid part; liveliness; living person.

   de vive force, by main force.

   de vive voix, viva voce, by word of mouth.

vif-argent, *m.* quicksilver, mercury.

vigilamment, *adv.* vigilantly, watchfully.

vigilant, *a.* vigilant, watchful.

vigne, *f.* vine, specif. grapevine; vineyard.

   — *vierge,* Virginia creeper.

vigneron, *m.* vine grower, wine grower.

vignette, *f.* revenue stamp; label; vignette; (*Bot.*) (1) meadowsweet, (2) clematis, (3) mercury.

vignoble, *m.* vineyard. — *a.* winegrowing, grape-growing.

vigogne, *f.* (*Zoöl.*) vicuña; vicuña wool.

vigorite, *f.* (*Expl.*) vigorite.

vigoureusement, *adv.* vigorously.

vigoureux, *a.* vigorous.

vigueur, *f.* vigor, vigour.

vil, *a.* (of prices) low; vile, low, base.

vilain, *a.* villainous, bad, vile, ugly.

vilebrequin, *m.* brace, wimble; crank shaft.

vileté, *f.* cheapness, lowness; insignificance.

village, *m.* village.

ville, *f.* town, city.

vin, *m.* wine.

   — *antimonié,* (*Pharm.*) wine of antimony, antimonial wine.

   — *aromatique,* aromatic wine.

   — *artificiel,* artificial wine.

   — *blanc,* white wine.

   — *bourru,* unfermented (or only slightly fermented) wine.

   — *chalybé,* (*Pharm.*) wine of iron.

   — *de Bordeaux,* Bordeaux wine, Bordeaux.

   — *de Bourgogne,* Burgundy wine, Burgundy.

   — *de bulbe de colchique,* (*Pharm.*) colchicum wine.

   — *de Champagne,* Champagne wine, champagne.

   — *de choix,* choice wine, fine wine.

   — *de coca,* (*Pharm.*) wine of coca.

   — *d'écorce d'orange amère,* (*Pharm.*) orange wine.

   — *de fruits,* fruit wine (includes cider, perry, etc.).

   — *de grappe,* — *de goutte,* wine from must which runs out before the grapes are crushed.

   — *de liqueur,* very sweet wine; liqueur; cordial.

   — *de marc(s),* wine made from the marc of grapes (cf. vin de seconde cuve, below, and piquette).

vin *de paille,* straw wine.

— *de palme,* palm wine.

— *de Portugal,* Portuguese wine, esp. port.

— *de presse,* wine made from expressed juice.

— *de quinine, (Pharm.)* quinine wine.

— *de quinquina ferrugineux, (Pharm.)* bitter wine of iron.

— *de raisins secs,* raisin wine.

— *de seconde* (or *troisième*) *cuve,* wine made from marc by gallization (i.e., addition of sugar and water).

— *de seigle ergoté, (Pharm.)* wine of ergot.

—' *de semence de colchique, (Pharm.)* colchicum wine.

— *d'Espagne,* Spanish wine, esp. sherry.

— *de sucre,* = vin gallisé.

— *de tire,* racked-off wine.

— *d'ipécacuanha, (Pharm.)* wine of ipecac.

— *d'opium composé, (Pharm.)* wine of opium, Sydenham's laudanum.

— *doux,* wine not completely fermented, sweet wine.

— *du cru,* native wine, home-made wine.

— *émétique, (Pharm.)* wine of antimony.

— *ferrugineux, (Pharm.)* wine of iron.

— *filant,* ropy wine.

— *fort,* strong wine.

— *gallisé,* wine made by gallization (addition of sugar and water).

— *médicinal, (Pharm.)* medicated wine.

— *mousseux,* sparkling wine.

— *non-mousseux,* still wine.

— *rouge,* red wine.

— *sec,* dry wine.

— *stibié, (Pharm.)* wine of antimony.

— *sucré,* sweet wine.

vinage, *m.* addition of alcohol to wine or must.

vinaigre, *m.* vinegar.

— *anhydre,* glacial acetic acid.

— *blanc,* white vinegar.

—'*cantharidé, (Pharm.)* vinegar of cantharides.

— *de bois,* wood vinegar, pyroligneous acid.

— *de cidre,* cider vinegar.

— *de plomb,* lead vinegar, solution of lead subacetate, Goulard's extract.

— *de Saturne, (Old Chem.)* vinegar of Saturn (lead vinegar).

— *de toilette,* aromatic vinegar.

— *de vin,* wine vinegar.

— *distillé,* distilled vinegar.

— *d'Orléans,* Orleans vinegar, wine vinegar made by the Orleans process.

— *glacial,* glacial acetic acid.

— *radical, (Old Chem.)* radical vinegar (acetic acid).

— *scillitique, (Pharm.)* vinegar of squill.

vinaigrerie, *f.* vinegar factory, vinegar works; vinegar making; vinegar trade.

vinaigrier, *m.* vinegar maker; vinegar cruet; *(Bot.)* tanner's sumac (*Rhus coriaria*).

vinaire, *a.* wine, of wine, relating to wine.

vinasse, *f.* vinasse, the residual liquid from the distillation of saccharine materials (as fruits, molasses; cf. drèche); poor wine.

— *de betteraves,* beet vinasse, beetroot vinasse.

vinée, *f.* vintage, wine crop.

vinelle, *f.* piquette (see piquette).

viner, *v.t.* add alcohol to (wine or must).

vinerie, *f.* wine making.

vinettier, vinetier, *m. (Bot.)* barberry.

vineux, *a.* vinous, of wine, winy; wine; (of wine) high in alcohol, strong; wine-producing, wine; wine-colored, wine.

vingt, *a. & m.* twenty; twentieth.

vingtaine, *f.* twenty (or about that), score.

vingtième, *a. & m.* twentieth.

vinicole, *a.* winegrowing, vinicultural.

viniculture, *f.* viniculture.

vinifère, *a.* wine-producing.

vinificateur, *m.* vinificator.

vinification, *f.* vinification, wine making.

— *en blanc,* white wine making.

— *en rouge,* red wine making.

vinique, *a.* vinic.

alcool —, vinic alcohol (ethyl alcohol).

vino-benzoïque, *a.* In the phrase *éther vino-benzoïque,* ethyl benzoate.

vinocolorimètre, *m.* an apparatus for determining the intensity of the color of wine.

vinomètre, *m.* vinometer, œnometer.

vinométrique, *a.* pertaining to the vinometer or to the determination of alcohol in wine.

vinosité, *f.* vinosity, vinous quality.

vin-pierre, *m.* wine stone, tartar.

vinrent, *p.def. 3 pl.* of venir (to come).

vint, *p.def. 3 sing.* of venir (to come).

vioforme, *m. (Pharm.)* vioform.

violacé, *a.* violet, violet-colored, violaceous.

violacées, *f.pl. (Bot.)* Violaceæ.

violacer, *v.i.* assume a violet tint.

violant, *a. (Dyeing)* violet, tending to violet.

violariacées, *f.pl. (Bot.)* Violaceæ.

violation, *f.* violation.

violâtre, *a.* tending to violet, violescent.

violemment, *adv.* violently.

violence, *f.* violence.

violent, *a.* violent.

violenter, *v.t.* force, do violence to.

violer, *v.t.* violate.

violet, *a. & m.* violet.

— *acide, (Dyes)* acid violet.

— *alcalin, (Dyes)* alkali violet.

— *au méthyle,* methyl violet.

— *bleu,* bluish violet.

— *crystallisé,* crystal violet.

— *d'aniline,* aniline violet.

— *de cobalt,* cobalt violet.

— *de Hesse,* Hessian violet.

— *de méthyle,* methyl violet.

violet *de Nuremberg,* mineral violet, manganese violet (impure manganese metaphosphate).
— *foncé,* dark violet.
— *formyle,* formyl violet.
— *Hofmann,* Hofmann's violet.
— *Lauth,* Lauth's violet.
— *minéral,* = violet de Nuremberg.
— *neutre,* neutral violet.
— *rouge,* reddish violet.
— *Van Dyck,* a violet pigment made by calcining precipitated iron hydroxide.
violet-chair, *m.* (*Glass*) a brownish-violet color produced with iron and manganese oxides.
violet-évêque, *m.* (*Glass*) a bluish-violet color produced with manganese dioxide.
violet-méthyle, *m.* methyl violet.
violette, *f.* (*Bot.*) violet. — *a. fem.* of violet.
— *odorante,* sweet violet.
violon, *m.* violin; violinist.
violoncelle, *m.* violoncello; violoncellist.
vipère, *f.* viper.
virage, *m.* turning; specif., turning of color, color change; (*Photog.*) toning (also toning bath); transfer of funds, adjustment of an account by transfer.
— *à l'or,* gold toning.
— *au platine,* platinum toning.
virée, *f.* turning, turn.
virement, *m.* turning, etc. (see virer and virage).
virent, *p.def. 3 pl.* of voir (to see). — *pr. 3 pl.* of virer.
virer, *v.i.* turn; specif., turn color, change color; (*Photog.*) tone, undergo toning; veer.
— *v.t.* turn; (*Photog.*) tone; transfer (a sum); clear (an account). — *v.r.* be turned; turn color, turn; (*Photog.*) be toned, tone; be transferred; be cleared.
— *au vert,* turn green.
vireux, *a.* poisonous, virose; malodorous, fetid, unpleasant, virose.
Virginie, *f.* Virginia.
virgule, *f. & a.* comma.
virole, *f.* ferrule; collar, sleeve, ring, hoop; (*Bot.*) Virola.
virtuel, *a.* virtual.
virtuellement, *adv.* virtually.
virulence, *f.* virulence, virulency.
virulent, *a.* virulent.
virus, *m.* (*Med.*) virus.
vis, *f.* screw, specif. male or external screw (cf. écrou). — *pr. 1 & 2 sing.* of vivre (to live). — *p.def. 1 & 2 sing.* of voir (to see).
à —, screw, fitted with or operating by a screw.
— *à ailettes,* — *ailée,* wing screw, thumbscrew.
— *à anneau,* screw ring.
— *à droite,* right-handed screw.

vis *à gauche,* left-handed screw.
— *à volant,* a screw turned by a hand wheel.
— *calante,* leveling screw; centering screw.
— *concave,* female screw.
— *d'ajustage,* adjusting screw.
— *d'Archimède,* Archimedean screw.
— *d'arrêt,* set screw; locking screw.
— *de distribution,* worm conveyor.
— *de pression,* set screw, binding screw.
— *de réglage,* adjusting screw, regulating screw.
— *de serrage,* set screw.
— *sans fin,* endless screw, worm.
visage, *m.* face, visage, aspect.
vis-à-vis, *prep.* opposite, facing; with respect to, in comparison with. — *adv.* opposite.
— *m.* state of being opposite, opposition; vis-à-vis.
— *de,* opposite; with respect to.
viscères, *m.pl.* viscera.
viscidité, *f.* viscidity.
viscine, *f.* viscin.
viscosimètre, *m.* viscosimeter.
viscosité, *f.* viscosity.
vis-écrou, *m.* (*Elec.*) binding post; (of a shell) fuse plug.
visée, *f.* aim; aiming; sighting; sight.
viser, *v.t.* aim at; sight on; visé. — *v.i.* aim.
viseur, *m.* sighting tube; (*Photog.*) finder; aimer. — *a.* sighting.
visibilité, *f.* visibility.
visible, *a.* visible; evident, manifest.
visiblement, *adv.* visibly; obviously.
visière, *f.* visor; (eye) shade; sight.
vision, *f.* vision.
visionnaire, *a. & m.* visionary.
visite, *f.* visit; inspection; search.
visiter, *v.t.* visit; inspect; search.
visiteur, *m.* visitor; inspector.
visqueux, *a.* viscous; viscid.
vis-robinet, *m.* screw faucet, screw cock.
vissage, *m.* screwing; (*Ceram.*) a defect consisting in spiral streaks or markings.
vissent, *pr. 3 pl.* of visser. — *imp. 3 pl. subj.* of voir (to see).
visser, *v.t.* screw, screw in, screw on, etc.
— *à fond,* screw tight.
visuel, *a.* visual.
visuellement, *adv.* visually.
vit, *pr. 3 sing.* of vivre (to live). — *p.def. 3 sing.* of voir (to see).
vital, *a.* vital.
vitalement, *adv.* vitally.
vitalisme, *m.* vitalism.
vitalité, *f.* vitality.
vitamine, *f.* vitamin, vitamine.
vitaux, *masc. pl.* of vital.
vite, *a.* quick, fast, rapid, speedy, swift; prompt; hasty. — *adv.* quickly, quick, fast, speedily, promptly, hastily. — *interj.* quick!

vitellin, *a.* (*Biol.*) vitelline.

vitelline, *f.* vitellin.

vitesse, *f.* velocity; rate, speed, rapidity, swiftness.

— *angulaire,* angular velocity.

— *de combustion,* rate of combustion.

— *de la lumière,* velocity of light.

— *de réaction,* reaction velocity.

— *de régime,* normal speed, working speed.

— *moyenne,* mean velocity.

— *uniforme,* uniform velocity, uniform rate, uniform speed.

viticole, *a.* vine-growing, grape-growing, viticultural.

viticulteur, *m.* vine grower, wine grower, viticulturist.

viticulture, *f.* viticulture, grape growing.

vitrage, *m.* glazing; glass; glass windows; glass partition; small curtain.

vitrail, *m.* stained-glass window.

vitrail-camée, *n.* enameled glass window.

vitraux, *pl.* of vitrail.

vitre, *f.* pane of glass, window pane.

vitré, *p.a.* glazed, glass, furnished with glass; vitreous; (of parchment) transparent.

vitréosil, *m.* a form of quartz glass.

vitrer, *v.t.* glaze, fit with glass; render (parchment) transparent.

vitrerie, *f.* window-glass making or trade; window glass; glazier's work or business.

vitrescibilité, *f.* vitrifiability.

vitrescible, *a.* vitrifiable, vitrescible.

vitreux, *a.* vitreous, glassy.

vitrier, *m.* maker or seller of window glass; glazier.

vitrifiabilité, *f.* vitrifiability.

vitrifiable, *a.* vitrifiable.

vitrificateur, *a.* vitrifying. — *m.* vitrifier.

— *sur verre,* glass enameler.

vitrificatif, *a.* vitrifying.

vitrification, *f.* vitrification, vitrifaction.

vitrifié, *p.a.* vitrified.

vitrifier, *v.t. & r.* vitrify.

vitrine, *f.* glass case (as a cabinet or showcase).

vitriol, *m.* vitriol.

— *blanc,* white vitriol (zinc sulfate).

— *bleu,* blue vitriol (copper sulfate).

— *de Chypre,* = vitriol bleu.

— *de cuivre,* copper vitriol, blue vitriol, copper sulfate.

— *de fer,* iron vitriol, iron sulfate.

— *de Goulard,* = vitriol blanc.

— *de plomb,* lead vitriol, lead sulfate.

— *de Vénus,* = vitriol bleu.

— *de zinc,* zinc vitriol, zinc sulfate.

— *martial,* = vitriol de fer.

— *rouge,* red vitriol, colcothar.

vitriolage, *m.* vitriolation; vitrioling; vitriolizing, vitriolization. (See vitrioler.)

vitrioler, *v.t.* vitriolate, convert into or treat with vitriol; (*Metal.*) vitriol, pickle with dilute sulfuric acid; (*Bleaching*) sour; vitriolize, injure with vitriol.

vitriolerie, *f.* vitriol making; vitriol factory.

vitriolique, *a.* vitriolic.

acide —, vitriolic acid (sulfuric acid).

vitriolisation, *f.* vitriolation, production of or conversion into vitriol.

vitrosité, *f.* vitreousness, glassiness.

vivace, *a.* long-lived, lasting; (*Bot.*) perennial.

vivacité, *f.* liveliness, briskness; quickness (as of an explosive); vividness, brilliancy (as of colors); vivacity, vivaciousness; vehemence, violence; acuteness; hastiness.

vivait, *imp. 3 sing.* of vivre (to live).

vivant, *p.a.* living, live, alive; lively. — *m.* living organism; living man; liver; life, lifetime.

vive, *a. fem.* of vif. — *pr. 1 & 3 sing. subj.* of vivre.

vivement, *adv.* briskly, vividly, sharply, greatly, quickly, etc. (see vif.)

vivent, *pr. 3 pl. indic. & subj.* of vivre.

vivifier, *v.t.* vivify, enliven, animate.

vivisection, *f.* vivisection.

vivre, *v.i. & t. vivre* — *m.* life; living, board, food; (*pl.*) provisions, victuals; victualling.

vivres de conserve, canned (or tinned) provisions.

vivres secs, dry provisions.

vocabulaire, *m.* vocabulary.

vocation, *f.* calling, call; vocation; talent.

vœu, *m.* vow; wish, desire; will, intention.

voici. here is, here are, here am, this is, these are, see here, behold.

— *qui,* here is one that, here is something that.

— *qu'il vient,* here he comes.

— *venir,* here comes, here come.

voie, *f.* way; method, process; road, track, line, path, trail; gage (the width); load, cartload; kerf (of a saw); (*Anat.*) passage, canal, duct. — *pr. 3 sing. subj.* of voir (to see).

La Voie lactée, The Milky Way.

les voies digestives, the digestive tract.

par la — *de,* by the way of, via.

par la — *interne,* (*Med.*) internally.

par — *de,* by means of.

— *aérienne,* air passage.

— *d'accès,* approach.

— *d'eau,* leak (in a boat); about 30 liters of water; waterway.

— *de bois,* about 2 cubic meters of wood.

— *de charbon,* a hectoliter of wood charcoal.

— *de charbon de terre,* about a cubic meter (1200 kg.) of coal.

— *de coke,* about 15 hectoliters of coke.

— *de fer,* = voie ferrée.

voie de Paris, 1.92 cubic meters.
— de terre, wagon road; land (as distinguished from sea).
— ferrée, railway, railroad, railway track.
— fluviale, waterway.
— humide, moist way, wet way.
— ignée, method in which high heat is employed (often synonymous with voie sèche).
— pâteuse, (Ceram.) a method of mixing or grinding intermediate between the wet and dry methods, the ingredients being in a pasty form.
— sèche, dry way, dry method, dry process.
voient, pr. 3 pl. indic. & subj. of voir.
voilà. there is, there are, that is, those are, such is, such are, see there, behold.
— que, all at once, suddenly.
— qui, that, it.
voile, m. dimness, obscurity, clouding; pellicle, scum; veil; (Biol.) velum; (fig.) screen, cover. — f. sail.
voiler, v.t. cloud, obscure, dim; hide, conceal; muffle; veil; rig with sails. — v.r. be clouded, etc.; cloud, become dim; (of wood, etc.) be warped, warp. — v.i. warp.
voir, v.t. see; look at; examine; find (e.g. in a writing); see to; overlook, face. — v.r. be seen; see oneself (or each other); find oneself. — v.i. see; look, face (on).
(The infinitive voir is used as imperative in references; as, voir page 305, see page 305.)
faire —, laisser —, show; let see.
voire, adv., **voire même.** even, nay even.
vois, pr. 1 & 2 sing. of voir (to see).
voirie, f. commission of public ways; public ways, means of communication; offal, garbage (or the place where it is deposited).
voisin, a. near, neighboring, adjacent, bordering (on); closely related, similar, alike; (Org. Chem.) vicinal, neighboring. — m. neighbor.
— de, near, near to, bordering on, approaching.
voisinage, m. neighborhood, vicinity.
voit, pr. 3 sing. of voir (to see).
voiturage, m. carriage, conveyance.
voiture, f. vehicle, cart, wagon, carriage, car, etc.; carriage, conveyance; fare; load, freight, burden.
— à marchandises, (Railways) freight car, goods van, goods wagon.
— de chemin de fer, railway car, railway coach, railway carriage.
voiturer, v.t. carry, convey, transport.
voix, f. voice; vote.
vol, m. flight; theft; robbery.
à — d'oiseau, in a straight line; (of views) bird's-eye.
volage, a. unsteady; top-heavy; fickle.
volaille, f. poultry, fowls; fowl.

volant, p.a. flying; movable; portable; (of leaves of paper, etc.) loose, detached; (of rockets) sky. — m. flywheel; handwheel; loose leaf; detachable part; interval (between supports); shuttlecock.
— à main, handwheel.
volatil, a. volatile.
volatilisable, a. volatilizable.
volatilisation, f. volatilization.
volatiliser, v.t. & r. volatilize.
volatilité, f. volatility.
volcan, m. volcano.
volcanique, a. volcanic.
volcaniser, v.t. volcanize.
volée, f. flight; volley, round; jib (of a crane); bar.
à la —, in the air, flying, on the fly; quickly, promptly; hastily, at random; (of brick-burning) in clamps; (of sowing) broadcast.
volémite, f. volemitol, volemite.
voler, v.i. fly; steal; rob. — v.t. fly after; steal; rob.
volet, m. shutter; damper; trapdoor; sorting board; vane; paddle.
voleter, v.i. flutter.
volettement, m. fluttering.
voleur, m. thief, robber, burglar; flier.
volontaire, a. voluntary; willful. — m. volunteer; willful person.
volontairement, adv. voluntarily; willfully.
volonté, f. will.
à —, at will, at pleasure.
volontiers, adv. readily, easily; willingly.
volt, m. (Elec.) volt.
voltaïque, a. voltaic.
voltamètre, m. voltameter.
voltigement, m. fluttering, etc. (see voltiger).
voltiger, v.i. flutter; flicker; fluctuate; hover; tumble; vault.
voltmètre, m. (Elec.) voltmeter.
volume, m. volume.
— critique, critical volume.
voluménomètre, m. volumenometer.
volumètre, m. volumeter.
volumétrie, f. volumetric analysis.
volumétrique, a. volumetric, volumetrical.
volumétriquement, adv. volumetrically.
volumineux, a. voluminous; large.
volvic, m. Volvic lava.
vomique, a. In the phrase noix vomique, nux vomica, vomit nut. — f. (Med.) vomica.
vomiquier, m. nux vomica (the tree).
vomir, v.t. vomit.
vomissement, m. vomiting; vomit.
vomitif, a. & m. vomitive, emetic.
vont, pr. 3 pl. of aller (to go).
vorace, a. voracious.
vos, a.pl. your.
vosgien, a. Vosges, of the Vosges.
votant, p.a. voting. — m. voter.

votation, *f.* voting.
vote, *m.* vote.
voter, *v.i. & t.* vote.
votre, *a.* your.
vôtre, *pron. & m.* yours.
voudra, *fut. 3 sing.* of vouloir.
voudrait, *cond. 3 sing.* of vouloir.
vouer, *v.t. & i.* vow.
voulant, *p.a.* willing, etc. (see vouloir).
vouloir, *v.t.* will; be willing; like; wish, desire; want; require; need; mean, intend; try, attempt; admit, grant. — *m.* will.
  en — à, long for; look for; have designs on; have a grudge against.
  — dire, mean.
voulu, *p.a.* willed, desired, etc. (see vouloir); needful, requisite; usual, due.
voulut, *p.def. 3 sing.* of vouloir.
vous, *pron.* you, to you.
vous-même, *pron.* yourself.
vous-mêmes, *pron.* yourselves.
voussoir, vousseau, *m.* voussoir, archstone.
voûte, *f.* vault; arch; vaulting; vaulted roof, arched roof.
voûter, *v.t.* vault, arch. — *v.r.* be bent, arch; (of persons) stoop.
voyage, *m.* journey, trip, run, voyage; travel, traveling; stay, sojourn; quantity carried at one time.
voyager, *v.i.* travel; (of things) be carried.
voyageur, *m.* traveler. — *a.* traveling.
voyait, *imp. 3 sing.* of voir (to see).
voyant, *p.a.* seeing, etc. (see voir); showy, conspicuous. — *m.* conspicuousness, showiness; eyepiece (of a level); vane (of a leveling rod); seer.

voyelle, *f.* vowel.
voyer, *v.t.* run, cause to flow (as lye); shake (to rid of dust). — *m.* road supervisor.
vrac, *m.* In the phrase *en vrac,* in bulk, not packed or arranged, loose.
vrai, *a.* true; proper, right; veritable, downright. — *m.* truth, true. — *adv.* truly, really, indeed.
  à — dire, to tell the truth.
  au —, dans le —, pour —, in truth, truly.
  pas —? is it not true? is it not so?
vraiment, *adv.* truly; in truth, really, indeed.
vraisemblable, *a.* probable, likely.
vraisemblablement, *adv.* probably, likely.
vraisemblance, *f.* probability, likelihood.
vrille, *f.* gimlet; (*Bot.*) tendril.
vriller, *v.t.* bore with a gimlet. — *v.i.* follow a spiral path (as a rocket).
vu, *p.a.* seen, etc. (see voir); looked upon, regarded. — *prep.* seeing, considering, in view of; according to. — *m.* examination, inspection; sight; (*Law*) preamble.
  — que, considering that, seeing that, whereas, since.
vue, *f.* view; sight; inspection; survey; intention; window, opening (thru which to look).
  — de côté, side view.
  — d'ensemble, general view.
vulcanisation, *f.* vulcanization, vulcanizing.
vulcaniser, *v.t.* vulcanize.
vulgaire, *a.* common; popular; vulgar.
vulgairement, *adv.* commonly; vulgarly.
vulgariser, *v.t.* popularize; render common.
vulnéraire, *a.* (*Med.*) vulnerary. — *m.* vulnerary. — *f.* (*Bot.*) kidney vetch (*Anthyllis vulneraria*).

W

wacke, *m.* (*Geol.*) wacke.
wad, *m.* (*Min.*) wad.
wagon, *m.* (railway) car, carriage; wagon; cart; flue tile of rectangular cross section.
— *à houille,* coal car.
— *de marchandises,* freight car, goods van.
— *plat,* flat car.
wagon-citerne, *n.* tank car; tank wagon; water cart.
wagon-lit, *m.* sleeping car.
wagonnet, *m.* small car, truck, trolley.
wagon-réservoir, *m.* tank car, reservoir car.
wallon, *a.* Walloon.

warranter, *v.t.* warrant.
watt, *m.* (*Elec.*) watt.
watt-heure, *m.* (*Elec.*) watt hour.
wattmètre, *m.* (*Elec.*) wattmeter.
wedro, *m.* vedro (Russian measure, 12.3 liters).
withérite, *f.* (*Min.*) witherite.
wolfram, *m.* wolfram (the mineral wolframite).
wolframine, *f.* (*Min.*) tungstite.
wolframocre, *m.* (*Min.*) wolfram ocher (tungstite).
woorari, *m.* curare, woorari.
wootz, *m.* (*Metal.*) wootz.

X

xanthéine, *f.* xanthein.
xanthine, *f.* xanthine; xanthin.
xanthinine, *f.* xanthinine.
xanthique, *a.* xanthic.
xanthochromique, *a.* xanthochromic.
xanthocréatine, *f.* xanthocreatinine.
xanthogène, *m.* xanthogen.
xanthone, *f.* xanthone.
xanthophylle, *f.* xanthophyll.
xanthoprotéique, *a.* xanthoproteic.
xanthoxyle, *m.* (*Bot.*) Zanthoxylum.

Xbre, *abbrev.* (décembre) December.
xénon, *m.* xenon.
xylème, *m.* (*Bot.*) xylem.
xylène, *m.* xylene.
xylidine, *f.* xylidine.
xylique, *a.* xylic.
xylite, *f.* xylitol, xylite (the alcohol).
xyloïdine, *f.* xyloidin.
xylol, *m.* xylene; xylole, xylol. (Cf. benzol.)
xylose, *m.* xylose.
xylotile, *f.* (*Min.*) xylotile.

Y

y, *adv.* there. — *pron.* to him, to her, to this, to that, to it, to them; in him, in her, in this, in that, in it, in them, therein; with him, with her, with them (and similarly with *by, for,* etc.); so.
yahourt, *m.* yogurt.
yeuse, *f.* holm oak (*Quercus ilex*).
yeux, *pl.* of œil.
-yle. -yl (in names of radicals).

yogourt, yoghourt, *m.* yogurt.
youfte, *m.* yufts, Russia leather.
ypérite, *f.* (*Mil.*) yperite (mustard gas).
ytterbium, *m.* ytterbium.
yttria, *m.* yttria.
yttrialite, *f.* (*Min.*) yttrialite.
yttrifère, *a.* yttriferous.
yttrique, *a.* yttrium, of yttrium, yttric.
yttrium, *m.* yttrium.

Z

zèbre, *m.* zebra.
zébrer, *v.t.* stripe (like a zebra).
zébrure, *f.* stripe, striping, stripes.
zédoaire, *f.* (*Pharm.*) zedoary.
zéine, *f.* zein.
Zélande, *f.* Zealand.
zèle, *m.* zeal.
zélé, *a.* zealous. — *m.* zealot.
zénith, *m.* zenith.
zéolite, zéolithe, *f.* (*Min.*) zeolite.
zéolitique, zéolithique, *a.* (*Min.*) zeolitic.
zero, *m.* zero; cipher; naught.
— *absolu,* absolute zero.
zérotage, *m.* determination of the zero point.
zeste, *m.* peel (of citrous fruits); dividing membrane (in nuts).
— *de citron,* lemon peel.
zester, *v.t.* cut the peel of (into strips).
zigzag, *m.* zigzag.
zigzaguer, *v.i.* zigzag.
zinc, *m.* zinc.
— *cuivré,* copper-plated zinc.
— *grenaillé,* granulated zinc.
— *laminé,* rolled zinc, sheet zinc.
zincage, *m.* = zingage.
zinc-éthyle, *m.* zinc ethyl.
zincifère, *a.* zinciferous.
zincique, *a.* zinc, of zinc, zincic.
zincographie, *f.* zincography.
zincographier, *v.t.* reproduce by zincography.
zingage, *m.* zincking, zincing; specif., galvanizing.
zingibéracées, *f.pl.* (*Bot.*) Zinziberaceæ.
zinguer, *v.t.* zinc; specif., galvanize.
zinguerie, *f.* zinc works; zinc ware; zinc trade.
zingueur, *m.* zinc worker.

zinquer, *v.t.* = zinguer.
zinquier, *m.* zinc founder.
zinzolin, *a. & m.* reddish violet.
zinzoliner, *v.t.* color reddish violet.
zircon, *m.* (*Min.*) zircon.
zircone, *f.* zirconia.
zirconien, *a.* containing zircon, zirconian.
zirconique, *a.* zirconium, of zirconium, zirconic.
zirconium, *m.* zirconium.
zone, *f.* zone.
zoné, *a.* zoned.
zoochimie, *f.* zoöchemistry.
zoochimique, *a.* zoöchemical.
zoologie, *f.* zoölogy.
zoologique, *a.* zoölogical.
zoologiquement, *adv.* zoölogically.
zoologiste, zoologue, *m.* zoölogist.
zooxanthine, *f.* zoöxanthin.
zumatique, *a.* (*Painting*) zumatic.
zumique, *a.* zymic.
zymase, *f.* zymase.
zymine, *f.* zymin.
zymique, *a.* zymic.
acide —, old name for lactic acid.
zymogène, *m.* (*Biol.*) zymogen, zymogene.
zymohydrolyse, *f.* zymohydrolysis (hydrolysis by fermentation).
zymologie, *f.* zymology.
zymologique, *a.* zymological, zymologic.
zymosimètre, *m.* zymosimeter, zymometer.
zymotechnie, *f.* zymotechnics.
zymotechnique, *a.* zymotechnical, zymotechnic.
zymotique, *a.* zymotic.